ETHOLOGY
The Biology of Behavior

ETHOLOGY
The Biology of Behavior

SECOND EDITION

IRENÄUS EIBL-EIBESFELDT
Max Planck Institute for Behavioral Physiology

translated by ERICH KLINGHAMMER
Purdue University

HOLT, RINEHART AND WINSTON, INC.
New York Chicago San Francisco Atlanta
Dallas Montreal Toronto London Sydney

Library of Congress Cataloging in Publication Data

Eibl-Eibesfeldt, Irenäus.
Ethology, the biology of behavior.

Translation of Grundriss der vergleichenden Verhaltensforschung, Ethologie.
Bibliography: p. 535
Includes indexes.
1. Animals, Habits and behavior of. I. Title.
[DNLM: 1. Behavior, Animal. QL751 E34g]
QL751.E3313 1975 591.5 74-23700
ISBN 0-03-091280-6

Printed in the United States of America
6 7 8 9 038 9 8 7 6 5 4 3 2

To my esteemed teacher and friend
Dr. Konrad Lorenz with gratitude

Preface

Ethology—the comparative study of behavior—has in recent years achieved a significance far beyond the realm of biology. The realization that phylogenetic adaptations determine the behavior of animals in a definable manner has increasingly led even those sciences which deal exclusively with man to search for the biological bases of human behavior. All behavioral sciences are based on the assumption that predictions about behavior can be made if a sufficient number of relevant variables are known. It is their goal to investigate such lawful relationships. Animals possess predictable behavioral programs. How these programs are acquired is a matter of dispute. There exist schools of thought which hold tenaciously to the view that animals and man learn all their behavior during the course of ontogenetic development. Ethology has refuted the generality of this assertion. It was first shown in animals that a basic repertoire of behavior patterns matures during the course of development. Emergence of these movement patterns, like that of cells and organs, is guided by phylogenetically acquired developmental blueprints. The impulse to study behavior anew came from biology, and the applicability—in principle—of working hypotheses derived from the study of animal behavior to human behavior is today an accepted fact. We know that even human behavior is determined to a certain extent by phylogenetic adaptations, and that these adaptations are of utmost practical as well as theoretical importance for the sciences dealing with man; consider, for example, the implications for education and sociology.

Phylogenetic adaptations determine the course of behavior in various ways. At hatching or at birth animals are fitted out with motor patterns that serve certain functions. Other motor patterns mature during ontogeny even if opportunities to copy the behavior from conspecifics or learn it by trial and error are withheld. Comparisons of closely related species frequently reveal behavior patterns that are identical in form, pointing to a common phylogenetic origin. Sometimes it is possible to reconstruct their evolution by **comparative methods.**[1]

[1] **Boldface** words or phrases invite the reader to consult the subject index, where further references to the subject will be listed.

In addition animals have detector devices tuned to certain environmental stimuli. These mechanisms enable the animal to react to specific, sometimes complex, stimuli at first encounter in a biologically appropriate way (see **innate releasing mechanisms**).

Furthermore, animals do not merely wait passively for stimuli to impinge upon them. Behavior is an active affair! Animals are motivated by inborn physiological mechanisms that act as drives and that cause an animal to seek actively for stimulus situations that will allow the performance of specific acts.

In addition, we have learned that modifications of behavior by learning are guided by inborn species-specific learning dispositions. Animals certainly do not learn everything with the same ease at all times but behave in this respect with considerable bias (see Chapter 14).

This ethological knowledge, based on animal studies, can contribute to a better understanding of human behavior, and K. Lorenz recognizes this as "essentially the most important task" of the branch of science he founded. This is so because in species that are mentally more advanced behavior to conspecifics is determined to a greater extent by innate components and less by learning than is their behavior toward the environment. "That this is unhappily so even in man," Lorenz (1950a, p. 431) writes, "is expressed drastically by the discrepancy between the enormous success in controlling the external environment and the crushing inability to solve intraspecific problems."

With disorders of our own social behavior so acute today, it would be of prime importance for education and the study of the psychology of man if we could discover which disorders of social behavior can easily be influenced by education and which less so. The latter is to be expected if phylogenetic adaptations determine the disposition which has to be counteracted by education. Of course, education has in any case to provide the means of correction, but strategies that make allowance for human nature are certainly more humane than environmentalist programs that take no such considerations into account. In this context I want to draw attention to the fact that B. F. Skinner (1971), in proposing to shape man for his survival, took as his only guideline what contributes to the survival of culture, assuming that no universal norms of human nature exist. Orwell's vision of *1984* is closer to reality than many of us would like to think. Because of its relevance to human behavior ethology extends into the areas of social sciences and philosophy. D. Ploog (1964a) has published an extensive review of the impact of ethology on psychiatry. Recent publications by S. J. Hutt and C. Hutt (1970), E. A. Tinbergen and N. Tinbergen (1972), and B. Hassenstein (1973) provide examples of the practical therapeutic applications of ethological concepts.

The subject matter of ethology was first presented in textbook form by N. Tinbergen (1951). In 1952 the German translation of the original

English version became available. To date no revised edition of this basic work has been published. However, especially during the last fifteen years, decisive advances have been made in the area of behavioral research, and a new review of the field is definitely needed.

Of those books that discuss the ethological point of view substantially, G. Tembrock (1964), W. H. Thorpe (1963), P. R. Marler and W. H. Hamilton (1966), R. A. Hinde (1966), and J. Altmann (1966) must be mentioned. Because ethology is a young area of research, no canon of subject matter as yet exists. It can, for example, be debated how extensively hormone physiology or sensory physiology should be covered. Depending on whether a student of behavior is oriented toward a comparative-morphological or comparative-physiological point of view, his study would be expected to reflect his particular preference. While basing this book on specifically *Lorenzian* views, I have nevertheless attempted to present all other current viewpoints as well.

As a colleague of Konrad Lorenz it has been my privilege to be a part of the development of ethology during one of its decisive phases. As early as 1946 I was introduced by Otto Koenig, at the Biological Station Wilhelminenberg near Vienna, to the problems and methodology of this research area.

Since 1949 I have worked with Konrad Lorenz, whom I followed in 1951 to the Max Planck Institute for Behavioral Physiology in Seewiesen, near Munich, Germany. My especial interests were in phylogenetic adaptations in the behavior of mammals and the ways in which, during the course of development in young animals, innate and acquired behavioral elements became linked into new functional units. In addition, I investigated, by comparative methods, the function and phylogeny of expressive movements of vertebrates and became acquainted with many different species. I received many stimulating new ideas from research expeditions into the tropics, to which I was repeatedly invited by my friend Hans Hass. The abundant animal life in the coral reefs led me to a deeper appreciation of the ecology of behavior. Ethological studies on various vertebrates then led to additional research projects concerning man. By the establishment of the Research Unit for Human Ethology the Max Planck Society provided a base to follow up research along this new line.

Research in the field of ethology has progressed rapidly in recent years. The studies of Willows (1971), Truman and Sokolove (1972), and others linked neuronal events to motor behavior; Ewert's (1973) investigations revealed how innate releasing mechanisms work at the neuronal level; the contributions of Gardner and Gardner (1971) and of Premack (1971) provided new and fascinating insights in the chimpanzee's way of conceptualizing and helped to narrow further the gap between ape and man, as far as the evolution of language is concerned. In numerous publications ethological concepts and methods have been examined critically. Progress can be reported from all disciplines of ethological research, and this made a thorough revision of the book necessary. Since the text had to be reset numerous illustrations and graphs, as well as chapter summaries, could be added. Thus brought up to date the book should be an aid in introducing those interested into a field of research that has proved most stimulating for interdisciplinary discussion and cooperation. The temperamental attacks of the early years have been replaced by sober appreciation and criticism from the related fields of anthropology, sociology, and psychology, which has found its most recent expression in the awarding of the Nobel prize to Konrad Lorenz, Nikolas Tinbergen, and Karl von Frisch.

While making plans for the first edition of this book I derived great value from the stimulating discussion with my colleagues at the Max Planck Institute for Behavioral Physiology, where practically all the important lines of behavior research are represented. I wish to thank all the members of this institution. This book is based on a previous book of mine which first appeared in the *Handbuch der Biologie* (Akademische Verlagsgesellschaft Athenaion, 1966a), and has here been expanded and adapted to meet the needs of students.

I wish to thank my esteemed teacher and friend Konrad Lorenz for all his encouragement and help. I want to thank especially my friend Dr. Hans Hass for the many hours of stimulating discussion as well as for inviting me to participate in the expeditions to the tropical seas which were filled with so many new impressions for me. I also want to thank Otto Koenig for his support while I was on the staff of the Biological Station Wilhelminenberg from 1946 to 1949; Otto Koehler, whose frank and encouraging criticism was of immense benefit to me; as well as my friends B. Hassenstein, E. H. Hess, E. Klinghammer, P. Leyhausen, P. Marler, J. Nicolai, E. S. Reese, H. Sielmann, H. Schöne, N. Tinbergen, and W. Wickler. I remember with great pleasure my collaboration with H. Sielmann while making films.

Special thanks are due all colleagues who contributed pictures and observations to this book, especially to the illustrator Hermann Kacher, the Piper Publishing Company for their care in the preparation of the original German edition, and Holt, Rinehart and Winston, Inc., for their help in preparing both this and the previous English edition.

As an Austrian citizen it is my duty as well as pleasure to thank Germany, the country where I am a guest, and the Max Planck Gesellschaft, the Deutsche Forschungsgemeinschaft, the A. v. Gwinner Foundation, and the Fritz-Thyssen Foundation for the generous manner in which they have supported my work.

Percha, West Germany I.E.-E.
1974

Contents

1

A Short History of Ethology

Behavior consists of patterns in time. Investigations of behavior deal with sequences that, in contrast to bodily characteristics, are not always visible. It is true that even the development of an organism can be thought of as a sequence and it is possible to investigate its growth as a kind of "behavior." However, in respect to our own time perception, bodily structures appear to be static, and they can always be fixed as an anatomical preparation. Behavior patterns, however, must be artificially transformed into spatial structures by means of motion picture film and sound tape if they are to become a preparation and a continuing document.

Behavior is usually expressed by the movement of muscles and sometimes also by the activity of glands or migration of pigments. Growth, swelling, and turgor movements, which are elicited and directed by means of specific stimuli, comprise the behavior repertoires of plants, of which C. Darwin (1872), to my knowledge, was the first investigator (see F. Gessner, 1942). In this book we will be concerned exclusively with the behavior of animals and man. Just as one can speak of the study of behavior of single-celled organisms, one may also study the behavior of the cells of an organism to determine the releasing stimuli for the movement of leucocytes, how scleroblasts deposit calcium, or how myxomycetes migrate to form a sporophore. Each cell possesses a behavior program and reacts to certain releasing and directing stimuli.

When an ethologist observes an animal performing a certain action, he may ask why the animal behaves in this and not some other manner. If we hear a bird sing, we may ask why it does so. In what way does its song contribute to the preservation of that particular species; that is, what selective advantage does the song confer upon the animal? With questions about the specific adaptation or function of behavior, the ethologist closely approaches the interests of the ecologist.

Once the function of a behavior is known, one can ask how it developed. During the course of development in the young animal the process of differentiation and integration of behavior patterns can frequently be observed directly. Ethologists employ the methods of comparative morphology (W. Wickler, 1961a, 1967b) to investigate behavioral phylogeny.

Behavior always has a cause. External sensory stimuli are just as responsible for its manifestation as internal drive mechanisms of the central nervous system, hormones, and internal sensory stimuli. These physiological causes of behavior are also the subject of ethological investigation, making in this way a bridge with physiology, whose methods are often employed. However, where physiologists generally try to get an understanding of the simplest behavior patterns (for example, heartbeat, respiration, muscular reflexes, and function of isolated muscle fibers) ethologists—who come mainly from zoology—investigate primarily the behavior of the total organism and its relationships with the organic and inorganic environment. They usually investigate the complex and well-integrated functioning of various muscle groups and less frequently the function of isolated parts that have been removed from the whole. They work, in other words, at a higher level of integration than the physiologists, although their areas of research are by no means sharply defined.

Students of the biology of behavior share this especial concern with the behavior of the whole organism with some schools of psychology. There is nothing mystical about the term "wholeness," a point that is made clear by the cybernetic study of behavior. Feedback systems operate as a whole, but are nevertheless subject to causal analysis (B. Hassenstein, 1966).

Psychologists and ethologists have developed their formulations about behavior from different points of view. Psychology is derived from philosophy. Relatively early it became involved in the mechanism–vitalism controversy, the results of which K. Lorenz (1950a, 1950b, 1957) has discussed at some length. Vitalists were as a rule excellent observers of animals, but they did not look for a causal analysis of behavior. They considered the behavior of the whole organism to be unanalyzable by mechanistic methods. Instead, they postulated as final causes entelechial soul-like factors and unfailing and inexplicable instincts. "We consider an instinct but we do not explain it," said J. A. Bierens de Haan (1940). The American school of purposive psychology also contains a strong vitalistic bias. The investigators of this group emphasize that behavior is purposefully directed toward a specific goal. In their view, this goal directs the activity as a specific motivation. The animal is motivated by expectancies that require no further explanation (E. C. Tolman, 1932; W. McDougall, 1936; E. S. Russell, 1938). Finally, a number of Gestalt psychologists employ concepts of wholeness (*Ganzheit*) in the manner in which Driesch uses "entelechy" (F. Krueger, 1948).

The mechanistic schools, on the other hand, have been convinced since Descartes that all behavior can be derived in the final analysis from physical laws. The concept of wholeness, in their view misused by the vitalists, is rejected by them. Furthermore, they ignore subjective phenomena and describe only that which is objectively observable, asserting that one cannot make reliable statements about the experiences of other organisms. They adhere to a "psychology without the mind." They all search for elements out of which they can build up even complex behavior. A. Bethe had already turned against a subjectivistic psychology by 1898. J. Loeb (1913) in his theory of tropisms endeavored to establish a purely mechanistic or machinelike explanation of the behavior of animals. Reflexology, founded and developed by W. Bechterew (1913) and I. P. Pavlov (1927), explained all behavior on the basis of conditioned and unconditioned reflexes and asserted that complex sequences of behavior are nothing but chain reflexes (J. Loeb, 1913; H. E. Ziegler, 1920). The concept of instinct as misapplied by the vitalists was rejected by the reflexologists. Similar is the position of some American behaviorists, who repudiate terms such as *feeling, attention,* and *will,* asserting instead that one can only determine stimuli and reactions and the laws governing their interactions.

On the whole this assertion has validity when we deal with an animal whose subjective "inner experience" is forever closed to us because of reasons implied in the theory of cognition. We may assume that subjective phenomena play a role in the behavior of animals, but we cannot say anything specific about them. The temptation to reason by analogy from our own experiences to similar ones in higher animals is constantly present, particularly in those who know animals well, but such reasoning can be no substitute for proof and becomes less valid the more dissimilar a species is from our own. About the subjective experiences of our fellow men, however, we obtain objective data not only through self-observation but also from the reports of others. In spite of the fact that definite statements cannot be made about an animal's subjective experiences, many concepts of a subjective nature are still used today. Physiologists speak of hunger and thirst. This kind of shorthand description is useful if for no other reason than that it is generally understood.

Of the representative American behaviorists only E. L. Thorndike (1911), J. B. Watson (1930), K. S. Lashley (1938), and B. F. Skinner

(1953) will be mentioned. Because the founder of behaviorism, J. B. Watson, stressed the influence of the environment, subsequent investigations tended to focus on learning phenomena and frequently overlooked the inherited, innate bases of behavior.

These mechanistic schools have been accused of a certain one-sidedness—justifiably so when an attempt was made to elevate one explanatory principle to the level of an all-inclusive one. Thus, J. Loeb wrongly generalized his discoveries into his theory of tropisms. His turbellaria and insects, which normally move toward or away from light, moved in circles when blinded in one eye. From this he concluded that normally the equally strong stimulation of sensory organs on the left and right would neutralize opposing muscle contractions on both sides of the body. On the basis of such simple reflexes he attempted to explain all oriented movements as tropisms, which in fact they are not.

For a long time reflexologists and behaviorists overlooked spontaneity of behavior, which was not readily observable in their particular experimental situations. For them all behavior consisted of reactions to stimulation. They clung overtenaciously to an experimental method once it was found successful (for example, the maze experiment), and this resulted in a certain one-sidedness. Some extreme proponents of behaviorism consider all behavior to be the result of learning processes. They state that the environment alone determines the behavior of the animal during the course of his ontogeny, a view which will be discussed in more detail later. Behaviorists and reflexologists have made contributions to the study of behavior through the rigorous application of the scientific method; and this is especially true in respect to the phenomena of learning.

Ethology emerged as another discipline exploring these research areas in the natural sciences. It developed out of zoology, especially through the work of K. Lorenz and N. Tinbergen, and is based on the discovery of phylogenetic adaptations in behavior. However, the knowledge that some behavior is relatively uninfluenced by individual experience is much older. As early as 1716 F. A. v. Pernau knew that animals possessed innate skills in addition to those they acquired, behavior patterns that they did not have to learn by imitation or other forms of training. He described the behavior patterns of various birds and showed which species had to learn their songs from their parents and which, upon becoming sexually mature, were able to sing their species-typical songs without prior exposure to them. H. S. Reimarus (1762) wrote in a similar vein:

> How do the spider and the ant lion go about finding means of supporting themselves? Both can do no other than to live by catching flying and creeping insects, although they are slower in their own movements than is the prey which they seek out. But the former perceives within herself the abil-

> ity and the drive to weave an artful net, before she had as much as seen or tasted a gnat, fly, or bee; and when one has been caught in her net she knows how to secure and devour it. . . . The ant lion, on the other hand, who can hardly move in the dry sand, mines a hollow funnel by burrowing backward, in expectation of ants and worms that tumble down, or it buries them with a rain of sand that it throws up in order to cover them and bring them into his reach. . . . Since these animals possess by nature such skills in their voluntary actions that serve the preservation of themselves and their kind, and that admit many variations; so they possess by nature certain innate skills. . . . A great number of their artistic drives are performed at birth without error and without external experience, education, or example and are thus inborn naturally and inherited. . . . One part of these artistic drives is not expressed until a certain age and condition has been reached, or is even performed only once in a lifetime, but even then it is done by all in a similar manner and with complete regularity. For these reasons these skills are not acquired by practice. . . . But not everything is determined completely in the drives of the animals, and frequently they adjust, of their own volition, their actions to meet various circumstances in various and extraordinary ways. . . . For if everything and all of their natural powers were to be determined completely, that is, would possess the highest degree of determination, they would be lifeless and mechanical rather than endowed with the powers of living animals.

D. A. Spalding (1873) demonstrated the maturation of innate behavior patterns when he raised swallows in cages so small that they could not flap their wings. In spite of this the birds flew excellently when first given the opportunity to do so. Innate behavior patterns were also reported by R. A. F. Réaumur (1734–1742), A. J. Rösel v. Rosenhof (1746–1761), B. Altum (1868), G. Peckham and E. Peckham (1904), and J. H. Fabré (1879–1910).

W. James (1890) presented a thoroughly mechanistic definition of instinct. He called instincts the correlates of the organs. Just as an animal has certain organs, so it also possesses an inborn ability to use them, and this is founded upon a given neural organization. C. Lloyd Morgan (1894, 1900) expressed himself in a similar manner when he said that the structure of the central nervous system, which underlies instincts, is the result of phylogenetic evolution.

As the direct forerunners of ethology, however, we nominate C. Darwin (1872), C. O. Whitman (1899, 1919), O. Heinroth (1910), and W. Craig (1918). In his work on the expressive movements of man and animals Darwin was the first to introduce the comparative phylogenetic method to the study of behavior. Heinroth investigated the systematics of closely related species of ducks and geese, and Whitman did the same for pigeons and doves. In their search for taxonomically useful characteristics, they encountered predictable inborn behavior patterns, which were characteristic for certain taxonomic categories in the same way as morphological characteristics. By their degree of similarity they could reveal the close or distant relationship of the taxa. Heinroth called these behavior patterns *arteigene Triebhandlungen.*

W. Craig (1918) was the first to distinguish the stereotyped *consummatory action* from the more variable initial *appetitive behavior,* which is a search for the appropriate releasing stimulus situation.

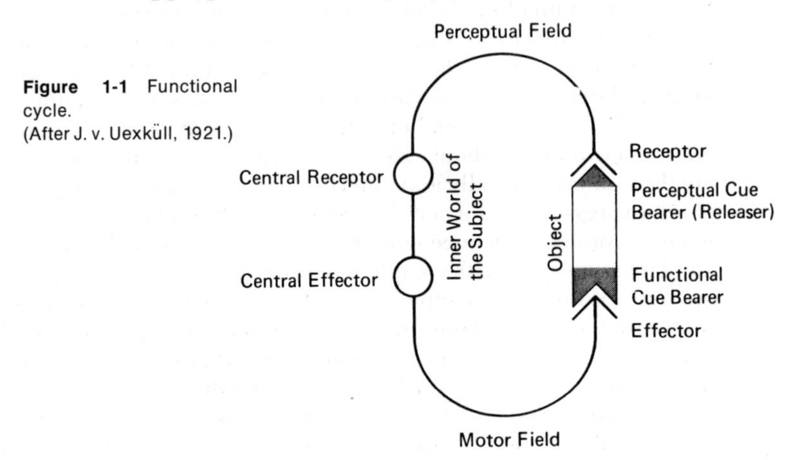

Figure 1-1 Functional cycle.
(After J. v. Uexküll, 1921.)

Great influence was exerted upon the development of ethology by J. v. Uexküll (1921), who conducted experiments to investigate the interrelations between organisms and their environment. He showed that an animal can perceive only a limited part of its potential environment with its sense organs. Some of these perceived characteristics of the environment serve as specific cues. According to Uexküll only those objects serve as cues that are of significance in the life of an animal, thereby becoming bearers of meaning for the subject, or **key stimuli.**[1]

> The appearance of an object (cue bearer) in the perceptual field of a subject always has an effect that imparts functional significance (effector cues) to the object. The effector cue or meaning extinguishes the receptor cue or meaning and brings the action to an end. The effector cue may be objectively extinguished (eaten if it is food), or it may be subjectively extinguished by satiation, when the stimulus filter of the sense organ is turned off. As soon as the effector cue of the object extinguishes its perceptual cue, the functional cycle passing from the object through the subject and again to the object is completed (J. v. Uexküll 1937:34).

Figuratively speaking, each animal subject grasps its object with the two arms of a forceps (receptor and effector) and impresses upon it receptor and effector cues (meaning) (Fig. 1-1). Uexküll illustrates this schema of a functional cycle with the example of the tick. Mated females climb bushes and wait until a mammal passes by. When they perceive

[1] **Boldface** words or phrases invite the reader to consult the subject index, where further references to the subject will be found.

the odor of butyric acid, which is secreted by the skin glands of all mammals, they release their grip and drop. If they should hit something warm, they begin to search for an area of skin free of hair, bore through the skin with their head, and suck blood until full. However, other warm liquids are also accepted, as was shown during experiments with artificial membranes. If we now fit into the schema of the functional cycle the tick as the subject and the mammal as the object, we can see that three functional cycles are completed:

> The skin glands of the mammal constitute the cue bearer for the first cycle, because the stimulus of butyric acid releases in the receptor organ specific receptor cues that are projected outward as an olfactory cue. These processes in the receptor organ induce (how is not known) in the effector organ appropriate impulses which cause the legs to release their hold and the tick to drop down. The falling tick imparts to the hair of the mammal the effector cue of the collision, which in turn releases a tactile cue by means of which the olfactory cue of butyric acid becomes extinguished. The new cue releases the search for a hair-free spot where the cue—warmth—takes over, which results in the boring through (J. v. Uexküll and G. Kriszat, 1934; new ed. 1963:28).

K. Lorenz (Fig. 1-2) was the first to appreciate the full significance these discoveries (1935, 1937). Working inductively from them and from his many personal observations he proposed a synthesis that forms the basis of ethology. The target of his investigations was at first the innate motor pattern. He recognized the spontaneity that underlies instinctive movements, a physiological fact of great importance that had been overlooked by the classical reflexologists. He investigated the **key stimuli** that release a specific behavior prior to all experience and studied the phylogenesis and ontogenesis of innate behavior patterns. In the "**instinct—learning intercalation**" he found a new mode of integration of innate and acquired components of behavior, and in the phenomenon of **imprinting** he discovered an inborn disposition to learn, of especial significance. He repeatedly emphasized the importance of these discoveries for the sciences dealing with man.

In 1937 the first volume of *Zeitschrift für Tierpsychologie*, consisting of 289 pages, appeared. By 1973 the yearly volume amounted to over a thousand pages. In 1948 the journal *Behaviour* (Brill, Holland) began publication, in 1953 the *British Journal for Animal Behaviour (Animal Behaviour* after 1958) and in 1966 the *Revue du comportment animal* first appeared. Today one generally talks about the comparative study of behavior (*vergleichende Verhaltensforschung*) or ethology (*ethos*, habit, manner). The term "ethology" had previously been used by biologists and covered what today falls under the heading "ecology" (L. Dollo, n.d., 1909). J. S. Mill (1843) understood by this term "the exact science of human nature." From 1907 to 1940 the *Zoological Record* carried a section on ethology—meaning the study of behavior—for each class of animals. Since 1951 (see N. Tinbergen, 1951, and Fig. 1-3) the term has been generally accepted as referring to this specific branch of the natural sciences. The main emphasis of ethological research initially dealt with the "study of instinct" but without being limited to this. Ethology is a natural science, a branch of biology, from which it took the comparative method for the study of behavioral morphology and the analytic method for the causal analysis of behavioral physiology. Its philosophical base is a critical realism. Its orientation is neo-Darwinistic, and it enjoys a fruitful exchange of ideas with other schools of behavior, especially with behaviorism, and increasingly with Russian schools (L. V. Krushinskii, 1962). **Human ethology** (I. Eibl-Eibesfeldt and H. Hass, 1966) is emerging as a new field of research.

An excellent report by J. K. Kovach (1971) deals with Russian ethology. In contrast to those American schools which assume that in the behavioral development of the phenotype everything is determined by experience, Russian scientists view such development as an integrative process involving both innate and acquired reflexes.

> In view of this, one may safely conclude that there is a more marked disagreement between some current epigenetic postulates in the West on the one hand and the postulates of Krushinskii and Promptov on the other, than between general ethological postulates and the postulates of Soviet eco-

Figure 1-2 Konrad Lorenz. (Photograph: H. Kacher.)

> logical physiologists. It is not the concept of innate but rather the separation between innate and learned elements in behavior . . . and the explanatory schemes of endogenous energy levels and the hierarchical organization of instincts . . . that are rejected by Soviet investigators (J. K. Kovach, 1971, p. 246).

The publications of L. V. Krushinskii (1962), in particular, demonstrate that differences in points of view between Western and Russian ethologists are not basic, and we may expect further contact between the two schools. The points of emphasis are differently distributed, however. According to Kovach, the comparative study of behavior on an evolutionary basis has not yet begun in Russia. On the other hand, much work is being done on the early development of behavior and on the releasing stimuli sufficient for the formation of conditioned reflexes. G. Razran (1971) offers a further introduction to Soviet ethology.

SUMMARY

*Ethology can best be defined as "the biology of behavior." In trying to understand *why* an animal behaves the way it does, ethologists search for the functions of the observed behavior patterns in order to learn what selection pressures have shaped their evolution. By applying the com-

Figure 1-3 Niko Tinbergen. (Photograph: B. Tschanz.)

parative technique developed in morphology, ethologists attempt to reconstruct the phylogeny of motor patterns. They explore the processes underlying the ontogenetic development and finally search for its causes, investigating the releasing stimuli and the underlying physiological processes. The discovery of phylogenetic adaptations preprogramming behavior has spurred the development of this new field of research and provoked lively discussions about the relevance of ethological concepts to students of human behavior.

2

The Ethogram
a behavioral inventory

Science begins with the description and categorization of the events it studies. The basis of each ethological investigation is the *ethogram*, the precise catalogue of all the behavior patterns of an animal. For this catalogue one selects functional units of behavior that are neither too small nor too large. In practice it is not too difficult to find easily recognizable functional units that are constant in form, such as scratching, chewing, and "head-up tail-up." The suggestion that each behavioral study of an animal species begin with the establishment of a behavior inventory was made as early as 1906 by H. S. Jennings (1906), who called them *action systems*.

It seems axiomatic that the species studied should be accurately classified. However, with some animals this is not easily done. W. Wickler (1960a) suggests that in such cases a specimen of the species studied be sent to a museum, and that the museum's name and address as well as the catalogue number of the specimen be included in the publication. Should the need arise, other investigators can then reexamine, and if necessary reclassify, the specimen. A. Seitz (1940) and L. R. Aronson (1949) originally were thought to be studying the same cichlid fish, *Tilapia macrocephala* Bleeker. Their results did not agree. Whether or not they actually worked with the same species can no longer be determined, because neither of the investigators preserved sample specimens.

The description of a behavior pattern should include each detail of the event. Such a *physical* description is never complete in actuality, because the observer usually omits what is not important to him. For this reason motion picture film has become the ethologist's most important means of documentation.

On film the behavior patterns become fixed and can be preserved for later comparison. In addition, fast and slow motion allow for the analysis of data that would not normally be accessible to direct observation. The slow-motion technique has frequently been used to make visible to the human eye events that run off too quickly to be perceived. The value of the **fast-motion** technique has received little recognition in ethology. This is especially true for human ethology. The Institute for Scientific Film (Institut für den Wissenschaftlichen Film) in Göttingen, Germany, has been assembling an archive, called the Encyclopaedia Cinematographica, of technically perfect motion picture films (16 mm) of behavior sequences (G. Wolf, 1957a, 1957b).[1] (In the United States these films and many others are now available from Audio-Visual Services, 6 Willard Building, The Pennsylvania State University, University Park, Pennsylvania 16802. A select group of titles of these films will be found on page 593.

Documents of human behavior are being collected systematically in the Film Archive for Human Ethology of the Max Planck Society, and are being published as separate films. Supplementary publications to films in the Film Archive for Human Ethology are printed in the anthropological journal *Homo* (I. Eibl-Eibesfeldt, 1971c).

Frequently behavior is labeled according to its *function*. In this case one focuses on the goal rather than on coordinated movements that lead to it. "Carrying in" or "nest building" are functional terms. However, they imply an interpretation by the observer, and therefore this procedure involves a certain risk (R. A. Hinde, 1959).

In order to record a sequence of movement patterns in terms of their duration, frequency, and relative positions in time, without interrupting the continuous observations, one may use multichannel event records. Each of a number of previously selected behavior patterns is represented by a particular key that activates a pen. Pressing down the key records the event on a paper roll, which moves at a constant rate of speed.

Most observations are made on captive animals, and this has certain disadvantages. Lack of opportunity to hunt, explore, and so on may lead to distortions in behavior—especially in mammals that normally are quite active. Pacing back and forth, retracing the same paths, swinging to and fro, and other movement stereotypes can often be observed (M. Hol-

[1] Since 1964, commentaries on these films, of great use to the student of behavior, have been published in *Publications to Scientific Films* (*Publikationen zu wissenschaftlichen Filmen*) by the Institut für den Wissenschaftlichen Film, Göttingen, Germany.

zapfel, 1938, 1939). Such behavior may have various causes. An armadillo in the Amsterdam Zoo stopped its stereotyped movements at once when the bottom of its hitherto bare cage was covered with a layer of earth, 20 cm deep, so that the animal could bury itself at night in order to sleep. The stereotyped movements reappeared when the earth was removed. H. Hediger (1942) discussed a large number of additional behavior aberrations found in captivity. Higher mammals frequently suffer from a lack of opportunity for various activities, which has led many zoos to initiate what might be termed "work therapy." Breeding frequently fails because the animals do not mate or do not raise their young. Observations in the natural environment often lead to a solution of the problems. O. Koenig (1951a, 1951b) was able to induce his bearded titmice to breed, but the parents threw their young out of the nest shortly after hatching. The cause was found in the overabundance of food he had provided: The parents stuffed their young full in a very short time, and when they returned to feed them more the young no longer gaped; this can never happen in the wild, because the parents need to spend time finding the food. In the wild, young titmice that do not gape are either sick or dead and are removed from the nest by the parents. The captive bearded titmice behaved toward their nongaping young as if they were dead. It was sufficient to offer food to the captive birds in small, infrequent portions to correct the problem.

Many other disturbances of behavior in captive animals have similar causes. If one knows the animal well they can be corrected. It is by no means true that captive animals always behave abnormally and that observations in captivity are therefore of little value, as is sometimes asserted. A number of excellent studies attest to the opposite (H. Kummer,

Figure 2-1 Konrad Lorenz among his free-ranging geese. (Photograph: I. Eibl-Eibesfeldt.)

Figure 2-2 Baroness Jane van Lawick-Goodall feeds one of the chimpanzee males she tamed in the wild. (Photograph: Baron van Lawick, with permission of the National Geographic Society.)

1957, 1968; H. Kummer and F. Kurt, 1965). Details of behavior can be observed only during close and continuing contact with a particular species. Distortions can be minimized if one maintains the animals relatively free in their natural environment. K. Lorenz chose this means when he settled his jackdaws near his home in Altenberg and permitted them to fly free. His graylag geese and ducks have been settled by raising and feeding them year-round near a small lake, where they move about freely (Fig. 2-1). C. R. Carpenter introduced a group of rhesus monkeys to the island of **Cayo Santiago** (Puerto Rico) in 1938. This colony has since been almost constantly observed. P. Krott and K. Krott (1963) raised European brown bears in the wild and observed them there. It is rarely possible to observe a large number of animals in the wild for long periods of time. J. van Lawick-Goodall (1968) has been camping for years in a valley near Kigoma, Tanzania (Tanganyika) that is inhabited by chimpanzees. Owing to her great patience, the chimpanzees gradually became used to her presence and today they move without fear around her campsite. She can feed or groom them and they even solicit her to play with them (Fig. 2-2). Additional examples that illustrate the value of observations in the wild are given by the works of

M. Altmann (1952), J. A. King (1955), J. Adamson (1960), S. L. Washburn and I. DeVore (1961), L. Crisler (1962), G. B. Schaller (1963, 1972), N. Tinbergen (1963), I. DeVore (1965), W. Kühme (1965), F. R. Walther (1965), D. Fossey (1970), V. Geist (1972), and H. Kruuk (1972).

Observations in the wild and in captivity complement each other, and a discussion of the relative merits of each method would be superfluous.

The unbiased observation and recording of behavior patterns is also a prerequisite for the scientific study of man from an ethological point of view. That the ethology of human behavior is only now beginning may be seen from the fact that today there exists hardly one **motion-picture** document of natural human behavior which is published and on file.

SUMMARY

Ethological studies start with the description of the behavioral repertoire of a species, the so-called *ethogram.* In naming behavior patterns one should preferably confine oneself to descriptive terms. Film and tapes are the most important tools for objective documentation of motor patterns. Observations in the wild and in captivity complement each other.

3

The Fixed Action Pattern (Inborn Skills)

THE FIXED ACTION PATTERN AND ITS TAXIS COMPONENT

In the behavioral repertoire of an animal one encounters recognizable and therefore "form-constant" movements that do not have to be learned by the animal and provide, like morphological characteristics, distinguishing features of a species. In a manner of speaking we are confronted with "innate skills." Such innate movement coordinations have been called *fixed action patterns* (fixed patterns) or *instinctive movements* (K. Lorenz and N. Tinbergen, 1938; K. Lorenz, 1953), and the German term *Erbkoordination* (inherited coordination) indicates that the innateness[1] of these sequences is the deciding criterion. This innateness is not, however, recognized by the stereotyped sequence of the pattern, as M. Konishi (1966) erroneously states of ethologists, but by means of specific experiments that will be discussed later (see **deprivation experiment**). Form constancy may be a strong indication that the movement is

[1] The term "innate" is meant to denote the fact that the neuromotor structures which underlie these movements develop in a process of self-differentiation, guided by developmental instructions in the genome.

The translation of *Erbkoordinationen* as *fixed* action patterns was unfortunate, since it implies a rigidity which is nonexistent, but since the term has been established I feel, for the moment at least, that we must abide by it.

inherited, especially when closely related species show similar movement patterns. We know, however—especially from the study of bird songs (J. Nicolai, 1959a, 1964)—that learned behavior patterns also possess a high degree of stereotypy, so that this criterion alone cannot serve as a definition of the fixed action pattern (W. Wickler, 1961c).

Innate behavior patterns may already be fully functional at the time of hatching or birth. A newly hatched chick is able to walk, peck at seeds, scratch on the ground, and drink. It flees to the mother hen when predators appear, it calls loudly when it has lost contact with its mother, and it shakes itself when it has become wet. These and many other behavior patterns are present from the time of hatching. The same holds true for a just-hatched duckling, but its behavior deviates in important details from that of the chicken. The duckling runs to the water, swims, dives, feeds below the surface, and oils its feathers. These differences in the behavior of chicken and duck must be rooted in inheritance, because even a chick hatched under a duck retains the chick characteristics, while a duckling hatched under a chicken runs to the water and feeds on the bottom with its bill, which is adapted for straining, in spite of the foster mother's efforts to entice it away.

Not all inborn behavior patterns are fully developed at the time of hatching. Some, such as the complex **courtship movements** (head-up tail-up, grunt whistle, and so on) **of ducks**, develop gradually as the animal grows older. Since these behavior patterns develop in each male even if raised in isolation from other ducks, so that it has no opportunity to imitate these complex behavior sequences, we assume that they, too, are inherited as phylogenetic adaptations and merely require a longer period of maturation.

We will discuss later in more detail what is meant by the **form constancy of fixed action patterns**. At this time it should be pointed out that a fixed action pattern can occur at several levels of intensity, ranging from mere intention movements, which indicate what an animal is about to do, to completely executed actions. However, the typical pattern is always recognizable, in the same manner in which a rhythm, whether repeated slowly or rapidly, can be recognized as long as the relative spacing of the sounds remains the same.

Fixed action patterns usually proceed without any indication of insight into the species-preserving function of the activity on the part of the animal, as is clearly shown by inappropriate actions. When the inner **readiness to act** coincides with the appropriate **releasing stimulus** situation then a particular fixed action pattern will run its course almost automatically. Thus a dog hiding a bone in the living room shows the movements of covering it as if earth were available; this is the way its behavior is genetically preprogrammed to be adaptive in nature. It will turn several times before lying down, although there is no grass to be trampled.

Fixed action patterns normally occur with orienting movements or taxes superimposed on them. In contrast to fixed action patterns, taxes require continuous directing stimuli to be noticeable. The unity of taxis and fixed action pattern is the basis of *instinctive activity*. Instinctive activities can be quite complex. By the incorporation of various taxis components they acquire a higher degree of adaptability and variability appropriate to a particular situation. However, it can be shown upon close examination that even here we are dealing with genetically programmed behavior sequences. While constructing a cocoon the spider *Cupiennius salei* first produces a base plate, then a raised rim that provides the opening into which the eggs are deposited. Having laid the eggs, the female closes this opening. If she is interrupted while spinning her cocoon, after the base plate has been completed, she will not produce a new base plate half an hour later when she builds a new cocoon, but instead spins only a few threads and then continues with construction of the rim, so that the bottom of the cocoon remains open. If one adds the number of spinning movements she performed for the previous base plate and for the new substitute cocoon, the number roughly equals the number normally used to build a complete cocoon. She has available, so to speak, a limited number of spinning movements—approximately 6400 dabbing movements. This number of movements is performed, even if, under abnormal circumstances, she is no longer able to secrete any threads. This has happened when the glands dry up as a result of the hot lights used during filming. In such instances the spider still produces her behavior program. After the appropriate number of ineffectual dabbing movements she will lay her eggs, which will then drop to the ground. Then she continues as if she were closing the rudimentary cocoon. Finally, the spider tries to remove this partial structure from the substratum (M. Melchers, 1964).

The spider is therefore not affected by the success of her efforts. This can also be seen when the same spider is placed upon a half-completed structure. The existing structure is not taken into account. Instead, she continues as if she were sitting on her own cocoon. In this manner structures may be produced that are unsuitable for receiving the eggs (M. Melchers, 1960, 1963).

The complex cocoon structure built by the caterpillar of the moth *Platysamia cecropia* is precisely programmed. The larva spins three layers of a cocoon. Displacement experiments showed that it is capable neither of continuing a cocoon that was already begun nor of producing a new outer layer in a new place when it had already begun its own elsewhere. It is unable to repeat any of the layers. In this species the behavior is determined by the amount of substance in the glands. If the spinning glands are filled to 60 percent of capacity, the animal will begin with the construction of the inner cocoon layer, with an accompanying change in the frequency of rotation about the body axis (W. G. van der Kloot and C. M. Williams, 1953).

The manner in which taxis and fixed action pattern are combined into one instinctive activity has been shown by K. Lorenz and N. Tinbergen (1938) in the egg-rolling movement of the graylag goose. If a graylag goose is presented with an egg outside its nest it will reach out with its bill over and beyond the egg and pull it in with the underside of the bill, balancing it carefully back into the nest (Fig. 3-1). This behavior may be broken down into two components. If one removes the egg after the rolling movement has been started, then the movement continues in vacuo. The bird behaves as if the egg were still there. However, the lateral balancing movements cease and the neck is pulled back in a straight line to the nest. This movement, which once released will continue in the absence of additional external stimuli, is the fixed action pattern. The lateral balancing movements are the orienting movements or taxis components, which are also inborn but are discontinued in the absence of the releasing stimuli. Taxis and fixed action pattern are related to each other in the same way as are the steering mechanism and the engine of a car. Each change of direction requires an external impulse, but the engine, once started, will continue without an external impulse.

While a male stickleback fans his eggs he is positioned head down in front and above them. When fanning spontaneously and without eggs he is horizontal to the ground. Fixed action pattern and taxis may mature at different times during ontogenesis. A newborn mouse will at first show scratching movements in the air with its hind legs, without touching its body.

In the original concept of a taxis (orienting movement), the movements of the animal were combined with orienting movements. N. Tinbergen (1951) suggested that only the simple turning movement be called a taxis. This separation of taxis and fixed action pattern becomes

Figure 3-1 Egg-rolling movement of the graylag goose. (After K. Lorenz and N. Tinbergen, 1938.)

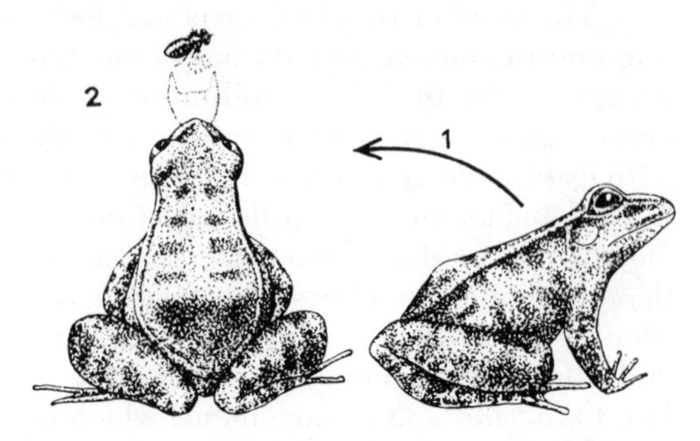

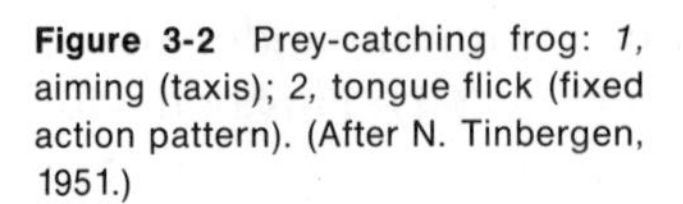
Figure 3-2 Prey-catching frog: *1*, aiming (taxis); *2*, tongue flick (fixed action pattern). (After N. Tinbergen, 1951.)

especially obvious when they are not coupled but occur one after the other. This occurs in a prey-catching frog, who will turn his body prior to the attack, with its snout pointing directly toward the prey (Fig. 3-2). This turning movement or alignment with the prey is the taxis, while the strike comprises the fixed action pattern (N. Tinbergen, 1951).

The original concept of taxis is different from Tinbergen's formulation in this important respect: Tinbergen uses the term merely to describe the event, while A. Kühn tried to characterize the physiological mechanisms involved.

In line with the distinction between the fixed action pattern and taxis one must also distinguish between stimuli that release and those that steer or orient fixed action patterns (N. Tinbergen and D. J. Kuenen, 1939).

THE DEPRIVATION EXPERIMENT—RAISING AN ANIMAL WHILE WITHHOLDING SPECIFIC INFORMATION

We have stated that behavior patterns may be inborn in an animal. How can the innateness of behavior be experimentally demonstrated? In recent years this question has fanned a controversy between ethologists and some behaviorists who subscribe to an environmentalist position.

Ethologists maintain that the question of innate and acquired components of animals' behavior can be answered if one raises an animal in isolation from all other conspecifics, thus preventing any possible imitation, and by making it impossible to learn the behavior in question by trial and error. The value of such *Kaspar-Hauser experiments* has been questioned by some investigators (D. S. Lehrman, 1953; T. C. Schneirla, 1956, 1966; R. A. Hinde, 1966). They argue that it is impossible to raise an animal without any experience, because it is always a part of an environment, even within the egg or the uterus, and is always experiencing something while interacting with its environment. A paper by Z. Y. Kuo

(1932) is frequently cited as an example of how a chick within the egg "learns" the movement coordination of pecking at food: The head of the 3-day-old embryo rests on the heart and is at first passively raised and lowered. At the same time the yolk sac is said to provide tactile stimulation of the head, because it is moved by amnion contractions that are synchronous with the heart beat. One day later the embryo bends its head actively when touched and opens and closes its beak. During these movements liquid is pressed into the mouth, which is swallowed from the tenth day on. In this way the initially isolated movements of swallowing, nodding, and beak opening become integrated into a stereotyped behavior pattern – as Kuo sees it, by "experience." The chick is able to peck at food immediately after hatching.

In response to these observations of Kuo, K. Lorenz (1961) raised the question why other species of birds, which have similar experiences with their heartbeat within the egg, do not peck but gape, while other species, ducks, for example, strain the mud for food and still others, such as doves, put their bill into the mouth of their parents.

R. A. Hinde (1966:327) accepts Kuo's interpretation up to a certain degree and accuses us of making fun of his hypotheses. He writes: "Such suggestions have been greeted with ridicule by a number of writers (for example, K. Lorenz, 1961; I. Eibl-Eibesfeldt, 1961a; W. H. Thorpe, 1963), but it is difficult to see why."

We must reject this reproach, because we have admitted the possibility of such learning (for example, I. Eibl-Eibesfeldt, 1963:706). Kuo's example, however, was rejected as a speculation that was not experimentally supported, especially because it has long been known (W. Preyer, 1885) that at the time the thorax is able to move the chick's head passively, the connection between sensory and motor neurons in the spinal cord is not yet made. This point escaped Kuo, as well as all others who uncritically quoted him in support of their hypotheses. It is further known as a result of the careful investigations of V. Hamburger (1963), R. Oppenheim (1966) and V. Hamburger, R. E. Wenger, and R. Oppenheim (1966) that this tactile self-stimulation is irrelevant for the development of the behavior. Thus, as a result of removal of the amniotic membrane there is no change in activity, and the hypotheses of Kuo are disproved.[2]

More recent investigations on monkeys (three rhesus monkeys, one baboon), deafferented on both forelimbs within four hours of birth, revealed that ambulation, climbing, and reaching developed spontaneously, evidently without the contribution of somatosensory feedback. Two rhesus monkeys that in addition to deafferentation of the forelimbs were blinded by sewing the eyelids together developed the patterns too,

[2] In a later publication Kuo denied that he ever believed in learning within the egg. He emphasized, instead, that self-stimulation processes undoubtedly play an important role in the ontogeny of behavior of the bird embryo, in line with the sentiment stated above, and that pecking has its behavioral precursors before actual hatching (Z. Y. Kuo, 1967:50, 108).

with a delay of only one to two weeks. "During infancy a great deal of motor learning takes place, but the basic programs for many patterns of movements do seem to be present in the primate central nervous system at birth" (E. Taub, P. Perella, and G. Barro, 1973:960).

These findings, of course, do not mean that in other species and at different stages of development animals may not learn even from processes of self-stimulation. However, even if this could be shown in a specific instance, it would by no means do away with the concept of innate behavior. The important fact that requires an explanation is the adaptation of behavior patterns to certain environmental situations. Behavior is molded toward these situations so as to fit or even duplicate models. As K. Lorenz (1961, 1965b) has explained, such an adaptation or copy can come about only when the organism has at some time obtained information about such environmental contingencies. When bird A and bird B sing the same song or when two people recite the same poem, concurrence by chance can be excluded. We have to assume, then, that the organisms somehow acquired information about the pattern which they copy or otherwise depict. A specific adaptation requires interaction with a specific pattern. The acquisition of information, for example, can take place in the course of the development of the young animal through an active dialogue with its environment, and we know today that some adaptations in animals are the result of traditions passed on by some model. In all these instances the experience that has been gained is stored in the central nervous system of the individual.

A specific adaptation, however, is quite often the result of phylogenetic evolution. In that case the species has come to terms with its environment. Natural selection is the "teacher," so to speak, and the acquired "experience" has been preserved in the genome of the species and become decoded during ontogenesis.

The expression "acquisition of information" as applied to phylogenetic adaptation may seem strange to those who think of the chance events that are involved in increasing the probability of survival for an animal. It is at first difficult to speak of acquisition of information when a change in phenotype has come about through the loss of a part of a chromosome, which then results in adaptiveness. The individual in question did not, in fact, acquire information about its environment. If one looks at the level of the species, however, we gain a new perspective. When the more advantageous genome begins to increase in frequency within a population, the increased fitness at the species level is comparable to an acquisition of information, because from that moment on an interaction with the environment begins by means of natural selection.

That adaptation always presupposes such an interaction has not been understood by those critics who have been concerned with these questions (for example, A. D. Blest, 1966).

The path by which a particular adaptation came about can be discovered with the aid of a deprivation experiment. All that needs to be done is to withhold from a growing animal a specific kind of information to which the behavior pattern being investigated is adapted. If the subject still shows appropriate, adaptive behavior, then we know that this specific adaptativeness is the result of phylogenetic evolution.

Such behavior is adaptive as a result of inheritance, as distinguished from adaptive modifications of behavior that are acquired. If the phylogenetic adaptation is on the motor level, we are dealing with fixed action patterns. It may, however, also lie on the receptor side as a selective stimulus filter, by means of which an animal reacts with specific behavior patterns to specific **key stimuli** or stimulus configurations prior to all necessary experience. In addition, there are **innate dispositions to learn** and motivating mechanisms that result from phylogenetic adaptations, causing the animal to act as a result of inner drives. Instead of using the term "phylogenetically adapted," we often say innate or "instinctive."

L. Carmichael (1926, 1927, 1928) raised tadpoles under permanent narcosis (acetone chloroform) until such time as control animals could swim well. When he withdrew the narcotic, the experimental animals swam equally well, although they had been unable to practice. When A. Fromme (1941) repeated these experiments, the experimental animals did not swim as well as the controls, but were nevertheless able to do so.

J. Grohmann (1939) raised pigeons in cages so small that they were unable to flap their wings. He released them when their control siblings were able to fly. In spite of having been prevented from practicing they flew well. The classical experiment of **D. A. Spalding** (1873) with swallows has already been mentioned.

In these instances the experiments showed that the behavior in question is present as a phylogenetic adaptation, probably in the form of a coordinating central mechanism. This can also be verified for more complex behavior sequences.

The squirrel *Sciurus vulgaris* L. buries nuts in the ground each fall, employing a stereotyped sequence of movements. It picks a nut, climbs down to the ground, and searches for a place at the bottom of a tree trunk or a large boulder. At the base of this conspicuous landmark it will scratch a hole by means of alternating movements of the forelimbs and place the nut in it. Then the nut is rammed into place with rapid thrusts of its snout, covered with dirt by sweeping motions, and tamped down with the forepaws.

One cannot decide by observation the degrees to which these behaviors are innate or acquired. However, one can easily withhold the relevant information a squirrel would require in order to learn how to collect its winter stores. The animal is hand raised in isolation with liquid food, and placed in a cage with a bare floor. It can neither observe another squirrel burying nuts nor can it practice the burying of food. In addition, it never experiences times of starvation; thus it is unable to learn that food hidden by chance can be useful in times of need.

If an animal so raised is tested when fully grown, one finds that it masters the complete hiding sequence on the first attempt. If one presents it with nuts, it will eat some first. Upon satiation, additional nuts are not dropped but are carried about in the mouth as if in search of something. Vertical structures seem to attract the squirrel. At a corner in a room, perhaps, the squirrel will deposit the nut, push it into the corner with its snout, and finally make the covering and tamping-down movements with the front legs, although it has not dug a hole (I. Eibl-Eibesfeldt, 1963). The entire behavior sequence therefore is preprogrammed as phylogenetic adaptation. Shrikes (*Lanius collurio*) remove the stings from hymenoptera before eating them, and this knowledge is just as innate as the recognition of which bees and wasps sting (E. Gwinner, 1961).

The ability of sexually mature ring doves (*Streptopelia roseogrisea*) to feed newly hatched squabs in a species-typical manner is not acquired individually by chance learning but is innate (E. Klinghammer and E. H. Hess, 1964), and in the same way **honeybees** will perform the complex **waggle dance** without the need to learn the code. Young bees that have been isolated in groups shortly after hatching develop the dance in seven days (M. Lindauer, 1952). The spider *Aranea diademata* will spin threads without a recognizable pattern shortly after hatching; after the first molt, however, it produces its artistic net. It can be shown that this behavior is not learned by placing the spiders into small glass tubes in which they can just turn around but cannot string threads. When released after their first molt they will spin their nets just as perfectly as previously unconfined conspecifics (G. Mayer, 1952).

The innateness of cricket song patterns was demonstrated by D. R. Bentley (1971), who conducted experiments in cross-breeding and also raised crickets under varying conditions.

> Despite being raised under different conditions of temperature, diet, light cycle, time of the year, and population density, individuals always produced calling patterns corresponding to genotype. The "correct" song for a genotype was produced even if an animal was the first of its type to mature and therefore had heard many "incorrect" songs, but none of its own. Individuals with different genotypes produced different song patterns even if raised under nearly identical environmental conditions. . . . The ultimate source of information for this programming network appears to be genetic. . . . (p. 1139).

Good examples of phylogenetically adapted behavior patterns are the songs of some birds. In these behavior patterns, serving the purpose of communication, it is clear that a complex store of information must be acquired to produce the appropriate song. This information can be acquired by listening to a sample of the song. Theoretically it would also be possible for the correct song to be rewarded by the appropriate behavior of a respondant. To date no such instance is known. If these two possibilities are excluded for the moment and a bird still produces its species-typical song, then the conclusion is unavoidable that a phylogenetic adaptation is present.

F. Sauer (1954) raised whitethroats (*Sylvia communis*) singly in soundproof chambers. They still developed all 25 species-typical songs. M. Konishi (1963) deafened chickens by surgical procedures. They could still produce the species-specific calls. Doves also produce their calls innately. Oregon juncos (*Junco oregonus* and *Junco phaeonotus*) and the blackheaded grosbeak (*Pheucticus melanocephalus*) sing in species-typical fashion provided they can hear themselves, even if raised in isolation (M. Konishi, 1964, 1965a). Sound-isolated chaffinches (*Fringilla coelebs*) develop a song that is similar to the species song in the number of syllables and total length but that lacks the characteristic patterning into three stanzas. This must be learned, but the animals possess an innate knowledge about which song to imitate. If offered tape recordings of several bird songs, they imitate only those that resemble their species song in tone quality and form of strophes. The sequence of the strophes is not preprogrammed, because species-typical songs with experimentally reversed strophes are imitated (W. H. Thorpe, 1954, 1958a, 1958b, 1961). These last two examples are especially interesting, because they show that the phylogenetic adaptation is not always present as a fixed action pattern. It may also be present as a specific "learning schedule"—here the innate knowledge of the song of the particular species (see also **imprinting-like learning schedules**).

In the bullfinch this learning schedule consists of the young birds' habit of imitating only their male parent. J. Nicolai (1959a) once had a bullfinch male raised by canaries. This bird sang like a canary and passed on this song to its young, and these in turn passed it on to their young. The widow birds (Viduinae), who are breeding parasites of the grass finches (Estrildidae), in whose nests they lay their eggs, possess both an innate territory song and an acquired courtship song which they learn from their foster-parent species. They imitate the latter so perfectly that the imitation cannot be distinguished from the original (J. Nicolai, 1964; see also Fig. 3-3).

A behavior sequence usually consists of parts, which in turn can be broken down into simpler **functional units,** so one must always specify the level of integration at which one is operating. Thus, an animal could possess a learning mechanism as a phylogenetic adaptation by means of which the coordination of two antagonistic muscles is learned. These learned units, however, could be integrated into new functional units on the basis of an innate program. In this case behavior at the higher level of

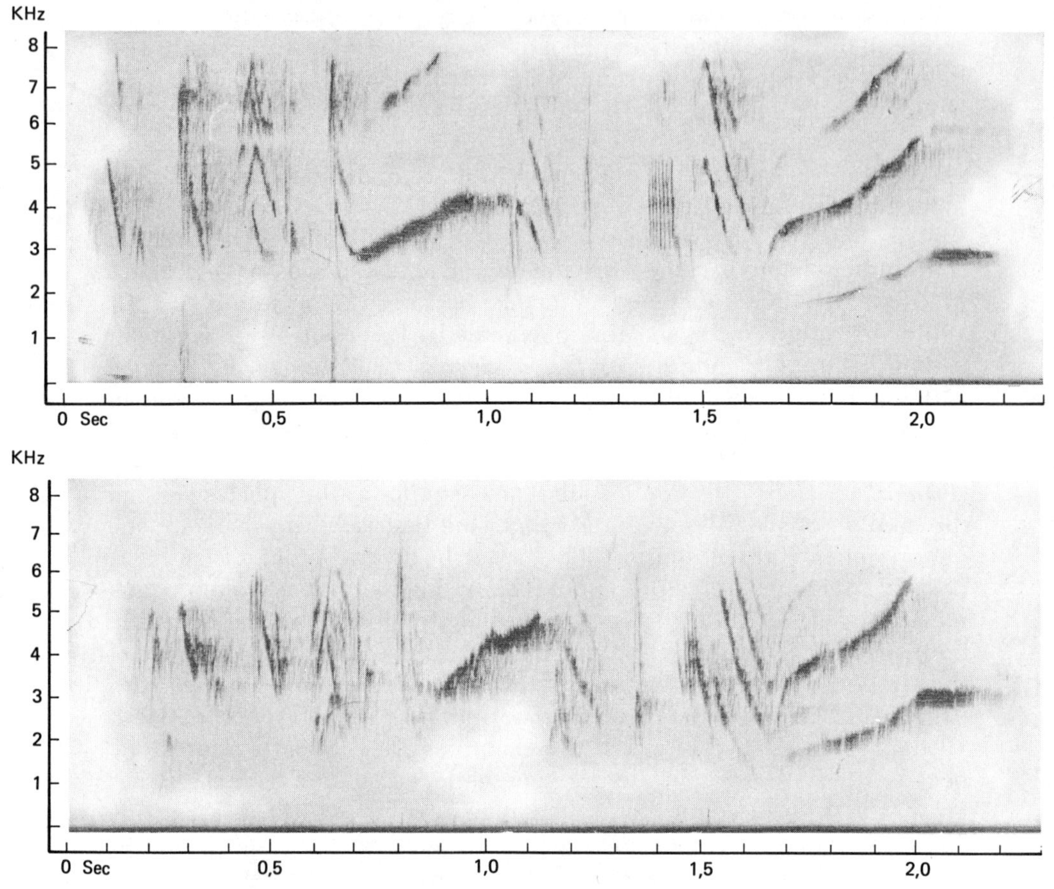

Figure 3-3 *Top:* Greeting song of the female estrildid finch *(Granatina granatina), bottom:* imitation of this song by a male widow bird *(Tetraenura regia).* (From J. Nicolai, 1964.)

integration would be considered as phylogenetic adaptation. In this way, learning processes could theoretically serve as a means of decoding phylogenetically acquired information that has been stored in the genome.

The strengths as well as the limitations of the deprivation experiment have been discussed by K. Lorenz (1961) and stated in some specific rules that are presented here with both additions and deletions:

1. The question asked deals with the origins of the adaptiveness of a behavior, so it is a prerequisite for each deprivation experiment that the species-preserving function of a behavior be known. This requires a thorough knowledge of the biology of the experimental animal.
2. The experiments must be set up in such a manner that only the adaptation to be investigated is disrupted (I. Eibl-Eibesfeldt and W. Wickler, 1962). If one

wants to know whether a particular strophe of a bird's song is inborn in its entirety, then one should prevent the bird only from hearing this particular strophe. If one wants to assess the ability of an organism to respond to optical stimuli or stimulus configurations, then the animal should not be reared in complete darkness, because this may lead to atrophy of the retina (A. H. Riesen, 1960) and to the disruption of all visual reactions. In other words, one must be aware that there are not only various levels of behavioral integration, but also levels of adaptation, and that during ontogenesis a diversity of factors may influence later behavior. Rats that were exposed during the first ten days of life to mechanical, electrical, and cold stimuli later showed not only increased resistance to the effects of food deprivation and cold stress, but they also learned faster (J. P. Scott, 1962). In this connection the experimental hypotheses are frequently stated too broadly and generally. Thus B. F. Riess (1954) asked whether the nest-building behavior in its entirety is inborn or learned. With such complex behavior sequences it is preferable to work at a lower level of integration. In reply to R. A. Hinde's (1966) recently repeated argument that one can never raise an animal devoid of all experience, it can be said that such an attempt would miss the point. The experimental design must be at the appropriate level.

3. The deprivation experiment informs us only about what does not have to be learned. Although we should always strive to disrupt only a particular adaptation, as stated in rule 2, it is not always possible to prevent a disruption of the total condition of an animal. We know that spontaneously occurring fixed action patterns are reduced in intensity, perhaps because the endogenous energy production is disrupted. This results in an increased threshold of responsiveness to the releasing stimuli. Furthermore, **innate releasing mechanisms** frequently lose their selectivity, and social inhibitions are often lost. As an example of such disruptions K. Lorenz reported that his hand-reared red-backed shrikes showed the movement of impaling prey but had to learn to direct this movement toward thorns. He cited this as an example of an instinct–learning intercalation until G. Kramer demonstrated that shrikes which had been raised on a more adequate diet aimed for the thorns innately. This has since been verified in additional experiments by K. Lorenz and U. v. Saint Paul (1968).
4. In the testing situation the experimental animal must be presented with all the relevant releasing stimuli for the behavior. This dictum was overlooked by B. F. Riess (1954) in his tests for nest-building behavior when he placed inexperienced rats in a strange experimental cage from whose walls paper strips were hanging. The rats, which had until then lived isolated in wire cages without nesting material and powdered food, did pull down the paper strips, but instead of building nests they scattered them. Riess concluded from their behavior that nest building must be learned as a result of experience with solid objects. They would discover that nest material gathered by chance protects against cold, and based on this experience they would then build nests. He overlooked the fact that even experienced rats will not build at first in a strange environment but will explore instead. Rats I (1963) raised according to Riess's method, but that were tested in their home cages, built nests in their sleeping corner and showed all the species-typical nest-

Figure 3-4 Nest building by a naive female rat: *a*, laying-back movement; *b*, pulling in; *c* and *d*, splitting. (From Scientific Film B757, I. Eibl-Eibesfeldt, 1958b.)

building movements (Figs. 3-4 and 3-5). The rat must therefore first have a nesting place before it can appropriately deposit material. A later experiment by F. Wehmer (1965) has confirmed this.

That even skilled and experienced experimenters may commit such errors is shown by the work of D. S. Lehrman (1955), who concluded on the basis of his experiments that ring doves do not recognize or feed their young innately but must learn to do so. He explained that they at first look down at their newly hatched young, which move under them and then touch the crop region of the parent. Under the influence of the hormone prolactin, the crop is swollen with crop milk and becomes sensitive. If the young succeed by chance in causing the vomiting of crop milk, the behavior of the parent would be reinforced and it would recognize the young as a stimulus leading to a reduction of tension in the crop. As a result parents would actively approach the young. Experiments seem to support this view: Of 12 experienced ring doves injected with prolactin, 10 fed the 7-day-old squabs. On the other hand, 12 similarly treated inexperienced ring doves did not feed young of the same age. E. Klinghammer and E. H. Hess (1964) obtained conflicting results when they placed newborn mourning doves under incubating experienced and inexperienced ring doves. All fed the young; inexperienced doves, who had as yet no crop milk in the crop, provided instead a clear liquid. Lehrman's 7-day-old squabs were already well feathered, and Klinghammer and Hess suggest that the down of the newly hatched young provided the more appropriate releaser for parental feeding behavior. This suggests that doves recognize and feed newly hatched young innately. Lehrman's conclusions are then limited to the statement that only experienced ring doves recognize 7-day-old squabs as young, and he has since accepted this interpretation.

5. In view of the fact that results obtained while working with one species are often checked with another species, K. Lorenz emphasizes the almost axiomatic rule that agreements in experimental results with respect to inherited behavior patterns can only be expected when genetically similar animals are used. If this precaution is taken, the deprivation experiment is the appropriate way to demonstrate phylogenetic adaptations in behavior. We often use the terms "inborn," "innate," or "instinctive" instead of "phylogenetically adapted" as a convenient shorthand description, even in cases in which the behavior matures during the course of development of the young animal. To

Figure 3-5 Nest building by naive female rats with crepe paper strips: *a*, pulling paper from the holder; *b*, pulling the paper in during nest building; *c* and *d*, one sleeping nest each built by naive female rats: *c* made from split straw; *d* made from crepe paper strips. (Parts *a* and *b* from Scientific Film B757; photographs: I. Eibl-Eibesfeldt, 1958b.)

be exact, it is not a certain behavior that is inborn but the developmental "blueprint," the norm of a reaction. Characteristics in themselves are not inherited but are developed within the bounds of inherited variations. The term "inborn" has frequently been understood only in a negative sense as "unlearned" (D. O. Hebb, 1953; R. A. Hinde, 1966). In line with our discussion it should be clear that we define the term on the basis of the origin of the adaptation.

It should also be emphasized that when ethologists claim that the development of a particular behavior does not require certain experiential influences, they certainly do not mean to imply that no experience at all is necessary. Occasionally we come upon authors who impute this view to ethologists (J. P. Kruijt, 1971). We undoubtedly owe our critics (especially D. S. Lehrman) a debt of gratitude for having compelled ethologists to clarify these concepts. In addition, fruitful discussions and a narrowing of the gap between points of view have taken place. Even T. C. Schneirla (1965, 1966), who argues against the possibility that inborn and acquired components of behavior can be distinguished, makes certain concessions. "A concept of interactionism, therefore," he writes, "that implies engagements of separate and disjunctive developmental entities seems invalid, *at least for the early stages*" (1966:283, italics mine). On the other hand, he (1965) feels compelled to distinguish "maturation" and "experience" during the embryonic development of behavior. But he does not look for the source of the developmental "blueprint." More recently he emphasized the point, as did R. A. Hinde (1966), that embryonic development ought to be studied and that one can then discover how environmental influences (always described as "experience") affect behavior during each stage of development. A similar position is taken by G. Gottlieb and Z. Y. Kuo (1965), who move away from an extreme environmentalist position when they propose the term "self-stimulation" instead of "experience." In that case all that remains of the old environmentalist position is the statement that self-stimulation is a factor in behavior development. How this position is incompatible with the concept of the inborn, as Gottlieb and Kuo still believe, is difficult to understand. Perhaps they did not understand us. No ethologist has ever asserted that diverse environmental influences, especially those of the internal environment or self-stimulation processes during embryonal development, play no role during development.

After all, we know from the pioneer work of H. Spemann in experimental embryology that organic substances secreted from tissues of newt embryos stimulate neighboring tissues to develop into specific organs. Thus the eye cup induces lens formation in the vertebrate epidermis above it. If one transplants the eye cup of a newt embryo into the ventral region, then lens formation is stimulated in the new location. In this manner genetically encoded information becomes decoded during the process of self-differentiation. This is also valid for the embryology of phylogenetically adapted behavior. In both cases one can activate by specific factors such as temperature or chemical stimulation the potentials provided by the specific genetic code. Thus, monkey and rat females can show irreversible male behavior characteristics (for example, increased aggressive behavior) if they are given male hormones during a sensitive phase of embryonal or early development (G.

W. Harris, 1964; W. C. Young, 1965). In rats the sensitive period, in which hormone treatment can exert this sex-reversing influence, ends one day after birth. Females that were treated with testosterone before this time showed no normal sexual behavior even after they were castrated and treated with estrogen and progesterone, which causes estrus in normal females. The hormonal influences then fixate a male (or in the reverse case the female) role during a sensitive phase, and later in life sexual hormones primarily activate the sexual behavior patterns preformed in the sensitive period.

Whether the cyclic feminine or the acyclic masculine pattern of gonadotropin secretion during the reproductive life of a mammal is manifested depends on the presence or absence of androgenic hormones during a sensitive period in development, which in most species is prior to birth. Androgen influence in female rats during that period leads to permanent suppression of cyclic secretion. On the other hand, the periodic pattern of gonadotropin secretion can be induced in male rats by antiandrogen treatment during the mother's pregnancy and immediately after birth (M. D. Neumann and H. Steinbeck, 1972).

Those who desire a better understanding of these relationships will find A. Kühn (1955) very helpful. The same phenotypic expression can come about in several ways and no one can deduce by appearance alone the course that development has taken. It has been possible, however, to distinguish the times and stages during which inherited and environmental factors act upon developmental events. The same is true for the development of behavior. Specific neural structures that underlie a behavior develop like other organs on the basis of a developmental code contained in the genome.

On this point D. S. Lehrman (1970:34) suggests that such a formulation contributes little to an understanding of the process and that one must approach the issue from a developmental point of view:

> Now, it may be comforting, in the sense that it gives us the feeling that we have increased our understanding of the problem, to say that a behaviour pattern (or a structure) is innate if it is "blueprinted in the genome" or, in a more modern vernacular, "encoded in the DNA." There are, of course, contexts in which such expressions are meaningful, but I believe that the comfort and satisfaction gained from disposing of the problem of ontogenetic development by the use of such concepts are misleading, and are based upon the evasion or dismissal of the most difficult and interesting problems of development.

Lehrman acknowledges our formulation of the problem and attempts merely to reduce its significance. On the other hand, Z. Y. Kuo (1967, 1970) rejects any kind of evolutionary interpretation. He denies the existence of species-specific behavior that would involve the concept of species survival value and represent the developmental results of natural selection. After all, he argues, in no animal does one stride exactly mirror another, and behavior and environment are in all cases quite variable.

But ethologists have never insisted on an absolute rigidity of the inherited movement patterns, merely on a constancy of form. Admittedly, the English translation of the term *Erbkoordination* as "fixed action pattern" is rather unfortunate, for it leads all too easily to misunderstanding. According to Kuo (1967), morphological structures determine an animal's behavior only in a negative, limiting sense:

> Morphological structures and their functional capabilities act as determining factors of behavior only in a negative way, that is, they merely set a limit to certain body movements (for example, a dog can only snarl at or bite its enemy but cannot throw a stone at him) (Z. Y. Kuo 1967:13).

In another place he writes:

> There are to be found some common factors in behavior such as those due to some common morphological characteristics of the species. For example, morphological structures of the limbs determine the modes of locomotion; the oral structure determines the modes of eating and drinking; the vocal apparatus determines the characteristics of voice and singing (p. 23).

He is even led to make such statements as:

> The fact that the human hand has a far greater flexibility in movement, dexterity and range of potential capacities than those of any other primate is sufficient, in our view, to explain why human beings became the most creative and the most resourceful creatures on earth even long before human language was developed. Some primates are almost human. But not quite. The hands tell the difference. I often speculate that if we could succeed in exchanging brains between a human neonate and a gorilla neonate and raise them in an identical environment with complete absence of human language and culture, the human child would grow up to behave with human characteristics and the gorilla with the characteristics of its own species because the skeletal framework of the body and the fine structures of the hands of the two different species are different (p. 188).

On page 195 Kuo writes:

> If the species known as *Homo sapiens* is so far superior to all the other species throughout the animal kingdom, it is not because it has a human brain per se, but because it possesses a pair of human hands and because the human vocal mechanisms have developed a most complex spoken and written language.

Phylogenetic adaptations that preprogram behavior apparently cannot exist in the central nervous system, according to Kuo. We would not deal with this unscientific speculation in such detail had it not appeared in publications that are widely read by students.

In a strangely self-contradictory manner Hailman (1967) refers to himself as an interactionist. Although he rejects the innate–learned dichotomy, he speaks of an interaction between the two components. Yet he does not believe it possible for these components to be distinguished in any way. Charlesworth (on press) has attempted to deal with this line of thought: He refers to an interesting difficulty that confronts those who speak of interactions and immediately insist on the fruitlessness of the nature–nurture problem. This type of self-contradiction is similar to that practiced by the individual Freud mentioned in his analysis of humor: "This person declared that he did not believe in ghosts, yet was quick to add that neither did he fear them. Thus, whoever rejects the innate–learned dichotomy and at the same time offers the solution of an interactionism displays a remarkable conceptual confusion."

We only need to ask Hailman, when he speaks of continual interaction, what it is that interacts—and once again we face the task of dealing with the problem: What influence is exerted by heredity, what by environment? Sometimes the objection is raised that while in theory it is possible to distinguish between innate and learned components, in practice this would be of little value, since at best only rare and extreme cases could be assigned to one or the other category; intermediate cases would in reality constitute the majority. This statement, however, is ill-founded. If we consider the courtship behavior of the mallard duck, for example, we find an array of highly specific, innate courtship movements, but not a single learned one or one that is substantially modified through learning. In many instances there seems to have been a strong selection for resistance to modifiability.

This is true, for example, for **expressive movements,** which when performed with typical intensity improve their signaling function by becoming less easily mistaken. The studies on the goldeneye ducks (*Bucephala clangula*) by B. Dane and W. G. van der Kloot (1964) revealed very little variability in the performance of most courtship displays, and recently R. H. Wiley (1973) has described even more stereotyped courtship movements in the male sage grouse (*Centrocercus urophasianus*).

Occasionally we hear the argument that only when differences are found in two individuals raised under identical conditions are we allowed to attribute these differences to differences in the genetic outfit. We agree that such an experiment proves genetic determination of the differences. However, we do not agree with the statement that this is the only way of proving genetic determination. The deprivation experiment certainly is another way of proving it. And certainly the often used phrase "only differences are inborn" must be considered as loose, since "differences" are an abstract.

PHYSIOLOGICAL CHARACTERISTICS OF THE FIXED ACTION PATTERN

For a long time the concept of the classical reflex influenced our thinking on the nature of a movement sequence. According to this concept each act is a response to external or internal stimuli. Afferent nerve endings are stimulated and pass the excitation on to the central nervous system. From there the excitation is carried, often via intermediate neurons, to an effector neuron that in turn excites efferent pathways leading to an effector, a muscle or a gland, which is then activated. This chain of excitation is called a *reflex arc,* and the process is called a *reflex.* In monosynaptic reflexes of mammals the wave of excitation is said to pass directly from the sensory to the motor neuron. The excitations activate the same muscle from whose proprioceptor they have been released. All other reflexes transverse additional intermediate neurons, and the excitation of many neurons can activate a specific organ, just as the excitation of a few receptors can activate many organs.

More complex movements are called *chain reflexes.* In these one reflex provides the stimulation for the release of the next. The release of one reflex may inhibit or facilitate others. For each reflex arc there are inherited "unconditioned" releasing stimuli. By means of learning processes new stimuli can become **conditioned stimuli** or releasers, or a given stimulus can become linked to new reaction sequences. These processes of stimulus and response selection frequently occur in combinations, and these newly acquired reaction sequences are called *conditioned reflexes.* The participation of such reflex processes in the structure of behavior cannot be denied. However, it is not true that each movement is the result of an afferent impulse. T. Graham Brown (1911, 1912) advanced the theory that the quadrupedal walk is a central mechanism after he had discovered that two completely deafferented antagonistic leg muscles of a cat showed rhythmic movement. E. v. Holst (1935, 1936) then demonstrated in a number of experiments that an inborn movement sequence can be centrally coordinated without participation of afferent stimulation.

According to the classical reflex theory the regular undulating movement of an eel occurs by the participation of internal sensory organs, the proprioceptors. The contraction of one muscle segment is said to release the contraction of the adjacent segments via these proprioceptors. If this were true, then an eel whose central nervous system does not receive impulses from the periphery would no longer show undulating movements. E. v. Holst showed that this is not the case. If one separates the spinal cord of an eel from the brain, by a cut behind the head, one obtains a spinal preparation that can be kept alive for a while by artificial respiration. If one now cuts all the dorsal roots of the spinal cord, which alone are capable of transmitting impulses from the sensory organs to the spinal cord, the eel will show undulations once the operative shock has subsided. A purely mechanical transmission of the undulating movement is also excluded. If the central third of the eel's body is mechanically restrained so that it can no longer move, an undulating movement starting in the first segment will appear in the posterior third after the same amount of time that it would take if the central segment had participated in the movement.

These experiments prove, first, that there is an endogenous production of excitatory potential in the central nervous system and, second, that these central impulses are also centrally coordinated. Such central automatisms also seem to be the basis of the respiratory movements of the gill covers in goldfish. E. D. Adrian and F. J. J. Buytendijk (1931) recorded rhythmic impulse patterns from the isolated respiratory center in the medulla, which corresponded to the frequency of movement of the gill covers. Rhythmic electrical impulses of the normal crawling rhythm of the earthworm were also found by E. v. Holst in the isolated ventral cord. The movement rhythm is apparently centrally produced, but proprioceptors also participate. It is generally known that the severed posterior portion of an earthworm continues to wriggle, while the anterior part proceeds to move forward normally. If the two severed edges are connected by means of two threads, then the rear part will follow the front part in the typical crawling rhythm (E. v. Holst, 1932, 1933).

In grasshoppers, peripheral feedback affects only the frequency of the otherwise central rhythm of the flying movements. If receptor impulses are excluded, the only result is the somewhat slower rate of rhythmic discharge of the thoracic ganglion. The resulting pattern of wing movements corresponds, however, to the normal wing beats (D. M. Wilson, 1961, 1964). It is also unnecessary to postulate adaptive reflexes to explain changes in the gait of insects after amputation of a leg (G. Wendler, 1965; D. M. Wilson, 1966). In intact insects, however, the system is much more influenced by peripheral reafferences: They control the phase relations between the six walking legs (E. v. Holst, 1943; G. Wendler, 1964, 1965; D. M. Wilson, 1968). This is clearly shown by the alternation of walking rhythms in leg-amputated arthropods, but we do not yet know the way in which reafferent signals influence the phase angle of the legs.

As we already know, the walking movements of an insect's legs are caused by a system of six self-sustained, mutually coupled oscillators. In his 1968 experiments Wendler tried to find the way the oscillators work on each other to produce the observed coordination of leg movement in *Carausius* and also the alternations of walking rhythm after amputation of legs. One hypothesis is shown in Fig. 3-6. The system consists of six oscillators of nearly the same spontaneous frequency. Oscillators of each segment show mutual influence on each other, while in those of ipsilateral legs—for instance, the hind and middle leg of one side—the influence travels from back to front. A nonrhythmic signal from the central nervous system controls the speed of walking. As Wendler showed by means of an analog computer model, the system holds for all observed phase relationships of an animal's legs. It also provides a framework for several hypotheses of leg-receptor influence on coordination: Leg receptors could measure the leg movement and directly influence the neighboring oscillator, thus determining the phase. An alternative hypothesis postulates central coupling of all oscillating systems, without the direct influence of afferent systems (receptors). In this case, receptor influence would be restricted to keeping the amplitude of the oscillator of the same leg high enough to keep the dependent oscillators in phase. This hypothesis receives support from experiments with partially leg-amputated stick insects whose leg stumps move with lesser amplitude than the intact legs (G. Wendler, 1965).

K. D. Roeder (1935, 1937) concluded that an endogenous automatism forms the basis of the mating behavior and locomotion of the praying mantis. Upon removal of the supra- or subesophagal ganglion, locomotor and mating behavior were disinhibited. The two behavior patterns occurred continuously, although releasing stimuli are normally required. Roeder postulated the existence of endogenous, self-activating systems that are responsible for the coordinated movements and are controlled by inhibitory centers. Additional experiments have supported this interpretation (K. D. Roeder, 1963a).

Endogenous activity of the central nervous system has also been demonstrated by P. Weiss (1941a), who implanted a section of embryonic spinal cord and a forelimb anlage into intact axolotls. The developing forelimb then became innervated from the implanted spinal cord, and the motor nerves grew faster than did the sensory nerves, which reached the limb much later. However, as soon as the efferent motor connection had become established the leg began to move. Although it did not show coordinated walking movements, an alternation between agonistic and antagonistic muscles could be observed in the irregular movement.

In the neural elements underlying spontaneous and reflex behavior only slight differences exist in their thresholds of excitability (K. D. Roeder, 1955). In nonautomatic cells the excitation remains at a constant resting potential, and a stimulus is necessary to elevate it above the

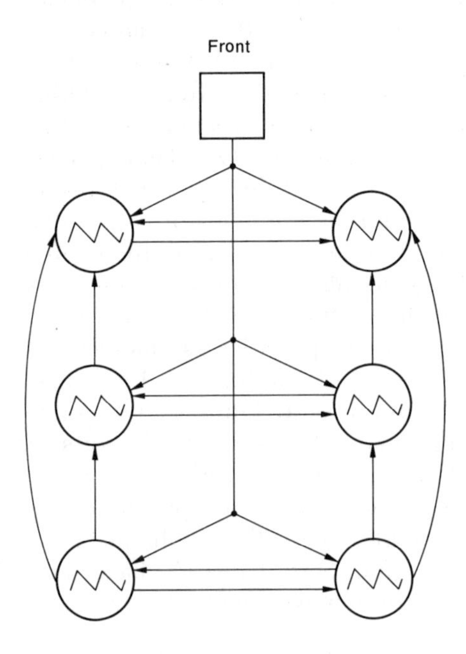

Figure 3-6 Diagram of a system of six self-sustained, mutually coupled oscillators (represented by circles). Oscillators belonging to the same segment (for example, hind legs) influence each other mutually, those of ipsilateral legs (for example, hind and middle leg of one side) in a direction from back to front. (Additional explanation in the text.) (After G. Wendler, 1968.)

threshold. After discharge it drops to zero but increases again until it rises above the level of the resting potential and then returns to it. For spontaneous elements, however, the readiness for discharge increases until the threshold has been reached, whereupon spontaneous discharge occurs. Between these two extremes there are intermediary cases. When the resting potential and discharge threshold are nearly equal a simple stimulus is able to release a sequence of discharges because the curve of excitability reaches the discharge threshold during the phase of increased irritability.

Chick embryos move their hind limbs spontaneously and rhythmically even when all sensory ganglia have been neutralized by the removal of the entire dorsal half of the spinal cord and when all influence of the brain is ruled out by the removal of a section of the spinal cord (Figs. 3-7–3-9). Turtles and fish will move before the reflex arcs have been closed. The oysterfish (*Opsanus tau*) hatches and is able to swim in this condition in a well-coordinated manner, before sensory input is possible (H. C. Tracy, 1926).

All these examples demonstrate the spontaneous, endogenous activity of motor neurons. In response to the objection that the neurons are still in an environment capable of influencing them, and that perhaps

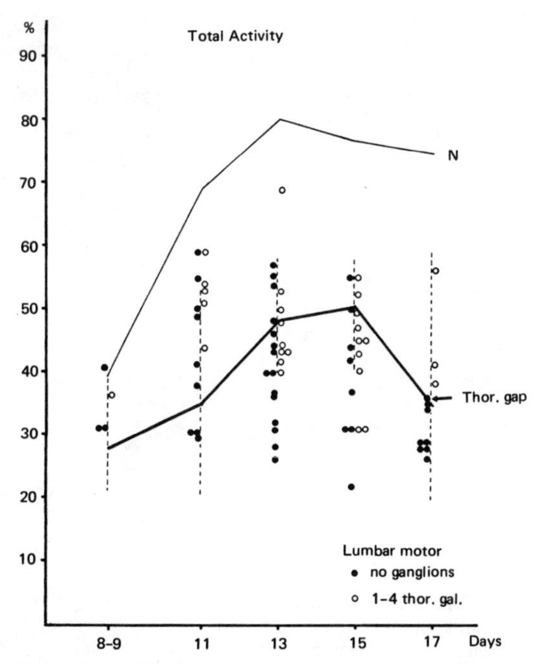

Figure 3-7 Activity during the standard observation period of 15 minutes. *N:* normal embryos. *Thor. gap:* embryos with thoracic section of the spinal cord (controls). The vertical dotted lines indicate the range for these controls. *Black dots:* embryos without any ganglia. *Circles* indicate embryos with several small ganglia in the lumbar region that do not innervate the extremity. (From V. Hamburger, R. E. Wenger, and R. Oppenheim, 1966.)

chemical stimuli from the blood could serve as releasing stimuli, Roeder (1963b:438) replies very aptly:

> In speaking of endogenous activity of the central nervous system I have used the term in the same sense—activity (in this case detected as nerve impulses) generated by mechanisms within the central nervous system. The criterion is that it continues to take place after all afferent nerve connections with the outside have been severed. Factors in the extracellular medium surrounding the nerve cell play an important part in determining whether a given nerve cell will remain inactive until stimulated or will regularly discharge impulses without stimulation. Nevertheless, the coupled regenerative system responsible for the sequence of nerve impulses must be considered to reside in the neurons themselves, and it would be misleading to think of the ambient medium bathing the cells as providing stimuli equivalent to those that normally reach it from the outside via afferent impulses.

A similar statement is made by T. H. Bullock and G. A. Horridge (1965:314):

> The term *spontaneous* means repetitive change of state of neurons without

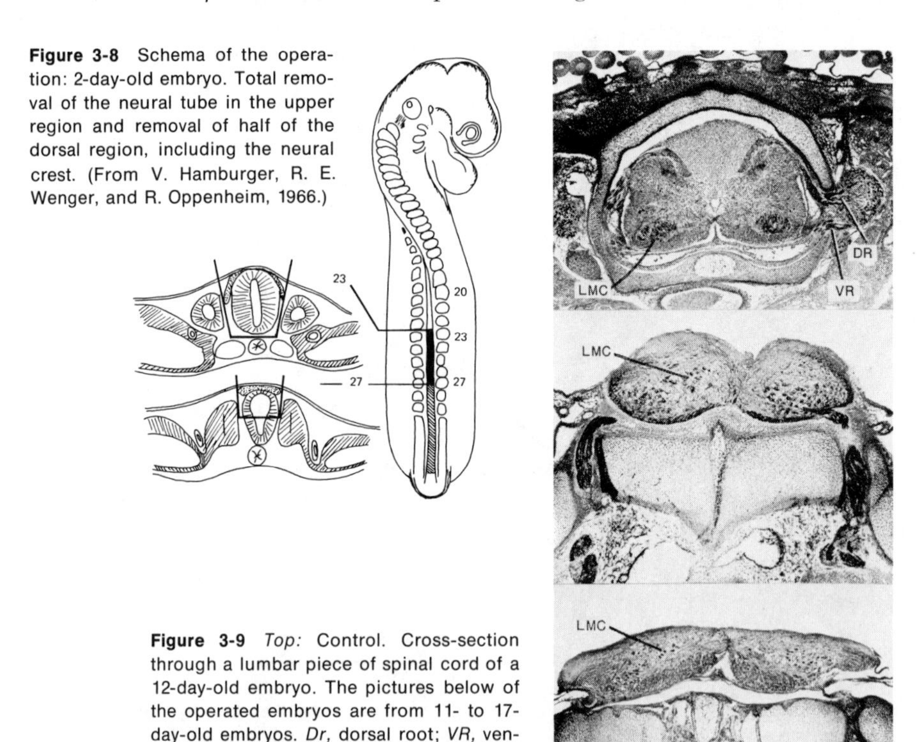

Figure 3-8 Schema of the operation: 2-day-old embryo. Total removal of the neural tube in the upper region and removal of half of the dorsal region, including the neural crest. (From V. Hamburger, R. E. Wenger, and R. Oppenheim, 1966.)

Figure 3-9 *Top:* Control. Cross-section through a lumbar piece of spinal cord of a 12-day-old embryo. The pictures below of the operated embryos are from 11- to 17-day-old embryos. *Dr,* dorsal root; *VR,* ventral root; *LMC,* lateral motor column. (From V. Hamburger, R. E. Wenger, and R. Oppenheim, 1966.)

change of state of the effective environment—that is, activity without stimulation other than the standing conditions. Of course the activity occurs only if many aspects of the milieu remain within certain limits—for example, the temperature and the ionic balance. These could be thought of as steady-state stimuli; but unless there is evidence of physiologically significant control of milieu, the term *stimulus* is not appropriate.

The fact that the oysterfish shows coordinated swimming before the reflex arcs are closed, and that a completely deafferented eel shows normal undulating movements, proves, in addition to the existence of a central nervous automatism (spontaneity), that the central coordination is independent of afferent impulses. E. v. Holst (1935, 1936) explains this by postulating that within the central nervous system there are groups of cells which are spontaneously active and which send their impulses to the musculature, if not prevented by inhibiting factors. These spontaneous cell groups influence one another, which results in certain specific movement coordinations. Holst demonstrated how these mutual influences work in fish that do not show undulating movements of the whole body but swim by means of rhythmically moving fins. He transected the medullas of these fish and provided artificial respiration; the fins were connected to recording pens (Fig. 3-10). When the operative shock had subsided, the fins showed rhythmic movements. If only one fin moved, a regular sine wave was obtained, but when several fins were moving this curve was more or less modified, which demonstrates the mutual influences of the rhythms. This must be a central influence because the passive movement of a resting fin did not affect the rhythm of another fin.

The fin rhythms may influence one another equally strongly. Frequently, however, one fin or pair of fins maintains its constant rhythm while the movement of the other fin changes rhythmically. In this instance a dominant, independent rhythm is superimposed on a dependent one. The influence of the dominant rhythm on the dependent rhythm can be discerned from the recordings (Fig. 3-11). Each fin has the tendency to maintain its own rhythm, but the dominant rhythm becomes superimposed upon the dependent one. If the dependent rhythm accelerates too fast it is slowed down, and if it lags behind it becomes accelerated. If the dominant rhythm is of sufficient strength it will superimpose its own rhythm completely (*absolute coordination*). If this is not possible, then the phase relationships of the rhythms change periodically in a lawful manner (*relative coordination*). An everyday example may serve to illustrate this. If we walk with our small daughter she tries to keep in step with us. Gradually she loses the synchronization, and the phase differential increases until the child corrects this by making a small jump that again brings her into phase. This attraction that two rhythms have for one another E. v. Holst calls the *magnet effect*. During *superposition,* finally, the dominant rhythm becomes superimposed on the dependent rhythm in an arithmetical relationship. Whenever the dependent fin moves in the same direction as the dominant one its amplitude becomes larger, and the reverse is true when their movements are opposite one another. This, then, is another way in which the independent rhythm can become imposed upon the dependent one and thus result in absolute coordination. Pure magnet effects or pure superposition are rarer than the intermediate forms.

In E. v. Holst's experiments the pectoral fin rhythm was always dominant over the rhythms of the dorsal and caudal fins. Their independent rhythms appeared only when the preparation emerged from operative shock or when the animal died. In the former case "coming to" moved from posterior to anterior, and death occurred in the opposite way, with the dominant rhythm dropping out first.

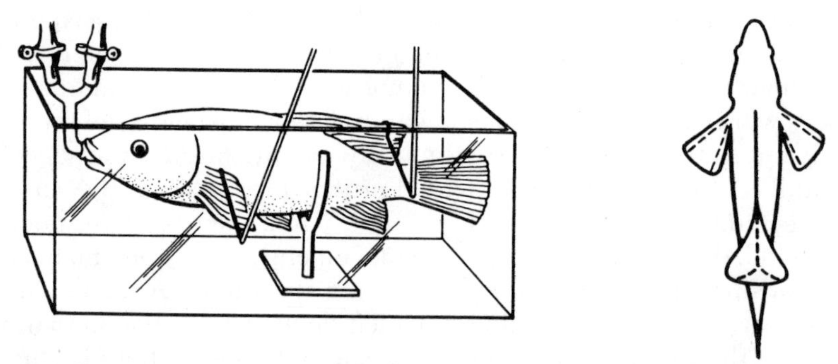

Figure 3-10 Experimental setup by E. v. Holst. The spinal lip fish receives artificial respiration. The fins are connected to recording pens. (After E. v. Holst, 1939.)

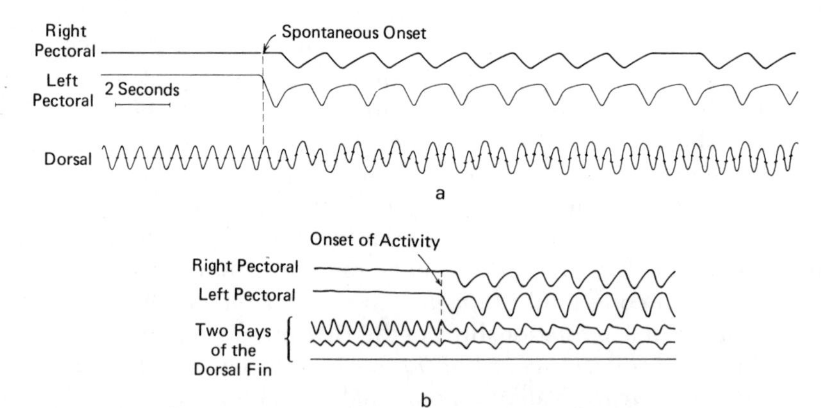

Figure 3-11 Recordings of fin movements of the spinal lip fish showing superposition and magnet effects. *a:* During the spontaneously appearing pectoral fin rhythms the rhythm of the dorsal fin, which until then had been regular, changes. This illustrates a purely central superposition. *b:* With the beginning of pectoral fin activity (dominant rhythm) the more rapid rhythm of the dorsal fin rays is slowed to coincide with the frequency of the pectoral fins. This illustrates the dependence of the dorsal fin rhythm and is an illustration of a strong magnet effect. (After E. v. Holst, 1938, 1939.)

What is true for the mutually influencing automatisms of the various fins is also true for each individual fin, which does not beat in the manner of a stiff board. Rather, wave movements pass over it, and each fin ray has its own automatism. These automatisms influence each other in such a way that the individual rays move in regular phase intervals. We can detect a hierarchical order of the automatisms. Each automatism can be broken down into subgroups, which in turn have a magnet effect upon one another, and the coupling is usually stronger or absolute at the lower levels of integration.

Theoretically there are a large number of possible interactions between various automatisms, which, however, do not occur randomly. Thus, E. v. Holst found several stages of stability in the phase relationships of different automatisms, which increase with the simplicity of the reciprocal frequency relations. The most stable is absolute coordination 1:1, followed by 1:2, 1:3, and 2:3 (or 1:2:2, 1:2:3, and 2:3:4 for three rhythms). For locomotion on land absolute coordination is undoubtedly the most efficient. The adaptive character of the various forms of relative coordination in water animals is not yet understood.

When their phase relationships are stable, the reciprocating automatisms form a transposable "gestalt," because a change in the frequency of one automatism influences the other in such a way that the original phase relationship is maintained. This is also true for stereotyped, learned movement sequences. Even if a person writes at different speeds, the same relationship is maintained for each letter, and the general character of the handwriting remains the same whether the person writes in capital or in lowercase letters. Even these learned movements are based on automatisms. Whereas in fixed action patterns the relationship between various automatisms is genetically programmed, in learned coordinations an initially unstable relationship between automatisms becomes fixed. The automatic groupings are led, in a manner of speaking, because of success into new patterns, whereby the automatic cells seek new stable relationships. The transition from the clumsy execution of a movement into a new and stable coordination occurs suddenly, as everyone knows who has learned to dance or to ski. This explains also why practicing parts of action patterns is not very useful. If one first learns the components until a fixed, automatic relation is established, these relations must be broken up and newly coordinated when a movement pattern of a higher integration is established. That this reordering can take place centrally without any aid from afferent input is shown by the experiments of E. Taub, S. J. Ellman, and A. J. Berman (1965). Their rhesus monkeys learned to grasp a cylinder with a deafferented hand from a fixed position without the aid of vision, to avoid a shock that followed an auditory stimulus. Preoperative training was not necessary.

The basic unit of the automatic movement is always the spontaneously rhythmic group of cells in the central nervous system. We have discussed earlier the form constancy of the fixed action pattern. If we now ask what it actually consists of, we will quickly find the answer, because whether a wave passes down a dorsal fin quickly or slowly, the phase distances of the muscle contractions that participate in the movement remain constant. And this constancy is, as P. Leyhausen (1954a) emphasized, to be understood by the stereotypy of the fixed action patterns and not as an unmodifiable absolute.

A central coordination was also demonstrated by J. Gray (1950), who deafferented toads completely, with the exception of the labyrinth. In spite of this the animal swam in a coordinated fashion. The walking pattern of the toad remains well coordinated even after deafferentation of all limbs, provided at least one spinal nerve remains intact (J. Gray and H. W. Lissmann, 1946a, 1946b). There must exist then a central movement coordination, because the coordination could not come from the afferent nerves of the labyrinth. The findings of H. W. Lissmann (1946) have to be interpreted in the same way; his largely deafferented sharks were still able to swim with well-coordinated undulating movements.

The studies of A. O. D. Willows (1971) on the flight reflex of the marine snail *Tritonia* provide an impressive example of central nervous coordination. When this animal touches a starfish it first retracts its anterior part, stretches the body, and broadens the anterior part into a kind of paddle. Then it swims away, employing a series of strong dorsal and vertical arching movements (Fig. 3-12a). Recordings of individual ganglion cells in the snail's brain have shown that the cells fired in precise coordination with the swimming pattern (Fig. 3-12b). By recording from single neurons Willows succeeded in relating several key cells to particular behavior patterns. When he finally experimented with the isolated brain, he discovered that a short electrical stimulation (for example, of the pedal ganglia) resulted in a series of ordered discharges. To the experimenter's surprise the isolated brain was capable of running through the entire flight response. That this coordinated firing of cells had its origin in neural circuits within the brain was shown by the complete isolation of the brain from any possible muscular or peripheral nervous feedback.

The butterfly *Platysamia cecropia* performs a series of stereotyped movements with its posterior part when hatching from the pupa. This prehatching behavior is released by neurosecretory hormones and consists of three phases, each with specific frequencies and types of movement. It has been demonstrated by electrical recordings of deafferented nerve fibers that this prehatching behavior is preprogrammed in the abdominal ganglia. Under the influence of hormones the sequence (which may last one, two, or three hours) can be released even without sensory feedback (J. W. Truman and P. G. Sokolove, 1972; Fig. 3-13).

Even reflex movements can be centrally organized. The coordination of the wiping reflex of a spinal frog remains if the leg that performs the movement is deafferented (E. Hering, 1896).

In mammals deafferentation of a limb usually results in the loss of complex movements. But it has been known for some time that the scratch reflex of the dog, in which 19 muscles cooperate in rhythmic coordination, remains well coordinated even following deafferentation (C. S. Sherrington, 1931). Recently it has been found that in bilaterally deafferented monkeys the function of the hands recovers to an almost completely normal level. The monkeys climbed and swung along with their hands even when their eyes were covered. They grasped and pointed toward an object although they could not see their hand. After complete bilateral deafferentation (C2–S5) of the spinal cord the monkeys remained capable of performing a large number of learned and

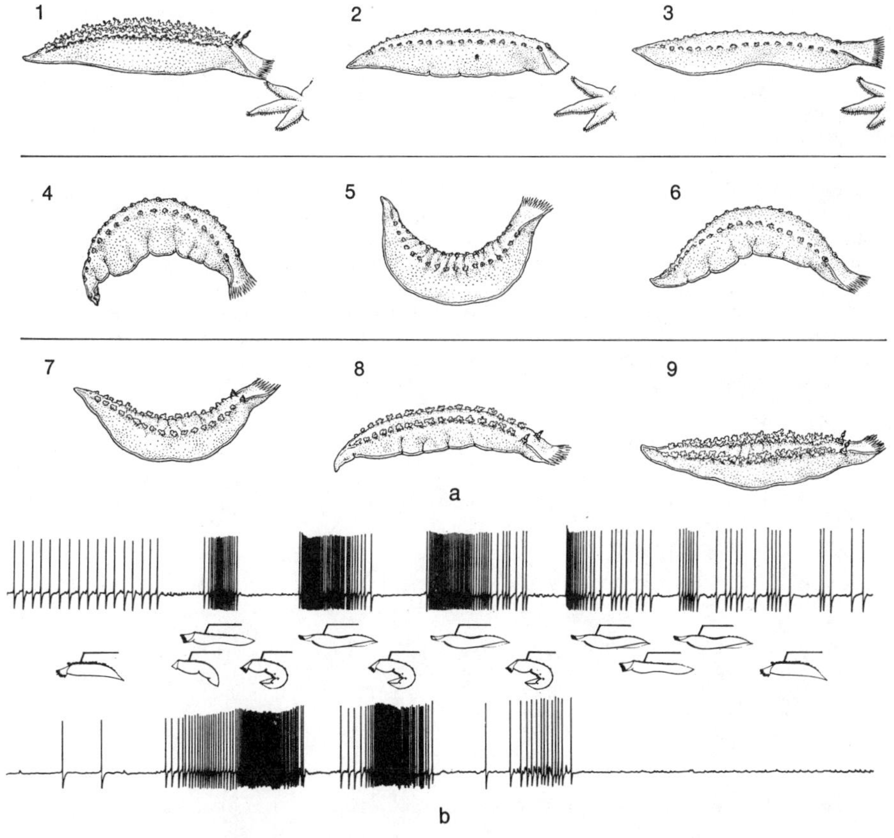

Figure 3-12 Flight reaction of *Tritonia: a:* Snail touches a starfish (*1*), then contracts the frontal part of body (*2*), broadens the head and tail region into a paddle shape (*3*), and swims away with a series of vertical (*4*) and dorsal (*5*) arching movements. The response ends with a series of gradually weakening movements (7–9). *b:* Concurrent recording of two ganglion cells that propel the arching movements, shown in relation to the movements. *Above:* Recording from the cell that propels the dorsal movement. *Below:* Recording of the cell that propels the ventral movement. (Willows, A.O.D., "Giant Brain Cells in Mollusks." Copyright © 1971 by Scientific American, Inc. All rights reserved.)

goal-directed movement patterns. However, unilaterally deafferented monkeys are unable to use the deafferented hand freely (E. Taub and A. J. Berman, 1964). E. Taub, P. Perella, and G. Barro (1973) demonstrated that ambulation, climbing, and reaching developed in monkeys deafferented in their forelimbs shortly after birth. The maturational process seems to be a central one. A detailed discussion of central coordination and central automatism is available in T. H. Bullock (1961, 1962), T. H. Bullock and G. A. Horridge (1965), and B. Hassenstein (1966).

Originally central coordination and automatism were used as cri-

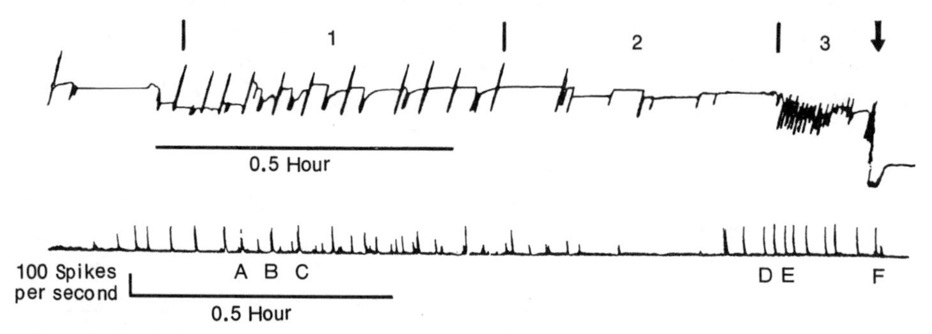

Figure 3-13*a* Behavioral and electrical activity during the three phases of prehatching behavior of the *Cercropia* moth. *Above:* Recording of the abdominal movements. Rotational movements produce upward movements of the recording pen. Ventral twists and peristaltic movements of the abdomen produce downward movements of the recording pen. *1,* first hyperactive period; *2,* rest period; *3,* second hyperactive period. The arrow indicates the moment of hatching. *Below:* Integrated representation of the efferent activity of the right nerve 1 of the A_2 ganglion after exposure to the hatching hormone of the semi-isolated, deafferented nerve fiber. The hormone was applied to the preparation 40 minutes prior to the first discharges. The letters refer to the discharges that are represented in *b* in a less crowded manner.

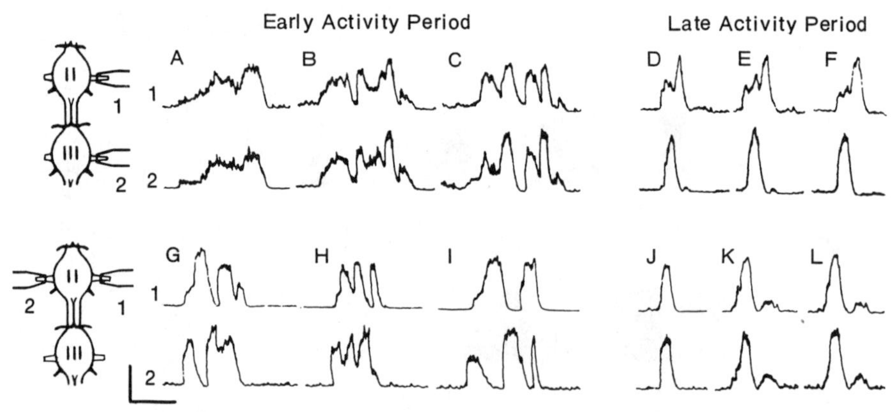

Figure 3-13*b* Examples of spontaneous discharges (integrated) that were recorded from a semi-isolated, deafferented nerve fiber after application of the hatching hormone. *Above:* Electrode discharge from the right roots of the ganglia A_2 and A_3. *Below:* Electrode discharges from the right and left roots of the ganglion A_2. The discharges *A–C* and *G–I* correspond with the rotational movements typical of the first hyperactive period. The discharges *D–F* and *J–L* correspond to the peristaltic movements of the second hyperactive period. The vertical line represents 100 discharges per second; the horizontal line, 10 seconds. (From J.W. Truman and P.G. Sokolove, 1972.)

teria of the fixed action pattern. In line with these criteria, behavior patterns that are coordinated by feedback from afferent stimulation (such as reflex movements) would not be considered fixed action patterns. But since the movement coordination via afferent pathways (proprioceptors) can occur as a phylogenetic adaptation, it seems useful to classify these behavior patterns as fixed action patterns as well (I. Eibl-Eibesfeldt, 1963, 1966a). Unconditioned reflexes and automatic movements represent extremes on a continuum.

When discussing earlier the automatisms of fins and fin rays we mentioned the hierarchical organization of the fixed action patterns. This holds for all complex fixed action patterns, whether they are birdsongs or sucking movements. They can be subdivided into elements that in turn are fixed action patterns. However, it is possible in principle to establish an upper limit of conjoined functional motor units, by calling fixed action patterns all those motor activities that, once activated by external stimuli or by spontaneous discharge, will then continue in the absence of additional external stimuli. The orderly sequence does not have to originate exclusively in an automatic, central pattern of excitation. Internal sensory stimulation frequently controls the discharge of a movement pattern in the manner of a chain reflex. As long as their regulation has been determined genetically and the movement pattern does not depend upon additional external stimuli, one can speak of fixed action patterns. If the concept were restricted to purely centrally coordinated automatic movements, then in my opinion a too-rigorous limitation of the concept would result, especially because the demonstration of fixed action patterns would then be difficult in many cases.

When **taxis** components are added, we speak of instinctive actions. If, on the other hand, an ordered movement sequence comes about when a behavior changes the releasing stimulus situation and activates new behavior via this new stimulus situation, then we are confronted by a **reaction chain** of fixed action patterns.

Such a chain exists, for example, when a falcon separates his victim from a flock of birds by a sham attack. Then, when he has been successful, he captures the single bird, plucks it, and eats it. In this case each succeeding stage presupposes a new and adequate stimulus situation, which is brought about by the previous activity of the animal. A fixed action pattern is always an inborn, internally coordinated sequence, which merely requires a releasing stimulus.

SUMMARY

Animals are fitted out with a repertoire of comparatively stereotyped motor patterns that mature during the ontogeny of the individual, the typical patterning not being a result of individual learning. These motor patterns have been named *Erbkoordinationen* or fixed action patterns.

They occur normally with orienting movements or taxes superimposed on them. As a shorthand description fixed action patterns are said to be inborn or innate, implying that the organic structures, serving as the base of the behavior, develop in a process of self-differentiation by means of blueprints contained within the genetic code.

This concept of the innate has been attacked on theoretical grounds. The main argument against its acceptance was that one could not possibly deprive an animal of all sources of information: even within the egg or the uterus it is exposed to environmental influences that can shape the behavior. This argument, however, misses the all-important fact that behavior patterns are molded to fit certain features of the environment. For this to occur information concerning the environmental structures toward which the behavior in question is adapted must have been fed into the organism. Whether this occurred during phylogeny or during ontogeny can be checked by the deprivation experiment. One needs only to withhold information from the organism about the environmental features with which the animal interacts in its behavioral adaptation. If it nonetheless shows the adaptive pattern in question (for example, sings the species-specific song) which we had observed to occur regularly in other members of the species, than we have proved that the information concerning the specific patterning was indeed contained in the genome.

There are certain rules to be observed when carrying out a deprivation experiment. In particular, we have to set up our experiment in such a way that only the adaptation to be investigated is disrupted and that the animals in the testing situations are presented with the releasing stimuli that normally set the behavior in motion. Neglect of these basic considerations causes experimental errors.

As a physiological characteristic fixed action patterns often show a spontaneity that in some cases has been traced back to the spontaneous activity of groups of motor neurons. These centrally created impulses are furthermore centrally coordinated. In the classical experiments of Erich von Holst deafferented spinal preparations of eels swam in a free and well-coordinated manner until death. The way in which spontaneously firing cell groups for different effectors influence each other has been studied. Dominant rhythms have been found to exert a "magnetic" effect on dependent ones, the latter being either slowed down if its own rhythm is too fast, or speeded up if too slow. The central coordination is absolute if the dominant rhythm forces the dependent one into its phase. In the case of relative coordination the dependent rhythm follows the dominant one only approximately, therefore periodically dropping out of phase. In the case of superposition the dominant rhythm superimposes itself on the dependent rhythm arithmetically.

More recent experiments with invertebrates have had basically the same results.

4

Motivating Factors

As I have tried to make clear in Chapter 3, a behavior is not merely a response to external stimuli. The animal is not simply an automaton into which one drops a coin and from which one then receives a response. Internal motivating mechanisms also activate it. This is convincingly shown by the study of intact animals. Animals that are maintained under constant conditions display, for example, a **circadian rhythm;** rest and activity alternate with a periodicity that coincides approximately with the day-night rhythm. The anemone (*Metridium*) shows spontaneous rhythmic contractions at 10-minute intervals (E. J. Batham and C. F. A. Pantin, 1950). Furthermore, animals that for a time have had no opportunity to perform a certain behavior pattern are in a state of specific readiness to perform it. W. Craig (1918) clearly recognized this specific drive state. The observer at first notes merely a general restlessness of the animal—"as if it were searching for something."

That this is not merely a general motor restlessness but the expression of a specific readiness to act can be recognized by the readiness to respond to specific releasing stimuli. The thirsty animal seeks water and passes up food objects. An animal in a hunting "mood" searches for a releasing situation that permits the discharge of hunting behavior patterns, and the sexually motivated animal searches for appropriate stimuli. If the motivated animal does not find adequate releasing objects, it may on occasion accept substitutes. Female rats are so ready to retrieve during the first few days following parturition that they will repeatedly grasp their own tails, retrieve them in their mouths, and deposit them in their nest. Sometimes they even grasp one of their own hind legs and limp into the nest on three legs (I. Eibl-Eibesfeldt, 1963; W. E. Wilsoncroft and D. U. Shupe, 1965). This search for a specific releasing situation, which W. Craig (1918) called *appetitive behavior,* is variable and adaptable to changing situations. The animal is capable of mastering detours that lie between it and a desired goal that is remembered, as when a dog in a mood to hunt proceeds toward a chicken yard known to him. Once he has found the releasing situation, the more automatically discharged fixed action patterns of prey-catching run off. Their occurrence not only changes the **releasing stimulus situation** but also results, as W. Craig pointed out, in a change of mood. One often speaks of a drive-reducing *consummatory act.* In cats the behavior patterns of prey-catching: lying in wait, creeping, catching, pouncing, and pawing, normally occur in a certain sequence, which is directed by the releasing situation. However, P. Leyhausen (1965a) has shown that each of these individual actions also has its own spontaneous motivation. If a cat has had no opportunity to perform one or the other behavior pattern, the pattern will develop its own appetitive behavior. The animal searches for a releasing stimulus situation, merely to paw or catch. The mouse alternately becomes an object to be caught, to be intently observed, killed, eaten, or pawed. The behavior patterns leading up to each action then become appetitive behavior for the desired consummatory act in each specific case.

Appetitive behavior can also be demonstrated by electrical brain stimulation. Cats that are eating stop when stimulated in certain parts of the hypothalamus and attack a rat that has until then been ignored. If they do not see prey they search, and they learn a maze when they find as a reward a rat they are allowed to attack. Rats stop eating if the appetitive behavior for gnawing is released by electrical stimulation and begin to search for objects that are suitable for gnawing. They also learn a maze under this motivation (W. W. Roberts and H. O. Kiess, 1964; W. W. Roberts and R. J. Carey, 1965).

The specific readiness to act—the mood—of an animal is also shown by a noticeable lowering of the threshold for certain releasing stimuli. A predator in a mood to hunt will react most readily to stimuli that release hunting-behavior patterns. If prevented from hunting for some time it will even accept substitute objects, and in some instances the **response** may occur **in vacuo** following a prolonged absence of the appropriate releasing stimuli. At the same time the thresholds for other behavior patterns, for example those belonging to the area of sexual behavior, are raised to a point where very strong releasing stimuli are required to distract the animal from its hunting and switch it to sexual behavior.

Observations on intact animals show that fixed action patterns often occur in sets, and then they show a common and identical fluctuation of the releasing thresholds. This points to a common physiological mechanism. The sets of behavior patterns are to some extent mutually exclusive. In male cichlids a readiness to flee suppresses a readiness to fight as well as to court. The readiness to attack and to court, however, are positively correlated. This is not the case in cichlid females, where a readiness to attack generally suppresses sexual readiness (B. Oehlert, 1958). In male sticklebacks the readiness to court suppresses the readiness to bite. P. Sevenster (1968) rewarded the males whenever they swam through a ring by briefly lifting an opaque screen in front of a glass pane so that the male could see a neighboring female and court it. The males learned their task slowly, because they rarely swam through the ring of their own accord. Once they had learned it, however, they traversed the ring several times in a session in order to see the female. If the task was to bite a rod for the same reward they learned this very fast, because they often spontaneously bite and nibble on rods. Even so, however, they never achieved a large number of completed tasks, for they seemed unable to repeat the biting immediately after the performance of the courting dance. Although they often stayed motionless in front of the bar, clearly intending to bite it, they could not bring themselves to do it, owing, obviously, to an inner inhibition. Biting and attacking, in contrast, do not inhibit each other. One male quickly learned to bite the bar if rewarded with the view of a rival whom he could fight through the glass pane. Whenever the opaque screen was lowered again, he would bite the rod without delay. Recognition of these kinds of relationships permits one to draw inferences about the mechanisms underlying the behavior. During continuous observations one records the sequence of several behavior patterns, computes the correlation, and sets up models that express these relationships.

Examples for this type of *motivation analysis* are found in the work of D. Morris (1958), P. R. Wiepkema (1961), and W. Heiligenberg (1964). In male ten-spined sticklebacks (*Pygosteus pungitius*), Morris showed that 1766 dances were followed by 1232 attacks (70.4 percent). On the other hand, 208 nest-showing actions were followed by 5.3 percent attacks, 1 percent nesting, and 93.7 percent sexual behavior. One can conclude from these results that aggression is dominant during the early stages of stickleback courtship and the sexual drive later.

W. Heiligenberg (1963) counted the frequency of swimming in schools and of biting in young of the chichlid *Pelmatochromis subocellatus*. The more frequently they school, the less they bite (Fig. 4-1). We may interpret this observation as evidence that the drive to school inhibits the drive to bite. Yet we cannot always make such clear determination of the temporal relationships between behaviors. In such cases, motivational factors may be elucidated by analyses of form and context. The **greeting ceremony of the graylag goose** involves two rituals: rolling and cackling; both are differently motivated. Ontogenetic studies throw light on the origin of cackling (H. Fischer, 1965). After separation from their parents, goslings utter a tiny one- or many-syllabled "wee" upon refinding them (Fig. 4-2). As the intensity of this "greeting call" increases it becomes louder and is accompanied by a distinct stretching out of the neck. We can observe a striking, formal similarity to adult behavior, which involves emitting a contact call without stretching out of the neck and a loud cackling with stretching out of the neck. On closer examination the contact call and cackling prove to be the same behavior at different levels of intensity (Figs. 4-2 and 4-3). Furthermore, the

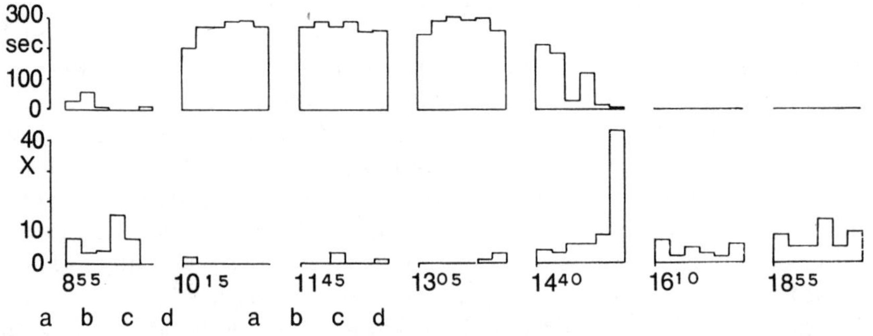

Figure 4-1 Two drives inhibit each other: For 25 young of the fish *Pelmatochromis* schooling activity was observed 7 times for 30 minutes each (*top*), and at the same time the attacks per 5-minute period were counted (*bottom*). It is evident from the block diagrams that the animals' biting behavior varies inversely with their schooling behavior. (From W. Heiligenberg, 1963.)

"wee" sound, the contact call, and the cackling call are all released by the sight of a conspecific. There is a congruence between the releasing stimulus situation and the resulting formal motor pattern in the young and the old that indicates a common root. The contact call and cackling occur temporarily independent of sexual and aggressive behavior, intake of food, and behavior concerned with care of young. For this reason we may assume for the former the existence of an independent motivational base. Loss of the partner is followed by a display of appetitive behavior that ends with a discharge of cackling. H. Fischer postulates a drive to establish bonds.

Appetitive behavior is always the first indication of a specific internal readiness to act. A great variety of physiological mechanisms act as underlying causes and they exert their influence at different levels of integration. Hormones, for example, can lower the thresholds of release for a great number of motor patterns, whereas neurogenic processes often cause the spontaneity of a few acts only. Motor patterns that are often used in the service of numerous function cycles (for example, locomotion or gnawing in rodents) have a particularly strong motivation. The variety of motivating mechanisms are subsumed under the functionally defined term "drive."

By means of brain stimulation with thin electrodes, E. v. Holst and U. v. Saint Paul (1960) activated various drives in intact chickens. Upon stimulation the chickens began, for example, to walk about restlessly. That this was not the activation of a general motor pattern but typical appetitive behavior was made clear when the animal was offered variously

Figure 4-2 "Wee"-sounding in the gosling with stretching out of the neck toward the mother. (From H. Fischer, 1965.)

a rival, a female, water, or food. The chickens always responded to a particular object when a particular brain area was stimulated, for the duration of the stimulus. The strength of the activated drive was measured by the level of stimulus voltage necessary to release the behavior. For instance, if they activated two opposing drives, such as those for sitting down and standing up, by means of two electrodes placed at different positions in the brain stem, they were able to measure (in volts) how strongly a chicken was motivated to sit and to stand, respectively (Fig. 4-4). They could measure changes in the specific readiness for each behavior. A chicken that was originally motivated to stand was brought into a sitting mood, following repeated activation of the sitting behavior; after this change a stronger stimulus was required to release standing up.

Figure 4-3 *a:* "Wee"-sounding with and without stretching out of the neck. *b:* Series showing the intermediate stages from contact call to cackling (top to bottom). (From H. Fischer, 1965.)

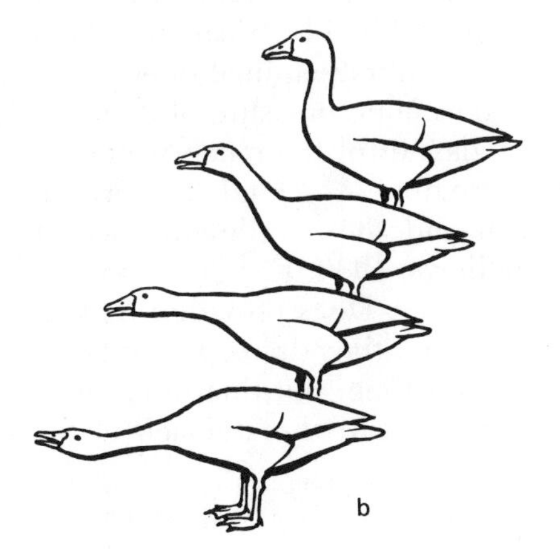

The observations and experiments with intact animals show clearly that a living organism is not a reflex automaton but is motivated from within to act in a specific manner. What are the motivating factors behind such a specific readiness to act? What are the mechanisms[1] that bring the animal into a state of specific restlessness, and what leads to the termination of the once-activated behavior? Let us examine these questions with some examples.

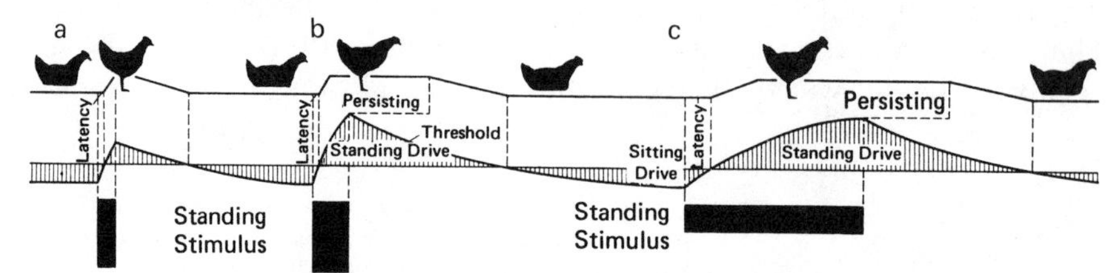

Figure 4-4 Measuring a sitting and standing drive. A chicken in a spontaneous mood to sit (*a*) is made to stand up by stimulating a central region that controls the standing-up drive. The stimulus is strong and short; after a short latency period the chicken stands up, but sits down again immediately upon cessation of the stimulus. The second standing-up stimulus (*b*) is of equal strength but lasts longer and the chicken stands up longer; latency and standing up are unchanged. Finally, if the duration of the stimulus is longer (*c*), but weaker, the animal rises more slowly but remains standing as long after the stimulus as in (*b*), because stimulus strength and duration compensate each other; that is, the total value remains constant. The central process that underlies these actions is designated by the middle curve. Below the zero line the physiological sitting drive is greater than the standing drive. The strength of these drives can be measured. As long as the animal sits, the threshold value that must be crossed to cause the animal to stand up is a measure of the strength of the sitting drive that must be overcome. (After E. v. Holst and U. v. Saint Paul, 1960.)

Many investigations have dealt with the phenomenon of "thirst" (summary by A. V. Wolf, 1958). It was found that the appetitive behavior of searching for water is released by osmoreceptors in the hypothalamus that respond to hypertonicity of the blood. By injecting saline solution into the veins of dogs and humans they can be made thirsty. On the other hand, it is possible to eliminate thirst by intravenous injection of water. If a minute quantity of a hypertonic saline solution is injected directly into the hypothalamus, the animal becomes thirsty; the same effect is achieved by electrical stimulation of this area in rats and goats.

A thirsty animal does not have to continue drinking until the normal osmotic pressure of its body fluids has been restored. There would be danger of too much water being taken in, because absorption takes some time. As a kind of safety measure, the amount of water in the stomach and the swallowing activity itself is taken into account (R. T. Bellows, 1939; E. J. Towbin, 1949).

Dogs provided with an esophageal fistula, through which all the water they drank drained outside, drank regularly and stopped after a certain time, but this satiation through the act of drinking did not last very long. If the stomach of the drinking animal was also filled with

[1] When speaking of motivating mechanisms, we refer to the totality of the physiological machinery involved in the specific activation of an animal.

water satiation was more lasting and correspondingly less water was taken in by mouth. If a rubber balloon was inserted into the stomach and inflated, the amount of sham drinking was also significantly decreased. This drinking activity is controlled by several mechanisms: the osmoreceptors in the hypothalamus release the appetitive behavior for drinking and finally terminate it, while a short-term "satiation of thirst" is achieved by filling the stomach and by the drinking activity itself. The last activity is especially interesting because it often appears as if the mere performance of the movements is "drive-reducing."

In this connection the observations of R. A. Spitz (1957) and D. W. Ploog (1964a) deserve especial attention. Both noted a clear correlation between the degree of satiation and the amount of sucking movements in infants. If the infants had taken in a certain quantity of food during 20 minutes by sucking, they were satisfied and slept. If the openings in the nipples were too large, so that the same amount or even half of it was consumed within 5 minutes, they remained dissatisfied. They continued to suck in vacuo and began to cry. If they were given the empty bottle, they sucked for another 10 to 15 minutes and then seemed satisfied. According to M. Mead (1937), infants of peoples who have not been influenced by Western ways and whose mothers started to nurse them shortly after birth, do not suck their thumbs. Puppies fed from nipples with large openings afterward sucked on their own bodies, even during sleep. On the other hand, when they were nursed with nipples containing small holes they did not do so (D. M. Levy, 1934; S. Ross, 1951). Calves that are fed from buckets, and hence drink their milk too fast, develop the habit of sucking on their steel chains or on other calves. Some become stunted as a result of this; they show the so-called tongue flick, which is perhaps a vacuum-sucking stereotypy. K. Zeeb (personal communication) was able to eliminate this behavior by letting even older heifers drink only from a bucket fitted with a rubber nipple. Ducks that have been fed grain on land dabble in vacuo (K. Lorenz, 1963a).

There are a number of comprehensive investigations dealing with the mechanisms that regulate the intake of food in mammals (L. de Ruiter, 1963; J. Mayer and D. W. Thomas, 1967). The glucose level of the blood is registered by glucose receptors in the hypothalamus. The motivating systems also are located in the hypothalamus: Electrical stimulation of the lateral part leads to an increased food intake, but there are also activating and inhibiting systems outside the hypothalamic region (P. Teitelbaum, 1961; B. G. Hoebel and P. Teitelbaum, 1962). The satiating mechanism is sensitive to an increase in the glucose level of the blood, but, as with drinking, filling of the stomach also leads to inhibition. Mechanical receptors report the volume to the ventromedial hypothalamus, and chemical receptors report the nature of the food that has been received.

In the blowfly the appetitive behavior for feeding depends on the amount of food in the foregut. When the foregut is filled, inhibiting impulses pass via the nervus recurrens to the central nervous system. If the nervus recurrens is cut, the feeding inhibition is removed, and the fly continues to suck up food until it becomes extremely distended and dies (V. G. Dethier and D. Bodenstein, 1958).

The interaction of external and internal factors affecting the fanning drive of the stickleback was investigated by J. v. Iersel (1953). After males had fertilized three to four clutches of eggs their sexual drive waned and they began to ventilate the eggs with fanning movements of their pectoral fins. Van Iersel measured the intensity of fanning and

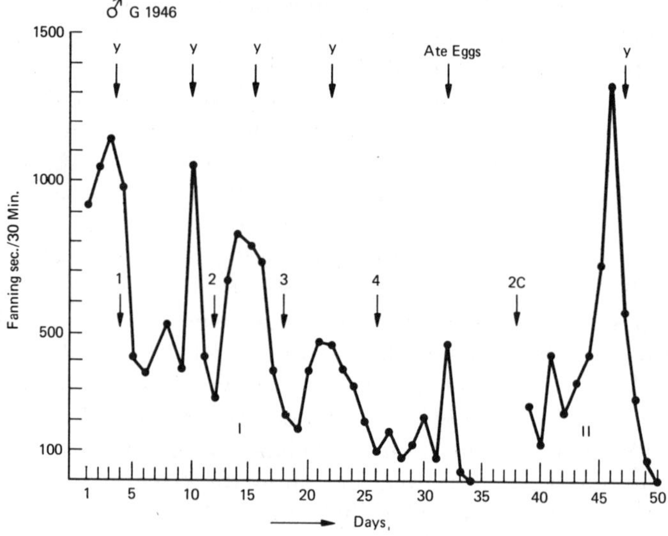

Figure 4-5 Series of fanning cycles in a stickleback male that were repeatedly induced by the exchange of clutches. Arrows *1–4* indicate presentations of new clutches. The *y* symbols mark the time of hatching. The eggs that induced the fifth cycle were eaten. *II*:Sixth fanning cycle, before which the male had courted for five days and then built a new nest. (From J. v. Iersel, 1953.)

found that the total fanning time increases from day to day until the eggs hatch, but drops sharply shortly thereafter. The increase of fanning activity is caused by the oxygen consumption of the eggs: If the CO_2 content of the water is artificially raised, the stickleback will fan more. However, the fanning activity does not depend exclusively upon external stimuli, as is shown by the following experiment: If one presents a fanning male with a new clutch just prior to the end of the fanning cycle and shortly before the activity stops completely, one can induce a new fanning cycle that is very similar to the previous one. The peak of fanning is lower, however, and if additional cycles are induced, the peaks will successively be lower (Fig. 4-5). Because the releasing stimulus situation is always the same, this change must be dependent on events within the stickleback. The mechanisms of drive reduction in the sexual behavior of male sticklebacks were investigated by A. C. A. Sevenster-Bol (1962). She found that the presence of eggs in the nest, and not the act of fertilization, functions as an inhibiting stimulus situation for the zigzag dance and for leading by the male.

D. S. Lehrman (1961) investigated the development of the reproductive behavior of the ring dove (*Streptopelia risoria*). Males or females that were placed alone in a cage containing a nest and eggs could not be induced to incubate them. A sexual partner had to be present. When pairs were placed in a cage with a nest and eggs, they courted, built a nest, and from the fifth day on some began to incubate; by the seventh day all did. It could be that the birds had to become used to the cage; Lehrman therefore placed a pair in a test cage but separated the partners by an opaque partition. When he removed the partition after seven days and gave them a nest and eggs, they still required seven days before they began to incubate. Becoming used to a cage was therefore not a factor. If, on the other hand, he kept the partners with nesting material in the cage and gave them eggs on the seventh day, they all incubated within two hours. If he did not give them nesting material first and presented them with a nest and eggs on the seventh day, they first built nests; some began to incubate the same day and the others by the end of the next day.

The birds seem to pass through two stages: First, courtship induces the readiness to build a nest, and this, in turn, induces the readiness to incubate eggs. It is sufficient for females to be able to see males through a glass partition, provided the males court. C. J. Erickson and D. S. Lehrman (1964) presented females with castrated males (which do not court) and this had no effect on ovarian development—an elegant demonstration of the significance of the courtship behavior. The changes induced by courtship behavior are hormonal. Lehrman injected 80 pairs of doves with progesterone 7 days prior to placing them together. If he then presented them with eggs, they incubated at once. If he injected them with estrogen instead, they began to build a nest first and incubated within 11 days. According to R. A. Hinde (1965), the courting canary male stimulates estrogen production in the female.

The interaction of various motivating factors in the courtship behavior of canaries is quite complex (Fig. 4-6). Changes in the amount of daylight induce growth of the gonads and estrogen production via a mechanism in the hypothalamus and the hypophysis. The latter is affected by stimuli coming from the male. The female responds to the courting male while under the influence of estrogen and builds a grass nest. This, in turn, further stimulates the female. The nest, together with the estrogen, stimulates the development of the brood patch and the oviducts, which are further stimulated in their development by secondary hormones initially activated by the estrogen. Then the eggs are laid. As a result of increased sensitivity to the nest, the bird stops building with grass and selects only feathers. Contact of the brood patch with eggs

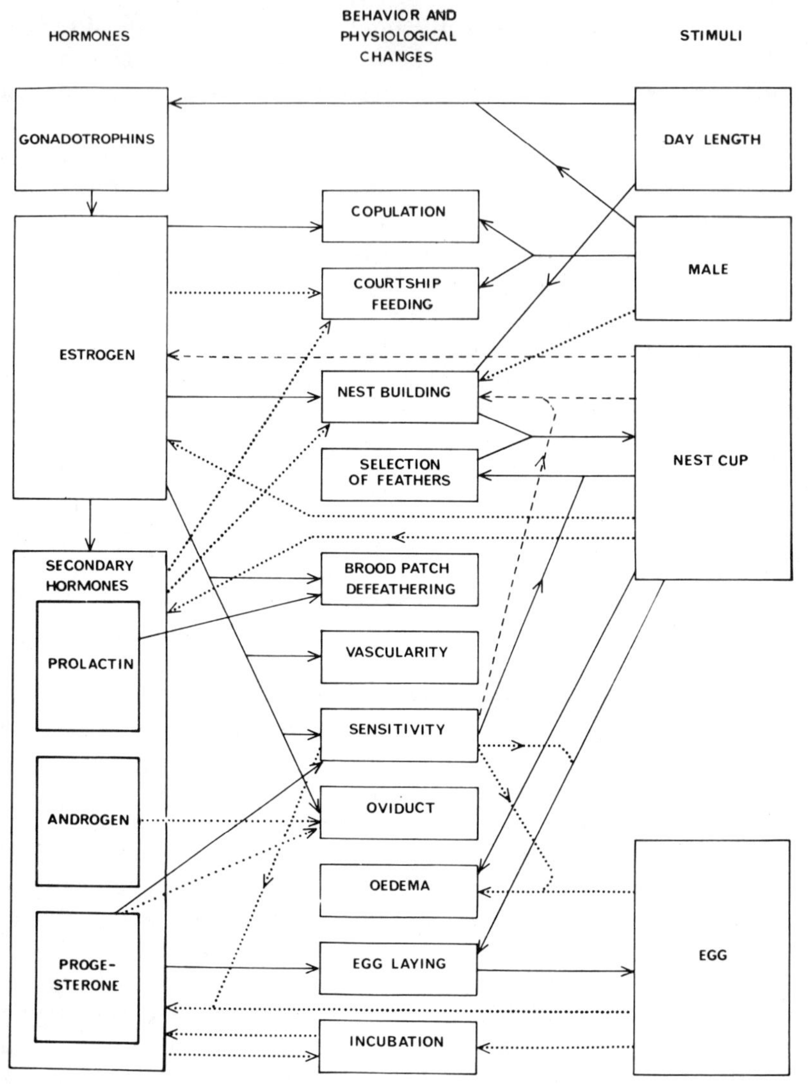

Figure 4-6 Interrelationships of various factors in the reproductive cycle of the canary, showing relations among hormones, external stimuli, behavioral changes, development of brood patch, and so on. *Solid lines* indicate facilitating influences; *dashed lines* indicate inhibiting influences; *dotted lines* indicate probable but not proved facilitating influences. (From *Animal Behavior* by R. A. Hinde. Copyright © 1970 by McGraw-Hill Book Company and used by their permission.)

and nest now induces incubation. R. A. Hinde (1965) emphasized four points about these interactions in his summary:

1. The causes and consequences of sexual behavior are closely intertwined with nest building and cannot be understood by themselves.
2. The external stimuli induce endocrine changes that are added to the immediate influences on the behavior.
3. Hormone production has multiple causes.
4. Hormones have multiple effects.

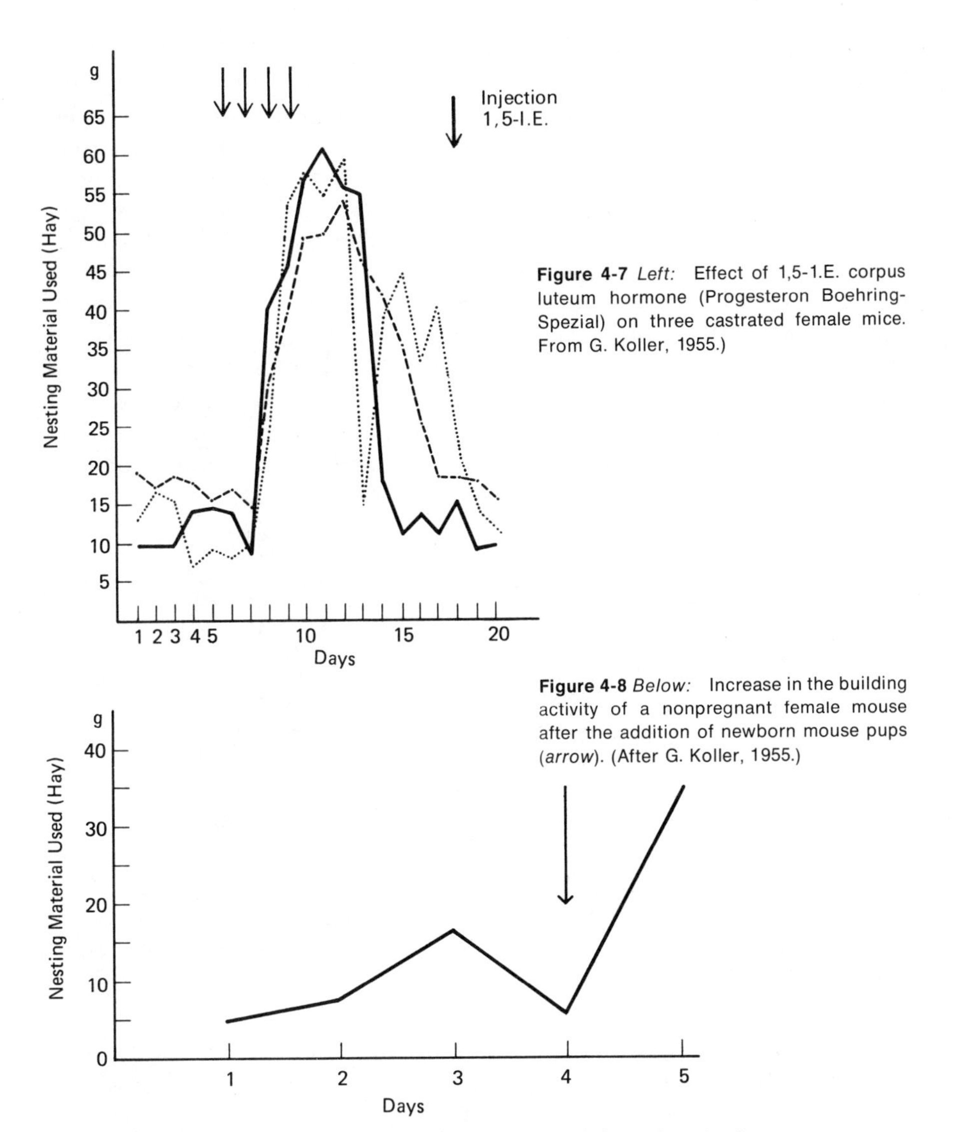

Figure 4-7 *Left:* Effect of 1,5-1.E. corpus luteum hormone (Progesteron Boehring-Spezial) on three castrated female mice. From G. Koller, 1955.)

Figure 4-8 *Below:* Increase in the building activity of a nonpregnant female mouse after the addition of newborn mouse pups (*arrow*). (After G. Koller, 1955.)

L. R. Aronson (1949) found that cichlid females (*Tilapia macrocephala*) prepared no typical nest pits when alone. If they see a male, even through a glass in the next tank, their ovaries begin to develop and they start building a nest.

Similarly, other external stimuli participate, via the hormonal system, in the buildup of a drive. The increase in length of day during the spring stimulates in many songbirds the growth and activity of the gonads and hormone production and in this way induces the reproductive drive.

Common house mice build a nest shortly before parturition, using up to four times as much nesting material as for their sleeping nests. This nest building can be released by the injection of the hormone progesterone, but not by prolactin. The hormonally stimulated nest-building drive normally wanes immediately following parturition, but the increased nest-building activity is maintained by the presence of the young in the nest. If the young are removed, this activity decreases. Virgin females can be stimulated to build nests if one presents them with very small young (G. Koller, 1955; see also Figs. 4-7, 4-8, and 4-9).

How decisively and specifically hormones are involved in the organization of drives can be seen in the numerous experiments on the sexual behavior of castrated and hormone-treated animals (F. A. Beach, 1948). Female dogs urinate in a squatting position, while males stand and raise one leg. The development of this behavior depends on the male sex hormone. Young males still urinate without raising the leg and continue to do so if they are castrated before they are four months old. If injected with testosterone, they later raise a hind leg while urinating.

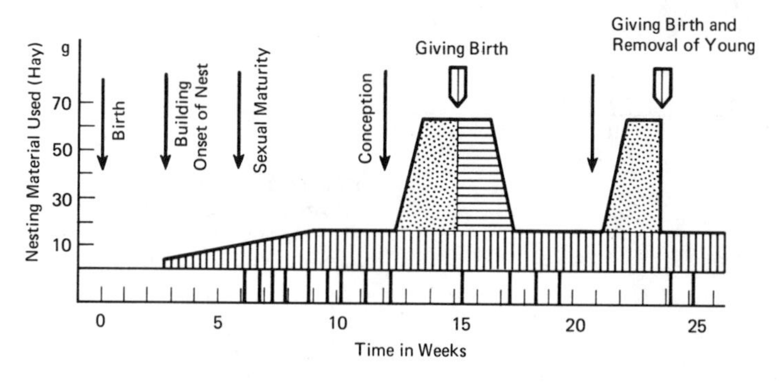

|| = Estrus |||| = Amount of Nest-Building Independent of Sex ⁘ = Amount of N.B. Due to Progesterone ☰ = Amount of N.B. Due to Presence of Young

Figure 4-9 Schematic representation of the nest-building activity of untreated house mice during the course of their lives. (After G. Koller, 1955.)

Females do the same if they are spayed shortly after birth and treated with progesterone but not when they are spayed as adults (T. Martins and J. R. Valle, 1948).

In the golden hamster there is a clear negative correlation between sexual receptivity and aggression, which is dependent on hormones. Golden hamsters in estrus show a decreased readiness to attack males in the rutting period (J. W. Kislack and F. A. Beach, 1955).

E. S. Valenstein, W. Riss, and W. C. Young (1955) present us with a fine example of interaction between hormonal and experiential influences. These authors compared the behavior of male guinea pigs that had been raised with females with the behavior of those raised in isolation. Subsequently the animals were castrated, and sexual behavior was markedly reduced. Finally the subjects were given testosterone, whereupon the sexual activity in both groups increased. However, the animals that were socially inexperienced remained at a lower level of sexual activity than those in the control group (Fig. 4-10).

From what has been said so far, it can be seen that a specific readiness to act can be released by many different factors and that it is usually activated by the interaction of several of them. We have discussed external stimuli (day length, sexual partner), internal stimuli, and hormones.

The main problem—which has been little investigated to date—in respect to the fixed action pattern, is the lawful fluctuation of the inner

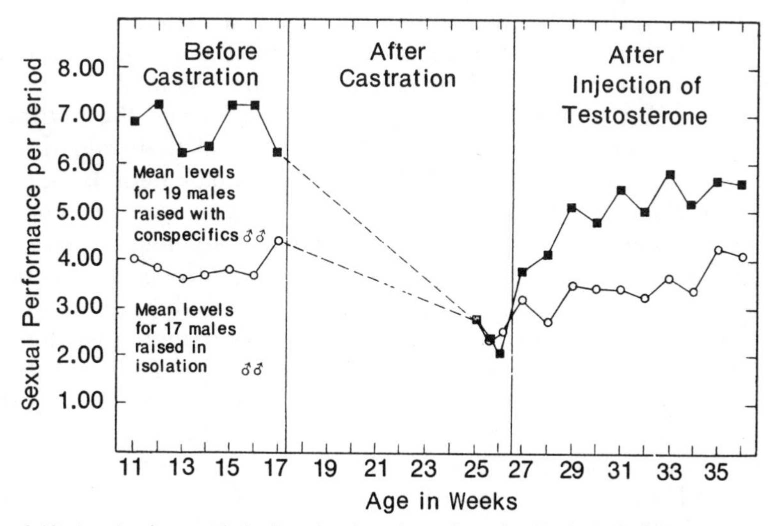

Figure 4-10 Levels of sexual behavior of male guinea pigs raised in isolation or with conspecifics before and after castration, and after injection of testosterone. (From A. Manning, 1967, after E. S. Valenstein, W. Riss, and W. C. Young, 1955.)

readiness to act, which cannot be explained on the basis of factors discussed so far. K. Lorenz (1937) writes about his well-fed starling, which had never had the opportunity to catch live insects but would nevertheless fly up from its perch, act as if it were catching something, return to its perch, perform killing movements, and finally swallow, although Lorenz assured himself that the bird had caught no prey—a good example of behavior in vacuo. H. N. Kluyver (1947) observed waxwings (*Bombycilla garrulus*) that showed insect prey-catching behavior in vacuo during hard frosts when insects are absent. While still in the nest young honey buzzards perform the movements associated with digging out wasp nests (K. Gentz, 1935). Sticklebacks court in vacuo (N. Tinbergen, 1952). Many more examples of such behaviors[2] in vacuo are now

[2] M. Bastock, D. Morris, and M. Moynihan (1953) have proposed to replace the term "vacuum activity" with the term "overflow activity," because one could never be certain of the complete absence of a releasing stimulus. But the term "vacuum activity" does not imply this, so we may as well retain the original term.

known (L. Koenig, 1951; P. Leyhausen, 1956a, 1956b; and others). It has also been shown that the mere performance of a movement can be rewarding in itself. We have cited the example of satisfying the **sucking drive.**

In the very aggressive cichlids *Etroplus maculatus* and *Geophagus brasiliensis* the males must fight with other males before successful pairings with females can take place. If they are not given the opportunity to fight they kill their females, because they discharge their aggression on them. To avoid this, it is only necessary to separate two pairs by a glass partition; then the males attack the glass and do not harm their females. The same occurs if other conspecifics are kept in a sufficiently large tank with them. If these "whipping boys" are removed, the male regularly attacks the female and finally kills her (K. Lorenz, 1963a). A. Rasa (1969) investigated this in *Etroplus maculatus* and confirmed the findings of Lorenz. Those males that could attack other fish in their tank directed few attacks against their females. Being able to attack neighbors, if only through a glass partition, they showed little aggression toward their females. If there was no stranger to attack, they fought with their females, and the number of attacks directed at them was markedly increased in comparison to the number in the two other groups (Fig. 4-11). This increase in aggressive behavior toward their own females, which was at first not understood, may be due to the female's continued efforts to seek contact with the male in spite of his attacks. An investigation of the spontaneity of aggression in cichlids that were raised in isolation from conspecifics is under way. In view of the great theoretical significance of this phenomenon, investigations of the behavior of other vertebrates are needed.

W. Heiligenberg (1964) has demonstrated that the readiness to fight wanes in male cichlids (*Pelmatochromis subocellatus*) if the animals fight briefly without damaging one another. They were usually not fatigued, as was demonstrated by their readiness to perform other behavior patterns. O. Drees (1952) gave salticid spiders an opportunity to exhaust prey-catching behavior, including approach running, stalking, and creeping, long before overall physical fatigue set in. He was able to rule out avoidance conditioning and adaptation of **afferent mechanisms,** so he interpreted these central damming-up and discharge processes in line with K. Lorenz's hypothesis.

If the gobbling call[3] of the male turkey is repeatedly released by a stimulus of constant amplitude and frequency, the threshold for this stimulus rises, and the animal no longer calls. This depends primarily upon adaptive processes of afferent **releasing mechanisms,** because the gobbling calls immediately reappear in response to a tone of different frequency and amplitude, even if the new stimulus is normally less ef-

[3] A call that can be released by various noises and tones.

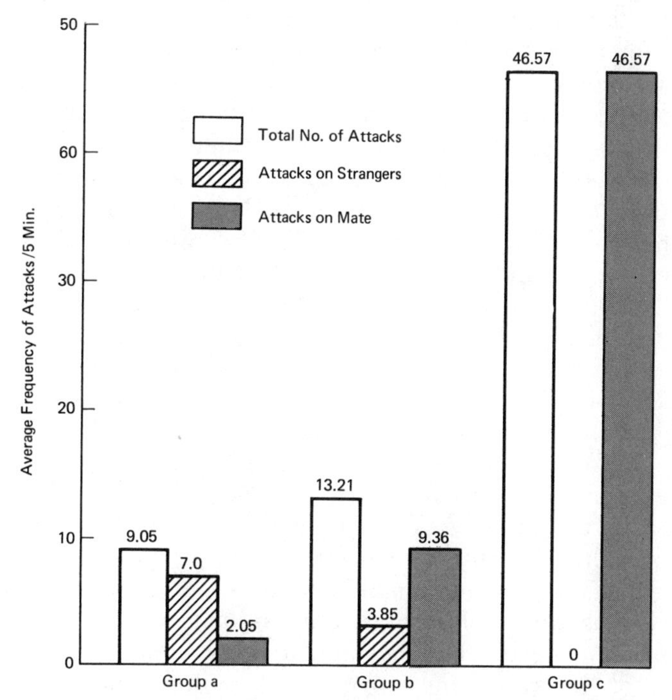

Figure 4-11 Average number of attacks of male cichlids (*Etroplus maculatus*) during the reproductive phase. In *Group a* the pair lived together with several unmated adult conspecifics and several young animals. Three such pairs were observed for a total of 83 hours and 15 minutes. Two of these pairs spawned twice and one three times. In *Group b* a glass partition separated the pair from their conspecifics. They were able to see their neighbors but could attack them only through the glass. Two pairs that raised five broods successfully were observed for a total of 62 hours and 40 minutes. In *Group c* the pair were completely isolated from all other conspecifics. In three pairs that were kept thus the pair bond disintegrated and the female had to be removed for her own protection. In the fourth pair the bond also disintegrated, but the animals mated once more briefly before spawning. They spawned, ate the eggs, and continued to fight. One day later they again mated and remained together until the eggs hatched. Then the male drove off the female and finally killed her three days after the young hatched. He successfully raised the young alone. The total observation time for *Group c* was 84 hours and 17 minutes. (After A. Rasa, 1969.)

fective as a releaser than the one in the preceding stimuli series. If the two differentially effective stimuli are presented alternately in a continuing sequence, the weaker of the two is soon not responded to, which points to a central fatiguing process in addition to adaptation (M. Schleidt, 1954).

In the cichlid *Pelmatochromis subocellatus kribensis* the readiness to fight, as measured by the number of bites, is increased following a short fight, but wanes increasingly as the fights continue; short pauses lead again to an increase. "Straining" (sifting), a movement associated with feeding, suppresses attack bites. Shortly before and during straining the attack readiness is markedly decreased; afterward it is clearly higher. It appears as if the fish had "saved up" the biting behavior, or as if the behavior had accumulated during the straining period. The digging movement of biting into the sand does not have the same influence on biting directed at conspecifics.

In physiology a similar accumulation phenomenon has been known since 1892 as *spinal contrast* (C. S. Sherrington, 1931). E. v. Holst (1937) investigated it in more detail in the sea horse. This fish swims about very little, usually fastening itself to some weeds with its prehensile tail; its dorsal fin is collapsed. The dorsal fin is raised only when the fish swims and it beats in an undulating movement. If the **spinal** cord is cut, this does not cause, as in most **fish,** uninhibited, spontaneous locomoting movements; instead the dorsal fin of the spinal sea horse remains in a half-raised position. If the animal is gently squeezed in the gill region the fin is folded down completely. When the fish is released the fin is raised a little higher than before. If this is repeated and the fish is held longer, upon release the dorsal fin is raised completely and begins to undulate, having been prevented from doing so by an inhibiting stimulus caused by the squeeze hold. The explanation for this peculiar phenomenon is that the **automatism** for swimming movement of the sea horse produces only a very small amount of endogenous excitatory potential, so that the potential has to accumulate before the swimming movements can occur. This accumulation is achieved by a reflexive inhibition of the swimming movements. In the spinal sea horse this inhibition is lacking, accumulation of excitatory potential does not take place, and the weak action-specific "excitation" is continuously discharged, causing the half-raised position of the dorsal fin.

In this connection we are reminded that E. v. Holst demonstrated the central spontaneity underlying swimming movements of some fish. The extent of the endogenous excitation production varies from species to species. In the **eel,** which swims a lot, a **spinal preparation** will show undulating movements until it dies. In the sea horse, which moves relatively little, the underlying automatism becomes visible only when continuous discharge is prevented experimentally in a spinal animal, and an accumulation of excitatory potential is achieved in this way. It is quite possible that the relative need for movement in higher vertebrates can be understood in terms of such differences in the accumulation of central excitatory potential. The lion, which stalks its prey, is a calm animal that may be kept in a small cage. Weasels and wolves, which run down their prey, have a great need to run and after feeding will continue to discharge this drive to satiation, by running up and down in their cages for hours.

K. Lorenz and E. v. Holst recognized the connection between these physiological states and observations of the spontaneity of instinctive behavior of intact animals. K. Lorenz hypothesized that each fixed action pattern—not only those of locomotion—is based upon the accumulation of central excitatory potential of E. v. Holst's automatisms. With this generalization he bridged the gap between behavior studies and physiology. The investigations of many ethologists and physiologists have since supported this view. K. D. Roeder (1955) noted that reflex movements possess an endogenous automatic base, which, however, is not sufficient for a spontaneous discharge.

We do not yet know what biochemical events within the central nervous system correspond to or are correlated with the observed fluctuations of the specific readiness to act (specific excitatory potentials). Perhaps the key to an understanding of these phenomena is in the catecholamine metabolism. Recent investigations indicate that animal and human behavior is dependent in some way on the catecholamine level (noradrenaline and dopamine) as well as on indolamine serotonin. Drugs that lower the central catecholamine level have a calming effect; those that raise this level stimulate motor activity and aggressiveness. In man they have an antidepressive effect. The manner in which these substances act is still unknown, but it has been suspected that they are involved in aiding synaptic transmission. Their accumulation and depletion at certain locations in the brain may explain the phenomena of central lowering as well as raising of thresholds (G. M. Everett, 1961; G. M. Everett and R. G. Wiegand, 1962; D. X. Freedman and N. J. Giarman, 1963; N. J. Giarman and D. X. Freedman, 1965; J. J. Schildkraut, 1965; J. J. Schildkraut and S. S. Kety, 1967).

I. J. Bak (1965) and R. Hassler and I. J. Bak (1966) were able to demonstrate the existence of submicroscopic catecholamine stores, which changed under the influence of drugs. Reserpine depletes these catecholamine stores, which possibly results in a loss of spontaneous activity. Following administration of iprozianid, catecholamine stores are increased and spontaneous movements increase at the same time.

The implications that follow if we postulate an accumulation of a central nervous excitatory potential as the basis for instinctive behavior are of great importance for human ethology. Many examples indicate that man is dependent upon an accumulation of central excitatory potential in some areas of his behavior, which is difficult to control because he is not conscious of it and which, together with other motivating factors, affects his inner readiness to act in a specific way. This may be true, for example, of the aggressive drive, which in present-day human society finds very few adequate opportunities for discharge. The constant endogenous accumulation of excitatory potential continues to lead man to seek a discharge for this drive, and in ignorance of his biological condition he projects his periodically occurring "anger" outward—in his personal daily life to those who are next to him (for example, the spouse) and in a larger context, perhaps against minorities or neighboring peoples. Only a clear understanding of the nature of these phenomena can help us in the search for reasonable solutions.

SUMMARY

Observations on intact animals have shown that fluctuations in their responsiveness to external stimuli are in part caused by built-in physiological mechanisms that act as "drives." These mechanisms motivate or cause an animal to actively seek, in what we call appetitive behavior, for stimuli situations that allow certain behavior patterns (according to the "mood" of the animal) to discharge. Behavioral analyses of motivation reveal that motor patterns occur in sets, sharing identical fluctuations of the releasing thresholds and thus pointing to underlying common physiological mechanisms. At the same time other behavior patterns are excluded.

A variety of motivating mechanisms are known. Inner sensory stimuli, hormones, and central nervous mechanisms are among these. There is no unitary drive concept. The same effect can be achieved by different mechanisms, as the study of "hunger" in flies and mammals demonstrates.

For a number of motor patterns neurogenic spontaneity has been proved and the hypothesis has been suggested that central excitation somehow accumulates, causing an increased readiness of the animal to act, sometimes even in the form of a "vacuum activity" (without the presence of releasing stimuli). The physiology of this process of damming up of "energy" is not known. Changes in the catecholamine level may in part be responsible.

5

Behavior as a Response to a Stimulus

THE INNATE RELEASING MECHANISM AS A BASIS FOR INNATE RECOGNITION

The concept of a central accumulation of excitatory potential whose discharge is inhibited by higher central controls requires the assumption of a special afferent mechanism that removes these inhibitions at the biologically appropriate moment. This neurosensory *innate releasing mechanism* (IRM) allows the central impulses to proceed to the effectors only when certain key stimuli are encountered. Key stimuli are usually simple. They can be discovered by means of experiments with models that are presented to inexperienced animals. Innate releasing mechanisms, which respond unselectively to the simplest stimuli, can become more selective through individual experience. The **toad,** which at first snaps unselectively at moving objects, soon learns to avoid noxious prey.

Unconditioned stimuli can also be inhibitory: The search automatism of infants—a rhythmic head movement when searching for the nipple—comes to an end as soon as the child touches the nipple with the mouth (H. F. R. Prechtl, 1958). Many precocial animals show an innate avoidance of a precipice, which they recognize visually before having had the adverse experience of falling off a cliff. Chicks, kids, lambs, and 4-week-old kittens that have never experienced a fall stop when they reach a cliff that is covered by the same glass plate on which they are standing, while the less visually oriented Norway rat walks without hesitation on the glass plate above the abyss (E. J. Gibson and R. D. Walk 1960). Changes in parallax while moving seem to be the effective stimuli. In order to exhibit the behavior the cats must merely have had visual pattern experience coupled with active movement. Cats that could walk in a rotating striped drum, but which were prevented from seeing their own feet, reacted to the visual cliff. Kittens that were carried passively through the same path traversed by the other kittens, but without the walking experience, did not avoid the visual cliff, although the visual impressions had been the same for both groups of animals (R. Held and A. Hein, 1963). Day-old chicks exhibit unlearned visual depth discrimination. The cues provided by focusing are critical for this discrimination, but not those from binocular and motion parallax (P. G. Shinkman, 1963). Three-day-old chicks that were never fed but grew up under normal lighting conditions preferred to peck at photographs of half-spheres that were illuminated from one side, provided they were "correctly" oriented (that is, with the bright side facing up). Three-dimensional objects normally are lighter on the upper side, because the light usually comes from above. The chicks prefer such photographs to those that are mounted upside down, even if they have been raised in cages illuminated from below, which points to an inborn capacity to utilize surface shadings as a stimulus parameter for three-dimensional objects (R. Dawkins, 1968). There undoubtedly exists, then, an innate ability to recognize releasing and inhibiting stimulus situations of a highly complex nature, as was also pointed out by W. McDougall (1936) and J. v. Uexküll (1921; see also Fig. 5-1).

Figure 5-1 Young and inexperienced cat at a visual cliff. Although the animal has had no experiences with a precipice, which is here also covered by a glass plate, it hesitates at the edge. (After E. Gibson and R. Walk in P. R. Marler and W. J. Hamilton, 1966.)

The innate releasing mechanism responsible for these actions is first defined in purely functional terms. It is a stimulus filter. Its seat within the brain is unknown in most cases. The investigations of H. R. Maturana and others (1960) of the leopard frog (*Rana pipiens*) show that analysis and integration of stimuli takes place as early as the retinal level. In the retina they found five types of ganglion cells, which respond only to different stimuli. One group fires briefly only when a light is turned on and off. They also respond to each moving edge and they fire during changes of illumination as well as during the passing of the leading and trailing edge of a stripe. If the image is stationary in the receptive area of the retina these cells do not respond; they only respond to changes in contrast—they are event detectors. Another group of cells does not respond to turning light on and off, only to the passing of a straight or curved edge. If the image stops the frequency of discharge drops to a lower level of continuous discharge. These cells inform the frog continually about the contours of objects; they are contour detectors.

One group of cells is of special interest, because it does not respond to a change of level of illumination but responds with vigorous discharges when a small object that is darker than its background passes over the receptive field. The authors call these "beetle detectors." Finally, there are special cells that measure the decrease in illumination and others that measure the light intensity. Here the selectivity of the stimulus filter exists in the retina. In the retina of the rabbit different nerve cells have been found that process the arriving impulses prior to entry into the central nervous system. There are cells that fire only when a dark object moves across the visual field in a certain direction (H. B. Barlow et al., 1964).

J. P. Ewert (1973, 1974a, 1974b) has investigated the process by which the visual stimuli characteristic of prey and predator are analyzed in the central nervous system of the toad (*Bufo bufo* L). Toads react to patterns that extend in the direction of movement ("wormlike objects") with prey-catching responses. Objects that extend perpendicular to the direction of movement release avoidance and flight.

Recordings from single fibers and single cells at different stages on the visual path showed that the processing takes place in three stages. In the retina three types of cells (Classes II, III, and IV) were identified with excitatory fields of 4° diameter for Class II, 8° for Class III, and 12° for Class IV. These three cell classes did not change their responses when the model was extended in the horizontal direction (Fig. 5-2*C*). Increase in the vertical plane brought about an increase of the discharge rate for Classes II and III until the diameter of the excitatory receptive field was reached. From then on the rate of discharge decreased. The discharge rate remained high, however, for Class IV cells (Fig. 5-2*A*,*B*). Prey normally excites cells of Classes II and III, enemies, Classes III and IV. For all the three classes the discharge rate increases with speed.

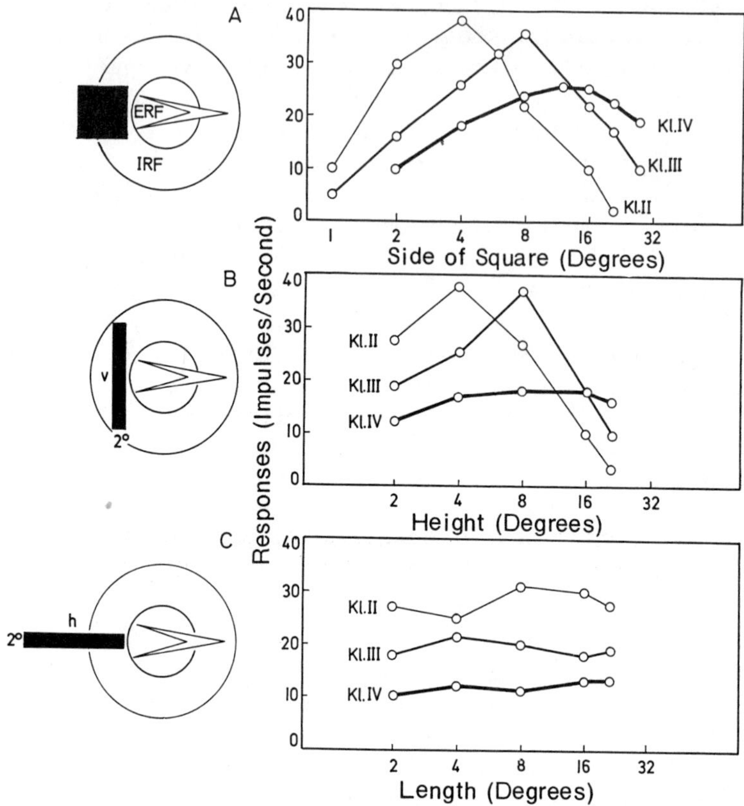

Figure 5-2 Responses of retinal ganglion cell Classes *(Kl)* II, III, and IV to black moving objects of different shapes. These objects were all moved at the same speed (7.6°/sec). *ERF,* excitatory receptive field; *IRF,* inhibitory receptive field. Further explanation in the text. (From J. P. Ewert, 1972.)

Although processing of the visual input data thus takes place in the retina the interpretation of the stimulus object in relation to behavior as a prey or a predator takes place more centrally.

In the mesencephalon (tectum opticum) each point of the retina is projected in a retinotopical order. Each location of the visual field is topographically represented (Fig. 5-3), as can be demonstrated by electrical recording; each projection field being activated only by the corresponding retinal area. Electrical stimulation of these tectal projection fields releases orientation movements which for each point are oriented to topographically defined locations in space and prey and also to prey-catching responses (Fig. 5-4 *left*). These tectal neurons undoubtedly represent a localization system. The tectum opticum also comprises a neuronal system that processes behaviorally relevant aspects of moving stimuli. Its cells are found in layers (Fig. 5-3), with an excitatory field about 27° in diameter. One type (Type I) is activated mainly by moving objects extended in the direction of movement. Extension perpendicular

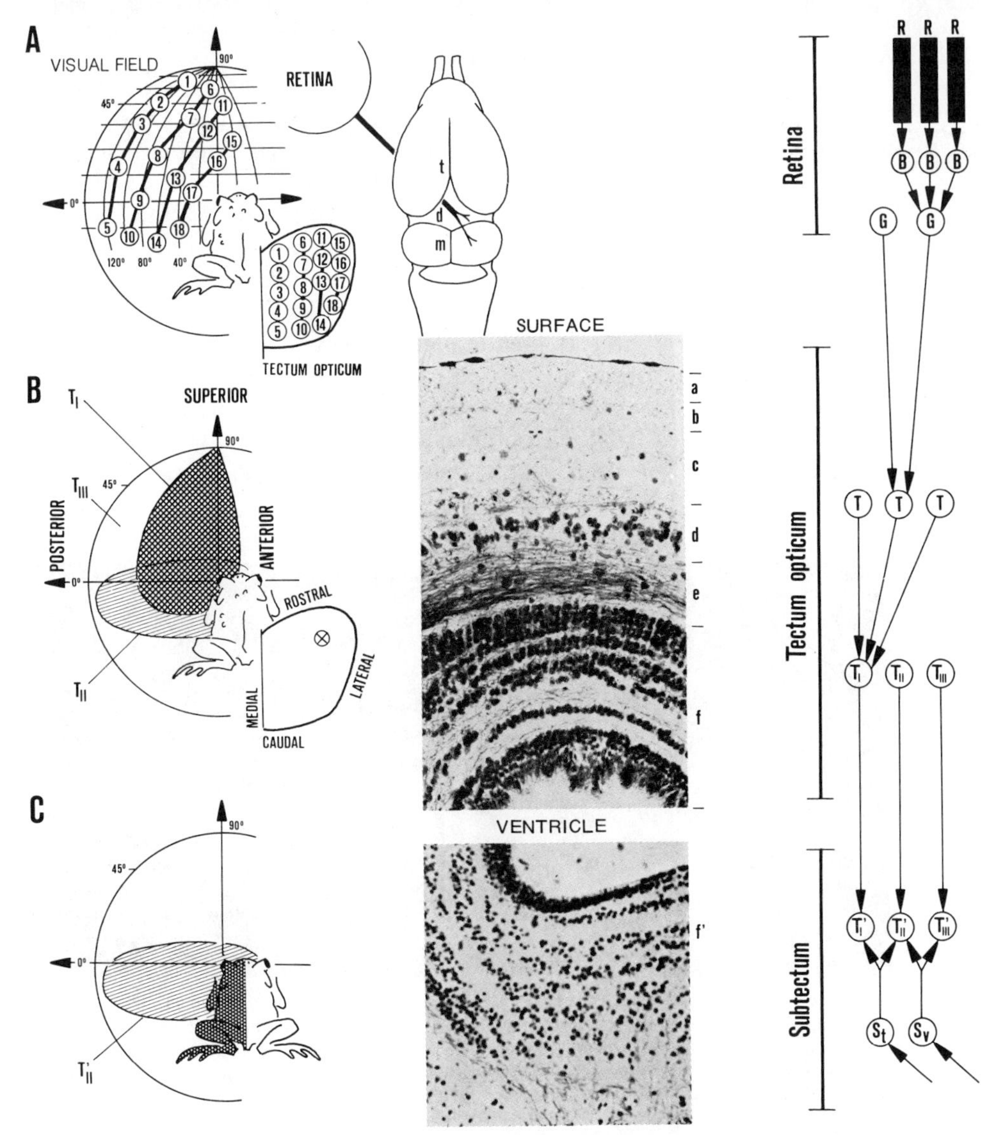

Figure 5-3 Location and size of the receptive fields of the retina (*A*), the tectum opticum (*B*), and the subtectum (*C*), and their correlation with certain tectal layers. The histological photograph (Klüver-Barrera) shows a transection of the tectum opticum and the subtectum.

A: Projection of the left visual field upon the surface of the contralateral (right) tectum opticum. Movement of an object in one part of the visual field (indicated by numbers) results in activations of retinal fibers in the correspondingly numbered parts of the projection field. Retinal Class II fibers end in tectum layers *a* and *b*, Class III fibers in *c*, and Class IV fibers in layer *d*. On the right: top view of the toad brain.

B: Location of the receptive fields of three different large-field units, T_I, T_{II}, and T_{III}, from the tectum layers *e* and *f*.

C: Visual (striped) and mechanoreceptive (dotted) fields of one of the visual- and tactile-sensitive units T'_{II} in the subtectum.

Diagram on the right: Supposed connections of the retinal, tectal, and subtectal neurons. *R*, receptor; *B*, bipolar cell; *G*, ganglion; *T*, movement-specific tectal small-field units; T_I, T_{II}, T_{III}, movement-specific tectal large-field units; T'_I, T'_{II}, T'_{III}, corresponding movement-sensitive large-field units from the subtectum with additional inputs from tactile- (S_t) and vibration-sensitive (S_v) neurons. (From J. P. Ewert, 1973.)

to the direction of movement does not have this effect. Other tectal neurons (Type II) differ from Type I in that their discharge rate actually diminishes with surface extension perpendicular to direction of movements (Figs. 5-5 and 5-6). They constitute the trigger system for the prey-catching response. In deeper tectal layers (Fig. 5-3 *e, f*) units with very large receptive fields are found. The receptive fields of these units cover the contralateral frontal (T_I), lower (T_{II}), and upper (T_{III}) visual field (Fig. 5-3). They probably help to localize large objects.

The thalamic pretectal region provides a "caution" system. The visually sensitive neurons (receptive field 40–90° diameter) in this region are activated by stimuli objects extended perpendicular to the direction of motion, movements of objects toward the toad, and large stationary objects, all situations that call for evasive movements. Brain stimulation of this region releases ducking, turning away, and even panicky flight.

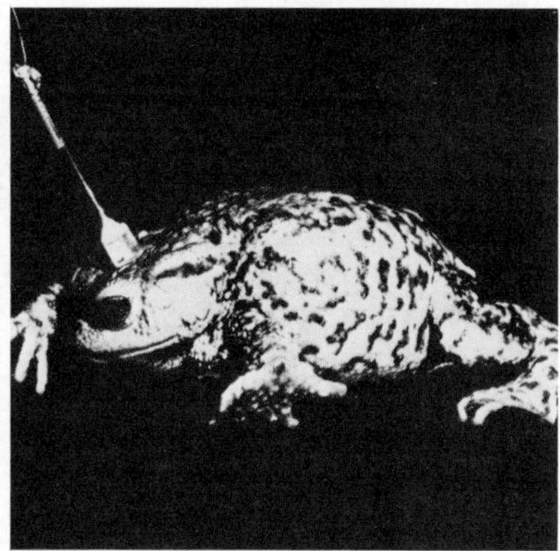

Figure 5-4 *Left:* Electrical stimulation of the tectum opticum with exclusion of optical stimuli releases a prey-capturing reaction in the freely moving animal. The toad draws itself up as if it saw an object of prey in the upper left part of the field of vision. *Right:* Stimulation of the thalamus praetectum, on the contrary, activates escape behavior. The toad cowers and turns away as if it saw an enemy in the upper left part of the field of vision. (From J. P. Ewert, 1973.)

It is assumed from these findings that the retinal projection fields in the mesencephalon (tectum opticum), which are responsible for turning toward the stimulus object, and those in the diencephalon (thalamus praetectum), which are responsible for turning away from a stimulus object, are both involved in the stimulus-identification process by subtractive interaction of their respective nerve nets. The thalamic signal is supposed to be inhibitory upon the trigger units in the tectum, since after removal of the thalamus the tectum cells respond even to vertical bars and other objects irrelevant for prey catching (Fig. 5-6). The interaction is schematically depicted in Figure 5-7. Every step of the analysis is, for reasons of clearness, depicted as if it would work like a window discrimi-

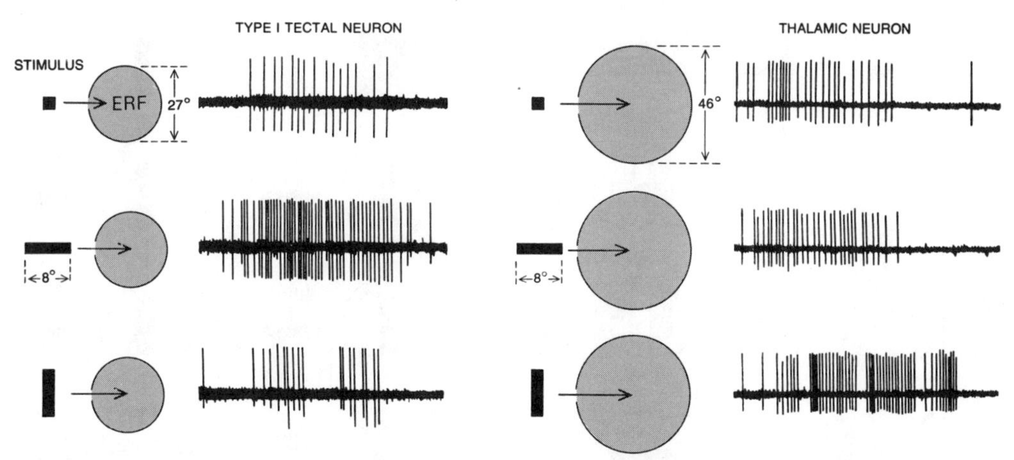

Figure 5-5 Feature detection beyond the retinal level is accomplished by cells in the tectum and the thalamus. Recordings from individual cells indicate that tectal Type I neurons (*left*) are most activated if the object moving through the field is extended in the direction of movement. The cells in the thalamic area (*right*) respond most to an object extended perpendicularly to the direction of movement. (Ewert, J. P. "The Neural Basis of Visually Guided Behavior." Copyright © 1974 by Scientific American, Inc. All rights reserved.)

nator. The retinal Class II and III neurons code extension of an object perpendicular to the direction of movements (vertical window). Tectal nerve nets (Type I) code object extension in the horizontal direction (horizontal window). In another tectal layer (II) impulses from the thalamic pretectal region inhibit, while impulses from the tectum I cells stimulate the units ("trigger units," II).

The visual world as perceived by the ganglion cells of the retina is thus projected in the retinal projection fields of the brain and by special

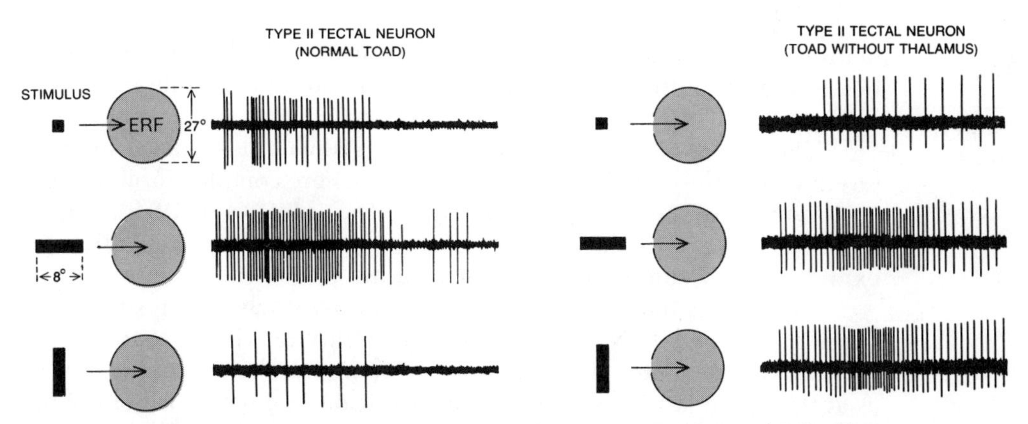

Figure 5-6 Trigger units for the entire prey-catching response seem to be the Type II tectile neurons. In the normal toad (*left*) the cells are most activated by wormlike objects (*horizontal bar*). They are less activated (and the decrease is greater than in the case of Type I tectal neurons) by stimuli that in behavioral experiments are irrelevant for prey catching (*vertical bar*). After removal of the thalamus, however (*right*), their response to these irrelevant stimuli is greatly increased, suggesting that the thalamic signal is inhibitory. (Ewert, J. P. "The Neural Basis of Visually Guided Behavior." Copyright © 1974 by Scientific American, Inc. All rights reserved.)

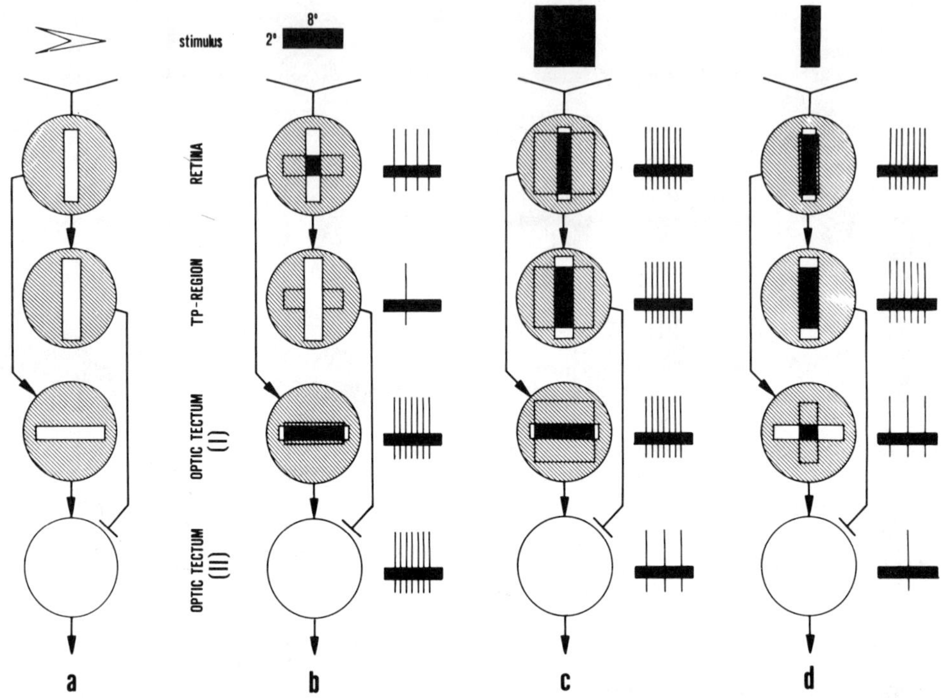

Figure 5-7 Logic linking of neuron networks (*circles*) in different sections of the visual pathway: retina, diencephalon, and mesencephalon. An arrow means stimulation, a line ending in a bar inhibiting influences. The shape parameter of a moving stimulus pattern (horizontal and vertical stripes or squares) can be seen in the corresponding window. The activity within each network is indicated by the inscribed pulse pattern. The "window discriminators" are a symbolic depiction to serve to understand the various steps of shape analysis. Object extension beyond the vertical window frame (retina and TP region) results in inhibiting influence. In *b* no such inhibiting influence occurs, while in *c* and *d* the prey-catching responses are inhibited. (From J. P. Ewert, 1973.)

neuronal circuits the visual information is screened, whereby genetically stored information is compared with patterns incoming from the filter passage.

The processing of visual information seems to take place in a different way in the visual cortex of mammals, where complex forms are not represented by special types of neurons (J. T. McIlwain, 1972). No object detectors are found, but objects are represented by populations of exited neurons, different cell classes coding different characteristics.

The events that occur in the retina of the cat have been studied by D. H. Hubel and T. N. Wiesel (1959). Contrary to what is found in the frog, where the center of an on-off area responds at the beginning as well as at the end of a light stimulus, the corresponding receptor field of the light-adapted cat is so arranged that an *on* area is surrounded by a peripheral *off* area. The investigations of D. H. Hubel and T. N. Wiesel (1959, 1962) dealt with information-processing beyond the retinal organization by recording the activity of single neurons in the cor-

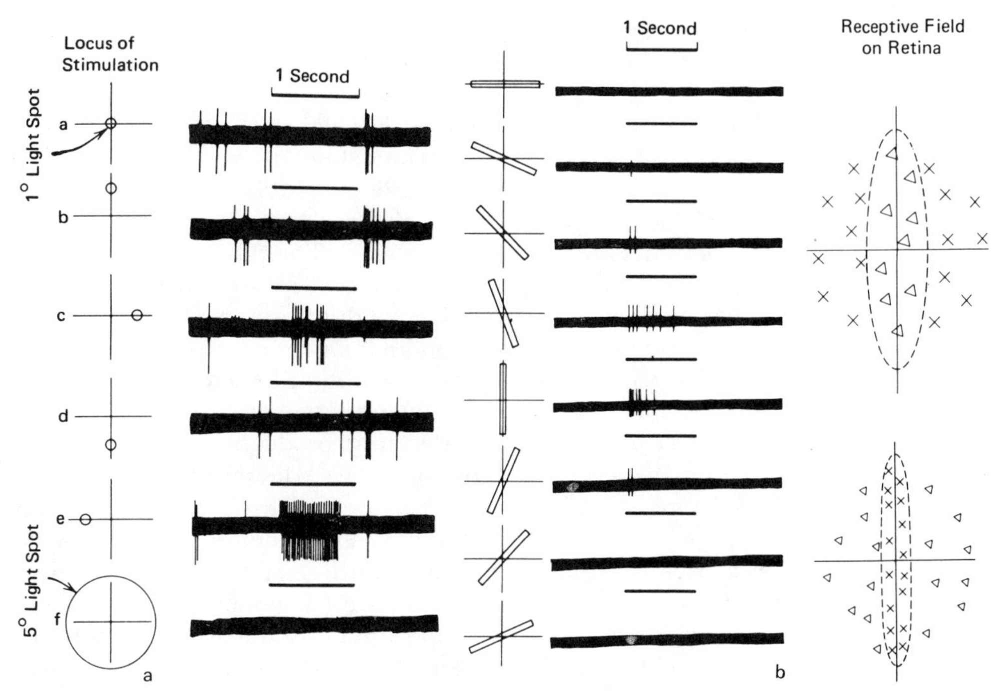

Figure 5-8 *a:* Responses of a cell in the striate cortex of a cat during stimulation of various points of the retina with a 1° light spot (*a*) and (*b*) with a bar of light (1° × 5°) that was rotated in seven different positions around the receptive area. The receptive areas of the retina are shown on the right (above for *a* and below for *b*); *triangles* denote inhibiting fields, *crosses* excitatory fields. (After Hubel and Wiesel, 1959, from P. R. Marler and W. J. Hamilton, 1966.)

pora geniculata, in the cortex striatum, the visual center of the cat brain, along with the retinal events. The cortical cells interact with the retinal receptor areas; they are connected with specific ganglion cells in such a way that a certain retinal stimulus area is projected onto a particular cortical cell. If the retina is stimulated with a narrow long band of light instead of with a spot of light, and if one records from a cell in the striate cortex, then one obtains responses of various amplitudes depending on the position and type of movement of the image in the light-stimulated area. Thus we encounter within the visual system data processing by steps that begin at the retinal level (Fig. 5-8).

The investigations of D. H. Hubel and T. N. Wiesel (1963) demonstrate that many of the complex physiological stimulus-processing functions within the cortex of the cat can be found in the newborn kitten. The neural connections that underlie this complex data processing must be present at birth, although the newborn animals appear not to be able to utilize this functioning visual system. Visual deprivation results in degeneration within the brain that can be demonstrated histologically, but this is secondary. The lack of visual impressions does not prevent the development of neural connections but leads to the degeneration of those present at birth (T. N. Wiesel and D. H. Hubel, 1963a, 1963b, 1965).

The work of C. Blakemore and G. F. Cooper (1970) and C. Blakemore and D. E. Mitchell (1973) is of interest here. They have shown by experiments on cats, in which young kittens were reared completely in the dark except for short periods each day in cylindrical chambers with only vertical or horizontal stripes, that almost all cortical visual neurones responded only to angles within 45° of the stripes to which they had been habituated. This was an active modification, as there were no "silent areas" or signs of degeneration. In the later paper they report that there is a particularly sensitive period at around 28 days of age, during which even one hour's exposure produces a strong preference for the relative field. This visual deficit is long-lasting, producing an effect similar to human astigmatism (D. W. Muir and D. E. Mitchell, 1973).

C. Blakemore and D. E. Mitchell state that these results seem to indicate a direct relation between learning and memory in this type of modification of cell response. As the infantile condition of little selectivity persists when the eyelids are sutured together, it would seem probable that it shows an innate disposition of the learning pattern. R. Mize and E. Murphy (1973) have attempted to repeat this work in the rabbit without success and therefore suggest that this animal lacks the neural plasticity seen in some other mammalian species.

Electrical recordings from the antennas of male silk moths led D. Schneider (1962) to the conclusion that the specificity for the sexual odor of the females is dependent upon the structure of the receptors.

There are a number of inborn mechanisms that process and integrate sensory data. The processing starts in the sensory receptor and we have learned that the integration of the data can occur at different levels of the central nervous system. Of course not all of these data-integrating mechanisms are releasing mechanisms.

On the whole we know much less about the localization and nature of releasing mechanisms than about the way they work. This will be examined in Chapter 6 (a comprehensive discussion can be found in W. M. Schleidt, 1962, 1964b).

KEY STIMULI AND RELEASERS

Behavior may be activated by an internal drive but is normally released by specific stimuli from the environment. During the breeding season a male stickback will attack male conspecifics, but court females. What are the cues by which these differential responses are triggered?

Before we can discuss this question we must recall that each animal can perceive with its sense organs only a limited portion of the total environment. Its external environment, or *Umwelt*, according to J. v. Uexküll, is made up of the particular stimuli that the animal perceives. Whoever studies the reactions of an animal must first be acquainted with its sensory-physiological capacities, because these differ from species to species. Some examples may illustrate this: Bees can see ultraviolet light and distinguish polarized from nonpolarized light—capacities that man does not possess. Bats can hear ultrasonic sounds that we cannot hear, but they do not see very well. But because they can establish for themselves an "image" (*Abbild*) of their environment by means of the echoes from their own calls, they are able to orient as effectively in flight as birds with good vision. Nile pike send out electrical impulses and react to small differences in the potential of the surrounding electrical field. Bees accept sugar substitutes that are tasteless to us. Pit vipers are very sensitive to infrared radiation and perceive temperature differences of 0.005°C. Red-breasted robins (*Erithacus rubecula*) utilize the earth's magnetic field for navigation.

The species differ qualitatively and quantitatively. Guinea pigs, being macrosmatic animals, are able to detect nitrobenzene in a 1/1000 dilution of the concentration we can sense. Some performances tax our comprehension. Eels respond to odorous substances in dilutions of 1:2.9 million billion. This, according to H. Autrum (1943, 1948), is equal to 1 ml of substance dissolved in a body of water 58 times the volume of Lake Constance. The olfactory cells of the silk moth respond with a neural impulse to a single odor molecule (K. E. Kaissling and E. Priesner, 1970). The minimum output requirement for the human eye is roughly 10^{-17} watt. To better illustrate this fact, Autrum offers the following comparison: According to physicists, the universe is 10 billion years old, equal to 3×10^{17} seconds. If from the beginning of the world's existence we had tapped, from some source of energy, a continuous output of 5×10^{-17} watt, from then to the present it would have expended no more than 15 watt-seconds. This would be the energy expended by burning a 15-watt lamp for one second.

More important than absolute thresholds are differential thresholds. In minnows the difference threshold for the pitch of a tone is half a note. Old World monkeys distinguish as many colors as we do. The critical distance between the wavelengths of two just-distinguishable colors in the red region is 10 mμ and in the blue region is approximately 9 mμ (W. F. Grether, 1939).

Finally, of utmost biological significance is the ability of an organism to locate a stimulus source with respect to distance and direction. Here the visual sense attains the greatest precision and range by means of the highly developed camera eyes of vertebrates and the compound eyes of insects. However, bats also locate their prey with the echoes of their ultrasonic cries. We shall discuss the capacities of the sense organs and their functioning in space in more detail in Chapter 16.

The functioning and capacities of the sense organs are studied by sense physiologists. The methods may vary. The conditioning method has been most successful: A certain reaction—for example, eating—becomes associated with a specific stimulus—for example, a whistle. When the animal has formed this association, one tries to determine, with the aid of appropriate controls, which sense mediated the relevant stimulus. For example, if one puts the whistle in the mouth without blowing it and the animal does not react, this indicates that the animal heard and reacted to the sound and has been conditioned to it. In this way K. v. Frisch (1923) demonstrated hearing in the dwarf sheatfish. The same author used the conditioning technique to demonstrate color vision in the honeybee. C. v. Hess (1913) placed bees into a darkened room with two different-colored lights of differential brightness. Bees approached the brighter of the lights regardless of whether it was red or green. They oriented to brightness, and v. Hess concluded that bees are color blind. K. v. Frisch (1914), who was unwilling to believe that an insect which searches out flowers would be color blind, offered food to bees on yellow paper which was placed between papers of various shades of gray. The bees did not confuse the yellow with a single shade of gray—hence they could see color. This indicates that the animals could behave differently in different functional systems. The investigations of N. Tinbergen and his collaborators (1943) give an additional, impressive illustration. Male grayling butterflies (*Eumenis semele*) fly randomly toward models of females of various colors as if they were color blind. When approaching blossoms, however, they prefer certain colors over others and distinguish them from shades of gray of equal brightness. In this functional system they do demonstrate color vision.

The capacities of sense organs can be elegantly investigated electrophysiologically by the recording of action potentials (H. Autrum, 1958; J. Schwartzkopff, 1962). The structure and functioning of sense organs are discussed in detail by H. Autrum (1952), W. v. Buddenbrock (1952), R. Granit (1955), D. Burkhardt (1960, 1961), L. J. and M. Milne (1963), D. Burkhardt, W. M. Schleidt, and H. Altner (1966), and H. Heran (1966).

Of all the sensory stimuli perceived by an animal, only relatively few innately release reactions. In dogs only stimuli from food objects initially release salivary secretion, and after appropriate **conditioning a dog** will show this reaction later to a bell or light stimulus.

We must therefore distinguish between the *perceived* and the *effective* stimuli (N. Tinbergen, 1951). The former are the subjects of sensory physiology; the latter are studied by ethologists.

It has been shown in numerous investigations that "unconditioned" stimuli and stimulus patterns exist to which an animal will react with appropriate actions prior to any experience with them. It has been demonstrated that the sense organs may serve quite different functional systems. The carnivorous water beetle (*Dytiscus marginalis*) reacts with prey-catching behavior, not to a moving tadpole in a glass vial, which it normally attacks readily, but to the odor of meat extract (Fig. 5-9).

Which specific stimuli or stimulus combinations release a specific reaction is determined by the use of models. Observation alone will often tell the ethologist something of their nature. Whenever a toad spies a moving insect, it fixates upon it and moves toward it. As soon as the insect becomes motionless, nothing further happens. The toad continues to stare at the spot where the insect was last moving, and will rarely snap at the motionless prey. After some time the toad seems to lose interest.

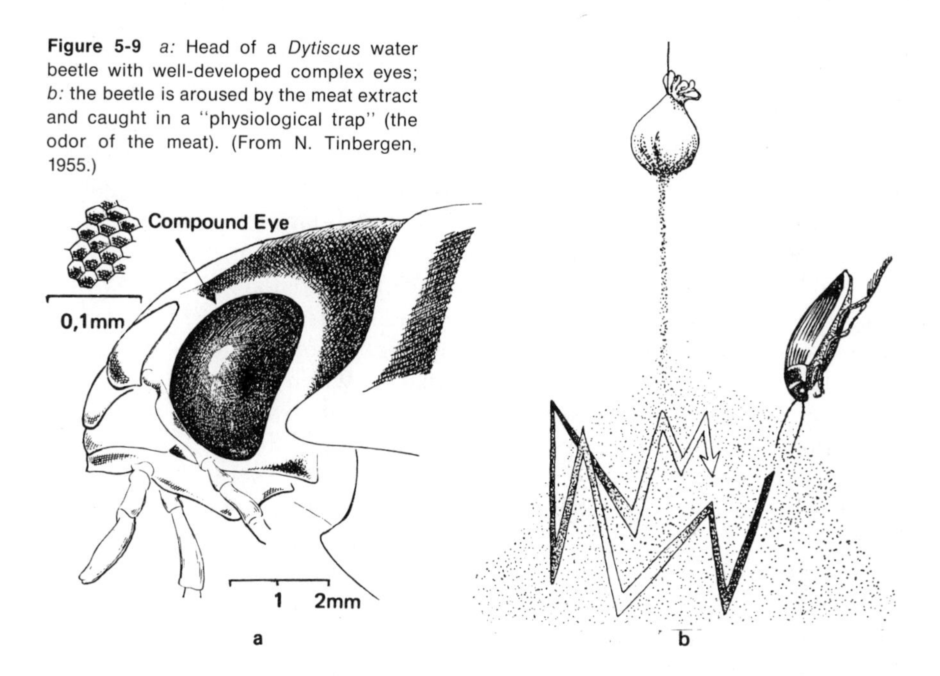

Figure 5-9 *a:* Head of a *Dytiscus* water beetle with well-developed complex eyes; *b:* the beetle is aroused by the meat extract and caught in a "physiological trap" (the odor of the meat). (From N. Tinbergen, 1955.)

Thus it appears as if the movement of a prey is a strong releasing stimulus for the prey-catching response. Experiments with models support this. If we move objects such as stones with a string, the toad will fixate and snap at them. However, if the object is over a certain size it will release escape (I. Eibl-Eibesfeldt, 1951a, 1962a). Toads react this way shortly after metamorphosis. Neurophysiological studies finally reveal the mechanisms underlying the processing of the visual data (see p. 67). In the clawed frog (*Xenopus laevis*) visual stimuli and vibrations in the water will release the prey-catching behavior, and will do this even in the inexperienced animal, as can be readily demonstrated. Tadpoles are plankton filterers that never snap at objects with an oriented movement. If they are isolated in clear water shortly before metamorphosis, so that they have never snapped at prey, and one then projects a light spot against the background of the container, the clawed frog will at once swim toward it, make the specific fanning movements with the forelimbs, and snap at it. A small jet of water from a pipette or slight touch of the legs also releases turning toward the stimulus and the prey-catching movements. The animal snaps at the stream of water.

This behavior will normally lead a frog to its prey, because in general small moving objects in its environment are prey animals. Toads and frogs can learn something new very quickly. If a toad has snapped at a leaf repeatedly without success it will refrain from doing so in the future. The same is true if it has caught a bad-tasting or even stinging insect. For avoidance to take place only one unpleasant experience is necessary. The innate knowledge is limited to snapping at small, moving objects. These characteristics sufficiently identify the natural prey objects. Similarly unselective is the response of a dragonfly larva to small moving objects; it snaps at small objects and flees from larger ones (M. Hoppenheit, 1964). When deprived of food for some time, larger objects are taken than before. Pike innately snap at moving objects. Whitefish (*Coregonus wartmanni*) fixate and snap even at floating, nonmoving particles (E. Braum, 1963).

Such releasing cues are called key stimuli. In analogy to a key that opens a lock, the key stimuli act upon a mechanism (the IRM) that normally prevents the release of central impulses when it is not appropriate and will only open the way to the musculature when the appropriate key stimuli are received. Each functional cycle has its own key stimuli, and the animal reacts differently to correspondingly different key stimuli. For a herring gull to roll an egg into its nest it must be spotted; when the gull is robbing eggs this characteristic is of no importance (J. P. Kruijt, 1958).

Key stimuli exist for almost all senses. The night moths of the families Noctuidae and Geometridae show escape reactions of dropping, flying downward, or other evasive maneuvers when they hear the ultrasonic cries of bats or artificially produced ultrasonic stimuli (K. D. Roeder and E. A. Treat, 1961). Crickets and grasshoppers react to their own specific song in a predictable fashion (J. Regen, 1924; A. S. Weih, 1951; A. Faber, 1953a, 1953b; W. Jacobs, 1953a, 1953b, A. C. Perdeck, 1958a). The males of the mosquito *Aëdes aegypti* react selectively to the whirring sound of the female's wings (L. M. Roth, 1948; H. Risler, 1953, 1955), and many frogs respond to the calls of their own species (C. M. Bogert, 1961).

The female turkey recognizes her chick only by its calls and will brood a stuffed polecat that is fitted with a loudspeaker uttering the call of newborn turkeys. She will kill her own young if she is deaf and cannot hear their calls (W. M. Schleidt, M. Schleidt, and M. Magg, 1960). Mallard ducklings (*Anas platyrhynchos*) and wood ducks (*Aix sponsa*) each prefer the calls of their own mothers (G. Gottlieb, 1965a). It might be that ducklings and chicks which were incubated could hear their own calls while still in the egg and would generalize from them to their mothers. If this were so, newly hatched chicks should prefer their own or calls from other chicks over the somewhat different calls of their mother. This, however, is not the case, as G. Gottlieb (1966) found in later experiments. The chicks always preferred the specific call of their mother, even if prior to hatching they were exposed to additional calls of other chicks. In this case they will later follow their mother's call even better, with a shorter latency and a higher proportion of followers, than those chicks that could only hear themselves in the egg. The ability to recognize the call note of their own species is therefore acquired as a phylogenetic adaptation. It is quite possible that the innate releasing mechanism which underlies this capacity to respond is facilitated in its embryonic development by auditory stimuli. A summary of investigations of acoustical key stimuli and releasers can be found in R. G. Busnel (1964).

Spiders react to slight vibrations of their net with prey-catching behavior. The ant lion throws sand at the ant if loose sand comes tumbling down into the funnel it has constructed. Water boatmen react to vibrations of the water surface. Disturbances in the surrounding water release prey catching or search in many fishes and clawed frogs (G. Kramer, 1933).

Special alarm substances warn fish swarms and the tadpoles of the toad; they all escape when they detect substances secreted by an injured member of their species (K. v. Frisch, 1941; I. Eibl-Eibesfeldt, 1949; F. Schutz, 1956; W. Pfeiffer, 1963). Alarmed honeybees secrete an odorous substance through their widely exposed cloacae which excites the others and makes them aggressive (K. v. Frisch, 1965).

Some gastropods of the tidal zone escape when they perceive substances that predatory starfish secrete from their feet, but they do not flee when a plant-eating starfish approaches (T. H. Bullock, 1953).

The male silk moth is very sensitive to the sexual odor of the females (I. Schwink, 1955), which was analyzed by A. Butenandt (1955) and A. Butenandt, R. Beckmann, D. Stamm, and E. Hecker (1959). In man, too, there are sex-specific reactions. Men, and women before sexual maturity and after menopause, can barely detect the odor of certain arousing substances derived from musk glands, whereas young women can smell these substances, which are widely used in perfumes, especially about two weeks after menstruation. Men can detect this substance following estrogen injection (J. LeMagnen, 1952). Additional examples of communication via the **sense of smell** can be found in E. O. Wilson (1963, 1965) and G. W. K. Cavill and P. L. Robertson (1965). Substances that are effective in communication are called *pheromones*.

About 14 days after giving birth the female albino rat emits a pheromone. This date coincides with the age when the young become responsive to the smell of the pheromone and are attracted by it. The mother ceases to emit the smell around 27 days postpartum, which again coincides with the age when the young cease to be attracted by the pheromone. The young around day 16 approach any lactating mother emitting this odor. The pheromone is under prolactin control but is also dependent upon external stimuli, since its emission can be prevented if the mother is made to experience neonatal pups only during the first two weeks (M. Leon and H. Moltz, 1971, 1972; H. Moltz and M. Leon 1973). More details about olfactory communication in mammals can be found in J. F. Eisenberg and D. G. Kleiman (1972).

Chemical stimuli often aid in the search for food. The moray eel hunts at night with the aid of its olfactory sense. The cuttlefish is able to neutralize the olfactory sense of this predator by secreting its "ink" as part of its defensive reaction (I. Bardach et al., 1959). Many parasites find their hosts by means of olfaction, and they selectively react to the specific odor of the host species (D. Davenport, 1955; G. Osche, 1962; M. Lindauer, 1963). Many marine polychaetes that live on shrimp and starfish react to the odorous substances that are diffused in the water around the host species. The ichneumon fly (*Pimpla bicolor*), which parasitizes the larvae of the South African night moth (*Euprotis terminalis*), is attracted by the latter's odor. The ichneumon fly (*Alysia manductor*), which parasitizes fly larvae, approaches only the odor of fresh flesh. *Nasonia vitripennis*, which only attacks pupae of flies, is only attracted by the odor of decaying flesh. The coddling moth (*Carpocapsa pomonella*) is only attracted by the odor of apples. Newly hatched garter snakes (*Thampnophis species*) react selectively to extracts of certain prey prior to all feeding experiences (G. M. Burghardt, 1966).

In sharks the odor of blood releases search for prey. They are able to locate bait with their olfactory sense (I. Eibl-Eibesfeldt and H. Hass, 1959; see also Fig. 5-10). For mosquitos, bedbugs, and mites, the key stimulus that attracts them is the heat radiated from the warm-blooded animals upon which they feed. L. J. Milne and M. Milne (1963) reported that an electric clock which radiated heat attracted mites each night, as did the chickens in whose house the clock was mounted. The clock, which because of this did not work, began to run again when the mites left in the morning. Later we shall cite additional examples of visual key stimuli.

It can be demonstrated that innate behavior is activated by key stimuli, so there must have evolved corresponding releasing mechanisms that serve as stimulus filters. Such an adaptation may be unilateral, in that only the recipient is adapted to a specific relevant environmental situation. This is true for the perceptual mechanisms of a predator, while its prey will not develop special signals that would make it more recognizable; on the contrary, the prey develops characteristics that make it as hard as possible for the predator to recognize. It is quite different where the contact between two organisms is of selective advantage for both, as in the interrelationships between mates, between the mother and young, or in a symbiotic relationship. In these instances receiver and sender of a signal are mutually adapted to one another. A fish is not apt to evolve a signal for its predator, but a female is apt to evolve a distinguishing characteristic for the male. These signals may consist of special morphological structures, odorous substances, calls, or conspicuous movements or postures. Such highly differentiated structures and behavior patterns that serve special signal functions have been called *Auslöser* (social releasers) by K. Lorenz (1935).

Figure 5-10 Sharks (*Carcharhinus menisorrah*) aroused by odors from a bait placed above the reef (Maldive Islands). In the lower picture one shark has grasped the bait. (Photographs: I. Eibl-Eibesfeldt.)

As man is especially adapted to the perception of visual signals, most investigations have been concerned with visual key stimuli and releasers, especially in fish and birds. For signaling we find bodily structures such as plumage patterns, color patches, and manes, as well as special behavior patterns that have been called **expressive movements.** The latter we shall discuss separately.

Various methods of analysis are available for the investigation of releasing stimuli. G. K. Noble and B. Curtis (1939) offered jewel fish females a choice between males exhibiting courtship coloration and others with inconspicuous coloration, by presenting one each on either side of an aquarium containing the female. The females always spawned on the side of the conspicuously colored males. If, however, a colorful male was blinded with eye cups, which resulted in lack of movements, the female spawned next to the colorless but active male.

Releasing stimuli can also be studied by making changes on the living animal. The schooling fish *Pristella riddlei* has a conspicuous dorsal fin with a black mark. A group of such fish with amputated dorsal fins are less attractive to an isolated fish than a group of intact fish, so one may assume that the black dorsal fin is a visual following signal (M. H. A. Keenleyside, 1955; see also Fig. 5-11). G. K. Noble (1934) investigated the stimuli that release fighting in the lizard *Sceloporus undulatus.* Only the males of this species possess a blue stripe at the border of the belly and a blue patch on the throat. When Noble painted such stripes on a female it was attacked. On the other hand, males courted males when he had painted over their markings. If one paints the black bills of juvenile zebra finches (*Taeniopygia castanotis*) red so that they are like those of the adults, they are not fed by the adults, despite their intense begging behavior (K. Immelmann, 1959).

Employing Skinner's instrumental conditioning method, G. P. Sackett (1966) demonstrated innate recognition of threat expressions in rhesus monkeys. Four male and four female monkeys were raised in isolation from conspecifics from birth until nine months. Controlled visual experience consisted of slides projected against the cage wall that showed monkeys and neutral objects (sunset, landscape with trees, geometric figures, and so on). After each presentation the monkeys could again project the slide they had just seen by pressing a lever. They could do this repeatedly during a 5-minute period and in each projection the picture was visible for 15 seconds. The monkeys soon learned how to operate the projector, and they viewed the pictures they preferred more

Figure 5-11 The fish *Pristella riddlei,* showing the dark spot on the dorsal fin that serves as a releaser for the following reaction by conspecifics. (After M. H. A. Keenleyside, 1955.)

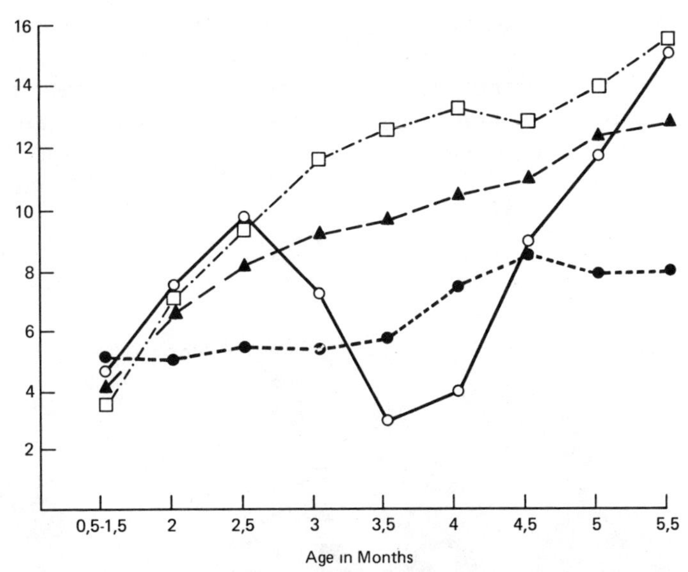

Figure 5-12 Frequency of bar presses for viewing of pictures by rhesus monkeys raised in isolation. The ordinate shows the average number of bar presses per five-minute test. The abscissa shows the age (in months). The curves show the reactions of the monkeys to various slides: *open circles,* a threatening conspecific; *squares,* a young monkey; *triangles,* other monkeys; and *filled circles,* control pictures. (After G. P. Sackett, 1966.)

often than others. It was found that they preferred pictures of conspecifics, especially pictures of a young monkey and of a threatening adult. These two pictures released also the most frequent social responses (vocalization, invitation to play, climbing about, and visual and manipulatory exploration of the pictures). At 2½ months of age the young monkeys suddenly reacted with fear to the picture of the threatening adult. They

Figure 5-13 Two robin models. *Left:* A stuffed young bird without red breast feathers; *right:* a bundle of red feathers. (Explanation in the text.) (After D. Lack, 1943.)

withdrew before the picture, crouched, clasped themselves, and showed fearful facial expressions. At the same time lever presses for this picture were markedly reduced (Fig. 5-12). Not until 2 months later was an increase noted for this picture. The animals recognized the threatening expression at 2½ months, although they had never seen a conspecific or their own mirror image. There must exist then an innate releasing mechanism that matures in the absence of social experience. Habituation may account for the increasing interest at a later time.

For the experimental analysis of the releasing stimuli the technique of adding components to a model is frequently used. First an attempt is made to release the behavior in question in inexperienced animals by the simplest of all possible stimuli. Prior observation aids in

Figure 5-14 *Above:* The male frog clasps the female; *below, left:* two fingers forms an acceptable model of a female; *right:* the male clasps the tip of a rubber boot. (Photographs: *above,* I. Eibl-Eibesfeldt; *below,* H. Sielmann in I. Eibl-Eibesfeldt, 1954.)

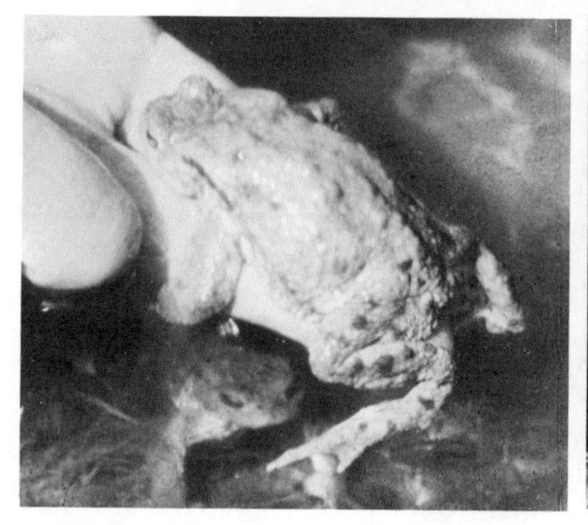

making the initial selection. In this way D. Lack (1943) released the most intense fighting behavior in the redbreast (*Erithacus rubecula*) by presenting a bundle of red breast feathers in the territory of the male. A stuffed juvenile without red feathers was ignored (Fig. 5-13). This leads to the assumption that the behavior of territorial defense is released in the redbreast by the red breast feathers. Similar results were obtained by V. A. Peiponen (1960) in the bluethroat, where the blue breast feathers are the releaser.

A male toad approaches all moving objects during the breeding season and tries to clasp them. The male releases its grip only if the clasped object gives the defensive call of another male. Females remain silent and continue to be clasped, as are carps, the human hand, and so on (Fig. 5-14). The innate releasing mechanism is quite unselective in this case, but suffices, because during the breeding season mainly toads of this species are encountered in ponds.

In comparing the effectiveness of different models they are usually presented consecutively. Only a few experiments are possible with a single subject because learning is quite rapid. This shortcoming of the method of successive presentation of stimuli can be overcome by presenting two different models simultaneously to the animal. D. Franck (1966) investigated the pecking reaction of the chicks of the common gull to beak models, comparing the methods of simultaneous and successive presentation. Both showed the same results; however, by simultaneous comparison Franck was able to measure differential preferences that could not be detected by successive presentation.

We owe an exemplary analysis of innate releasing mechanisms to E. Kuenzer and P. Kuenzer (1962). The following reaction of the young of cichlids that spawn on the seabed is released by the movement and coloration of the mother. Form and size have no effect. The key stimuli for the following reaction are species-specific and correspond to the reproductive colors of the females. Young of *Apistogramma reitzigi* approach yellow models, while the young of *A. borelli* approach models that are painted with contrasting black and yellow. In a similar manner, in *Nannacara anomala* the selectivity with which the young react to various models with the following reaction corresponds exactly to the behavior and appearance of the parents. The young in these experiments were without experience, so the development of the innate releasing mechanism and releaser must have occurred during phylogenesis (P. Kuenzer, 1968).

In the stickleback the red belly releases fighting; a crude wax model with a red underside, but lacking all other fish characteristics, is attached at once, while models resembling a stickleback but without the red markings rarely release fighting (Fig. 5-15). It is important, however, that the underside is red; if the model is turned upside down it loses its fight-releasing qualities. The cues release fighting only when they are presented in a particular relation to others, in this case "red below" (N. Tinbergen, 1948).

A male stickleback recognizes the female by her swollen abdomen, which she presents in a definite manner. One can imitate the swollen abdomen and the posture with a simple dummy and elicit mating behavior. Even animals raised in complete isolation react correctly to the signals of the same or opposite sex of their own species. No differences can be observed between the latter and animals that have developed normally (E. Cullen, 1960).

A relationship between two characteristics (cues) constitutes a *configurational stimulus*. For example, a red dot at the tip of the bill releases the food-begging response of the herring gull (N. Tinbergen and A. C. Perdeck, 1950). The blackbird (*Turdus merula*) gapes toward a simple

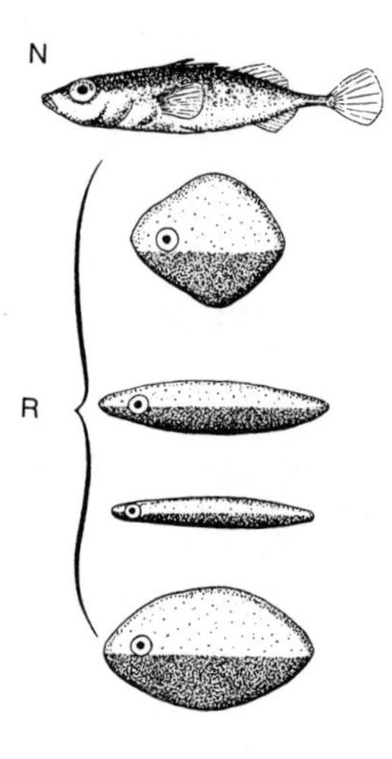

Figure 5-15 Stickleback models. *N*, a carefully made form- and color-true imitation of a stickleback without a red belly. It is less frequently attacked than the four simple red-bellied models of series *R*. (After N. Tinbergen, 1951.)

model of an adult where the head and the body are represented by two black disks of different sizes. The smaller of the two disks is gaped at as if it were the head of the adult; here the key stimulus is the size of the head in proportion to the body. If two-headed models are offered, the birds will prefer one of them and orient to a certain size of the head. The two pictures of the models (Fig. 5-16) show in each case heads of the same size but bodies of different sizes. When the model shown in Figure 5-16*a* is presented the smaller of the two heads is responded to; in *b*, the larger is gaped at. The animals respond to the key stimulus of a certain size relationship between head and body. As in gestalt perception, relations between stimuli are attended to and this seems to be true for perception in general: If a bird is trained to respond to the lighter of two gray stimuli, and then the darker of the two is exchanged for one lighter than the other (which had been the positive stimulus) the bird will prefer this new, lighter stimulus. This is also true of key stimuli. In the mouth-breeding cichlid *Haplochromis multicolor*, the young disappear into the mouth cavity of the mother when danger approaches. They also try to enter a simple model of the mother's head and orient to the position of the eyes and a point between them. If the eye spots are positioned on a horizontal plane, the model is more effective than when they are placed vertically (H. M. Peters, 1937a). In *Tilapia mossambica* the young gather predominantly near the undersides of disk-shaped models. They also approach dark spots and try to dig into any depression. This response would normally lead them into the mouth of the mother (G. P. Baerends, 1957; see also Fig. 5-17).

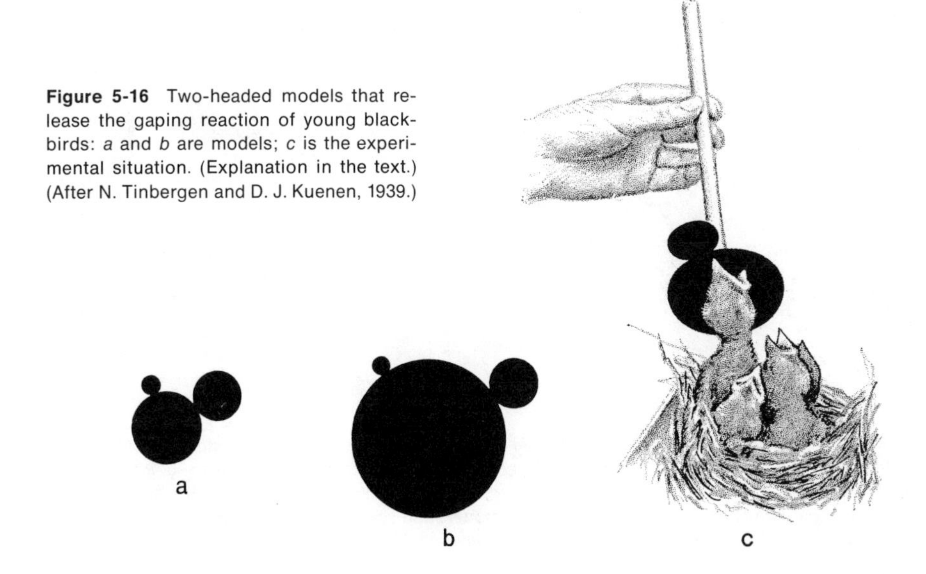

Figure 5-16 Two-headed models that release the gaping reaction of young blackbirds: *a* and *b* are models; *c* is the experimental situation. (Explanation in the text.) (After N. Tinbergen and D. J. Kuenen, 1939.)

In response to a silhouette of a model pulled across the sky, turkeys, geese, and ducks show specific escape reactions. K. Lorenz (1939) found that in geese and ducks the important cue is the "relative speed" of the moving silhouette (expressed in diameters per time unit) and that the form of the object is without relevance. However, in his experiments with turkeys he found that a hawk-shaped model was more effective than one in the shape of a goose. A model shaped like a cross with two arms of even length, with a long rear end and a short head end, frightened the turkeys when it was pulled so that the short end pointed forward; the same model pulled in the opposite direction, so that the long post pointed forward like the outstretched neck of a goose, proved relatively ineffective. Lorenz concluded that the "short-neckedness" of the escape-releasing model was an innate characteristic. N. Tinbergen (1948, 1951) generalized this opinion to "gallinaceous birds, ducks, and geese." W. M. Schleidt (1961a, 1961b), in an attempt to repeat the original experiments with turkeys, found that the relative speed is also of importance in this species. However, his turkeys, without previous experience of flying objects, responded equally well to shapes with long or short necks and to silhouettes of birds of prey as well as to simple disks. Presenting such models at different frequencies (for example, the "long-neck" 10 times more often than the "shortneck"), the more frequently shown model soon decreased in effectiveness, while the rarely shown ones lost little of their frightening effect. In his pilot study Lorenz had used turkeys that had been frequently exposed to ducks and geese flying overhead, so we can expect that they were already habituated to long-necked flying objects before the experiments were started, and therefore only the short-necked models appeared sufficiently different to elicit escape responses. D. Müller's (1961) investigations of the escape-releasing

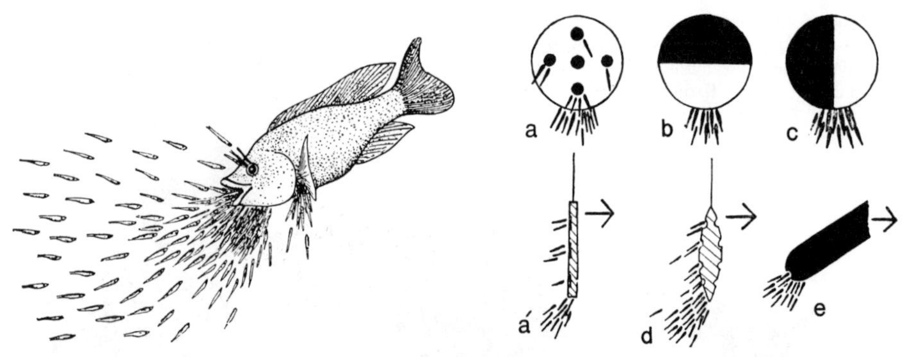

Figure 5-17 Model experiments with *Tilapia mossambica,* a mouthbreeding cichlid. *Left:* Mother with young, which move toward her mouth; *right:* Models in the form of a flat disk: *a–c*; *a*, a disk with indentations; *d*, a disk with pits; and *e*, a dark-colored test tube with an opening at the bottom. When the young fish are startled they approach these models and seek to enter them at the underside. They are also attracted by dark spots. (From G. P. Baerends, 1957.)

stimuli in inexperienced capercaillies also failed to show differences in response to various shapes or to provide evidence for an optimal effective range of "relative speed." Recent experiments by Schleidt indicate that turkeys, exposed to a variety of shapes of equal size but all shown at the same frequency, habituate more to some than to others. This would indicate some "unlearned" preference for certain shapes over others; however, no models clearly resembled a bird of prey.

The male lightning bug (*Lampyris noctiluca*) reacts specifically to the configuration of the species-specific stimulus pattern of the female's light organ, which consists of two parallel bars, one behind the other, and two dots. Stencils of this pattern illuminated with a flashlight from behind elicit approach by the males in preference over other patterns. The males of *Phausis splendidula* possess a less selective innate releasing mechanism, in that they always approach models of larger than na-

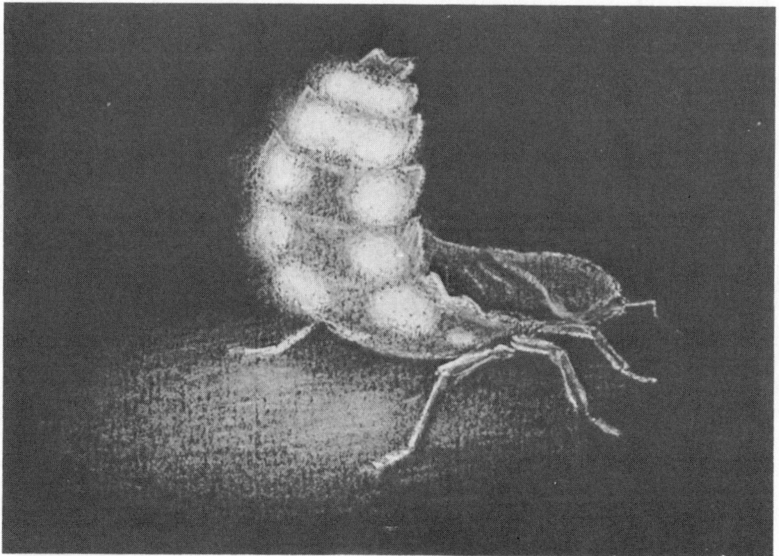

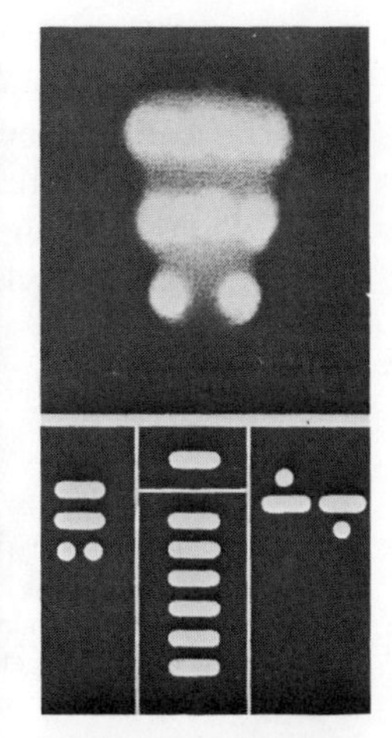

Figure 5-18 *Left: Phausis* female in the typical lighting position. *Right, top:* light organ of *Lampyris noctiluca; bottom:* light organ of *Phausis splendidula* (both females). Below in three columns each: models ordered by effectiveness. The real signal of the species is shown above each of the sets of three columns. On the left the most effective model, on the right the least effective one. (After F. Schaller and H. Schwalb, 1961, in W. Wickler 1967a; H. Kacher, artist.)

tural size (up to four times), even when they showed great departures from the species-specific pattern (F. Schaller and H. Schwalb, 1961; see also Fig. 5-18).

The pattern on the head of the cichlid ***Haplochromis burtoni*** includes a black stripe running over the eye. The male of this species reacts with increased aggression to a model carrying this trait. Consequently, W. Heiligenberg, U. Kramer, and V. Schulz (1972) constructed a model whose eye stripe could be rotated to different positions (Fig. 5-19). If the whole fish was rotated without a change in the relative position of the stripe, the releasing effect of the model remained constant in most positions. On the other hand, if the stripe itself was rotated a clear dependence of the aggression-releasing effect on the angle of the stripe was observed. The experimenters measured the biting rate among young conspecifics living in the same aquarium, which increased markedly after a short viewing of the model. Figure 5-20 shows the biting rate as a function of stripe angle. It is evident that the stripe achieves its greatest aggression-releasing effect when positioned at 0° or 180°. We may ask why the eye stripe did not develop in this position as an optimal

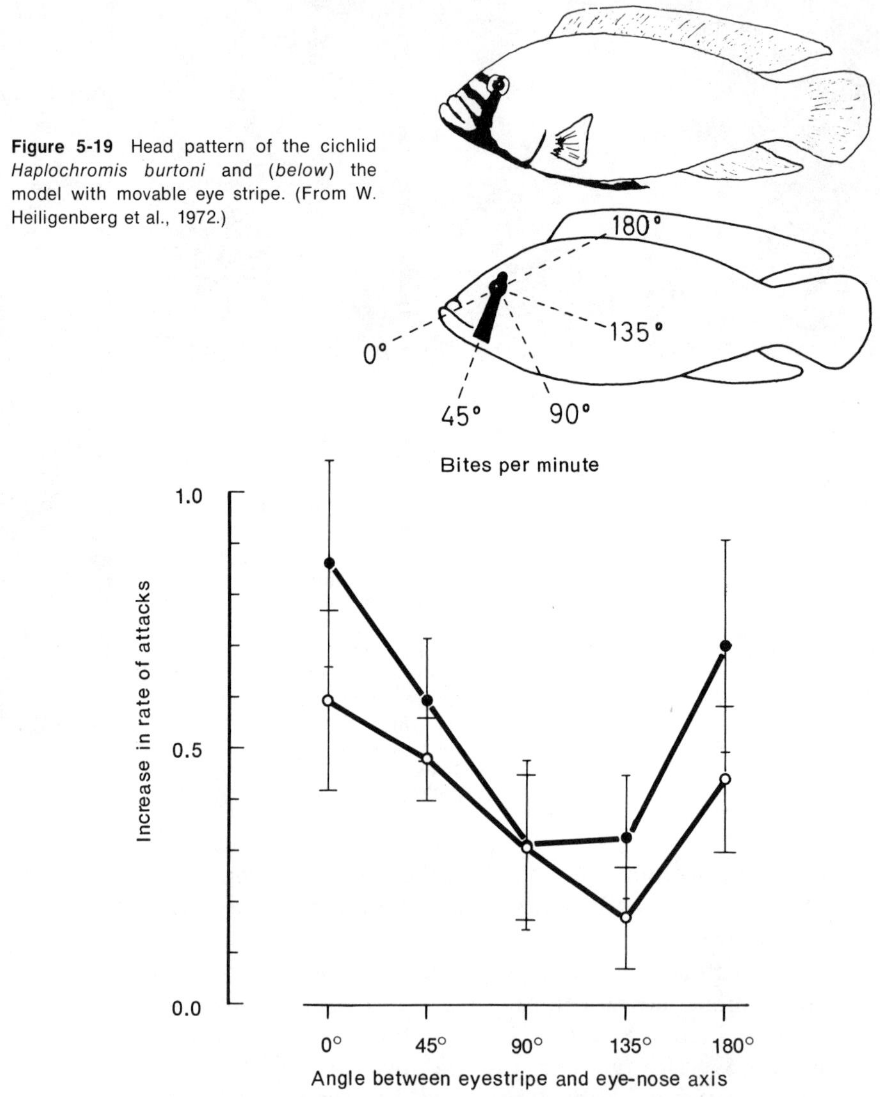

Figure 5-19 Head pattern of the cichlid *Haplochromis burtoni* and (*below*) the model with movable eye stripe. (From W. Heiligenberg et al., 1972.)

Figure 5-20 Mean increase of rate of attack as a function of the angle between eye stripe and eye–nose axis of models that were presented either horizontally (*open circles*) or vertically (*solid circles*). (From W. Heiligenberg et al, 1972.)

aggression releaser. The same characteristic is revealed when the model is stood on its head, however, and an upward shift of the curve can be observed. This species stands on its head when its threatening becomes intense.

The same behavior is often elicited by several different key stimuli. These stimuli, which can also be presented separately, become additive in their effectiveness if they are combined. We have already mentioned that the red belly of a stickleback is a strong fight-releasing stimulus. Sticklebacks threatening one another will each assume a head-down position. Models without red undersides, which rarely release reactions when presented in a horizontal position, release fighting when presented in a head-down posture. The head-down position, a behavioral cue, is therefore a fight-releasing stimulus. A red-bellied model, which releases fighting in the horizontal position, becomes increasingly effective if shown head down (N. Tinbergen, 1951). This *law of heterogeneous summation* was first described by A. Seitz (1940).

The males of the cichlid *Astatotilapia strigigena*[1] are blue with black marks on the dorsal and ventral fins. These characteristics are displayed during threat and also constitute the lowest intensity of fighting behavior. The displaying fish shows his lateral side with erected fins to the opponent. What happens next depends on the behavior of the other. If it responds similarly and shows breeding coloration, the opponents will stand parallel to one another and spread the skin that covers the branchiostegal rays. This leads to the second stage of the hostile encounter. The males exchange tail beats. They stand parallel to each other, facing in opposite directions, and hit the opponent in its face. This is followed by the third stage, the actual fight, in which one male opens its mouth and rams the other wherever it can reach it. The other fish will always try to escape and attempts in turn to ram his opponent, so a circling ("merry-go-round") results.

Seitz found that the blue coloration, the black marks at the fins, as well as the behavior patterns of lateral position, spreading fins, tail beats, and ramming thrusts, each by itself released threat behavior of varied intensity. The stimulus components are exchangeable up to a certain degree. The tail beating by a model without reproductive coloration is as effective as a model that shows only spreading of fins and the reproductive coloration. If all these characteristics are combined a stronger response is obtained.

U. Weidmann (1959) investigated the stimulus-summation phenomenon quantitatively. He released the peck reaction in young black-headed gulls by simple cardboard models and counted the number of

[1] The species has not been clearly identified. It is probable that this is a species of *Hemihaplochromis*.

pecks directed at them. In this way he found that a gray, round cardboard disk, for example, received y number of pecks as opposed to y' number of pecks for a square model. If he painted both models red,[2] the number of pecks increased by the same amount x for each model. The effects of various releasing stimuli can be additive, but they do not add up in such a simple manner in all cases, as was shown by E. Curio (1961, 1963).

In the cichlid *Haplochromis burtoni,* two pattern designs of the male may influence the aggressiveness of conspecifics of the same sex.

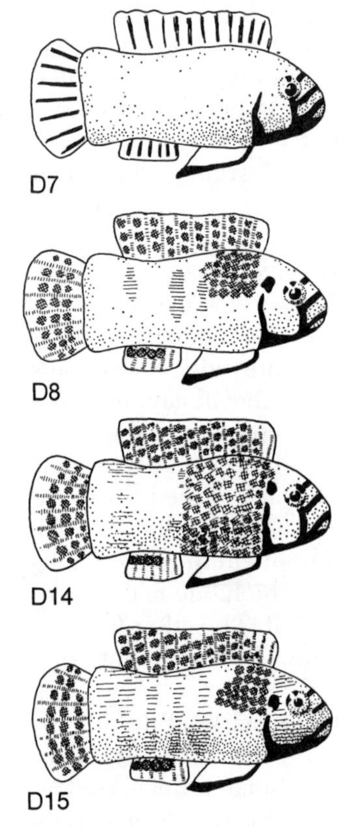

Figure 5-21 Four of the models employed by Leong: *solid black,* black; *cross-hatched,* orange-red; *horizontal lines,* blue. (From C. Y. Leong, 1969.)

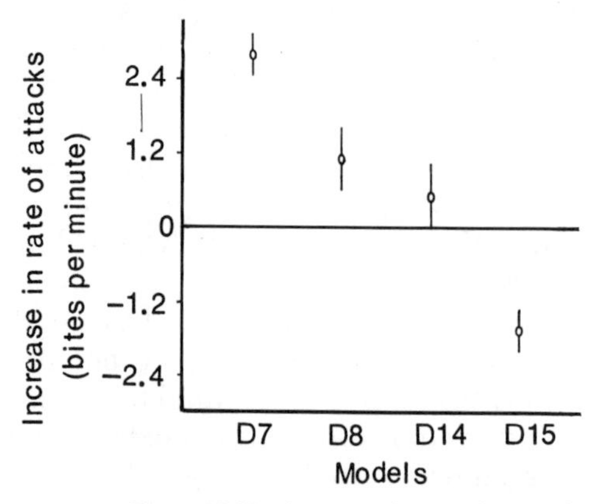

Figure 5-22 Average changes in attack rates after presentation of the models D7, D8, D14, and D15. The vertical stripes indicate deviations from the mean. (From C. Y. Leong, 1969.)

Compared to the established basic value a vertical stripe in the head pattern of a model effected an increase in biting rate of 2.79 bites per minute against young conspecifics continuously present in the same aquarium. An orange-red patch above the pectoral fin effected a decline of 1.77 bites per minute. A model combining both color traits effected an increase of 1.08 bites per minute, equal to the sum of the values effected by each component when presenting individually (C. Y. Leong, 1969). We have here the finest example to date of Seitz's rule of summation of stimuli (Figs. 5-21 and 5-22).

A behavior is dependent not only upon the strength of the releasing

[2] Adult gulls have red bills, and red releases pecking.

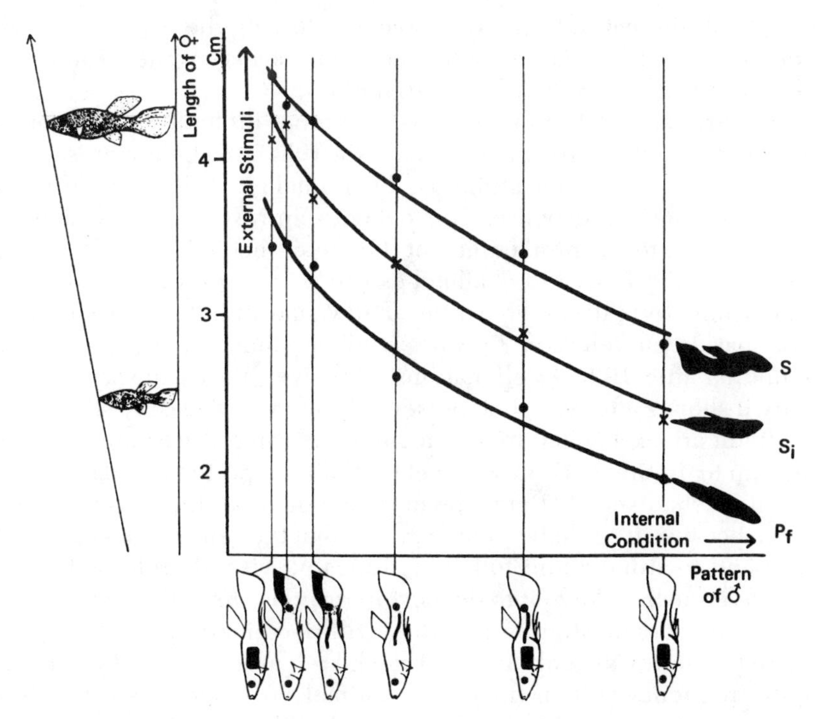

Figure 5-23 Interaction of internal and external factors, illustrated in the courting behavior of *Lebistes*. The strength of the external stimulus is indicated by the size of the female (ordinate). The internal state can be read from the melanophore pattern (abscissa). The relative positions of these patterns on the abscissa were derived from recording the readiness to perform sexual behavior patterns. The curves show the combinations that release the same behavior with respect to the behavior patterns of swimming after (P_f) and *S* forms of the male (two intensities, *S* and S_i). (From G. P. Baerends, 1956.)

stimuli, but, as I have said earlier, also upon the internal readiness to respond. This has to be considered in experiments with models. When the readiness to respond is high, even a weak stimulus can release the complete behavior with full intensity. On the other hand, an animal with low motivation may show strong reactions only when the releasing situation is especially effective. The interaction of internal and external factors has been studied by G. P. Baerends, R. Brower, and H. Waterbolk (1955). First, they determined how the various conspicuous markings of male guppies are correlated with their specific sexual motivation. This gave them good indicators of the inner readiness to respond with sexual behavior. They were able to present releasing stimuli of varying degrees of sexual motivation and found that both factors compensate one another in a lawful relationship (Fig. 5-23).

In order to assess the effectiveness of a model one has to know whether the animal is ready to react at all. This is done by testing the animal with the natural releasing object following the experiment with models—but not before, because the behavior may wane in intensity. This is the *method of double quantification.*

Experiments with models are made more difficult by a phenomenon called *afferent throttling* (M. Schleidt, 1954). If a behavior is repeatedly released by the same stimulus, the animal reacts less and less until there is no longer a response. This decrease in responsiveness does not have to be due to a central fatigue of the motor mechanisms. The gaping reaction of 5- to 7-day-old chaffinches can be released by vibrations of the nest, imitated calls of the parents, and visual stimuli. If the gaping reaction has been released by one kind of stimulus, the young stop responding after 10 to 13 elicitations. However, they gape with full intensity if shortly afterward one presents a different releasing stimulus. If the stimuli are exchanged in this manner one can release gaping in such a bird up to 46 times. H. F. R. Prechtl (1953a) reported an adaptation of afferent mechanisms that must be more central than the sense organ involved, because it could be demonstrated that the sense organ continued to respond to stimulation: following repeated visually released gaping the birds crouched down into the nest in response to further repetition of the stimulus, demonstrating that they still perceived it.

In the turkey the gobbling call can be released by sound stimuli of a certain frequency. When finally the animal no longer responds, it remains ready to respond fully if stimulated with a new frequency.

It is remarkable that artificial stimulus situations can be developed which surpass natural releasing objects in effectiveness. This was discovered by O. Koehler and A. Zagarus (1937). The ringed plover, for example, will roll white eggs with black spots into his nest in preference to his own, which have dark-brown spots. Even more surprising is its preference for large eggs. An egg four times the size of its own is preferred, although the bird is unable to sit and incubate it properly (Fig. 5-24). Male grayling butterflies (*Eumenis semele* L.) approach black models more frequently than those with natural colors (N. Tinbergen et al., 1943). Male butterflies (*Argynnis paphia* L.) prefer models that have the species-typical brown coloration, but the illumination, size of the colored area, and number of stimulus changes per unit of time can be exaggerated. A horizontal rotating cylinder with brown horizontal bars was preferred by males over an actual female. The greatest number of approach flights was obtained when the cylinder rotated so quickly that the males could just barely perceive the change between brown and dark stripes (D. Magnus, 1954, 1958).

Firefly males (*Lampyris noctiluca*) always prefer the pattern of light of their own species, but they prefer, as reported earlier, a model with a larger illuminated area. They also prefer a model that contains a larger amount of yellow than is contained in the light produced by their own females (F. Schaller and H. Schwalb, 1961).

A herring gull chick will respond to a sharp red rod with three white rings with more pecking responses per time unit than it will to the realistic three-dimensional model of a gull's head (N. Tinbergen, 1963; Fig. 5-25).

This responsiveness to "supernormal" releasers is exploited by some parasites. O. Heinroth referred to the European cuckoo as the scourge of songbirds because the gaping mouth of the young cuckoo releases in the foster parents stronger reactions than do their own young.

Figure 5-24 Supernormal models: the oyster catcher tries to roll a giant egg into its nest. The bird prefers it to its own normal-sized one. (After N. Tinbergen, 1951.)

This exaggeration of the releasing stimuli also shows that the evolution of the existing releasers is not necessarily completed. This may be due to counteracting selection pressures. A signal should be as conspicuous and unique as possible; that is, it should not be confused with other signals and thus lead to errors. Thus, from the receiver comes a selection pressure in the direction of conspicuousness and uniqueness with a corresponding consistency on the part of the sender of the signal. But whatever is conspicuous is also more readily seen by a predator; hence a selection pressure in the opposite direction exists. The result frequently is a compromise. Many bony fishes, for example, carry their releasers on fins that can be folded. During courtship they spread these fins and wave them in a manner that exposes their signals. Other fish can quickly change their colors. The unicorn fish *Naso tapeinosoma* Bleeker, which lives above coral reefs in the open water, normally has an inconspicuous,

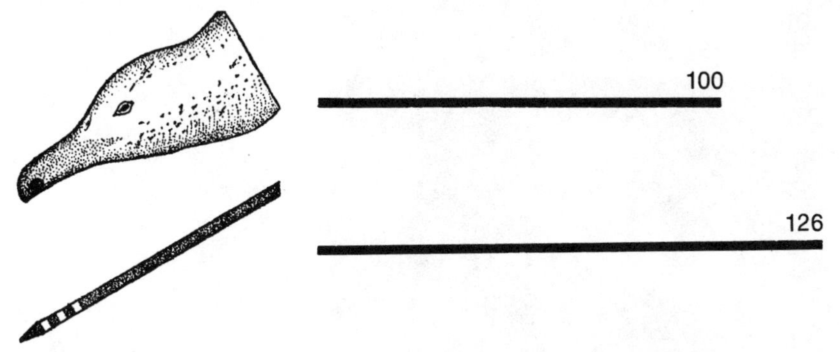

Figure 5-25 A thin red rod with three white rings releases more pecking responses per time unit in a herring gull chick than a realistic plaster head model. (From N. Tinbergen, 1963.)

darkish coloration. However, when a male courts a female he develops within seconds a light-blue saddlelike spot on his back, similar vertical stripes on the sides, blue lips, and a blue caudal fin (Fig. 5-26). As quickly as these brilliant colors appear, the former darkish colors reappear when the fish ceases to court (I. Eibl-Eibesfeldt, 1962b). Many cichlids, which normally possess stripes that camouflage the outlines of their bodies, are also capable of these sudden physiological color changes. During fights and courtship they acquire very conspicuous patterns and colors. They are even capable of displaying several such colorful dresses and are thus able to present several signals (Fig. 5-27 and Plate I).

That these various colorations and patterns are appropriately reacted to by other fish was shown by H. Albrecht (1966b) in *Haplochromis wingatii*, which when seeking to escape shows horizontal stripes, and vertical stripes in an aggressive mood. A mother with young does not attack them when they have horizontal stripes but will do so when one establishes a territory and shows vertical stripes. She also attacks models with vertical stripes but not those with horizontal stripes.

In the territorial fish of the coral reef, which in their adult state are rarely captured by predators, releasers have frequently been developed without apparent compromise. Many coral fish exhibit their optical signals continuously. They appear to be moving advertising signs. In other words, they are quite conspicuous, and they share with posters the char-

Figure 5-26 *Above:* Swarm of *Naso tapeinosoma* Bleeker (Maldive Islands) with inconspicuous swarm coloration; *below:* coloration of the courting male (light-blue lips, light-blue saddle spot, cross-stripes, and light-blue tailfin). This is an example of color changes for the purpose of signaling. (Photographs: I. Eibl-Eibesfeldt.)

acteristic of not being readily confused with others despite the simplicity of their signals (Plate II). The patterns possess a high degree of improbability; that is, it would be highly unlikely that another fish would evolve the same pattern, unless it were a **mimic.** Fish primarily fight with conspecifics or, if they have no opportunity to do so, with species similar in appearance (K. Lorenz, 1962; D. Zumpe, 1965; see also Fig 5-28). If a mirror is placed on a coral reef many fish will fight their mirror images (Fig. 5-29). The gaping mouths of many altricial birds present conspicuous visual signals. The young of the cave-nesting gouldian finch develop light-reflecting papillae (Plate II). Many grass finches recognize their young by their species-specific gape markings in the mouth. The **widow birds** (Viduinae), which parasitize these species by laying eggs in the nests of the grass finches, imitate these gape markings exactly. Finally, there is the example of a confusing signal that has the function of misleading predators. In the tropical seas there exist several species of saber-toothed blennies (*Runula*, *Aspidontus*, and so on) that

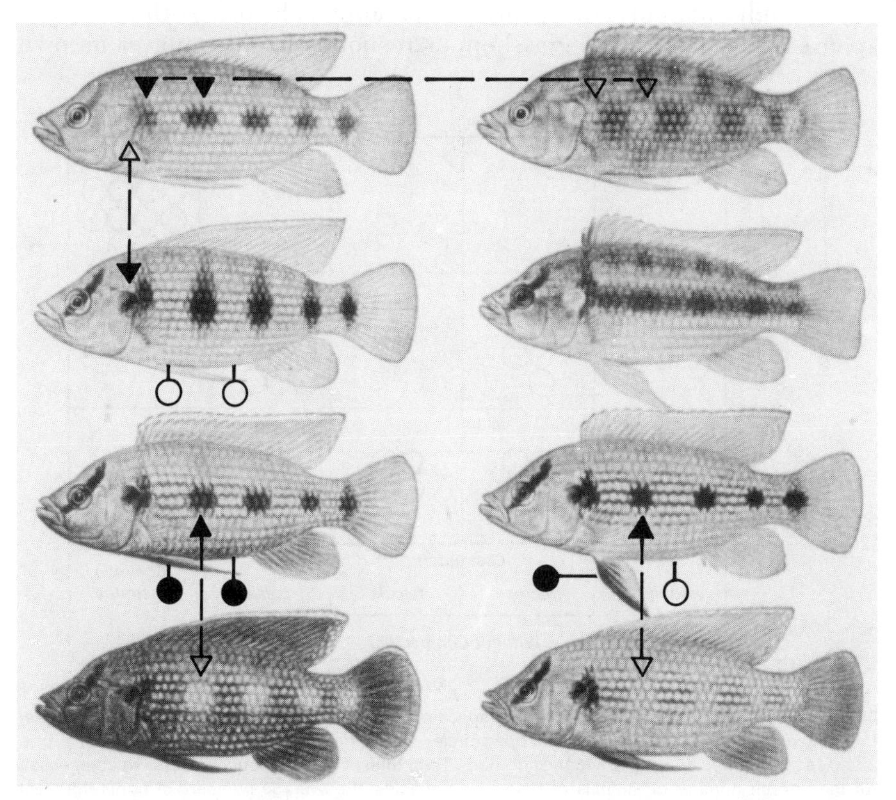

Figure 5-27 Explanation of Plate I. The color patterns of *Hemichromis fasciatus.* The symbols denote pattern components that vary independently of one another. They can be light (*empty symbols*) or dark (*filled symbols*). This illustration is simplified; all phases are connected by smooth intergradations. (From W. Wickler, 1965e.)

specialize in biting off pieces of skin and fins from other fishes for food (I. Eibl-Eibesfeldt, 1955a, 1959). They attack the eyes preferentially, and in many coral fish these are camouflaged by a dark eyeband. In addition, some species have developed an eyespot elsewhere on their bodies, and these are reported to divert attacks from the eyes (W. Wickler, 1961b; see also Fig. 5-30).

T. C. Schneirla (1965) has advanced the hypothesis that all reactions can be explained by a simple principle of approach and withdrawal. Weak stimuli or stimuli decreasing in intensity would activate a system that results in approach, whereas strong stimuli or stimuli increasing in intensity would result in withdrawal. This is sometimes true. With increasing voltage stimulation E. v. Holst and U. v. Saint Paul (1960) observed a change from attack to escape in their brain-stimulated chickens. It is also known that toads approach small worms with prey-catching responses but flee from larger ones, which could be explained by Schneirla's hypothesis. But on the basis of the observed facts we cannot accept his view that there are no a priori differential reactions to effective key stimuli but only the principle of approach and withdrawal responses. The fact that a grasshopper responds to the song of its own

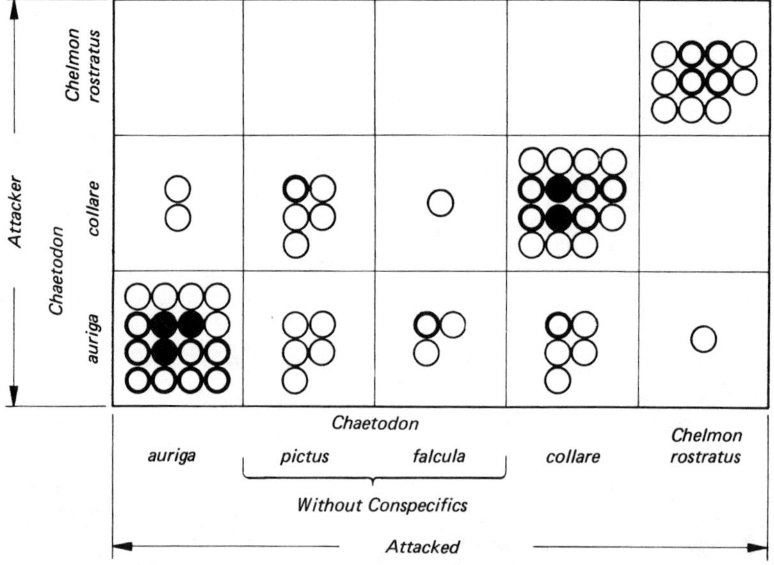

Figure 5-28 Distribution of intra- and interspecific fights within a group of butterfly fish (Chaetodontidae) that were kept together: *light open circles,* fights that lasted 2 seconds; *heavy open circles,* 3 to 9 seconds; *filled circles,* 10 seconds or more. Total time of observation, 13 hours. To compensate for the uneven number of animals of the various species, the average numbers of fights that each animal would have had with another of a different species were computed. For example, 6 *Chaetodon collare* had 36 fights with 3 individuals of *C. auriga.* One *C. collare* had on the average only 2 fights with 1 *C. auriga. C. pictus* and *C. falcula* had no species members in their tank and fought only occasionally with very similar species. (From D. Zumpe, 1965.)

Figure 5-29 *Above:* Two parrot fish fight their mirror images on a coral reef. (Photograph: H. Hass, 1957.)

Figure 5-30 *Left:* Example of a deceptive signal (Maldive Islands). The butterfly fish *(Chaetodon auriga)* protects itself from saber-toothed blennies *(Runula)* by camouflaging its eyes with a black band and by an additional eye spot near the end of the dorsal fin. (Photograph: I. Eibl-Eibesfeldt.)

species but not to another species song, that a chick within the egg, before it could hear or follow a hen's call, responds to a call of a hen instead of the call of a duck, whereas ducklings respond the opposite way, that firefly males prefer the signal pattern of their females—these facts can be explained only by assuming the existence of specific releasing mechanisms.

After all I have said it must be obvious that the animal does not possess a "picture" of the conspecific or of nonspecies members. The conspecific seems to be an object that emits various releasing stimuli for various responses. However, there are instances where the conspecific has few species-specific markings, and in this case the partner learns to recognize the other individually. K. Lorenz (1935) showed this in mallard ducks, which display a well-developed sexual dimorphism. The males are conspicuously colored (for example, a green head with a white neckband); the females, on the other hand, are drably marked. Lorenz raised one male and one female mallard with pintail ducks, without any contact with conspecifics until sexual maturity. At this time the female mallard did not react at all to the courting pintail males with which she had been raised. But she responded immediately to the courting behavior of a male mallard which she saw for the first time through a crack in the wall of her cage. The male mallard, however, indiscriminately courted the male and female pintails with which it had been raised. More recent investigations by F. Schutz (1965a, 1965b) support the view that male mallard ducks cannot distinguish conspecific females from those of other species. They court all ducks, with the exception of male mallards, which they recognize innately. Male mallards raised with other species congregate in groups for communal courting of their own kind.

A. Seitz (1940) tried in vain to release the courtship behavior of the cichlid *Astatotilapia*. The same animals that responded with fighting to crude dummies of other males refused to court detailed models of females. All had already had experiences with real females. When he was able to raise males in complete isolation they courted, much to his surprise, even very simple models (for example, a glass rod with a round wax ball). Apparently these males normally learn the appearance of their females, and are then not deceived even by a good imitation of the real fish. This phenomenon has been known to gestalt psychologists for a long time. The complex quality of a learned gestalt is made up of many details. A small change of one detail can so change the appearance of the whole that it is no longer recognized. Thus, a tame redbreast became very frightened by Lorenz when he wore glasses for the first time. K. Lorenz (1943) discussed questions of this kind at length and has pointed out that it is also possible for an animal to abstract secondarily, as it were, simple cues from a complex gestalt, to which the animal then reacts as to a releasing signal. He cited the example of a sea lion reported by M. Spindler and E. Bluhm (1934), which responded in one instance to the cockade on the cap of the caretaker, and in another time to the feeding bucket as a releasing characteristic of the total feeding situation.

For a long time the gestalt principle was seen as the criterion of the acquired releasing situation, while the stimulus-summation phenomenon was seen as incompatible with it and as characteristic of an innate releasing situation. However, G. P. Baerends, K. Brill, and P. Bult (1965) reported that the stimulus-summation phenomenon is also applicable to acquired releasing situations, so that a difference in principle between innate and acquired integration of stimulus situations is perhaps not warranted. It appears, however, that an innate situation is usually simpler than an acquired one, and as a rule of thumb we may use K. Lorenz's (1945b:29) statement: "Where an animal can be 'tricked' into responding to simple models, we have a response by an innate releasing mechanism; where it cannot be thus confused, we have an acquired recognition of a gestalt."

Following this principle K. Lorenz (1943) concludes that man, too, responds on the basis of innate releasing mechanisms to certain **stimulus configurations.** Releasing signals evolve from already existing structures, the course of evolution being determined by the receiver of the signal. Some examples may serve to illustrate this. Mouthbreeding cichlids pick up their eggs after spawning. The eggs act as releasing stimulus upon a releasing mechanism of the adult fish. A strong selection pressure, caused by predators, favors a fast acquisition of the eggs. In some species of the genera *Haplochromis* and *Tilapia* the females pick their eggs up immediately after laying, thus leaving their males no time to fertilize the eggs on the ground. Fertilization is achieved by a special adaptation. After the female has taken the eggs in its mouth, the male presents its anal fin, which bears yellow spots resembling eggs. The female tries to pick up these egg dummies and thereby inhales the sperm. Comparative studies reveal that the egg spots have evolved from much less conspicuous pearly spots on the vertical fins, which have no signaling function. It can be shown, however, that even such pearly spots can attract the attention of females. When a female *Haplochromis wingatii* was spawning with a male that had lost its anal fin, she picked at the pearly spots on his caudal fin. The existence of pearly spots in the ancestral form of the egg-spot cichlids was, however, only one of the prerequisites for the evolution of the egg dummies; a behavioral precondition was also necessary. Before the signal evolved, the males were presenting their side to the females during courtship and spawning. This is still characteristic of other species of *Haplochromis* that still fertilize their eggs on the ground (W. Wickler, 1962a).

That this behavioral characteristic was decisive for the evolution of egg dummies on the anal fin can be inferred from another development that has taken place in the genus *Tilapia*. Here, too, the females of some species pick up their eggs immediately after spawning. Although the males have pearly spots on their vertical fins, no *Tilapia* species has evolved spots simulating eggs. During courtship the males guide the females to the spawning site, swimming in front of her with the head bent down and the ventral side exposed to the female's view. In those species in which the females take the eggs in their mouth immediately after spawning, the males have evolved signals around their cloaca that release a snapping response in the female (see Fig. 7-5).

To evolve the egg spots as a signal, the following preadaptations had to exist:

1. The readiness of the females to pick up eggs.
2. The existence of pearly spots on the vertical fins of the males.
3. A lateral display by the males during courtship and spawning.

The progress of evolution of signals can be studied in some species of gulls (N. Griffith Smith, 1966). The arctic gulls *Larus argentatus, L. hyperboreus, L. glaucoides kumlieni,* and *L. thayeri* are very similar to each other. Cross-breeding, however, is rare, even where the ranges of several of the species overlap. Experiments reveal that the gulls recognize their species mates mainly by the coloration of a small ring around the eye (yellow to orange) and by the color of the iris. If one paints the eye ring of a male with the color of another species, he gets no female of his own species. If one changes the eye-ring color of a female that is already mated, no copulation takes place and the pair finally breaks up. Thus the eye-ring coloration is a signal for both males and females.

Only in the Kumlien's gull does the coloration of the iris vary considerably. Where this species overlaps in its distribution with other species there is less variation. At the south coast of Baffin Island, where Kumlien's gull occurs together with the light-eyed herring gull, dark-eyed Kumlien's gulls are preponderant. In areas where this species is sympatric with the dark-eyed *Larus thayeri,* bright-eyed Kumlien's gulls prevail.

Assortative mating takes place only where this gull lives with other gull species. Light-eyed females pick light-eyed males and dark-eyed females pick dark-eyed males. The assortative mating system may be maintained by a fixation of the chicks on the iris type of their parents. The evolution of distinguishing signals of these gull species is evidently still in progress.

SUMMARY

Animals are equipped with data-processing mechanisms or detectors that are tuned to specific environmental stimulus situations. These mechanisms enable the animal to react in a biologically adequate—that is, adaptive—way on the first encounter with the unconditioned stimulus. The capacity to react, often to highly specific stimuli, prior to any individual experience with specific behaviors presupposes an afferent mechanism as a phylogenetic adaptation. This processes the incoming data in such a way that only when they meet the characteristics of a key stimulus do they trigger responses. These have been called innate releasing mechanisms. Many social behavior patterns are released by means of such releasing mechanisms. Since this situation is of survival value for both the sender of the signal and the receiver, mutual adaptation has acted on the receiver as well as on the sender, who has developed signalling devices (conspicuous color patterns, postures, odors, sounds, and so on) that are called releasers. By experiments with decoys a number of characteristics of releasers have been discovered. Quite simple, but sometimes pattern-dependent, stimuli may act as releasers.

Often one movement pattern can be released independently by any one of several sign stimuli. If these independent stimuli are presented together the strength of response increases arithmetically (law of heterogeneous summation). Finally, models can be made that surpass the natural object in their releasing quality. From these experiments it becomes clear that an animal does not possess a "picture" of its conspecific or symbiont. Rather, these appear to be objects bearing or emitting a number of releasing stimuli for various responses. Releasing and directing stimuli can be distinguished.

6

Releasers
expressive movements and other social signals

ORIGIN OF EXPRESSIVE MOVEMENTS AND OTHER RELEASERS

A courting bird behaves conspicuously. It spreads its feathers, assumes certain postures, sings, and frequently offers the female food and other gifts. One dog greets another, wagging its tail, or growls and bares its teeth at a stranger. A threatening cat humps its back and hisses but purrs when in a friendly mood.

Behavior patterns of this kind have a communicative function. Their effectiveness is often enhanced by conspicuous morphological structures (feathers, manes). The behavior patterns that have become differentiated into signals are called *expressive movements*. They have evolved in the service of coordinating social behavior and are therefore releasers like the morphological structures that have evolved as signals.

Certainly one animal can understand many behavior patterns of another. If someone shivers, he communicates something. It is best to distinguish such undifferentiated expressive behavior from differentiated expressive movements that have become signals, although the former may become transformed into the latter, in a process called ritualization. Expressive movements may be innate or learned. They may be quite simple. Frequently several expressions are superimposed on one another, which leads to an apparent multiplicity and variability of expressive behavior; nevertheless, the complexity can be broken down to a few invariables (fixed action patterns).

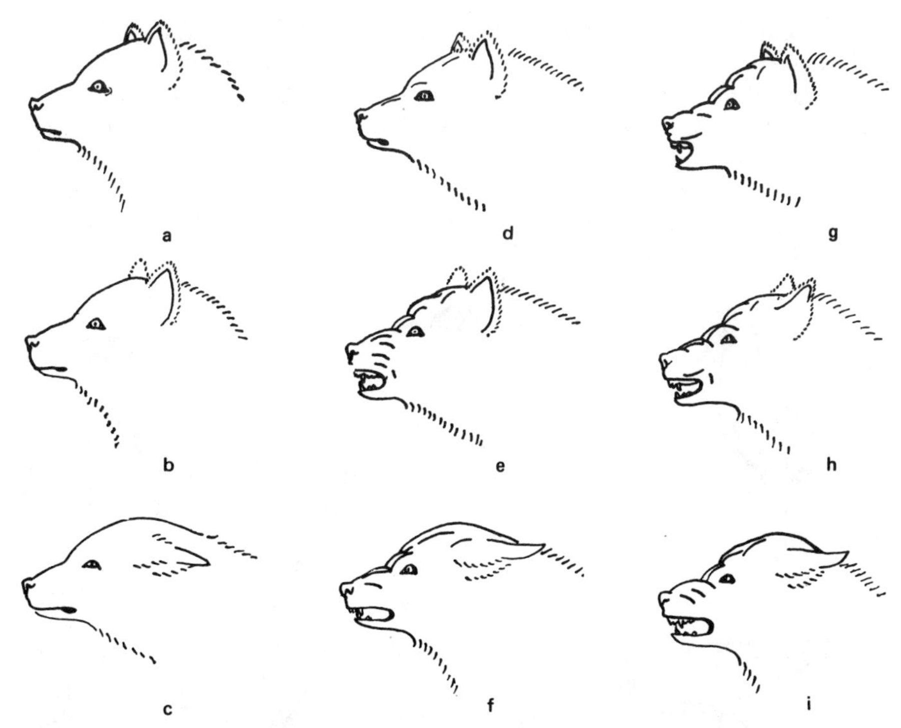

Figure 6-1 Facial expressions of the dog that result from a superposition of various intensities of fighting and flight intentions. *a–c:* Increasing readiness to flee; *a–g:* increasing aggression and the corresponding superpositions. (From K. Lorenz, 1953.)

This fact has misled some investigators. R. Schenkel (1947) stated that the richness and variability of the facial expressions in the wolf argue against the validity of the concept of fixed action patterns in mammals. In a reply K. Lorenz (1953) showed that in the dog's facial expressions the combination of the intention movements to flee with those to fight leads to a great variety of expressions. The intention to flee is characterized by pulling back the corners of the mouth, retracting the upper lip, and wrinkling the muzzle and forehead. Elements of both expressions can be superposed in varying degrees. Fighting and fleeing are often activated simultaneously, so one usually sees a combination of the two expressions, rarely a pure expression of one or the other (K. Lorenz, 1953; see also Fig. 6-1). P. Leyhausen (1956b) confirmed this in the facial expressions of the cat. Depending on the relative strength of the simultaneously activated drives to attack and to flee, the black-headed gull shows various display postures (Fig. 6-2), which can be interpreted as superpositions or combinations of various intention movements (M. Moynihan, 1955).

Species members understand such combinations, and some mammals communicate complex messages through the combination of various expressive movements. Pregnant polecats threaten the male by defensive biting, but simultaneously pacify him by uttering a call that indicates readiness for contact (*Muckern*) (I. Eibl-Eibesfeldt, 1956c). M. R. A. Chance (1963) has observed how Indian macaque females arouse the interest of the males by threatening them, but at the same time neutralize the released aggression by showing submissive behavior. The night monkey (*Aotes*) produces complex messages by arranging stereotyped elements into a sequence (M. Moynihan, 1964).

Figure 6-2 Various display postures of the black-headed gull, in each case with the appropriate strength of the simultaneously activated attack and escape drives. The block diagrams next to the figures: *crosshatched*, attack drive; *open*, escape drive. (After M. Moynihan, 1955.)

When two hamadryas baboons fight, one often flees to a higher-ranking animal. He appeases the latter by conspicuously presenting to him while at the same time threatening his opponent, thus directing the aggression of the higher-ranking individual against his opponent ("protected threat," H. Kummer, 1957). Protected threat is also found in chacma and olive baboons (K. R. L. Hall and I. DeVore, 1965) and in many macaque species (M. R. A. Chance, 1956; K. Imanishi, 1957; S. A. Altmann, 1962; P. Jay, 1968). However, in these species it is usually the males rather than the females that are involved. Indeed, Kummer states that subadult hamadryas males do not use this strategy with adults; instead they grab nearby infants and behave very like mothers, while moving away from the adult (see also J. M. Deag and J. H. Crook, 1971).

The richness of expressive movements is quite different from species to species even in closely related ones. The wolf possesses a much more varied expressive repertoire than the fox (G. Tembrock, 1954). This is correlated with the fact that wolves hunt in packs and in coordinating this activity, their need to communicate is greater than it is for the fox, which is a solitary hunter.

By comparing the behavior of related species it has been possible in some instances to reconstruct the phylogenetic history of various expressive movements. Expressive movements are often derived from other behavior patterns when these have accompanied a certain state of arousal or activity of the partner frequently enough to serve as cues to others. Social grooming behavior, for example, is always an expression of readiness for social contact and peaceful intentions. A dog in a friendly mood greets by licking and nibbling the same way as does a tame badger (I. Eibl-Eibesfeldt, 1950a) and many other mammals. The companion understands the "friendly" meaning of this gesture, which is also shown by the mother in caring for the young. Such a gesture can calm an aggressive animal; O. Antonius (1947) cites an impressive example. He kept a wild onager stallion (*Equus hemionus*) that was very aggressive and always attacked him when they met. Because the stallion could not reach Antonius, it redirected its aggression by biting the fence or attacking animals in adjacent enclosures. Once, while trying to attack its neighbor, it had its back to the fence so that Antonius could reach out and touch the animal with a bundle of keys. The result of this contact was dramatic. The stallion acted as though he had received an electric shock; he stood still, turned around for a second, showing intention to bite but refraining from doing so. Instead, he continued to permit this obviously pleasing stimulation. From that moment on the animal was tame. Whenever Antonius appeared the animal no longer theatened, but approached, showed greeting behavior, turned around, and presented his rump in order to be scratched. In a similar manner I was able to tame a galago (*Galago crassicaudatus*). The animal seemed to like being scratched behind the ears and in the armpits. Soon it showed that it wanted to be scratched there by raising an arm.

Since the behavior patterns of social care of skin and fur already express contact willingness, it is understandable that they sometimes became ritualized into expressive movements. The lemur (*Lemur mongoz*) greets others with a movement that is used to comb the fur, a behavior common to this group. This combing movement with the lower mandible is made without touching the fur and is accompanied by rhythmic calls and even intense licking movements. *Macaca speciosa* makes similar licking movements and rapidly opens and closes the lips. By imitating these movements one can pacify aggressive animals. Many lower monkeys groom only after performing the intention movements of licking (R. J. Andrew, 1963a, 1963b). Vervet monkeys smack their lips before cleaning one another and also gnash their teeth. This generally expresses a peaceful mood (T. T. Struhsaker, 1967). During the courtship of many birds and mammals preening and grooming behaviors play a great role. They aid in pacifying the aggressiveness of the partner. Even the behavior of an attacker can be transformed into "friendly" grooming. In herons, cormorants, guillemots, and other birds the bird that is attacked pacifies the attacker by presenting its head. The attacking behavior then leads into preening behavior (C. J. O. Harrison, 1965). If one rat bites another accidentally during play, the one that was bitten will squeak and is at once groomed by the one that bit it (I. Eibl-Eibesfeldt, 1957a).

Frequently behavior patterns that lead to attacks evolve into threat gestures. Thus the opening of the mouth as an intention movement preceding biting has evolved into baring of the teeth in many mammals (carnivores, rodents, and so on).

In crabs, threats with the main weapon, the claws, are ritualized in several ways. In only a few instances are the claws raised and lowered in a slow rhythm. This is done by the shore crab (*Grapsus grapsus*) in threats against conspecifics as well as against others (H. Schöne and I. Eibl-Eibesfeldt, 1965). The mangrove crab (*Goniopsis cruentata*) threatens similarly, and in a slightly modified form this claw movement is used in a waving motion during courtship (H. Schöne and H. Schöne, 1963). Ritualization has evolved furthest in the fiddler crabs (*Uca*): one enlarged claw of the male is waved, and each species has evolved its own mode of waving (H. Hediger, 1933; J. Crane, 1943, 1957; R. Altevogt, 1955, 1957; see also Figs. 6-3 and 6-4). Not all movements are derived from threats. In a number of fiddler crabs the feeding movements of the claws have become ritualized into the waving movement (J. Crane, 1966).

Many threat postures seem to have evolved out of movements of preparing to jump at the opponent. We may see only an intention movement of rising, or the animal may actually jump toward the opponent, and then stop short in an exaggerated manner, thumping the feet hard on the ground, frequently erecting the hair at the same time (for example, in badgers and squirrels as a "putting-on-the-brakes" display). In man, stamping the feet in anger seems also to be a ritualized attack movement (I. Eibl-Eibesfeldt, 1957a). Movements to protect parts of the body, such as pulling back the ears to protect the pinna and the inner ear, have evolved into expressive movements. So have epiphenomena accompanying general arousal, such as displacement activities (N. Tinbergen, 1952) or autonomic events (blushing, paling, glandular secretions), but only if they characterize the physiological state of the animal unambiguously (D. Morris, 1956; I. Eibl-Eibesfeldt, 1956b, 1957a). Thus, many snakes vibrate the ends of their tails. In some species this has become a threatening gesture and special rattling devices have evolved (rattlesnakes). In many rodents and other mammals, tail movements and ear movements become ritualized into expressive movements (R. Schenkel, 1947; P. Bopp, 1954; H. A. Freye and H. Geissler, 1966).

Movements of embarrassment and other epiphenomena of arousal do not have additional functions that could counteract selection in a certain direction, so they seem to be especially suitable for modification into signals.

Porcupines possess spines on their tails that have become modified into sound-producing organs, as was noted by Darwin (1872; Fig. 6-5). Similarly, the raising of hair and feathers may have led to the evolution of manes and conspicuous feathers. Blushing may have led to the development of bare skin areas that are strongly vascularized, which can be exhibited as tumescent bodies, and so on. The habit of marking with

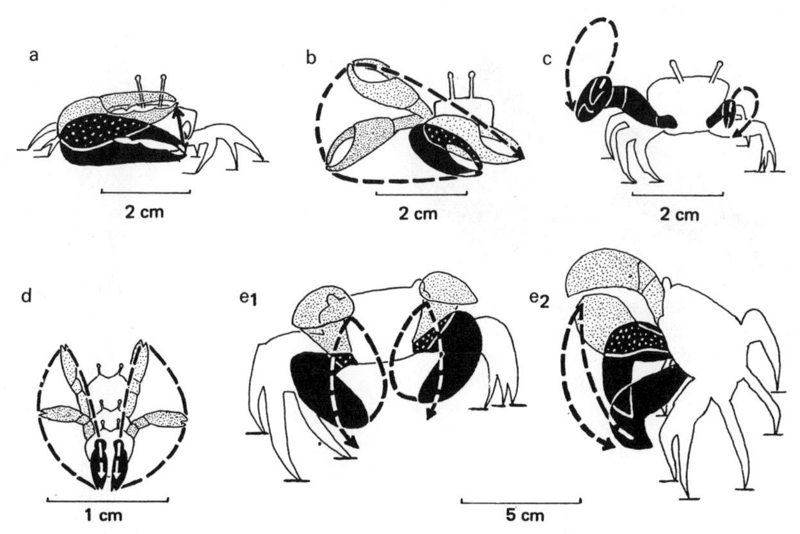

Figure 6-3 Various types of claw waving in crabs; (*a*), *Uca rhizophorae* (vertical waving); (*b*), *Uca annulipes* (waving sideways); (*c*), *Uca pugilator* (waving with the claw stretched far out); (*d*), *Dotilla blanfordi;* (*e*), *Goniopsis cruentata.* (After H. Schöne and H. Schöne, 1963.)

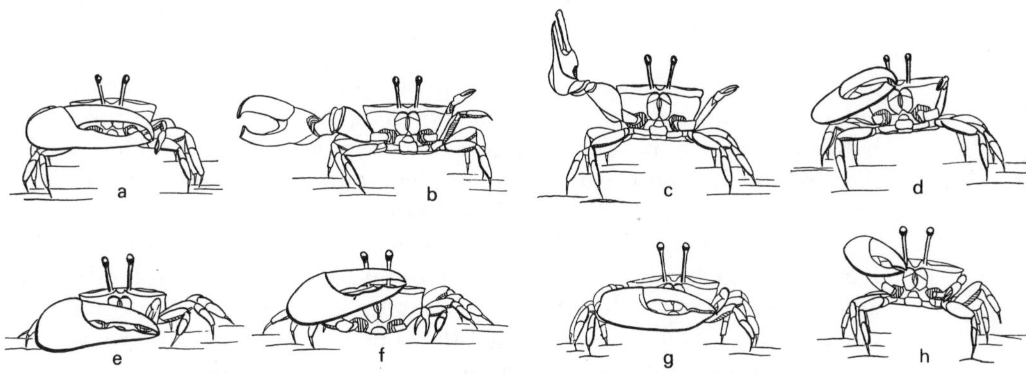

Figure 6-4 Various types of claw waving in the genus *Uca. Top: Uca lactea* (Fiji Islands), lateral waving. The claw, which is initially pulled in *a,* is stretched out in a sideways movement *b;* then it is raised (*c*) and returned in an arc into the original position. *e–h:* Examples of vertical waving; *e* and *f, U. rhizophorae* (Malaya); *g* and *h, U. signata* (Philippines). (After J. Crane, 1957.)

urine, found in so many mammals, may have been derived from urination when the animal was frightened (I. Eibl-Eibesfeldt, 1957a). In a strange environment rats leave trails by regularly excreting drops of urine. They also do this when crawling over conspecifics, thus marking them. In porcupines (*Erethizon*), agoutis (*Dasyprocta*), and maras (*Dolichotis*) the males rise up and spray urine with an erect penis on females as well as on male rivals (R. Kirchshofer, 1960). In maras the mere presentation of an erect penis serves as a threat. W. Wickler (1966c) suggests that the **genital presentations of male primates** have evolved in a similar manner.

Figure 6-6 *Emblemaria pandionis* courting before the cave in which it lives by waving with the dorsal fin, which becomes enlarged as it assumes the signal function. (From W. Wickler, 1966a.)

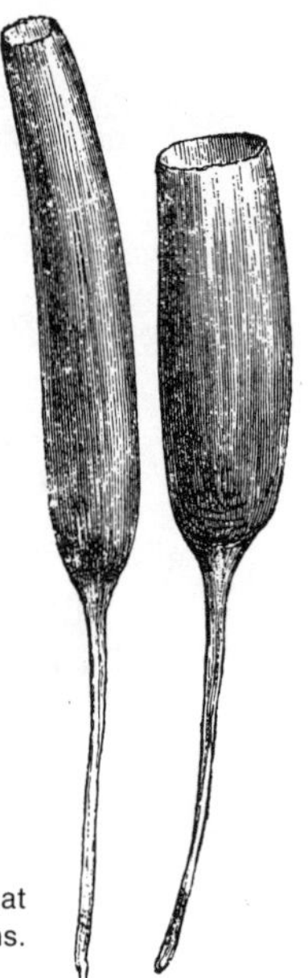

Figure 6-5 Tail quills of a porcupine that have been modified into sounding organs. (From C. Darwin, 1872.)

Whenever it is of advantage for an animal that some of its incidental behavior be understood by another, selection operates to transform the behavior pattern in question into a conspicuous signal. This modification of a behavior pattern to serve communicative function is called *ritualization* (J. S. Huxley, 1923; see also Fig. 6-6).

On the basis of homologous patterns such as grooming or parental feeding, expressive movements have often evolved independently and along similar lines in different animal groups as **homologies** especially in cases where it is not essential that the species are clearly differentiated from one another in their expressive movements. At times similarities develop because certain characteristics are required for the effectiveness of a signal and then selection pressure leads to converging evolution. This is true for many threat vocalizations, such as hissing, roaring, and spitting, as well as for those that must be heard above the roar of breakers at the seacoast or that are difficult to detect by predators (P. R. Marler, 1956b; I. Eibl-Eibesfeldt, 1957a).

Figure 6-7 shows several quite similar calls that five species of songbirds give when a predator passes overhead. The high-pitched, thin, and long-drawn-out calls ("siit" note of the chaffinch) can almost never be sufficiently localized to orient a predator. They are too high to be useful in detecting binaural phase differences, according to P. R. Marler (1956b), and are too low for the detection of appreciable differences in intensity. In addition, they begin and end unnoticeably, so that binaural comparisons with respect to time differences are also of no help. The call serves to warn conspecifics of danger; they, in turn, seek the nearest hiding place without trying to locate the caller. Chaffinches and many other songbirds also react to the quite similar warning calls of other species.

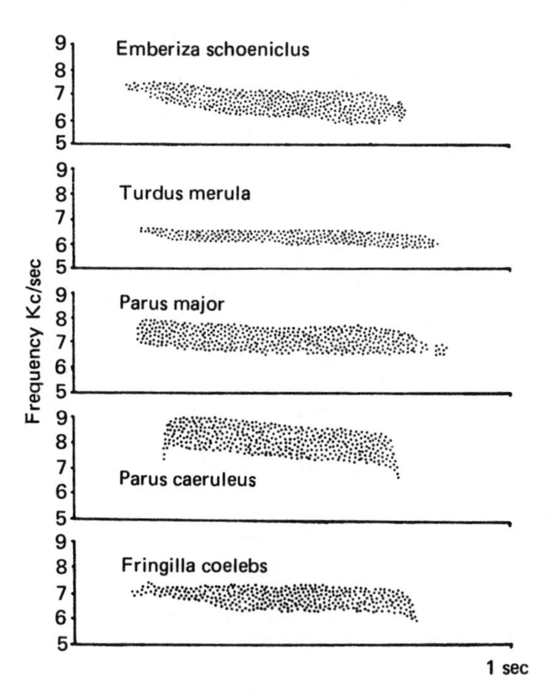

Figure 6-7 Calls of five different species of songbirds while a diurnal predator flies overhead. The calls range between 6000 and 9000 cycles per second. Note the narrow frequency range and the lack of irregularities. (From P. R. Marler, 1956b.)

On the other hand, when the signals distinguish species they are substantially different, especially in closely related species where the chance of hybridization exists (P. R. Marler, 1957a). Thus the territorial and courtship songs of the three warbler species, the willow warbler, chiffchaff, and wood warbler (*Phylloscopus trochilus, P. collybita,* and *P. sibilatrix*) are very well differentiated (Fig. 6-8). They can also be easily located because of the break between frequencies in their song patterns. On small islands where fewer birds live together, the calls are much more variable within a species than are the calls of conspecifics that live on the continent. This is true for the calls of the blue tit of Tenerife

Island and for the goldcrest on the Azores (P. R. Marler and D. J. Boatsman, 1951). Closely related species of grasshoppers that live sympatrically have very distinct calls (Fig. 6-9), and sympatric lightning bugs have different blink signals (F. A. McDermott, 1917). Frog species with partly overlapping ranges have calls more clearly distinguishable in areas where they occur side by side than where only one species is found (W. F. Blair, 1958). Contrast is emphasized only where it has utility. This phenomenon is called *character displacement*. Such selection pressure toward uniqueness with the resulting lack of confusion of signals exists also when expressive movements have evolved in opposite directions within a species (see Darwin's **principle of antithesis**).

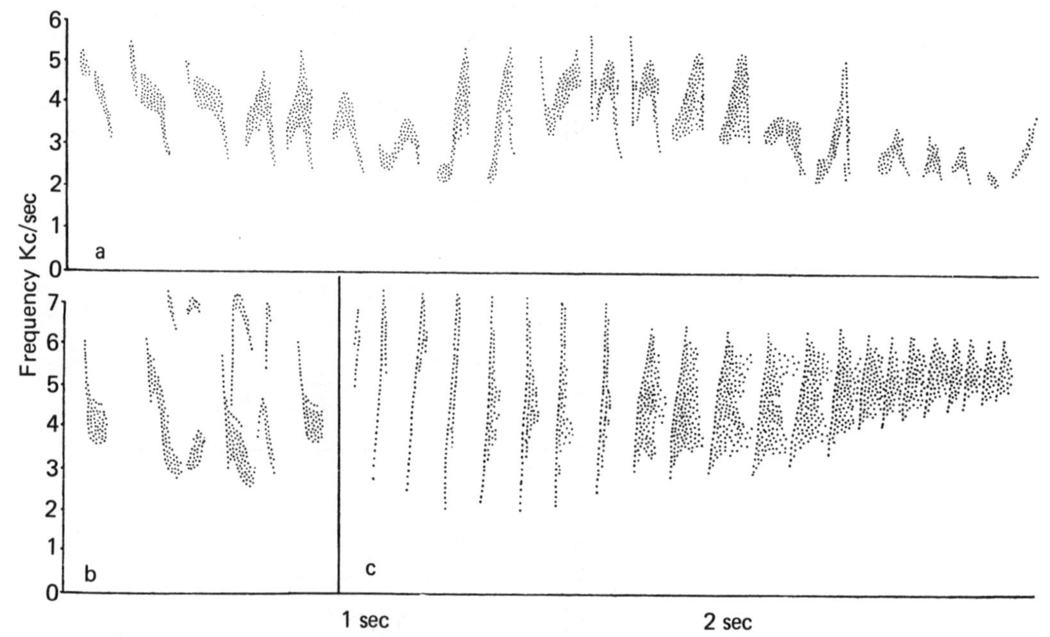

Figure 6-8 Songs of the willow warbler (*a*), chiffchaff (*b*), and wood warbler (*c*). *b* consists of four parts: *tsip-tsap-tsap-tsip*. Note how greatly the songs of these closely related species differ with respect to length, tempo, and structure. (From P. R. Marler, 1956b.)

All the changes that accompany ritualization seem directed to meet the requirements of signaling to an optimal degree. This means that in most instances the signal is conspicuous, precise, and not easily confused. Selection takes place via the receiver, who selectively perceives and responds to the signal. From many comparative studies we know that the understanding of a certain behavior pattern precedes its development into a signal. This is valid for the attracting and mood-inducing effect of the feeding movements of chickenlike birds on conspecifics, which in mother hens—in its ritualized form—serves to attract other hens. Within the Phasianidae this food calling has become ritualized into an important courtship movement, as will be shown below.

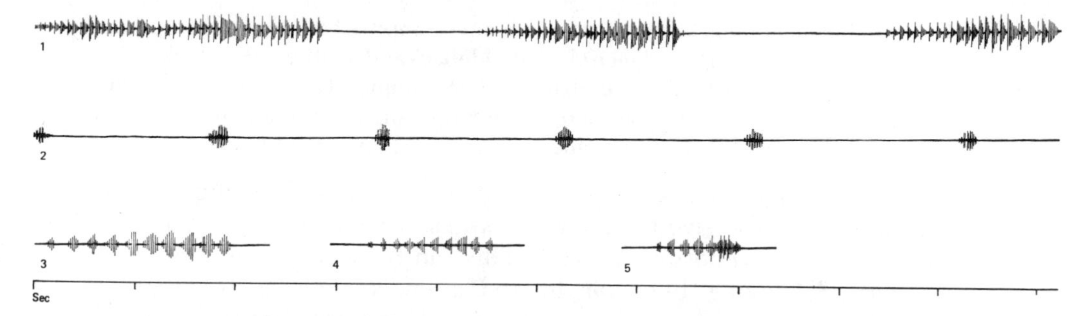

Figure 6-9 Songs of five field grasshoppers of the genus *Chorthippus* (after oscillograms by A. Faber): *1, C. biguttullus; 2, C. brunneus;* these species are closely related to each other, occur side by side, and differ greatly in their songs; *3, C. montanus; 4, C. longicornis; 5, C. dorsatus.* The close relationship is reflected in a certain similarity of the songs; *3* lives on very wet, *5* on moderately wet, *4* on fairly dry ground. These three species do not live sympatrically, so no selection pressure favored the development of divergent songs. (From W. Jacobs, 1966.)

The following specific changes can take place during ritualization:

1. The behavior undergoes a *change of function.* The original food enticing in various **phasianids** became a **courtship behavior.** The same holds for the **inciting of ducks.**
2. The ritualized movement can become independent from its original motivation and develop its own motivating mechanisms. Such a *change in motivation* was illustrated by W. Wickler (1966c) with the example of the "presenting behavior" of female baboons, which has become a "greeting gesture." A similar autonomy in respect to a specific innate movement can be traced during the course of the increasing ritualization of the inciting behavior of ducks (K. Lorenz, 1941).
3. The movements are frequently *exaggerated* in respect to frequency and amplitude, but they are at the same time simplified, by the dropping out of some components while others become exaggerated. We have already mentioned the waving of claws in fiddler crabs as an example. The visual effectiveness of this movement was strengthened not only through the exaggeration of the *amplitude of movement* but also by the frequent *rhythmic repetition* (K. Lorenz, 1941; A. Daanje, 1950; W. Wickler, 1963).
4. The *threshold values* for releasing stimuli often change to such a degree that the more ritualized behavior pattern, in general, is more easily released (A. Daanje, 1950; B. Oehlert, 1958).
5. Movements frequently "freeze" into *postures.* Many threat postures, for example, have developed out of the opposing motivations of attack and fleeing, which are usually activated simultaneously during encounters with enemies (K. Lorenz, 1951).
6. Components of *orientation* are changed. (An example is **inciting in ducks.**)
7. A behavior pattern that had previously varied in response to the intensity of motivation and stimulus can become stereotyped, with constant intensity (frequency and amplitude) even if the animal is strongly motivated (*typical intensity,* D. Morris, 1957). In this manner the behavior becomes unambiguous (B. Dane and W. G. van der Kloot, 1964). As one example we may cite **drumming in woodpeckers.**

8. Variable movement sequences can become compressed into stereotyped and simpler ones (see **zigzag dance** of the **stickleback** and inciting in the duck).
9. Along with these behavioral changes there frequently occurs the development of very conspicuous body structures, such as ornamental plumes, enlarged claws for waving, manes, sailfins, and tumescent bodies.

All these changes can occur during ontogenesis as well as during phylogenesis, because even a learned expressive movement can be improved in effectiveness as a signal. This can be observed in the development of **begging behavior** in zoo animals. This is learning by reward, in that the zoo visitor preferentially rewards behavior or postures that are most attractive or pleasing to him, so that the animals exaggerate the desired postures and leave out the unnecessary ones.

The "polishing" of ceremonies, which at times occur between two married partners, is not included under the process of ontogenetic ritualization.

The cultural ritualizations of man follow the pattern of phylogenetic ritualization. Lifting the hat as a greeting developed from removal of the helmet. The military form of greeting, the salute, in which the hand is brought to the edge of the cap, developed from the movement of raising the visor. Both are gestures expressing confidence. These traditional **greeting forms** are widely ritualized, and hardly anyone using the behavior knows the origin of these gestures. This will be discussed further in Chapter 18.

Similarities between the development of individually acquired expressive movements and the phylogenetic process of ritualization have at times led to Lamarckian interpretations. However, the principle of selection alone is sufficient to achieve a directed evolution or development. Only upon a superficial consideration does it appear unreasonable that nature should provide the raw material for selection only through "blind" mutation rather than by striving towards adaptation with the aid of directed mutations. Upon further reflection one can see that organisms would be in danger of running into evolutionary culs-de-sac in this case. Only through blindly random mutations is it possible to try out all possibilities that could be of use in meeting the challenges of a changing environment. This method of evolution, which at first appears so unwieldy, actually turns out to be the most appropriate one—that is, the most advantageous in terms of selection. Random changes in the environment can only be successfully matched by a random exploration of all possibilities.

As can already be seen from the summary above, the concept of ritualization implies the improvement of a signal—the development of releasers. Behavior patterns and changes in structures in respect to signal function also fall within this concept. The direction of evolution is determined by the perceptual mechanism of the receiver. We have also discussed ontogenetic ritualization and contrasted it with phylogenetic ritualization. The term "stylization" has been proposed for the concept of ontogenetic ritualization (D. Morris, 1956), but it is possible to avoid adding yet another concept. It is important to recognize that the development of a signal does not always mean that it becomes more conspicuous. Many animals have ritualized behavior that makes them less conspicuous. We have in mind the waving motions of leaf-dwelling insects, with many examples of convergence. These are also signals, but we are dealing with deceptive signals, which hide the animals in the moving leaves. Frequently animals develop conspicuous patterns that serve to facilitate specific learning processes. One is reminded of the wasp pattern, whose meaning has to be learned by many song birds. Not all releasers have their counterparts in innate releasing mechanisms.

Not included in the concept of ritualization are all those changes that are concerned with the improvement of the reception of signals—that is, the development of innate or acquired releasing mechanisms (W. M. Schleidt, 1962). W. Wickler (1967b) has proposed that all processes which lead to an improvement of communication be collectively termed *semantization.* This can take place unilaterally from the sender ("semantization from the sender"—German, *senderseitige Semantisierung*) and is then called *ritualization.* Often sender and receiver develop together by mutual adaptation, and finally a semantization on the part of the receiver can take place. This includes all development of releasing mechanisms. An example is the case when a night moth evolves an innate releasing mechanism in response to the ultrasonic calls of a bat that releases escape responses, or when a toad develops an innate releasing mechanism for a prey that "does not want to be noticed." The imparting of meaning, so to speak, occurs from the receiver's direction. For this reason the development of the sender does not mirror exactly the development of the receiver.

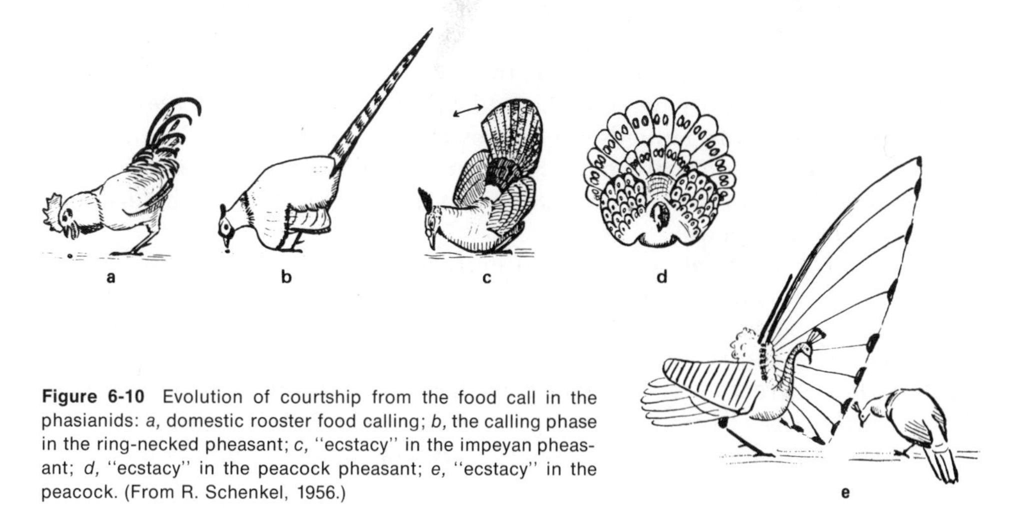

Figure 6-10 Evolution of courtship from the food call in the phasianids: *a,* domestic rooster food calling; *b,* the calling phase in the ring-necked pheasant; *c,* "ecstacy" in the impeyan pheasant; *d,* "ecstacy" in the peacock pheasant; *e,* "ecstacy" in the peacock. (From R. Schenkel, 1956.)

After this discussion we now return to the concept of ritualization, illustrating the process with some examples.

R. Schenkel (1956, 1958) was able to trace the progressive ritualization of food enticing during courtship by a comparison of various phasianid birds. The least ritualized form is shown by the male chicken (*Gallus*). It scratches several times with its feet, steps back, and pecks at the ground while uttering calls to attract, even if there is no food present. It will, however, pick up small stones as if they were food objects; the hen comes running (Fig. 6-10*a*). If the rooster has actually found food, she will eat it; if not, she will seek in vain.

The ring-necked pheasant (*Phasianus colchicus*) attracts hens in a similar manner (Fig. 6-10*b*). The courting impeyan pheasant (*Lophophorus impejanus*) bows low with a slightly spread tail before the hen and pecks vigorously at the ground.[1] The hen approaches and searches in front of him, while he spreads his wings and tail feathers to their full extent and keeps his head lowered. Only his spreadout tail bows rhythmically up and down during this stage of "ecstasy" (Fig. 6-10*c*).

The peacock pheasant (*Polyplectron bicalcaratum*) behaves similarly. After scratching on the ground in the manner of a food-enticing male domestic chicken, he will bow with raised wings and spread tail feathers (Fig. 6-10*d*). If the female approaches he will quickly move his head back and forth in her direction. If one gives him food, he will offer it to the female, thus revealing the original motivation. The movements seem to be ritualized feeding movements. Normally the male does not feed the female at this time, although she pecks on the ground in front of him. The courtship movements of the peacock (*Pavo*) have become so ritualized that one cannot recognize their origin without a knowledge of the intermediate forms. The male peacock spreads his tail feathers, shakes them, and moves back several steps. Then he bends the spreadout tail forward and points downward with his beak, while his head is still upright (Fig. 6-10*e*). As a result, the female runs in front of him and pecks in a searching manner on the ground in the focal point of the concave mirrorlike shape of the fanned tail. The male peacock points, so to speak, with his fanned-out tail toward imaginary food. Young male peacocks, incidentally, food entice in the original unevolved form with scratching and pecking. Thus there is a recapitulation of phylogenesis during ontogenesis.

This food enticing (food calling) also appears in the functional cycle of care of young. A hen attracts her chicks by scratching, pecking, and giving special calls, and this is undoubtedly the less ritualized form of the behavior. We can often observe that expressive movements which indicate a readiness for contact are derived from actions performed during the care of young. W. Wickler (1967b) and I. Eibl-Eibesfeldt (1966a) have compiled a series of examples. Among many songbirds the adults

[1] The impeyan pheasant does not scratch for its food but pecks it free with its beak.

feed one another during courtship, as if the partner were a young bird, while the other "begs" with the infantile "wing trembling" (Fig. 6-11). In the cuckoo (*Clamator jacobinus*), which no longer feeds its own young, we can only observe the derived feeding of the pair during courtship. Many carnivores feed their young. Young of the jackal (*Thos mesomelas*) of Africa cause their parents to regurgitate food by pushing their parents with their muzzles (Fig. 6-12). Between adults this pushing into the corners of the mouth is part of a greeting ceremony (W. Wickler, 1966c). The "affectionate mouthing" of wolves probably developed in a similar manner (R. Schenkel, 1947). In some seals, which no longer regurgitate food, pushing and rubbing with muzzles is used exclusively as a greeting between mother and child and among adults as well (Figs. 6-13, 14). Sea lion bulls also use this greeting ceremony to stop fights between their females (I. Eibl-Eibesfeldt, 1955b). In anthropoid apes and man mouth-to-mouth feeding of young by their mothers has been observed (Fig. 6-15*a*),[2] but adult chimpanzees also greet one another with a kiss (J. v. Lawick-Goodall, 1968; Fig. 6-15*b*), and as early as 1915 M. Rothmann and E. Teuber (1915) suggested that kissing in man is ritualized feeding. This will be discussed further. S. Freud recognized the similarity between care behavior for the young and sexual behavior in man, but misinterpreted the direction of development by saying that the mother experienced sexual feelings when stroking, kissing, and rocking the child, which became a substitute for the sexual object (S. Freud, 1950).

In analogy, we find similar ritualization in insects, in which mutual feeding also has an important role in keeping the group together. E. Roubaud (1916) proposed the hypothesis that in wasps the larval saliva is

[2] A photograph of a Papuan mother mouth-to-mouth feeding her baby was published by A. Duprerat (1963:128). And see Fig. 18-28.

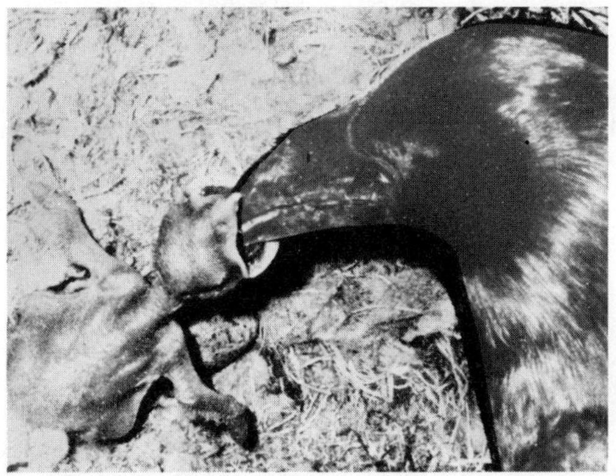

Figure 6-11 *Left:* Raven feeding its young; *right:* raven pair during courtship feeding. (Photographs: E. Gwinner.)

Figure 6-12 Jackal *(Thos mesomelas)* feeding young (Tanzania). (Photographs: W. Wickler.)

the means of establishing a bond between mother and larvae and resulted in the raising of many young. The feeding is mutual. W. M. Wheeler (1928) coined the term *trophallaxis* for this phenomenon. Exchange of food is also believed to be the basic cohesive factor among ants and termites, where behavior in respect to care of young is continued into adulthood (T. C. Schneirla, 1946; W. Wickler, 1967b). Bees

Figure 6-13 *Above:* Seal mother (*Arctocephalus galapagoensis*) greets her young by rubbing snouts. (Photograph: I. Eibl-Eibesfeldt.) **Figure 6-14** *Below:* Sea lion male (*Zalophus wollebaeki*) greeting a female (Galápagos Islands). (Photograph: I. Eibl-Eibesfeldt.)

Figure 6-15 *Left:* The 1-year-old gorilla child Jambo is fed a cherry by her mother. The mother holds a piece between her lips and offers it to her young. *Right:* Chimpanzees greeting one another by kissing. (Photographs: *Left:* P. Steinemann [Zoological Garden, Basel] from E. M. Land, 1964; *right,* Baron and Baroness H. van Lawick-Goodall [Tanzania], with permission of the National Geographic Society.)

that want to enter a strange hive appease the guards by offering them food.

By comparing the courting behavior of many species of ducks, K. Lorenz (1941) was able to reconstruct the evolution of some highly specialized courtship movements (Fig. 6-16). In some species, for example, the up-down movement is lacking. In its place in the sequence there is a "drinking toward" the partner, which has the same appeasing function. The similarity in form and of position within the sequence, and, in this instance, the similarity in function as well, indicate that the up-down movement is probably a ritualized form of the drinking movement.

Many drakes preen their wings during courtship, which has been interpreted as conflict behavior (**displacement activity**). The primary feathers are only touched in passing with their bill, as if they were pointing at their colorful wing speculum. In the mandarin drake this behavior has become an actual demonstrative movement. In the mandarin drake special conspicuous feathers have evolved, which are exposed during this apparent preening (Fig. 6-17). Other comfort movements, such as shaking or bathing, have also become ritualized within the duck family. We owe an extensive presentation and discussion to F. McKinney (1965).

Another behavior pattern whose evolution we can trace step by step is inciting behavior in ducks. This movement patterns is part of the behavior repertoire of females and serves the function of separating a particular male out of the group of communally courting males, by aggression against other males. In sheldrakes this behavior is still found in its least ritualized form. If one pair meets another, the female attacks with threatening gestures. But as soon as she approaches the other pair, her escape drive is activated and the female returns to her drake. There her aggressive impulses become stronger again. She stops and threatens back toward the strange pair, but she does not turn directly toward them. Her body remains turned in the direction of her mate, and she threatens with her head over her shoulder at the other pair (Fig. 6-18). This is not a fixed position but is the result of the conflict between attack and escape. In mallard ducks this movement sequence has become a fixed action pattern: The female always threatens over her shoulder, even if the bird toward which she threatens is in front of her and she then points away from him. She continues to look at him and does not turn the head back

Figure 6-16 Courtship movements of the mallard drake. *Two top rows,* the basic movements of courtship: *1,* bill shake; *2,* shake and stretch; *3,* tail shake; *4,* grunt whistle; *5,* head-up tail-up; *6,* looking toward the female; *7,* nod swimming; *8,* showing the back of the head; *9,* pull up; *10,* up-down movement. The movement patterns *1–4* and *10* appear during group courtship of the drakes; *5–9,* on the other hand, appear during the sexual courtship before the female in a coupled sequence from *5* onward. In the lower four rows complete movement protocols are represented. (Lorenz, K., "The Evolution of Behaviour." Copyright © 1958 by Scientific American, Inc. All rights reserved.)

as far as if the opponent were actually behind her (Fig. 6-18). In low-intensity inciting she may point her head directly toward him, but with increasing excitement her head is forced back.

The oscillation between two antagonistic intention movements has become ritualized into an expressive movement in the stickleback. The dance of the courting male consists of movement components that are directed alternately toward the female and toward the nest. The former are activated by a drive to fight, the latter by a drive to mate (J. v. Iersel, 1953).

W. Wickler (1963, 1964b) showed that the nod swimming of the mimic (*Aspidontus taeniatus*) of the cleaner wrasse is derived from a conflict behavior that is widespread in the Blennidae. He observed and

Figure 6-17 *a:* Sham preening of the garganey drake. It preens the exposed side of the wings, exposing the blue shoulder feathers in the process. *b:* Sham preening of the mandarin drake. It touches a sail-like arm feather which is turned upward. This movement is firmly coupled with "drinking toward" (*c* and *d*.) (After K. Lorenz, 1941.)

filmed a large number of species and noted that all these fish raise their heads when emerging from their hiding place. When frightened and moving backward, however, they press their heads down toward the ground. When both tendencies are equally strong, they nod. They also nod while swimming when they are in a conflict between continuing to swim on and turning back. If the disturbance is caused by a conspecific, this nodding occurs in a rhythmic pattern, that is, is clearly ritualized, but at other times the pattern is irregular (Fig. 6-19). In the mimic of the cleaner fish the nod swimming is very similar to the ritualized nodding and nod swimming of its relatives, except that this species exhibits the ritualized form of the behavior in intra- as well as in interspecific encounters. W. Wickler's comparative investigation not only explains for the first time the evolution of mimic behavior patterns but also documents it completely with motion picture film. According to U. Weidmann (1955) the head flagging of the black-headed gull should be interpreted as intention to flee. If the two partners of a pair are acquainted with one another, the female begs from the male and is fed. Out of nest building and nest repair come many expressive movements of birds—for example, in several species of Pelicaniformes, where they have greeting and threat function (G. P. v. Tets, 1965; see also Fig. 6-20).

In the black woodpecker two expressive movements are derived from chipping out a nest cavity (H. Sielmann, 1955, 1958). Drumming against dry branches, with quick rhythmic repetitions, signals to other males that the territory is occupied. Translated it would mean "someone is working (chipping) here." Females are attracted. A second expressive movement is less ritualized and therefore more easily recognized as derived from the original chipping behavior. If a woodpecker is building a nest cavity and wants to be relieved it will fly to the entrance of the nest cavity, where it will peck with deliberate slowness at the edge of the entrance hole. The partner approaches to relieve the bird and continues the work. This "relief drumming" is also shown by incubating birds, when the partner is asked, so to speak, to take over. Then the gentle pecking is directed against the wall of the nest cavity. This behavior seems to be a general signal for demands to be relieved from

Figure 6-18 Evolution of "inciting" in ducks. The sheldrake (*above*) threatens toward a neighboring pair. The mallard duck (*below*) always threatens back over her shoulder. When the threatened opponent is positioned to her side and behind, her neck is bent back farther than when the opponent is in front of her. (Additional explanation in the text.) (From K. Lorenz, 1963a.)

other tasks as well as chipping. In one of our films adult woodpecker finches (*Cactospiza pallida*) can be seen using as gestures during pair formation the food-begging behavior shown by the young. The females actually beg from the males and are fed. The male begs with the same behavior patterns at the entrance to the nest and invites the female to follow (I. Eibl-Eibesfeldt and H. Sielmann, 1965). This shows that in the course of evolution expressive movements experience changes (generalization) similar to those traceable in the evolution of language (K. Lorenz, 1941).

Cichlids generally lead their young with "follow-me" signals, which can be thought of as ritualized swimming movements. Normally, a fish that swims away will collapse its dorsal fin and make undulating

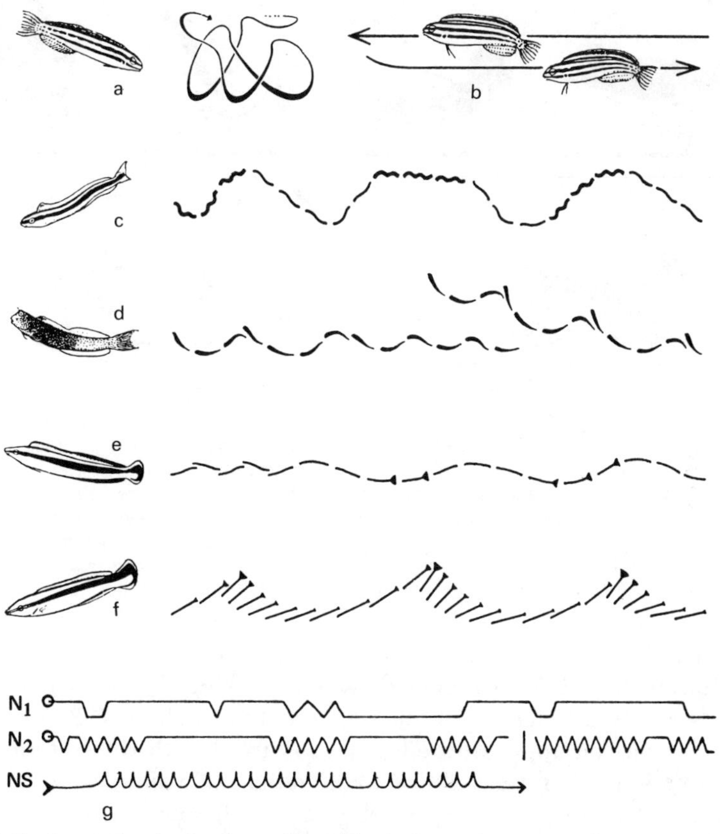

Figure 6-19 Forms of nod swimming. *a:* Courtship of a *Petroscirtes* male. *b:* Threatening *Petroscirtes* female. When swimming forward the fish slightly raises its head; when swimming backward it lowers the head. Out of these intention movements nodding can be derived. *c: Runula,* threat swimming. *d: Ecsenius* male, courtship. *e: Aspidontus,* nod swimming. *f: Labroides,* dance. *g:* Comparison of movement frequencies in *Ecsenius:* N_1, irregular nodding when disturbed (unritualized); N_2, ritualized nodding toward conspecifics; *NS,* ritualized nod swimming. *e, g,* blennies; *f,* cleaner wrasse. (From W. Wickler, 1964.)

body movements. The cichlid *Aequidens* swims a short distance with exaggerated undulating movements and collapsed fins and waits for its young. In the dwarf cichlids, on the other hand, we observe these follow-me signs only in the form of an exaggerated head shaking as a last remnant of the swimming movement. In the cichlid (*Herichthys cyanoguttatus*) this follow-me movement has become a warning signal during danger; the same movement occurs in the jewel fish (*Hemichromis bimaculatus*), in which, in addition, a quick, repeated raising and collapsing of the dorsal fin serves as a signal to call the young to the nest pit—a movement that K. Lorenz (1951) called a releaser for "putting the young to bed." In *Hemichromis*, then, two expressive movements with separate meaning have been derived from the movement of leading young.

Figure 6-20 Probable evolutionary course of postures or movements with slight threat functions toward strangers and greetings toward the partner, which probably were derived from nest-building or repairing movements in the Pelicaniformes. *a: Pelecanus erythrorhynchos,* reduced grasping of nest material; *b: Phalacrocorax auritus,* nest repair; *c: Morus bassanus,* bowing with lowering of the bend of wing; *d: Sula leucogaster,* trembling bow; *e: Phalacrocorax aristotelis,* bowing forward; *f: Sula sula,* bowing forward with raising of the wings; *g: Anhinga anhinga,* bowing with bill clapping in vacuo or toward branches. (After G. P. v. Tets, 1965, from W. Wickler, 1967.)

Several interesting examples of a change of meaning in the course of evolution have been described by W. Wickler (1965b, 1965e). The sexual presenting behavior of female hamadryas baboons (*Papio hamadryas*) also has an appeasing effect and is used with this intention by males, who in further assimilation to the females have imitated their red tumescent anogenital regions. In the spotted hyena (*Crocuta crocuta*) the presenting of the male genital region is an appeasement greeting ceremony. The sexes cannot be externally distinguished; the females have penislike organs capable of erection, with which they greet the males.

Innate expressive behavior may be brought under willful control in higher animals and man. Everyone knows that children who have been only slightly injured will wait until nearing home before they begin to cry loudly. From the time she was 4 months old a female gorilla in the Basel Zoo whimpered only when she was sure that she could be heard (R. Schenkel, 1964).

Of the well-established *learned expressive movements* we mention first the begging movements of many domestic and zoo animals, which usually are derived from intention movements of grasping and approach and sometimes from imitation of gestures (K. H. Winkelsträter, 1960). The animals learn by trial and error, and the movements become stereotyped, rhythmic, and frequently exaggerated, like the innate expressive movements.

Higher mammals can also use expressive movements deceptively. A chow chow female that disliked to accompany her owner by following his bicycle limped badly when called to come along. On the return leg of the trip, however, she bounded ahead in a lively fashion (K. Lorenz, 1950c). A gorilla female in the Basel Zoo, which disliked being alone, sought to call the keeper to her by a number of staged accidents. A new woman caretaker fell for this ruse, when the gorilla pretended to have caught her hand between the cage bars. When the woman entered the cage, the contact-starved animal rushed toward her and held her in her arms throughout the night (E. M. Lang, 1961).

Dogs use the learned movement of "shaking hands" as an appeasement gesture, which comes close to an ability to talk, as K. Lorenz (1950c:178) emphasized.

> Who does not know the dog who has done some mischief and now approaches his master on its belly, sits up in front of him, ears back, and with a most convincing "don't-hit-me" face attempts to shake hands? I once saw a poodle perform this movement before another dog of whom he was afraid.

Dogs also are able to make themselves understood through simple gestures appropriate to the situation at hand.

> If your dog bumps you with his nose, whines, runs to the door, scratches there, or places its paws on the edge of the sink and looks around in a questioning manner, then he does something which is much more like human language than anything a jackdaw or graylag goose can ever say . . . (K. Lorenz 1949:125).

The female gorilla child Goma began at age 2 to communicate her desires to her keeper by pointing at what she wanted. If she wanted to have the door opened she put her finger into the keyhole. Later she would pull a person to the door by his hand (R. Schenkel, 1964).

Even closer to human speech was the behavior of a tame raven (*Corvus corax*) called Roa that was hand-raised by K. Lorenz. Ravens have a special way of inviting conspecifics to fly with them. They fly from behind closely over the other raven, wag their folded tails, and call "crack-crack-crack." Roa did this with his caretaker, especially when he noticed the latter in a place where he had been frightened previously. He then approached the caretaker from behind as he would have done with another raven, but instead of the innate call he now called "roa-roa-roa," in imitation of the human voice. At the same time, however, the bird retained its innate warning call and used it appropriately with his conspecific sexual partner.

E. Gwinner (1964) had a raven that was called to the wire of its cage with the German word "*Komm.*" Later he called his female in the same manner. B. Grzimek's (1951) raven called all children Gregor, after the first child he had come to know. The learned sounds are used in communication between ravens who are friends, but the sounds used in immediate, species-preserving situations such as begging and position calls of young, food, attack and threat calls, and calls preceding copulation cannot be replaced by learned ones (E. Gwinner, 1964).

O. zur Strassen (1952) reported on a young grey parrot that had learned to say "*Bitte*" (please) when given something to eat. Once the bird said "*Bitte*" continuously when its teacup was empty and the bird was obviously thirsty. After receiving its tea the bird talked continuously, but used only other words. E. Gwinner and J. Kneutgen (1962) found that paired ravens and shama thrushes (*Copsychus malabaricus*) call one another with the song strophe each has learned from its partner. They name, so to speak, the partner with the strophe that is its characteristic and use this strophe alone to call it.

Crows of the United States have alarm, fright, and collective call notes that differ from those in France. If tape recordings of American crows are played to French crows, they either do not react or interpret them wrongly. For example, they gather instead of flying away when they hear the American alarm call. French herring gulls do not react at all to the calls of American conspecifics (H. Frings and M. Frings, 1959). They do not understand their "language."

The highest level is reached by learned expressive movements in human language. What in animals is commonly referred to as "language" is with the exception of the two last examples, nothing more than interjections, noninsightful sound productions. Human language can be said to be a taxonomic characteristic of man. We have here a largely open system; the concept-carrying units are combined according to specific, apparently universal rules into complex signals with new meanings. According to N. Chomsky (1970) and E. H. Lenneberg (1964, 1967) man is endowed with a genetically determined ability to speak. Human speech consists of a variety of phonetic systems that have been built upon a relatively small number of sounds. In all languages these sounds are combined into words; in all cases we find the construction of syntactic systems that serve to organize the larger conceptual associations, on the basis of universal ground rules of grammar, as N. Chomsky hypothesizes. In learning to speak, children everywhere demonstrate the same chronological peculiarities appropriate to their stage of development. The learning sequence is determined by ontogeny, just like the process of learning to walk, but the speech of children by no means represents an automatic reproduction of adult language. Children articulate differently, combine words uniquely, and invent new words. Past tenses like "winned," comparative words like "gooder," and so on, show that the child has grasped the principles of language—one might say "precociously," when looking at his other capacities. Simply the fact that the 1-year-old is able to adapt an auditorially perceived word to his own lingual motor capacity, demonstrates a specific, inborn learning disposition. The same child is far from capable of tracing a circle with its hand, although this constitutes a comparatively simple motor task. In addition we may hypothesize an actual drive to speak.

In the first three months of life the children of deaf and dumb parents cannot be distinguished from those of normal parents (E. H. Lenneberg et al., 1965). Children born deaf begin to babble but cease to do so after a time, apparently because feedback, necessary for future development, is lacking. Because of this motor ability for language it is conceivable that children would develop a language of their own. O. J. Jespersen (1925) describes the case of two Danish children who grew up neglected and were cared for by their deaf-mute grandmother. They conversed fluently in a language no one else could understand and that had no similarity to Danish.

Even the content of speech often seems to develop according to predetermined patterns. For instance, an examination of bonding dialogues points to some universal appeals (I. Eibl-Eibesfeldt, 1970, 1972a, 1972b, 1972c).

A number of investigators have dealt with the question of whether chimpanzees can be taught to use learned symbols to communicate in a way similar to the verbal communication of man. The chimpanzee that was raised by C. Hayes (1951) babbled the syllables *pu, pwa, bra, bu, wa, io, aho, baho,* and gurgled *k-k* with saliva until the fourth month of its life. With great difficulty and repeated physical guidance it was possible to teach the animal the four words "Mama," "Papa," "cup," and "up." However, the animal did not always use them appropriately, although it was able to obey 50 verbal orders correctly. The chimpanzee failed when presented with new combinations ("Kiss the cup," Kiss the dog."). Attempts at teaching chimpanzees a language using words have failed. But other methods have shown that, with human aid, they can develop a system of communication which is to a great extent "linguistic." R. A. and B. T. Gardner (1967, 1969, 1971) taught their female chimp Washoe a manual sign language, similar to that used by American deaf-mutes (Table 6-1) without ever conversing orally. First, Washoe learned simple signs for *come, come hug,* and *come swing.* She responded to the signs *look, stay, no, more, sweet* and the movement of showing or pointing. She soon began to signal, for example she **begged** with outstretched arm and open hand when seeking help. (This behavior, however, is perhaps inborn since it can also be observed in **chimpanzees** living in the wild, and in addition the behavior occurs spontaneously as early as three weeks.) These movements then led to the greeting gesture of handshaking. Washoe also pointed with extended index finger and guided the caretaker's hand to the spot she wanted cleaned. Angry behavior with foot stamping, swinging the raised clenched fist, and shaking large objects is surely innate. By waving towards herself with open right hand, Washoe invites humans to throw her a ball; recently she has begun to include the arbitrarily acquired signs for *more* and *sweet.* In all, Washoe had learned 30 signs in a period of 22 months and was able to use them spontaneously and correctly. The first signs were simple demands such as *come, tickle, out.* Later, object names followed such as *flower, blanket,* and *dog.* These descriptions were used as demands as well as answers to questions. Noun descriptions were also employed for pictures of familiar objects. At the point where the chimpanzee had learned eight to ten signs, she began to use them spontaneously in sequences of two or more—*come open, gimme sweet.* Often this occurred in free and original combination. Thus, *open flower* was used to describe a garden gate and *listen eat* for an alarm clock that signaled meal time. When the chimpanzee's doll dropped behind a wall out of her sight and reach, she signaled to the Gardners in a spontaneous new combination, *open baby.* In this case, the concept *open* was generalized to denote the wish *please make it available.* She would also engage in "monologues" when looking at picture books: On seeing the picture of a dog, she would signal *dog,* and likewise would make the sign for *cat* on coming upon the picture of a cat. On such occasions she preferred to be alone and would move away with her book if anyone was watching. Once, upon entering a part of the garden which was off limits to her, she signaled to herself *silent.*

Three years after her training began, Washoe knew 85 signs for words. She understood the signs *you* and *me,* and expressed wishes such as *you me out.* Even persons whom no one before had addressed as *you* were indicated with this sign—evidence that Washoe was generalizing correctly. The "sentences" were correct even in a grammatical sense. For example, Washoe would combine the signs *Greg (person) tickle* or *Naomi (person) tickle*—but never *Greg–Naomi,* which would have been meaningless. And the sentences could be varied: *you me go out, you Roger Washoe out, you me go out hurry.* Washoe signalized *key open food* when wanting the cooler opened, and *open key clean* when wanting the cabinet opened. Frequently she would use the word *sorry* in varied combinations and always in the proper context, meaning *excuse: please sorry, sorry dirty, sorry hurt, please sorry good, come hug sorry.* A request for soda water ("sweet drink") was expressed as : *Please sweet drink, more sweet drink, gimme sweet drink, hurry sweet drink, please hurry sweet drink, please gimme sweet drink,* and other combinations of these words. If the caretaker (Susan) stepped on Washoe's doll—which she often did as a test in various situations—then Washoe would use one of the following alternatives to indicate her wish that the doll should be released: *Up Susan, Susan up, mine please up, gimme baby, please shoe, more mine, up please, please up, more up, baby down, shoe up, baby up, please more up,* and *you up.* On encountering a closed gate she would signal one of the list: *Gimme key, more key, gimme key more, open key, key open, open more, more open, key in, open key please, open gimme key, in open help, help key in,* and *open key help hurry.*

Furthermore, Washoe's vocabulary allowed her to carry on simple dialogues:

Washoe: Please
Person: What do you want?
Washoe: Out

Washoe: Come
Person: What do you want?
Washoe: Open

Washoe: More
Person: What more?
Washoe: Tickle

Washoe: Out, out
Person: Who out?
Washoe: You
Person: Who more?
Washoe: Me

Another remarkable dialogue took place one day late in the evening (B. T. Gardner and R. A. Gardner, on press). Washoe and her companion were in her housetrailer. He peered out of the window and coming back initiated the following interchange:

Person: Washoe, there is a black dog outside—with big teeth. It is a dog that eats little chimps. You want to go out now?
Washoe: (prolonged and emphatic) "Nooooooo."

TABLE 6-1 The First 59 of the 85 Signs Used Reliably by Chimpanzee Washoe within 3 Years of the Beginning of Training

		Form		
Gloss	Usage	*Place where sign is made (P)*	*Configuration of active hand (C)*	*Movement (M)*
1. Come–gimme	For a person or an animal to approach, and also for objects out of reach. Often combined: *come tickle, gimme sweet*	At arm's length, in front of body	Flat hand, palm up	Bending action, toward signer
2. More	For continuation or repetition of activities, and for further portion of food, etc. Often combined: *more go, more fruit*	Fingertips	Flat hands, palm toward signer	Repeated contact, both hands active
3. Up	Indicating the position of an object, and for a person to raise an object, to lift Washoe, or to rise	At arm's length, above body	Flat hand, sometimes with index finger extended	Repeated thrusts upward
4. Sweet	For dessert or candy	Tip of tongue and lower lip	Index and second finger extended from spread hand, fingertips toward signer	Simple contact
5. Open	For help or permission in opening various doors (house, room, car, cupboard, etc.), various containers (bottles, jars, suitcases), and faucets	Index-finger edge of flat hand	Flat hands, palm down	The hands contact, then both hands move up and apart
6. Tickle (Touch)	For tickling and chasing games	Back of flat hand, palm down	Index finger extended from compact hand	C drawn across P

7. Go	For a person to move, while walking hand-in-hand with Washoe, or while carrying her. Also, while riding in her toy wagon, or in a car that has stopped at a stoplight. Washoe usually indicates the direction of movement desired	At arm's length, in front of body	Compact hand	Away from signer
8. Out	For going outdoors, and occasionally for leaving a place, whether indoors or out. For help in removing objects from containers. Also, indicating location, as in *look out*	Palm of curved hand, palm toward signer	Flat hand, palm toward signer	Hands contact, C inside P, then flat hand is drawn upward
9. Hurry	Impatience with delays, as during meal preparation. Often combined: *open hurry, blanket hurry*	Shoulder height, in front of body	Spread hand, bent at wrist. (ASL[a]: index and second finger extended side by side)	Vigorous shaking
10. Listen (Hear)	For sounds and for objects that produce sounds: bells, radios, clocks, watches, etc.	Ear	Index finger extended from compact hand	Simple contact
11. Toothbrush	For toothbrushes and for toothbrushing	Upper teeth	Index finger extended from compact hand, side of finger toward signer	Back and forth, across P
12. Drink	Used for the object and the action, as in *you drink, me drink,* and *sweet drink,* the usual phrase for soda pop. Also used for containers, such as cups and bottles	Lips	Thumb extended from compact hand	Simple contact
13. Hurt (Pain)	For cuts and bruises on herself or on others. Can also be elicited by small red stains on skin, or by small tears in a person's clothing	As site of injury	Index fingers extended from compact hands	Fingertips converge, both hands active

[a] American Sign Language

TABLE 6-1 *(Continued)*

Gloss	Usage	Form		
		Place where sign is made (P)	*Configuration of active hand (C)*	*Movement (M)*
14. Sorry	For apology and appeasement, e.g., after biting someone or when someone has been hurt in some other way, not necessarily by Washoe. Also when ordered to *ask pardon*	Chest, near shoulder	Compact hand, palm toward signer	Repeated clasping
15. Funny	During social interaction games, such as tickling and chasing. Occasionally, while being pursued after mischief	Nostrils (Washoe added a distinctive sound component to this sign by snorting while contacting nostrils)	Index finger extended from compact hand	Simple contact
16. Please	Asking for objects and activities. Frequently combined: *please open, please flower.* Also, when ordered to *ask politely*	Chest, near shoulder	Flat hand, palm toward signer	C drawn across P
17. Food	Used for the object and the action, as in *open food, you food, me food*	Lips	Flat hand, fingertips toward signer	Repeated contact
18. Flower	For flowers	Nostrils	Flat hand, fingertips toward signer	Contact one or both nostrils
19. Cover	For blankets, and other large pieces of fabric, and for games in which Washoe covers herself with these	Back of flat hand, palm down	Flat hand, palm down	C drawn across P and toward signer
20. Dog	For dogs and for barking	Thigh	Flat hand, palm down	Repeated contact

21.	You	Indicates any one of Washoe's companions. Often combined: *you drink, you tickle, you Susan*	On a person's chest	Index finger extended from compact hand	Simple contact. (ASL[a]: pointing at the person, without touching)
22.	Bib (Napkin)	For bib, washcloth, facial tissue, etc.	Mouth region	Flat hand, palm toward signer	C drawn across P
23.	In	For going indoors, and for indicating locations, as for objects placed inside containers	Palm of curved hand, palm toward signer	Flat hand, palm toward signer	C enters P
24.	Brush (Rub)	For hairbrush, paintbrush, etc., and for being brushed	Back of flat hand, palm down	Compact hand, palm down	C drawn across P, from fingers to wrist, repeatedly
25.	Hat	For hats, caps, bandeaux, etc.	Top of head	Flat hand, palm down	Repeated contact
26.	Me	Indicates herself. Often combined: *me drink, me Washoe, tickle me*	Chest	Index finger extended from compact hand	Simple contact
27.	Shoes	For shoes and boots	Index-finger edge of compact hands	Compact hands, palm down	The hands contact and move down together, both hands active
28.	Roger	For Roger S. Fouts, research assistant	Earlobe	Thumb and index finger touch	Grasp, then pull down
29.	Smell	For odors, and for scented objects, such as tobacco, sage, perfume bottles	Nose	Flat hand, palm toward signer	C drawn across P

TABLE 6-1 *(Continued)*

Gloss	Usage	Form		
		Place where sign is made (P)	*Configuration of active hand (C)*	*Movement (M)*
30. Good (Thanks)	For goodbye, and as part of apology and appeasement: *me good, sorry good*	Protruded lips (Sometimes accompanied by a kissing sound, a common mannerism of her human companions)	Flat hand, palm toward signer	Contact, then move away from signer
31. Washoe	Indicate herself	Ear	Flat hand, held upright	Forward, brushing, sometimes bending ear
32. Pants (Trousers)	For diapers, rubber pants, and trousers	Hips	Flat hands, palms toward signer	Upward, hands in contact with body, both hands active
33. Clothes	For jackets, shirts, dresses, nightgown, etc.	Chest	Flat hands, fingertips toward signer	Downward, fingertips in contact with P, both hands active
34. Cat	For cats and for mewing	Side of face	Thumb and index finger touch	Grasp, then pull sideways
35. Key	For keys and locks	Palm of flat hand	Index finger extended from compact hand	Repeated contact
36. Baby	For human infants, and for dolls and figurines, including animal dolls, such as a toy duck, and miniatures, such as a toy car	Elbows	Curved hands	Cross arms and grasp elbows, both hands active

37.	Clean	For washing and for soap	Palm of flat hand, palm up	Flat hand, palm down	C drawn across P, toward fingertips
38.	Catch	For games of chase and for games in which objects are thrown and caught	Back of compact hand	Flat hand, palm down	Contact and grasp, repeatedly
39.	Down	Indicating the position of an object, and for a person to lower an object, or to lie down	The ground below Washoe	Flat hand, sometimes with index finger extended	Simple contact. (ASL[a]: pointing without touching)
40.	Look	For the act of looking or peeking, and for optical devices such as glasses, binoculars, and magnifying lenses	Side of eye	Index finger extended from compact hand	Simple contact
41.	Susan	For Susan Nichols, research assistant	Forehead	Flat hand, palm toward signer	C drawn upward in contact with P
42.	Book	For books, magazines, and notebooks	Palm of flat hand, palm up (ASL[a]: the flat hands are held palm to palm, then opened as if opening a book)	Flat hand, palm down	C clasps P
43.	Oil	For lotion, for the bottles containing lotions and medicines, and for rubbing a lotion onto the skin	Little-finger edge of flat hand	Thumb and index finger touch	Grasp, then pull down
44.	Mine	Claiming objects held out to her, such as toys, and in response to questions about such objects and her possessions, *Whose that?*	Chest	Flat hand, palm toward signer	Repeated contact
45.	Bed	For beds and for going to bed	Side of face	Flat hand	Simple contact
46.	Banana	For bananas	Tip of index finger, extended from compact hand	Hooked index finger, extended from compact hand	Grasp, then pull toward signer

TABLE 6-1 *(Continued)*

	Gloss	Usage	Form		
			Place where sign is made (P)	*Configuration of active hand (C)*	*Movement (M)*
47.	Hug (Love)	For hugging	Waist and shoulder. (ASL[a]: both shoulders)	Curved hands	Cross arms and grasp body, both hands active
48.	Bird	For birds, and for bird-calls	Lips	Thumb and index finger touch, pointing toward signer	Repeated grasping
			(ASL[a]: point hand away from the signer's face, imitating the bird's beak)		
49.	Pencil (Write)	For pencils, pens, chalk, etc.	Palm of flat hand	Index finger extended from compact hand	C drawn across P, toward wrist
50.	Mrs. G.	For Beatrice T. Gardner	Side of face	Index finger extended from compact hand	C drawn downward, touching P
51.	Quiet (Hush)	For hushing, as when a person makes a loud or anguished sound, and in hide-and-seek games	Protruded lips	Index finger extended from compact hand	Simple contact
			(Often includes auditory component produced by breathing in, imitating our "shhh" sound)		
52.	Greg	For Gregory R. Gaustad, research assistant	Chest, near shoulders	Index finger extended from compact hand	Raise forearm, C contacts contralateral shoulder

53.	Help	For assistance with tasks that are difficult: operating locks and keys, looping a rope around a rafter, etc.	Palm of flat hand, palm up	Compact hand	Repeated contact
			(ASL[a]: the open hand is the active one and moves up to contact fist)		
54.	Wende	For Wende Sharrock, research assistant	Nape of neck	Flat hand, fingertips toward signer	Simple contact
55.	Dr. G.	For R. Allen Gardner	Brow-ridge region	Index finger extended from compact hand	Draw finger sideways, outlining part of one brow
56.	Naomi	For Naomi W. Rhodes, research assistant	Side of face	Thumb extended from spread hand	Repeated contact
57.	Fruit (Apple)	For fruits	Side of face	Compact hand, palm toward signer	Downward, little-finger edge of C touching P
58.	Comb	For combs, for combing, and for grooming	Side of head	Flat hand, palm toward signer	C drawn down P
59.	Dirty	For defecating, voiding the bladder, or their products, for the toilet, and for items that are soiled	Underside of chin	Spread hand, back of wrist toward P	Repeated forceful contact with wrist joint
			(Usually accompanied by sound of teeth clacking together)		

From B. T. Gardner and R. A. Gardner, 1971. Reproduced by permission.

Structurally, Washoe's language shows striking similarities with the early language of children (B. T. Gardner and R. A. Gardner, 1971; Table 6-2). R. S. Fouts (in print) trained several chimpanzees in the silent language. Again, the invention of new sign combinations was observed. The female, Lucy called a radish *food* for the first three days. On the fourth day she took a bite and called it *cry hurt food.* A watermelon was called first *drink fruit* and later *candy fruit.*

Once while learning the sign *monkey* she was observed exchanging threats with a mature male rhesus monkey. Fouts interfered and showed her monkeys in other cages. Upon being asked Lucy correctly named siamangs and squirrel monkeys with the monkey sign. The rhesus monkey, however, was described in answer to each of Fouts' several questions as *dirty monkey.* Since then she has been observed to use the *dirty* sign as an adjective to describe experimenters who refused to grant her requests. Prior to this time the sign was used to describe soiled items and feces only. This seems to be the genesis of an insult.

D. Premack (1971) trained a chimpanzee (Sarah) to use plastic pieces as word symbols. For example, if the chimpanzee wanted an apple or a banana, she had to stick the appropriate plastic piece to the board, where it was held in place by a magnet. This she grasped quickly. At the next level she learned to associate the gift with the giver. If Mary was there to present an apple, the chimp had to stick *Mary apple* on the board. Here attention was paid to the correct order of the words, since the experimenters wanted to teach *Mary give apple Sarah* as the first sentence. This stage too, was completed successfully. Early in the training, Premack taught his chimpanzee the signs for *same* and *different.* For this he presented two identical objects with the sign for *same* and two dissimilar ones with the sign for *different.* Sarah managed to grasp this association too, and thereafter generalized it so that any two unfamiliar but identical objects were recognized as being *same.*

Now the experimenters could introduce the question mark. They presented, for example, two objects A, below them the signs for *same* and *different,* and between the signs a new sign denoting a question mark:

A A
same ? *different*

After she had come to understand this procedure, the experimenters could, with the aid of the question mark, introduce *yes* and *no.* For this they presented the familiar sign sequence *X same X* and added the question mark. In a similar way they could ask *X different from Y?* and other questions. In this way the new signs *yes* and *no* were introduced. In further trials the chimpanzee was taught the sign for *word.* Premack placed an apple beside the sign for apple and between them the sign for

TABLE 6-2 Parallel Descriptive Schemes for the Earliest Combinations of Children and Washoe

Brown's (1970) scheme for children		Our scheme for Washoe	
Types	*Examples*	*Types*	*Examples*
Attributive: Ad + N	*big train, red book*	Object–attribute[a] Agent–attribute	*drink red, comb black* *Washoe sorry, Naomi good*
Possessive: N + N	*Adam checker, mommy lunch*	Agent–object Object–attribute[a]	*clothes Mrs. G., you hat* *baby mine, clothes yours*
Locative: N + V	*walk street, go store*	Action–location Action–object[b] Object–location	*go in, look out* *go flower, pants tickle*[c] *baby down, in hat*[d]
Locative: N + N	*sweater chair, book table*	(not applicable)	
Agent–action: N + V	*Adam put, Eve read*	Agent–action	*Roger tickle, you drink*
Action–object: V + N	*put book, hit ball*	Action–object[b]	*tickle Washoe, open blanket*
Agent–object: N + N	*mommy sock, mommy lunch*	(not applicable)	
(not applicable)		Appeal–action	*please tickle, hug rurry*
(not applicable)		Appeal–object	*gimme flower, more fruit*

[a, b] Indicates types classified two ways in Brown's scheme and only one way in the Gardners' scheme.
[c] Answer to question, 'Where tickle?''
[d] Answer to question, ''Where brush?''
From B. T. Gardner and R. A. Gardner, 1971. Reproduced by permission.

word (name), and thus Sarah learned the combinations *word for* and *not word for*. Her capacity was tested, for example, by presenting the signs *? banana word apple* ("Is banana the word for apple?"). The chimpanzee had the choice of answering *yes* or *no*, and in fact chose the proper sign in each case. Next, she was asked: *? word for key* ("What is the word for key?"), to which she answered correctly by using her acquired repertoire.

The symbol for *word* could be introduced only after the chimpanzee had become able to name several things. In order to acquire the class concepts *color*, *form*, and *size*, Sarah first had to learn several colors and the concepts *round*, *angled*, *large*, and *small*. The concepts *red* and *yellow* were learned with the aid of diverse objects, whose only similarity lay in their color, red or yellow. Then Sarah was handed one of the red objects and the words *give Mary Sarah*, and the new plastic piece which meant red. On this basis the class concepts were taught. Sarah was presented with the words *red ? apple* ("What is the relationship between *red* and *apple?*") or *yellow ? banana*. The only familiar word at her disposal in both cases was *color*. She combined *red color apple*. The experimenters could ascertain that Sarah had really understood this concept by posing reverse question, such as *? red color feather* ("Is *red* the color of the feather?"). The answers could be *yes* or *no*.

Premack taught the chimpanzee the concept *on* by placing several differently colored cards on top of one another. The green card was placed on the red one, then the combination *green on red* was stuck on the board, and the process repeated for *red on green*. After she had also grasped the sign for *is*, she was taught the *plural (pl): red, yellow is pl. color* ("Red and yellow are colors"). In this order Sarah learned object classes (*fruit*, *sweets*, etc.) and collective nouns (*all*, *none*, *one*, *several*). For example, she could answer the question *? apple is pl. green* ("How many apples are green?") correctly with *several*.

She even managed to connect *if* and *then*. She was rewarded with a piece of chocolate when she chose an apple, but received no reward when she took a banana, and at the same time was shown on the board: *If Sarah takes apple, then Mary gives Sarah chocolate*. Here, *if* and *then* were the only new words. After a while she came to understand this concept and responded correctly to the instructions: *If Sarah takes apple, then Mary gives no chocolate*, or *If Mary takes red, then Sarah takes apple*.

The experiments conducted by the Gardners and Premack demonstrate that chimpanzees are capable of communication not unlike our communication by language (see also D. Ploog and T. Melnechuk, 1971). As far as we know, however, this ability remains undeveloped in the wild. A. Kortlandt (1967b) did mention that free-living chimpanzees communicate through simple hand movements, which differ from one group to another and which he referred to as local conventions. But they obviously do not make use of the complex mental concepts revealed by the experiments. We are faced with the question of whether this ability mirrors a particular stage prior to the rise of human language, or whether we are tapping the remnants of capacities that had been further developed at the time of divergence of hominids and pongids—a suggestion brought forth by the "**dehumanization concept**" of A. Kortlandt and M. Kooij (1963).

CLASSIFICATION OF RELEASING SIGNALS ACCORDING TO FUNCTION

Expressive movements can be grouped according to their function. It seems appropriate to make a distinction between those expressive movements that are used in intraspecific communication and those used between species. R. A. Stamm (1964) raised the valid objection that certain contacts (threats) with other species are similar to those with conspecifics. Frequently, however, they are not. The predator is often threatened in a different manner than is the conspecific, something frequently overlooked, and this is the reason for emphasizing these differences. In both categories we can distinguish expressive movements signifying a willingness for social contact from those indicating avoidance. Often one and the same signal has a different effect on different perceivers. The "long call" of the black-headed gull attracts unmated females but repels rivals. The "head-to-ground" posture repels neighbors and attracts the female (G. Manley, 1960; see also **drumming in woodpeckers**).

Releasers for Intraspecific Behavior

Signals That Promote Group Cohesion

Introductory Remarks Between members of one species there exist attractive as well as repelling forces. Frequently, the conspecific is the bearer of aggression-releasing signals, which erect a barrier, in a manner of speaking, that discourages approach. At certain times, however, this barrier must be overcome, for example, when males and females have to come together for mating, or when aggressive animals of a species are to live temporarily or permanently as a group. In such instances a multitude of behavior patterns and signals play the role of buffers against aggression, when new contacts are first attempted as well as during continued contacts. The signals that regulate intraspecific interaction and serve to isolate species from one another as a rule are so specific that they are understood only by members of the same species.

This is especially true where closely related species occur sympatrically. Such species are clearly distinguishable in the **calls** with which they attract the sexual partner. This is true for grasshoppers (A. C. Perdeck, 1958a) as well as for frogs (C. M. Bogert, 1961), songbirds (P. R. Marler, 1957a), and other animals. We have already given some examples of this. Here is an additional example. Male fence lizards exhibit a specific head-nodding pattern that differs from species to species (Fig. 6-21). D. Hunsaker (1962) imitated such patterns of head nodding with plastic models of lizards and presented them simultaneously to females of *Sceloporus torquatus* and *S. mucronatus*. The females turned toward the model that nodded with the species-specific rhythm.

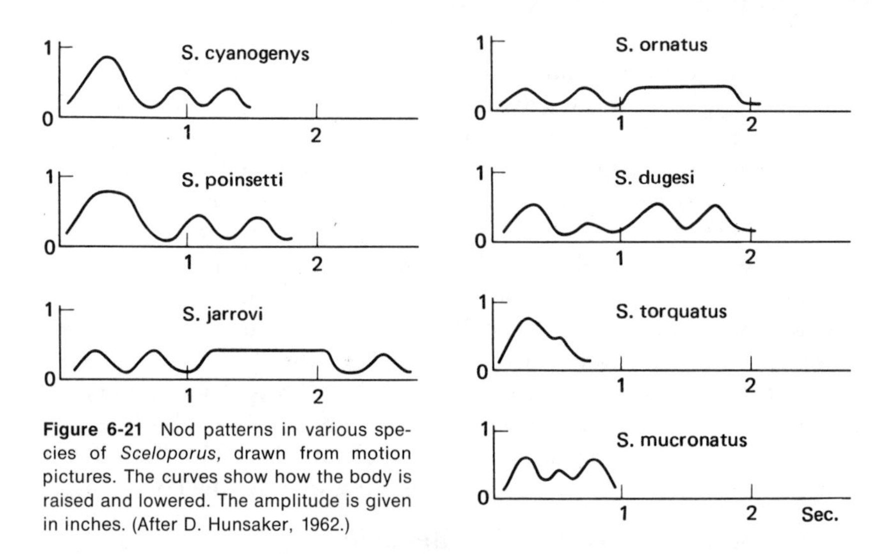

Figure 6-21 Nod patterns in various species of *Sceloporus*, drawn from motion pictures. The curves show how the body is raised and lowered. The amplitude is given in inches. (After D. Hunsaker, 1962.)

Courtship Behavior Many courtship ceremonies serve to establish contact between animals. The sexual partner is first attracted by special behavior patterns and signals. Its hesitancy in making contact is then reduced, and the behavior patterns of the partners become synchronized to one another so that fertilization is possible. Frequently both males and females court, but one partner is usually more active; as a rule this is the male. An exception to this, for example, is that the female pipefish courts the male (K. Fiedler, 1954).

Female insects use odorous substances, and both bird and locust males use courtship songs in order to attract partners at great distances. In addition, male birds frequently make themselves conspicuous by special display behavior. Male frigate birds (*Fregata*) inflate their reddish throat pouches and wait for passing females (Plate II). Birds of paradise display their magnificent feathers, often accompanying this with unusual postures. The bluebird of paradise (*Paradisaea rudolphi*) and the white bird of paradise (*P. guilielmi*) hang upside down on branches during courtship (Plate III). In the latter species, which court in groups, two males form a symmetrical figure: One sits on a branch and the other hangs below with its back to the ground (H. O. Wagner, 1938). The superb lyrebird (*Menura novaehollandiae*) cleans a courting place. There he courts, calling loudly, placing his fanned out tail on his back so that the patterned underside becomes visible (Fig. 6-22).

The most unusual courting behavior is úndoubtedly that of the bowerbirds of Australia and New Guinea. They clean special areas for courting and build bowers and decorate them, so to speak, with a "removable courtship plumage" (E. T. Gilliard, 1963). The crested gardener bowerbird (*Amblyornis macgregoriae*) selects a straight young tree and decorates it with twigs and lichens (Fig. 6-23). He dances

Figure 6-22 Courtship of the superb lyrebird (*Menura novaehollandia*): *a*, seen from the side; *b*, seen from the front. The bird folds its spread tail over its back, exposing the conspicuously marked underside. (Photographs: H. Sielmann.)

around this "maypole" displaying his erected nape feathers. The orange-crested gardener bowerbird (*Amblyornis subalaris*) builds a tentlike stage, which he decorates with fruit, flowers, snail shells, and colorful beetles (Fig. 6-24). H. Sielmann (1967) observed these birds closely and for the first time filmed them in the wild. He found that the males are very selective about the objects they choose. They put a flower into the wall of moss that they have erected, stand back, and look it over. It frequently happens that they remove it and place it elsewhere.

Figure 6-23 Pair of crested gardener bowerbirds (*Amblyornis macgregoria*) before a tower made of branches and covered by lichens. (Photographs: H. Sielmann.)

Lauterbach's bowerbird (*Chlamydera lauterbachi*) builds a bower with a central basketlike cabin. On either side of this cabin he erects two more walls, creating two more passages. The central cabin is decorated with bluish ficus fruits. In front of the bower he keeps red fruits. Should a female approach, he picks one up in his bill and presents it to her (Plate III).

Figure 6-24 Male of the orange-crested gardener bowerbird (*Amblyornis subalaris*) inserting a flower into a wall of moss. (Photograph: H. Sielmann.)

Another bowerbird (*Chlamydera nuchalis*) builds a bower that is decorated with bones and other bright objects (Fig. 6-25). This bird brings one of these objects as a kind of symbolic present to the female, then turns his nape, which has conspicuous feathers when they are spread, toward her. Some bowerbirds – for example, the satin bowerbird (*Ptilonorhynchus violaceus*) – paint the inside of the bower with saliva mixed with crushed berries and charcoal; frequently they use a leaf or piece of bark as if it were a brush.

An interesting parallel to the bowers of these birds is found in the males of the cichlid *Tilapia macrochir*, which decorate their spawning pits with furrows radiating outward from the center so that the entire structure appears like a star. Thus the spawning pit becomes a signal that attracts females (M. Huet, 1952).

Once the sexual partner has been called or approached, its fear of close contact must be overcome. Even gregarious animals, and to a greater extent solitary ones, often maintain a certain distance between conspecifics and react defensively toward transgressions of their individual distance. This barrier of aggressiveness must be overcome and is done with special appeasing gestures. Terns (*Sterna hirundo*) offer fish to females that they court (H. Rittinghaus, 1963; see also Fig. 6-26).

Some insects behave similarly: for example, the predatory flies (Empididae), where the males are in danger of being devoured by the females during mating. The females of *Empis trigramma*, for example, will attempt to do this if they have nothing else to eat. Males of *Empis borealis* and *Empis tessellata* avoid this danger by catching a prey before mating, which they pass to the female. This is the beginning of a very in-

Figure 6-25 Courting bowerbird (*Chlamydera nuchalis*). *a:* Presenting the object. The area before the bower is decorated with bones and other bright objects; *b*, the male displays the colored feathers of its nape to the female, whom he has enticed into the bower. He still holds in his bill an object he had presented to her earlier. (Photographs: H. Sielmann.)

Figure 6-26 Courting terns. The male offers a fish. (Photograph: N. Tinbergen.)

teresting sequence of ritualizations. The males of *Empis poplita* and *Hilaria quadrivittata* before presenting the prey object, spin a cocoon around it with the aid of spinning glands on the fore legs, so that the females are kept busy longer. In *Hilaria maura* the present is merely a symbol: the males spin a cocoon around an inedible object, for example, a leaf. At the extreme of this sequence we find *Hilaria sartor*, whose males spin balloonlike structures that serve as visual attracting signals (O. M. Reuter, 1913; see also Fig. 6-27).

This giving of food may be of different origin from the same behavior in birds, despite a similar appearance. The "bill flirt" and **courtship feeding** of the bullfinch (J. Nicolai, 1956) and many other birds is ritualized feeding derived from the functional system of parental care. This may also be true for the mutual feeding of champanzees, who pass food to one another with their mouths (M. Rothmann and E. Teuber, 1915). An orangutan mother nursed her infant with pablum by mouth feeding, as did a gorilla mother in the Basel Zoo. Mouth-to-mouth feeding was also practiced until recently in the German province of Holstein (D. W. Ploog, 1964a). The children react appropriately when approached

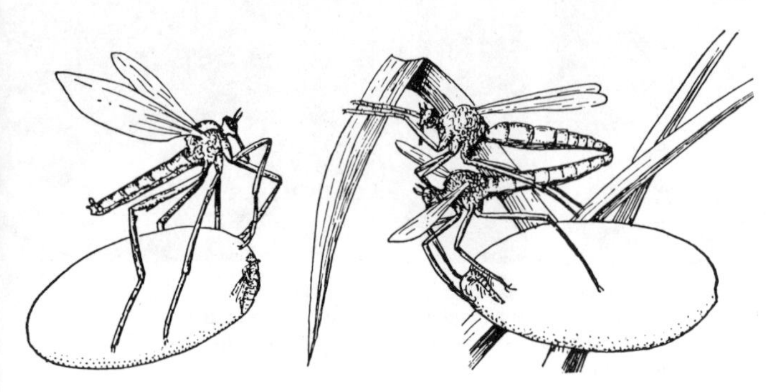

Figure 6-27 Ritualized passing of food during the courtship of *Empis*. The male presents a balloonlike cocoon that contains a fly. (From J. Meisenheimer, 1921.)

Figure 6-28 Courting diamond doves (*Geopelia cuneata*). The male presents a twig to the female. (Photograph: J. Nicolai.)

by protruding their lips and making licking movements upon contact with the mouth. Kissing in man could be derived from this. Upon closer observation one can see pushing movements with the tongue which remind us of movements used in passing food, and the noticeable sucking movements may be interpreted as associated with receiving food. However, in the flies discussed above, food passing almost certainly is not derived from the brood-care system.

Other gifts besides food are brought. Diamond dove males give nesting material (Fig. 6-28) to their females. **Courting exotic finches** and many other birds (P. Kunkel, 1959) behave similarly. Infantile behavior inhibits aggression just as effectively as does gift bringing. Behavior patterns usually shown by the young, most frequently used by males, seem to have this effect (I. Eibl-Eibesfeldt, 1957a; D. Burkhardt, 1958). The courting bearded titmouse shows begging movements with the wings in the manner of a young bird (O. Koenig, 1951a, 1951b). The woodpecker finch male (*Cactospiza pallida*) entices the female to the nest with this behavior. At the same time she begs for food from him with the same movements (I. Eibl-Eibesfeldt and H. Sielmann, 1965). Male hamsters call like nestling young when they court. When a man is courting he often speaks tenderly with childlike words that emphasize diminutives.

Both mating partners of the African bird *Trachyphonus d'arnaudii* sing a melody in duet, in which each takes a turn and sings only certain parts. They harmonize so well that a listener has great difficulty in discerning two birds instead of one. At a particular place in this song duet, the male adds a *shräh* call that is derived from the food-begging calls of young (Fig. 6-29; W. Wickler and D. Uhrig, 1969). An interesting counterpart can be seen in the bill clacking of the grass finch species *Lonchura*

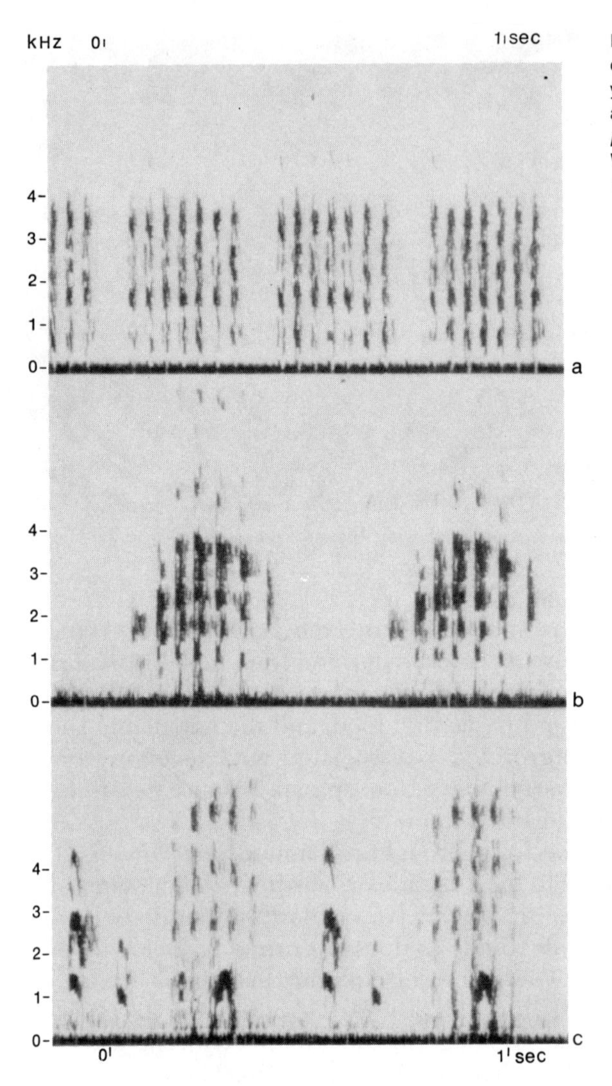

Figure 6-29 *a:* Food call of the young; *b:* call of a young male; *c:* part of a duet sung by *Trachyphonus d'arnaudii.* (From W. Wickler and D. Uhrig, 1969.)

and *Spermestes.* This gesture is a ritualized feeding behavior that occurs during courtship, and with *Lonchura* constitutes part of the courting song (H. R. Güttinger, 1970). The same author discovered that most signal movements which grass finches employ in courtship are derived from the behavior patterns of caring for young: two originate in the functional cycle of nest building, three from that of raising young, and two can be interpreted as ritualized food-begging behavior.

Yellow baboons (*Papio cynocephalus*) greet with "lip smacking." These movements, which are made into the air, can be explained as rapidly performed sucking movements and also as being derived from the sucking movements of the young during ontogeny. Pink parts of the body, such as the nipple, the penis, the female reproductive region, and the face of the child all act as strong releasers of smacking. As these parts are also attractive to other baboons they probably contribute to the cohesion of the group (T. R. Anthoney, 1968). With this the **female breast** received an additional signal function that aids the cohesion of the group. This is interesting, because we can point to a certain parallel development in **humans.**

Finally, many animals appease during courtship by covering up signals or weapons that ordinarily release aggression. According to N. Tinbergen (1959) head flagging in the black-headed gull serves to hide the black face mask that releases aggression (Fig. 6-30). Terns hide the tops of their heads with a stretch posture, and the raven appeases by looking away and raising the bill (E. Gwinner, 1964).

Figure 6-30 Presenting the back of the head (head flagging), an appeasement gesture of the black-headed gull. (Photograph: N. Tinbergen.)

Several of these expressive movements often occur in a single courtship sequence. In courting albatrosses we can observe behavior patterns that can be interpreted as ritualized food begging, appeasement gestures, showing of nest, preening behavior, as well as others that are not yet understood (Fig. 6-31). The courting ritual, which is repeated many times, begins with a dance. The male walks around the female with his neck pulled in and wobbles conspicuously from one side to the other and in rhythm with his steps. Still keeping time, both birds turn their heads in turn to the side so that the bill touches the raised shoulder. The dance is followed by fencing with the bills (Fig. 6-31*a*). The birds stand facing one another, stretch their necks forward, and hit their bills together with rapid sideways motions of the head; at the same time they make nibbling motions with their bills. Young birds beg for food in this manner, and perhaps this behavior is ritualized begging. Other behavior patterns may follow—for example, clapping the bill, which is done by standing up, opening the bill widely, and closing it with a loud noise. This is frequently done by both birds at the same time (Fig. 6-31*b*). This can also be observed when albatrosses threaten one another. More bill fencing or display movements may follow, the birds (Fig. 6-31*d*) raising their bills straight up and calling.

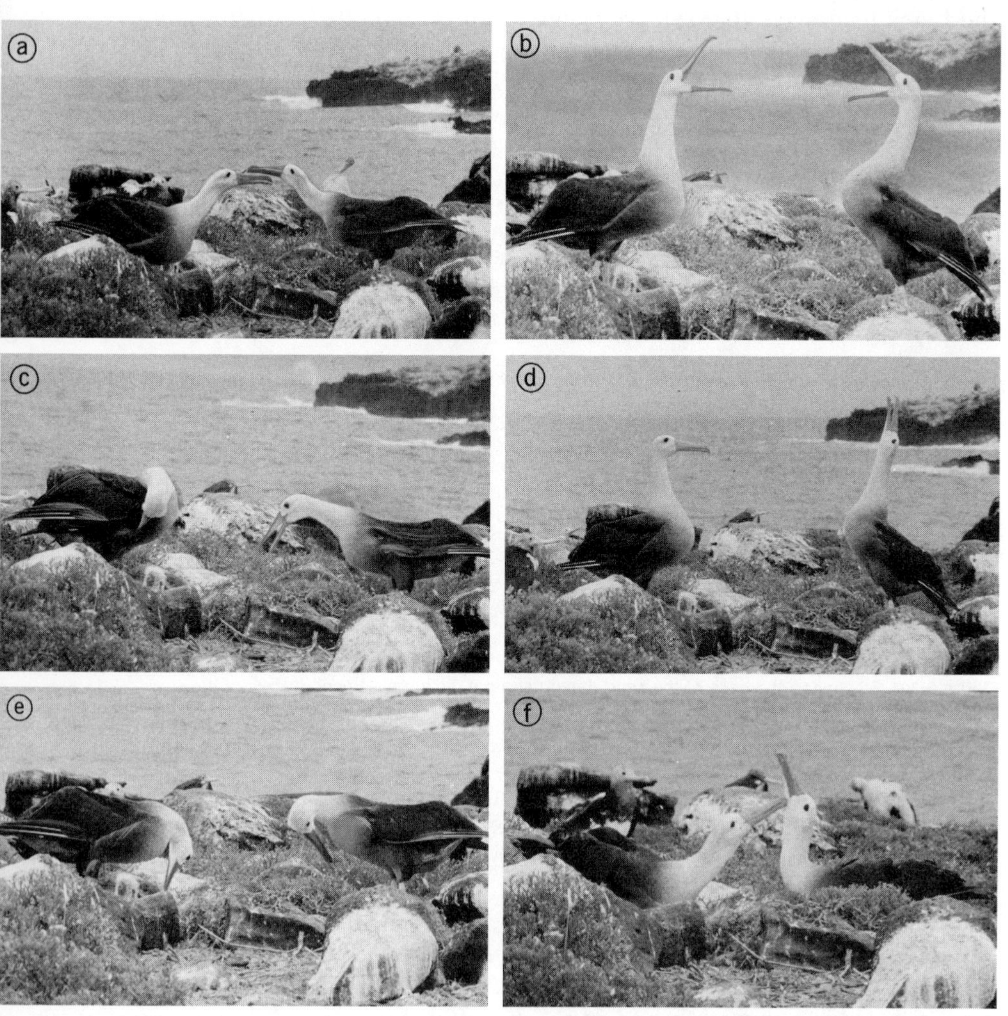

Figure 6-31 Courtship of the Galápagos albatross (*Diomedea irrorata*) (Hood, Galápagos Islands): *a,* Bill fencing; *b,* clapping the beaks; *c,* preening the shoulder (left) and clapping the beak (right); *d,* "presenting" the beak (right); *e,* showing the nest site; *f,* social preening. (Photographs: I. Eibl-Eibesfeldt.)

This reminds one of the appeasement gesture of boobies that N. Tinbergen (1959) described. Sometimes one bird claps its bill—stretched forward (Fig. 6-31*c*)—and at this time the partner always preens its shoulder feathers. At the moment the bird stops clacking and raises himself up, the other throws up his bill in display, clapping once loudly. These behavior patterns may follow one another in various sequences. At the end of such a courtship sequence the birds bow to each other, pointing their bills to the ground (Fig. 6-31*e*); while doing this they utter two-syllable calls. This could be symbolic pointing at the nest site. Usually they then both sit down and begin preening each other's neck feathers (Fig. 6-31*f*). After a short interval the entire sequence may be repeated.

Some mammals take symbolic possession of the courted female by marking her with scents, creating an odor bond. For example, porcupines (*Erethizon dorsatus*), agouties (*Dasyprocta aguti*), and maras (*Dolichotis*) approach the females on their hind legs and spray them with urine. Similar urine ceremonies are known in guinea pigs and rabbits (I. Eibl-Eibesfeldt, 1958a). A gesture of contact willingness in females is the special movement that **presents** the genital region, frequently conspicuously altered. An animal can form a bond with a conspecific by activating its aggression against a third, which is the case with the **inciting behavior of ducks.** The "**triumph**" **calls of geese,** which are greeting gestures in the wider sense, may have a similar origin.

Submissive Gestures, Greeting, and Other Appeasement Behavior

Many of the courtship behaviors I have discussed are appeasement gestures that also play an important role in other situations. Frequently the loser of a fight appeases the victor by so-called submissive gestures that are usually the opposite of threat postures. The **marine iguana** submits by prostrating itself before the winner, who then ceases fighting but waits in a threat position until the vanquished retreats from the area. The appeasing function of behavior has been shown in chaffinches by P. R. Marler (1956a). Animals that behave submissively are permitted to come closer to conspecifics than those who show a threat posture. Some vocalizations have appeasing effects, such as the **muttering (*Muckern*) calls of polecats** and the squealing of young rats. If one rat bites another too hard during play, the latter will squeal, whereupon the former will gently groom its fur. Many postcopulatory displays of birds can be interpreted as appeasement ceremonies, and many, but not all, greeting ceremonies have this function. In general, contact between conspecifics, frequently of different sexes, is established and maintained by greeting ceremonies.

When the flightless cormorant (*Nannopterum harrisi*) returns to its nest and mate, it will bring a gift of a sea star or a bundle of seaweed, which is presented to the bird on the nest. The latter often pulls it away

Figure 6-32 Appeasing gesture: passing of nest material to the mate while exchanging places at the nest by the flightless cormorant (*Nannopterum harrisi*) (Narborough, Galápagos Islands). (Photographs: I. Eibl-Eibesfeldt.)

aggressively (Fig. 6-32). One can recognize from the vehemence of the behavior that aggressiveness is directed toward the gift. A simple experiment shows that this is indeed the case. If one takes the gift away from a bird returning to its nest, which is possible because the birds are quite tame, the bird is driven away from the nest by its mate (I. Eibl-Eibesfeldt, 1965b).

The male sea lion employs appeasing greeting gestures to keep his herd together. If two females fight, the ruling bull approaches at once and pushes himself between the two combatants, extending greeting toward both, which has a calming effect (I. Eibl-Eibesfeldt, 1955b).

Storks greet their mates by placing their heads over their backs and clapping their bills. This can be interpreted as a pronounced turning away of the weapon, because during a threat the tip of the bill points at the other bird. In principle the head flagging of gulls is also a turning away of organs used in fighting. Here it is primarily a threat signal, while in storks the weapon is demonstratively turned away. This is also found in other birds.

In the so-called contact species, in which individuals know one another well, touching the body serves a greeting function. Cats greet by "presenting the head" (O. Antonius, 1939; P. Leyhausen, 1956b). Grooming behavior patterns clearly express a readiness for social contact, and they have frequently evolved into greeting ceremonies. We remember the lemur *Lemur mongoz* discussed earlier. According to J. v. Lawick-Goodall (1965, 1968) **chimpanzees** have several **greeting gestures.** They embrace one another and kiss with a touching of lips when they meet someone they know. The embrace can probably be derived from infantile clasping, which now serves the function of maintaining group cohesion. The gesture calms both animals. Goodall reports that even large males, when frightened, clasp young chimpanzees and calm down as a result. Another appeasing greeting behavior is the **sexual presenting** of females (Fig. 6-33*a–c*), whereby the greeting animal turns its posterior toward the other, a behavior found in baboons and other **apes.**

Males also use this originally female behavior. Chimpanzees also shake hands the way people do. The initiative is taken by the lower-ranking animal, who reaches toward the higher-ranking animal, palm up, in a kind of **begging gesture** (Fig. 6-34). In response to this gesture, probably derived from the infantile search for contact, the higher-ranking animal gives his hand, which in turn calms the other. Lower-ranking animals solicit approval in this way from higher-ranking members when they attempt to obtain food at a common feeding place. They also bow when greeting others (J. Goodall, 1965, and van Lawick-Goodall, 1968). The very aggressive cichlids (*Tropheus moorii*), which live in groups, appease others by presenting a yellow band to the attacker, which is also shown during spawning and courting (W. Wickler, 1965e; see also Fig. 6-35).

Interestingly enough, some greeting gestures can also contain elements of threat behavior. According to E. Trumler (1959), this is true for the "greeting face" of horses; opening the mouth and exposing the corners of the mouth are clearly aggressive. But this is "canceled out" by also raising the ears (Fig. 6-36). The facial expression of a mare in estrus (*Rossigkeitsgesicht*) originated out of the mimic expression of threat. Redirected threat movements can be observed in graylag geese, which threaten with outstretched necks past one another as if they were confronting a common enemy (K. Lorenz, 1963a). The "triumph ceremony" (*Triumphgeschrei*) plays a special role during pair formation. The male at first makes sham attacks toward objects that are normally avoided. Following such an attack he "triumphantly" returns to his intended mate and threatens beyond her (Fig. 6-37). If she joins the "triumph ceremony," a defensive alliance has been formed, which is a

Figure 6-33 Greeting chimpanzees (Tanzania). *Above, left:* arriving female presents and is touched on the genital region by the male; after she has turned around (*above, right*) he touches her face; *right:* she bows "laughingly" and the male begins social grooming, which remains, however, a mere symbolic gesture. (Photographs: I. Eibl-Eibesfeldt.)

Figure 6-34 Chimpanzees greeting by shaking hands. The lower-ranking animal presents the hand to the higher-ranking one by holding the palm up in a begging fashion. (Photographs: Baron and Baroness H. van Lawick-Goodall, with permission of the National Geographic Society.)

Figure 6-35 Quiver display of *Tropheus moorii*, an appeasement movement probably of sexual origin. (Photograph: W. Wickler.)

prerequisite for the successful rearing of a brood (K. Lorenz, 1943). This behavior continues to maintain the pair bond by functioning as a greeting ceremony. The neck movements of a greeting goose have the same form as the threat movement and undoubtedly were derived from it. The "cackling" (*Schnattern*) of the "triumph ceremony" has developed from the contact call of the young (H. Fischer, 1965). For this reason and by the orientation of the threatening neck posture past the other, the gesture has become neutralized, similarly to the threat expressions of horses, which become "greeting faces" by the erection of the ears. Although the "triumph ceremony" of the graylag geese is undoubtedly derived in part from aggressive behavior patterns, it has its own motivation that is independent of aggression (H. Fischer, 1965). The *rab-rab* palaver of ducks, as well as the greeting ceremony of the female bullfinch, still contain much aggression to this day (K. Lorenz, 1941; J. Nicolai, 1956).

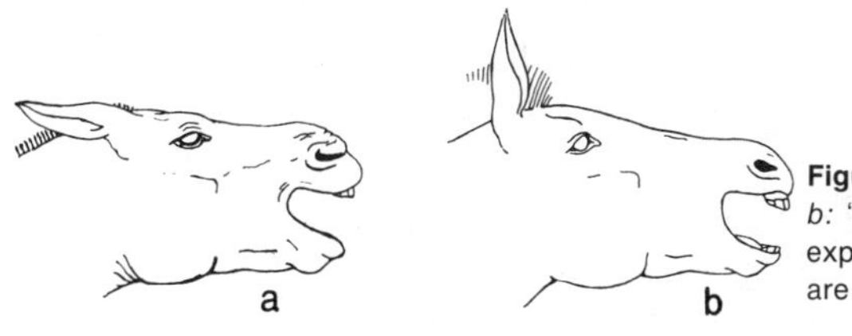

Figure 6-36 *a:* Threatening zebra colt; *b:* "greeting face" in the same zebra. The expression is as in the threat but the ears are erected. (From E. Trumler, 1959.)

Figure 6-37 Sequence of behavior patterns of the "triumph ceremony" in the graylag goose. The male threatens toward the opponent (*F*) (*1* and *2*) and drives him off (*3*), turns around and returns in an impressive display posture to the female (*4*), who approaches him with rolling calls. They roll and cackle together (*5* and *6*). (From H. Fischer, 1965.)

Two threat postures, the upright and forward, are a part of the greeting ceremony of the mated black-headed gull pair (Fig. 6-38). At first these postures have a distinctively aggressive function. The female is attacked as an intruder and chased off. Soon this picture changes; she is allowed to remain, although he continues to threaten, but now with a new orientation. When alighting, he greets with the "long call"; she approaches with her neck stretched forward, and he reacts likewise. Finally, they no longer point at each other, but stand parallel, similar to geese during the "triumph ceremony." This ritual is reduced to a bare minimum as the birds get to know each other better (G. Manley, 1960).

In rhesus monkeys the mounting by the males of others of the same sex not only means aggressive threat or assertion of rank but is also an expression of an accepted order within the group, strengthening the bond between the two individuals. The higher-ranking animal usually mounts first, but is frequently mounted in turn by the lower-ranking one. C. B. Koford (1963a) compares this kind of greeting with a military salute. In most cases now reported the order of such mounts seems to reflect the dominance hierarchy of the troop (see T. E. Rowell, 1967). The turn-around mounting sequence is seen mainly in juveniles and subadult males, rarely in adult males unless the pair are of nearly equal status.

In the raven, the ceremony of mutual feeding between members of a pair becomes increasingly superficial. Initially both partners actually feed one another; finally this is done only rarely, the ceremony consisting mostly of a brief grasping of the bill. This has been interpreted as an ontogenetic ritualization but I do not agree. For this to be true one would have to demonstrate that better communication is achieved by simplifying the behavior, that is, that the signal has become more understandable and effective. This kind of ritualization does not seem to have occurred. Instead it appears as if the ceremony has become more superficial, deritualized—perhaps because the animals know one another better and the need to appease aggressive behavior has diminished.

The smile of man is an important buffer against aggression. A smile can be disarming, and reports from warfronts contain examples of how aggression may be inhibited by a smile. Every traveler has experienced the release of tension between strangers that is provided by a smile. People usually smile politely when making a refusal to someone, and they smile when excusing themselves. However, a smile not only inhibits the aggression of another person, it also frequently brings about a friendly reply. Infants smile, and this increases the bond with the parents. In adults a smile will build a bridge to total strangers, and people smile at each other when flirting as well as during friendly greeting.

A smile frequently changes into laughing, which has therefore often been seen as a higher level of the smile, but it is not exclusively so. J. A. Ambrose (1963) interprets smiling as an ambivalent behavior that is

Figure 6-38 Upright and forward (standing parallel) postures in the threatening greeting of the black-headed gull. (Photographs: N. Tinbergen.)

derived from the simultaneously present tendencies of turning to and turning away from someone. Light tickling releases turning toward in human babies, and strong tickling releases turning away. This is also true for other stimuli—sudden surprise, saying "Boo!" and so on—which have to be presented at just the right intensity and frequency if they are to be successful in releasing laughter in a child. Ambrose sees certain similarities between smiling and crying—a rejecting gesture. K. Lorenz (1963a) interprets laughing as a greeting ceremony derived from a threat movement, a view made plausible by the exposure of the teeth. N.

Bolwig (1964) interprets it as ritualized biting (a playful intention to bite).

It is quite certain that there is some aggression involved in laughing. The rhythmic vocalizations remind one of similar sounds made by primate groups when they threaten in unison against an enemy. Such a combined threatening unites the members of a group, and it has been observed during investigations of laughing behavior that a strong bond is established between members of the group in a similar way. Those outside such a group are quite uncomfortable in the presence of the laughter, especially when it has the character of "laughing at," when it is definitely aggressive and challenging. In its original form laughing

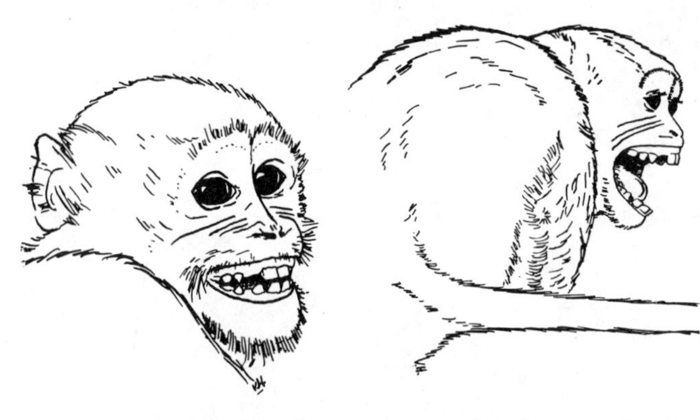

Figure 6-39 Silent bared-teeth display and bared-teeth scream of a submissive *Macaca irus.* (From J. A. van Hooff, 1971.)

seems to unite *against* a third party. In the smile, however, the aggressive component (the sound utterance) is lacking. It furthermore must be emphasized that the baring of teeth—which strengthens the idea that there might be a phylogenetic connection between smiling and threat behavior—differs markedly from the teeth exposure during **rage**, in which the corners of the mouth are opened and drawn downwards. In some primates (Gelada baboon) this leads to the full exposure of the upper canines. During a smile only the front teeth are exposed, and these are used in social grooming and play.

Studies by J. A. van Hooff (1971) have revealed homologies for smiling and laughing among several different primates. He also distinguishes between the two expressive movements as representing more than merely differences in intensity. The "silent bared-teeth display" (sbt), which indicates submission and, during intense fear occurs as "vocalized bared-teeth display" (Fig. 6-39) would be homologous to the smile. The latter expression has a long phylogenetic history, for we can observe this behavior even in lower mammals during defense. Spitting and hissing add vividness to this facial expression. In chimpanzees Hooff discoverred three kinds of bared-teeth display (sbt). In a friendly mood one animal will show an equally amiable other its teeth with an "open mouth" (sbt) (Figs. 6-40 and 6-41). Vertical sbt (see Fig. 6-43) is displayed by high-ranking individuals to reassure smaller ones. Horizontal sbt is motivated by fear. In this way low-ranking chimpanzees appease high-ranking ones who are showing aggressive tendencies.

The development of laughter begins with the play face, also termed the "relaxed open-mouth display" (Fig. 6-42*a*, *b*). We are not certain about the origin of this expression. It could be derived from friendly bite intentions that signal the playful character of a fight. In such a case the origin would involve an element of aggression.

But the intention to bite playfully is only one component of laughter. As I have mentioned, the rhythmic vocalizations may be derived from mobbing. In J. A. van Hooff's summary (Fig. 6-43), the initially separate lines for smiling and laughing in humans eventually converge. I

Figure 6-41 Interspecific play. The chimpanzee, playing the active part, shows the relaxed open-mouth display, accompanied by *ah-ah* sounds. The more passive boy laughs. (From J. A. van Hooff, 1971.)

Figure 6-40 A female chimpanzee shows the open-mouth bared-teeth display before hugging her child. (From J. A. van Hooff, 1971.)

believe, however, that because of the numerous possible superpositions of expressions this merely appears to be the case. For surely we can distinguish a pure, intense smile from pure, intense laughter, and the two expressions are then very different indeed.

A larger number of greeting ceremonies in man have their functional analogies in animals, serving in the establishment and maintenance of a bond between individuals who are acquainted with one another. The reader may convince himself of their appeasing function by not greeting his closest relatives and friends for one week. It is surprising to find out how quickly aggression that is not buffered by these appeasing gestures is turned against him. Human greeting gestures include, besides the smile, gestures of symbolic submission. People bow or nod their heads, in Europe in the same way as in Japan. One bares the head and removes weapons, thus demonstrating trust by giving up one's protection.

The **greeting** with the **raised, open right hand** is widespread. The Chom Pen of the Great Nicobar Island in the Indian Ocean, who had had no contact with Europeans, greeted us with this gesture in the same way as the Karamojong tribesmen in East Africa.

Sometimes weapons are used in greeting. They are demonstratively turned away to a position that is not dangerous, as when presenting arms. When someone greets us with a spear he does not point its tip against our stomach. The ever-present cultural differences do not affect the basic principles. However, greeting ceremonies have become

Figure 6-42 *Left:* Two *Macaca irus* engaged in playful fighting. The one facing the observer shows the relaxed open-mouth display. *Right:* The same expression from the side. (From J. A. van Hooff, 1971.)

changed in characteristic fashion depending on rank and sex, a problem that needs still further investigation.

Frequently presents are given, and this seems especially appropriate if one enters another's territory, such as a gift of flowers when we visit someone's apartment or house. This encounter seems to require stronger, more effective gestures, and if it is omitted in cultures where it is customary, the omission is regarded as impolite, and the unappeased aggression is experienced as annoyance.

In man threat gestures are sometimes changed into greetings. As in the **inciting of the duck,** laughing in its *original function* seems to serve as a common and thus uniting threat against an enemy. One demonstrates the willingness to attack together. The greeting with a raised fist

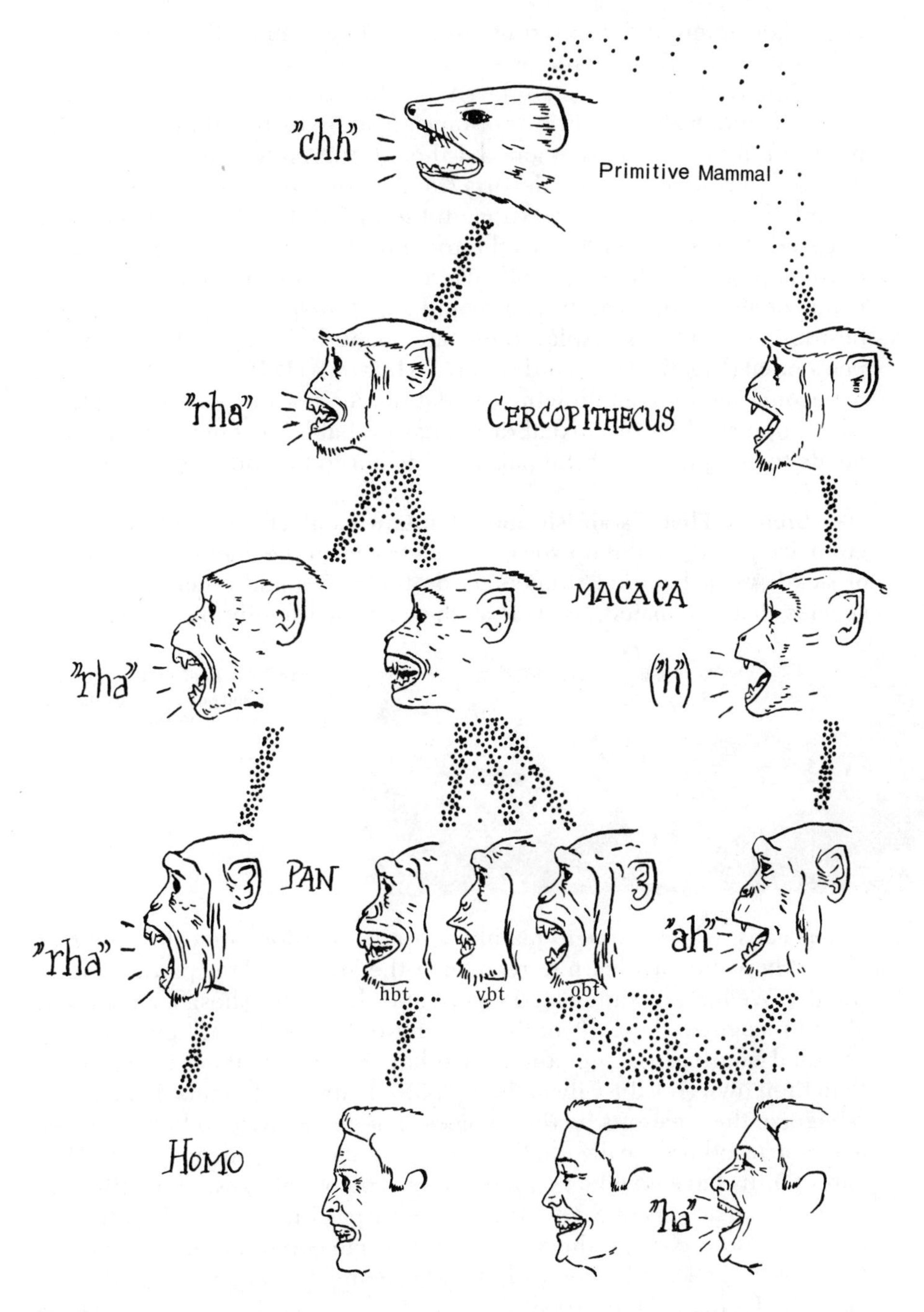

Figure 6-43 Evolution of smiling and laughter according to J. A. van Hooff (1971). *Left:* The line of development that leads to the silent bared-teeth display and to the bared-teeth scream. The *sbt* display starts as a submissive gesture but develops into a friendly one. *hbt,* horizontal silent bared-teeth display; *vbt,* vertical sbt display; *obt,* open-mouth sbt display. *Right:* line of development leading from relaxed open-mouth face (play face) as signal for play to laughter. (From J. A. van Hooff, 1971.)

is another example. The uniting function of communal threats becomes obvious in all military parades.

Finally, we must include under the heading of greeting the *farewell*, which also needs further investigation. Its function seems to lie in the strengthening of the bond for the future. Another component may possibly play a part. If one departs from another person there exists the potential danger that aggression, hitherto inhibited, would become released. A person who leaves the room backward and bowing continuously is probably afraid as well. We find a functional parallel to this in many of the complicated "postcopulatory" displays in birds, where besides elements of display there are many behavior patterns of appeasement (Fig. 6-44). According to O. Heinroth (1910) eider ducks tend to become aggressive following copulation. Aggression, which until then was suppressed by the sexual drive, has probably become released and needs to be appeased if the pair bond is not to be endangered.

Signals That Establish and Maintain Contact Animals within a group frequently maintain vocal contact (for example, members of a flock of jackdaws or bearded titmice) as do the males and females of many species that are mated, as well as the maternal family of the squirrel.

Figure 6-44 Postcopulatory display in the striped goose *(Anser indicus)*. (Photograph: H. Kacher.)

These calls, which serve to maintain group cohesion, are called *contact calls*. Many animals learn to recognize the voice of their partner individually. We have already listed ravens and shama thrushes. Sea lions and sheep recognize the calls of their young individually, and only this individual familiarity keeps them together. Sea lions attack young other than their own (I. Eibl-Eibesfeldt, 1955b). Young guillemots (*Uria aalge*) recognize their parents by their voices. They react only to the call notes and not to other vocalizations of the parents. The call notes of other parent birds have no effect on the young. The parents respond with calls to the vocalizations of the young while it is still in the egg. Whether or not they also recognize the young at this stage could not be determined because they recognize the individual egg and no longer accept strange eggs (B. Tschanz, 1965, 1968).

Some birds have developed alternating or duet songs, which maintain the pair bond. This is found in some Australian honeyeaters (Meliphagidae) and some African shrikes. In the honeyeater (*Acanthagenys rufogularis*), both members of a pair sit next to one another; one sings the strophe and as soon as it ends the other continues. Other species sing their duets together and in surprising synchrony. The highest level of development is reached in the duets of antiphonal songs of African shrikes. In *Laniarius aethiopicus* both birds have one song which each is able to sing alone. Members of a pair, however, frequently sing only certain parts of a strophe in alternation, but so perfectly adjusted to one another that one does not recognize at once that two birds are singing one melody. These duets are found primarily in birds that live in dense forests (K. Immelmann, 1961; W. H. Thorpe and M. E. W. North, 1965).[3] There exists then one group of common signals that keep a group of animals together but that does not presuppose individual recognition, and another group that depends on it. This also holds for olfactory signals, which in many mammals serve as cues that maintain group cohesion. Norway rats, bees, and many other gregarious animals recognize one another by odors common to the group or hive, without, however, recognizing one another individually (see **anonymous groups**). Sea lions, on the other hand, recognize each other individually.

If a group is to be kept together, it is helpful for its members to do the same things at the same time. A flock of birds could never keep together if each bird did something different; one would sleep, another eat, and others might want to fly. We can often observe that feeding is contagious: if one bird eats, others follow. Frequently specific expressive movements have evolved that facilitate the synchronization of moods. Graylag geese ready to fly off begin to walk, shake their heads with stretched-out necks, and call. If some members of a flock begin with this activity, others follow, and within a short time, they all leave together. In man yawning seems to have a similar contagious effect: it makes everyone sleepy.

This was described among others by K. v. Steinen, who was the first European to come into contact with the Bakairi of Central Brazil.

> If they seemed to have had enough of all the talk, they began to yawn unabashedly and without placing their hands before their mouths. That the pleasant reflex was contagious could not be denied. One after the other got up and left until I remained with my dujour (K. von den Steinen, 1894; new ed. 1917:183).

The maintenance of contact is also achieved by all those signals that stimulate a conspecific to flee, such as the conspicuous spot on the rear end of deer and antelopes.

[3] Duet singing evolved in convergence in the genus *Trachyphonus*, a monogamous bird that does not belong to the song birds (H. Albrecht and W. Wickler, 1968).

Communication about the External Environment

Warning and Distress Calls Many warning calls have evolved that alert conspecifics to the presence of a predator. The ground squirrel (*Citellus citellus*) and the marmot (*Marmota marmota*) utter a call before fleeing from a predator, and many birds do the same. It has been shown to be an effective warning to conspecifics.

Chicks fall silent while still in the egg and cease scratching movements when they hear the warning call of chickens (E. Baeumer, 1955), but herring gull chicks eagerly respond prior to hatching to all calls from the outside, including the gull's own distress call (F. Goethe, 1955). The chicks of some wading birds (Scolopacidae) also do not know the meaning of the alarm call and must first learn to associate it with the silhouette of the aerial predator (O. v. Frisch, 1958).

Gregarious aquatic animals warn conspecifics chemically by alarm substances. For example, if the snail *Heliosoma nigricans* perceives the body juice of an injured conspecific, it will bury itself in the mud (W. Kempendorff, 1942). Minnows and many other schooling fishes flee when they perceive a substance that is released from the skin of injured conspecifics (K. v. Frisch, 1941; F. Schutz, 1956; W. Pfeiffer, 1960, 1963), as is true for tadpoles of the common toad (*Bufo bufo*), which move about in swarms (I. Eibl-Eibesfeldt, 1949; E. Kulzer, 1954). The death cry of many animals may be analogous to these warning signals and bring about an association with the dangerous situation. Systematic studies are not available to my knowledge. I have personally caught many rats, one after another, in traps of the type that, when tripped, breaks the rat's back. The animals stepped on the trigger and even ate from the dead conspecifics, until one rat was merely injured and did not die at once. From that time on, no more rats were caught (I. Eibl-Eibesfeldt, 1953b).

Often animals react to the distress call of a conspecific that has been caught by an enemy. Many apes and monkeys who know their keeper well will attack blindly when a conspecific, after being grasped by the keeper, utters the distress call (W. Köhler, 1921; S. Zuckermann, 1932). This happens so automatically that, for example, terns will come to the aid of a strange chick, which they will then attack themselves (K. S. Lashley, 1915).

Language of Bees We know from the pioneering and careful investigations of K. v. Frisch (last summary of his work in 1965) that honeybees communicate the direction and distance of a food source to their fellow hive members by means of special dances.

The returning forager bee begins to dance on the comb in a very specific manner. When the feeding place is near the hive, the bee performs a round dance that contains no directional information. New foragers become aroused by this dance and then search around the hive in all directions, seeking the odor the dancer brought back from the food source. When the food source is more than 70 meters away from the hive, the bee waggle dances. Wagging her abdomen, she runs straight for a short distance while accentuating this wagging distance indicator by a rasping sound produced with her wings. Then she turns to one side and returns to the original position without wagging. She then repeats the wagging dance over the same route but turns to the other side to return again to the point of origin, and so on (Fig. 6-45). A number of bees in the hive become excited by this dance and follow the dancer. They perceive the odor of the blossoms visited by the forager and they learn in which distance and direction they have to search. If the food source is close to the hive, then the straight wagging run is short, and the wagging dances follow one another more rapidly. This enables the bees to compute the distance to the feeding place. An experienced observer can do this with a stopwatch.

Wind velocity and direction are also reflected in the dancing rhythm. When head winds prevail the bees dance more slowly; thus they report a greater distance. They do this also when they have flown to the food up a steep slope. Thus the dance does not indicate either actual distance or the flight time but rather the energy that the bees will expend in reaching their goal. This information is passed on by the amount of time they wag, emphasized by the rasping sounds mentioned above.

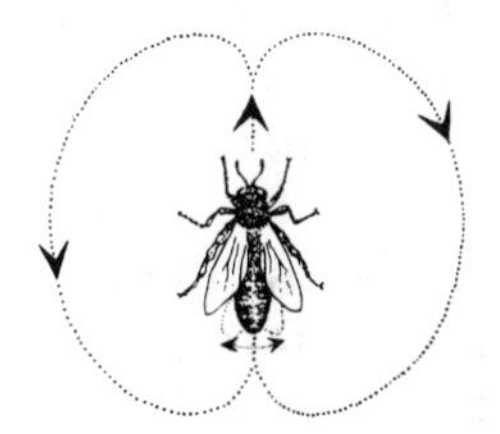

Figure 6-45 Waggle dance of the honeybee. (After K. v. Frisch, 1959.)

More recent investigations by H. Esch (1967) indicate that these sounds are of more importance than had been assumed. Sometimes returning foragers dance on the comb without making sounds. Esch observed 15,000 of these dances, and in no instance did this silent dance lead other bees to the food source.

The direction is given with reference to the position of the sun. If a bee dances in front of the hive, which rarely occurs, then the method of transmission of information can be observed. The straight line of the wagging dance runs at the same angle to the sun that the bee had maintained on a straight-line course to the food source (Fig. 6-46). The bee also does this when the combs of the hive are in a horizontal position and when they can see the sun. Then the straight portion of the traversed path leads also directly toward the target. If one hides the sun the bees become disoriented, and the aroused new foragers find the feeding place only by chance. When v. Frisch placed four odor plates in all four compass directions, they were visited equally often by the newcomers in this shading experiment. However, when the foragers danced in the sunlight on the horizontal combs they could indicate the direction, and the new foragers visited one of the odor plates in preference over the others. Normally, however, the bee dances on vertical combs within the dark hive. In this case, the angle to the sun is translated into an angle with respect to gravity (Fig. 6-47). If the feeding place lies in the direction of the sun, the straight part of the wagging dance points straight up.

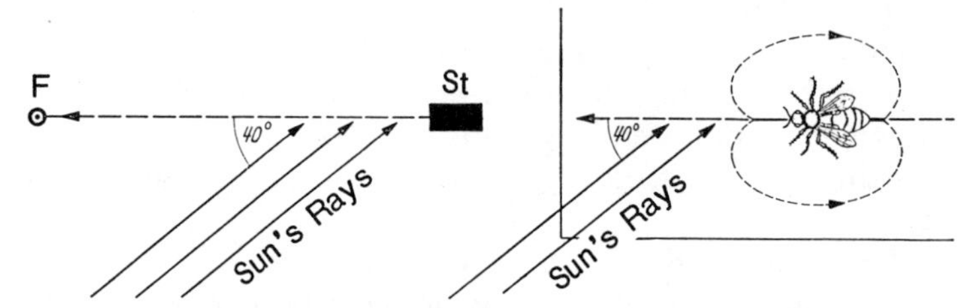

Figure 6-46 Indication of direction with respect to the position of the sun during the dance on a horizontal plane. *Left: broken line,* direction of flight to the feeding place; *right:* the waggle dance on a horizontal plane. (From K. v. Frisch, 1959.)

If the location is 60° left of the sun, the wagging dance deflects 60° from the vertical to the left. If the food source is away from the sun, the wagging dance is executed straight down. This ability to transpose is also shown by the dung beetle, which does not dance. If allowed to walk on a level surface, it will maintain a certain angle to an artificial light source. If one then raises the surface to a vertical position and illuminates the area from above with diffuse light, the beetle transposes the angle of his previous path with respect to gravity just as the bees do (G. Birukow, 1953). It is not known of what use this ability is to the dung beetle. Other insects also show this capacity to transpose. Ants transpose the angle to the light source into an angle with respect to gravity when the horizontal experimental table is tipped vertically and the light is turned off at the same time. However, ants are not as precise as dung beetles or bees. If the angle of direction was 20° to the right of the sun, they will maintain 20° to the perpendicular line, but it may be either to the right or the left, and they may walk up or down with respect to gravity. The ladybug, which equates the direction to the light with a vertical upwards direction, may, however, deviate right or left; the dung beetle transposes in the manner of the bee except that the direction toward the light coincides with a downward direction. G. Birukow (1956) attempted to explain these phenomena. An integration of light and gravity orientation can be found in bumblebees. When they leave their nest, they are positively phototactic in respect to their orientation to the sun. If one places them on a vertical plane in the dark when they are on the way to leave the nest, they will go upwards, the same way they would within their dark nest in the ground when they were ready to leave. On the other hand, they are negatively phototactic on the return flight and crawl into darkness. If placed on a vertical plane in darkness they move downward. On the ground they will find their way to their nest in this way (U. F. Jacobs-Jessen, 1959).

If the bees were to indicate the angle to the sun when they leave the hive, this would ultimately lead to errors, because the position of the sun changes and with it the angle to the sun, although in a lawful manner. K. v. Frisch's experiments have shown that the bees take this movement of the sun into account, although it is not known how they accomplish this. A prerequisite for this ability is that a bee experience this movement of the sun herself, and from this they learn what is necessary. Bees that had seen the movement of the sun for only a few afternoons in their lives, spending the remaining time in the hive in a dark basement,

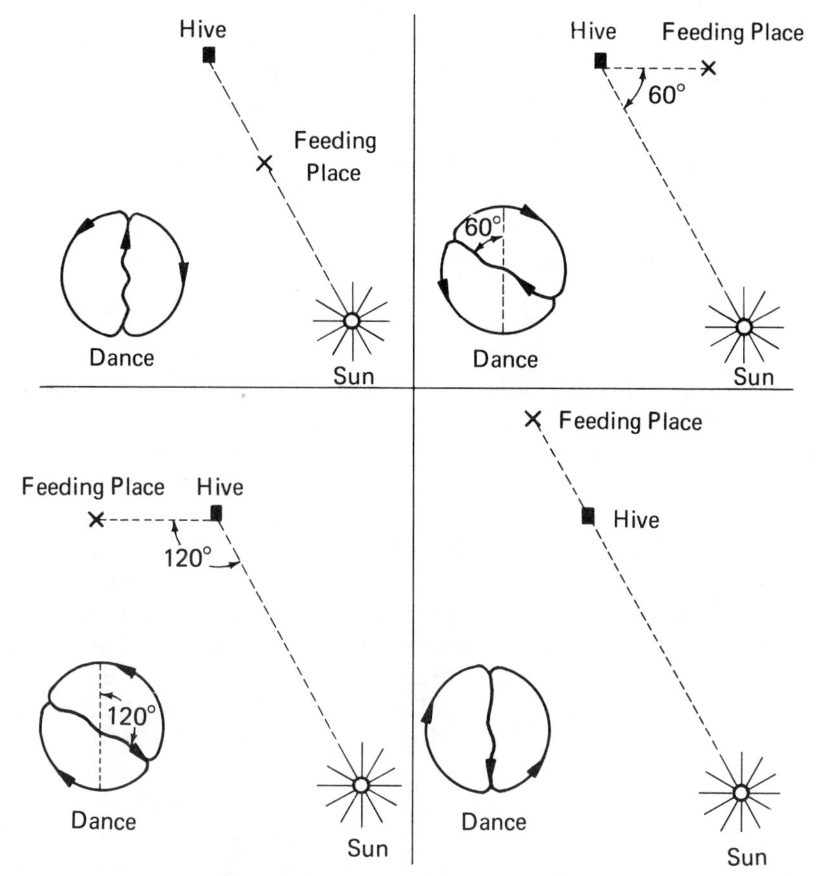

Figure 6-47 Indication of direction according to the position of the sun during the dance on the vertical plane of the comb. On the left is shown in each case how the bee dances on the comb to indicate a given location of the feeding place. (Additional explanation in the text.) (From K. v. Frisch, 1959.)

acquired the knowledge about the daily course of the sun. If they were subsequently tested in the morning during daylight, they did not make errors in the direction in which they had been trained.

When making detours bees indicate the distance in a straight line, but they also communicate the length of the detour. Dancing is innate in bees, and there are several dialects. The Egyptian honeybee begins wagging dances when the food is more than 10 meters from the hive; the Krainer race only beyond 50 to 100 meters from the hive. The latter also has the fastest wagging dance.

D. L. Johnson (1967) and A. M. Wenner (1967) proposed that bees are able to communicate direction and distance of a goal by means of the dance, but that this is not of any use to other worker bees. These are interested in the dancers but do not utilize their information. Instead, the odors of food, hive, and other bees are said to aid them in finding their goals.

These statements cannot be brought into agreement with the experiments we have discussed so far. During the detour experiments returning foragers danced in a straight line to the food source, and later foragers flew in a straight line toward the goal and over the obstacles, while the previous foragers had made the detour. Thus it is inconceivable that the recruits could have followed an odor trail.

If one raises the question as to how Johnson and Wenner came to their contrary opinion, one discovers several errors of experimental design that the authors overlooked. K. v. Frisch set up his fan experiments in such a manner that he trained marked bees to a food source 250 meters from the hive. The food source consisted of a weak sugar solution, which did not result in much dancing in the hive and which only barely excited other bees. During the actual tests 7 odor plates without food were placed 15° apart at a distance of 200 meters. At the same time the original feeding place was marked with the same odor as that on the other odor plates, and a strong sugar solution was offered. Now the trained bees danced and the number of new recruits on the various odor plates indicated in which direction the recruiters were searching. The number of arrivals at the food source was not counted, because the marked bees also made use of their odor glands, which are an additional attractant.

Johnson performed similar experiments, but he then added a "control experiment." To create identical conditions at all locations he placed another bee hive, containing light-colored bees, next to the first hive. These lighter bees visited three locations that were closer to the hive, while the darker bees continued to visit the food source more distant from the hive, which was located on the same line of flight as that of the central plate of the three closer feeding places. Only marked bees of both hives were fed; other arrivals were killed until the actual tests began, when Johnson began counting recruits. As one would have predicted from the findings of K. v. Frisch, the recruited dark bees preferably visited the central feeding stations. Alas, they did so even when Johnson shifted their feeding place to the north, again on a line with one of the closer feeding places. From this finding he concluded that the bees were not guided by the dance but merely visited the geometric center of all the feeding stations. However, during the experiment a strong southeast wind prevailed, and because the bees in the nearer hive used their odor glands, the bees on their way to the northern food location were detracted by the more intense odor reaching them from the southern feeding place. Just such distracting effects had already been described by v. Frisch, and Johnson might have avoided this error by a more careful reading of the literature. Furthermore, Johnson tested in only three directions, while v. Frisch covered a much larger area with seven odor plates. He also tested bees over greater distances (600 and 1250 meters).

Wenner's experiments were just as uncritical as are those of Johnson. He fed bees, using Frisch's gradient-experiment procedure, 400 meters from the hive. The nearest and farthest feeding places (200, 300, and 500 meters) were visited by only a few new bees. The original food source was visited by 74 percent of all new bees, although all feeding places were supplied equally well with food. Then Wenner placed another hive with lighter bees directly north of the line of flight and fed one group each of these bees at locations 200 and 300 meters, while the trained dark bees continued to fly to the 400-meter location. The recruits of the darker bees no longer responded in accordance with the distance message received from the trained bees, since 18 percent alighted at the 200-meter place, 48 percent at the 300-meter place of the light bees, and only 33 percent at the 400-meter food source of their own bees. One percent visited the feeding place at 500 meters. These results also could be predicted on the basis of v. Frisch's experiments. In a strong wind, which prevailed during Wenner's experiments, bees fly close to the ground. Under these conditions many of them visited the closer food sources, to which the learned odor and the odor secretions of other bees attracted them (K. v. Frisch, 1968).

Meanwhile, experiments by J. L. Gould et al. (1970) have confirmed once again that the dance of bees serves to impart directional information to other bees.

Observations of other insects hint at likely ways along which dancing might have evolved. V. G. Dethier (1957) fed sugar to blowflies (*Phormia regina*). If he withheld it from them, they performed a kind of round dance, running in circles to the right and left. They also did this when they were taken to another location. They "danced" more rapidly when the sugar solution was of higher concentration. These dances were oriented in respect to light and gravity. The flies approached the light source and ran up and down on a vertical plane, without, however, indicating direction.

These flies also regurgitate food when they meet another fly, and the latter then begins to search. The "dance" here is always a searching, but one can imagine that the "round dance" of the bees might have evolved from such initial stages; in other words, that it is a ritualized searching which stimulates others to search also. With increasing delays, the searching behavior of the fly decreases in intensity.

A parallel to the waggle dance was discovered by A. D. Blest (1960) in some New World saturnid moths. These moths shake their body after landing by alternately bending and straightening their legs. The duration of this shaking increases linearly with the length of flight time.

A similar physiological mechanism may have been possessed by the ancestors of bees, and which could then have become useful in indicating distance. The indication of direction can be interpreted as derived from an intention movement of flying toward the feeding place. This interpretation can be seen as probable when we consider the behavior of primitive bees. The stingless bee *Trigona postica* leads her fellow hive members. On the way home she marks various landmarks with secretions of her mandibular glands. Upon returning to the hive she runs about on the comb with fluttering wings and bumps other bees. She dispenses food samples, and when a number of bees have gathered around she flies off along the odor-marked trail that she prepared earlier (M. Lindauer, 1961). This leading behavior is shortened in the more advanced stingless bee species. Several species of the genus *Melipona* use a kind of Morse code on the comb to indicate distances: short bursts of sound indicate close goals; longer bursts signal greater distances (H. Esch, 1967). The bees indicate direction by flying toward the goal, first in a zigzag course and then in a straight line. After repeated indications of direction in this way, some worker bees fly off in the same direction, as if they had understood the message. The dwarf honeybee (*Apis florea*) performs a complete dance, but only in sunlight and on a horizontal plane. Here the wagging dance can be recognized as a repeated intention to fly off. This fits with the observation of the kinds of sounds made during wagging dances in the advanced bees. In this way it becomes plausible to suggest that the dance evolved from a more general invitation to search to the wagging dance, which indicates distance and direction of a goal. The ability to transpose orientations from the sun to gravity was probably present as a preadaptation, because it also exists in many insects that do not dance.

The dance language shows some similarities to human language. It is a means of communication between conspecifics, and relations between things are communicated. However, in contrast to human language the system is a stereotyped, innate coding system. **Human language** is also based on an inborn potential for specific sound production and perhaps on the drive to speak, but the language symbols are individually learned and passed on by tradition. Individual experiences can be described by words and passed on, and abstract thinking permits communication about relationships between relationships. The bee dance is similar to human language in so far as it is also a symbolic language by which inexperienced individuals acquire knowledge without the object in question being present. The transfer of knowledge, however, is only made by animals that have experienced the location of the feeding place in a previous flight. No bee will communicate a message just received by another bee to a third unless it has itself visited the feeding place. This language is "rumorproof" in a manner of speaking (W. Wickler, 1967b). O. Koehler (1949, 1952, 1954b, 1955) has repeatedly commented on this subject, most recently in a review (1966b) of the work of C. F. Hockett (1960). He presents one of Hockett's tables amended by him, in which he compares six methods of communication among animals (including bee dances) with human language and instrumental music. This he does in respect to each of 13 characteristics of human language.

Figure 6-48 Play-fighting young gorillas. After roughhousing, the victor drums with the hands on his breast while the loser climbs away. (Photographs: Johnston, from the film "African Apes," F. 96 of the former R. St. f. d. Unterrichtsfilm.)

Intraspecific Threat Signals

By intraspecific threat signals we mean all those behavioral and morphological characteristics that serve to reject a conspecific. Colorful plumage, for example, has this effect. K. Immelmann (1959) showed that zebra finches with colorful plumage maintain a certain distance from one another, while all-white birds of the same species perch much closer together. Many sounds and vocalizations produced by insects and vertebrates have the same function. Many fishes threaten one another with calls, for example, cichlids (A. A. Myrberg, 1965) and damselfish (I. Eibl-Eibesfeldt, 1960a; H. Schneider, 1963). The territorial song of birds is well known and is directed against rivals. Male sea lions roar at the owner of the neighboring territory. Many rodents utter ultrasonic threat calls. These are only a few examples. Threatening sounds may also be produced by other means. Many monkeys display anger by shaking the branches of their trees. Japanese monkeys, in addition, beat against resounding objects, behavior that is considered to be derived from tree shaking (S. Kawamura, 1963).

Gorillas and sometimes chimpanzees pound their chests (Fig. 6-48); chimpanzees beat against "drumming trees" and, in captivity, against other resonating objects (B. Grzimek, 1954; G. B. Schaller, 1963; J. van Lawick-Goodall, 1965, 1968). Human beings do the same. Macrosmatic animals try to intimidate their opponents with odors.

In addition, special postures and movements serve to keep others at bay; these often consist of ritualized elements of attack behavior (for example, biting and rush attacks). Generally the threatening animal will make itself larger and more impressive, and may display its weapons (Figs. 6-49–6-52). The animals may rise up high and spread manes, skinfolds, fins, and feathers, which frequently exhibit conspicuous patterns or colors. On occasion it is possible to distinguish aggressive from defensive threats. An aggressive squirrel pulls its ears back and chatters with its front teeth, but if cornered and defensive, it threatens by raising its ears, which appear still larger because of tufts of hair on their tips; at the same time it squeals (I. Eibl-Eibesfeldt, 1957a; see also Fig. 6-53). Threatening calls are very widespread. Many species threaten predators in the same way as they threaten conspecifics, for example by displaying their weapons.

Signals Effective in Interspecific Communication

Signals Indicating Contact Readiness between Species

Animals belonging to different species often come together to their mutual advantage. Some shrimps (*Alpheus*) live together with gobies on the sea bottom, which is bare of all cover. The shrimps dig a hole in

Figure 6-49 *Above:* Lateral enlargement of the threatening Andes anolis: *top,* a threatening male; *bottom,* normal posture. (After W. Kästle, 1963.)

Figure 6-50 *Left:* Threatening marine iguana (bite threat and lateral display) (Narborough, Galápagos Islands). (Photograph: I. Eibl-Eibesfeldt.)

Figure 6-52 A coconut crab (*Birgus latro*) threatens by raising its opened claws toward the photographer. In intraspecific disputes crabs often threaten with their claws and this has led to a widespread ritualization of the gesture (see Figs. 6-3 and 6-4). (Photograph: I. Eibl-Eibesfeldt.)

Figure 6-51 Galápagos sea lion (*Zalophus wollebaeki*) threatening near the border of his territory. (Photograph: I. Eibl-Eibesfeldt.)

which they live; the gobies, which do not dig, profit well from this, and in turn warn the shrimp of approaching danger (W. Luther, 1958; W. Klausewitz, 1961; see also Fig. 15-34). Anemone fish (*Amphiprion*) live in certain specific anemones without being injured by their poisonous tentacles (I. Eibl-Eibesfeldt, 1960a; E. Abel, 1960a). In symbiotic relationships of this kind a problem of communication between species exists. This has been investigated in more detail in the "cleaning symbioses" (I. Eibl-Eibesfeldt, 1955a, 1959).

Figure 6-53 *Left,* Defensive and, *right,* aggressive threat in the European red squirrel. (From I. Eibl-Eibesfeldt, 1957a.)

A number of marine fishes are specialists in freeing other fish of parasites (I. Eibl-Eibesfeldt, 1955a, 1959; J. E. Randall, 1958; C. Limbaugh, 1961), for instance, the cleaner fish (*Labroides dimidiatus;* Fig. 6-54). This fish entices its hosts to permit themselves to be cleaned by means of a signal consisting of a special nod swimming (cleaner dance). It butts with its snout against their fins or opercula so that they spread them, and against their mouths so that they open them, so that it can get inside. While the cleaner fish inspects its host, it continuously vibrates its ventral fins against the latter's body so that the host knows at all times where it is being cleaned. It can be clearly seen that the host reacts to this procedure, because it stops moving the fins that have been so touched.

On the other hand, host fish invite the cleaners to do their job by opening their mouths, and they signal them when to leave by closing their mouths half-way with a jerky motion and opening them again. Following this signal the cleaners leave the mouth cavity in which they are cleaning. The host signals its intention to swim on to the fish cleaning it on the outside by shaking its entire body. In this way the cleaner and the fish being cleaned communicate with a few expressive movements. The **saber-toothed blenny** (*Aspidontus taeniatus*) imitates the cleaner fish and sneaks up on its victims. It looks like the cleaner and imitates the nod-swimming behavior in all its details, although this style of swimming is not typical for this group of fish. In this way the host is deceived and the mimic bites chunks from the fins and gills.

The honey guide, a bird of the savannahs of Africa south of the Sahara (*Indicator indicator* and *Indicator variegatus*) leads honey badgers and men to beehives that they themselves cannot open but on whose honeycombs they feed. They produce conspicuous calls, spread their tails when one approaches them exposing a white pattern, and fly off some distance.

The begging movements of young brood parasites such as the European cuckoo and the learned begging movements of zoo and domesticated animals are all directed toward members of other species.

Figure 6-54 *Left:* Cleaner wrasse cleaning *Plectorhynchus diagrammus* (Maldive Islands). The cleaner fish swims above the mouth of the larger fish; others wait nearby to be cleaned. *Right:* The cleaner fish is just disappearing into the mouth of the larger fish. (Photographs: I. Eibl-Eibesfeldt.)

Threat Postures and Other Signals for Warding off Nonspecies Members

A large number of threat postures and gestures, which are often similar to those used in intraspecific disputes, are used to repel members of other species. Many animals threaten each other with their weapons. Carnivores show their teeth; **crabs threaten** with raised, open claws. Specific postures adapted for use against predators are found in many butterfly larvae. The caterpillar of *Dicranura vinula* raises itself up when touched or when the leaf it sits on vibrates and displays a conspicuous face mask, which if touched is turned toward the stimulus (Plate IV). At the same time the animal projects two long, red threads, which spring from the last pair of modified legs that are raised into the air, and which are repeatedly twisted into a spiral before being withdrawn. In general, an animal defending itself makes itself larger and more conspicuous. Some species imitate stronger species (see **mimicry**). One category of defensive behavior consists of the so-called mobbing reactions with which many songbirds attack birds of prey. They have special mobbing calls and often make sham attacks from all sides against the enemy, who usually departs. In addition to being bothersome, the advantage seems to be that the detected predator is unable to surprise a prey (E. Curio, 1963). Some fish species also mob predators. While diving in the waters near the Maldive Islands I observed fusiliers (*Caesio*) that repeatedly swam toward a moray eel and passed closely overhead until it disappeared (I. Eibl-Eibesfeldt, 1964c).

A *distraction display* is behavior that serves to mislead a predator. When a mammalian predator appears many incubating and brooding birds flutter to the ground and run away limping as if they were injured. This behavior is also shown by the Galápagos dove, which lives in an environment totally lacking predatory mammals. This behavior seems to be a remnant from those times when the ancestors of this dove had encounters with mammalian predators (I. Eibl-Eibesfeldt, 1964b).

SUMMARY

Movement patterns that function as releasers have been termed expressive movements. Their function and origin have been investigated in many instances. Many, for example, derive from intention movements of attack and flight. By superposition of several expressive movements a rich variety of expressions can result. Expressive movements derive from functional patterns (biting, grooming, nestbuilding) and epiphenomena or displacement activities (blushing, trembling), provided they characterize the readiness of the animal to act (its mood) distinctively enough as to serve another animal as an indicator.

The process by which movement patterns change into signals is called ritualization. In the course of ritualization behavior patterns may change function and also motivation. The pattern itself becomes simplified but at the same time amplified, and often rhythmic repetition makes the signal more conspicuous. Threshold changes may occur, movements may freeze into postures, orientation components may change, and behavior patterns that formerly varied according to strength of motivation and stimulus now occur in a stereotyped manner at a constant level (typical intensity). Variable movement sequences can become condensed into simpler stereotyped ones. Often morphological structures develop along with the ritualization of the movements, underlining and emphasizing the movement.

Phylogenetic and ontogenetic ritualization in principle follow along the same lines. Human language is unique insofar as man can communicate about objects without their presence, with the help of a learned code. Recent studies of chimpanzees have proved the ability of these animals to communicate by means of learned symbols, with a grammar and syntax like human language.

Signals can be classified according to their functions. Courtship patterns, greeting behaviors, and acts of submission have in common the function of appeasing and often of establishing a bond. Appeasement is achieved by the withdrawal of signals that provoke aggression (facing away of the blackheaded gull) and by signaling behaviors that release "friendly" responses, such as feeding and other altruistic behaviors. Often patterns of infantile behavior serve this function in a ritualized way. Thus the mates in many adult birds imitate the begging behavior of the young during courtship and indeed get fed as a result.

Dogs in submission prostrate themselves on the back, expose their genitals, and urinate, thus imitating puppies that present for being cleaned, with the result that the attacker changes from attack to sniffing and licking. In greeting, also, patterns from the repertoire of threat behavior are used in some species, as in geese, where both mates threaten jointly a third, often imaginary, enemy. Communication about the external environment occurs in the form of the warning and distress calls and in a very elaborate form in the dances of the honey bee, by means of which the bees communicate to others the direction of a food source with respect to the position of the sun. Threat signals have generally evolved from patterns of attack. Sometimes signals have evolved for interspecific communication, as in cleaning symbioses between fishes.

7

Natural Models and Mimicry

The simplicity of key stimuli permits model making not only by the ethologist and the fisherman but also by many animals. They, too, are able to imitate specific releasing stimuli with which to release behavior patterns in others to their own advantage. This is above all true for a number of predators. The alligator snapping turtle (*Macroclemys temminckii*) rests at the bottom of rivers with its mouth open; the inside of the mouth and tongue are darkly colored. At the tip of the tongue are two thin, red processes; these are dangled in the water and move like small worms. They attract fish, which are caught as soon as they begin to nibble at the protuberances. The large-mouthed catfish (*Chaca chaca*) has two small moving barbels that serve the same function (H. Schifter, 1965). The anglerfish (*Antennarius* and others) has a movable first ray on its dorsal fin with skin attachments at its tip that serve as lures. These animals are camouflaged and lie quietly on the bottom with only their lures moving (W. Wickler, 1964a, 1964c). The same author has recently shown that different species of anglerfish possess different lures which are adapted to specific prey animals (Fig. 7-1).

The snail (*Succinea*) is the intermediary host for the sporocysts of the liver fluke (*Leucochloridium*), which parasitizes songbirds. The sporocysts develop extensions that penetrate into the tentacles of the snails and attract birds with their pulsating movements. Their conspicuousness is enhanced by yellow-green rings and the considerable swelling of the antennae. The bird is deceived by this model of an insect larva and eats the antenna including its contents, and thus becomes infested with the parasite (C. Wesenberg-Lund, 1939).

The small swordtail characin (*Corynopoma riisei*) from Venezuela attracts the female with a model of a daphnia when he wants to copulate. The male's gill operculum is modified into a long extension with a dark knob at its end, which is moved in a trembling fashion before the female.

Figure 7-1 Various anglers. *Top left:* the alligator snapping turtle (*Macroclemys temminckii*) fishing with its tongue; *below:* the large-mouthed catfish (*Chaca chaca*) angling with its barbels; *top right:* the anglerfish (*Phrynelox scaber*) fishing with its wormlike lure. The lure is not only moved by the stem on which it is fastened, but also wriggles by itself. *Below: Ogcocephalus,* which fishes for prey hidden in the sand with a lure that points downward. (From W. Wickler, 1968a; H. Kacher, artist.)

She may actually bite at it (K. Nelson, 1964). The male then takes advantage of the closeness of the female and copulates with her (Fig. 7-2).

Female fireflies of the genus *Photurus* attract the males of another firefly species, *Photinus,* by imitating their flashing code. The deceived males are then eaten by the females (J. E. Lloyd, 1965).

The fly orchids of the genus *Ophrys* have a modified lower lip on their flowers that resembles the female of certain wasp species as well as its sexual odor. The pollen sticks to the males when they attempt to copulate with these models and is carried to the next flower (B. Kullenberg, 1956; F. Schremmer, 1960; see also Fig. 7-3). In many species, such as *Ophrys fusca*, the surface of the lip sports a furlike structure resembling the dorsal hairs of the female insects they are mimicking. The "hair" growth is pointed up towards the pistil, and thus the male insects, after landing, are forced to turn around and insert their abdomen into the flower. In this way contact is made with the pollen carriers. H. Baumann and G. Halx (1972) have published an excellent series of illustrations of this procedure.

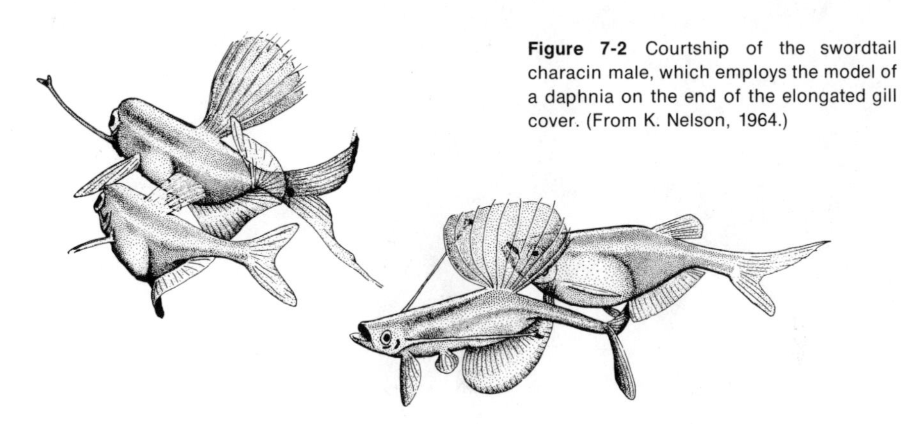

Figure 7-2 Courtship of the swordtail characin male, which employs the model of a daphnia on the end of the elongated gill cover. (From K. Nelson, 1964.)

The African devil's flower (*Idolum diabolicum*), a predatory mantid, mimics a flower that attracts insects. Because flies are attracted by other flies, a part of the flowerlike body of the mantid is dotted to resemble flies sitting there—a fact to which W. Wickler (1968a) called attention (Fig. 7-4).

The **saber-toothed blenny** (*Aspidontus taeniatus*) looks so similar to the cleaner fish (*Labroides dimidiatus*) that it is mistaken for a cleaner by other fish and can approach them easily (Plate V). It rushes at them and bites pieces out of their fins (I. Eibl-Eibesfeldt, 1959). J. E. Randall and H. E. Randall (1960) discovered that the mimic even resembles differences in the races of various cleaner fish species. In the

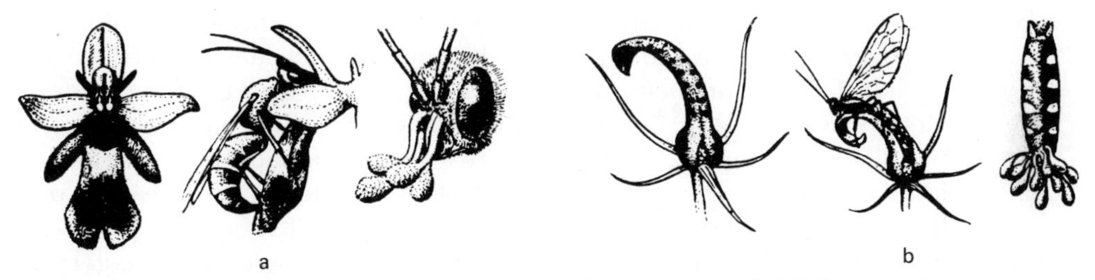

Figure 7-3 *a:* Fly orchid (*Ophrys insectifera*) with a wasp (*Gorytes mystaceus*); *left:* flower; *center:* wasp visiting the flower; *right:* head of the wasp with attached anthers. *b:* The wasp (*Lissopimpla semipunctata*) on the orchid (*Chryptostelis leptochila*); *left:* flower; *center:* wasp visiting the flower; *right:* abdomen of the wasp with adhering anthers. (From W. Wickler, 1966d.)

Figure 7-4 African devil's flower (*Edolum diabolicum*) in the prey-catching position. (After a color plate by P. Flanderky in Brehm's *Tierleben*, ed. 4, vol. 2, p. 80.)

Tuomotus region the cleaner has an orange-red color around the middle; so does the mimic. In other areas cleaner and mimic have a dark stripe at the base of the pectoral fins, which is not found in other members of these species.

The African cichlids of the genus *Haplochromis* are such highly developed mouthbreeders that the females take the eggs they have just laid into their mouths even before the male has an opportunity to fertilize them. However, the male has "imitations" of the eggs on his ventral fin. Once the female has taken the eggs into her mouth, the male spreads his ventral fin before the female, exposing the dummy eggs. She tries to take them into her mouth; the male milts and in this way the eggs already in her mouth become fertilized. In this example of intraspecific mimicry, discovered by W. Wickler (1962a), the conspecific is deceived (Plate V).

In *Tilapia macrochir* the female also picks up the eggs immediately after spawning. Here the male fertilizes the eggs in another way. He produces filamentlike spermatophores that the female picks up if she finds them; but many of them are lost. Again, fertilization is ensured through the existence of another deceptive signal. The males possess long, filamentlike spermatophore models that protrude from the genital region. These are even stronger releasers for the females than the actual spermatophores, just as in the case of the dummy eggs. The male presents these dummies to the female; she takes them into her mouth

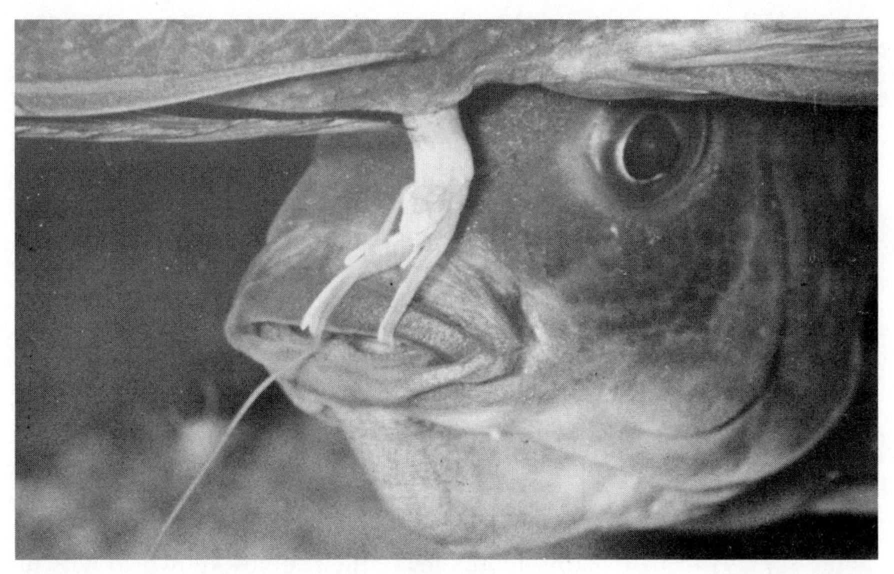

Figure 7-5 Spermatophore model and spermatophore of *Tilapia macrochir*. The female has just taken the spermatophore-like attachments of the male genitals, including a spermatophore (long filament), into her mouth. (Photograph: W. Wickler, 1966b.)

and in this way receives the spermatophores that are between the dummies (R. Apfelbach, 1967b; see also Fig. 7-5).

W. Wickler (1965b) was able to explain some releasers that function in keeping groups together as examples of intraspecific mimicry. In hamadryas baboons signs of females in estrus become appeasing signals that are also present in the males. Males have strongly vascularized skin areas that are similar to the swollen parts of females in estrus, but in specific instances the areas in question are not homologous. Males present their posteriors to other group members in the manner of females. This has a definite appeasing effect. The swelling in these instances no longer has a sexual "meaning." W. Wickler (1965e, 1966b) discovered this same principle in other monkeys, also in carnivores and fish. Males of the **spotted hyena** (*Crocuta*) present their slightly erected penis during each greeting encounter with others. Females do the same and possess a pseudopenis that looks deceptively like a real penis. This makes it almost impossible for even a trained observer to distinguish males and females by external signs.

What has been said so far indicates that the concept of mimicry must be considerably broadened. Not only protective similarities, but all similarities that involve the falsification of signals, should be included (W. Wickler, 1964c).

An interesting ethological example of mimicry is discussed by J. Nicolai (1964). The **widow birds** (Viduinae) are breeding parasites of various species of grass finches (Estrildidae). Their young so closely resemble the host species with respect to plumage as well as to the gape markings in the mouths that the young are raised by the host parents along with their own (Plate VI). The Viduinae also imitate the courtship song of the host species in every detail, thus attracting their own females to the correct host-species pairs. The larvae of the *Lomechusa* bettle, as guests of ants (*Lomechusa strumosa* and *Atemeles pubicollis*), mimic the begging behavior of ant larvae as well as their attractive odors. Thus they are fed by the hosts in the manner in which the latter feed their own larvae (B. Hölldobler, 1967).

Finally, there are many examples of mimicry in the traditional sense of protective resemblance. Songbirds that have had an unpleasant experience with wasps will avoid them and their harmless mimics as well, for example, those from the group of Diptera and Lepidoptera (see Plate IV). Toads learn quickly to distinguish a mealworm from a bee, and henceforth will also avoid mimics of bees (L. P. Brower and J. v. Zandt-Brower, 1962). Some mimics of wasps imitate not only the appearance but also the buzzing sound of the poisonous model (A. T. Gaul, 1952). Some species of bad-tasting moths make themselves known to bats by warning sounds: they click in a special manner when they are hit by a burst of sounds from a bat. Some species of edible butterflies also do this, and they are probably mimics (D. C. Cunning, 1968). In experiments bats avoided edible prey when it was presented in conjunction with warning sounds. It is fascinating to observe that a mimic adopts several forms, thus imitating different models (M. Tweedie, 1966; see also Plate IV).

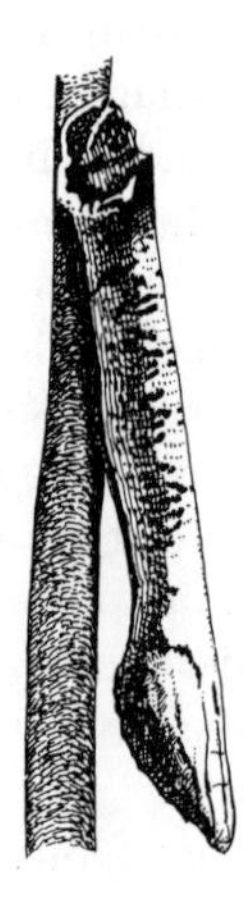

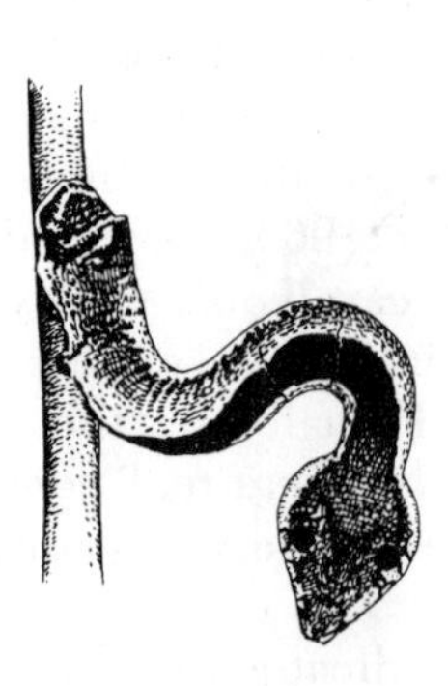

Figure 7-6 The caterpillar of *Leucorampha ornata* (Sphingidae) mimicking a snake; *left:* resting; *right:* when alarmed, turning the underside toward the observer. (After A. Moss, 1920.)

The caterpillar of the moth *Leucorampha ornata* raises its posterior end, imitates the head of a snake, and curves its body into an **S** when it is alarmed (Fig. 7-6). The snake-head model has two eye spots on its underside that are turned toward the predator. The caterpillar of the moth *Pholus labruscae* imitates a snake with the anterior and posterior ends.

The anterior portion reminds one of the snake head. A small black "tongue" which protrudes from a dark-colored area at the posterior end wiggles like the tongue of a snake (E. Curio, 1965a).

A remarkable instance of behavioral mimicry has been described by T. Eisner and J. Meinwald (1966). The ground beetle (*Eleodes longicollis*) defends itself against predators by standing on its head and spraying an irritating substance from the tip of its abdomen. Another darkling beetle (*Megasida obliterata*), which lives in the same desert region of Arizona, imitates this behavior, although it does not have a defensive excretion.

Many insects have conspicuous dark eye spots on their wings that they expose suddenly when in danger. These patterns are more feared by several species of songbirds than patterns that are unknown to them (A. D. Blest, 1957). The same principle applies to the threat posture of the spectacled cobra (*Naja*). The argument could be made that the eye spots on the wings of butterflies which are suddenly exposed when danger appears should not be considered under the category of mimicry but merely as warning signals which present a sudden optical stimulus. This optical stimulus could then drive off the predator, independently of a similarity with another structure. The observation that songbirds fear "eye spots" more than other unknown patterns could be a result of the greater effectiveness of concentric circles as a pattern on a given plane. They would thus be the strongest stimulus for a mechanism attuned to gestalt perception. There are, however, butterflies whose eye spots also have unsymmetrical "reflections" (for example, *Caligo eurylachus*), which produce a most deceptive similarity with a vertebrate eye. This is perhaps the strongest argument for considering these spots as examples of mimicry (Plate IV). An excellent treatment of these problems, which are only outlined here, has been published by W. Wickler (1968a).

SUMMARY

Many animals mimic the signals of other species, or patterns that serve other species as signals (such as certain characteristics of their prey). They do it mainly to lure other species in order to prey upon them. We know also, however, of instances of intraspecific mimicry. In hamadryas baboons males mimic the red swellings of the females, using their presenting behaviors in appeasing greeting rituals. Furthermore, animals often mimic others that are armored or protected by unpleasant smells or tastes in order to profit from the resemblance (protective mimicry).

8

Reaction Chains

When a living organism responds to a stimulus, the releasing-stimulus situation is frequently changed because the animal then comes into a new position in which additional stimuli become effective. We know, for example, that a bee is visually attracted to a piece of colored paper, but will rarely alight on it. On closer approach she knows from odor cues that no nectar will be found here. If an appropriate odor is added she will land and continue to search. Other stimuli are then needed to release the sucking movements. The bee-hunting digger wasp (*Philantus triangulum*) flies from flower to flower in search of bees and reacts first only optically to moving objects, including small flies, upon which they do not prey. If the wasp perceives a moving object it positions itself leeward from it at a distance of 10 to 15 cm in the air and tests the wind. If the appropriate odor is present—such as a model with bee odor—the attack is made. However, the wasp only stings if a real bee is present. This reaction could not be released by the model (N. Tinbergen, 1935). In all these instances the animal comes into new releasing-stimulus situations by its own actions.

The same is true for the behavior sequence of hermit crabs during the selection of snail shells. E. S. Reese (1963a) distinguished eight different fixed action patterns that occur in a specific sequence. This sequence is dictated exclusively by the releasing stimulus situation, and

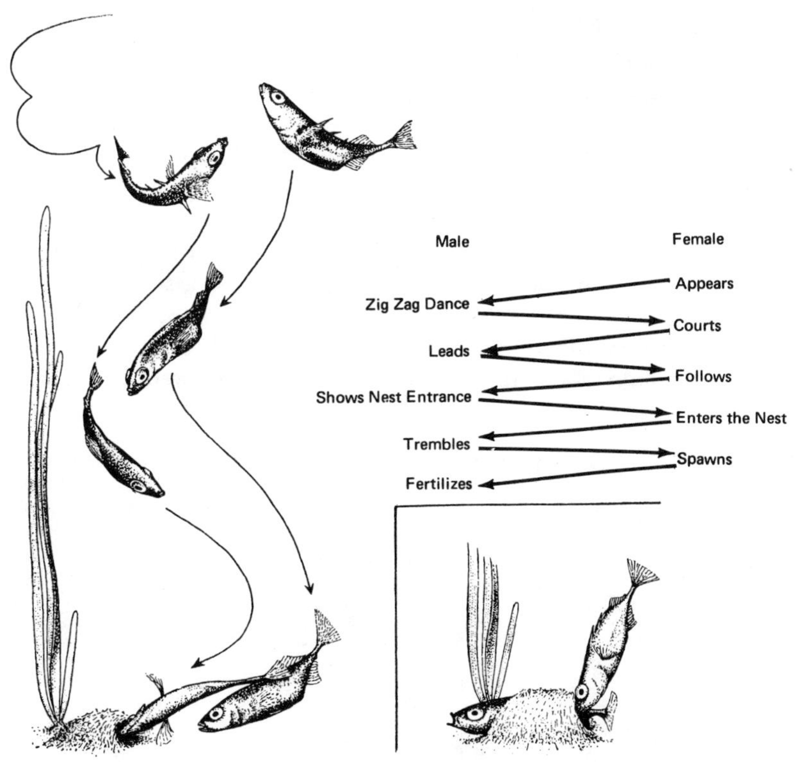

Figure 8-1 Courtship behavior of the three-spined stickleback, illustrating the mutually releasing actions of the male and female. (After N. Tinbergen, 1951.)

at times a specific movement may not take place. Then the releasing stimuli are not present. Thus, if a hermit crab finds the opening of a snail-shell when first encountering it, all those behavior patterns with which the exterior of the shell is normally investigated are omitted. The animal proceeds at once with the investigation of the interior of the shell with its claws and the first pair of legs. Only then will it slip into the shell and raise it up. Whether or not the hermit crab will show additional appetitive behavior depends on the suitability of the encountered shell (E. S. Reese, 1962b, 1963b).

In these cases the behavior is not normally terminated because of **action-specific fatigue,** but by a stimulus situation that cuts it off. A single individual of a schooling fish species becomes calm when it swims within the group and stops searching for a school after it has joined one. In squirrels and agoutis the food-hiding behavior is terminated when the food object is actually buried. If the expected success is not achieved, then renewed efforts to bury it are made or **conflict behavior** occurs.

In cases where two mutually attuned partners are present—for example, sexual partners—they reciprocally release certain reactions from one another, which in turn are themselves releasing stimuli. An especially good example is presented by N. Tinbergen (1951). If a stickleback female appears in the territory of a male, he at once begins a zigzag dance. This in turn releases a special display movement of the female. He then leads her to the nest; she follows; he shows the nest entrance and she slips into it. Then he butts her repeatedly at the base of the tail, which still protrudes from the nest; in response to this she spawns. She then swims off; he enters and milts. Each of these stimuli can be imitated by a model. For example, one can remove the male after the female has entered the nest and release spawning by drumming against the base of her tail. If this stimulus is omitted, the behavior chain breaks at this point; the female does not spawn. Whenever the behavior sequence

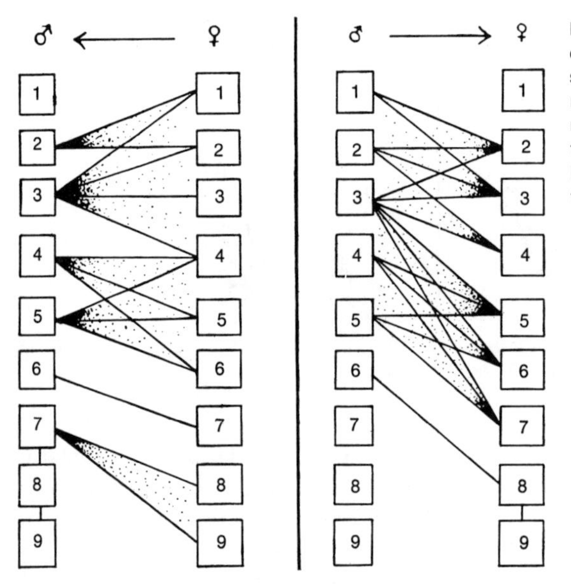

Figure 8-2 Observed behavior sequences in mating behavior of the stickleback. *Left:* Reactions of the male to actions of the female; *right:* reactions of the female to actions of the male. Numbers represent behaviors. For key and details see text. (From D. Morris, 1958.)

depends on releasing stimuli, segments of the chain of behavior can be skipped. Even a behavior that has already taken place can be recapitulated by means of the appropriate stimulus configuration.

Figure 8-1 shows the two chains of reciprocally releasing actions. The sequence of actions of male and female partners has been simplified; in the natural situation it is not so precisely determined. There are many deviations from it, but the sequences are by no means random (G. P. Baerends et al., 1955; D. Morris, 1958; G. W. Barlow, 1962).

D. Morris observed which reactions of the sexual partner released behavior in the other. Deviations from the ideal sequence are shown in Fig. 8-2. The following numerical key specifies the various behavior patterns of the males and females.

Behavior Pattern	Male	Female
1	Appears	Appears
2	Zig-zag dance	Presents
3	Leads	Orients towards male
4	Shows entrance to nest	Follows
5	Ritualized fanning	Swims beneath the male to the entrance of the nest
6	Snout tremolo	Buries head in nest
7	Slips into nest	Slips into nest
8	Milts	Spawns
9	Swims away	Swims away

On the left are the male's reactions to actions of the female; on the right the responses of the female to actions of the male. Evidently most reactions can be released by more than one action of the partner. Furthermore, we can see from the diagram that the various actions that can release a particular behavior are grouped within the complete sequence.

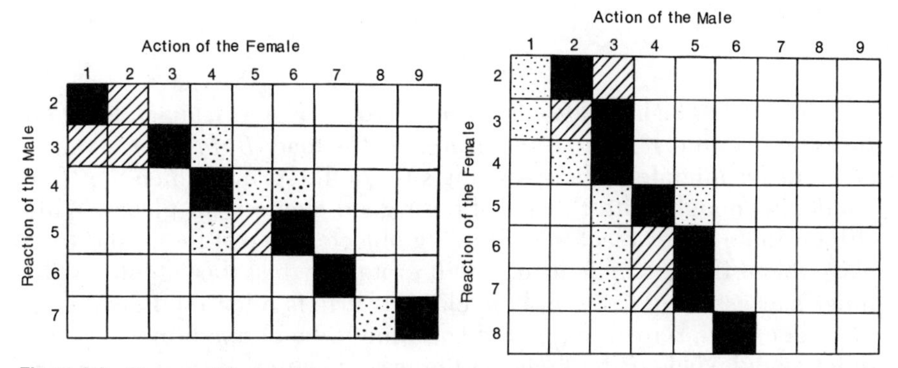

Figure 8-3 Diagram of the behavior sequences in the mating behavior of the stickleback. The patterning of the squares demonstrates the frequency with which one action releases certain others. Numbers represent actions as in Fig. 8-2. For details see text. (From D. Morris, 1958.)

Reactions and responses overlap. One reason for this is differential motivations of the partners: male and female often are not in perfect sexual synchrony. Some stages can even be skipped if the organism is highly motivated. Yet none of the overlaps spreads over more than a few stages. Furthermore, quantitative analysis shows that actions may release others in the partner with varying probabilities. In Fig. 8-3 the actions are once again represented by numbers. Black squares represent behaviors that in a typical case release a specific response; dotted squares represent actions that do this only occasionally; and lined squares refer to actions that seldom release a specific response. G. W. Barlow (1962) found similar stimulus response overlaps in *Badis badis*.

Figure 8-4 *Above:* Toad pair shortly before spawning; *below:* signal posture of the female and the "basket" position of the hind legs of the male. (After I. Eibl-Eibesfeldt, 1950b.)

The relationships are also very clear in water salamanders (*Triturus*) and anurans (*Bufo, Rana, Hyla*). In the **toad** (*Bufo bufo* L.) males and females migrate to the spawning sites in the spring, which they find, strangely enough, with the aid of their good memory for places. While still en route males react to all moving objects, jump at them, and try to clasp them. If it is a male toad, it will protest against this attempt with a quick succession of calls, and the clasping grip is released. Females, on the other hand, remain quiet and continue to be clasped, the same as a quiet model, such as the finger of the experimenter. If the male arrives unmated at the pond and if nothing stirs nearby, it begins to call, attracting females. If it perceives a movement nearby, it again approaches indiscriminately everything that moves and clasps it. As on land, the further behavior of the clasped object determines the subsequent reactions of the male. He clasps the female behind her front legs; his hind legs assume a position from which he can kick against all rivals to keep them away. The pair remains together until the female gives a signal by assuming a pronounced lordotic position. The male then slips back and forms a "basket" with its hind legs in front of the cloacal opening of the female, where the discharged spawn is collected and fertilized (Fig. 8-4). In this manner several spawnings take place, interrupted by intervals during which the female swims about with the male, again in the former clasping position. Finally there is a signal posture that is not followed by spawning. The male again forms the "basket" posture, but since no spawn appears, the clasp is soon released and the male dismounts. If the

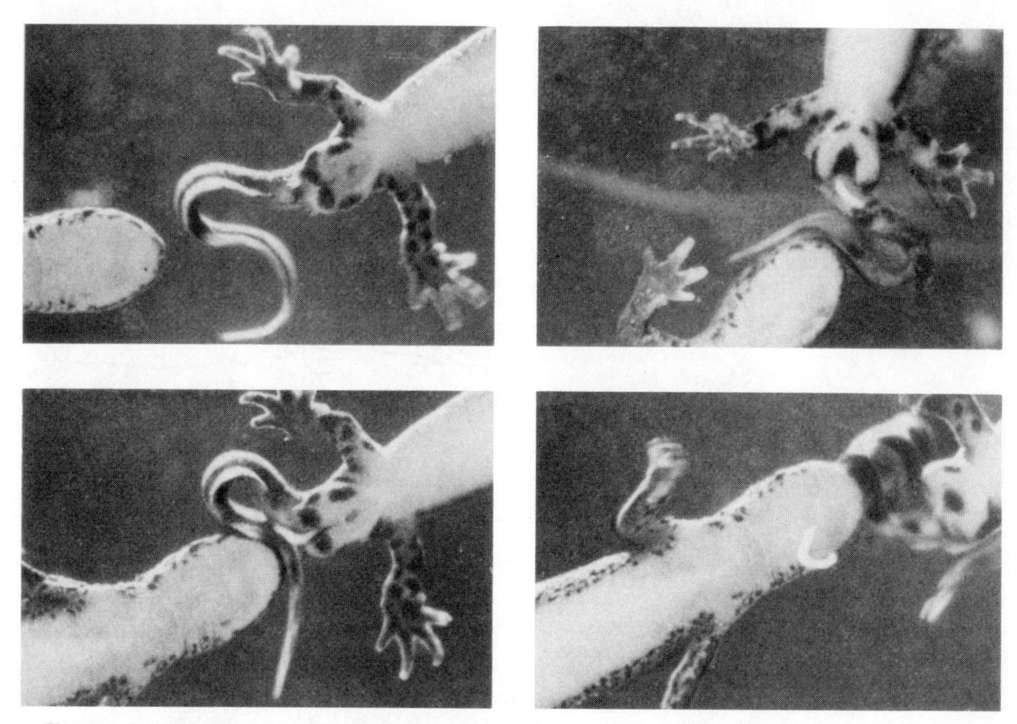

Figure 8-5 Deposition of the spermatophore and leading seen from below. After depositing the spermatophore the male opens his cloaca wide and is followed by the female. (From Scientific Film C698 (Göttingen), I. Eibl-Eibesfeldt, 1955, photographs: H. Sielmann.)

female should then again be clasped by a male, she will behave like a male, making the appropriate rejection movements with her hind legs but remaining silent. The rhythmic movements are sufficient to repulse a clasping male (I. Eibl-Eibesfeldt, 1950b, 1954, 1956a; H. Heusser, 1960).

In water salamanders the males must be excited by the species-specific odor substance of the females before they will react further. One has only to put water from a container holding a female into one holding a male, and the latter will then react to a simple, moving model which until then has not been attended to (H. M. Zippelius, 1949; H. F. R. Prechtl, 1951). The male makes an olfactory examination and then blocks the path of a conspecific female and begins to fan odor substances toward her with tail movements. If she approaches him, he turns around and waddles off slowly with his tail bent. The female follows and butts her snout at his tail. This is the signal to deposit the spermatophore: the male raises the tail, deposits the sperm packet, opens his cloaca, and leads the female in a straight line across the spermatophore, which then becomes attached at the female cloacal opening (J. Marquenie, 1950; H. F. R. Prechtl, 1951; I. Eibl-Eibesfeldt, 1955c; see also Figs. 8-5 and 8-6).

Botanists know of comparable reaction chains. In the mushroom

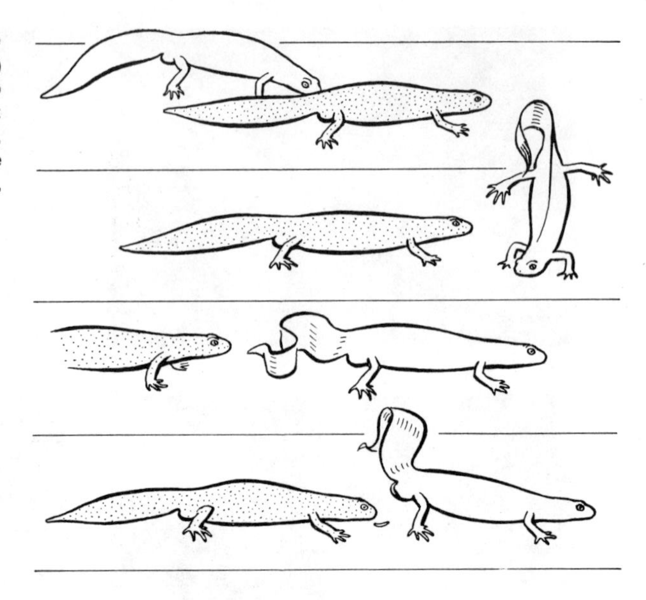

Figure 8-6 Courtship of the Alpine water salamander (*Triturus alpestris*) (female stippled). The sequence is to be viewed from top to bottom: olfactory investigation of the female, blocking her way and wagging the tail, releasing the spermatophore, and leading. (H. Kacher, artist.)

(*Achlya ambysexualis*) male mycelia begin to form antheridial cells when they come into contact with a substance (*A*) which is produced by female mycelia. The antheridial cells in turn produce a substance (*B*) which stimulates the female mycelium to produce oogonia. The oogonia in turn produce a substance (*C*) which attracts the antheridial cells chemotropically and which causes the closing off of the antheridia once the union with the oogonia has taken place. A substance (*D*), which is produced by the fully developed antheridia, stimulates the closing off of the oogonia, but only after direct contact with the antheridia (J. R. Raper, cited in M. Hartmann, 1956).

SUMMARY

The orderly sequence of behaviors in interactive encounters is often brought about by animal A performing a pattern, which releases another behavior pattern in B, which in turn changes the stimulus situation and releases yet another action in A, and so forth. The best-studied reaction chain is the courtship sequence of the three-spined stickleback, where both sexual partners reciprocally release certain reactions from one another, which in turn act as releasing stimuli, until a consummatory act in each concludes the sequence. The old scheme of Craig: A releasing stimulus–appetitive behavior–consummatory act is certainly much too simplified. Behavior is not normally terminated because of action-specific fatigue but because of a stimulus situation that cuts it off.

9

The Hierarchical Organization of Behavior

Behavior patterns occur in a specific order. In the reaction chains discussed in Chapter 8, the orderly sequence of the various actions was dictated by the relevant releasing-stimulus situation. However, as was shown by the examples of the squirrel hiding its nuts and the spider building its net, there are also endogenously programmed action sequences. Scratching is followed by placing of the nut, butting with the snout, covering, and stamping down, even though the squirrel has not previously dug up the ground. Each individual movement has its fixed position in the total sequence, and each individual movement is in turn an internally programmed sequence of various muscle contractions.

The orderliness of behavior does not consist only of a temporal sequence but also of temporal parallelisms. Behavior patterns may be more or less coupled with one another, or they may exclude one another. We can also observe that behavior patterns are ordered in sets, in which each set is distinguished by a common fluctuation of the threshold for releasing stimuli. In an animal that is in a fighting mood, for example, we can see that the behavior patterns of threat, attack, biting, and so on, can be more readily released than at other times—for example, when the animal is eating. Other behavior patterns, such as eating and nest-building behavior, are inhibited at this time. This is an indication that the behavior patterns are grouped and depend on higher, coordinating organizations, which influence one another mutually in specific ways.

Thus, a male squirrel is not only more ready to court during the rutting season but is clearly more aggressive as well. Which behavior patterns are activated at a specific time, whether those of courting or of fighting, depends on the releasing situation. However, the male has a lowered threshold for both kinds of behaviors. We know that this is due to the influence of male hormones. In many birds the behavior patterns of nest building, courting, and fighting are similarly organized during the breeding season.

This order with respect to sequence and simultaneity reflects at the same time an hierarchical order of behavior in which several levels of integration can be recognized. An example may illustrate this: In the early summer, hatching **digger wasps** (*Ammophila campestris*) come into a reproductive condition and are then ready to mate and care for their broods. Caring for the brood involves a number of specific drives: nest-site selection, nest building, hunting for caterpillars, egg laying, feeding of larvae, and opening and closing of the nests. Each of these specific drives in turn consists of chains of individual actions that are controlled by specific releasing stimuli. Thus, the digger wasp will first search for a nest and will begin scratching and biting in order to build the nest only when she has found a suitable place. The loose sand that she digs up is carried away. When the nest chamber is completed the entrance is closed off with a clump of dirt of the proper size. Then the wasp begins a new activity. She searches for and kills a caterpillar. After that the mood to retrieve supplants the mood to hunt, and the innate actions of transporting, dropping the caterpillar before the nest, opening the nest, entering, turning within, grasping and pulling in the caterpillar follow. Finally, the digger wasp will deposite an egg and close the nest. Thereafter she will repeatedly visit the nest, and once the larva has hatched she will feed it at first with small and later with larger caterpillars; when the larva pupates, the wasp closes the nest for the last time. She is able to adjust her behavior according to the demands of the situation; she will bring fewer caterpillars if some are left in the nest, and she will bring small ones when the larva itself is still small (G. P. Baerends, 1941).

Observations show that there are dominant and subordinate instincts. N. Tinbergen (1951) has illustrated this in his schema of the "hierarchy of instincts." He developed his ideas during his studies of the reproductive behavior of sticklebacks.

In the spring, as the length of day increases, the male stickleback comes into a reproductive mood. But the change to reproductive coloration does not take place suddenly, and he shows no courting or fighting behavior. Instead, the fish migrate together peacefully in swarms from their winter habitat in deeper water to warmer and more shallow water. There each male establishes a territory, an area containing some water plants. Only when a territory has been selected does the fish acquire reproductive coloration and become receptive to new stimuli. He fights or threatens when another male appears; he courts females and builds a nest if he can find suitable material. What, specifically, he will do depends on the releasing-stimulus situation, but he is in a state of internal readiness to perform all these activities. Fighting behavior is activated by the appearance of a red-bellied male, but which specific fighting actions take place again depends upon still more specific stimuli. If the intruder flees, he is pursued. If he beats his tail, the territory owner reacts likewise. The red male releases the readiness to fight but not the actual fight itself. One can recognize several levels of integration that lead from a more general to a more specific behavior, and certain key stimuli activate the next more specific action system or mood. For example, if one collects several migrating sticklebacks and places them into a bare tank, they remain in a group without change of color, because no territorial borders can be defined with respect to landmarks. If one adds some plants in a corner, one male will remain there and change color, and he is then in a reproductive condition. He is ready to court, fight, or build a nest in response to the appropriate stimuli.

N. Tinbergen assumes that this observed system of behaviors reflects an order of functional organizations within the central nervous system (Fig. 9-1). He speaks also of a hierarchy of centers. So the term "center" is defined purely functionally. Hormones affect the highest center responsible for reproduction—the migrating center—and cause appetitive behavior in the form of migration. There seems to be no specific key stimulus for this. The appetitive behavior of migration ends when the fish perceives the key stimuli in a specific biotope. These, in turn, affect a specific innate releasing mechanism that frees the next main center for propagation that was blocked until then. Impulses can now pass to the lower centers, such as care of brood, courting, nesting, and fighting, but each of these centers is blocked until specific key stimuli release the behavior—for example, when a rival appears. The rival must then provide still more specific stimuli before the specific fighting behavior patterns are released.

Later investigations by P. Guiton (1960) compel us to modify this model somewhat. According to Tinbergen, migration, establishment of territory, and reproduction follow in a sequence, but fighting, nesting, courting, and care of young are parallel in time. According to Guiton, establishment of territory, digging a pit, nest building, courting, and care of young follow one after the other. After a territory has been established, the male digs a pit, and only this releases the carrying in of nesting material and gluing. If the pit is covered, digging is gradually reactivated, but the fish does not dig as long as before and soon begins to carry in nesting material. He is not fully ready to mate until he has dug a tunnel into the nest. These new findings do not, however, change the basic

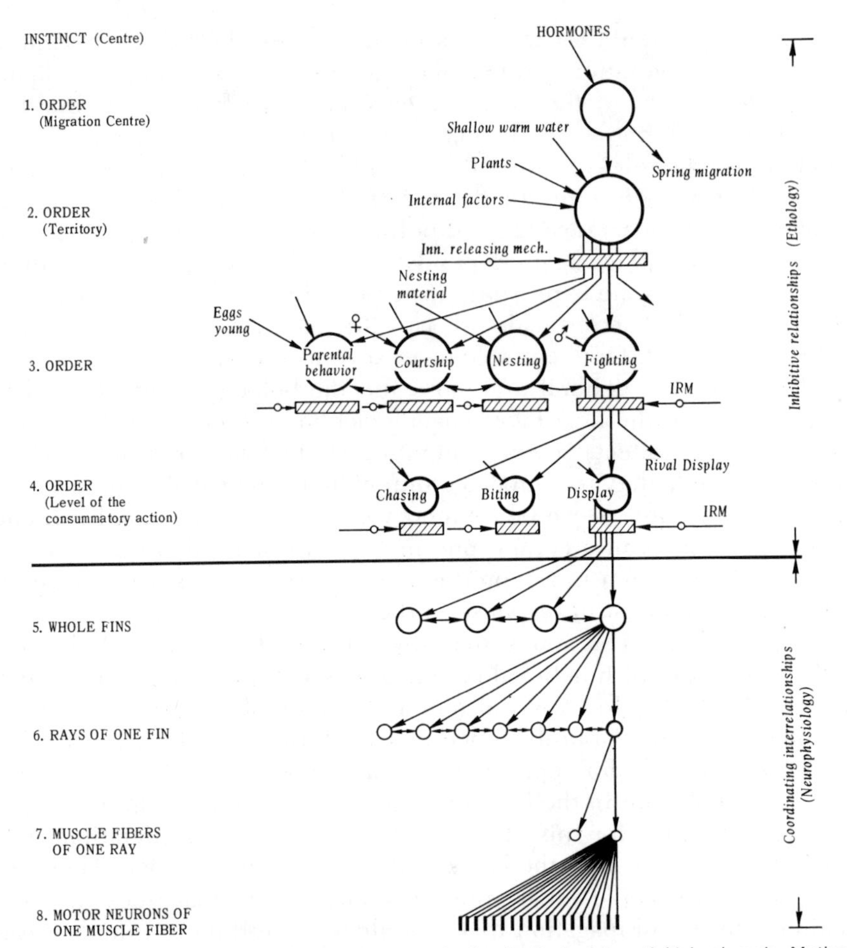

Figure 9-1 Hierarchical centers of the major reproductive instinct of the stickleback male. Motivational impulses are represented by straight arrows which "load" the centers (shown here as circles). These impulses may come from the external environment as well as from superordinated centers, or they may occur spontaneously within a center (this is not considered in the schema). The shaded rectangles indicate inhibiting influences, which prevent a continuous discharge of motor impulses. These blocks are removed by innate releasing mechanisms. When this has occurred the animal will show a specific appetitive behavior until more specific releasing stimuli activate the next subordinated instinct and the still more specific appetitive behavior. The concave, two-headed arrows between centers of the same level indicate mutually inhibiting relationships and the existence of **displacement activities.** Below the level of the consummatory acts a number of centers come into action simultaneously. The relation between subconsummatory centers of the same level is indicated by horizontal lines. (Additional explanation in the text.) (After N. Tinbergen, 1951.)

principle of Tinbergen's schema, and for this reason the schema is reproduced here unchanged, if only for historical reasons.

Basing his work on neurophysiological considerations, P. Weiss (1941a) independently developed the theory of the hierarchy of the central nervous system. He distinguished six levels of integration. The lowest level represents the single motor unit (1). This is followed by all the motor units of one muscle (2); next the coordinated function of muscle groups that move a joint (3); followed by the coordinated movement of one extremity (4); then the coordinated interaction of several appendages (5); and finally the movement of the entire animal (6). The sixth level includes, as Tinbergen shows, several levels of integration. In Tinbergen's schema three of Weiss' levels are represented. The horizontal line is meant to separate the fixed action patterns (consummatory actions) from the more simple and subordinated movement coordinations.

Tinbergen (1951:112) defines *instinct as*

> . . . a hierarchically organized nervous mechanism which is susceptible to certain priming, releasing, and directing impulses of internal as well as of external origin, and which responds to these impulses by coordinated movements that contribute to the maintenance of the individual and the species.

He distinguishes between major and subordinated instincts.

In complete agreement with this statement is W. H. Thorpe (1951: 3), who speaks of an instinct as

> . . . an inherited and adapted system of coordination within the nervous system as a whole, which when activated finds expression in behavior culminating in a fixed action pattern. It is organised on a hierarchical basis, both on the afferent and efferent sides.
>
> When charged, it shows evidence of action-specific-potential and a readiness for release by an environmental releaser.

The hierarchical scheme of G. P. Baerends (1956) is comparable to that of Tinbergen, except that Baerends demonstrates more clearly that higher centers very often control several lower ones (Fig. 9-2). This view is supported by the findings of E. v. Holst, which have shown that running, pecking, and so on occur in diverse functional relationships. These experiments will be discussed in more detail later.

There are several ways of discerning the functional organization of the various behavior patterns. Whether a behavior sequence depends upon external stimuli or is programmed within the system can be determined by manipulation of the releasing situation, as was illustrated by the examples given earlier. The study of fatigue phenomena also illuminates these relationships. For example, if a behavior pattern *A* repeatedly occurs and a behavior *B* shows fatigue but the performance of *B* does not influence the occurrence of behavior *A*, we have an indication that the actions are ordered in a sequence *A–B*, in which *B* is dependent on *A*. If on the other hand, an increase in the threshold of *A* is accompanied by an increase in the threshold for *B*, and if the performance of *B* is followed by an increase in *A*, then there is no strict hierarchical organization between both actions, but both are dependent upon a common higher stage (A. Kortlandt, 1955). Further discussion of the hierarchical organization of behavior can be found in G. P. Baerends (1956) and R. A. Hinde (1953). Hinde makes the point that the hierarchical order is not only manifested in simple, linear relations but also in a network of relations. How this is to be understood will be discussed in the experiments by E. v. Holst and U. v. Saint Paul (1960).

The conclusion that the organization of behavior of intact animals is a reflection of an organization within the central nervous system should be obvious. In attempts to understand this organization, experimentors

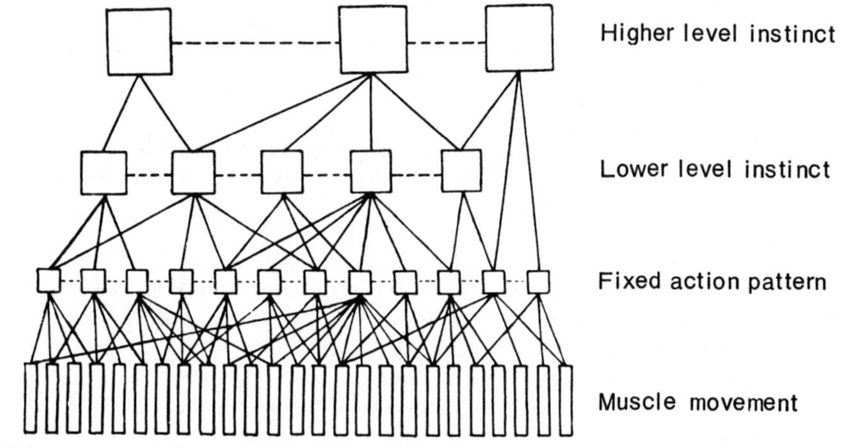

Figure 9-2 The hierarchical scheme of G. P. Baerends (1956), clearly showing the interrelationships of the centers. Especially noteworthy is the fact that lower centers are often controlled by several higher ones. Dotted lines represent inhibitory relationships between mechanisms of the same order.

at first used surgical techniques. However, the effects of lesions allow only a rough approximation in localizing certain behaviors within specific areas in the brain, and contradictory results lead to unrealistic interpretations. While localization theories assume the existence of strictly defined anatomical centers for certain functions, the theory of plasticity (mass action) does not contain such a strict localization of function. For those favoring a localization hypothesis, the brain is a more or less fixed mosaic of individual centers, each responsible for specific activities. According to the mass-action theory the brain functions as a whole, whose parts have partially overlapping functions. Both groups can base their conclusions on experimental results (see the discussion by J. Dembowski, 1955). K. S. Lashley (1929, 1931), for example, found that the learning performance was proportional to the extent of cortical lesions. Thus, the location of removed cortical tissue was less important than the amount, and the learning abilities depend less on a localized anatomical structure than upon a more widespread cortical mechanism. F. A. Beach (1937, 1938, 1940) obtained similar results in the study of reproductive behavior of rats. The percentage of male rats that continued copulating decreased with an increase in the size of brain lesions. There were other findings, however. Small lesions in the cingulum, for example, disrupted the parental-care behavior of rats quite substantially. The same is true for food hoarding (J. S. Stamm, 1954, 1955). H. F. Harlow (1953) disagrees with Lashley's concept of mass action. According to O. L. Zangwill (1961), there is not necessarily any contradiction. Undoubtedly there exist specific brain areas where largely instinctive behavior (emotional behavior) is localized in specific structures in the brain stem. In

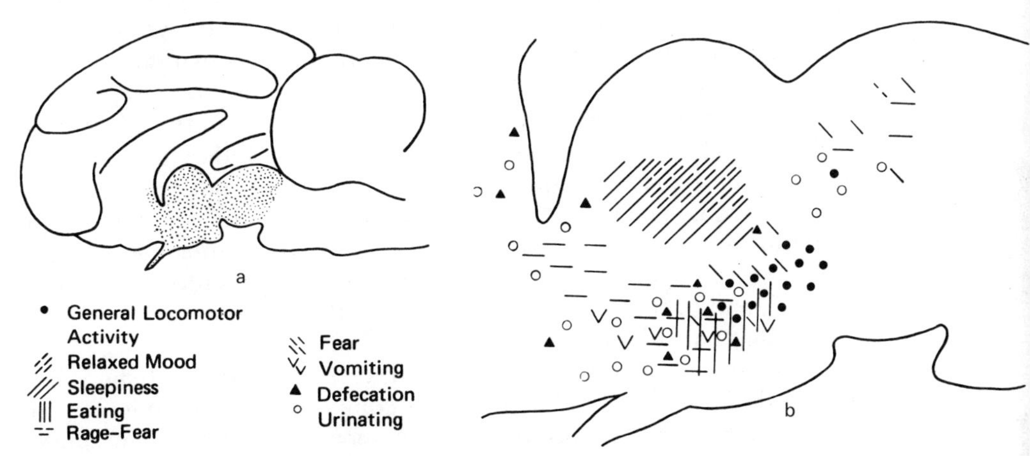

Figure 9-3 *Left:* Schematic representation of the cat brain showing the stippled area of the brain stem that was stimulated in experiments by W. R. Hess. *Right:* The same region enlarged. The various symbols mark the areas in which different behaviors were released. (After W. R. Hess, simplified by E. v. Holst, 1957.)

respect to other functions the number of ganglion cells is important. This is true with certain exceptions for the relationship of the cortex to learning and intelligent behavior.

An entirely new impetus was given to brain research by the investigations of W. R. Hess (1954, 1957), who activated many behavior patterns in the brains of largely intact cats by means of electrical stimulation with fixed electrodes. He was able to localize points in the midbrain of the cat where he could activate threat, escape, eating, and other behavior patterns. These points were concentrated in certain areas but were also widely distributed, as could be expected in such a complicated "wiring system" (Fig. 9-3). Cats raised in social isolation displayed upon electrical hypothalamic stimulation all the patterns of attack behavior that normally raised cats demonstrate when so stimulated. The brain mechanisms underlying these patterns are probably innately organized (W. W. Roberts and E. H. Bergquist, 1968). F. Huber (1955) was able to determine the nerve centers of some instinctive behavior patterns in crickets by brain stimulation.

By recording activity from single neurons B. R. Komisaruk and J. Olds (1968) demonstrated a direct relationship between neural activity and observed behavior in rats. The discharge of single neurons was always coupled to definite activities, such as feeding, twitching of whiskers, sniffing, and exploring.

An entirely new road was taken in pursuit of the question of the organization of the central nervous system by E. v. Holst and U. v. Saint Paul (1960), who used the method of electrical brain stimulation to investigate the hierarchical organization of behavior. They used electrodes that could be lowered into the brain in discrete small steps. They studied tame domestic chickens that were habituated to the experimental situation. By trial and error they placed several electrodes into the brain stem of a chicken and were able to stimulate several separate points at the same time. One group of behavior patterns, out of the many they were able to elicit, is of special interest and had the following characteristics: (1) The behavior sequence, in which various individual components occurred, remained the same even during long-lasting electrical stimulation or when higher voltages were used; (2) persons who were familiar with the species considered the movement coordinations to be "natural"; and (3) the behavior that was elicited consisted of functional patterns that had adaptive value.

It appears as if they had activated an instinct that satisfied Tinbergen's definition: individual behavior components appearing in an orderly sequence that seems dependent upon the different thresholds required to release them. Thus, the stimulation of a certain area of the brain stem releases blinking of the left eye in a rooster. If the stimulus is continued or its strength increased, the rooster will shake his head, and if stimulation is continued further, will rub his head on his shoulder. Finally he scratches his left cheek with his foot. Head shaking and scratching are then repeated as long as stimulation lasts; it appears as if the rooster is bothered by an invisible fly.

In a similar manner the authors were able to release the entire "disgust reaction" in chickens. The neck is at first stretched forward and the head is bent so that the beak points toward the ground. The beak is opened slightly, and the tongue moves. During continued stimulation saliva is secreted, as if an unpalatable object had to be washed away. Finally the chicken shakes its head and scratches itself. When the stimulus is terminated the chicken makes a final wiping motion on the ground (Fig. 9-4*a*). It was possible to release complicated functional behavior sequences such as escape from an aerial or ground predator as well as simple reactions such as cackling. If only the cackling drive was activated, the animal continued to perform this behavior even when prolonged

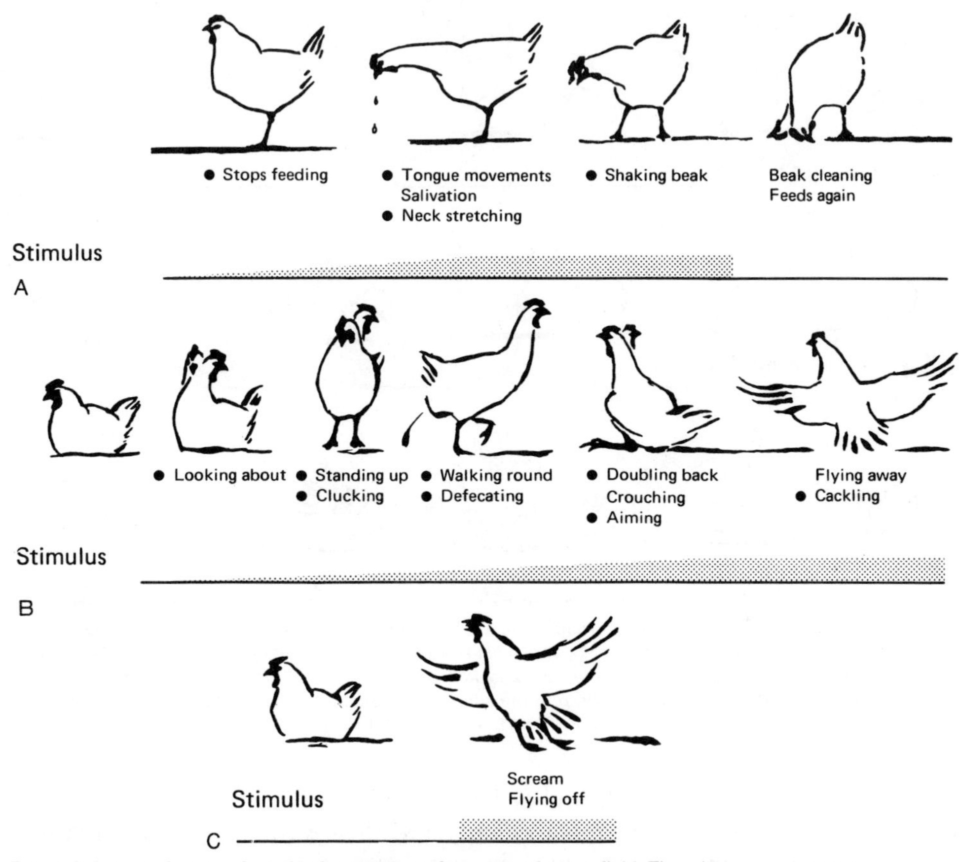

Figure 9-4 *a:* Activation of a behavior sequence from one stimulus field. The whole complex serves to remove something unpleasant from the beak. The individual acts marked with black dots can also be obtained in isolation by stimulating other areas. The increasingly deeper stippled area denotes the strength of the stimulus and its duration. Several seconds after cessation of the stimulus the chicken wipes its beak. *b:* Behavior sequence of fleeing from a ground enemy with slowly increasing stimulation of the field. *c:* Reaction to a sudden strong stimulus. (From E. v. Holst and U. v. Saint Paul, 1960.)

or stronger stimulation was applied. Finally a kind of fatigue set in, and a stronger stimulus was required to continue releasing the behavior.

During other experiments this cackling was merely a component of a more complicated behavior. The animal began to cackle when the stimulus was turned on; with an increase in strength the animal became restless, walked around, showed orienting head movements, and finally flew off with calls of fear, just like a chicken that has been frightened by a ground predator (Fig. 9-4*b* and *c*). In this case the escape drive had been activated with its individual parts appearing in the proper order as their respective thresholds were reached. If stimulation began with high voltage the animal flew off at once. This escape behavior, in turn, can be a

Figure 9-5 Centrally released ground predator behavior. Without a suitable object the stimulated hen shows only locomotive unrest. *a:* Toward a fist she shows only slight threatening. *b:* A stuffed, motionless polecat is vigorously threatened and attacked; if the stimulus ends at this moment the hen remains standing and threatening slightly. *c:* If the stimulation continues, she checks and flees screeching, unless the polecat drops to the floor. (From E. v. Holst and U. v. Saint Paul, 1960.)

part of a drive of a still higher level of integration. E. v. Holst and U. v. Saint Paul sometimes released only a restless wandering about. In these cases they tried to find out, through presentation of various releasing stimuli, whether they had activated a general motor restlessness or a specific appetitive behavior. In one experiment they found that the chicken was not ready to eat, drink, or court, but when presented with a fist it threatened lightly. If presented with a stuffed predator, a polecat, the chicken at once threatened and attacked. When the polecat was knocked to the ground, the chicken retained the threat posture (Fig. 9-5).

The conclusion to be drawn is that an "attack drive" against ground predators had been activated. An additional experiment, however, showed that this was not the case. If the stuffed polecat was fixed to the table so that it could not be knocked down, the chicken behaved quite differently. After the initial attack, which failed to dislodge the predator, the animal then called and flew off if brain stimulation was continued. Thus, "behavior directed against ground predators" had been activated, including attack as well as escape. The alternatives were ordered according to their different releasing thresholds. A very strong brain stimulus could also cause an attack to be switched to escape.

In the last three experiments cackling had been activated once by itself, another time as part of an escape drive, and finally as part of the ac-

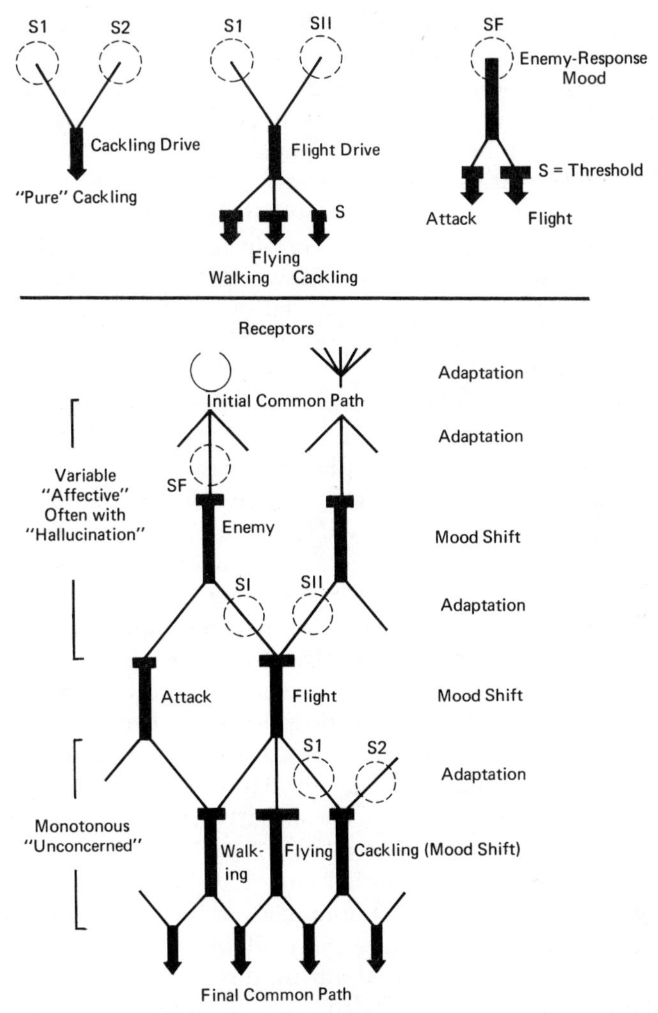

Figure 9-6 Sketch of a fragment from the functional organization of some behavior patterns in the fowl. *Above:* The behavior patterns referred to in the text, arranged side by side, which can be activated from one or more stimulus points. *Below:* The partial actions have been arranged in some kind of order. This order results in a diagram that illustrates how several paths lead to the same final action. Walking may be at one time a part of the activated escape drive, at another time part of the feeding drive, and so on; just as sitting may be pure or part of the sleeping or incubation drive. Change of mood and adaptation can occur at various levels of the central nervous system. When the behavior patterns of a higher level of integration are activated, which are closer to the sensory areas, then the animal behaves "affectively," for example, it behaves as if it were afraid. When actions of a lower level of integration are activated, the animal behaves with indifference. (After E. v. Holst and U. v. Saint Paul, 1960.)

tivated ground-predator behavior complex. These levels seem to represent three stages of integration, which if arranged in a logical order, express a hierarchical organization (Fig. 9-6). This can be represented in the form of a diagram in which additional experiments are included to reveal details about the particular interrelationships. One and the same behavior can often be activated from two different stimulus points. By appropriate experiments one can discover how these points are interrelated within the functional system. At first one tries to find out if there is a connection between the two points by stimulating both points simultaneously at intermediate strength, which from one point alone releases only a moderately strong reaction. If the simultaneous stimulation results in a stronger response, one can conclude that the excitations from the two points flow together somewhere in the central nervous system. A behavior pattern can also be fatigued by repeated stimulation from a particular point; in order to release the behavior, stronger stimuli are required. If stimulus point 1 no longer releases a reaction and point 2 is stimulated, then the behavior can again be elicited with full intensity. In this instance a pathway leading to the center solely from point 1 has been locally affected, but the stimulation did not affect the motor center.

The "fatigue" phenomenon at the point of stimulation was called a "central local adaptation" by E. v. Holst and U. v. Saint Paul, and they distinguished it from a "change of mood," which is a different process. Repeated stimulation can result in a change of the basic mood of an animal. An animal that spontaneously cackled "angrily" could be induced to sit down after repeated activation of the drive to sit, so that it did not get up again after stimulation ceased. Mood shifts, local adaptation, and the possibility of activating behavior from several stimulus points were observed by the authors at various levels of integration. This is represented in the diagram (Fig. 9-6). Here the three experiments that I have discussed above (cackling, escape drive, and reactions to a ground predator) have been represented in a specific order. The resulting diagram—the "functional organization of drives" (*Wirkungsgefüge*)—shows not only the hierarchical organization but also a network of connections. The same final action is the result of several initial approaches. We know from observations that sitting down, for example, can be activated at one time as part of the drive to sleep and at another time from a drive to brood, just as running or flying can be a part of several drives. K. Lorenz called behavior patterns that are involved in several drives "behavioral tools."

E. v. Holst's organization of drives emphasizes the interrelationships more than Tinbergen's hierarchical schema, which is primarily linear, although even here, cross connections are shown by arrows pointing to different levels of integration. Holst was also concerned with the histological mapping of the various stimulus points, but his untimely death interrupted this work.

Brain stimulation studies on the opossum (*Didelphis virginiana*) achieved in principle the same results (W. W. Roberts et al., 1967). The hierarchical organization, however, was much less pronounced, which fact might reflect the organizational difference between the avian and mammalian brain.

Hierarchical organization of behavior can be recognized in many invertebrates and in all classes of vertebrates. As one ascends the phylogenetic scale, this organization becomes less and less linear. In the prey-catching behavior of the cat a linear arrangement of behavior components occurs only if the animal is hungry: lying-in-wait, stalking, catching, killing, and eating. But a satiated cat will also capture prey without eating it, and as mentioned earlier, each partial component of a sequence can become a final consummatory act because it has its own specific motivation. The other behavior components then become appetitive behavior in the service of this final action. P. Leyhausen therefore called this a "relative hierarchy of drives" (moods).

SUMMARY

In the case of a reaction chain the orderly sequence of the various acts is dictated by releasing stimuli. There are, however, also endogenously programmed action sequences. Behavior patterns may furthermore be coupled in sequence as well as simultaneously, and while certain behavior patterns thus occur together, others are mutually exclusive. Tinbergen and Baerends explained these facts by hypothesizing a hierarchical organization of (functional) centers within the central nervous system, analogous to the schema Weiss had developed for motor patterns on a lower level of integration. Baerends' schema is less linear than Tinbergen's, indicating that higher centers are often linked with several lower ones, an observation which was confirmed by the elegant brain stimulation experiments of E. v. Holst and U. v. Saint Paul.

10

Conflict Behavior

Sometimes a stimulus situation will activate several drives simultaneously – for example, the drives to attack and to flee. Such opposing behavior patterns come into conflict with one another and the resolution can take various forms. By using their method of electrical brain stimulation E. v. Holst and U. v. Saint Paul (1960) have examined this problem in domestic chickens by simultaneously activating two opposing drives from two different stimulus points. In the most simple cases the activated behavior patterns are superimposed on one another (*superposition*): for example, pecking and head turning. We have already discussed such a superposition in normal animals with the example of **expressive movements and facial expressions in dogs.** In *averaging*, two behavior patterns are also superimposed but their intensity changes. If watching with a stretched-out neck and looking around with widely sweeping head movements are activated together, the result is watching with a still more stretched-out neck and looking around with less extensive sweeping head movements. The simultaneously activated behavior patterns may also be expressed in *alternation* according to a pattern *a-b-a-b-a-b*. We know such ambivalent behavior from the zigzag dance of the stickleback. These instances have also been referred to as successively ambivalent behavior, in contrast to examples of simultaneous ambivalence, where both activated systems are simultaneously expressed. I do not think that this is an accurate choice of terminology, because several inappropriate things are combined. "Successive" versus "simultaneous" refers to the expression of the behavior; whereas "ambivalence" refers to the two simultaneously activated motivational systems. Someone might interpret the term "successive ambivalence" as referring to successive internal drives. E. v. Holst and U. v. Saint Paul obtained alternation when they activated watching out and eating at the same time. Opposing behavior patterns such as turning right and turning left *cancel* one another.

An especially interesting aspect of behavior has been called *transformation.* If attack and escape are simultaneously activated by electrical stimulation an entirely new behavior pattern appears: The chicken runs about with fluffed feathers, calling loudly, and this is the behavior of an incubating hen if one approaches her nest. However, the term "transformation" must be used with caution. What might appear upon superficial consideration to be transformation may have other physiological causes. For instance, if in a hungry animal the escape drive is activated and at the same time the sleeping drive, which normally suppresses the drive to escape, the result is that the chicken will eat before falling asleep. Formally expressed, $a + b = c$, but in actuality c is suppressed by a and a is suppressed by b; hence c is liberated.

We speak of *masking* if one behavior suppresses another without preventing it from occurring altogether. If a cackling chicken receives a brain stimulus that releases the drive to sit, it will sit down and cease cackling. If both stimuli are turned off, the chicken cackles briefly, the suppressed action thus reappearing for a moment. The suppressed drive exists latently but had been blocked somewhere before the motor areas had been affected. If such an afterdischarge does not occur, E. v. Holst and U. v. Saint Paul speak of *preventing* (Fig. 10-1).

Which behavior pattern will suppress another depends generally on the strength of the stimulus, but not exclusively so. Some dominant activities suppress others, even if they are expressed only with low intensity. This is true for several escape reactions, especially watching out and freezing.

The effect of some behavior patterns on one another has been measured quantitatively. It was found that the "disgust reaction" only barely raised the threshold for head turning, but the threshold for pecking, which is part of the feeding system, was sharply raised. Thus, head turning is more readily combined with the "disgust reaction" than with pecking behavior. Finally, an activated behavior can influence another behavior in such a way as to make it disappear altogether. If a spontaneously scolding hen is stimulated long enough to sit down, she will eventually calm down (Fig. 10-2).

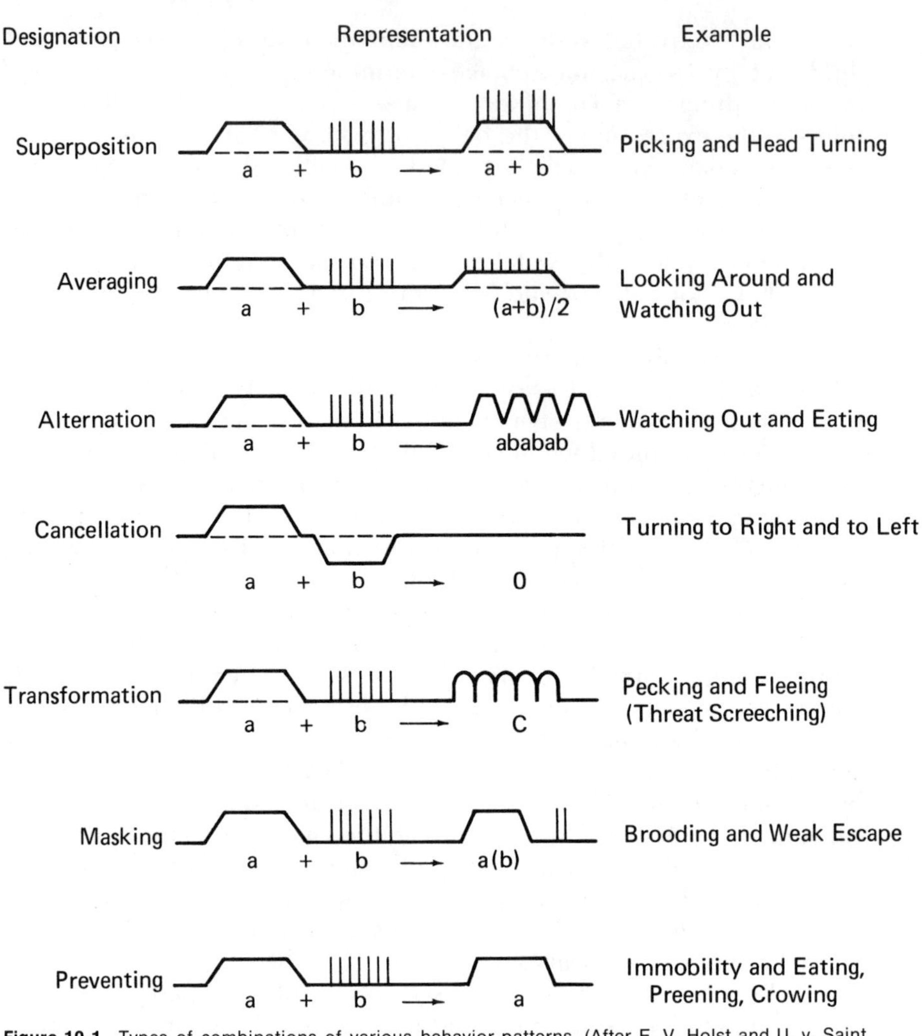

Figure 10-1 Types of combinations of various behavior patterns. (After E. V. Holst and U. v. Saint Paul, 1960.)

Ethologists have observed very similar phenomena in intact animals and have interpreted the behavior correctly. One example that did not occur in the experiments discussed above will be considered. A. Kortlandt (1940) and N. Tinbergen (1940, 1952) discovered independently that animals frequently show behavior patterns in conflict situations which cannot be attributed to competing drives. They called these behavior patterns *displacement activities* (*Übersprungbewegungen*). This concept is based on the hypothesis that the observed behavior patterns are not activated from their normal source, that is, autochthonously, but receive their excitatory potential allochthonously from those drives whose expression has been prevented by the conflictful situation in which the animal finds itself. The dammed-up excitation spills over, so to speak, into another channel and there finds its discharge. We shall return to this hypothesis, but first the phenomenon itself will be described. Fighting roosters, which are inhibited from attacking by the simultaneously activated tendency to flee, both behaviors being released by the presence of their opponent, begin to peck at the ground. The same behavior is seen in male turkeys (H. Räber, 1948). Deer engage in sham feeding when the tendency to remain conflicts with a tendency to flee (D. Müller-Using, 1952). Cleaning movements, shaking, bill wiping, bathing movements, and others dealing with the care of the body often appear as displacement activities. Fighting starlings preen themselves vigorously between bouts of fighting. Similarly, many mammals scratch themselves in conflict situations. Sticklebacks show displacement digging when they threaten one another head down near the boundary of

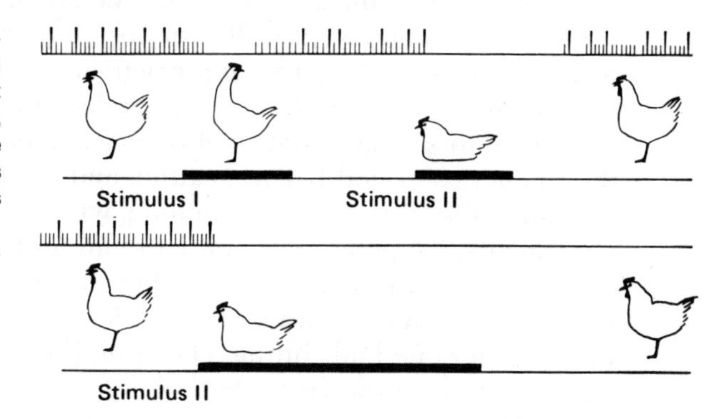

Figure 10-2 As a result of a preceding fleeing stimulus a fowl cackles spontaneously and without stopping (*kut, kut kudaw, kut, kut, kut kudaw! . . .*, suggested by the upper graph). *Stimulus I* produces watching out; *stimulus II* produces sitting down. When the second stimulus is applied long enough, a change of mood takes place. (From E. v. Holst and U. v. Saint Paul, 1960.)

their territories, until deep pits result. It has been shown that the fish are in a conflict between attack and escape and that the drives are in balance. Each stickleback attacks more readily the nearer he is to the center of his territory. He is more ready to flee outside it. However, it is possible to subdue a stickleback inside his territory with a model that can resist his attacks. He then hides between the plants. If one does not move the model for some time, he will again attack as his aggressive drive becomes stronger. He will come out of his hiding place and before attacking again will show displacement digging (N. Tinbergen, 1952).

Cormorants exhibit sham incubation during intervals of fighting (A. Kortlandt, 1940). Avocets even assume a sleeping posture by placing their bill between the feathers in a conflict situation (G. F. Makkink, 1936). Herons show prey-catching behavior when they are sexually excited (J. Verwey, 1930). Since specific displacement activities occur in particular conflict situations, they have frequently become ritualized as expressive movements (**displacement preening of ducks**). N. Tinbergen (1952) interprets the relieving ceremony at the nest in black-headed gulls and herring gulls as a displacement activity. The birds bring nest material to their partner. They are in an incubating mood, but the place where they could incubate is occupied. In this conflict the animal picks up nest material.

Man also shows comparable conflict behavior. R. Seiss (1965) extensively studied the behavior of speakers. A speaker exposes himself before his listeners, so he feels isolated. Escape reactions that cannot be discharged are also activated, because the speaker is in a situation that does not permit withdrawal. This motivation to flee can in extreme cases lead to behavior reminiscent of neuroses, such as sweating, trembling, and restlessness. In most cases the speaker will adjust to the situation, and several paths are open to him. He can be strictly formal and stick to his subject matter, and in this way his interaction with the environment is greatly reduced, or he may edge into his lecture by speaking to himself. He can modify his exposed position by belittling his achievements, by appearing humble and exhibiting behavior that indicates a readiness for contact, such as friendly smiles and submissive behavior. Autochthonous behavior patterns are activated—for example, those of seeking comfort (clutching the lectern, and so on). Finally, the conflict tensions lead to a large number of displacement activities. These fall into the categories of bodily care, eating, and sleeping behavior.

Into the first category belong wiping, rubbing, and scratching movements, bringing the open hand around the neck and nape of the neck, stroking the beard, even when no beard is present (!), stroking back the hair on the head, or scratching the head. In the category of eating behavior one finds biting, chewing, sucking on objects (pens), spontaneous chewing movements, licking, and swallowing. Finally, many learned behavior patterns appear in man as displacement activities, such as fondling of the tie, rhythmic pushing of the button on a ball-point pen, and many more.

Displacement activities occur not only when antagonistic drives are activated, but also when the "goal" of a behavior sequence has been reached too quickly, such as when a rival with whom the animal has been fighting leaves prematurely, or when an expected stimulus does not appear, or when a female does not follow a leading male. Tinbergen explained displacement activities with the concept of **central excitatory potential**; an energy surplus that cannot be discharged into its normal channel will then flow over into another channel and discharge itself in an irrelevant activity. The dammed-up energy "spills over," so to speak, from one center to another, according to Tinbergen. P. Svenster (1961) pointed out that in the stickleback, displacement fanning can come about by disinhibition analogous to the example of the cackling chicken described earlier in this chapter. Two drives, each independently inhibiting a third one, mutually inhibit each other in a conflict situation and lose their capacity to inhibit the third, which is then free to be discharged.

It appears as if the choice of the term "displacement" was premature; thus for a long time all preening behavior that occurred in conflict situations was called "displacement preening." But when threat postures are exhibited in conflict situations feathers or hair becomes disarrayed, which presents the proper stimuli for grooming or preening behavior.

In principle all displacement activities can be explained by the disinhibition hypothesis. Tinbergen's overflow hypothesis as a basis for displacement activities has not been proved or disproved to date. Displacement activities deserve our special attention because they are interesting in themselves, independent of any interpretation of their internal mechanism.

When a releasing stimulus simultaneously activates and inhibits a behavior pattern, the result need not always be a conflict movement (B. Grzimek, 1949a). An animal that is attacked or threatened by a higher-ranking one does not necessarily challenge the dominant animal but redirects his aggression against a still lower-ranking one, who in turn can pass it on. M. Bastock and others (1953) proposed the term *redirection activities*. Another example of redirected behavior is the grass pulling of herring gulls in border disputes, which Tinbergen originally interpreted as displacement activity.

SUMMARY

If two drives (such as flight and attack) are activated simultaneously conflict behavior often results. Conflict behavior was studied under controlled conditions by electrical brain stimulation in the chicken. If two drives were activated from two different stimulus points the patterns were superimposed in most cases, but other types of interactions were also observed, such as alternation, cancellation, and transformation. In the latter case the simultaneous stimulation of two patterns results in the occurrence of a third one. This resembles the phenomenon of displacement activities which was discovered by Tinbergen and explained as a central spilling over of the energy of an inhibited drive, to find its outlet in other motor pathways. However, another explanation is proposed by the brain-stimulation experiment. By inhibiting each other the two activated drives lose their inhibitory influence upon a third one, which is then freed.

11

Genetics of Behavior Patterns

What is meant by inheritance of behavior has been discussed in detail in Chapter 3. Developmental "blueprints" are passed on from generation to generation in the genome. They determine the prospective potentials, not all of which may be actually realized. What develops can be influenced by the environment up to a certain degree but not in all directions with equal magnitude. Instead, ranges of modifiability are inherited. The biologist can demonstrate inheritance in several ways. One of these is the tracing of the course of inheritance, but the genetics of fixed action patterns has not been thoroughly studied. More is known about the genetics of movement anomalies in dancing mice, shyness and aggressiveness in mice and dogs, the fighting prowess of rats, and so on. The reason for this dearth of studies of fixed action patterns may be that species and subspecies that are readily crossed are generally not qualitatively distinct in their inherited movements. The observed differences are often quantitative. One rat population may be slightly more ready to fight than another; another may learn a little faster (R. C. Tryon, 1940); and so on. These examples are presented especially in the books by J. L. Fuller and W. R. Thompson (1960), as well as J. P. Scott and J. L. Fuller (1965), each of which deals with behavior genetics. More oriented toward ethology is the excellent contribution of J. Hirsch (1967). The study of human behavior has also so far revealed only quantitative differences. Chinese-American newborn differ from European-American newborn, the latter being more restless, more apt to cry, and less easily comforted (D. G. Freedman and N. C. Freedman, 1969). The faster motor development of African babies in the first year of life in comparison to those of Europeans, demonstrated by the study of M. Geber (1958), might be another example of a genetic difference. The much-discussed IQ differences are probably genetically determined (A. R. Jensen, 1969). (For a critical discussion see W. F. Bodmer and L. L. Cavalli-Sforza, 1970.) It would indeed be difficult to see what selection pressure could operate to keep the potential capacities alike, considering the enormous differences in environmental conditions, that have caused so many other local adaptations.

These differences, however, merely reflect adaptations to different environments and do not at all imply differential values between people. My personal opinion is that the existence of these differences is an expression of the variety of endowments, which constitutes a value by itself, since they contribute to what we call the richness of human diversity. Furthermore, we have to keep in mind that IQ differences based upon a comparison of populations cannot be used to classify individuals along a scale. The curves of the two populations overlap extensively and individuals with what we call high IQ scores can be found in all human populations. Finally, I feel that the one-sided emphasis and value put on IQ differences reflects a typically ethnocentric viewpoint.

Chromosome abnormalities (for example, males with an extra Y chromosome) have been correlated with certain behavioral predispositions, but again there exists a considerable controversy (L. Jarvik, V. Klodin, and S. S. Matsuyama, 1973). Less controversial are the studies on monozygotic twins, which have demonstrated great similarities in many aspects of behavior, even when the twins are raised in different environments (J. Shields, 1962).

There is a certain tendency to trace behavioral effects to genetic effects on sensory organs and adaptive morphological characteristics. This certainly holds true in many instances. The preferred temperature ranges of various races of mice were attributed by K. Herter and K. Sgonina (1938) to differences in skin characteristics of these animals. Undoubtedly there exist many pleiotropic effects of sense and motor organs, but from the point of view of behavior genetics such findings are rather "trivial," in the words of E. Caspari (1964). It is more interesting to trace the inheritance of qualitatively different fixed action patterns in closely related species.

W. C. Dilger (1962) crossed the parrots *Agapornis roseicollis* and *A. fischeri,* which are well distinguished by the manner in which each species transports nesting material. *A. roseicollis* tucks strips of nesting material cut from leaves or paper under the rump feathers, which have small hooks and can hold them in place. *A. fischeri* carries the nesting material in the bill. The F_1 hybrids cut strips from leaves in the manner of their parents and try to tuck them under their rump feathers, usually failing. They show the usual tucking movements, but do not let go of the strip. After repeated attempts they finally drop the nesting material and cut a new strip. Often they perform the tucking movements at the wrong place, against their breast, for example, or they do not press the feathers down tightly enough against the tucked-in strips. Finally, the movements of tucking in nesting material often change into those of preening the feathers, or the animal carries the strips in its bill, and eventually gives up all attempts to carry nesting material by tucking it in between its feathers. The hybrids show a mixture of behavior patterns; unfortunately, they could not be paired, perhaps because they are sterile.

G. Osche (1952) crossed two races of Nematoda (*Rhabditis inermis inermis* and *R. i. inermoides*). Only the latter show the so-called "waving" above the floor with the raised anterior part of the body, a behavior pattern that results in contact with the insects that it parasitizes. In the F_1 generation all animals show this behavior; thus this "waving" is dominant. Backcrossing with the recessive parent resulted in some animals that "waved" and others that did not, which indicates monofactorial inheritance.

E. Clark, L. R. Aronson, and M. Gordon (1954) crossed the platy (*Xiphophorus maculatus*) with the green swordtail (*Xiphophorus helleri*) – two fish that differ in some respects in their reproductive behavior. The results indicate a polygenic inheritance. The same seems to be the case in the finch hybrids of R. A. Hinde (1956).

Of special interest are the investigations of S. v. Hörmann-Heck (1957), which can serve as a model for ethological genetics. She was able to cross two closely related species of crickets (*Gryllus campestris* and *G. bimaculatus*), which differ in several behavior patterns quantitatively as well as qualitatively, and to trace the inheritance of behavior patterns through the F_1 and F_2 generations and through backcrosses. Four behavior patterns were investigated. Antennal vibration during the post-courtship period and larval fights are only quantitatively different. *Gryllus bimaculatus* fights little or not at all during the juvenile period. *Gryllus campestris*, on the other hand, fights very intensively. This characteristic shows monofactorial inheritance, as does the antennal vibration in the postmating courtship, where both species are only quantitatively different in this respect. The pendulumlike movements of the thorax during mating are only seen in *campestris*, and in respect to this character backcrosses indicate polygenic inheritance. However, the stridulating sounds preceding courtship, which only occur in *bimaculatus*, seem to depend on only one pair of alleles.

D. R. Bentley (1971) crossed the cricket species *Teleogrillus commodus* with *Teleogrillus oceanus*, afterwards examining the hybrids as well as the products of backcrossing. Cricket songs are especially well suited for such a study. The movements that produce the songs are controlled by a relatively small neural nexus and are virtually independent of external stimuli. Even upon deafferentation, the nervous system of crickets produces a song pattern indistinguishable from normal songs. These experiments support the view that this neurally determined behavior is controlled almost totally by the genotype—indeed polygenetically and multichromosomally. Individual song traits are determined by genes localized in the X chromosome (see also D. R. Bentley and R. R. Hoy, 1972).

Also of special interest are the investigations of W. C. Rothenbuhler (1964). He crossed two different races of bees that differed clearly in their "hygienic" behavior. The hygienic bees opened cells that contained dead pupae and removed the dead. The nonhygienic bees left the dead pupae in the closed cells. When Rothenbuhler crossed the two races he obtained an F_1 generation that contained only nonhygienic bees. One F_1 queen produced four different kinds of drones. When the F_1 generation was backcrossed with the hygienic form Rothenbuhler obtained an F_2 generation that contained four groups of bees. One group was hygienic. Another group opened cells, but did not remove the dead pupae. Another group did not open the cells, but removed the pupae when the cells were opened. The last group was nonhygienic. These groups occurred in approximately equal proportions. The inheritance of the behavior patterns of uncapping (u) and removal of pupae (r) should then each depend on the homozygotic occurrence of a recessive gene:

P	♀ (queen) hygienic (uncapping and removing) uu rr	×	♂ (drone) nonhygienic (pupae rot) U R
F_1		nonhygienic Uu Rr	

The backcross of the four kinds of drones obtained (UR, ur, Ur, uR) with a queen of the hygienic race (uu rr) resulted in the following F_2:

F_2	hygienic	uncapping, but no removal	no uncapping, but removal	nonhygienic
	1:	1:	1:	1:
	uu rr	uu Rr	Uu rr	Uu Rr

It is not to be expected that the complicated neuronal mechanism underlying the behavior patterns of uncapping and removing derives from a single gene. Indeed, unhygienic workers in rare cases perform the hygienic activities, when the stimulus situation is very powerful. But the threshold of the uncapping pattern is virtually determined in an all-or-nothing fashion by the alleles U and u.

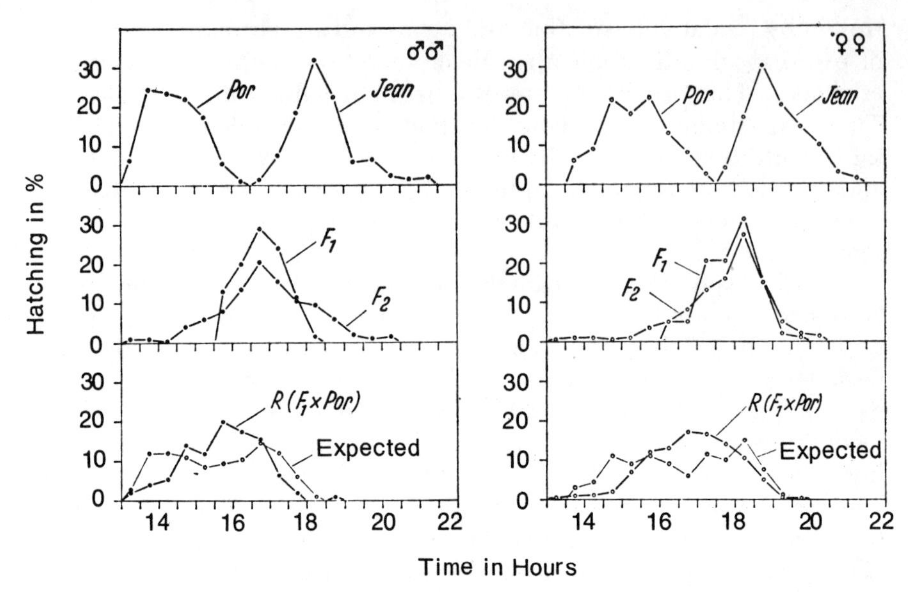

Figure 11-1 Cross-breeding of two populations (*Por* and *Jean*) of the marine fly *Clunio* that are local species (Normandy and the Bay of Biscay), differing in their genetically fixed hatching periods. Curves show the daily hatching distribution of the F_1 and F_2 generations, also the products of back-crossing, compared to the parental stock. Beside the backcross curve is the expected curve for the case of a monohybrid inheritance. Conditions: light–dark alternation, 12 + 12 hrs: light 6–18 hrs at 20° C. (From D. Neumann, 1966.)

The investigation of F_2 male hybrids between mallards (*Anas platyrhynchos*) and pintails (*Anas acuta*) revealed a significant positive correlation between the inheritance of behavioral and of plumage characteristics, indicating a genetic control for both groups of characteristics depending on relatively few genes (R. S. Sharpe and P. A. Johnsgard, 1966).

R. J. Konopka and S. Benzer (1971) isolated three mutants of ***Drosophila melanogaster,*** in whom the 24-hour cycle had been altered. One mutant was arhythmic, another had a period of 19 hours, and the third a period of 28 hours (see Fig. 17-5). All these mutations seem to be based on the same gene in the X chromosome. The same authors managed to determine the gene locality of the mutation.

The pupae of the marine fly ***Clunio*** hatch on spring days when the water is low. The time of ebb tide varies with locality, and with it the time of hatching. D. Neumann (1966) found that this process is based on genetic adaptation. When he crossbred populations that normally did not overlap in their hatching periods, the result was an F_1 with an intermediate hatching time. The majority of F_2 were also intermediate, but the author found relatively large deviations towards the ranges of the parent generations. This result indicates that the differences in time of hatching have their origin in a few genes (Fig. 11-1).

SUMMARY

Inheritance has been traced by crossbreeding for motor patterns, orientation movements, activity rhythms, and other behavioral traits. Behavioral genetic studies encounter the difficulty that closely related species which are easily crossbred do not normally differ qualitatively in behavior patterns. Nonetheless, the few exceptions studied so far prove that the pattern of inheritance follows in principle that found for morphological traits, which after all was to be expected. Mono and polyfactorial inheritance was found as well as patterns dependent on both dominant and recessive genes.

12

Phylogenetic Development of Behavior Patterns

GENERAL REMARK.

Before evolution of behavior patterns can take place, there must first exist a genetic variation in behavior on which natural selection can act. A. Manning (1961) was able by artificial selection to change the time between the first meeting of male and female fruit flies (*Drosophila melanogaster*) and copulation. He produced population-specific courtship periods of 80 minutes and 3 minutes. By rigorously destroying intermediaries between these two populations, he created a reproductive barrier between the two groups, with members of each group preferring one another. In *Drosophila obscura* positively or negatively geotactic populations can be obtained by appropriate selection (additional examples can be found in A. Manning, 1965, 1967; J. Hirsch, 1967). J. Hirsch and J. C. Boudreau (1958) selected from a population of heterozygous *Drosophila* the strongly and weakly phototactic flies and these they continued to breed separately. After 29 generations of careful selection, the authors produced two lines that could be clearly distinguished in terms of their phototactic reactions.

Each behavior pattern that changes the selective value of a species can initiate a phylogenetic development and be subject to adaptive changes. It is not necessary for them to be always new mutations of behavior patterns. Many behavior patterns that are at first neutral with respect to selection, that exist as pleiotropic effects, may become subject to selection when the animal changes its habitat or when the environment changes. In retrospect this is often called "preadaptation."

B. F. Skinner (1966) writes that adaptedness is not always an irrefutable proof that a process of adaptation has in fact taken place. Behavior patterns may be advantageous by chance without being selected for. This argument seems to be based on the premise that the process of adaptation takes place in many adaptive steps and moves toward a specific adaptation. Adaptations are present, however, whenever a selective advantage, however small, results from their presence. Measured in respect to such adaptations the result may be more or less advantageous, and when they first appear they always occur randomly. Adaptiveness is defined and measured by its selective advantage. How this came about is irrelevant.

J. Nicolai (1964) has advanced an hypothesis that might explain how the bond between widow birds and their hosts, which is based on traditions and maintained by imitations of the song of the host species by the widow species, has led to the evolution of different races in this group. It seems likely that the song dialects that can be demonstrated in various songbirds led to a certain ethological isolation of the various bird populations, which then led to an evolution of subspecies (C. W. Benson, 1948; G. Thielcke, 1961, 1965; P. R. Marler and M. Tamura, 1964). These questions must still be tested experimentally. In a similar manner, imprinting to a certain biotope or to a specific host plant may lead to a new development. The ichneumon fly *Nemeritis canescens* normally deposits her eggs on the caterpillars of *Ephestia* moths, which it recognizes by odor. If one artificially raises ichneumon fly larvae on caterpillars of *Meliphora,* wasps that develop from these larvae will respond to *Meliphora* odor when they are ready to lay eggs, preferring this odor to others, although they still respond most strongly to *Ephestia* odor (W. H. Thorpe and F. H. W. Jones, 1937; W. H. Thorpe, 1938). The fruit fly can similarly be imprinted to peppermint odor during certain developmental periods (W. H. Thorpe, 1939).

We have already described how selection differentiates already existing movements into expressive movements. Behavior patterns probably often initiated a new line of development, thus functioning as "key characteristics" (G. v. Wahlert, 1957). This can be seen from the fact that in many instances closely related species are more conspicuously distinguished from one another in their behavior than in their morphology, so that sometimes behavioral characteristics are used in the classification of the species. The dragonflies of the genus *Orthetrum,* for example, can easily be distinguished because they sit only on the two pairs of hind legs, with the front legs folded against their prothorax (K. F. Buchholz, 1957). Two very similar *Nereis* species can easily be distinguished by their reproductive behavior (R. I. Smith, 1958). Two species of butterflies differ in the manner in which their caterpillars spin their cocoons, the time of mating, and the selection of food (C. P. Haskins and E. F. Haskins, 1958). And two gallflies are distinguished by their food plants (B. Stokes, 1955). All these species can be identified morphologically only with great difficulty. Such species, which are primarily distinguished by their behavior, are called "ethospecies" (A. E. Emerson, 1956). Behavior patterns frequently seem to be the "pacemakers" of certain characteristics. The rattles of the rattlesnake and the porcupine evolved on the basis of an existing movement. It is also possible that different behavior in separate populations of a species can lead to a diverging development, even if these behavior patterns are initially learned and are maintained by tradition within the group. G. v. Wahlert (1962) described the differential behavior of some Mediterranean fish of the same species that function as cleaners in one area but not in another. In the Red Sea the white-spotted damselfish *Dascyllus trimaculatus* lives as an anemone fish between the tentacles of the giant anemones. Near the Maldive Islands and the Nicobar Islands only young fish of this species live near anemones, but they avoid contact with the tentacles. The morphological differences between populations that behave so differently are minimal, but the two populations clearly differ as ethospecies (I. Eibl-Eibesfeldt, 1964c).

In this connection the formation of rites in man deserves attention. Certain clan and tribal habits separate groups of people very effectively, so that E. H. Erikson (1966) actually speaks of "pseudospecies." Thus, the tribal tattoos bind the individual African to his group for the rest of his life. It is very difficult for him to emigrate, because an individual will always be recognized as a stranger in another group (P. Fuchs, 1967).

With reference to the evolution of larger taxa, behavior patterns relating to courtship, care of young, and other intraspecific relationships are of limited significance. But as isolation mechanisms they do play an important part in the development of species. Macroevolution, leading to the exploitation of new ecological niches (transition from water to land, from the ground to the trees, and so on), begins with changes in behavior that alter the choice of biotope or diet (E. Mayr, 1970:334–335):

> It is probably no coincidence that—to name only one example—classification of mammals is so closely tied to diet specialization: all hoofed animals are exclusively herbivorous, almost all carnivores are predominantly or exclusively meat-eating, rodents are herbivores that gnaw, and so on. It is this uniformity of environmental exploitation within the larger groups of mammals which seems to favor the typification indulged in by idealistic morphologists. Readaptation—here to a particular kind of diet—in all cases leads to a specialization that has significantly left its mark on structure, family life, migration, hibernation, and many other dimensions of the biology of these animals. Particularly interesting are the relatively rare exceptions:

> The large panda bear among the Carnivora has switched to a purely vegetarian diet (bamboo). Almost all bats (Chiroptera) are insectiverous, but the fruitbats (Megachiroptera) feed off the fruit of trees . . . When looking at such readaptations within relatively homogeneous groups, we are faced with new questions: How often do they occur? How does such behavior spread from the individual who initially employs it to other members of the populations? . . . How does such behavioral readaptation lead to an extension of niche, in the sense of Ludwig's annidation?

CONCEPT OF HOMOLOGY

Behavior patterns can be compared to each other like morphological characteristics, and in this way one obtains similarity gradients that can be used to reconstruct their phylogenetic development. Earlier in this book we tried to trace the phylogenetic history of several expressive movements. But to do this one must be able to distinguish analogies from homologies, so it seems advisable here to discuss the criteria of homology. Behavior patterns in general do not leave fossils, so we are dependent on the comparison of living species when we try to reconstruct the evolution of behavior. Only in very rare instances is it possible to order the products of animals' activities into a phylogenetic sequence. R. S. Schmidt (1957, 1958) was able to do this with various termite nests.

In general those structures are called homologous that owe their similarity to a common origin. Descent in most cases implies a direct genetic relationship, where the information that relates to the adaptiveness of the behavior pattern in question is passed on through the genome. The criteria of homology that are given below, however, allow us to conclude only that information has been passed on. They can do no more, and they are especially unsuitable for making the distinction between innate and acquired characteristics, as is shown by language homologies. Homologies that are passed on via memory have been called *homologies of tradition* (W. Wickler, 1965a), as opposed to *phyletic homologies*, which are passed on by the genome as the transmitter of the information. As no instances of homology of tradition had previously been known in the animal kingdom, this possibility had not been considered until then. J. Nicolai (1964) discovered song traditions some of which even cross species boundaries (**song mimicry**). For research dealing with homologies it is only necessary that information emanating from one common source is passed on. It is not necessary for reproductive relationships to be involved. The song that is imitated by widow birds is just as homologous to the host song as is Chinese learned by a European to that learned by Chinese themselves. For the assessment of homology it is important only that one source of information is tapped and that the animal did not individually acquire the information during its interactions with the environment. Let us assume that a predator innately possesses the neck bite with which to kill its prey; then it can be homologized with the neck bite of its mother and its siblings. This is also possible if the mother communicates this behavior in some way to her offspring. On the other hand, if each young animal acquires this information of its own, without tapping one common source of information, then we can speak only of individually acquired adaptations. In the latter case the similarities would be called analogies. The terms "homologous/convergent" overlap with "innate/acquired" in the manner depicted in Fig. 12-1. Not until a behavior pattern can be shown to be **homologous** and **inborn** can we infer common ancestry. In the use of the terms "analogous" and "homologous" we follow primarily G. P. Baerends (1958) and W. Wickler (1961a, 1967b), who took the criteria of homology from morphology (A. Remane, 1952) for use in ethology (see also K. Günther, 1956).

Remane distinguishes three main criteria. In the *criterion of position* the same relative position within a structural system indicates homology, as, for example, with cranial bones. With respect to behavior patterns the position in a temporal sequence is an important criterion. If we find a regular sequence of similar movement patterns – a–b–c–d–e–f–g – in two closely related species, and if in one species one of the elements appears somewhat more modified, then the specific location within a temporal sequence indicates its homology.

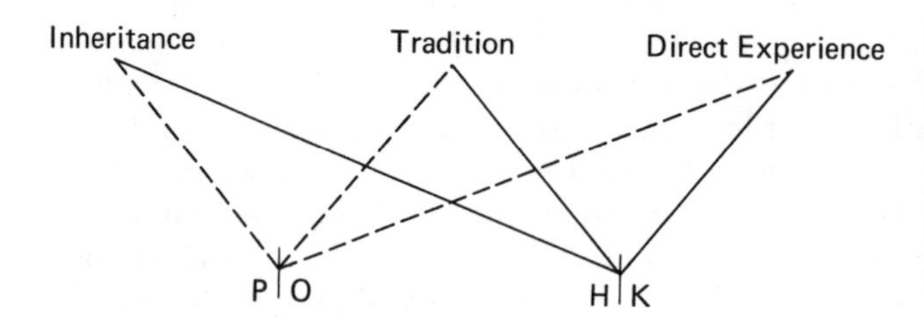

Figure 12-1 Schematic representation of the relationship between the concept pairs phylogeny/ontogeny (P/O) and homology/convergence (H/K), showing the source of the information. (From W. Wickler 1965a.)

The criterion of *specific* quality deals with formal similarities. Homology is indicated when agreement is found in many details of the character compared in two individuals. However, we already know from morphology that similar adaptations to specific environmental conditions can develop independently, which then are analogies – such as the fishlike form of fishes, marine reptiles, and marine mammals. This criterion will not suffice alone, but it is more likely to be valid in the study of expressive movements because their specific adaptations are mostly not a response to the nonliving environment. Here the ethologist argues similarly to the ethnologist, who does not necessarily interpret the similarity of stone axes from European, African, or Asiatic stone age peoples as evidence of a common cultural heritage. The form depends on the function; hence convergence can hardly be excluded.

However, if he finds the words *mère, mater, mother, Mutter, matka,* and *madre* with the same meaning in the languages of different people, then this indicates a common root. In a similar fashion one can consider the expressive movements of many animals as phylogenetically developed "conventions." However, simple expressive movements frequently come about by **convergence,** as, for example, was shown by the investigations of Wickler with respect to the nod swimming of **cleaner fishes and their mimics.**

The criterion of *linkage by intermediate forms* is of use whenever such connecting transitions are found. In this way quite dissimilar behavior patterns can be homologized. The intermediate forms can occur during ontogenesis, and then it is possible to trace the gradual change of a behavior pattern. Were this not the case, these intermediate forms must have been derived from taxonomically closely related forms. R. Schenkel (1956) used this criterion in his interpretation of the **courtship behavior of pheasants.**

As an *auxiliary criterion* one can use the statement that even simple behavior patterns are probably homologous if they occur in a large number of closely related species, and they are probably not homologous if they occur in a large number of unrelated species.

Serial organs can also be compared, such as the mandibles and walking legs of shrimp. The gradient of similarity indicates that the mandibles are actually modified legs. Such instances of serial homology are called *homonomy.* In respect to behavior there are two possibilities for homonomy, according to W. Wickler (1961a): (1) the serial homologous movements of the legs and of the mandibles derived from them, and (2) the various movements of one organ which have a common origin. The "carpentering" activity, drumming, and tapping of a woodpecker that wants to be relieved from the nest are all homonomous movements.

Analogies exist when a behavior pattern is found in animals that share a specific way of life (carrion eaters, predators) or in inhabitants of a specific biotope (cliff-dwellers, tree or desert inhabitants), independently of their taxonomic relationships. Analogy is also indicated when the ancestors of the various species whose similarities are being compared and which live similarly today led different ways of life in earlier times and did not show the similarities in question.

A number of bottom-dwelling fish from several families developed converging adaptations in their behavior (W. Wickler, 1957, 1958, 1959, 1965c). The convergent adaptations of fishes that live in rapid waters are most impressive (Gastromyzonidae and Homalopteridae). The former are descended from the Cobitidae; the latter from the Cyprinidae in the narrower sense. But the convergences in shape and behavior are so close that the members of both families were at one time combined in one category. Both groups of animals have two separate pectoral fins of large size that enable them to stay on the bottom. When they breathe, water is pressed under the head by the strong countercurrent, which the fish pump away by very rapid fanning of the pectoral fins or by the rhythmically beating posterior section of the pectoral fin, with the result that the water flows faster beneath the fish. This reduces the relative pressure, which enables them to remain in place without being swept away. Several species of Cyprinidae, Homalopteridae, Gastromyzonidae, and Siluridae have "invented" this mechanism independently, although they descend from fish that did not possess the rhythmic movement of the pectoral fin (W. Wickler, 1960b). The manner of drinking by immersing the beak and sucking up water in pigeons, sandgrouse (*Pterocles*), and pin-tailed sandgrouse (*Syrrhaptes*), which had been thought of as homologous, was shown to be an analogous adaptation to life in arid regions (W. Wickler, 1961c). This same "invention" was also made by other groups of birds, such as the grassfinches. It is not correct to consider sandgrouse and pin-tailed sandgrouse as relatives of pigeons because of their manner of drinking.

Homoiologies are analogies that have developed on the basis of an homologous structure. Thus, the flipper of a whale is homologous with the wing of a penguin with respect to vertebrate extremities. The adaptation as a flipper, however, is an analogy, because it was independently developed.

Desert mice (Gerbillidae), desert kangaroo mice (Dipodidae), hares (Leporidae), several species of mice (Muridae), and other rodents drum on the ground with their hind legs when they are excited (aggression or escape), which is probably a ritualized intention movement to jump (I. Eibl-Eibesfeldt, 1951b, 1957a). These animals evolved this behavior independently of one another, but certainly through the homologous basis of the jumping-off movement. Many other threat postures owe their similarities to convergent evolution. Tenrecs (*Echinops telfairi*), tree shrews (*Tupaia glis*), squirrels (*Sciurus vulgaris*), dormice (*Glis glis*), and hedgehogs (*Erinaceus europaeus*) threaten when they are disturbed from their sleep by hitting at the disturbing object with sudden stretching movements of their forelimbs, accompanied by shrill hissing and screeching. In tree shrews and squirrels the young already behave this way while still naked. The similarity of the movements can probably be explained from the function of frightening the attacker as well as by the fact that identical homologous origins from defensive and breathing movements served as the basis for ritualization.

Hissing threat sounds have evolved convergently in many vertebrates on the basis of their homologous breathing patterns. In an analogous manner behavior patterns of **care for young** (social grooming, feeding, and so on) and **infantilisms** repeatedly were changed in the service of group-uniting functions.

If the criteria of homologies discussed above are considered, then we can see that behavior patterns are of great taxonomic value and can help to elucidate the natural relationships among animals. We refer in this connection to the studies of the Anatidae by K. Lorenz (1941) and the investigations of A. Faber (1953a, 1953b) and W. Jacobs (1953a, 1953b) on locusts, U. Weidmann (1951) with fruit flies (*Drosophila*), J. Crane (1949, 1952) with New World mantids and jumping spiders (Salticidae), W. F. Blair (1957a, 1957b, 1958) with frogs, G. K. Noble (1927, 1931) with amphibia, J. Nicolai (1959b) with serins, G. P. v. Tets (1965) with pelicans, and P. Leyhausen (1956a, 1956b) with cats. The detailed investigation of tree shrews (*Tupaia*) showed that they should be considered a separate order (Tupaioidea) rather than included among the primates, as they have been (R. Martin, 1966a, 1966b). They share characteristics with both the primates and the insectivora. However, the classification is still disputed. It would be difficult to assign them to any other existing order because they share many characteristics with rodents, rabbits, and marsupials. Additional examples can be found in W. Wickler (1961b, 1967b). With respect to the assessment of body form and behavior there exist opposing viewpoints. D. Starck (1959:47) writes:

> To assume evolutionary relationships on the basis of behavior patterns is not justifiable when such findings clearly contradict morphological considerations. The methods of morphology will therefore remain the basis of the natural system. Its fundamental significance is based on the fact that it is the only method applicable to fossil material.

On the other hand, E. Mayr (1958:345) states:

> If there is a conflict between the evidence provided by morphological characters and that of behavior the taxonomist is increasingly inclined to give greater weight to the ethological evidence.

It seems advisable to take a position somewhere between these. In principle, taxonomic relationships based on morphology should agree with those based on behavior; otherwise one or both are apt to be wrong (N. Tinbergen, 1951).

The Study of Analogies

Whereas the study of homologies allows us to trace the phylogeny of behavior patterns and at the same time by inspection of its adaptive radiation provides information as to the potentiality of a taxonomic group, the study of analogies reveals the laws, dependent on function, that determine the development of a pattern.

The importance of the exploration of analogies is often not clearly understood, since it is often heard that, for example, the similarities in social behavior of geese, fish, and man are "just" analogies and therefore of little importance to us. It should be emphasized therefore that the study of analogies is sometimes even more illuminating than the study of homologies. If one wants to learn the principles after which a wing is constructed one may well turn to study the wings of insects and of birds, even though the one is made from an epidermal fold and the other from a vertebrate extremity. One may even compare these organs with the artificial "organ" of an aeroplane wing, which, after all, is also shaped in response to the laws of aerodynamics. Thus cultural and biological constructs can be compared because they share the same function. In the same way one may look at behavioral structures and ask what laws of function determine pair bonding or hierarchy on a biological or cultural level. Indeed, one would be ill advised to try to understand these phenomena by studying chimpanzees and other closely related primates, which have not evolved pair bonding at all, since they adapted to a different ecological niche. The study of geese or even monogamous fish may help in a better way to develop working hypotheses concerning the similar phenomena in man. And we may well compare culturally and biologically determined behavior patterns, since the material from which a structure ("wings") is made seems in this context to be of lesser importance than the identity of its function.

HISTORICAL RUDIMENTS

The original function of behavior patterns can change if there is a basic change in the way of life of an animal. They may either assume a new function or may be retained in the old or little changed form as behavioral *rudiments*, as long as this is not of an immediate disadvantage for the species. The group of very short tailed macaques (*Macaca speciosa*, *M. arctoides*, *M. fuscata*, and *M. maura*) perform balancing movements with their small tail stump, which of course are not effective. It is possible to release flying movements in ostriches (*Nandus*), although their wings are rudimentary and therefore unfit for flying (I. Krumbiegel, 1940).

Rusa and Dybowski deer and elk threaten by displaying a rudimentary organ. The oldest deer (lower Oligocene) possessed no antlers, as with *Moschus*, the most primitive living deer today. All of these primitive deer species, including the muntjac (*Muntiacus*), possess elongated upper canine teeth for slashing that are used in fighting. They display these weapons when making threats, walking to and fro in front of an opponent, nodding slowly, head raised high, gritting their teeth, and retracting their lips so that the daggerlike teeth are clearly visible. In the same manner Rusa deer, Dybowski deer, and the European elk threaten, although their canine teeth are reduced to small structures and they use their antlers when fighting (O. Antonius, 1939; see also Fig. 12-2). In the process of ritualization of the courtship of grass finches one can trace the

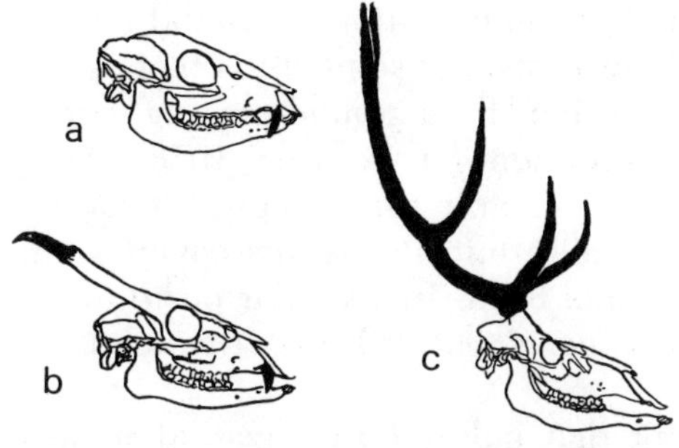

Figure 12-2 Morphological sequence of the recent elk, showing the stepwise replacement of the canine teeth by the antlers as weapons. (*a*): *Moschus;* (*b*): *Muntiacus;* (*c*): *Cervus.* (Explanation in the text.) (From G. Tembrock, 1964.)

change of function and rudimentation (M. F. Hall, 1962; K. Immelmann, 1962b). Carrying nesting material for nest building evolved into the male courtship actions using grass stems. This was again secondarily reduced in some species and became rudimentary, while at the same time the song, which originally served the function of staking out territories, also underwent a change in function. These animals are gregarious and are not really territorial. Instead of courting with grass stems, these males sing softly while sitting next to the females. In the genera *Bathilda* and *Aegintha* males are unable to court without a grass blade in their bills. They hold it continuously and perform various derived nest-building activities without actually building. Eventually they give the blade of grass to the female. The males of *Neochmia* use different material while courting and in the actual construction of the nest. This demonstrates convincingly that courtship with grass stems has acquired independent motivation. Grass finches of the *Lonchura* genus carry a stem around for some time before commencing to court, although they actually court without it, while *Aidemosyne* still uses a stem in the initial phases of courtship. *Emblema* merely pecks at stems of grass; *Poephila* pays no attention to grass blades, although on occasion courting with grass stems occurs as a behavioral rudiment, as well as in young males. Marine iguanas make threats by opening their mouths wide, although they normally do not bite one another during fights. Many birds threaten with widely gaping mouths, even species that actually attack with closed bills. They use the phylogenetically older intention to bite as a threat display. The blue-footed boobies of the Galápagos Islands pass nest material in the form of small stones between each other during the pair-formation ceremonies although they no longer build nests. In all these instances a behavior pattern has survived the disappearance of an original function in the form of an expressive movement. Rudiments that no longer have any function also exist.

The redheaded finch (*Amadina erythrocephala*) is a nest parasite that uses the nests of other birds and no longer builds its own. When sitting on a nest the bird still performs all the nest-building movements but in a randomized sequence. It reaches over the edge of the nest and "grasps" and pulls in nonexisting nest material as if the bird were actually building (J. Nicolai, personal communication). Ground-nesting birds roll in eggs that lie outside their nests, but some birds that nest in trees also do this. These birds were derived from ground-nesting species and still possess this behavior as a rudiment (H. Poulsen, 1953). Wingless *Drosophila* mutants perform wing-cleaning movements, as do the wild forms (H. J. Heinz, 1949). Several termites construct covered tunnels that end in culs-de-sac that are not used. In a similar position one can find passages in closely related species that still serve a function (R. S. Schmidt, 1957, 1958).

We have already pointed out that cultural **ritualization in man** shows remarkable similarities to its phylogenetic parallels. Thus we observe that certain objects lose their original function and acquire new functions or are kept on as mere rudiments. We have decorative buttons on the sleeves of suits that were originally used to button the sleeves. In the same way ribbons on hats originally were used to fasten the hat (Fig. 12-3). Today they are mere decorations (L. Schmidt, 1952). O. Koenig (1968) has collected material that documents this kind of development.

The Galápagos dove (*Nesopelia galapagoensis*) shows a distraction display near the nest, although this behavior, which is useful in deceiving predatory mammals, is no longer of use in the islands, which until very recently were free of such predators. In the same islands E. Curio (1965b) was able to demonstrate that Darwin finches from islands that are free of predators responded appropriately when presented with them, behaving like finches from islands where predators (snakes, raptors) are present. These examples illustrate a kind of natural **deprivation experiment.**

Sometimes we can gain an understanding of phylogenesis by studying the ontogenesis of behavior, although it is true even in morphology that the biogenetic law—that ontogenesis recapitulates phylogenesis—is true in only about 60 percent of the cases. There are some good examples from behavior. The young bearded titmouse at first crawls with all four limbs, which are moved in alternating diagonal pairs (O. Koenig, 1951a,

Figure 12-3 Change of the attachment cord on the headdress of Hungarian hussars. (*a*): Kalpak with cap bag and cord for attachment (1700). (*b*): felt cap dating back to approximately 1760. The cord has lost its original function and is purely decorative. (*c*): Tschako of a hussar before 1914. The cord is solely ornamental and developed to fit this new purpose. A new chinstrap is developed. (From O. Koenig, 1968.)

1951b). The larks, which are derived from hopping birds, hop when they are young but run when they are older. The young **peacock** performs his tail display and food calling, but when older only fans out his tail. Young marine iguanas bite one another when fighting, but when older they butt their heads.

The freshly metamorphosed glaucothoë of the coconut crab (*Birgus latro*) searches for snail shells in the manner of the hermit crab and performs the same fixed action patterns of testing shells and entering them as do the glaucothoë of the hermit crab *Pagurus longicarpus* (E. S. Reese, 1968). The shell protects the glaucothoë from dessication during their migration on land. Individuals that do not find a shell die. The pattern of shell selection still fulfills a function during a short period of life. Older crabs do not need the protection of a shell (E. S. Reese, 1968). Additional examples of processes of behavioral rudimentation are presented by E. Curio (1960) and W. Wickler (1960a).

During artificial manipulations in bumblebee nests, A. Haas (1962, 1965) was able to elicit older behavior patterns that are no longer performed but that in some form or other are common to all members of the genus ("generic behavior").

Evolutionary remnants can be observed in other areas of behavior as well. For instance. innate releasing mechanisms can "outlive" the signal to which they were adapted. Among other things our observations of human behavior bring evidence for this phenomenon. We have already mentioned the masculine habit of emphasizing the shoulders and the feminine one of emphasizing the buttocks in **clothing.**

An interesting example was described by J. D. McPahl (1969) for one species of stickleback. In the area of the Cephalis River (western North America) the predatory fish *Novumbra hubbsi* feeds on sticklebacks. In response, the male of the latter discarded its striking red coloring and instead acquired a black underbelly. Now, if we offer females of the species – which normally mate with black-bellied males – a choice of red-bellied or black-bellied males, we discover that they prefer the red-bellied ones five to one. The original scheme has been maintained by the black stickleback population over the 6,000- to 8,000-year period of their existence.

RESEARCH ON DOMESTICATED ANIMALS AND THE PROCESS OF DOMESTICATION

Research on domesticated animals gives us interesting leads about the phylogenetic development of fixed action patterns. The ancestor of our various races of domesticated pigeons – the rock pigeon (*Columba livia*) – possesses several characteristic courtship behavior patterns. During the display flight the male claps its wings loudly above its back. After that he glides several meters through the air with his wings elevated above the horizontal line of his body. This behavior pattern has been developed and changed further by the artificial selection of man (J. Nicolai, 1965b). In several races of pouters, especially in Steller's pouter, this clapping is so extensive and vigorous—30 instead of 4 to 5 claps in the ancestral form—that the tips of the primary feathers wear away progressively from spring through the summer. Shortly before the molt only a third of the vane remains on the wing feathers, so that the bird's flight is quite impeded. The position of the wings during the glide becomes exaggerated: the wing tips touch and the bird loses altitude quickly. In these instances a behavior pattern has become ritualized through selection by man, because no predator is selecting against this behavior. In roller pigeons, such as the oriental roller or the Birmingham roller, the gliding display flight has been changed through selection into a continuous smooth series of backward somersaults. These somersaults follow each other so quickly in some individuals that the bird seems to be falling from several hundred meters high like a swirling ball of feathers. The fall is broken just above the ground when the bird pulls out of its dive. This is an entirely new behavior pattern that does not occur in the rock pigeons. During courtship on the ground the cooing male turns about its vertical axis and performs a small jump when the female walks away. He may then clap his wings once or twice. In the German ringbeater this behavior pattern is hypertrophied; the cock flies up clapping noisily, flies in a tight circle about one meter above the hen three to four times, and returns to the ground. This flying up has become a new display flight. Turning and following has been combined into a new behavior pattern.

New behavior patterns probably evolved under natural conditions in the same manner. In the diamond firetail finch the behavior of bringing a gift to the female is combined with the infantile begging movement (Fig. 12-4) into a courtship movement. The male at first courts sitting upright, flicking its tail and holding a blade of grass in its bill. When the female approaches he will bend his head down in the position characteristic of the food-begging behavior of this group without letting go of the blade (J. Nicolai, 1965a; see also Fig. 12-5).

Research on domesticated animals supplies still more examples of changes in behavior patterns. Fighting cocks have been selected for aggressiveness; dogs have been selected for various characteristics.

Our domesticated animals have undergone a large number of changes in behavior and appearance under the changed conditions of captivity. These domestication characteristics in animals are similar to those of man (K. Lorenz, 1940, 1943, 1950a) and are the result of similarly changed selective conditions.

With the protection of pens, stalls, and other fenced-in structures, acuity of the senses and physical fitness are less important than a high

Figure 12-4 *Left:* Finches (*Carduelinae*) gape directly toward the adults in an upright body position. They are fed in small portions. Here a 21-day-old West African *Ochrospiza leucopygia* gapes toward its father, who already has regurgitated some seeds. *Right:* Most estrildine finches turn their heads when begging in the manner shown and make pendulumlike movements. The adult birds pump larger quantities into their gullets at one time without removing their bills. Here a 19-day-old estrildine finch (*Granatina granatina*) begs from its father. (After J. Nicolai, 1965a; H. Kacher, artist.)

Figure 12-5 Courtship of the diamond finch male (*Staganopleura guttata*). The male sits with a stem of grass on a branch and attracts a female by his song and courtship dance. The stem is then offered in the position of a begging young (see Fig. 12-4, *right*). (After J. Nicolai, 1965a; H. Kacher, artist.)

rate of reproduction. One result has been a breakdown of the finely differentiated social behavior patterns. Monogamy and high selectivity in the choice of partners impede breeding and are a selective disadvantage under conditions of domestication.

Where in wild animals—for example, graylag geese—a large number of conditions must be fulfilled before their highly differentiated sexual and family life can be developed, it is sufficient in the domesticated form to lock up together any two individuals of different sexes to ensure breeding (K. Lorenz, 1950a).

The selectivity of innate releasing mechanisms has been markedly reduced in domesticated animals. Domesticated zebra finches feed nestlings that do not show the species-specific gape markings, whereas wild birds are extremely selective. The tame finches also court very simple models (K. Immelmann, 1962a).

The wild form of our domesticated chicken, the jungle fowl, responds with brood-care behavior only to chicks with a very specific color pattern on the head and back. Chicks with deviations in their patterns are killed. We find such selective behavior occasionally in Phoenix fighting cocks and dwarf breeds that are relatively close to the wild form. Our domesticated country chickens, on the other hand, accept chicks of all colors, but react selectively to the calls of chicks of their own species, so that it is relatively difficult to give them young ducks to incubate and hatch. However, the most domesticated breeds, such as the Plymouth rock, will accept ducklings without difficulty (K. Lorenz, 1950a).

Graylag geese will mate only after a prolonged courtship period and remain monogamous thereafter. Domesticated geese pair off without showing preferences and are not monogamous. While the aggressive drive was often selected against because it was disruptive, the reproductive and eating drives became hypertrophied. The behavior becomes simpler. These behavior changes are accompanied by a large number of physical signs of domestication. K. Lorenz pointed out that there is a tendency to shorten the extremities and the muzzle, to become fat as well as a general weakening of the muscles and connective tissues (Fig. 12-6). All these physical characteristics of domestication can also be observed in many members of civilized peoples. Furthermore, in civilized man many of the "cardinal" virtues, such as loyalty to the family, courage, and moral behavior, are in danger. If someone commits unscrupulous acts that are barely permissible, he will gain a certain advantage. He who produces offspring recklessly has a higher rate of reproduction, but a society can only exist as long as the number of social responsible individuals prevails. Perhaps some of the old civilizations were in the final analysis the victims of such a degeneration. These degenerative signs, which are the result of domestication, are opposed by our innate esthetic and ethical value judgments, according to which we disapprove the degenerative symptoms discussed above.

It must be made clear that the changes brought about by domestication do not result in a less adapted species but only in a more specialized one. Biologists tend to value, however—without being able to reason why—what is called "creative evolution," by which an animal adapts to many aspects of its environment—for example, by developing finely differentiated sense organs, elaborate mechanisms of coordination, or a rich

Figure 12-6 Characteristics of domestication. In each case the wild form is on the left, the domesticated form derived from it on the right. *a:* Wild chicken – domestic chicken; *b:* wild goose–domestic goose; *c:* wolf–domestic dog. In the figure the relative sizes are not accurate. The domestic goose is larger than the wild form, and the pug is smaller than the wolf. (From K. Lorenz, 1965a)

repertoire of behavior patterns – whereas the processes of involution are felt to be a step backward. The process is best demonstrated in the evolution of parasites. When the parasitic crustacea *Sacculina* is in its larval stage it demonstrates a highly developed organization with complex eyes and elaborate motor organs, but after attachment to its host it "degenerates" into a myceliumlike creature, invading the victim's body. We call this "involution" and for some reason value it less highly. It is nevertheless a perfectly adapted organism. Domestication is in part such a process of involution, but only in part.

One must be extremely careful to distinguish between the degenerative appearances that are caused by domestication and domesticated characteristics. The former have a negative selective value for the preservation of the species. The latter have a positive selective value and are adaptations to specific environmental conditions. This is true, for example, for the **hypersexuality of man** as a result of domestication, which has sometimes been regarded as a sign of degeneration. There is much evidence that this hypersexuality serves an important function in the service of preserving the bond between man and woman. By reducing instinctive behavior patterns through domestication it also became possible to clear the path for learning and education. This undoubtedly was a precondition for the development of *Homo sapiens* (K. Lorenz, 1943).

Commensals of man experience changes in behavior similar to those caused by domestication. The various subspecies of the house mouse (*Mus musculus*) can barely be discriminated on morphological grounds, but can by their behavior. In their country of origin between the Caspian Sea and Lake Neusiedl, the subspecies *Mus musculus spicilegus* (harvest mouse) lives all year outdoors and shows a hoarding instinct. Two to six mice collect five to seven kilograms of seeds, which they store in mounds above the ground. They cover the mounds with earth and live in nests beneath. These harvest mice are herbivorous and lack the odor of the western European house mouse (*Mus musculus domesticus*), which lives as a commensal of man all year and lacks the hoarding instinct. Furthermore, these mice are polyphasic in their activity and, similarly to domestic animals, their estrous cycles continue all year. The half-commensal central European harvest mouse (*Mus musculus musculus*) links both the other races: in summer this subspecies lives out in the fields, in winter as a commensal of man (A. Festetics, 1961).

Figure 12-7 Feeding tracks of the sediment-feeding *Dictyodora* can be read as reflecting changes in behavior. *a:* Tracks from the Cambrian age, loosely meandering paths immediately below the surface; *b:* later they ate their way, in corkscrew fashion, deeper and more closely drawn round the initial spiral; *c:* during the lower Carboniferous Age they went deeply into the sediment. (Seilacher, A., "Fossil Behavior." Copyright © 1967 by Scientific American, Inc. All rights reserved.)

BEHAVIOR FOSSILS

It is possible to draw inferences regarding the behavior of organisms by studying the fossil tracks left by those which produced them. Certain behavior patterns can be deduced from tracks left by feeding (R. Richter, 1927; W. Schäfer, 1965). If one studies the feeding tracks of marine organisms during geologic time, progress in the technique can be observed (Fig. 12-7). The tracks left by snails and trilobites of the Cambrian period as well as those of worms of the early Silurian period remind one of the scribbling of children. The animals seem to have obeyed an impulse to move sideways, thus avoiding a retracing of previously covered ground. A more efficient use of the grazing area is achieved when grazing occurs in a tight spiral or in a tight meandering pattern. In order that such a spiral can be followed it is necessary for the animal to maintain contact with the previously made track. If it is to meander, the direction must be changed regularly. We can now observe that the scribbling patterns disappear completely during the course of the earth's history and that they are supplanted by spiral and meandering feeding patterns. More complicated meander patterns and tight double spirals do not appear until the end of the Mesozoic age (A. Seilacher, 1967).

The feeding tracks of the sediment-feeding *Dictyodora* show changes in the feeding technique through the ages that are correlated with morphological changes. From the Cambrian age to the Devonian age (600 to 350 million years) *Dictyodora* fed immediately below the sediment surface in a loose meandering pattern. Probably as a result of numerous competitors this species escaped into deeper layers of the sediment, while the breathing tube, which can be seen in the sediment, became longer. At first they drilled into the substratum and only then did they follow their usual meandering pattern. Later they began to eat their way downward, leaving a corkscrew pattern. When they reached the right depth they began to meander, and their pattern became changed in the course of time so that the meandering paths were placed tightly around the initial corkscrew spiral. Finally they gave up meandering completely and bored their screwlike spiral deeper into the bottom (A. Seilacher, 1967; Fig. 12-7).

SUMMARY

Behavioral evolution starts with genetic variation in behavior upon which selection exerts its influence. How selection shapes behavior has been studied experimentally in *Drosophila*. Behavior patterns are often the pacemakers of evolution. Song dialects promote isolation of a breeding pool and thus initiate subspeciation. Imprinting to hosts may in a similar way bring about reproductive isolation. In man "pseudospeciation" on a cultural basis may well have led to the rapid racial differentiation.

Macroevolution leading to the exploitation of new ecological niches begins with behavioral changes that lead to changed preferences in diet or biotope. In tracing the evolution of behavioral characteristics the criteria of homology as elaborated by morphologists are used. Homologies can be divided into homologies of tradition and phyletic homologies. The study of analogies throws light upon the laws of function that rule the evolution of a behavior pattern. This should be particularly noted, since many misconceptions derive from the fact that the value of this research has not been adequately appreciated.

As with morphological traits, a behavior pattern may be subject to a process of rudimentation, if the selection pressure that shaped it no longer exerts its influence. Rudiments can be discovered in motor patterns, innate releasing mechanisms, and even in cultural activities of man.

The study of domestic animals provides an interesting opportunity to study behavioral evolution.

13

The Ontogeny of Behavior Patterns

EMBRYOLOGY OF BEHAVIOR

In general an animal has its behavior patterns ready for use when they are needed. As organs grow and mature, so does their capacity to perform their function. It is much rarer for a behavior pattern to mature before the organ that is involved in its performance. Young fighting graylag geese will show wing-boxing behavior with wings that are as yet nothing more than tiny stumps with which they are unable to hit their opponent (K. Lorenz, 1943). During ontogenesis, behavior patterns develop gradually and overlap one another in time. Sense organs, coordination centers, and effectors can mature independently of each other and at different rates, as the last example demonstrated. The interlocking of fixed action patterns with the appropriate taxes may occur only after the basic patterns are established. Thus, newborn mice and rats scratch at first spontaneously in the air without touching their skin (I. Eibl-Eibesfeldt, 1950c). Some behavior patterns mature as organs do, while others owe their specific adaptiveness to an interaction of the young animal with its environment or its own body. The development of behavior begins in the embryonic stage, so that by the time of birth or hatching a number of actions are fully functional.

Little is known about the embryology of behavior. Although W.

Preyer wrote a pioneer work on this subject as early as 1885, there was no followup for a long time. G. E. Coghill (1929) came to the conclusion, following numerous, careful studies, that behavior always seems to appear in well-organized patterns and that it depends on the spontaneous activity of the central nervous system. In the larva of the salamander *Ambystoma,* the undulating movement first occurs as a turning of the head to one side only. In fishes it is frequently the same. In bird and mammal embryos we can also observe an increasing maturation of movements from the head to the posterior parts of the body; the first behavior that can be recognized is a bending of the head. Soon thereafter the entire body, legs, wings, head, and trunk become active. seemingly independently of one another. In this "mass action" repeated bouts of activity follow short intervals of rest. How such activity becomes integrated into well-coordinated movements needs to be investigated specifically. It appears certain, however, that the behavior of an animal is not (as W. F. Windle [1940, 1944], among others, thought) built up of larval reflexes, which become integrated as primary units of behavior into secondary, higher functional units, because the initial movements are always of a spontaneous nature (V. Hamburger, 1963; V. Hamburger et al., 1966). A chick already moves when it is still a 3-day-old embryo, but not until the seventh day of incubation can a response to tactile stimulation be obtained. This interval between movement and sensitivity was noted as early as 1885 by W. Preyer (1885).

It is possible that simple learning processes play a role in the **developmental physiology of innate behavior.** According to H. F. R. Prechtl and A. R. Knol (1958) later behavior in man is influenced by the position of the embryo in the uterus as a function of the relative freedom of movement of the embryo. Children who develop in a head-down position show good flexing and extension reflexes. If they are scratched on the soles of the feet, they pull up the legs. Following a breech position the same stimulus will extend their legs, and extension movements will later predominate in the movement repertoire of the infant. There are no noticeable differences in the musculature, so these differences in behavior must depend on habituation.

EARLY ONTOGENETIC ADAPTATIONS (KAINOGENESES)

When a living organism is born or hatched from an egg it may be more or less completely developed. Sometimes the newborn animal is a miniature version of the parents and lives in much the same way as they do. This is true for most reptiles and to a certain degree for several precocial birds and mammals (Fig. 13-1). In many other instances the young does not resemble the parents and its mode of living is quite different. The different ways of life of many larval and imaginal insects or frogs are well known. Many of these larvae possess highly specific behavior patterns, serving functions such as a particular manner of feeding or avoidance of predators, which are completely lost after metamorphosis. I will cite a few examples here. Ant lions build pits and throw sand at their prey (H. J. Nieboer, 1960). Larvae of the fly *Arachnocampa luminosa* live on the ceilings of caves in New Zealand. They are luminous and produce long,

Figure 13-1 *Left:* Precocial animal: a newborn hare; *right:* altricial animal: a newborn rabbit. (From F. Bourliere, 1955.)

dangling silky threads that have drops of a sticky substance attached at short intervals. Insects attracted by the light are caught in these traps and are consumed along with the threads (J. B. Gatenby and S. Cotton, 1960; V. B. Wigglesworth, 1964; see also Fig. 13-2). The butterfly caterpillar of *Aethria carnicauda* builds several fences consisting of hairs arranged in whorls around the branch, on either side of itself, which protect the pupa from predators (Fig. 13-3). Larvae of the long-horned woodborer beetles

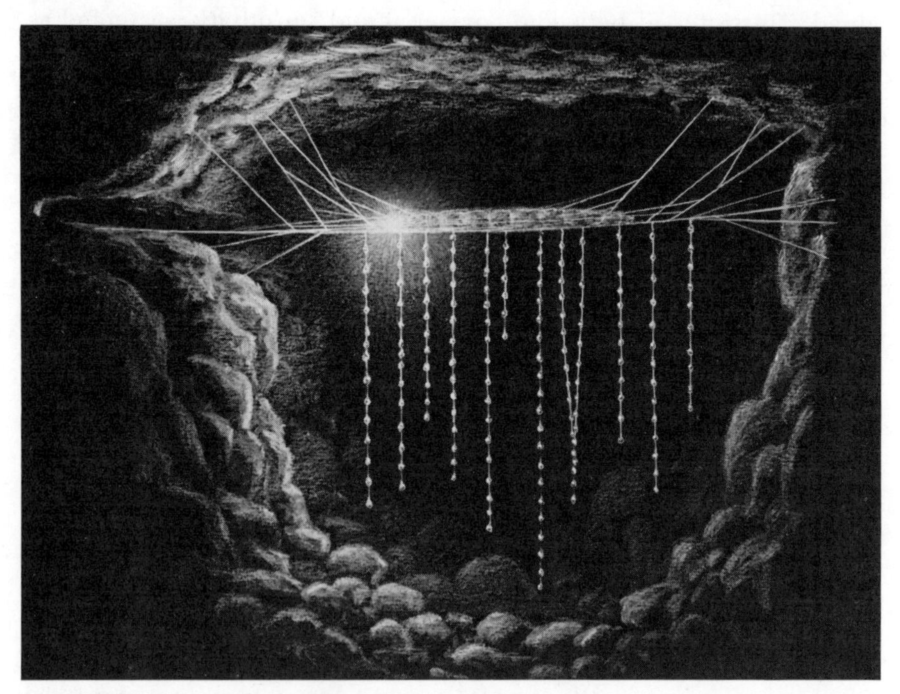

Figure 13-2 Larva of *Arachnocampa (Bolithophila) luminosa* rests on a horizontal web from which are suspended catching filaments with attached drops of a sticky substance. From the abdomen of the larva light is emitted that attracts prey (insects). These become entangled in the filaments and are eaten. When in danger the larva retreats into a hiding place in the wall of the cave. (After J. B. Gatenby, 1960, from W. Wickler, 1967a; H. Kacher, artist.)

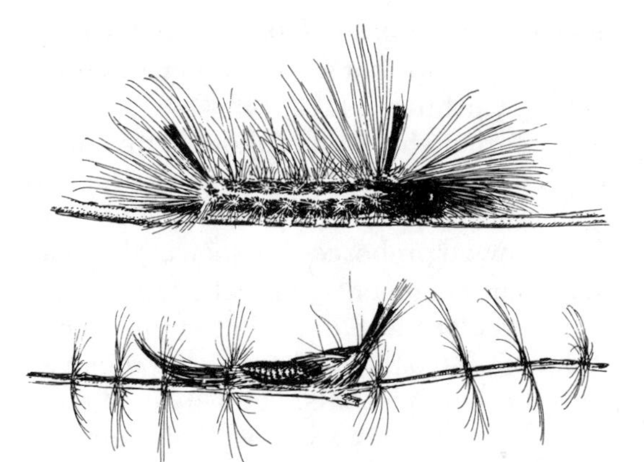

Figure 13-3 Caterpillar, and below it the pupa, of *Aethria carnicauda* Butler with the protective fences that were built by the larva before pupation. (After W. Beebe, 1953.)

(Cerambycidae) position themselves within their passages close to the surface of the wood just prior to the pupal stage. The adult beetle is unable to drill through the wood and would perish if the larvae pupated deep within the wood. The pea weevil (*Bruchus pisi*) metamorphoses within the pea. The developed beetle cannot eat its way to freedom; this is done by the larva, which eats a passage up to the thin peeling of the pea. Here it gnaws a round furrow so that the beetle merely has to push the cover open from the inside (J. H. Fabré, 1879–1910).

Even the larvae of primitive invertebrates can demonstrate surprising behaviors. The nematode (*Dictyocaulus viviparus*) lives in the lungs of cattle. The larvae, which hatch in the bronchial tubes, emerge via the intestinal tract onto the pasture. Because cows avoid grazing in the proximity of feces and the larvae are unable to migrate, further transfer takes place in the following manner. The larvae climb, positively phototactically oriented, to the tops of pilobolus fungi, which grow in great abundance on the feces. These fungi project a spore packet into the air when exposed to light. The larvae wait on top of these spore packets, from which they are shot out to land some distance away. Over 50 larvae have been found on one spore packet (J. Robinson, 1962).

The cercarian larvae of the fluke (*Dicrocoelium dendriticum*) control the behavior of their last intermediary host, the ants. Whereas most of the cercaria are in the abdominal cavity of the ant, one of them, with a different appearance from the others, is always found in the subesophageal ganglion near the nerves that innervate the mouth parts. This "brain worm" seems to change the behavior of the ants so that they climb upward on grass stems and there hold on with their mandibles. This ensures that grazing sheep will eat them so that the parasite is safely delivered to its host (W. Hohorst and G. Graefe, 1961).

Larval adaptations in respect to behavior are of various types and are lost after metamorphosis. Whether some of these abilities are retained in some changed form we do not know, but it is known that what is learned in the larval stage can be retained through metamorphosis. Mealworm larvae (*Tenebrio*) that were trained by W. v. Borell du Vernay

(1942) in a **T** maze retained the learned habit when they were tested as beetles. Fruit flies (*Drosophila*) prefer odors that were added to their larval food (W. H. Thorpe, 1939; J. E. Cushing, 1941).

Another intriguing observation that should be followed up comes from E. Fischer (cited by R. Fletcher, 1948). The pupae of *Hoplitis milhauseri* are enclosed in very hard cocoons which they dissolve with a silk-splitting enzyme from their narrowed probosces. As soon as the head is free of the pupa it performs circular movements, whereby the enzyme is excreted. In this way a roughly circular piece drops out of the cocoon and the butterfly can crawl to freedom. Two pupae that Fischer removed from the cocoon did not perform these movements, yet displayed them as butterflies. They stood with their heads against the wall of their container and performed the circular movements as if they were going to dissolve part of it. Normally, these butterflies unfold their wings right after hatching.

In vertebrates we often find early ontogenetic adaptations in the behavior of fish and amphibian larvae whose way of life is quite different from that of the adult animals. An early ontogenetic adaptation seems to be the specific movement coordinations involved in the hatching of birds (V. Hamburger and R. Oppenheim, 1967), as are the gaping reactions and begging movements of young birds, which sometimes reappear in the behavior repertoire of the adults as so-called **infantilisms**. Specific adaptations prevent soiling of the nest. In the black woodpecker defecation is released when the adult bird stimulates the anal region of the young, after which the adult simply eats it (H. Sielmann, 1955). Young bee-eaters walk backward to the nest wall and defecate there, thus keeping the center of the nest clean (L. Koenig, 1951). Other young birds defecate over the edge of the nest. The young cuckoo pushes its nestmates out of the nest. This behavior pattern becomes manifest ten hours after hatching and is no longer present four days later.

Young mammals also possess a number of early ontogenetic adaptations; some of these animals are often born in very early stages of their development. The young of the great gray kangaroo is born as a 2-cm-long embryo but is able to climb into the mother's pouch unaided by means of its powerful front legs. Young mammals often possess a searching mechanism, a rhythmic head movement, which leads to the teat. When they suck, they show treading movements with the forepaws or butting of the udder with the head (H. F. R. Prechtl and W. M. Schleidt, 1950, 1951; H. F. R. Prechtl, 1958; J. Adler, G. Linn, and A. V. Moore, 1958). Young mammals often have special alarm calls, and can hold on when in danger. Prematurely born human babies are able to hold on to a rope with all fours as well as with the hands alone without any aid. This ability is later lost and may again develop secondarily (A. Peiper, 1951, 1961).

Frequently young animals can produce sounds that enable them to maintain contact with their mother in a kind of dialogue. A young graylag gosling will give a two-syllable "wi-wi" call from time to time even when asleep under its mother, which she will answer. If a gosling is kept alone the "wi-wi" calls become more and more urgent. Such "lost calls" are widespread, and human infants also have them. Thus, nighttime restlessness of the child comes from a phylogenetically very old need to be reassured by the presence of the mother. To be left alone signifies the greatest danger for the breast-fed infant, and this crying is a contact call that alarms the mother and aids her in finding her child. Today we place our children in beds, but the old mechanism is still operating, and the child seeks with its crying to bring about the calming contact with the mother. The child can be calmed down by rocking or by providing it with a pacifier, which is a model of the mother's breast (A. Peiper, 1951).

MATURATION OF BEHAVIOR PATTERNS AND "INSTINCT–LEARNING INTERCALATION"

New behavior patterns are built up as the result of maturation and learning processes, while infantile and larval behaviors are completely or partially fragmented. They can disappear completely or they can reappear as expressive behavior in the repertoire of the adult (see **food begging during courtship**). Sometimes infantile behavior patterns appear as *regressions* in adult animals and man (M. Holzapfel, 1949; J. Adler, G. Linn, and A. V. Moore, 1958; D. W. Ploog, 1964a). This proves that they have persisted in latent form, in other words, that the mechanisms on which they were based have been retained. In older people who suffer from degenerative processes of the central nervous system a recurrence of the infantile searching mechanism, oral orientation, and sucking movements can be observed (H. F. R. Prechtl and W. M. Schleidt, 1950; S. Wieser and T. Itil, 1954; S. Wieser, 1955; G. Pilleri, 1960a, 1960b, 1961; D. W. Ploog, 1964a, 1964b).

The maturation of new actions and the breaking down of old fixed action patterns can overlap. Young nestling sparrows peck when they are still gaping. They do this especially when they are satiated after feeding. If they become hungry they gape again. Gaping initially inhibits pecking. K. Lorenz (1935) reported that young hand-raised starlings, which had been picking up food on their own while their caretaker was away for several days, again gaped continuously after he returned. They had eaten on their own until then, so he did not think it necessary to feed them until he noticed that they were becoming weak. The drive to gape, which was released by the return of the caretaker, had blocked the pecking reaction (see also M. Holzapfel, 1949).

Study of the behavior patterns of food intake in cormorants reveals a gradual transformation from infantile behavior patterns to an increasing integration of the individual acts with the appropriate motivations. Until the third week of life the young birds beg and gape. Between the third and fifth week they begin to eat fish by themselves, and from the sixth week on they begin to catch their own. Begging drops out when they are six months old. During the ontogenesis of nest-building behavior a similar picture emerges. The final action of "trembling" occurs first; it is followed by the fastening of the twig, bringing of the twig, and so on. This integration is not the result of individual learning. In this connection it is remarkable that a disintegration in reverse order occurs within the annual cycle toward the end of the breeding season (A. Kortlandt, 1940, 1955).

In species that metamorphose there is at the same time a complete reorganization of the behavior, with the appearance of behavior patterns for which there were no previous indications. As A. J. Rösel v. Rosenhof (1746–1761) pointed out long ago, butterflies do not have to learn to fly after hatching. In species with incomplete changes the differences are less startling and behavior patterns of the imago appear during the larval stage (W. Jacobs, 1953a, 1953b). The larvae of grasshoppers perform the leg movements necessary to produce songs, although their "song" is at first silent, because the morphological structure that is necessary to produce the song develops much later (A. S. Weih, 1951). The fully developed cricket, on the other hand, does not begin to sing immediately after the hardening of the elytra but several days later. In this instance the organ has matured before the behavior (S. v. Hörmann-Heck, 1957). We have already seen the opposite in the wing boxing of the graylag gosling.

Behavior patterns can be activated prematurely by the injection of hormones. Following an injection of testosterone, young male dogs raise their hind legs like adults when they urinate (J. Freud and J. E. Uylert, 1948). Fourteen-day-old male rats begin to mount females after testosterone injections, and in 21-day-old female rats the copulatory posture (lordosis) can be released following injection of follicle hormones (F. A. Beach, 1947). According to these results the neuromotor mechanisms underlying these behavior patterns are readily available long before the behavior normally occurs. Immelmann's observation that **zebra finch** females sing the **songs** they have heard when they were young, following treatment with male sex hormones, is also very interesting in this respect.

In the white butterfly (*Pieris napi*) the ability to fly improves as the wings harden. Until this occurs, spontaneous flight seems to be suppressed (B. Petersen, L. Lundgren, and L. Wilson, 1957). The locust *Schistocerca gregaria* is unable to fly properly until several days after reaching the adult stage. Only then is the cuticle of the wings and thorax sufficiently hardened (J. S. Kennedy, 1951).

In many instances the developing behavior patterns appear very clumsy. Only gradually does the initial lack of coordination disappear, which may be the result of maturational processes, learning processes, or a combination of both. Whether improvement depends upon learning or upon maturation can only be decided on the basis of experiments. No distinction can be made by mere observation of the phenomena. Newly hatched chicks peck at small objects, but their aim is not very accurate. When allowed to peck at the head of a nail embedded in soft clay, the initial impressions of the beak are more scattered around the nail than they are later on. On the fourth day they are clustered very closely around the head of the nail. E. H. Hess (1956) was able to show that learning was not involved in this improvement. He fitted chicks with hoods containing prisms that displaced the object to the right. The initial peck marks on day 1 were scattered to the left of the target. On day 4 they were tightly clustered but still displaced from the target so that it was outside the peck marks (Figs. 13-4 and 13-5). In this case the improvement is not due to learning but to a maturation of the aiming mechanism. The animals never learned to hit the nail on the head. In some cichlids (Cichlidae) the selectivity of the innate releasing mechanism for the following reaction is also a function of maturation (E. Kuenzer and P. Kuenzer, 1962).

K. Lorenz (1935) spoke of "instinct–learning intercalation" when innate and acquired components become integrated into one behavior sequence. I have described how these components can be distinguished experimentally (see **deprivation experiment**).

Objections that are based on the assertion that a behavior generally can be modified down to its smallest units (T. C. Schneirla, 1956) cannot be upheld. Behavior cannot be indiscriminately shaped, and the numerous examples that show great resistance to all attempts at modification disprove the hypothesis about the general modifiability of even phylogenetically adapted behavior mechanisms. P. Weiss (1941a) severed the leg muscles of newts from their tendons and grafted them onto their antagonists while keeping their original innervation intact. It was found that each muscle continued to react according to its original function, which led to inappropriate leg movements. The newts were unable to learn the new coordination. Corresponding muscle transpositions and artificial crossing of nerve fibers in rats and monkeys also did not result in corrections. In man partial reorganization has been observed following muscle transposition (R. W. Sperry, 1958). When the right and left limb buds were exchanged in newts at the time the anterior-posterior axis was already determined, this resulted in the front legs, which were now facing to the rear, walking toward the hind legs so that the animal was unable to move. No change occurred throughout the year following the operation. The newts were not able to learn to reverse the movement of the front legs and thus become able to move forward. Afferent impulses from the periphery were unable to change the central coordination and the legs moved in the manner they would have in their original position. Many of the functions of the central nervous system therefore

Figure 13-4 Goggle experiments of E. H. Hess: a chick with prism goggles. (Photograph: E. H. Hess.)

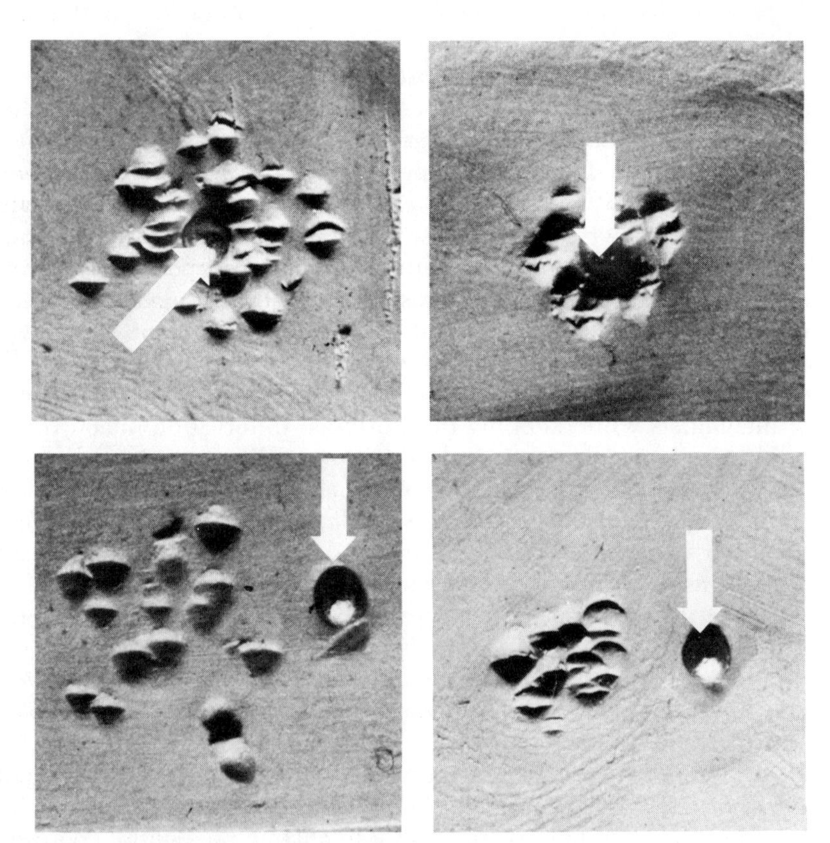

Figure 13-5 Space-perception studies of E. H. Hess. (*Above, left*). In the 1-day-old chick the impressions of the pecks are more scattered around a target (nailhead) embedded in soft clay, than (*right*) in the 4-day-old animal. A similar scattering, although displaced some distance away from the target due to the effect of the prisms, in a 1-day-old (*below, left*) and 4-day-old (*right*) chick. (Additional explanation in the text.) (Photographs: E. H. Hess.)

are predetermined as **phylogenetic adaptations** and are quite resistant to change. P. Weiss (1939:558) expresses this in the following comparison:

> The nervous system reminds us of an industrial complex, not only because of the multiplicity of cooperating agencies and the harmonic coordinations of their activities, but also because the organization is already present in its basic outline without prior experience and indeed before the beginning of its functions. The condition of the central nervous system, at the time it is first functionally activated, can be compared to a ship at the time of its launching. While it still remains to be completely outfitted it is already capable of floating, moving, and being steered. In the same way the central nervous system, when activated, can transmit impulses, coordinate them, and control the musculature in a general way.

In relation to the question of the predetermination of the functions of the central nervous system, it is interesting to note the changes in opinion in the field of neurophysiology. In the 1930s the dominant belief among neurophysiologists was that there existed a practically unlimited functional plasticity of the vertebrate brain. Neurosurgeons operated on the belief that the wiring diagram of the nervous system could be rearranged in any desired way. As long as any connections were established functional readjustment would take place (for a review see R. W. Sperry, 1971). It was supposed that one could even cross-connect arm and leg nerves to reinnervate a paralyzed leg. "Having the wires crossed was no challenge to the brain of the 1920's and 1930's. In his review of 1939 Kurt Goldstein concluded that it seemed immaterial what particular nerve connection exist; so long as any connections are present correct function follows. . . . Such was the thinking of the 1930's (if you hadn't been there, you wouldn't believe it!). On the above terms, where the nerve fiber connections seemed to be functionally plastic and interchangeable, there was no problem in the developmental prewiring of the nervous system to provide for selective growth of proper connections. The general motto of the day was 'Let Function do it' " (R. W. Sperry, 1971:28).

The experiments of R. W. Sperry (1940, 1943a, 1943b), however, challenged this view. After surgical transplanation of nerve–muscle and other nerve–end-organ relations, the predicted readjustment simply failed to occur; the concept of plasticity could not be experimentally verified (R. W. Sperry, 1945a). The interchange of nerve connections in fact caused directly corresponding disturbances in function that were resistant to re-education and often impossible to correct.

"The new picture that we emerged with implied a functionally specified system of wired-in behavioral nerve circuits, relatively implastic to rearrangement by function. The whole problem of developmental organisation and the question of how a brain gets itself wired for adaptive function was, of course, markedly changed. We were now confronted with the question of how the proper nerve connections get established correctly in the first place: was this achieved through selective fiber growth or by early irreversible training, or by some combination of the two?" (R. W. Sperry 1971:29).

In the 1930s the growth and termination of nerve fibers was thought to be nonselective. This conclusion was derived from studies of nerve growth in tissue cultures and of the growth and regeneration of nerves by P. Weiss (1936, 1937, 1938). However, when R. W. Sperry tested experimentally the selectivity of growth in central brain tracts of amphibians and fishes he arrived at the opposite conclusion (Sperry 1945b, 1951a, 1951b). Nerve growth and connections with the brain seemed to proceed with the utmost precision and selectivity. In one crucial experiment rotation of the eyes by 180° (which results in an inversion of the visual perception) was followed by section and scrambling of the optical fibers. The following regrowth of the brain connections did restore an orderly type of vision, but it was inverted. This proves prefunctional regulation by inherent growth mechanisms. If functional adaptation had occurred right-side-up vision should have been the result. The behaviorally highly maladaptive inverted vision persisted and was resistant to re-education. It reverted promptly to normal upon surgical rerotation of the eyeball into the normal position.

It is now assumed that the precisely and prefunctionally ordered growth of brain circuits takes place on the bases of chemical codes under genetic control. Selective chemical affinities on the molecular level are believed to operate in intercellular contacts to determine which cell connects with which. The growing fibers are selectively matched by inherent chemical affinities (R. W. Sperry, 1963, 1965).

> Each fiber in the brain pathways has its own preferential affinity for particular prescribed trails in the differentiating surround. Both pushed and pulled along these trails, the probing fiber tip eventually locates and connects with certain other neurons, often far distant, that have the appropriate molecular labels. The potential pathways and terminal connection zones have their own individual biochemical constitution by which each is recognized and distinguished from all others in the same half of the brain and cord. Indications are that right and left halves are chemical mirror maps. In general outline at least, one could now see how it would be entirely possible for behavioral nerve circuits of extreme intricacy and precision to be inherited and organized prefunctionally solely by the mechanisms of embryonic growth and differentiation (R. W. Sperry, 1971:31–32).

This, of course, is in complete agreement with the ethological concept of inborn or inherited behavior traits, and R. W. Sperry (1971) indeed pointed out that in the biological and closely related sciences polar revisions in our basic thinking relating to the nature/nurture problem have taken place, and he adds: "As yet the meaning and impact of these changes has only begun to permeate into areas outside biology and ethology. In the more human areas of behavioral science like clinical psychology, psychiatry, anthropology, education, and the social sciences generally, the prevailing conceptual approach on this subject remains today essentially unchanged or very little changed from where it stood 30 years ago" (R. W. Sperry, 1971:32).

Just how rigidly the various movement patterns are determined depends on the specific species and is different even within the same species in different functional cycles. Usually additional learning takes place in the sense of Lorenz's instinct–learning intercalation.

Squirrels (*Sciurus vulgaris*) possess the movements of gnawing and prying, but they must learn how to employ these behavior patterns effectively when opening a nut. Experienced squirrels can do this with a minimum of wasted effort. They gnaw a furrow on the broad side of a nut from base to tip, possibly a second one, wedge their lower incisors into the crack, and break the nut open into two halves (Fig. 13-6). Inexperi-

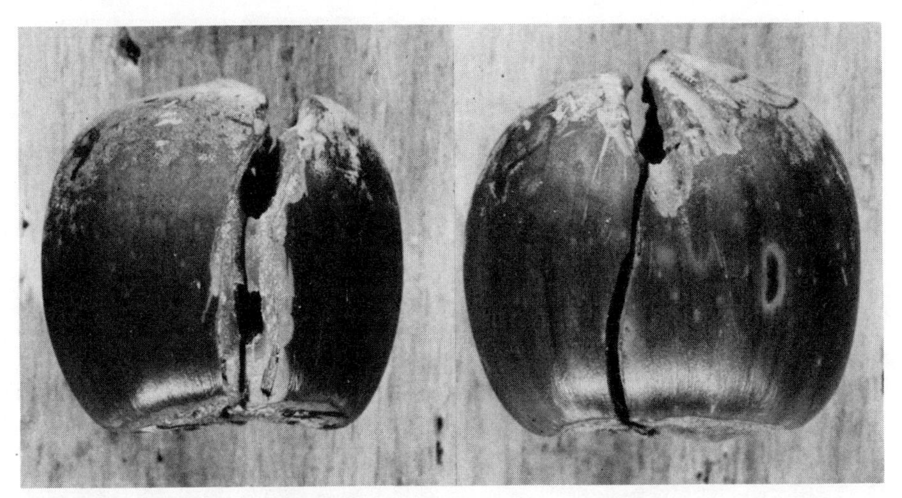

Figure 13-6 Hazelnut opened by an experienced squirrel using the wedging technique. (Photograph: I. Eibl-Eibesfeldt.)

enced squirrels, on the other hand, gnaw without purpose, cutting random furrows until the nut breaks at one place or the other (Fig. 13-7*a* and *b*). They try to wedge their teeth into the opening in attempts to pry the shell open; but this leads to success only if the furrows have been gnawed in the proper way. The first improvements in the technique can be seen when the furrows run parallel to the grain of the nut and are concentrated on the broad side of the nut (Fig. 13-7*c* and *d*). The squirrel follows the path of least resistance, and in this way the activity of the squirrel is guided in a specific direction by the very structure of the nut. The squirrel continues with its attempts to pry, and it keeps repeating those actions that have led to success. In this way most squirrels acquire the most efficient prying technique (I. Eibl-Eibesfeldt, 1963). There are, however, individual deviations. Some squirrels learn to open nuts by gnawing a hole through a few closely spaced furrows (Fig. 13-8). One squirrel achieved almost instant success by gnawing a hole into the base of the nut, and continued to use this technique. Eventually it learned to make a few closely spaced furrows and then gnaw the hole (Fig. 13-9). Finally it changed to attacking the thin-walled tip of the nut in the same fashion. Similarly, other rodents develop techniques to open hard-shelled fruits (E. Petersen, 1965).

I have referred in this connection to the **food-hiding** attempts of **squirrels,** which show that a fairly long chain of innate behavior patterns matures predominantly in the absence of experience. The animal adds little to this technique by learning. In this functional cycle the animal possesses a primarily phylogenetically acquired behavior program, and

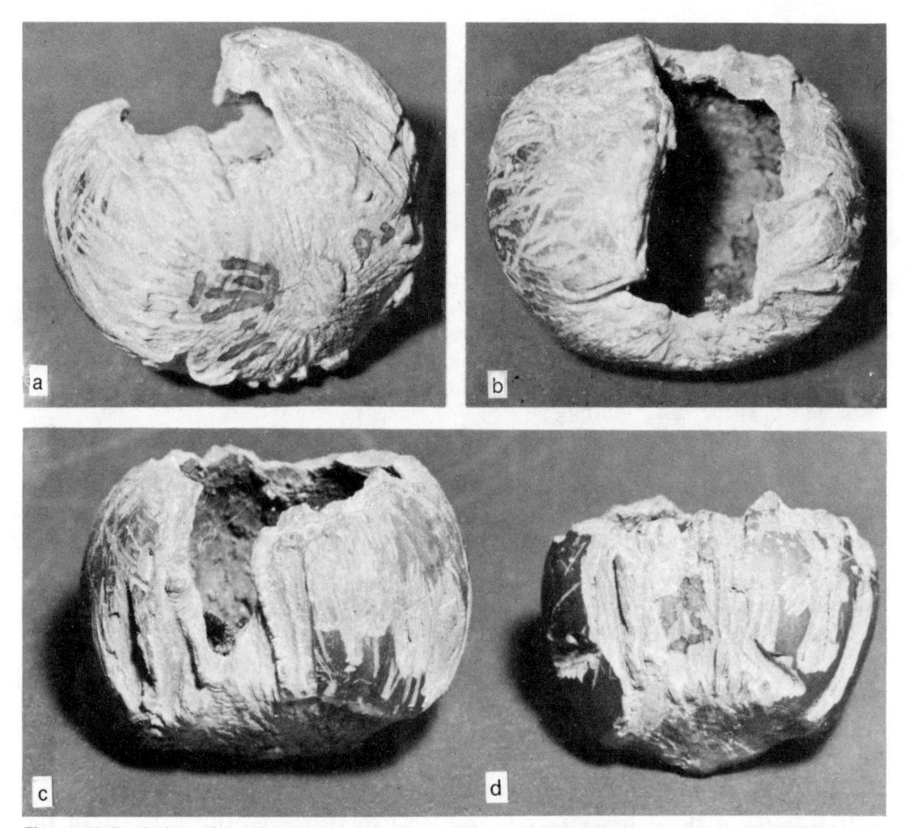

Figure 13-7 A view of (*a*) the base and (*b*) the tip of a hazelnut opened by an inexperienced 66-day-old squirrel, the fourth nut tried. Numerous gnawing marks have produced many furrows on the nut in a randon pattern. (*c*), the thirteenth and (*d*), the fourteenth nuts opened by the same animal, showing parallel gnawing tracks in line with the grain of the wood. (Photographs: I. Eibl-Eibesfeldt.)

Figure 13-8 Hole-cracking technique of experienced squirrels. (Photographs: I. Eibl-Eibesfeldt.)

it is quite obvious that in this case selection pressures work against adaptive modifiability, which is unnecessary because one behavior blueprint is sufficient for the performance of the task. Any modification of the tried and successful contains a certain risk. This is quite different in the case of nut opening. The variability of the available nuts requires various techniques and hence individual adaptation.

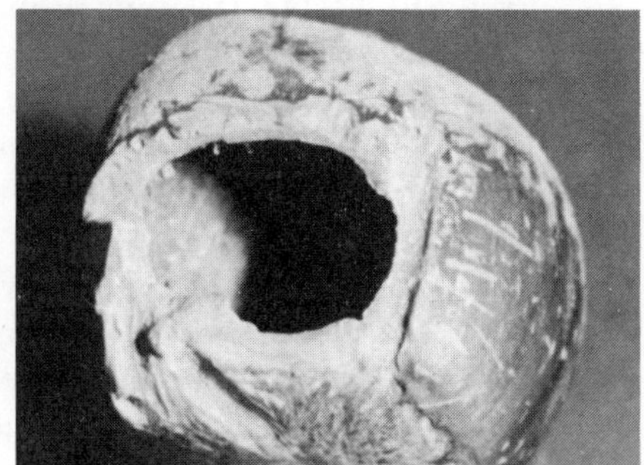

Figure 13-9 *Left:* Hole-gnawing technique; *Right:* the hole-cracking technique derived from it. (Photographs: I. Eibl-Eibesfeldt.)

In England titmice learned to open milk bottles. By a recombination of innate behavior patterns of food getting they developed various methods of solving the problem. The acquired habit of bottle opening spread geographically, which supports the hypothesis that the skill is passed on by tradition (J. Fisher and R. A. Hinde, 1949). Such learned behavior sequences can be called *acquired coordinations,* although they contain fixed action patterns as elements. They have been integrated into new functional units. Additional examples of acquired coordinations were discussed in connection with **bird songs** and **begging movements.** In higher mammals we can generally observe that fixed action patterns consist of very short movement sequences, which are combined by learning into acquired coordinations.

The rat possesses all nest-building behavior patterns innately but learns the appropriate sequence of the individual components (I. Eibl-Eibesfeldt, 1963). In a similar manner the inexperienced canary shows all the behavior patterns of nest building, but it must learn to integrate them into one functional whole (R. A. Hinde, 1958).

An animal may possess an innate behavior pattern, but its application must be learned. The woodpecker finch of the Galápagos Islands (*Cactospiza pallida*) uses a tool to probe insects out of wood. Once it has opened a tunnel made by an insect, the bird picks up a cactus spine or a straight, thin piece of wood, breaks it into the proper length, and proceeds to impale the insect in attempts to remove it (I. Eibl-Eibesfeldt and H. Sielmann, 1962, 1965; see also Fig. 15-11). A male bird that I acquired when it was quite young did not fully master the technique. It searched for sticks, probed in crevices and holes, but only in a playful manner after feeding. When it saw an insect in a crack it dropped the stick and attempted to catch it with its bill. The bird learned only gradually to use the tool for this purpose.

D. Morris told me that chimpanzees that were born in captivity in the London Zoo poke sticks into cracks and holes in a playful manner. In the wild they use thin twigs to fish termites out of holes in the ground (see Fig. 14-11). It seems probable that this behavior is based upon an innate disposition to use a tool, but that the specific application must be learned.

The inexperienced raven possesses a specific nest-building behavior, but it must learn what to use as nesting material. The bird tries everything—broken glass, cans, twigs, pieces of ice, and so on. These objects are pushed into place with sideways, trembling movements over the base at the prospective nest site. The frequency of trembling increases when the object meets with some resistance. If the object becomes wedged or caught the raven stops. Glass and cans do not work as well as twigs, so the raven quickly learns which to choose. The night heron, on the other hand, innately knows the right kind of nesting material, but must learn the best place for building a nest (K. Lorenz, 1954b).

Frequently an orienting component of behavior is learned. Polecats (*Putorius putorius*) and other Mustelidae kill rodents (such as rats) that are quite able to defend themselves by grasping them by the neck (Fig. 13-10). They learn this orientation of the killing bite toward the neck of the prey by trial and error. Isolated polecats, which have never killed prey, attack a rat when it runs away, but they bite it in any part of the body they can get hold of. If the rat defends itself the polecat lets go and tries to grab again. They learn very quickly how the prey must be grasped by the neck so that it cannot bite back. Polecats that have had an opportunity to play with litter mates learn this much faster (I. Eibl-Eibesfeldt, 1955e, 1963). The polecat also learns to recognize its prey. At first it pursues any fleeing animal. A quietly sitting rat is approached and

Figure 13-10 Female polecat killing a rat (neck bite). (Photograph: H. Sielmann from Scientific Film C697, Göttingen.)

sniffed inquisitively. If the rat runs toward the polecat, the latter runs away. Only after having killed a rat does the polecat recognize it as prey, whether it stays still or approaches. When one raises **polecats** from early life with **rats,** they are clearly inhibited from biting them; they accept the member of the other species as a **social companion.** In all other respects, however, the repertoire of prey-catching behavior remains unchanged; only different objects are able to activate them. Z. Y Kuo (1967) observed the same in cats that he raised with rats. However, he interprets this observation by saying that undoubtedly no prey-catching instinct exists, or it would not be possible to change the "nature" of the cat in such an obvious way. He concludes that the body structure alone explains why a cat behaves like a cat and that it is not necessary to postulate additional instincts in the form of structures in the central nervous system: "The behavior of an organism is a passive affair. How an animal or man will behave in a given situation depends on how it has been brought up and how it is stimulated" (Z. Y. Kuo, 1932:37). However, this can be seen as only partially true if one observes carefully. Even the cats that were friendly to rats could still capture prey, according to Kuo's account. Their behavioral repertoire was changed apparently as little as that of my polecats. The only thing that was changed was the object that released the behavior.

During mating polecat males grasp the females by the neck region, which immobilizes them. Animals raised in isolation grasp the female anywhere and have to learn as a result of the female's defensive behavior to grasp her at the right place. Polecats that were raised with siblings until they were two months old and were isolated after that time grasped the females correctly. They learned during play fighting that a conspecific will remain still when grasped by the neck (Fig. 13-11). The behavior patterns of mating itself are identical in both experienced and inexperienced animals (I. Eibl-Eibesfeldt, 1963). R. R. Maclennan and E. D. Bailey (1972) report comparable findings in the mink. F. A. Beach (1958) and K. Larsson (1959) found that inexperienced male rats copulated as successfully as experienced ones. This is also true for guinea pigs (*Cavia*), hamsters (*Cricetus*), golden hamsters (*Mesocricetus*), and rats (*Rattus*). The behavior patterns of mating are innate, but inexperienced animals often mount incorrectly oriented and, for example, clasp the head of the female (F. A. Beach 1942; I. Eibl-Eibesfeldt, 1953a, 1963; E. S. Valenstein, W. Riss, and W. C. Young, 1955; F. Dieterlen, 1959).

Rhesus monkeys raised in isolation are excited by females in estrus and try to copulate with them but they do not mount them correctly (Fig. 13-12). They are unable to learn this later in their lives (H. F. Harlow and M. K. Harlow, 1962a, 1962b; W. A. Mason, 1965), in contrast to hand-raised chimpanzee males that can achieve intromission with the aid of experienced females who help them to overcome initial difficulties. Here social isolation has less permanent effects, perhaps because normal mating behavior is more variable to begin with than in the lower monkeys (R. M. Yerkes and J. H. Elder, 1936).

During normal ontogenesis more than learning and maturation with respect to motor performance is observed. Innate releasing mechanisms mature as well and they become increasingly more selective through learning, as I have shown by the example of the **prey-catching behavior of toads and clawed frogs.** New releasing mechanisms can be acquired by learning, as is discussed by W. M. Schleidt (1962).

From all the examples presented so far, it is clear that various

Figure 13-11 *Left:* Polecats mating. The male holds the female by the neck. *Right:* Polecats wrestling playfully; one holds the other by the neck. During play the animals learn the neck bite, which later becomes a normal part of the mating behavior. (Photographs: *Left:* H. Sielmann, Scientific Film C697, Göttingen; *right:* I. Eibl-Eibesfeldt.)

species of animals are equipped with various innate dispositions to learn. These innate capacities to learn will be discussed in more detail in Chapter 14. Learning is not the result of the passive reception of stimuli by an organism. All observations support the view that there are often quite specific learning dispositions and internal motivating mechanisms, the latter being expressed in **curiosity** and **play** behavior.

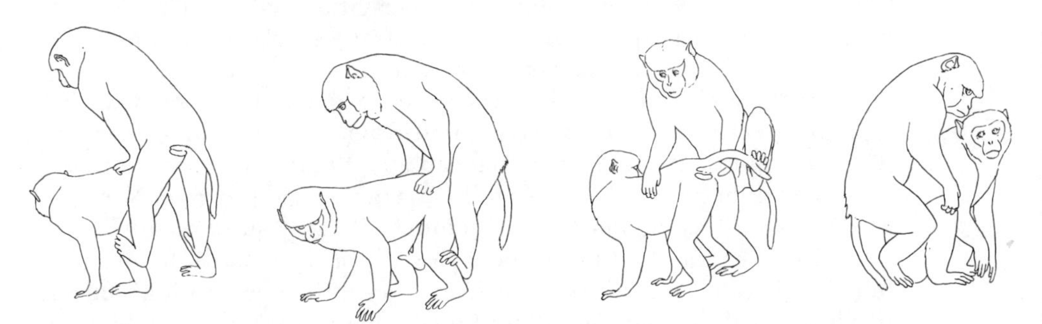

Figure 13-12 The sexual behavior of male rhesus monkeys who (*left*) grew up normally with others; and (*right*) who had no such social experience. These isolated animals become excited in the presence of a female in estrus and try to mount her, but intromission fails because they do not hold themselves properly on the hind legs of the female. (After W. A. Mason, 1965.)

INNATE DISPOSITIONS TO LEARN

Species-specific Learning Capacities

Different species of animals possess different learning capacities. In many species the behavior is largely determined by phylogenetic adaptations in the form of fixed action patterns and innate releasing mechanisms, and very little is left to learning. Animals that are so equipped have the advantage that they do not have to acquire adaptations through time-consuming, risky learning experiences. The Yucca moth (*Pronuba yuccasella*) "knows" as a result of its phylogenetic construction that it must collect pollen prior to laying its eggs in the seeds of the yucca plant and rub the pollen onto the stigma of the flower. Only in this way will the seeds develop on which the larva will feed. Adaptations of this nature are of advantage when the environmental conditions to which these adaptations are adjusted do not change appreciably. The more variable the environment is, the less precisely can the behavior be adjusted in advance. Changing environmental conditions require individual capacities for adjustment. Stenotopic forms can afford to "run on tracks" like trains, but eurytopic forms are specialized for adaptive modifiability of behavior; they are the "specialists in nonspecialization" as K. Lorenz (1959) observed appropriately.

When adaptations of behavior are precisely determined, any

change may be detrimental. J. H. Fabré showed this long ago in many experiments with insects. The digger wasp (*Ammophila*) opens and inspects the cavity it has dug before it deposits caterpillars in it to serve as food for its larva. It arrives with the caterpillar, drops it near the entrance, enters the cavity, inspects it, reappears head first, and pulls the caterpillar inside. If one removes the caterpillar to a place some distance from the nest while the wasp inspects the cavity, it will search for it until it has found the caterpillar, bring it back to the entrance, and the entire sequence of dropping, inspecting, and so on is repeated. This can be repeated 30 to 40 times, at which time the wasp will finally carry the caterpillar directly into the cavity without prior dropping and inspection (G. P. Baerends, 1941). The animal can adapt to the new situation only with great difficulty; its behavior follows a quite rigid program. Normally no disruptions occur, so the wasp achieves its goal quite readily.

However, this rigidity of behavior is not necessarily found in all functional cycles. Just as **squirrels** learn very little when **hiding food** and much when they **open nuts,** there are behavior systems in which the digger wasp can perform astounding learning tasks. For example, it learns the way home with its prey on its flight away from the nest. These **digger wasps** take care of several nests at the same time during the phases of caring for the brood; they supply each of their larvae, which differ developmentally from each other, with the appropriate amount of food. What the digger wasp will do at a particular nest throughout the day is decided during the first inspection visit in the morning. Before beginning with hunting activities, the wasp visits all nests not yet permanently sealed and checks contents. G. P. Baerends (1941) was able to induce wasps to bring more caterpillars than they normally would have by removing caterpillars from a particular nest. When he added caterpillars the wasp would bring less food. These manipulations influenced the behavior only when they were performed before the morning inspection by the wasp. Later manipulations were unsuccessful. This means that the wasp's behavior is determined for the rest of the day at the time of the morning inspection, and that the wasp is able to remember the condition of up to 15 nests for the entire day.

This is a memory task that could not be detected with the usual methods of delayed-response experiments. In this method, which was originally developed by W. S. Hunter (1913) and is still widely used today, an animal is trained in a multiple-choice apparatus to select one of several doors that are marked by specific stimuli (for example, a light that is turned on briefly). Once the animal has mastered the basic task, it is prevented from responding while the positive stimulus is presented. The animal is allowed to respond only after various intervals of time, and the longest possible delay is considered a measure of the memory capacity of an animal. N. R. F. Maier and T. C. Schneirla (1935) questioned the value of this method, according to which a gorilla capable of 48-hour delays would have a memory 576 times as good as an orangutan who mastered only a 5-minute delay. N. Tinbergen (1951), who cited this example, agrees with this criticism.

These observations teach us that standardized learning methods are not always appropriate for the comparative study of **learning capacities** in many species. If one wants to test the learning capacities of a heron, a rat, and a frog, one would be ill advised to run each of these animals through a maze. The rat, which normally lives in burrows, will do better than the heron or the frog. If we tèst the frog instead on bad-tasting prey models, we will discover a very rapid learning capacity.

There exist species-specific learning capacities for various function. Predators are very intelligent in the functional cycles dealing with capture of prey, and animals that live in a specific area of their habitat are very good at learning paths. Some social lemurs show great social intelligence (A. Jolly, 1966), which is in marked contrast to their otherwise lower level of intelligence. If one wants to obtain knowledge about the degree of adaptive modifiability of a species, one should first observe the animals under natural conditions. As stated earlier, learning capacities are adapted to the demands of the ecological niche of the species and its other inhabitants.

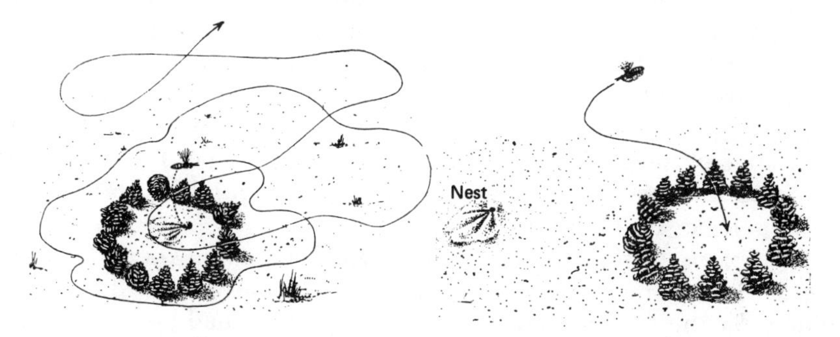

Figure 13-13 *Left:* After leaving the nest the digger wasp makes an orienting flight and learns the location of the nest entrance; *right:* after the circle of pine cones has been moved from around the entrance the returning digger wasp searches incorrectly in the center of the circle, now in a new place. (From N. Tinbergen, 1951.)

Most species whose members live in a specific place or repeatedly visit a landmark acquire knowledge of their surroundings. The limpet *Patella* follows its own mucous track back to its original resting place like a tracked vehicle (W. Funke, 1965). The bee-hunting digger wasp (*Philanthus triangulum*) remembers the landmarks surrounding its nest and is thus enabled to find its way back. If one surrounds the nest with a circle of pine cones or stones while the wasp is in the nest and then transposes this configuration by one meter after the wasp has flown away, it will search upon return within the circle and not find the nest entrance (N. Tinbergen and W. Kruyt, 1938; I. A. Chmurzynski, 1967; see also Fig. 13-13).

Bees remember the location of good foraging grounds for a long time. Continuously dancing honeybees reproduce from memory the distance and angle to the sun, making allowance for changes in the sun's position, although they have not visited the place whose position they indicate with their dances for days. M. Lindauer (1963) cited the case of a continuously dancing bee, which had visited a food source for the last time on November 1, 1959, and which indicated the position correctly in the closed hive on December 8. When such continuously dancing bees had visited two food sources at different times of the day, they indicated within the closed hive the position of the feeding place whose visiting time was closest to the time of dancing (M. Lindauer, 1957).

A young mammal which leaves its den for the first time begins by learning the return path to the nest. If the marine gobi (*Bathygobius soporator*), which lives in the tidal zone, is surprised in a tidal pool, it will jump out of the pool in the direction of the sea, from one pool to the next, until it reaches the sea. They cover distances of 10 meters, using up to 11 pools. If placed into an aquarium the fish will not jump, but they will jump if returned to the tidal pool, even after a stay in captivity of 14 days. If changes have been made in the environment, such as emptying a pool, they jump into the dry pool. They learn the path during high tide, as was demonstrated in elegant experiments by L. R. Aronson (1951). The path can be learned in darkness and may be remembered for 40 days. Equally extraordinary learning performances are shown by animals that return to their spawning grounds (for example, **salmon**).

Many species learn while obtaining food, both with respect to the knowledge of the object and the manner of handling it. I will add some examples to those already cited (**nut opening of squirrels; prey catching in the polecat; in the frog**).

Snapping turtles (*Chelydra serpentina*) which had been fed for 12 days after hatching with worms or fish, subsequently showed clear preferences for the specific food, fish or worms, they had been fed (G. M. Burghardt and E. H. Hess, 1966). Many mammals and some birds learn from the mother what to eat, and in this way traditions become established that are restricted to groups of animals in a specific locality. Mother squirrels, rats, and other rodents tolerate the seizing of food from their mouths by their young (I. Eibl-Eibesfeldt 1958a; see also Fig. 13-14) and so do the parents of marmosets (*Callithrix*). Wood pigeons must learn that acorns can be eaten. The picking-off movement is innate, but they do not attempt to eat large fruit without the presence of adult pigeons they have observed eating acorns (O. Heinroth and M. Heinroth, 1928). Rats quickly learn as a result of adversive experiences to avoid poisonous bait. If new bait is offered, at first only a few animals of the group will try sublethal amounts of it. Usually they are not poisoned but merely become ill, and the bait is avoided thereafter. It is interesting that other members of the group also avoid the bait, taking their cues from those who have had the bad experiences. In this way local traditions come about, so that specific types of bait are rejected in certain districts of a city for several generations. On the North Sea Island of Norderoog, Norway rats specialize in catching birds, which mainland rats do not normally do. They overcame their innate fear of fluttering birds and learned to creep up to them and ambush them (F. Steiniger, 1950).

Lions at Lake Manyara (Tanzania) and at Ishasha in Queen Elizabeth National Park, Uganda, climb trees. This is primarily a female activity; males do not climb nearly as much. It has been said that this activity is related to avoidance of biting flies at Manyara and to escape the heat and flies at Ishasha. Female meerkats (*Suricata*) actually present food to their young, which snap it out of their mouths. In this way they learn to eat the same foods eaten by the mother and thus develop food preferences. The young of one litter did not accept bananas which the caretaker offered them but snapped at them when presented by their mother in her mouth. From this moment on, they accepted bananas readily (R.

Figure 13-14 Young Norway rat (albino), pulling a piece of food from the mother's mouth. (Photograph: I. Eibl-Eibesfeldt.)

F. Ewer, 1963). Teaching seems to occur in the behavior of the moutain gorilla. G. B. Schaller (1963) observed a female removing an inedible hagenia leaf from the mouth of her young. Another animal helped a juvenile dig up a root. The gorilla mother Achilla in the Basel Zoo also taught her son to move by retreating gradually. When he was older she placed his hand on the cage bars so that he could hold on. She also enticed him actively to climb the cage bars and watched over him. By pulling him in the small of his back, she stopped him from doing certain things. She was definitely interested in various activities of her child and either encouraged or limited the direction in which they could develop (R. Schenkel, 1964).

The origin and spread of new feeding habits through groups of Japanese macaques (*Macaca fuscata*) are well documented. A troop of these monkeys on Kôshima Island was regularly fed with sweet potatoes beginning in 1952. In 1953 Imo, a 1½-year-old female, was first seen washing the potatoes at the edge of a freshwater brook. She held the potato in one hand and cleaned off the sand in the water with the other.

Figure 13-15 Japanese macaques of Koshima Island washing sweet potatoes. On the right the young is observing the mother. (Photographs: Japanese Monkey Center, Aichi.)

This new habit spread in the group during the course of the next several years, at first within the closer families and within the groups of playmates. Later it was passed on from mother to her children. In 1962 three quarters of all monkeys 2 years and older were washing potatoes (Fig. 13-15). At first the monkeys washed their potatoes only in fresh water. Later they also used seawater, which some monkeys seemed to prefer because of its salty taste. They gradually began to salt their potatoes by repeatedly dipping them into the water before the next bite.

At the same time these animals were fed wheat spread along the shoreline in the same area. At first the monkeys carefully picked up each individual kernel until the same female, Imo, who had invented the potato washing and who was now 4 years old, began to gather the wheat with the sand and throw it all into the water, where the sand quickly separated from the lighter wheat. By the time of the last report 19 of 49 monkeys were using this invention (J. Itani, 1958; S. Kawamura, 1963; M. Kawai, 1965). The habit of swimming in the sea and begging with specific postures developed as patterns specific for the group and were passed on by tradition. Near Kyoto the macaques learned to keep warm at an open fire after the fashion of their keepers. In 1958 a female began joining the keepers at a fireplace and now all have the same habit. Group-specific habits are known from free-living macaque populations. The monkeys of Mt. Takasaki in Kyushu eat the fruits of the Aphananthe trees but spit out the kernels. The monkeys of Mt. Arashi, however, crush them to get the seedling. The monkeys of Mt. Minoo feed on eggs; those of Shodoshima do not eat eggs. Even social behavior patterns are influenced by traditions. High-ranking males of the Takasaki group carry young monkeys around, which are already weaned but still need protection. This is not observed in other groups (D. Miyadi, 1965, 1967). It is only necessary to show a chimpanzee once how to operate a push-button water faucet (R. M. Yerkes, 1948). This behavior is then passed on by social tradition to others.

Knowledge of possible danger is also passed on by tradition. K. R. L. Hall (1965) frightened a monkey (*Erythrocebus*) mother by popping open a box containing a snake. Her young were present but could not see what was in the box. The mother was startled when she saw the snake and her response was observed by the young; from then on they avoided this box, although they had previously opened it frequently and without hesitation.

P. H. Klopfer (1957) trained ducks with electric punishing stimuli to avoid a bowl filled with water. Other ducks that were in the same cage and observed the procedure avoided the bowl without being punished themselves. Other examples of learning through observation can be found in P. H. Klopfer (1962).

Many animals learn about matters related to social behavior. Birds learn their songs, for example. Gregarious animals often learn to know their group members personally and form attachments or bonds with other individuals. They also learn their **place** within the **rank order** of the group and often attachment to a particular locality which they defend as their territory. J. P. Scott (1963) in this context spoke of a process of "primary socialization," which serves to organize and control agonistic behavior.

Such habits are often tenaciously adhered to once they have been formed, and this is advantageous to a certain degree, because in this way "tried-and-tested" behavior persists. Retention of such a habit is probably facilitated by the fact that each deviation from it is accompanied by feelings of displeasure and fear. K. Lorenz (1963a) cites some interesting examples in this respect. One of his graylag geese, which lived in his room, had become used to a certain detour. At first she always walked past the bottom of the staircase toward a window in the hallway before returning to the steps, which she then ascended to get into the room on the upper floor. Gradually she shortened this detour, but persisted in initially orienting toward the window, without, however, going all the way to it. Instead she turned at a 90° angle once she was parallel with the stairs. Once Lorenz forgot to let the goose into the house at the usual time. It was beginning to get dark and the goose ran, contrary to her usual habit, directly toward the staircase as soon as the door was opened and began to climb up:

> Upon this something shattering happened: Arrived at the fifth step, she suddenly stopped, made a long neck, in geese a sign of fear, and spread her wings as for flight. Then she uttered a warning cry and very nearly took off. Now she hesitated a moment, turned around, ran hurriedly down the five steps and set forth resolutely, like someone on a very important mission, on

> her original path to the window and back. This time she mounted the steps according to her former custom from the left side. On the fifth step she stopped again, looked around, shook herself, and performed a greeting display behavior regularly seen in graylags when anxious tension has given place to relief. I hardly believed my eyes To me there is no doubt about the interpretation of this occurrence: The habit had become a custom which the goose could not break without being stricken by fear (p. 112).

Margaret Altmann (cited by K. Lorenz) was obliged to unload and reload her old horse symbolically at a location where they had previously camped on several occasions. Unless this was done the horse would not continue. A dog belonging to an acquaintance once scared up a mouse while he accompanied his master to the shed to fetch coals. Since that time the dog performs a "mouse jump" each time he comes to the shed. At another time he had seen a rabbit running ahead in the car's headlight. Since then he always chases ahead wildly when the car returns at night. The force of habit is no less strong in man.

These examples should suffice to illustrate the learning dispositions of different animal species. The basic assumption underlying all classical learning theories that all responses are about equally conditionable to all stimuli is no longer tenable. Further support for this view is presented in K. Breland and M. Breland (1966) and J. Garcia et al. (1969). One of these innate learning dispositions is illustrated by the imprinting phenomenon, which occurs under special conditions and is most likely the result of specific learning mechanisms. It will be discussed in the following sections.

Imprinting and Imprinting-like Learning Processes

The course of learning may be determined in various ways by phylogenetic adaptations. We learned earlier that chaffinches in a choice situation prefer the song of their own species, the preference being based on innate knowledge as to the type of song to be learned. Chimpanzees know innately that threats are made by producing noise, but they must learn the method of producing it.

It is somewhat different in the zebra finch, which learns its song from those who feed it. If a society finch feeds zebra finch young, they will learn the society finch song, although zebra finches are singing in the adjacent cage. If they are fed by both society finches and zebra finches, they will learn the song of the zebra finch. Thus a preference for the song of the species as the model becomes evident even here (K. Immelmann, 1967).

In many cases learning is genetically programmed so that the animals are able to learn in specific sensitive periods and possess specific learning capacities only at this time. If this period is restricted to a developmental period after which the animal can no longer learn, we speak of a critical period. Such periods come to an end, even when no learning has taken place, as the result of internal changes. Zebra finches that have been isolated from species members before they are 35 days old are unable to distinguish males and females of their own species. They are chased by their own father when they are between 35 and 38 days old, and from then on males become the object of their aggressions. If they have not had this particular learning experience, they are unable to learn it at a later date (K. Immelmann, personal communication).

After an animal has learned something during a sensitive or critical period of its life, the readiness to learn ceases, and in general the animal firmly adheres to that which it has learned. Many examples are known of animals learning particular details of an object of a fixed action pattern during a specific developmental period. After this they seem to be fixed on the particular object in respect to the particular drive (object imprinting). In later research on specific sensitive periods for learning, motor patterns were discovered. These phenomena will be discussed in the following sections. For reviews, see W. Sluckin (1965) and P. P. B. Bateson (1966).

Object Imprinting

Many innate behavior patterns can be released by unspecific key stimuli. Rhythmic calls and the most diverse moving objects release the following reaction in a young graylag gosling shortly after hatching. It follows a man as readily as it would a goose or a moving box. If it follows such an object even for a short time, it will remain with it. Once a gosling follows a person it later cannot be induced to follow its own mother (K. Lorenz, 1935). In respect to the following reaction it has become imprinted to man. The same is true for chicks and ducklings. The reaction to auditory stimuli, while modifiable, tends to be more selective. A number of species clearly prefer the call notes of their own species (G. Gottlieb, 1965a, 1965b). This innate preference reduces the chance of imprinting to the wrong object.

The statement that an animal is imprinted on something always refers to a specific reaction determined by the releasing stimulus situation. In the example just cited this reaction was following. Cichlids frequently distinguish their own young from those of other species and eat the latter. *Hemichromis bimaculatus* pairs prefer young of another species if they were given eggs of that species instead of their own during their first breeding cycle (A. A. Myrberg, 1964). In this case the reactions involved in parental care become imprinted to an object. This is also true for behavior patterns of other functional cycles, such as those of reproduction and food getting. A jackdaw that has been raised by human hand from the nestling stage will join a flock of jackdaws when becoming fledged if an opportunity is available. In the following year, however, it will court humans during the reproductive season, even if other jackdaws are available. In respect to its sexual reactions the bird is imprinted to man and prefers him over its own species (K. Lorenz, 1931). This is all the more remarkable when you consider that this imprinting appears to take place at a time when the animal does not yet show sexual behavior (Fig. 13-16).

The fixed action patterns themselves do not undergo any noticeable changes in imprinted animals. A human-imprinted ring dove (*Streptopelia risoria*) courts the human hand with the same behavior patterns with which he would normally have courted a female conspecific, and human-imprinted female doves invite courtship feeding from the human hand and squat before it in the copulation posture (E. Klinghammer, 1967).

Figure 13-16 Examples of sexually imprinted animals: *Left:* The rooster, which was imprinted on mallard ducks, waded regularly into the water to join them. He did this most frequently in this duck cage because it was possible for him to approach the ducks very closely. Otherwise they would usually swim away when he approached. *Right:* Even when 7 years old these three male wood ducks still behaved homosexually, although numerous females were present. They inspect a nesting box in search of a nest site. (Photographs: F. Schutz, 1965a, 1965b.)

K. Lorenz (1935) discovered the phenomenon of imprinting and emphasized several criteria that distinguish this learning process from normal association learning. E. H. Hess (1959, 1973) has further clarified these differences.

1. Imprinting takes place only during a specific *sensitive* period. If this time passes the animal can no longer be imprinted. This sensitive period is not necessarily restricted to the first few days or weeks of life. The time and duration of this period can vary depending on the reaction even within the same animal. For the following reaction the sensitive period is between 13 and 16 hours after hatching. E. H. Hess (1959) determined this by imprinting ducklings in a special apparatus after keeping them in darkness for varying periods of time from 1 to 35 hours after hatching (Fig. 13-17). For 1 hour they were allowed to follow the model of a mallard male duck, which called by means of a built-in loud speaker. Then they were returned to darkness. For the test they were once again placed into the apparatus, where they could choose between a male and a female model. Both models were at first silent; after 1

Figure 13-17 Imprinting experiments of E. H. Hess. The experimental situation: Chicks or ducklings that were maintained in a dark box from 1 to 35 hours after hatching until they were to be used in the experiment were introduced onto the circular runway. Here a male mallard model with a loudspeaker inside was moved in a circle around the runway. Each animal remained for 1 hour with the model. After a certain time interval, during which they were kept in the dark, they were tested as described in the text. (From E. H. Hess, 1959.)

minute both began calling, the male model with an artificial "gock, gock, gock, gock" call, the female model with the recorded natural call of a mother duck. During the third phase of the test the female model alone called, and in the fourth test the female model also moved while the male model remained stationary. If a duckling that was imprinted on the male model approached the male model in all four tests the imprinting was scored as 100 percent. The results can be seen in the graphs (Fig. 13-18). The strength of imprinting increases proportionally to the distance followed. This could account for the increase and decrease of imprintability with age. Initially the young animal is too weak to follow. The older it becomes, the better it is able

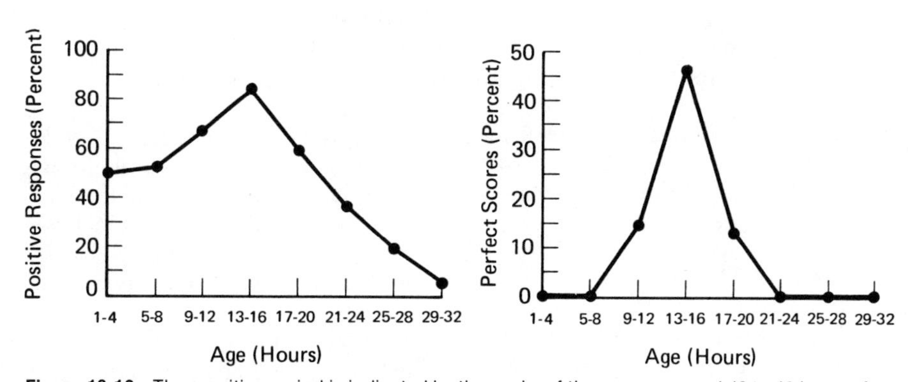

Figure 13-18 The sensitive period is indicated by the peaks of the curves around 13 to 16 hours after hatching. *Left:* Average of positive responses; *right:* percentage of animals that had errorless scores in all four test situations. (From E. H. Hess, 1959.)

to do so. The decrease of imprintability could be due to maturing avoidance responses (E. Fabricus, 1951; E. H. Hess, 1959).

2. The acquired knowledge of the releasing object is retained for life, while forgetting is normally common after learning. What is learned is not only retained but the imprinting object is also preferred during the rest of the animal's life; imprinting is then *irreversible*. It is possible to induce two hand-raised parakeets to mate and incubate eggs. But if they are again exposed to man both will court the human, the pair bond is broken, and the brood is neglected (G. Hellmann, cited by K. Lorenz, 1954). F. Schutz (1965a, 1965b) raised male ducks of various species with chickens, ducks, and males of other duck species and confined them all together for several weeks. Then they were released onto a lake on which conspecifics they had not seen until then were also present. Now they had a choice. In the following spring most of the males courted members of the species with which they had been raised, but normally not those individuals with which they were raised. In numerous instances the effects of imprinting were shown after the animals had mated with females of their own species and had lost them through accidental death or because of separation by the experimenter. The imprinting effects of early experience in such cases showed up even in males that were normally paired to females for a period of two years (F. Schutz, 1968). This shows quite clearly the difference between imprinting and association learning by means of the usual reinforcement contingencies. In additional experiments F. Schutz (1965b, 1968) imprinted male ducks on other males of the same species and then induced them to mate with females of their own species. The imprinted homosexual preference remained, however. W. M. Schein (1963) imprinted three male turkeys on people and three to conspecifics. All were equally tame. In the absence of other turkeys they all courted people, and they courted turkeys when no humans were present and mated with them. However, when humans and turkeys were present at the same time the human-imprinted males courted humans, and turkey-imprinted turkeys courted their own kind. The turkeys still showed these preferences when they were five years old. K. Immelmann (1966) had male zebra finches (*Taeniopygia guttata castanotis*) raised by female society finches (*Lonchura striata* var. *domestica*). All males courted only society finches. Then each male was kept isolated with zebra finch females. They courted, bred and raised young. When their preference was tested later in the free-choice situation, they preferred society finches to conspecific females.
3. During imprinting the animal learns only supraindividual species-specific characteristics. A male mallard duck that was imprinted to a sheldrake courted all sheldrakes. A human-imprinted graylag goose follows all humans.
4. Only specific reactions become imprinted to a particular object. A jackdaw raised by K. Lorenz (1935) regarded humans as parents and sexual companions.[1] The bird flew about with hooded crows as flight companions and accepted young jackdaws as child companions.
5. The determination of the object for an instinctive activity can take place, as in the case of sexual imprinting cited above, at a time when the appropriate

[1] "Companion" (*Kumpan*) is a partner in a specific functional cycle.

behavior pattern has not yet matured and thus has not yet been performed by the animal.

6. During imprinting of the following reaction of ducks, there was a greater effect when the imprinting experience of following a model took place in massed trials than in distributed trials, which are usually more effective for association learning. Painful stimuli strengthen imprinting, whereas punishment usually results in the avoidance of the associated stimulus (E. H. Hess, 1959).

Of the criteria for imprinting cited above some—such as the sensitive period and the learning of supraindividual characteristics—also pertain to other forms of learning, which are by no means present in all imprinting-like learning processes. Imprinting is principally recognized by the criteria listed under the second point, especially those pointed out by E. H. Hess: primacy of the imprinting experience is more important than recency, while the reverse is true for normal associative learning. From this fact derives the observation that imprinting is irreversible. The work of Blakemore et al. (1970, 1973) on the modifiability of visual fields in striate cortex neurones also shows the first two features, those of a specific sensitive period and of irreversibility.

The assertion made in point 5, that imprinting of an object can take place for an action that is not manifested until after later maturation, cannot be regarded as proved. In the example of sexual imprinting it is possible that the initial orientation of a young animal to an object actually constitutes the first link in a chain of sexual behavior patterns. Such an association could be rewarding in itself while being quite independent of the actual mating behavior. The motivation of such an initial orientation to another must be studied before we can make statements about the presence or absence of sexually motivated behavior in young animals.

As is the case with all "injunctive" concepts (B. Hassenstein, 1955), it is not always possible to draw a sharp distinction between imprinting and other learning processes. Transitions to normal association learning can be expected (E. Klinghammer, 1967).

Object imprinting takes place especially when the object of an instinctive action shows a lack of releasing signals. Female ducks of different species are often very similar to one another, while the males have very conspicuous releasers (green head and white neck ring in the mallard). Thus the imprintability of the various sexes is also different. The male **mallard** must learn to recognize the plain-looking conspecific female, and this knowledge is acquired during the imprinting experience. If raised with other species it will accept them as sexual companions. The female, on the other hand, cannot be imprinted with respect to her sexual behavior. She knows innately the releasing, species-specific signals of the male. Even if she is raised with other species she will only court males of her own species when adult (K. Lorenz, 1935; F. Schutz, 1964).

Imprinting on sexual objects creates barriers to crossbreeding between closely related species. In the case of species that undergo rapid development, this phenomenon is a relatively important one. Since more characteristics are involved in learned object recognition than in an IRM, there is in the case of the former less chance for confusion between individuals belonging to related species with a high degree of resemblance. This applies, for example, to African and Australian Estrillidae (see K. Immelmann, 1970, who also offers further references on isolation mechanisms based on imprinting.).

In a study entitled "Mother Love: What Turns It On?" P. H. Klopfer (1971) drew attention to the interesting fact that mother goats establish the individual bond with their young only during a five-minute period after giving birth. If the young are removed immediately after birth and returned to the mothers two hours later, the mothers attack them. If the young are allowed to stay with their mothers for five minutes after birth and then removed for two hours, the mothers accept them, but reject any strange young. P. H. Klopfer gave an interesting hint as to a possible physiological mechanism responsible for this short sensitive period. He points out that according to S. J. Folley and G. S. Knaggs (1965) the oxytocin level in the blood of the goat is high, but drops within a five-minute period to a low level. The release of the hormone is induced by the dilatation of the cervix. Manual dilatation will also cause oxytocin release (G. Peeters et al., 1965). P. H. Klopfer's model suggests that the cervical dilatation accompanying birth induces the release of a substance with a time course similar to that of oxytocin, which may either stimulate specific central ganglia or alter the sensitivity of peripheral receptors so as to make the animal react to stimuli to which it is normally unresponsive.

Imprinting of Motor Patterns

Imprinting-like learning processes also exist for the acquisition of motor patterns. Chaffinches learn their song only during the first 13 months of their lives. Toward the end of their sensitive period their learning capacity increases for several weeks. It is sufficient, however, that they are exposed to the song at a time when they do not as yet sing themselves. Chaffinches that were isolated from their parents during the month of September, long before they had begun to sing, produced a normal song during the following spring. What they had heard was not forgotten (W. H. Thorpe, 1958a). O. Heinroth (cited by K. Lorenz, 1954b) once recorded the song of blackcaps (*Sylvia atricapilla*). In the same room he kept 12-day-old nightingales (*Luscinia megarhynchos*) which only uttered the begging call at that time. The total time these birds were exposed to the blackcaps was about one week. When they began to sing the following spring they surprised Heinroth with the complete blackcap song, which was identical with the recorded song. The white-crowned sparrow (*Zonotrichia leucophrys*) possesses differing local song

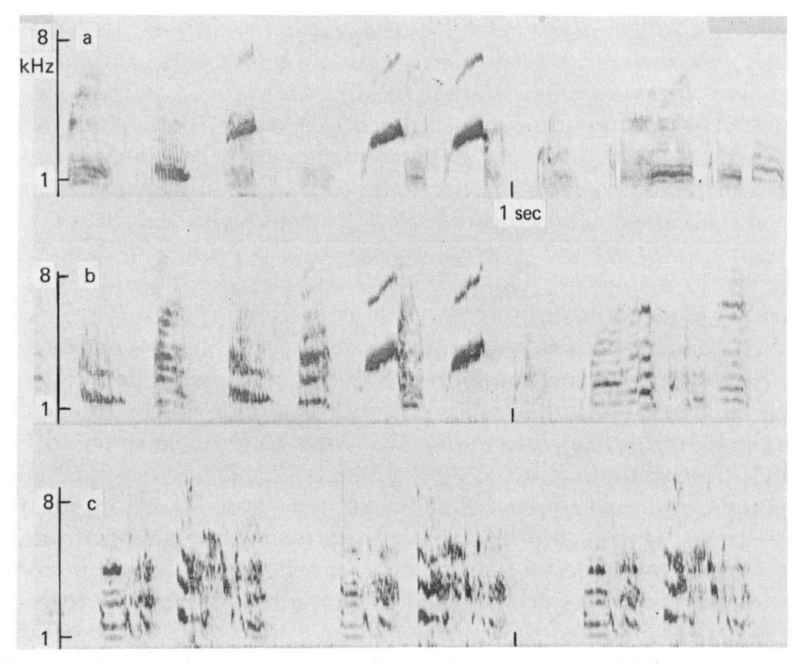

Figure 13-19 Sound spectrograms: (*a*) of a society finch; (*b*), of a zebra finch raised by (*a*); (*c*), of a normally raised zebra finch [natural father of (*B*)]. Abscissa: time in seconds; ordinate, frequency range, in cycles per second. (From K. Immelmann, 1965.)

dialects that are learned during a sensitive learning period. Several months separate this critical period and the time when the birds begin to sing. The birds therefore must have retained what they had heard (P. R. Marler and M. Tamura, 1964). Zebra finch males that have been raised by society finches learn the song of their foster father. This acquisition is completed by the time they begin to sing on their own. When a zebra finch has heard the song of a society finch for the first 35 days of its life, and from then on only hears the song of zebra finches, it will always sing like a society finch (Fig. 13-19). The females, too, remember what they have heard when they were young, although they do not normally sing. If, on the other hand, they are given male hormones they will sing the song of the species on which they have become imprinted (K. Immelmann, 1965). According to M. Konishi (1965b) the same is true for the white-crowned sparrow. In motor "imprinting" apparently the relevant information is acquired at a time when the behavior that is learned, the song, does not even exist in rudimentary form.

Imprinting-like Processes in Mammals

Only a few observations have been made about imprinting in mammals (U. Grabowski, 1941; G. Grzimek, 1949b). Hoofed animals such as horses and sheep can be imprinted to man in respect to their following reaction in the same way that graylag geese can, if one assumes the parent role shortly after birth. Dogs pass through a critical period with respect to the development of social relationships during weeks 4 and 6. During this time they form a close social bond to conspecifics or to man as a substitute, regardless of whether they are punished, fed, or treated indifferently. J. P. Scott and J. L. Fuller (1965) have emphasized that the internal process on which this readiness for contact is based seems to be more important than the external factors. It is possible that imprinting exists also in man (P. H. Gray, 1958; W. H. Thorpe, 1961a), but as yet there is no experimental investigation.

Psychoanalysts were perhaps the first[2] to recognize the existence of sensitive periods during the course of individual human development. During the early childhood years certain environmental influences can affect the behavior of a child in decisive ways and can cause possibly irreversible disruptions. The well-being of an infant is not dependent only on hygiene; personal contact provides an important stimulus for further development. During the first year of life a short separation from the mother can result in serious retardation. A separation of several months can often lead to irreparable damage, and infant mortality is increased in such instances (W. Goldfarb, 1943; R. A. Spitz, 1945, 1946, 1951; J. Bowlby, 1952, 1969; A. Dührssen 1960). It is especially during the second half of the first year of its life that the child develops a personal bond with its own or a foster mother. This bond is the basis for the development of "original trust" (E. H. Erikson, 1953), the basic attitude toward one's self and the world. The child learns that it can depend on another person and this positive attitude is a cornerstone of a healthy personality. If this relationship is disrupted, a basic attitude of distrust develops.

Such a development can also be caused by a prolonged stay in a hospital in the second half of the first year of life. The child will attempt even there to achieve close contact with a substitute mother, but no nurse can bestow enough intensive care on an infant for such a bond to be established. The nurses change so frequently that growing relationships are continually broken off. A child that has been repeatedly disappointed in its efforts to make lasting contacts becomes apathetic after a short rebellion. During the first month in the hospital it will cling to a caretaker and whine often. During the second month there is much crying and loss of weight. During the third month of separation it only whimpers a little and at last becomes completely apathetic. If such children return home after three or four months they recover, but after longer periods of separation the damage is irreversible. These children remain retarded in their development despite the best hygienic care and nutrition; no one plays with them and no one carries them around. Of 91

[2] M. Montessori, in her description of the human development, also employed the concept of a sensitive period: "One is here concerned with a special receptiveness which occurs during development, that is, during childhood. These periods are transitory and serve to acquire certain abilities" (M. Montessori, 1952:61).

children in one orphanage who were studied by R. A. Spitz (1945, 1965) and who were separated from their mothers from the third month of life, 34 had died by the end of their second year. The developmental index of the surviving ones was 45 percent of normal. The children functioned almost at the level of idiots. Even after four years many could not stand, walk, or speak. In this orphanage one nurse took care of ten children. In another home where the mothers were able to care for their children much of the time, none died and the children developed normally.

When these socially deprived children survive, they are frequently lacking in affect and avoid contact. They are capable only of superficial contacts and may completely shun all close relationships. Certainly these observations of R. A. Spitz need further examination, especially as to whether the retardation is primarily due to emotional factors and whether the damage is indeed irreversible. Contradictory evidence has been published by W. Dennis (1960).

J. Bowlby (1969) describes the effects of separation from the mother upon children aged 15 to 30 months. In an initial phase of protest the child appears acutely distressed at having lost his mother and tries to establish contact by crying. He behaves as if he expects the return of the mother any time and normally rejects other persons. This period may last for over a week. It is followed by the phase of despair during which the child is quieting down. He is withdrawn and inactive. The final phase of detachment is characterized by an increased interest in the environment. The child accepts nurses and is sociable, with, however, no strong emotional tie to anyone. When the mother visits him, the child seems hardly to know her. "He will cease to show feelings when his parents come and go on visiting day; and it may cause them pain when they realize that, although he has an avid interest in the presents they bring, he has little interest in them as special people. He will appear cheerful and adapted to his unusual situation and apparently easy and unafraid of anyone. But this sociability is superficial; he appears no longer to care for anyone" (J. Bowlby 1969:28). It seems that the human child is innately biased to attaching himself to one figure and if this attachment is prevented, serious disturbances of social behavior are the consequence.

In this connection the investigations of H. F. and M. K. Harlow (1962a, 1962b) deserve attention. They raised rhesus monkeys without their mothers. The animals had only surrogate mothers, which were either covered with terrycloth or consisted of bare wire mesh. Attached to these models were bottles with nipples, from which the baby monkeys could suck their milk. The monkeys that were raised in this way later proved to be poor mothers. They allowed their young to be removed without protest, did not nurse their young or only did so after some time, and even mistreated them (Fig. 13-20). Here, too, an early childhood experience led to substantial disruptions of later social behavior.

In human beings the case of hospitalism that was cited above is un-

Figure 13-20 *Above:* Normal rhesus monkey mothers care for their young even in captivity; *below:* rhesus monkeys that have grown up in isolation remain indifferent toward their own young; sometimes they are even rejecting or aggressive. (Photographs: Sponholz, Primate Laboratory, University of Wisconsin; H. F. Harlow and M. K. Harlow, 1962a, 1962b.)

doubtedly the crassest case of loss of love, but during the normal course of development there are various degrees of early childhood experiences of this nature that show their effects only much later. Infants are certainly adapted to a close contact with the mother, and children experience such contact among all peoples in which children are carried about. The child that is placed in a carriage or in bed by itself misses this feeling of security in our type of culture. We do not know what the consequences are for later development, but it is entirely feasible that our characteristically sober, detached, and critical attitudes, as well as many neuroses, have their root in these practices. M. Mead (1965) has attempted to interpret some idiosyncracies of various cultures in line with their differential early childhood experiences. However, with the current incomplete state of our knowledge, one must be careful not to equate a plausible interpretation with a causal explanation. M. Mead writes, for example, that the Mundugumur of New Guinea nurse their children only reluctantly, remove them abruptly from their breasts, allow them to cry, and in general treat them roughly. This, she feels, is the root of their later aggressiveness. The Arapesh of New Guinea, on the other hand, have a very close, warm, and permissive relationship with their children, and they are very docile as adults. Seen in this light the explanation seems to fit, but whether this different treatment of infants is actually the cause of later behavioral differences has not been proved. There exist a number of warlike tribes who nevertheless treat their children with great love and kindness. Included here are the Nilotic tribes, where the Masai are especially aggressive. They in no way mistreat their children; on the contrary, children are carried on the mother's body, spoiled, and cared for, by the fathers as well as the mothers.

Further sensitive periods can be found in the development of the human child. Between the ages of two and three years European children usually begin actively to explore and interact with their environment. If this activity is excessively suppressed and the child punished for its often destructive experimentation, and if too little guidance is given, then the impulses for independent actions soon become suppressed. Creative activity is impeded, and such children may lose the capacity to live according to their own perceptions and ideas, and may become disrupted in their initiative and work attitudes (A. Dührssen, 1960).

Around the fifth year of life the child experiences a critical period (the *oedipal stage*, S. Freud), which is important for later sexual life. The sexual drives begin to mature, and sex-specific behavior is determined. Members of the opposite sex become the object on which appropriate behavior directed to the other sex is practiced. A boy may seek physical contact with his mother during this period, crawl into bed with her, and be very tender. At the same time he identifies with his father, who is his model. If the mother is too rejecting, homosexual preferences may result. On the other hand, permissive or even seductive behavior of the mother may occasionally lead to behavior disorders. A boy may become strongly sexually attached to the mother, and this may lead to feelings of guilt related to the father. In this way at least the demonstrable sexual aberrations have been interpreted by psychoanalysts.

Girls seek contact with the father, who then becomes the object for practicing behavior toward male partners. At the same time they begin to assume the role of the mother. If the mother does not readily accept her female role or acts in a pronounced masculine way, identification with her is more difficult and female homosexuality can develop. A rejecting father can influence the behavior of a girl in the same homosexual direction.

The experiences of the oedipal stage affect the process of self-identification. Object fixations for the sexual motivations occur probably much later. Sexual pathologies such as handkerchief or shoe fetishisms can often be traced to earlier sexual experiences (R. v. Krafft-Ebing, 1924). Homosexuality may also be traceable to an object imprinting in the final analysis; a predisposition for a specific sex role may have been acquired during the oedipal phase. The parallels to imprinted homosexual animals are striking. This seems to be supported by the fact that such preferences are highly resistant to therapy. To what degree a true cure is possible could be tested with the pupillometric tests of E. H. Hess (1965).[3]

During puberty there is probably another sensitive period in which the young person is especially receptive to new values. Young people are apt to identify with a social group, and often the final political and religious attitudes are determined for the rest of life.

Knowing about the strong fixations caused by imprinting, one is confronted with the question as to whether it is right to expose a child during the respective sensitive period with ethical concepts which are not universal. Any fixations of that sort hamper the development of free will. This is certainly dangerous in modern times.

In discussing this point H. Hass (1968:206) wrote:

> To be fair towards our children one should expose them up to the age of 16 only with those ethical concepts that are agreed upon by all men. One should furthermore point out to them all the dangers of early fixations and emphasize that they have a right and even the duty to judge on their own, even if they should come into opposition with the parents or the commu-

[3] E. H. Hess found that the pupils of a test subject are dilated when some interesting material is viewed but contract when something unpleasant is seen. If a normal man is shown pictures of girls, men, and other objects, the pupil response is positive, especially to girls. Homosexual men, on the other hand, react positively to pictures of men. One could test how cured homosexuals respond. Although this work has been criticized (by, for example, I. E. Loewenfeld, 1966; J. J. Woodmansee, 1966; M. P. Janisse, 1973; and M. P. Janisse and W. S. Peavler, 1974), a specific validation of the findings on sex-differentiated responses was conducted by N. Bernick, A. Kling, and G. Borowitz (1971), who reported that male subjects showed greatest dilation to a heterosexual sex film than to homosexual or thriller films and that the degree of dilation was significantly correlated to the subjects verbal report of their own sexual arousal. Subjects in this test who were aroused by the homosexual film also tended to show increased pupil size to inocuous male slides.

It does, however, seem possible that the contraction response observed to negative stimuli may be an artifact due to lag in the return to baseline or to prior dilation in anticipation. This is not yet clear, as the positive–negative scales used by various experimenters are neither obviously clear nor comparable.

nity. This is, of course, a utopian concept for the moment. But such development is perhaps already indicated in our young people.

Curiosity Behavior and Play

A conspicuous disposition of animals for learning is expressed in curiosity and play. If we place a new object into a room inhabited by rats, all will explore it very soon. They will at first repeatedly but cautiously approach the new object. Finally, they will sniff, chew, climb, and urinate on it—hesitatingly at first, later with more confidence. Once they have explored the object in this way they lose all further interest. In a similar manner many higher mammals investigate a new object by sniffing, looking, chewing, biting, scratching, or manipulating it in some other way. What are the typical characteristics of this exploratory behavior?

Close observation reveals that the exploring animal alternately approaches and withdraws from the object of its interest. It makes contact with the object by means of its sense and effector organs, withdraws, and approaches again, with a new view of the situation, so to speak, in order to become acquainted with all aspects of the object. The animal is attracted by the object, but it does not get interlocked with it rigidly by a particular behavior, retaining the ability to withdraw again. This ability to withdraw to a distance is the basis for any dyadic interaction. It is typical for curiosity behavior and play. During the ontogenesis of the human child development of this ability can be readily observed. When the child first reaches for an object, the first actions are fairly stereotyped. The child grasps the object, brings it to its mouth, and begins to suck it; at first this is all that is done. Soon afterward the child is able to remove the object from its mouth again, to look at it, suck it some more, perhaps drop it, and reach for it with the other hand. By now the rigid action sequence has become more flexible and the child is able to explore. I will show later that this ability to withdraw to a distance is one of the roots of human freedom of action.

At least during their youth, most mammals are pronounced "creatures of curiosity," actively seeking out new situations for exploration under the compulsion of some internal drive. Various groups of animals behave differently—both qualitatively and quantitatively—in respect to curiosity. Primates and carnivores are more curious than rodents, and within the rodents porcupines are more curious than mice. Squirrels are somewhere in between. Rodents gnaw at new objects and occasionally hoard them. Monkeys look at the objects and try various things with their hands, the Cercopithecinae more than the Colobinae. Some fish and birds are also curious. Once a new object is explored, all interest gradually wanes (A. Wünschmann, 1963; S. E. Glickmann and R. W. Sroges, 1966; see Fig. 13-21).

There seems to be a drive to learn that can be called a "curiosity drive." It can be demonstrated that there is a drive to learn new motor skills as well as to receive new perceptual impressions and thus to acquire knowledge. Some experiments have supported this view. Rhesus monkeys learned a puzzle game without any other reward than the performance of the task itself (H. F. Harlow, M. K. Harlow, and D. R. Meyer, 1950). They also learned a task when as a reward they were allowed to look out through a window of their cage (R. A. Butler, 1953; additional examples in E. R. Hilgard, 1956).

In exploring rats, neurons in the lateral hypothalamus and in the preoptic region are active. Electrical stimulation at these locations is rewarding and rats learn readily to stimulate themselves by pressing levers (B. R. Komisaruk and J. Olds, 1968).

This drive to learn seems also to underlie the play behavior of animals. It is usually easy to recognize when an animal plays and when it is more seriously occupied; nevertheless it is not easy to give a definition of play. "The animal works," writes F. Schiller, "when the behavior

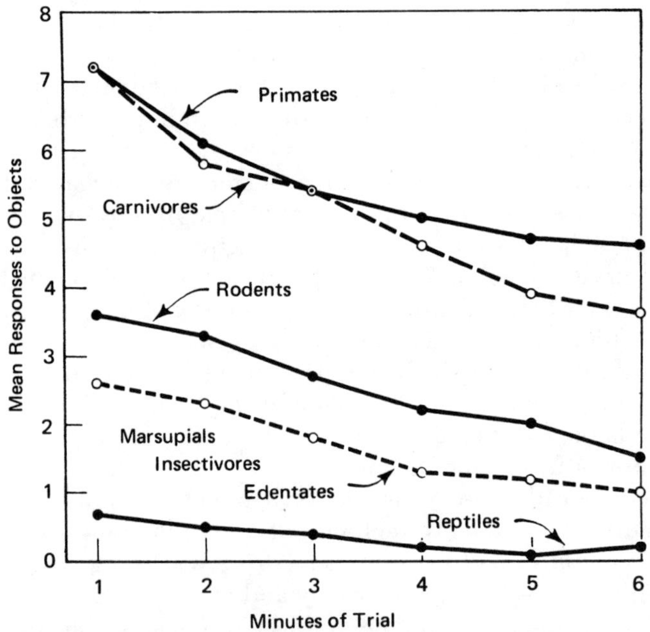

Figure 13-21 Average readiness to react to test objects during the course of a 6-minute period. To measure the approach to an object the authors divided the 6 minutes into 72 5-second intervals. When an animal reacted to an object within such a 5-second period this was noted. In this way a value scale was obtained that represents the ordinate. The curves show the average values of the length of contact. They were computed from four presentations each with different objects. Two conclusions can be drawn: (1) higher mammals interact more intensively with new objects than do lower mammals and reptiles; (2) within one 6-minute period the interest in a new object already begins to wane. (From S. E. Glickmann and R. W. Sroges, 1966.)

is motivated by some need, and it plays when an excess of energy provides the motivation." This statement aptly characterizes one of the roots of play. An animal only plays when it is satiated and when it is not occupied with other tasks. Play is not dictated by any immediate need, but it is nevertheless of enormous importance for the normal development of an animal. The nature and value of play becomes clear when we study how widespread it is and when we look more closely at the various forms it can take. Only the most highly developed "learning animals"—those that seek new situations out of their own initiative—play in the true sense. They are curious and try out new behavior patterns, learning in the process. Insects do not play; neither do fish and amphibians. However, many mammals do, especially when young, and so do some birds. The taxonomic distribution of play seems to indicate that play is related to learning, and we shall see that an animal interacts with its environment during play. It experiments with things in its environment and acquires knowledge about their characteristics. The animal collects experiences during play with conspecifics and learns the possible range of its own movements. Play always implies a dialogue with the environment, and this dialogue is always the result of an internal drive. One could even assume a separate drive to play, but I am inclined to believe that the drive to learn, which is the basis of all curiosity behavior, coupled with an excess of motoric motivation, will suffice to account for the phenomenon of play. Learning takes place during playful experimentation in the same way as during endlessly repeated movement games. In this connection it is interesting to observe that animals develop special modes of action during play; that is, they play at particular times and only until they fully master what they have practiced; then they lose interest and try something new (I. Eibl-Eibesfeldt, 1950a).

An animal may fight with a companion during play or flee before a predator, to mention only two examples. But in most instances the play version is clearly distinguishable from the actual behavior: The serious aspects of the behavior are lacking. An animal that plays at escaping does not actually flee. A rat that actually flees underground will not reappear as quickly and then only with hesitation. The rat playing escape reappears at once. When two polecats pursue one another in play they frequently change roles of the follower and the pursued, which does not happen when the chase is in earnest. When engaging in rough and tumble play they have an inhibition against biting, and the threat behavior that accompanies a real fight is missing. This is true for play-fighting rats as it is for squirrels or any other play-fighting predator. Squirrel monkeys squeak continuously when playing. If the squeaking stops, a serious fight has started (P. Winter, D. W. Ploog, and J. Latta, 1966).

Play behavior may also be repeated many times over, which is especially noteworthy because usually instinctive actions are quickly fatigued. A dog will retrieve a stick many times. It may play fight with another dog until sheer exhaustion or a new exciting stimulus diverts its attention.

How can this be interpreted? At first, one might think that in play we are dealing with instinctive behavior which has not as yet matured. Frequently this is not the case. Animals may still play fight with all the accompanying social inhibitions at an age when they are already able to fight seriously; this is readily observed in dogs. During play a number of fixed action patterns appear which are in fact part of serious fighting behavior, but these behavior patterns occur independently of their usual function. They do not appear in the sequence in which they appear in a real fight and sometimes behavior patterns from different functional cycles are combined that are mutually exclusive in an actual fight, such as when sexual behavior is mingled with fighting and prey-catching behavior. It appears as if these behavior patterns are activated in something other than their normal fashion. The movements of lower levels of integration are more or less independently activated and not by the arousal of an entire instinct (in Tinbergen's sense) (M. Meyer-Holzapfel, 1956; I. Eibl-Eibesfeldt, 1951c, 1953a). This could explain why subjective correlates that accompany the real fight are lacking during play. These subjective correlates are perhaps aroused only when the behavior is activated from a higher level of integration, as was implied by the **brain-stimulation experiments** of E. v. Holst and U. v. Saint Paul. Only when behavior patterns are thus independent of the normally involved superior centers can an animal freely manipulate its various movements, recombine them in new ways, and thus experiment with its movement repertoire. It has been shown that the animal actually learns many things which are indispensable for the acquisition of behavior needed in later life.

Polecats, for example, learn during play the orientation of the neck bite which is necessary for successful **mating,** and the **woodpecker finch** learns the use of the tool. Chimpanzees who have not handled sticks do not discover that they can be used to pull in a banana into their cage. After playing with sticks for three days they can solve this problem within twenty minutes (H. G. Birch, 1945). A prerequisite for this freedom to play is the condition that an animal is free of stimulation to fight, flee, or hunt. All this takes place, as G. Bally (1945) so aptly says, in a "relaxed field." True, innate behavior patterns are used, but the "instinctive need" is to be relaxed (G. Bally). The relaxed field is at first provided by the parental care and protection that frees the animal from the need to find its own food and shields it from predators. Adult animals only play when they feel safe, and especially when they are satiated. My own woodpecker finches regularly played after feeding. They took leftover mealworms, placed them in holes and crevices, and poked them out again. They often placed a mealworm within a split branch and pushed it to and fro with their sticks. Many songbirds sing their most variable and

PLATE I

Color patterns of *Hemichromis fasciatus* in eight different moods. Top left: without territory (neutral mood); below: three stages of increasing territoriality and readiness to fight; bottom left shows the greatest readiness to fight. Right bottom: spawning animal; above that, while caring for young; the two top right pictures show the fright coloration of a fish that would like to hide. The top picture shows a fish that is hiding in a plant; the one below is one that has no hiding place and must remain in the open. Note how the melanophores on various locations of the body sometimes function together and sometimes in opposite ways. The coloration of the very aggressive fish (bottom left) is actually a negative of the neutral mood. (From W. Wickler 1964a; H. Kacher, artist.)

PLATE II

Top: Pygoplites diacanthus (Red Sea) as an illustration of a reef fish that is marked like a poster. (Photograph: I. Eibl-Eibesfeldt.) *Center:* courting frigate bird *(Fregata minor)* (Galápagos Islands). (Photograph: I. Eibl-Eibesfeldt.) *Bottom:* gape releaser of a young Gouldian finch. (Photograph: I. Eibl-Eibesfeldt.)

PLATE III

Top: bluebird of paradise *(Paradisornis rudolphi),* courting with head down (New Guinea). (Photograph: H. Sielmann.) *Center:* Lauterbach's bowerbird *(Chlamydera lauterbachi)* (New Guinea). The male courts with a red fruit in its bill. (Photograph: H. Sielmann.) *Bottom:* Satin bowerbird *Ptilonoryhnchus violaceus)* building its bower (New Guinea). (Photograph: H. Sielmann.)

PLATE IV

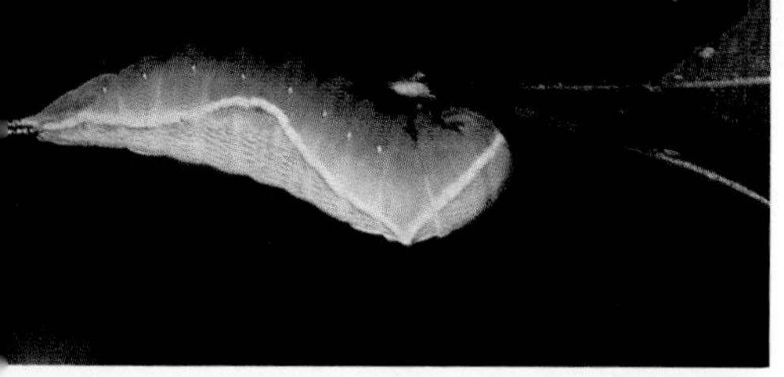

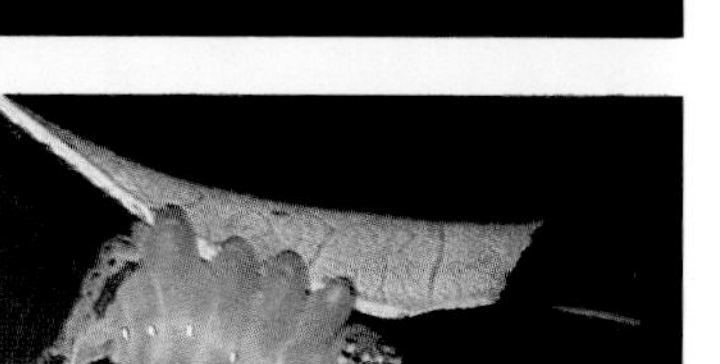

Top right: Automeris acutissima exhibiting the eye spots when frightened. (Photograph: E. Priesner.) *Top left:* larvae of *Dicranura vinula.* Top, resting position; below, frightened. The larvae has ejected two red threads from the raised abdominal pair of legs and exposed its face mask. (Photograph: I. Eibl-Eibesfeldt.) *Center left:* hornets *(Vespa crabo)* and the moth mimic *(Aegeria apiformis).* (H. Kacher, artist.) *Center and bottom right:* top row, poisonous Malayan butterflies whose larvae feed on poisonous aristolochiacees; left to right: *Atrophanura varuna, A. coon, A. nox.* Bottom row, three different morphotypes of the nonpoisonous butterfly *Papilio memmon;* these are females of the same species; each morphotype mimics another model. In this plate the various morphotypes of *Papilio memmon* were placed under the models they mimic. (After examples by M. Tweedie [1966]; H. Kacher, artist.)

beautiful songs outside the reproductive season. They "compose," as bird lovers say. My young polecats after feeding regularly engaged in rough-and-tumble play, as did my badger (I. Eibl-Eibesfeldt, 1950a, 1956c).

The playful actions are performed as a result of an inner drive, but this drive does not seem to be identical with that which motivates the behavior patterns when they are performed seriously. The playful actions seem to have become "unhooked." This "independence of actions from the underlying motivation" is, according to A. Gehlen (1940), especially characteristic of human behavior. In this way a hiatus between needs and their fulfillment is created in which planned human thought can function rationally without the disruptive influence of drives. The root of this specifically human freedom of action exists in the play of animals.

Play versions and their counterparts differ then in the basic foundations of their drive motivation. In the pure case play is undirected at first. The animal may engage in play fighting, hunting games, or it may experiment in play, each according to the nature of the environmental condition prevailing at the time. There are exceptions to this rule, however. Frequently play is directed toward certain functional cycles, so that there are transitions between play and nonplay behavior. My hand-raised badger often changed from play fighting to actual fighting, including threatening calls, especially during the time when he was beginning to become independent. When no play partner was available he often sought out a substitute object. I observed several times how he threatened a bush, butting and pulling at its branches. He invited the bush to play fight in typical badger manner. In young rabbits an initial escape game may begin with several cross-jumps and end in "serious escape," where the animal presses itself to the ground in a dark corner.

The expressions of play are quite variable and we find many examples in the summaries of F. J. J. Buytendijk (1933), K. Groos (1933), E. Inhelder (1955) and M. Meyer-Holzapfel (1956). The various theories of play are treated in detail by C. Allemann (1951), and M. Meyer-Holzapfel (1956).

We may state the following criteria for play behavior:

1. During play we see instinctive behavior patterns along with acquired behavior; neither, however, are performed in earnest (that is, with their specific function), although there are transitions between play and nonplay activities.
2. Behavior patterns that appear during play seem to be devoid of the motivating mechanisms to which the behavior is normally subject. I interpret this to mean that the entire instinct, in Tinbergen's sense, has not been activated. Instead, the behavior patterns have been individually activated at a lower level of integration.
3. This results in random combinations of behavior patterns that can include actions from different functional cycles and that allow a random change of ac-

tivities or roles. This seems to be the root of the ability to *withdraw*, which is a prerequisite for any kind of dialogue. This ability is lost at once when actual appetitive behavior becomes aroused.

4. The animal seems to learn things that are of use to it in its later life, and the development of some behavior patterns seems to include "planned-in" play activities with siblings (for example, **prey catching in the polecat**). This aspect of learning will be further discussed when we discuss fads (**movement games**).
5. There exists a specific motivation for play which is based on a curiosity drive, that is, a mechanism which moves the animal to seek new situations and to experiment with new objects. Added to this is a strong motor motivation. The urges to play and to learn have a common source. Play is an active form of learning.

If one applies these criteria, then it will be seen that real play can be found only in some higher mammals and some birds, such as the raven and the kestrel (E. Gwinner, 1966; O. Koehler, 1966a)). The fish *Gnathonemus* balances objects on the tip of its snout, which has been interpreted as play activity, an interpretation I hesitate to accept (M. Meyer-Holzapfel, 1956). No other observations of "play" in fish have been reported. It seems safe to say that fish do not play. Not all mammals play, however. In most species of mice I have never observed play behavior (I. Eibl-Eibesfeldt, 1958a).

Animals that fight with members of their own species as adults practice the actions during play fighting (Figs. 13-22 and 13-23). These are clearly distinguished from serious fighting by an inhibition against biting and by the lack of threat behavior, as well as the repeated interchanges of roles. If one polecat or rat unintentionally bites another too hard, then the bitten animal will utter a cry and thus inhibit the partner. Between bouts of play there are periods of chasing, where the roles of pursuer and pursued change often. Polecats and badgers jump to and fro with humped backs facing the partner, which may be interpreted as an invitation to play. Dogs play passing and cutting-off games (J. Ludwig, 1965). Here it is difficult to distinguish between play fighting and play prey catching. In predatory species the pursuer seems to be more in earnest when trying to catch up with the partner during the chase, which is in contrast to many vegetarian animals, where the pursued animal tries to escape. Squirrels run up trees during escape games in which each tries to get the tree trunk between itself and its pursuer. Neither animal makes an effort to actually catch the other. Instead it runs into hiding as soon as the other appears around a corner. Games of hide and seek are widespread and so is playful defense of a particular place. My tame polecats liked to hide under blankets and peek out. If another animal came close it was attacked. In this way they defended wastebaskets and other objects they occupied. Young deer and goats play king of the castle (F. F. Darling, 1937).

Figure 13-22 Play-fighting young polecats. (Photographs: I. Eibl-Eibesfeldt.)

Figure 13-23 Young lions play with one another and with adults. (Amboseli, East Africa.) (Photographs: I. Eibl-Eibesfeldt.)

During hunting games movements of prey catching are practiced, such as closing in, throwing down prey, shaking, and creeping. Often the animals use substitute objects that are treated like prey: the cat with a ball of yarn, a lion with its siblings, the dog with its ball. Sometimes impressive feats are performed. I knew a poodle that would carry a ball to the top of an embankment, push it down, and chase after it. Sea lions of the Galápagos Islands dive for stones, throw them into the air, and catch them again. These "hunting" games can be readily distinguished by their over-all appearance from the usual rough-and-tumble games and chases. Usually there are no invitation gestures, although transitions exist.

Of especial interest are the movement games and playful experimentation with objects. In the former the animal experiments with its own abilities. It jumps, frequently changes the direction, rolls on the ground, and invents new movement coordinations. Wild sea lions ride the waves. My young badger discovered forward somersaults accidentally, and he practiced them until he was able to roll down a long incline in one continuous series. On another occasion he discovered that he could slide down an icy road, after which he practiced this tirelessly. Under these circumstances animals behave like children, who also practice all manner of locomotory actions: walking on heels, balancing, and so on, learning many new things thereby (Fig. 13-24). It is also possible that animals acquire knowledge about their own body during play. They play with their limbs as well as with their own shadows.

Figure 13-24 Young gorillas playing. (From G. B. Schaller, 1963.) In zoos orang-utans engage in a rich variety of similar movements and object plays (see F. Jantsche, 1972).

New inventions may be retained as fads in both animals and people. W. Köhler (1921) reported that his chimpanzees retained newly invented games for some time. At one time they fished for objects with long sticks; after they had discovered that chickens could be attracted and chased away, the habits changed. After my hand-raised badger had invented the forward somersault he concentrated exclusively on this game for awhile. This animal always sought contact with his caretaker and attempted with great persistence to get to him when he had been locked up. Once he accidentally discovered that he could enter the room by climbing to the low windowstill. After this discovery he lost interest in his caretaker for awhile. Instead he practiced playfully leaving through the open door and climbing in again through the window. Similar behavior was observed in a beech marten. When we ate dinner on an elevated platform the marten used to visit us, to beg for food and to play. When he accidentally discovered a shorter route he forgot us and continued to practice climbing his new path.

During play an animal manipulates various objects. It chews and throws them about until it has acquired mastery over them. In most animals, objects are more commonly taken apart than put together in a constructive way, but among the apes games involving construction exist. As I have already mentioned, new objects are especially likely to arouse curiosity, but gradually interest wanes and the animal plays less (E. Inhelder, 1955; A. Wünschmann, 1963).

During play animals make inventions that later prove useful. The classical example is provided by W. Köhler (1921), whose chimpanzee Sultan had the task of reaching for a banana beyond arm's length by means of two sticks that could be joined. After several unsuccessful attempts he withdrew from the task, began to play with the sticks, and succeeded in putting one end into the other. Then he returned to the former task and reached for the banana. E. Inhelder (1955) described how a monkey (*Macaca silenus*) discovered that a ball could be put into a pail, how the animal repeated the game, and how others imitated it.

In man, too, play is an experimentation with one's own abilities, by interaction with the environment as well as with other people. A new element is that play is often constructive and based on models (Fig. 13-25). One should study in more detail the degree to which construction games of humans are based on specific innate predispositions. For example, children of a certain age especially like to build tree houses and other structures. City children who have never had an opportunity to observe adults performing the behavior begin to build tree nests or leaf shelters when they go to the country during vacation. Is there a bias in favor of this behavior?

In humans imagination plays an important role in all play activities. Children attach different meanings to objects and assume changing roles during play. Phantasy becomes superimposed on reality (A. Gehlen, 1940). We know from introspection that we play with our imaginations in our daydreams. We will discuss the biological significance of this capacity in Chapter 18.

Although innate dispositions of various kinds provide a framework in which play fighting, hunting games, and so on, occur, play is primarily liberated action (G. Bally, 1945; J. Huizinga, 1956). The independence from the basic motivation permits an exploratory change from approach to withdrawal, as I mentioned earlier. It is interesting to note that during play new traits are revealed which normally cannot be observed. Chimpanzees produce paintings during play that are quite pleasing to the eye, illustrating a basic esthetic appreciation for symmetry and balance (D. Morris, 1963). In addition to this, the pictures reflect individual expressions, and in three chimpanzees of different social rank within a group, this ranking was also expressed in the individual styles. The high-ranking male covered the entire canvass and began turning the brush after each stroke, including curved lines in the picture. The high-ranking female also filled all available space with bold strokes from the center to the periphery. When given several colors she painted something reminiscent of a rainbow. The low-ranking female, on the other hand, limited herself to painting a small spot at the lower edge of the canvas. She pressed down so hard that the paper became roughened (Plate VII). When given additional colors she did not fill the still vacant space but instead filled in the center of the spots she had painted earlier. This

Figure 13-25 Construction game of El-Molo children. *Top:* Scenery near Lake Rudolf (Kenya) with one of the typical huts; *bottom:* children playing "house." (Photographs: I. Eibl-Eibesfeldt.)

reminds one of the tree test that psychologists use: healthy children draw trees that spread out their branches in all directions, whereas children with psychological problems produce inharmonious or distorted pictures.

SUMMARY

At every stage of development functional patterns are available to the animal. These often differ considerably at different stages of development, particularly in animals that start their lives as larvae, or that are born or hatched at an altricial stage. Many behavior patterns mature during ontogeny, others are acquired by learning. How both processes interact has been studied in many instances. Innate learning dispositions ensure the animals modifying their behavior adaptively and learning the right things (skills, knowledge) at the right times. This can be achieved in a variety of ways, as the studies of bird songs have demonstrated. Chaffinches, for example, know as an innate learning disposition what song to imitate, while zebra finches learn any song they hear during a critical sensitive period.

In the case of object imprinting the releasing stimuli for a reaction are specified during a short critical period. Once an animal becomes attached to an object it does not easily change its preference. Sometimes the fixation seems to be irreversible. In the case of sexual imprinting the reaction becomes imprinted to an object even long before sexual maturation. Critical periods seem to play an important role in mammalian and human development.

Curiosity is a motivation evolved in the service of learning as a mechanism that urges an animal to seek the new in order to acquire information actively. The German word *Neugier* (greed for the new) is a good descriptive term for this motivation. *Neugier* is very pronounced in the young mammal. In some—particularly in man—this juvenile feature persists during adulthood.

The phenomenon of play is to be understood similarly. During play animals are experimenting and learning. Only higher mammals and a small group of birds play, and although fixed action patterns occur during play, play activity can be distinguished from serious performance by a number of characteristics. For example, whereas in a serious fight the motor patterns follow in a fairly rigid order and a consummatory situation or act ends the sequence, this order does not appear in play. The motor patterns occur fairly independently from one another, and patterns from different functional cycles, which normally exclude each other, may be combined. It seems as if the actions were liberated or unhooked from the physiological concomitants by which they are normally governed. It is assumed that the whole "instinct" with its hierarchical organization is not activated, but instead, by a special mechanism, only the motor patterns of lower integrational levels are involved, and thus available to the animal for experimentation.

14

Mechanisms of Learning

In everyday usage learning refers to the acquisition of new skills and knowledge. Scientifically speaking, this term can be used whenever the probability of the occurrence of certain behavior patterns in specific stimulus situations has been changed as a direct consequence of encounters with these or similar stimulus situations and not as a result of maturational or fatigue processes (E. R. Hilgard, 1956; P. R. Hofstätter, 1959). As a rule the resulting changes in behavior are such that they contribute to the survival of the individual. Learning in the broader sense therefore can be defined as adaptive modification of behavior (K. Lorenz, 1973, 1969). The modifications range from simple processes of sensitization and habituation to more complex patterns of reinforcement learning. This latter type of learning is based on the "invention" of feeding back the final success of a chain of actions to the behaviors initiating it. As K. Lorenz (1973) has emphasized, the evolution of learning took place in convergent development independently in the different classes of animals.

This modifiability of behavior presupposes special adaptations of the organism. First, an organism must have the ability to "remember"—to retain the effects of past experience. Furthermore, it must be programmed in such a way as to be able to distinguish experiences that are positive from ones that are negative with respect to the preservation of the species. This means, among other things, that an animal does not indiscriminately associate each environmental stimulus with specific perceptions. This is actually not the case. A rat that has tasted poisoned bait will henceforth avoid the bait but not the place where it was found (F. Steiniger, 1950). That taste and olfactory impressions are associated with visceral conditions, on the one hand, and pain stimuli are associated with auditory and visual stimuli, was demonstrated in the experiments of J. Garcia and F. R. Ervin (1968) and J. Garcia et al. (1968). When people have become seasick they tend to associate the nausea with specific odors or foods but not with the ship itself. Differences in this kind of programming from species to species have been discussed in Chapter 13. In accordance with these differential dispositions many species behave differently even in artificially uniform experimental situations (K. E. Grossmann, 1967). According to the theory of classical conditioning any behavior pattern can be extinguished by the application of punishing stimuli. H. A. Euler (on press and in preparation) applied electrical shocks to cockerels whenever they showed aggressive displays or attacked a conspecific. And indeed these cockerels finally stopped showing aggressive behaviors and accordingly assumed a low-ranking position. In another series of experiments he tried to extinguish submissive behavior by the same method, but this did not stop them showing submissive behavior whenever they met a high-ranking cockerel.

As we have seen in Chapter 13, an organism can be so constituted that it not only learns passively from events taking place, but also actively searches out the unknown. It is *curious*. This requires the existence of appropriate motivating mechanisms. In the following sections we will discuss the motivations underlying learning processes and the nature of memory.

THE EXPERIMENTAL ANALYSIS OF LEARNING AND ITS MOTIVATION

The simplest learning process is called *habituation*. The animal learns passively to refrain from responding further to repeated stimuli which are not accompanied by positive or negative reinforcement—that is, to those events which biologically are meaningless to the animal. A clawed frog is startled when one taps the side of its container. If this stimulus is repeated several times, the animal will no longer show escape behavior. An additional example is the **habituation** of **graylag geese** to silhouettes of birds flying overhead.

In addition to the process of habituation, H. Thorpe (1963) distinguishes the following kinds of learning:

1. The formation of conditioned reactions or reflexes by the standard process of conditioning;
2. Reinforcement learning as a consequence of the organism's own activity (instrumental conditioning, trial-and-error learning, or conditioned reflex, type II);
3. Latent learning;
4. Insight learning.

Complex innate behavior patterns can be released by conditioned stimuli after appropriate training. Japanese quail (*Coturnix coturnix japonica*) will show species-typical courtship behavior in response to a buzzer when it has previously been paired with the appearance of a female. The various components of the behavior become linked in a specific sequence (which approximately follows their appearance during ontogeny) with the conditioned stimulus, and they are extinguished in the reverse order (H. E. Farris, 1967).

The formation of conditioned reactions has been studied in great detail by I. P. Pavlov and his school. I have cited the case of a dog that salivates in response to the sound of a bell as an example of such a **conditioned reaction.** An originally neutral stimulus that repeatedly precedes the presentation of food will eventually elicit the unconditioned response to the food (salivation), even when the food itself is not presented. The initially neutral stimulus becomes coupled with the unconditioned stimulus for the response. If one illuminates the eye of an animal, the pupil will contract. If this light is accompanied by the sound of a bell, then the stimulus becomes associated with the unconditioned stimulus, and eventually the pupil constricts in response to the sound alone. From time to time the unconditioned stimulus must again be paired with the conditioned stimulus, or the conditioned reaction will gradually extinguish, after which the animal will no longer react to the conditioned stimulus. Negative associations are retained longer without reinforcement than positive ones.

In circus training an unconditioned reaction is frequently linked up with new conditioned stimuli. In this way circus horses can be brought to perform innate behavior patterns in a reliable fashion by presenting conditioned stimuli. Rearing up on the hind legs, for example, occurs naturally during conspecific fighting in horses (K. Zeeb, 1964).

Behaviorists in the United States, in contrast to Pavlov, base their investigations of the learning processes upon the spontaneous actions of the animal. Their focus is on trial-and-error learning (reinforcement). Typically, an animal is placed into a closed cage from which it can escape by pressing a certain lever (K. S. Lashley, 1935) or where it can obtain food or water by pressing a bar (B. F. Skinner, 1938; see also Fig. 14-1). However, it is possible to reward any other behavior of the animal—for example, when the animal turns to the right. Thus a pigeon can be taught a sequence of steps by reinforcing each successive step with food, so that within a short time the pigeon will walk in a circle, as was demonstrated by Skinner. If a pigeon is rewarded each time it raises its head above a certain line, it will soon run about with its head raised. If a pigeon is to be trained to peck at a marble, one initially rewards the mere turning toward the marble, then looking at the marble, approaching the marble, and finally the pigeon is rewarded only when it pecks at it. It has been possible to train whales (*Globicephala*) at Marineland, California,

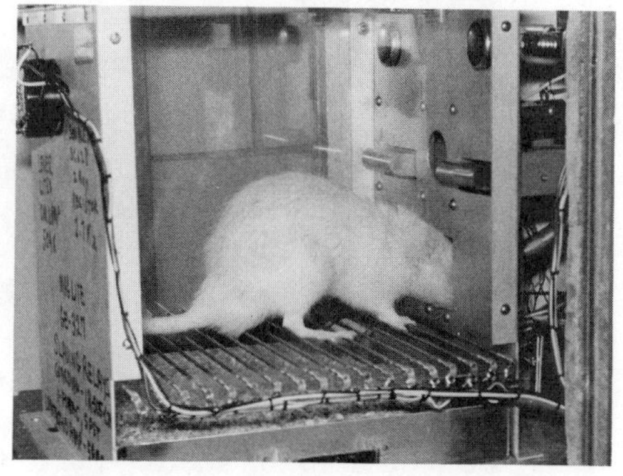

Figure 14-1 One form of Skinner box. When the rat presses one of the two levers a food pellet will drop into the food cup. The rat learns, for example, to activate the levers in response to various signals —here light signals. (Photographs: I. Eibl-Eibesfeldt, laboratory of T. I. Thompson.)

in the same way. For example, they will slap their fluke on the water surface in response to a command and continue until given a signal to stop. Initially the animals were rewarded whenever they accidentally hit the water surface with their tail. They very quickly learned on what the reward depended. Other toothed whales (*Delphinidae*) have also learned in a similar way to jump together out of the water, to dive down head first, propel themselves backward while standing on their flukes, and many other tricks (H. Hediger, 1963; see also Fig. 14-2).

Looking at a learning curve of, say, bar pressing in a rat, one can generally see that the first few successful responses seem to have had no effect. However, after a few successful performances the rate of bar pressing increases sharply (Fig. 14-3). B. F. Skinner (1953) developed a widely used method to lead school children through learning programs that have been broken into small steps, where the reward consists of the student being allowed to progress to the next step in the program after successfully completing the preceding ones. The immediate reward of success strengthens the learning process. The results of such programmed learning depend, of course, on the abilities of the programmer. Furthermore, one has to realize that in this way of teaching the forming of individual opinions has been precluded.

A favorite method for the study of learning processes is the maze

Figure 14-2 Trained blackfish whale (*Globicephala*) during a performance in Marineland (California). (Photograph: I. Eibl-Eibesfeldt.)

experiment, which was first used by W. S. Small (1900) to train rats. In such an experiment the animal must learn the path to a goal that cannot be seen from the starting point. Complex mazes have many dead ends, and in time the animal learns the shortest route to the reward in the goal box. Very simple mazes are of Y or T form. The animal may run in open, closed, or on elevated runways. Figure 14-4 shows three of the types of

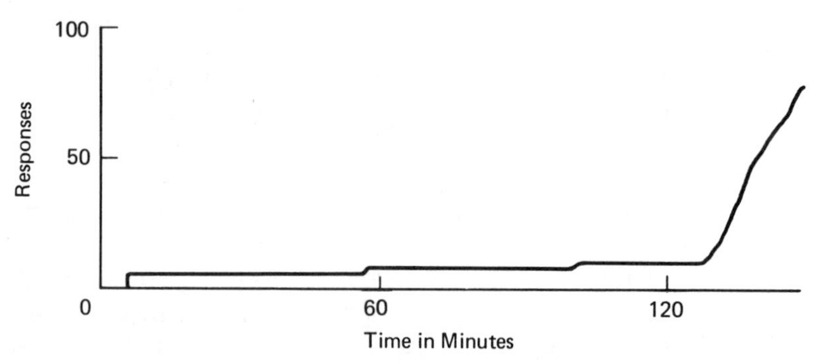

Figure 14-3 Learning lever presses in a Skinner box setup. Each correct response is indicated by a rise in the level of the curve. It can be seen that the first three correct responses apparently were without effect. After the fourth success, however, the number of correct responses (bar presses) increases sharply. Ordinate, number of responses; abscissa, time in minutes. (After B. F. Skinner from N. L. Munn, 1950.)

mazes that are frequently used (see also N. L. Munn, 1950). In the evaluation of maze experiments, the biology of the animals used should be considered. Rats, which normally live in tunnel systems, bring a specific learning disposition to the experimental situation that many other species do not have.

Maze experiments have shown that such reinforcement learning is not limited to the specific situation in which the learning has taken place. If mice have mastered one maze, then they can find their way, without additional learning, if all angles of paths leading away from the choice points have been changed from the original 90° to 45° or 135°, or if the lengths of the runways have been doubled. Even the mirror image of an originally learned maze can be mastered. At the first choice points they make errors but then seem to understand the principle and are able to transpose. In the house mouse this ability to transpose is maintained even after the mice are blinded, but if the visual cortex is destroyed, the ability is lost (W. Dinger and N. Heimburger, cited by O. Koehler, 1953). Rats that have learned to run through a maze can still find their way when they must swim it, and no additional learning seems required (D. A. MacFarlane, 1930). E. C. Tolman (1932) has pointed out in this connection that the animal does not learn movements but meanings. An animal that runs through a maze, he said, does not learn movement patterns but a concept of the path to the goal that it seeks. That higher animals are guided by conceptions of the goal can be demonstrated by the fact that rhesus monkeys, who have seen a banana hidden under a cover, will continue to search for it after they have discovered a leaf of lettuce that has been surreptitiously substituted for the banana. The animals are quite agitated during the search, undoubtedly filled with

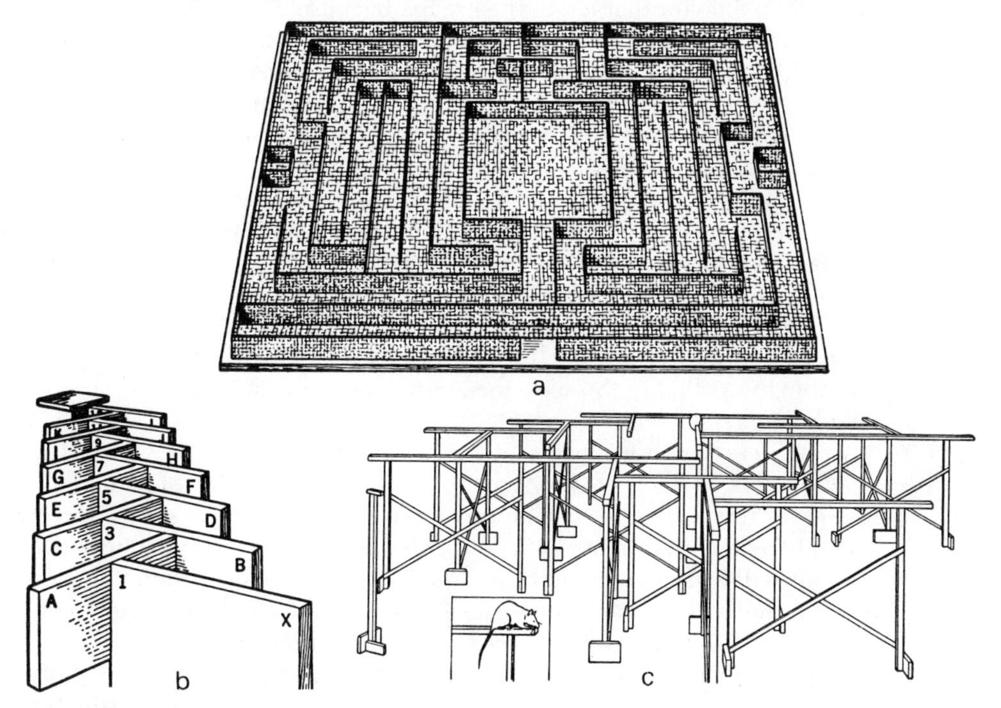

Figure 14-4 (*a*): Hampton Court maze of Small (depth maze); (*b*): elevated maze made of narrow boards; (*c*): elevated maze made up of T-maze components. (From N. L. Munn, 1950.)

some expectation. When learning a new task animals often try out what brought success last time. Rats in a new maze will, for example, repeatedly turn right or left, depending on what has most recently led to success, or they may alternate. Thus at the beginning of training there is no random trial; instead, the animals behave as if they were trying out "hypotheses" (I. Krechevsky, 1932). Improvements are made only after the rats abandon their hypotheses.

A number of learning theories have been developed on the basis of the various experiments. The preconditions necessary for an animal to learn has been the subject of much discussion. C. L. Hull (1943) emphasized his theory of the strengthening of reaction tendencies along the lines of E. L. Thorndike (1911), asserting that learning will take place only when the correct response is rewarded in some way, where the reward leads to a reduction of a specific drive—for example, hunger. Such a reward leads to reinforcement of the response that has brought it about.

In contrast to this, the theory of contiguity of E. R. Guthrie (1952) stated that such a reinforcement is not a prerequisite for learning. Associations are formed by the close contiguity of stimulus and response. The behavior is not strengthened by the reward but is prevented from decaying. For example, if an animal escapes from a puzzle box, it will not forget the behavior that led to the escape because it is removed from the environment and has no opportunity to form new associations. F. D. Sheffield and T. B. Roby (1950) claim to have demonstrated that a reduction of a physiological need is not necessary for learning. They rewarded rats with saccharine. Furthermore, it was found that rats will learn a maze not only because they receive a special reward. If they are allowed to run about in a maze without any reward, they will later learn the maze much more quickly when the goal contains food than will control animals that have not had this prior experience. During exploration, learning is said to be "latent" (H. C. Blodgett, 1929; M. H. Elliot, 1930; E. C. Tolman and C. H. Honzik, 1930b). This latent learning has been cited as evidence against Hull's reinforcement theory. However, it is possible to say that in this case the "curiosity" of an animal becomes satisfied.

It is certain, however, that the opinions expressed by C. L. Hull (1943) and B. F. Skinner (1938) that hunger, sex, avoidance of pain, and a few other drives are the only motivations for learning are surely incorrect. My own experiments with squirrels have shown that inexperienced animals will learn to open nuts, although the kernels have been removed and the shells glued together again (I. Eibl-Eibesfeldt, 1967). If gnawing is activated in rats as a result of brain stimulation, they learn a task if their only reward is to be able to gnaw at wood or cardboard (W. W. Roberts and R. J. Carey, 1965). If the rats happened to be eating during the experiments, they stopped in response to the brain stimulation and sought out objects to gnaw, which clearly demonstrated that the drive to gnaw was activated rather than the feeding drive. It seems then that the mere performance of fixed action patterns is rewarding, a point that has been made by ethologists.

In a Skinner box the rats learn to press a lever if they can administer an electrical brain stimulus to themselves by means of implanted electrodes. When the electrodes are in a certain location the frequency of bar presses increases rapidly. The self-applied stimulus seems to be rewarding. The animals will even cross an electrified grid to gain access to the lever by which they can obtain the brain stimulus. The frequency of self-stimulation of one specific location decreased until testosterone was injected, after which it increased. In other areas the frequency of self-stimulation increased when the animal was hungry. Undoubtedly there is an activation of mechanisms resulting in pleasurable sensations that are normally associated with mating or feeding (J. Olds, 1958).

Finally there are acquired secondary motivations. One can become used to a certain kind of food and have a very specific desire for it. One becomes attached to one specific environment and is homesick when one is away from it. If all these possibilities are included under the heading of motivation, then the theory of reinforcement by rewards is undoubtedly correct in respect to learning. Simple conditioned reactions and some other learning can also be explained according to Guthrie's principles: a temporal or spatial contiguity of objects or events, without any demonstrable reward or punishment, is remembered. Sometimes animals are taught unusual postures or movements by passively forcing them to perform these forms of kinesthetic learning. Circus animals are taught in this way to stand on their heads. Usually progress in learning is facilitated by additional rewards (H. Hediger, 1954). Balinese girls learn their complex dances through guidance from a teacher by means of kinesthetic training (G. Bateson and M. Mead, 1942; see also Fig. 14-5).

Up to a certain point one can use visual paired-comparison experiments to compare the learning achievements of various animal groups. The animal is presented with two stimuli side by side (for example, a circle and a cross) and allowed to choose. Choice of one stimulus leads to food; the other leads to no reward or to electric shock. The stimuli are randomly switched in respect to position. In this situation the animals learn to choose the rewarded stimulus. An octopus mastered 3 tasks, and was able to discriminate 6 different stimuli (J. Z. Young, 1961). Trout *(Trutta iridea)* mastered up to 6 tasks, iguanas *(Iguana iguana)* 5, large chickens up to 7, and an Indian elephant and a horse up to 20 (B. Rensch, 1962). Memory has been tested in many species; the octopus made 83 percent correct choices after an interval of 27 days. A carp was able to distinguish a circle from a cross after 1 year and 8½ months and

Figure 14-5 On Bali girls learn the complicated dances by watching examples and by direct leading by the instructor (kinesthetic learning). (Photographs: I. Eibl-Eibesfeldt.)

selected the positive stimulus significantly more often. A trout still retained a task after 150 days, a rat after 1 year and 3 months; an elephant retained 12 of 13 visual discriminations after 1 year, and a horse remembered 19 of 20. The optimal performances of these animal groups are quite similar and do not reflect the great differences commensurate with their phylogenetic level of organization (B. Rensch, 1962). M. E. Bittermann (1965) believes that he has demonstrated qualitative differences in learning abilities in fish, reptiles, birds, and mammals. Fish do not learn a habit reversal, a task in which after learning that one of two stimuli is positive, the previously negative one now becomes positive; then when the new task is learned it is again reversed, and so on. Monkeys, rats, and pigeons quickly learn reversals. Turtles did not learn a visual reversal but were able to make spatial reversals, where they had to chose between two identical signals on two different sides. Bittermann has only tested a few species, so that generalizations are not well founded. Gobies and blennies (Gobiidae and Blenniidae) probably would surpass many a reptile in their learning abilities.

NATURE OF THE ENGRAM

A basis for all higher accomplishments is memory, and this has been demonstrated in all animals with a central nervous system, including planaria. Opinions about the learning capacity of the Protista diverge (W. H. Thorpe, 1963). According to B. Gelber (1965) paramecia gather about a platinum wire which had previously been baited repeatedly. H. Machemer's (1966) experiments in training hypotrichous ciliates, on the other hand, yielded negative results. However, he agrees that in principle protozoans can learn.

In vertebrates learning performance is clearly correlated with the size of the brain, whereby the actual size of the brain and the number of ganglion cells seems to be more important than the taxonomic position (B. Rensch, 1962). It seems certain that several levels of the central nervous system are capable of learning (R. Hernandez-Peon and H. Brust-Carmona, 1961). The cuttlefish has a visual and tactile memory, each localized in different parts of the brain (J. Z. Young, 1965). Frogs are able to learn with only their spinal cord (L. Franzisket, 1955). In mammals, most, but not all, experience is stored in the neocortex. By electrical stimulation of various regions of the temporal neocortex, W. Penfield (1952) was able to elicit acoustical and optical hallucinatory images in epileptic patients. The patients also remembered these sensations after excision of the stimulated area. One can conclude, therefore, that the memory trace is also present in the temporal region of the other half of the brain. That memory traces of one brain hemisphere are also projected to the other has been demonstrated by the experiments of R. W. Sperry (1964) and R. E. Myers (1956).

If the optic chiasm of a cat or monkey is cut in the sagittal plane, then the stimuli impinging on one eye are transmitted only to the homolateral half of the brain. After this operation it is possible to teach the animal a simple discrimination (between a circle and a square, for example) with one eye. If it has learned with the right eye, this eye is covered and the animal is then tested for discrimination with the left eye. Such tests of transfer are successful, which proves that during learning a projection of the information from one hemisphere to the other has taken place. However, if before the training one also cuts the corpus callosum (Fig. 14-6), then the animal can perform the task only with the trained eye and completely fails with the other; in fact, the animal behaves as if it had two brains. Each eye learns independently, so it is possible to teach each eye something different. If, on the other hand, one transects the corpus callosum after training with one eye, the animal remembers what it has learned when the other eye is tested. This memory copy does not seem to be as sharp, however. For difficult problems this projection of the memory trace is not sufficient; direct sensory information is more effective. In humans one hemisphere is more specialized for the storage of memories than the other. This is especially true for the memory of words.

Several split-brain cases have been studied in man. It was found that each hemisphere received its own messages and that the subjects behaved in the test situation very similarly to the animals just described. They behaved as if they had two minds, which did not know each other (M. S. Gazzaniga, 1967; R. W. Sperry and B. Preilowsky, 1972).

The nature of a memory trace or engram is not really known today. According to J. C. Eccles (1953) memory consists of electrical reverberating circuits that, once activated by a specific excitation, continue. This reverberating-circuit hypothesis is opposed by theories that assume structural changes at and in the ganglion cells. Morphological changes have frequently been demonstrated at the cellular level. Changes occur in the number of microsomes, apical dendrites of pyramidal neurons swell, and changes take place at the synapses themselves (R. W. Gerard, 1961). In recent years biochemical hypotheses have also been advanced.

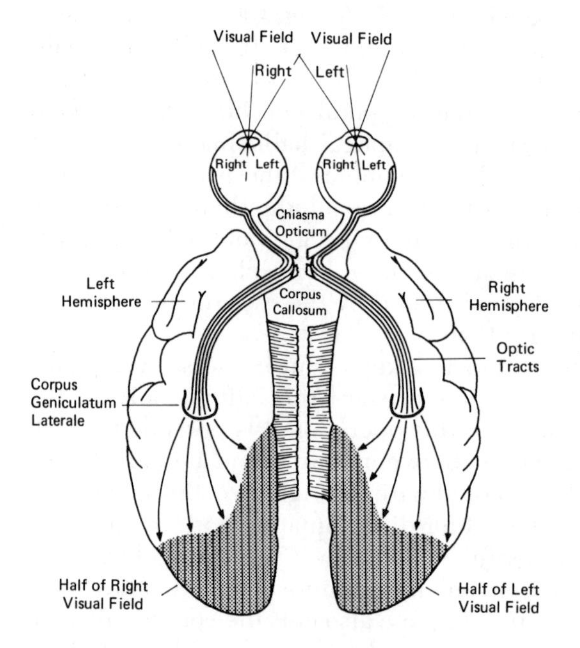

Figure 14-6 After cutting the optic chiasm and the corpus callosum the information obtained through one eye is transmitted to only one cerebral hemisphere. (See explanation in the text.) (After R. W. Sperry, 1964.)

Thus, experiments with planaria (*Dugesia*) suggest a basis of memory in some substance. Planaria were cut in half behind the pharyngeal region after they had learned a task. When the brainless posterior part had regenerated a head, this animal was said to have retained the task (J. V. McConnell, A. L. Jacobson, and D. P. Kimble, 1959; J. B. Best, 1963). The changes that accompany learning must have been distributed throughout the tissues of the entire animal and not restricted to the central nervous system. The biochemical hypothesis for the nature of the engram seems to be supported by experiments which showed that if trained flatworms were fed to untrained ones, the latter learned faster (J. V. McConnell, 1962). A. L. Hartry, P. Keith-Lee, and W. D. Morton (1964) found, however, that planaria in general learn better after cannibalism, even when they ate untrained planaria.

There are indications that ribonucleic acids (RNA) are the carriers of information. If planaria were trained, then cut in half, and their tails were allowed to regenerate in a ribonuclease solution, then no memory was found after regeneration; this was in contrast to the tails of control animals that regenerated in normal pond water (W. C. Corning and E. R. John, 1961).

A. L. Jacobson et al. (1965) and F. R. Babich et al. (1965) trained rats to search for food when a clicking sound was presented. When they had learned the task they ran toward the food dish upon hearing the click, even when it was empty. The trained rats were killed and RNA was extracted from their brains and injected into the peritoneal cavity of 7 untrained rats; 8 control animals received the extract of untrained rats. In 25 tests per animal, conducted by experimenters who did not know to which group an animal belonged, those injected with the brain extract of trained rats approached the food dish on the average of 6.86 times when the click stimulus was presented, while the control animals did so only once on the average. The effect took place 5 hours after the injection and lasted over 24 hours. In a second experiment the conclusion is probable that specific information was transferred. Inexperienced rats, which had been injected with extracts from rats that were trained to respond to the clicks with approach to the food magazine, responded in a critical test (during a random series of light and click stimuli) significantly more to click stimulation. In contrast, animals that had been injected with RNA from rats trained to respond to light stimuli approached food significantly more when a light stimulus was presented. F. R. Babich et al. (1965) were able to transfer memory contents from hamster brains into rats. All these experiments have been criticized, however, and are not yet conclusive.

There seems to be a correlation between an inhibition of RNA metabolism and a decrease in learning ability. An increase of this metabolism by drugs strongly enhances the learning capacity of rats. It has also been shown that the RNA content of nerve cells in rats increases during the first year of life as a result of stimulation. In animals that grew up with a lack of this stimulation, the RNA content remained low (H. Hydén, 1961; H. Hydén and E. Egyhazi, 1962).

A mechanism for these changes is proposed by Hydén as follows: impulse series from motor and sensory cells change the ion balance of the cytoplasm of the affected cells. A given impulse could produce a permanent change in an RNA molecule. This new RNA molecule, although only slightly changed, would direct the synthesis of a protein molecule differing slightly but significantly from that previously produced. Hydén assumes that the new protein has the property of responding to the same electrical pattern that created the change in the RNA. When the same electrical pattern does occur again, the new protein dissociates rapidly, causing an explosive release of the transmitter substance at the synapse. This allows the electrical pattern to bridge the synapse and be passed along by the second cell, then by a third cell, and so on. The nerve cells then respond differentially, depending on whether or not the arriving electrical pattern is new or has occurred previously.

More recently doubts have arisen as to whether RNA is the transmitter of information. The synthesis of RNA depends on the code that is present in deoxyribonucleic acid (DNA). It is possible that DNA changes during learning and that this causes the changes in RNA synthesis (J. Bonner, 1964; J. Gaito, 1964). RNA would then only be a link in a chain of events and not the storage substance for the engram. This role would be assigned to DNA, which is also the carrier of the "inherited memory." As early as the 1890s E. Hering (1921) suspected fundamental similarities between memory and inheritance, which he referred to as organic memory. Whether or not the ability, acquired during phylogenesis, to store information on a molecularly programmed basis, is used during learning has to be shown by additional experiments. There is much in favor of this view, but some caution is advisable in the interpretation of the work cited so far, especially because a number of investigators who repeated the rat experiments of Jacobson and Babich could not confirm their results (W. L. Byrne et al., 1966). For more recent discussions of this controversial subject see S. P. R. Rose (1970) and E. J. Fjerdingstad (1971).

The structural hypothesis and the reverberating circuit hypothesis earlier discussed supplement one another. It has been found, in the meantime, that a distinction can be made between short-term and long-term memory, which, as was shown in experiments with the octopus, are located in different parts of the brain (B. Boycott, 1965).

Short-term memory could depend on reverberating circuits. This hypothesis is supported by the observation that this type of memory can be eradicated by supercooling to the point of cessation of all electrical brain activity and also by electroshock. This short-term memory precedes long-term memory; the more time that has passed after training, the more difficult it is to eradicate what has been learned.

Rats and hamsters show a completely normal learning curve if they are subjected to an electroshock or supercooling 4 hours after a set of trials. If the same is done 1 hour after training, learning is slower. At an interval of 15 minutes between training and shock, learning is significantly disrupted, and if the animals are shocked 5 minutes after each training session, no learning improvement occurs. If a goldfish brain is injected with a drug that inhibits protein synthesis following a learning trial, then the fish forgets what it has learned; no long-term memory develops (B. W. Agranoff, 1967).

It is quite possible that excitation initially activates reverberating circuits, whose activity over a longer period of time leads to structural and biochemical changes (D. O. Hebb, 1949; R. W. Gerard, 1961). In line with this theory a dynamic principle is postulated for short-term memory, while structural changes are considered to be the basis for long-term memory.

ABSTRACTION, NONVERBAL CONCEPTS, AND INSIGHT BEHAVIOR

A notable result of training experiments is the realization that animals are capable of achievements usually placed in the category of human "higher brain functions."

This is true primarily for the ability to abstract and to generalize. In the previously discussed **visual discrimination experiments**, an animal learns to distinguish two simultaneously presented figures, patterns, or colors. The approach to one pattern is rewarded (positive stimulus) and to the other is punished (negative stimulus). During the course of this training it was found that the animals can recognize similarities between figures. They abstract and draw, so to speak, the nonverbal conclusion "This is similar to that" (B. Rensch, 1965). Even fish are capable of forming such nonverbal concepts. Minnows that were trained to distinguish a triangle as the positive stimulus from a square as the negative stimulus reacted positively to an acute angle versus a straight line, in line with the previous training. They apparently had "abstracted" the cue "pointedness" as the positive stimulus. An elephant that had learned to differentiate an X as a positive stimulus over a circle, later responded positively to any stimulus that contained crossed lines. In these cases the animals seem to learn only the most characteristic cues contained by the positive stimulus and ignore the rest.

This also seems to take place in the natural situation. Toads that have been taught to avoid inedible prey models "abstract" various differential cues. Initially they react quite specifically. Only later do they begin to generalize. Toads that I trained recognized negative models at first only in a specific location and when they were moved in a certain way. They had to learn to recognize the models in other situations. Some learned to distinguish the models according to color, and they recognized this color on other models. Other animals avoided everything presented after only a few negative experiences, everything, that is, except mealworms. They even rejected grasshoppers. One female that responded in this way later learned to distinguish a moving model that was pulled past her from prey that moved on its own. She eventually ate grasshoppers, but avoided mealworms that were pulled on a string. Mealworms that moved by themselves were eaten (I. Eibl-Eibesfeldt, 1951a). Rhesus and capuchin monkeys are able to recognize pictures of insects and flowers, implying a capacity to generalize (E. Lehr, 1967).

Animals do not necessarily learn all the characteristics of the positive or negative stimulus. Sometimes they learn the relation of the patterns to each other. If wider stripes are presented versus narrower ones, the animal does not learn the specific width of a stripe but instead recognizes the wider or the narrower of the two as the training stimulus. If the narrower stripe has been learned as the positive stimulus, and a new stimulus pair is now presented which contains a still-narrower stripe, then this new stimulus and not the original positive one (which now is negative) is responded to. In a similar manner, animals learn to choose the lighter of two shades of gray, where the positive stimulus is always lighter than the negative stimulus.

Abstractions are at the same time generalizations. B. Rensch and G. Dücker (1959) have shown how far this "nonverbal" concept formation can be carried. They trained a civet cat to develop the concept "bent/straight" by training the animal to respond to two parallel half-circles as the positive stimulus and to two parallel vertical stripes as the negative stimulus. When the animal had learned this, the patterns were rotated 90°, then in a different arrangement of the stimulus components,

Figure 14-7 Civet cat choosing between two patterns. (Photograph: B. Rensch.)

so that the animal was forced to pay attention to the one constant feature "bent in a circle." Finally, the animal preferred any bent lines over straight ones even when they appeared against an entirely different background. In a similar manner, Rensch and Dücker were able to teach the civet cat the concept "equal/unequal" (Fig. 14-7).

In this connection investigations dealing with "value concepts" in primates are important. Chimpanzees learned the differential symbolic value of chips of various colors and sizes (J. B. Wolfe, 1936; J. T. Cowles,

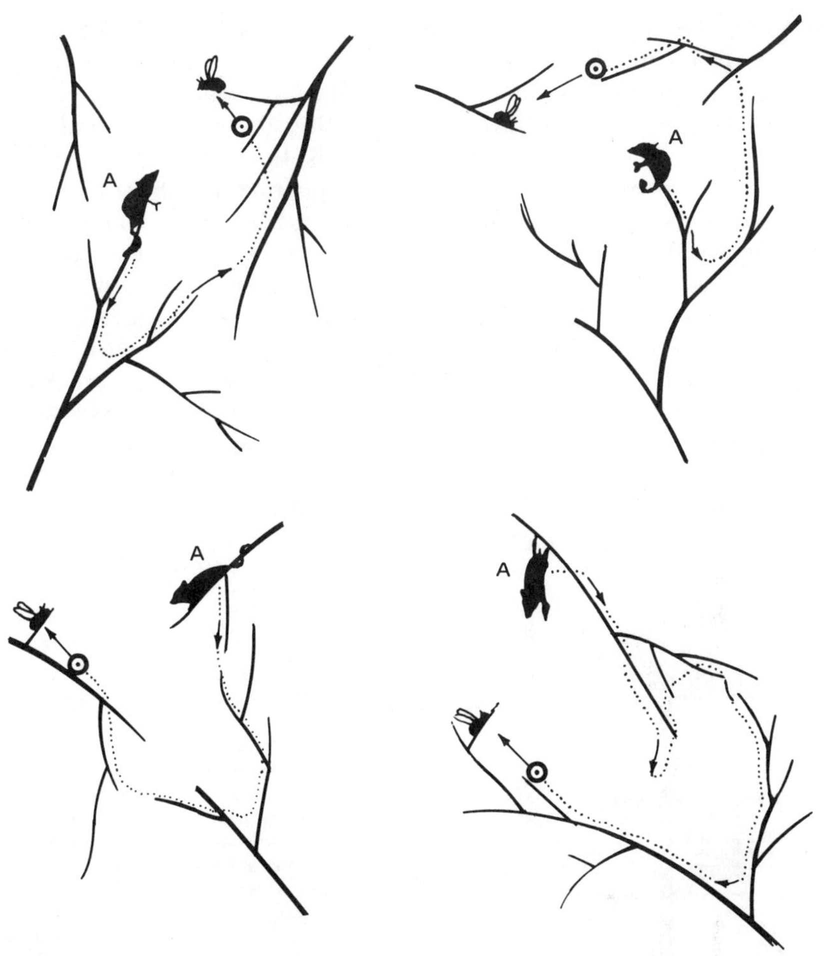

Figure 14-8 Examples of solutions to detour problems in the chameleon while stalking prey. *A:* Beginning of detour; ⊙ : position at time of tongue flick. (From O. v. Frisch, 1962.)

1937). The animals were able to insert blue, white, and brass discs into a food dispenser and receive one or two grapes or none at all. Soon they preferred the blue chips, which brought the biggest reward. The animals could be trained to perform certain tasks by rewarding them with chips which they then exchanged for food. These experiments were further varied by teaching the animals that some chips were for food, others for getting out of the cage, and others for playing with the keeper. Some animals utilized the chips according to their needs. One female would drop light blue chips into a slot in the door so that it would open whenever the cameraman, of whom she was afraid, arrived to take her picture. The chimpanzees undoubtedly had learned the value of the various chips. This same ability has been demonstrated in some other lower monkeys by T. Kapune (1966; see also for additional references). A female rhesus monkey finally mastered six different value concepts.

The highest achievements of generalization seem to have been discovered during the investigations of counting abilities of several animal species by O. Koehler (1943, 1949, 1952, 1954b, 1955). Pigeons, parrots, ravens, and squirrels learned to take only a certain number of kernels or pieces of food from a larger number. Ravens were taught to take only as many as were indicated on a sign. If the sign contained two dots, they chose the food dish whose cover also contained two dots. After further training it was no longer necessary for the size and arrangement of the dots on the sign to coincide with those on the cover. The birds selected solely on the basis of number. One gray parrot learned to open as many food dishes out of seven and take out as many kernels as was indicated by the number of times a bell rang. The bird also learned to take food kernels in response to two or three light signals, and then transpose, without further training, to one acoustic signal.

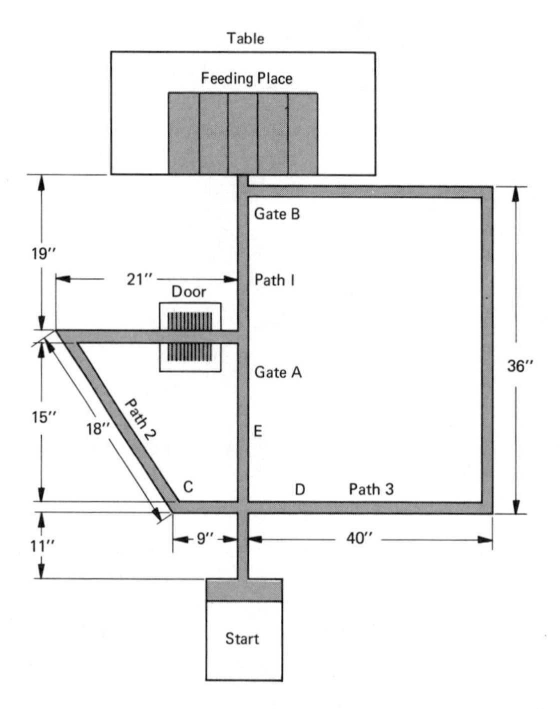

Figure 14-9 Maze used by Tolman and Honzik to demonstrate insightful behavior. Rats learned each of the three paths that led to the goal. When the straight path *1*, which led directly to the goal, was blocked at *A*, the animals could reach it via paths *2* or *3*. When path *1* was blocked by gate *B*, the rats at once selected the only correct path, *3*, which normally they did not prefer, and this was considered evidence of insight. Only a few rats were able to solve this problem in the elevated maze shown here. (After E. C. Tolman and C. H. Honzik, 1930a.)

The spontaneous development of nonverbal concepts without trial-and-error learning is already a hint that *insightful learning* is taking place. This is also the case when an animal masters a detour and combines several independently acquired experiences (Fig. 14-8). When I once would not allow my badger to enter my living quarters, it soon gave up its useless scratching at the door, and ran around the building and entered through a window on the other side. Such spontaneous insightful behavior was also shown by a dog who entered a garden that it had avoided until then (by opening the door leading to it) after discovering that its rival had been tied up there (W. Gnadenberg, 1962).

H. H. Hsiao (1929) built a maze in which a rat could reach the food box by three different routes. Two shorter paths led to a common entrance that could be blocked by a gate; the third path was much longer. All three paths were well known to the animals and they learned to use the shortest route. If the common entrance was now blocked, the rats at once used the third, longer, path without trying the second, shorter, route.

Figure 14-10 (*a*): Chimpanzee putting sticks together; (*b*): chimpanzees in an experiment to try to reach a suspended piece of food by placing boxes on top of each other and using a stick. (After W. Köhler, 1921.)

They behaved as if they possessed a spatial picture of the maze and knew that with the closing of the gate the second route was also blocked. A similar maze was used by E. C. Tolman and C. H. Honzik (1930a; see also Fig. 14-9).

Correct insightful behavior in the sense of understanding relationships is often observed in tool using. Here W. Köhler's (1921) experiments with anthropoid apes showed the way. His chimpanzees used sticks to pull in bananas that were outside their cages. They could put two short sticks together to make a longer one. They could put boxes on top of one another to reach a banana suspended from the ceiling (Fig. 14-10). From the descriptions one can see that the behavior sequences were not learned by trial and error. A chimpanzee would sit and look around, at the box, at the place under the banana, and at the banana, until the solution had been found. In these examples the behavior sequence is thought out as though the trial is internalized. The ability to plan mentally was demonstrated by chimpanzees that had even learned to visual-

Figure 14-11 Chimpanzee fishing for termites. *Left:* Selection of the tool; *right:* insertion of the tool into an opened passage of a termite hill. (Photographs: Baron and Baroness H. van Lawick-Goodall, with permission of the National Geographic Society.)

ize complex mazes covered by Plexiglas. The animals discovered the shortest path to the goal solely by looking at the maze and were later able to maneuver a piece of iron to the goal with a magnet (B. Rensch and J. Döhl, 1968).

Planning with foresight was also described by M. P. Crawford (1937). Two chimpanzees had to pull in a food box by ropes. The animals had to cooperate, because the box was too heavy for one alone. At first one animal tried to pull the box alone. When this failed, it directed the other's attention to the rope, gesturing at the rope. Additional examples about insightful behavior and tool using in captive chimpanzees can be found in N. Kohts (1935) and R. M. Yerkes (1948).

Observations from the wild are available (J. van Lawick-Goodall, 1968, 1970). The free-ranging chimpanzees she observed pulled termites from their tunnels with thin twigs or grass stems. They opened a tunnel the termites use when swarming, pushed the tool into the tunnel, and pulled it out, with the termites clinging to it (Fig. 14-11). This was repeated over and over. They were very careful in their choice of tools. In that same area chimpanzees use leaves to soak up water from holes in trees, which they cannot otherwise reach with their lips. They use the leaves to clean themselves. In one of the films made by Baron and Baroness van Lawick-Goodall, a chimpanzee suffering from diarrhea picks leaves and cleans itself. In captivity, one female chimpanzee cleaned her young with a cloth after each elimination (K. Heinroth-Berger, 1965). Tool using alone is no criterion for intelligence, but a varied, individually modifiable utilization of tools, as found in the chimpanzee, certainly is.

Interesting observations about tool using in chimpanzees were made by A. Kortlandt (1962, 1965, 1967a, 1967b). Chimpanzees that were captured in the savannah and kept in a large enclosure in Guinea directed well-aimed blows from above toward a stuffed leopard while they were standing upright (Fig. 14-12). Chimpanzees of the forests in the Congo were much clumsier in the same situation. They also beat about with sticks, even threw them in the direction of the predator, but their aim was bad and they never directed a blow from above at the predator, and they never hit the stuffed leopard. Kortlandt believes that the use of sticks as weapons originated in the open savannah. When the ancestors of the forest chimpanzees were pushed into the forest through competition with man's ancestors, their skill in using weapons atrophied.

H. Albrecht and S. C. Dunnet (1971) have made further observations relevant to this topic. Both draw attention to local peculiarities. For instance, the chimpanzees they studied did not possess the habit of fishing for termites by using sticks. Furthermore, T. T. Struhsaker and P. Hunkeler (1971) have reported that chimpanzees on the Ivory Coast crack open nuts with clubs and stones, a habit not observed with any other group.

Figure 14-12 Savannah chimpanzee beating a stuffed leopard with a stick. (Photograph: A. Kortlandt.)

Intelligent behavior implies the capacity of the individual to solve problems by recognizing relations between objects in a context not encountered before ("insight"). By learning processes during ontogeny the animal acquires knowledge that contributes to finding new solutions of problems. The capacity to perform intelligent operations was acquired during phylogeny, particularly in richly differentiated environments (K. Lorenz, 1973). Thus, the high capacity for insightful behavior in man developed together with binocular vision during his arboreal primate phase (see Chapter 18). With the capacity to internalize trial and error behavior—to think or act in imagination—a decisive step in the evolution of higher brain capacities was taken.

SUMMARY

The ways in which skills and knowledge are acquired have been studied experimentally. The laws of acquisition and extinction of conditioned reflexes and the processes of instrumental conditioning are well known. How acquired information is stored, however, is as yet little understood. Storage certainly takes place within the central nervous system, and by electrical stimulation memories can be recalled, but the physiological basis of the memory trace is as yet unknown. Changes on the synaptic level, the establishment of new connections between neurons, changes on the molecular level, and reverberating electrical circuits are hypothesized. The experimental evidence for the molecular hypotheses seems to be controversial. There are indications that long-term and short-term memory involve different physiological mechanisms. Higher vertebrates are able to abstract, generalize, and solve problems not encountered before by insight.

15

Ecology and Behavior

In previous chapters I have discussed the functions of various behavior patterns, but primarily the causal and historical aspects have been considered. In this section I want to discuss the factors with which an organism has to deal in its environment and how its behavior must adapt to the various contingencies. Each organism, first of all, must maintain a constant internal milieu and defend this homeostatic balance against various disruptive influences. The organism must grow and reproduce itself. We will discuss the dependence of these functions on diverse environmental factors. The function of behavior in preserving the species will be illustrated with some selected examples.[1] Much will inevitably be omitted. The presentation of the various locomotor behavior patterns alone, such as running, swimming, jumping, climbing, and flying, would go beyond the scope of this book.

For the purpose of clarity I will group the relationships to the nonspecies environment separately from the relations to species members. We must remember, however, that many behavior patterns are not really restricted to one or the other grouping. Many, such as the behavior patterns of locomotion, serve in several functional cycles. An animal may run to catch up with a rival or to escape from a predator (see **functional organization of behavior**).

[1] Additional literature on this topic is the unsurpassed work of R. Hesse and F. Doflein (1943) and the summaries by G. Tembrock (1954), W. Kühnelt (1965), and A. Kaestner (1965).

RELATIONSHIPS TO THE ENVIRONMENT OTHER THAN THE SPECIES

Adaptations to Nonbiological Factors

That each animal species has its own preferred temperature was demonstrated by K. Herter (1943, 1952, 1953) in numerous investigations. Animals in a temperature-choice apparatus were allowed to choose a room with a specific temperature. They gathered in a room whose temperature matched the optimal temperature for the species. In the same way they selected a certain humidity, they avoided dryness or sought it, and they sought exposure to the sun or avoided it. In short, a number of behavior mechanisms that enable animals to select their appropriate biotope (cliffs, loess walls, meadows, brush, and so on) are inborn. The New World mouse *Peromyscus maniculatus bairdi* selects grassland; *Peromyscus maniculatus gracilis*, on the other hand, selects woodland. Both subspecies occur in adjacent areas, but they are so strictly separated with respect to their biotopes that they do not hybridize, although in captivity they do so readily. T. v. Harris (1950) raised mice of both subspecies in captivity and allowed them to choose between terraria that were planted differently. In one he planted small trees; in the other he imitated grassland with thin paper strips. *Peromyscus maniculatus bairdi* innately preferred this "grassland" while *P. m. gracilis* selected the woodland habitat. In a room half of which contained pine branches and the other half oak branches, sparrows (*Spizella passerina*) selected the side with pine branches, although there were an equal number of perches on each side. The choice matched the natural preference of their biotope (P. H. Klopfer, 1963; P. H. Klopfer and J. P. Hailman, 1965).

Environmental adaptations also include those behavior patterns with which an animal creates a shelter against bad weather—structures and nests that protect it against heat as well as cold. It is known that many desert mice can survive the midday heat only by retreating to their dens (F. Bourliere, 1955). The earth mounds around the dens of prairie dogs protect them against flooding. Beavers dig dens into the riverbanks, but if the bank is so low that no cave can be dug, then the beaver will build its den in a large mound of twigs. The animal will gather branches up to four meters long, heap them on top of one another, plug the holes with mud, earth, and weeds, and finally will make a den inside. The entrance to the den is underwater. The beaver ensures itself adequate protection and a reliable means of transporting his food by damming up running streams. A drop in the water level immediately produces dam-building activity (G. Hinze, 1950). When building a dam the beaver cleverly utilizes natural elevations as supports. He piles twigs and branches on top of each other with the thick end facing upstream. The side branches catch on each other and mud collects at the steeper and upstream side of the dam. The dam may be anchored with trees that have been felled across the stream and by poles that have been pushed vertically into the bottom (P. B. Richard 1955, 1964; see also Fig. 15-1). The dams are rarely higher than 1.50 meters. They can be several hundred meters long and are then the work of many generations of beavers. In this way the beaver goes a considerable way in creating his own appropriate environment.

Termites provide yet another example. Some regulate the humidity and temperature within their mound so that it remains around the optimum of 30° C and 98 to 99 percent humidity. To accomplish this, water bearers carry water from tunnels that lead down to the groundwater level. The termites protect themselves against extreme temperature changes by building hard, thick walls that shield them against the environment. The high mounds, which have ridges, are made from a clay-saliva mixture. The ridges are used for ventilation. Air ducts lead up and down within them. The used-up air rises from the center of the nest to a central roof chamber and from there returns to the bottom via the air ducts. Carbon dioxide is given off through the pores of the outer wall and oxygen is taken up. The fresh air collects in a chamber below the nest and from there rises upward. The termites regulate the ventilation by widening or narrowing the air ducts in accordance with the need for oxygen and for warmth or cooling (Fig. 15-2).

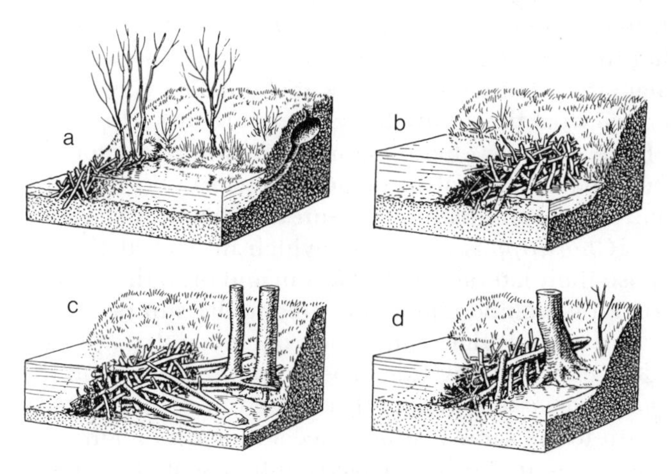

Figure 15-1 (*a*): Beaver den in the bank of a river with the entrance below the water surface. A dam raises the water level. (*b*)–(*d*): Various types of dam with trees along the banks used as supports and with tree trunks rammed into the bottom for support. (After P. B. Richard, 1955.)

Special behavior patterns for fur and feather care maintain their insulating and water-repellent qualities. Waterbirds whose feathers are caked together quickly perish. Behavior patterns for bodily care are therefore adaptations to climatic factors (M. Bürger, 1959; H. Dathe, 1964). Emperor penguins (*Aptenodytes forsteri*) court and incubate during the continuous night of the antarctic winter. Storms rage with

Figure 15-2 Termite hill of *Macrotermes natalensis.* One sector of the nest has been cut away to allow a view of the ventilation system. (P. Farb, *Insects,* Silver Burdett, 1964.)

average velocities of 80 km per hour and reach as high as 140 km per hour, coupled with temperatures of minus 60°C. The penguins survive by huddling together in V formations that effectively break the wind. In this way they maintain their body temperature at about 35.7°C, whereas in individuals separated from the group the temperature drops to 27.9°C (J. Prévost, 1961). Here extreme environmental conditions are met by adaptations of social behavior (see also **contact behavior**).

Reptiles assume special sun-bathing postures so that they warm up faster. Grasshoppers (*Chorthippus dorsatus*), which in central Europe court in the fall, expose their lateral side to the sun and drop their thigh, thus exposing the dark-pigmented abdominal area to the sun rays.

Many animals of the intertidal region seek out areas that are suitable for surviving desiccation at low tide. The limpet (*Patella*) and the false limpet (*Siphonaria*) have independently evolved an attachment to a home base. Both return to the same resting place after each foraging excursion. The outlines of their shells fit exactly into a seat (Fig. 15-3). There are also behavior patterns that are adapted specifically for survival in a particular ecological niche. Whenever an animal consistently seeks out a specific habitat, a number of changes in body structure and behavior evolve. We mention here only the parallel adaptations of fish in rapidly flowing waters, of bottom-dwelling fish, and of birds that breed on cliffs (E. Cullen, 1957; W. Wickler, 1958, 1959, 1965c; additional examples in Wickler, 1961a). Fish that live in the high seas, where they normally never meet with an obstacle, bump against the walls of the aquarium without ever learning not to, just as some birds of the plains bump into cage walls (K. Lorenz, 1959). By comparison, animals that live in structured biotopes such as woods or coral reefs show much more **intelligence.**

The Norway rat and the house rat (*Rattus norvegicus* and *R. rattus*), both followers of man, come from different biotopes, which are clearly reflected in their behaviors. The house rat, which originated from tree-dwelling forms and which in southern countries today still nests in trees, prefers to settle in the upper floors of buildings—hence the name roof rat. It climbs well, and when panicked tries to climb upward even if there is no hiding place there. This animal is primarily a vegetarian. The Norway rat, which in the wild lives near water and on river banks, prefers to live in the lower floors of human dwellings—the cellar rat. It also lives in sewage systems, where it fishes for food objects in the water with special straining behavior patterns. On occasion the animal leads a predatory way of life (I. Eibl-Eibesfeldt, 1953d).

Finally, some nonbiotic adaptations to environmental conditions are related to periodic environmental changes of the tides, diurnal rhythms, moon phases, and annual cycles. These will be discussed in Chapter 17. I will also discuss several questions concerning the spatial orientation of animals in Chapter 16.

Figure 15-3 Place loyalty of *Siphonaria gigas.* The young fit exactly into the seats, which will be eaten into to fit the shells of the adults. *Left:* an animal in its resting place; *right:* the empty seat of another snail, showing the shape of the seat and the impression of the foot. (Photograph: I. Eibl-Eibesfeldt.)

Procurement of Food

Adaptations to the procurement of food are just as numerous as adaptations to the nonliving environment. I will illustrate this by a few examples. In general, the specialists predominate. The struggle for existence of various species often leads to surprisingly specific adaptations. The small leatherjacket, the filefish (*Oxymonacanthus longirostris*), specializes in picking off individual coral polyps. Many coral fish have developed long snouts that enable them to pick off small animals hiding between crevices in the coral (Fig. 15-4). They not only have the appropriate morphological adaptations, but in their food-getting behavior shows a strong preference for seeking food in crevices and clefts, even when kept in an aquarium where such crevices do not contain food. Bottom feeders are often so rigidly specialized for foraging on rocks that it is very difficult to get them to accept food that lies loose on the bottom. This is true for the Moorish idol (*Zanclus cornutus*), which Lorenz was successful in keeping after many unsuccessful attempts to feed it. Finally, he placed chopped clam meat on stones, let it dry enough so it would stick, and placed the stones into the water. The fish at once took to this food and began to forage.

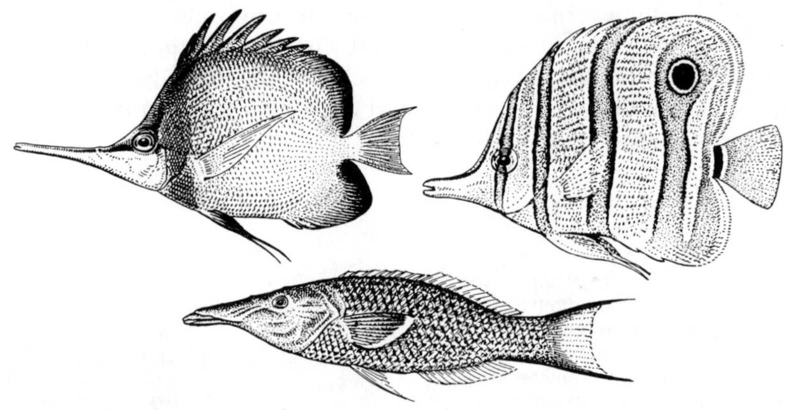

Figure 15-4 Pipefish, which pick their food out of the branches of corals. *Top:* The two butterfly fish *Forcippiger longirostris* and *Chelmon rostratus; bottom:* the lipfish *Gomphosus.* (From I. Eibl-Eibesfeldt, 1964c.)

The **saber-toothed blennies** *Runula* and *Aspidontus* tear chunks of flesh from the fins of other fish. Trumpet fish (*Aulostomus maculatus*) approach their prey by hovering above peaceful species such as parrot fish. When the latter feed, small fish approach in search of food particles that are broken off by them. At the opportune moment the trumpet fish will glide from its position above the parrot fish and catch the smaller ones (H. Hass, 1951; I. Eibl-Eibesfeldt, 1955a; see also Fig. 15-5). The archer fish (*Toxotes jaculatrix*) ejects a stream of water at insects that rest on leaves above the water surface. Remaining just below the surface the fish aims, and toward the end of the shot it jerks up its head, scattering the stream of water, which increases the probability of hitting the prey (H. Lüling, 1958; see also Fig. 15-6). The various methods used by spiders to catch their prey are as diverse as they are surprising (A. Kaestner, 1965); the lasso spider *Mastophora* throws a thread with a sticky ball at its end at passing insects. Other lasso spiders, such as *Cladomelea* and *Dicrostichus*, sit on a horizontal thread from which they suspend another thread with a sticky ball on the end. This pendulum is swung about with the claws of one leg. Insects that bump into this contraption

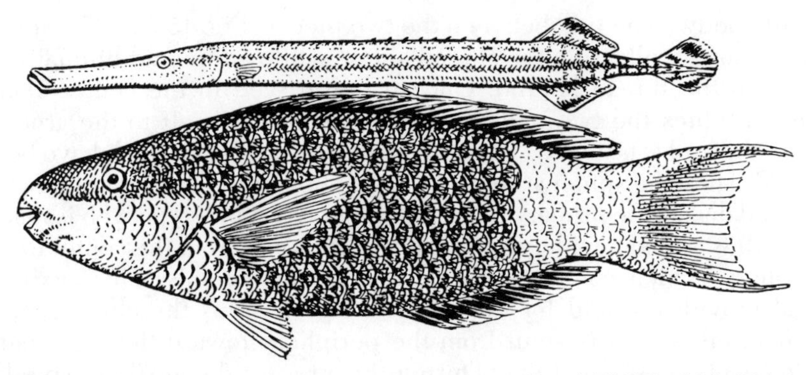

Figure 15-5 "Mounted" trumpet fish (*Aulostomus maculatus*). (From I. Eibl-Eibesfeldt, 1964c.)

stick to it and are caught (Fig. 15-7). The complex nets of the spoke-wheel web spiders (Fig. 15-8), which have been studied in detail by H. M. Peters (1939, 1953) and G. Mayer (1952), are well known. The spider begins by stretching a thread from one object to another. This is accomplished by raising its hindquarters in the air and ejecting a long thread that is caught by the wind. If it does not stick anywhere, the spider pulls it in and tries again. If it should stick, for example on another branch, the spider will also fasten its end, then move out onto this bridge, biting the thread in half but holding the two ends with the legs so

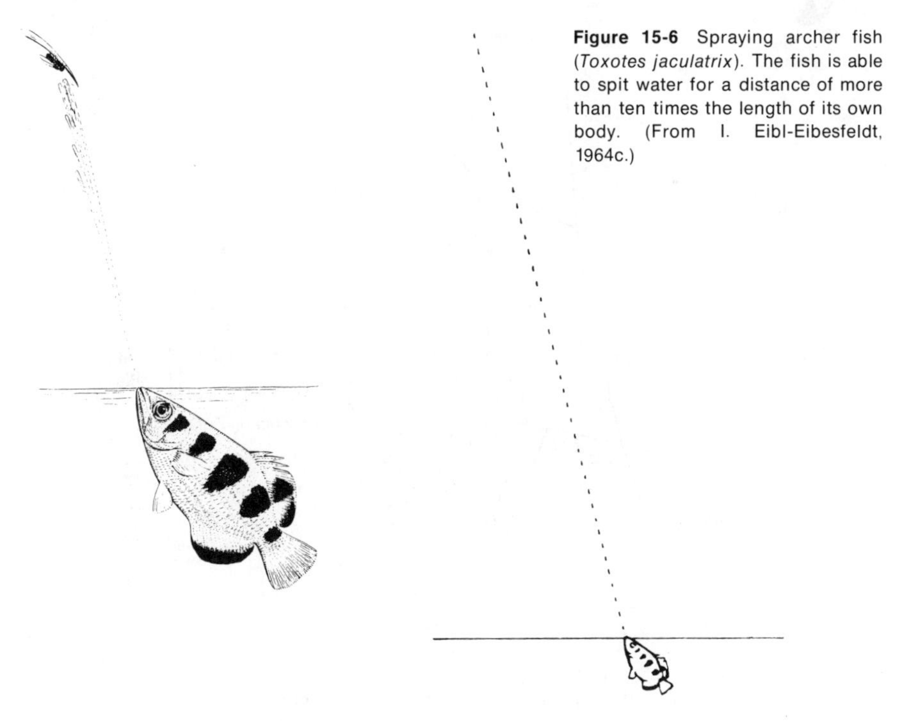

Figure 15-6 Spraying archer fish (*Toxotes jaculatrix*). The fish is able to spit water for a distance of more than ten times the length of its own body. (From I. Eibl-Eibesfeldt, 1964c.)

that the body is the link between the two pieces (Fig. 15-9*a*). The spider moves along, collecting the thread ahead while adding additional substance to the thread behind. Once it has arrived in the center of the thread it glues the two ends together and lowers itself to the ground, where the end is fastened. Thus the first three radii of the web have been formed. Then the spider spins new spokes beginning in the center of the web and makes the primary frame. In this way a spoke-type web is made, to which is added a widely spaced auxiliary spiral from the center to the periphery, which connects the radii at large intervals. This auxiliary spiral provides a hold for the spider when it spins the closer-spaced catch spiral, which is spun from the periphery toward the center and which contains sticky drops. During this process the auxiliary spiral is dismantled (Fig. 15-9*d*).

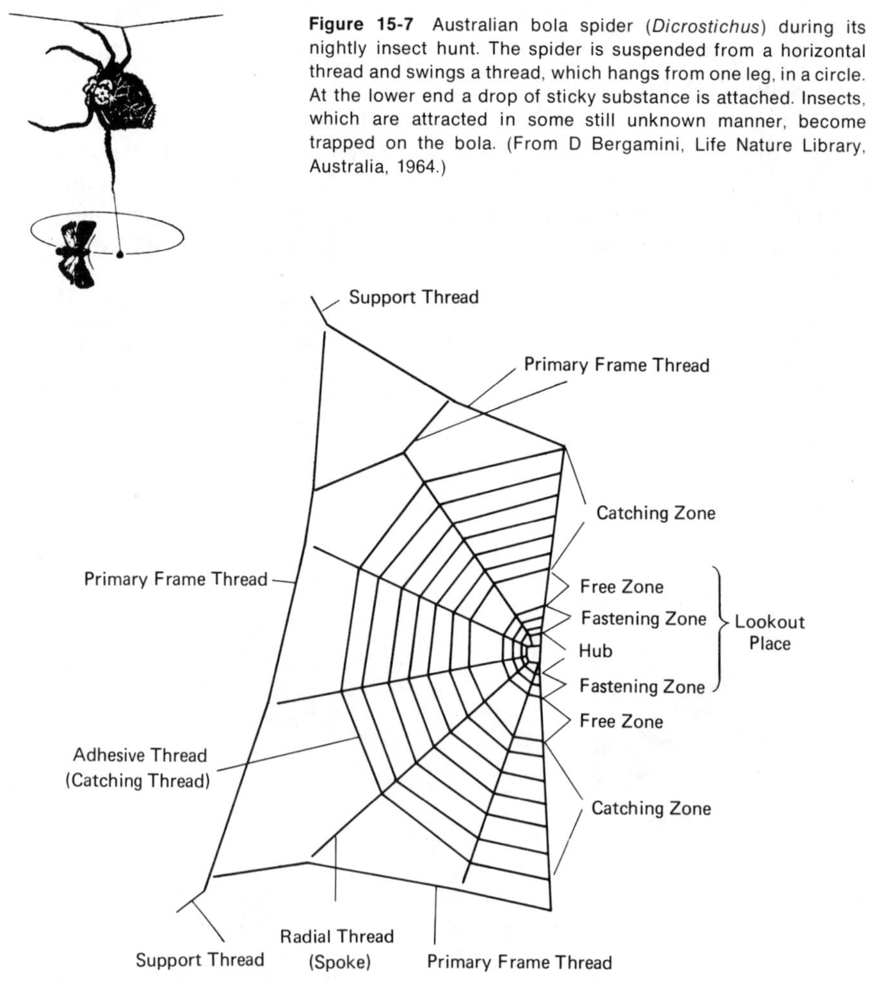

Figure 15-7 Australian bola spider (*Dicrostichus*) during its nightly insect hunt. The spider is suspended from a horizontal thread and swings a thread, which hangs from one leg, in a circle. At the lower end a drop of sticky substance is attached. Insects, which are attracted in some still unknown manner, become trapped on the bola. (From D Bergamini, Life Nature Library, Australia, 1964.)

Figure 15-8 Outline of a spider web. (After H. M. Peters, 1939.)

Hunters that capture quick-moving prey frequently stalk it. My tame iguanas (*Tropidurus*) stalked flies which they took from my hand, and they never learned that stalking was not necessary. When they saw the fly in my hand they quickly approached to within 30 cm; then they stalked closer, body pressed to the ground, until they had come very close, after which they made a final lightning-quick dash to grasp the prey. I have already discussed the technique of capturing dangerous prey by **polecats.** Other predatory mammals behave in a similar way (see P. Leyhausen, 1956a, 1956b).

The California sea otter (*Enhydra*) opens *Mytilus* clams by beating them against a stone, which is balanced on the stomach while the animal is floating on its back (K. R. L. Hall and G. B. Schaller, 1964). The Egyptian vulture (*Neophron percnopterus*) opens ostrich eggs with stones that are collected nearby. The bird positions itself next to the egg, raises its head as high as possible, and flings a stone down on the egg. Some stones weigh as much as 300 grams. This is repeated until the egg is cracked (J. van Lawick-Goodall and H. van Lawick, 1966). The North American nuthatch (*Sitta pusilla*) uses bits of scotch pine as a lever to loosen pieces of bark that hide insects (D. W. Morse, 1968).

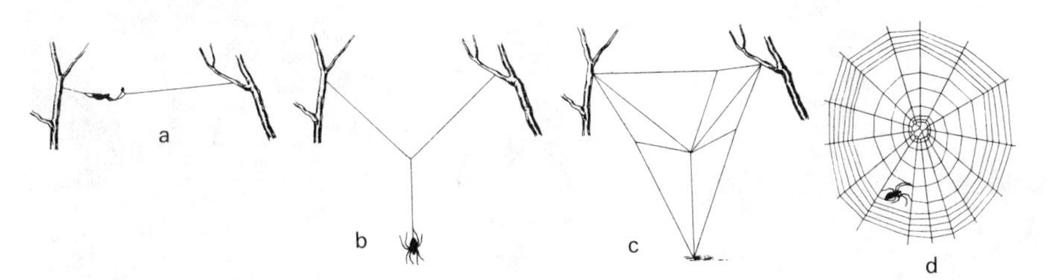

Figure 15-9 Construction of a spider web. (Explanation in the text.)

From these descriptions it should be clear that animals that live on other animals have developed the most varied adaptations. Fruit- and plant-eating animals have fewer adaptations, partly because prey animals evolve various forms of defense. This is correlated with the general observation that grazing and browsing animals are somewhat less intelligent than predators, which show more varied appetitive behavior, learn more, and perform spatially **insightful behavior.**

These varied specializations are the result of competition. The adaptive radiation of bony fishes shows this clearly, as does that of the Darwin finches, which, originating from one basic form, eventually filled the most varied ecological niches. Their different ways of feeding are expressed in the shapes of their bills (Fig. 15-10) as well as in their behavior. The cactus ground finch (*Cactospiza scandens*) has a pointed bill which it pokes into the flowers and seeds of cacti. The small insectivorous tree finch (*Camarhynchus parvulus*) searches for insects, and the woodpecker finch (*Cactospiza pallida*), lacking a long, sticky tongue, uses a long pointed twig or cactus spine for probing insects out of holes (Fig. 15-11). In Hawaii, where a similar adaptive radiation took place, the Akiapolaau (*Heterorhynchus*) fills the ecological niche of a small woodpecker. This bird gets at insects in a still different way. It chips at the wood with the straight lower part of the bill and retrieves the prey with its curved upper bill. In a bird on New Zealand, *Heteralocha acutirostris*, the male has a short, straight bill for chipping, the female a longer, curved probelike bill. Here both sexes collaborate when in search for food (D. Lack, 1947; see also Fig. 15-12). Many animals lay in stores; I have given the example of the **squirrel.** The various feeding patterns of insects are fascinating. The fungi-growing ants (*Atta*) make fertilizer from leaves they have carried in. They cultivate fungi and live off their bulbous swellings. One could fill an impressive volume with descriptions of ways of feeding in hymenoptera (H. Bischoff, 1927). Special adaptations with respect to body structure and behavior are found in those animals that live as **parasites** of other animals. I will discuss this separately.

Figure 5-10 Three Darwin finches as examples of adaptive radiation (Indefatigable, Galápagos). Notice the different shapes of the bills. *Top left:* Small ground finch (*Geospiza fuliginosa*); *top right:* medium ground finch (*Geospiza fortis*); *bottom:* cactus ground finch (*Cactospiza scandens*). The first two species are primarily seed eaters, specialized to feed on various seeds. The cactus finch with its sharp bill feeds on cactus fruits and drills into the juicy cacti. The species live side by side in he same biotope. (Photographs: I Eibl-Eibesfeldt.)

Figure 15-11 Tool using in the woodpecker finch (*Cactospiza pallida*). *Top left:* Insertion of the tool (cactus spine); *top right and bottom:* probing for and lifting a larva. (Photographs: I. Eibl-Eibesfeldt.)

Figure 15-12 Adaptations of birds that search for insects hidden in wood. The woodpecker (*1*) probes with its long tongue. *Heterorhynchus* (*2*) chips wood with the lower bill and probes with the upper bill. The woodpecker finch (*Cactospiza*) (*3*) uses a cactus spine or a piece of wood as a tool for probing. In *Heteralocha* (*4a*), the male does the chipping and the female does the probing (*4b*). (From D. Lack, 1947.)

Defense against Predators and Interspecific Competition

In competition with the pursuer, the pursued continues to evolve new adaptations in body structure and behavior that serve their defense and escape. The commonest response, of course, is to flee from the predator. There is commonly a species-specific flight distance that can be varied through individual experience (H. Hediger, 1934). In general small animals have a shorter flight distance than larger ones. The less protection a species has through other means, the longer is the flight distance. A protectively colored grouper which is hard to see may be approached very closely. The same is true of a hare that hugs a depression in the ground. Conspicuously colored fish, on the other hand, will soon flee into a hiding place when they are approached (Plate VII), but the poisonous dragonfish and other well-armored species permit a person to approach quite close. In these fishes conspicuous coloration serves as a warning. Fish that are thus protected flee less readily and are more easily kept in captivity because of their tameness. The three-spined stickleback (*Gasterosteus*) is better protected by its long spines than the ten-spined stickeback (*Pygosteus*) and is correspondingly less shy (R. Hoogland, D. Morris, and N. Tinbergen, 1957).

The direction and the goal of flight are frequently fixed as a phylogenetic adaptation in the same way as responsiveness to flight-releasing stimuli. A squirrel flees to the tops of trees, a mouse into its hole, a beaver dives, and a pheasant flies up. In two closely related gecko species on New Britain H. Hediger (1934) observed that one species always fled up the tree trunks while the other fled downward and hid in a crevice.

In a coral reef many fish have specific hiding places. Above the large reefs in the Indian Ocean large swarms of blue triggerfish (*Odonus niger*) swim about. When approached each flees into a particular hole in the reef. Many damselfishes, such as members of the genus *Chromis*, hover above clumps of coral and flee between the stalks when alarmed. A diver can break off such a clump and bring the entire swarm to the surface, so strong is their attachment. Many fish have evolved special wedging devices that enable them to remain fixed in their holes. An example is the velvet fish (*Caracanthus*), which lodges itself in place with its spine-armored operculae (Fig. 15-13). Fish that live above open sand have special escape adaptations; they can withdraw quickly into burrows they have made or bury themselves (W. Klausewitz and I. Eibl-Eibesfeldt, 1959; I. Eibl-Eibesfeldt, 1964c; Fig. 15-14).

Lizard fish (*Synodus*), which rest on the sand, scoop sand from beneath themselves with their pectoral and pelvic fins when danger approaches so that within seconds they disappear into the sand with only their eyes showing. Garden eels (Heterocongridae) remain in tubes in the sand that are held up by a gluelike substance secreted from their bodies (Fig. 15-15). They can withdraw into the sand within seconds if one tries to pull them out. Fish that live in the open water where there is no cover frequently jump above the water surface to escape from the pursuer's view. Many of these fish, such as the Mugilidae, dive back into the water head first, gather speed, and again jump into the air. A number of other fish return to the water tail first. Then they move the tail fin, whose lower part is extended, so rapidly that they are able to propel

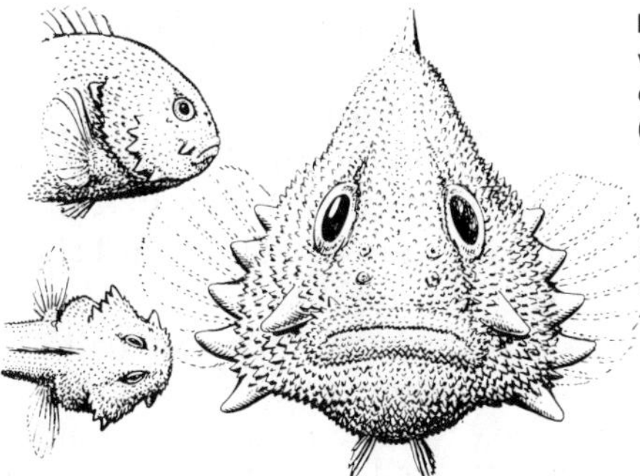

Figure 15-13 Armored spine operculae of the velvet fish (*Caracanthus maculatus*) as an example of a wedging mechanism for protection in corals. (From I. Eibl-Eibesfeldt, 1964c.)

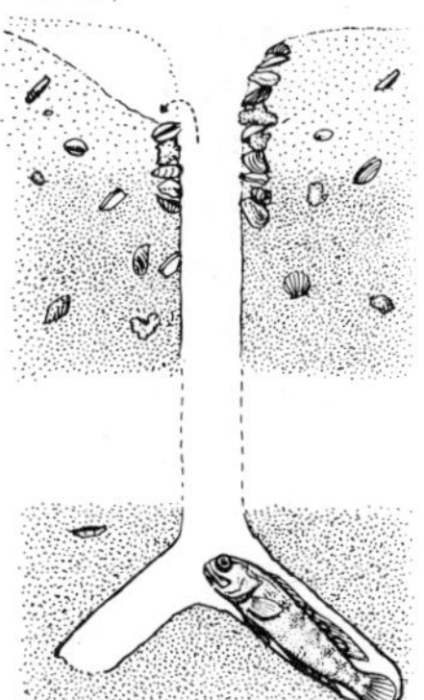

Figure 15-14 The jawfish (*Gnathypops rosenbergi*) builds a vertical tube into the sandy bottom in which it lives. It may be up to 1 meter deep and has branches at the bottom. The rim is reinforced with built-in pieces of coral, clam shells, and sea urchin pieces. (From I. Eibl-Eibesfeldt and W. Klausewitz, 1961.)

themselves along above the surface with the body pointing upward out of the water at an angle. As a result, the fish skims along the surface. Such surface skimmers were the evolutionary point of departure for the flying fish, as can clearly be shown by the increasingly differentiated forms within the Synenthognathi (K. Lorenz, 1963b; see also Fig. 15-16). One extreme is the garfish (*Belone belone*); at the other are various genera of flying fishes (Fig. 15-17), that can glide through the air using their pectoral fins, which are considerably broadened into the shape of wings. They increase their speed on the water surface with their tail immersed in the water until they have attained enough speed to rise up into the air. They are able to sail along for many meters. The head is higher than the tail, which eventually dips back into the water as the fish slows

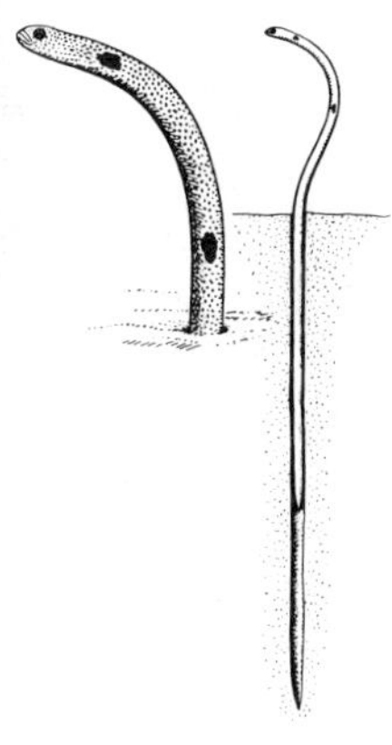

Figure 15-15 Garden eels. *Left: Gorgasia maculata* (Nicobar Islands); *right: Xarifania hassi.* (Photograph: I. Eibl-Eibesfeldt, 1964c.)

down. The fish then can propel itself into the air again after having attained sufficient speed.

The South American hatchet fish (*Carnegiella vesca*), a freshwater fish, is said to be capable of an actual whirring flight. Breder and Eisenmann of the American Museum of Natural History observed that these fish raised themselves into the air when they were driven toward shore with a net (cited by K. Lorenz, 1963b; see Fig. 15-18).

Animals that are pursued often show irregular behavior patterns that make it difficult for a predator to predict their behavior and thus to maintain contact with them: rabbits make sharp turns; some night moths fly zigzag courses or perform unpredictable loopings; pheasants scatter into all directions and then hide. Fish in a school react as a group. Sometimes there are deceptive maneuvers. A lizard that is caught by a predator often drops part of its tail, which continues to writhe on the ground, thus attracting attention while the lizard hides. These various deceptive behaviors have been called "protean behavior" (M. R. A. Chance and W. M. S. Russell, 1959), after Proteus of Greek mythology, who escaped his pursuers by assuming different forms.

Flight or escape reactions are often adapted to a specific category of predators. We have already mentioned that certain gastropods react to the odor of certain **starfish** with escape reactions. Domestic chickens show one set of reactions to aerial predators and another to ground predators, each with its own specific warning calls. They take cover from birds of prey and fly up into trees before ground predators such as polecats and cats. In brain-stimulation experiments these behavior systems can be activated separately from different stimulus points (E. v. Holst and U. v. Saint Paul, 1960). If one category of predators is no longer present, one set of appropriate reactions against predators may drop out. On the Galápagos Islands, where predatory mammals are lacking, the

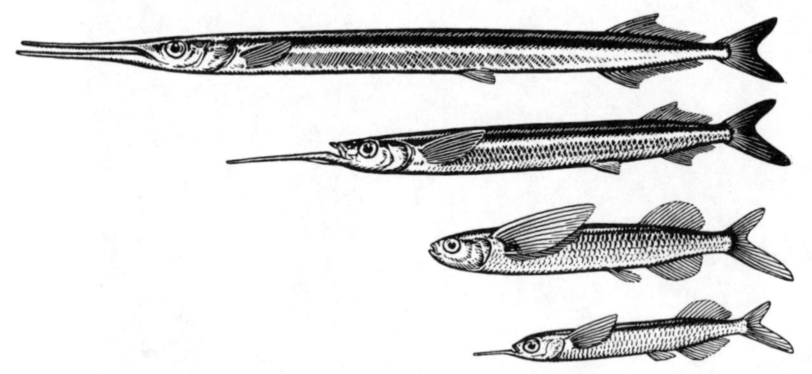

Figure 15-16 Differentiation of marine fish that skim along the surface. *Top to bottom:* garfish (*Belone belone*) approximately 70 cm; halfbeak (*Hemirhamphus*), 40 cm; *Oxyrhamphus micropterus, adult* animal, 138 mm; young of the same species, 40 mm. None of these animals can glide through the air, but they are intermediaries that lead to the flying fish, which "fly" in that manner. (From K. Lorenz, 1963b; H. Kacher, artist.)

Figure 15-17 California flying fish (*Cypselurus californicus*) taking off from the surface and gliding. (From K. Lorenz, 1963b; H. Kacher, artist.)

Figure 15-18 Freshwater fish capable of active wing beating. *Left: Carnegiella vesca,* the hatchet fish of South America; *right:* an intermediate form, *Triportheus elongatus,* that might possibly be capable of flight and lead into *Carnegiella.* (From K. Lorenz, 1963b; H. Kacher, artist.)

hawk (*Buteo galapagoensis*) allows itself to be touched by humans (Fig. 15-19). Similarly tame are the marine iguanas (*Amblyrhynchus cristatus*) and the Galápagos penguins (*Spheniscus mendiculus*) when they are on land. In the water, where they are threatened by sharks, they flee even from a swimming man (I. Eibl-Eibesfeldt, 1960b, 1964b). The Kittiwake gull (*Rissa tridactyla*) does not flee from humans when it is on the cliffs where it breeds, but it will flee when meeting man on land while gathering nesting material (E. Cullen, 1957).

Special adaptations to escape behavior are shown by the small crab *Dotilla.* After sifting sand pellets for food, it deposits them in a specific pattern, so that ring-shaped mounds are formed around the entrance to its hole. In addition, several radial "streets" are kept free of these pellets. In this way the crab maintains free escape routes by which it can reach its hole either directly or via a detour (H. Hass and I. Eibl-Eibesfeldt, 1964; see also Fig. 15-20).

Escape behavior is certainly activated by external stimuli in most instances, but K. Lorenz (1943) has shown that this behavior may also be based on internal motivation. Ducks "escape dive" frequently in vacuo, and animals that have not been frightened for some time are inclined to flee in response to stimuli that would normally be ineffective; they show a definite threshold reduction for escape reactions.

If an animal is prevented from escaping by being driven into a corner, it may attack once its assailant has reached a critical distance (H. Hediger, 1942). A circus trainer must at all times be alert to this reaction. An animal will also attack if it is suddenly surprised and the critical distance has been inadvertently violated. Frequently flight and defense are combined in behavior—for example, when an octopus ejects an inky

Figure 15-19 Example of "island tameness" (Duncan Island, Galápagos). The Galápagos hawk allows itself to be touched by man. (Photograph: I. Eibl-Eibesfeldt.)

substance when beginning its escape. The same occurs in an alarmed bombardier beetle (*Brachynus*), which excretes a volatile substance from its anal glands that quickly diffuses into the air.

Many animals seek protection near others that are better armed. I will discuss this later. Animals that do not possess their own mechanical protection frequently obtain it by building containers that they carry with them. The tubelike structures made by caddisfly larvae are well known, and their construction and repair has been described by C. Wesenberg-Lund (1943). The caterpillars of the Psychidae build a similar structure, an example of behavioral convergence. Empty snail shells are used by hermit crabs after careful inspection. The octopus (*Octopus aegina*) enters empty clam shells whose two halves it can open and close as needed (Fig. 15-21). The caterpillar of the wasp *Lygaeonematus compressicornis* surrounds its feeding place on an aspen leaf with a

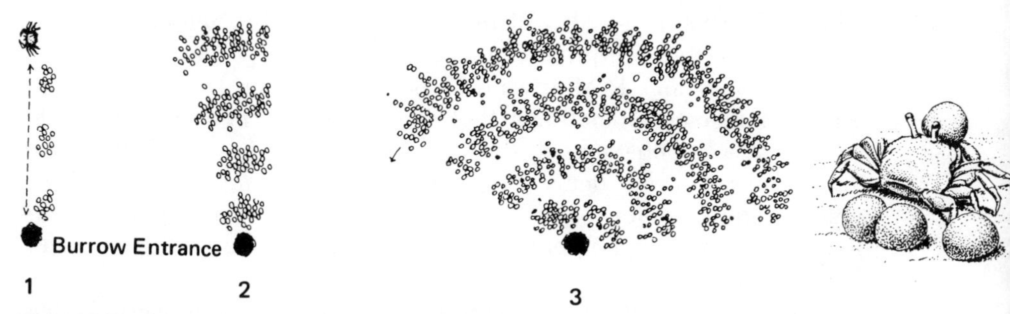

Figure 15-20 Sand pellets that have been deposited in ringed walls around the hole entrance by the pellet crab (*Dotilla sulcata*), *1*, *2*, and *3* depict different stages as the animal continues to feed. On the right the feeding crab. (From H. Hass and I. Eibl-Eibesfeldt, 1964.)

fence of foamy palisades, and sometimes it prevents access to a leaf by erecting such an obstruction at the stem. This foam is sticky and contains salicylic acid.

Many animals protect themselves through camouflage. Some are already adapted by their appearance to a specific background. However, they must also possess the appropriate behavior for selecting the correct background and maintaining specific postures if this protection is to be effective. In this way, caterpillars that are patterned not only select the environment that fits this coloration, but they also assume the postures that make them least conspicuous (H. B. Cott, 1957; W. M. Herrebout et al. 1963). E. Curio (1966a) found one species of moth that had three different forms of caterpillars. Each form had a different color and showed the appropriate behavior in selecting the most fitting resting place (Fig. 15-22). Some looper moth caterpillars effectively mimic dead twigs. Many animals camouflage themselves by covering their bodies with foreign objects; thus sponge crabs (*Dromia*) use their modified third and fourth pair of legs to hold clams and sponges on their backs. In majids the carapace contains special bristles for attaching foreign objects. All these examples show that camouflage is the result of a complicated interaction between structure and behavior.

Widespread methods of defense against predators are those of warning and deception (O. M. Reuter, 1913). The forktail caterpillar (*Dicranura vinula*) will stop eating when it is disturbed and remain in a stretched-out position with the head slightly drawn in. In this position the green caterpillar is well camouflaged. If one touches the animal or moves the leaf on which it sits, the caterpillar raises its front end and turns a very conspicuous "face mask" toward the attacker. The brown head, framed by red and yellow edges, carries two dark pigment spots, which seem to be imitations of eyes. At the same time the animal ejects two red glandular filaments out of the last pair of modified abdominal legs, which twirl for several seconds before they are withdrawn (I. Eibl-

Eibesfeldt, 1966b; see also Plate IV). From a well-developed gland in the prothorax the caterpillar can eject a bad-smelling secretion.

T. Eisner and J. Meinwald (1966) reported on the chemical defenses of insects. Sometimes these defenses are neutralized by the special attack behavior of the predators. In this way the mouse *Onychomys torridus* deals with the beetles *Eleodes* and *Chlaenius*, which secrete a defensive substance from the posterior tip of their abdomen, by grasping the beetles and ramming them into the ground, abdomen first.

Often bad-tasting insects are very conspicuously marked. A good example is provided by the wasps. Once a songbird has been stung by a wasp it will remember this for months and avoid all conspicuously ringed objects from then on. This protection is also extended to a number of insects that are marked like wasps. Many bad-tasting butterflies are mimicked by those that are edible (Plate IV). They deceive their pursuers. Some unusual examples of mimicry have been collected by E. Curio (1966c), and I have mentioned additional examples in Chapter 7 (see also O. J. Sexton, 1960).

Many animals are on the lookout for potential danger. All our native mammals raise their heads from time to time and look around, taking samples from various strata of air by sniffing while raising and lowering their head ("taking wind"). In addition, they interrupt other activities such as eating or digging with great regularity. A digging hamster repeatedly looks up and around. While analyzing films of people who were eating but were unaware of the observer, H. Hass (1968) discovered that they stopped at regular intervals and seemingly automatically looked up and around. Additional observations of this behavior have convinced us that it is a form of reconnoitering.

A conspicuous adaptation for protection from predators is the schooling of fish. First, it is easier for animals living in a swarm to detect

Figure 15-21 Octopus (*Octopus aegina*) in the clam shell in which it lives and protects its spawn. (Photograph: I. Eibl-Eibesfeldt.)

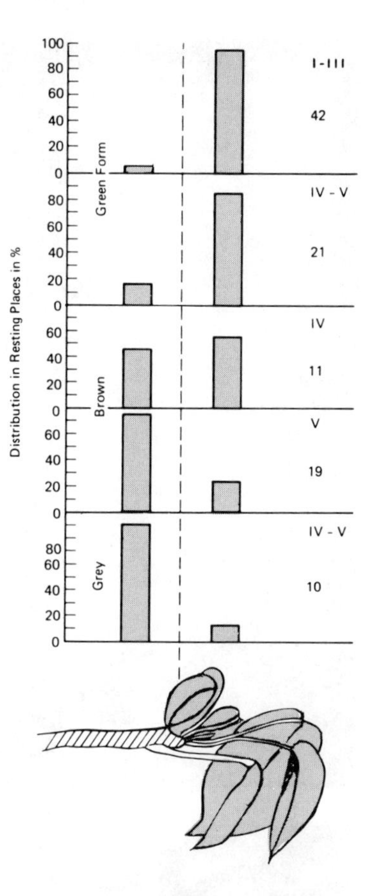

Figure 15-22 Resting places (branch or leaf) of three caterpillars of *Erinnyis ello* in all five stages (*I–V*). Arabic numerals: number of observations. (From E. Curio, 1966a.)

danger; more eyes can see more. This, however, is by no means the most important factor favoring the aggregation. It can be shown that a single fish is much safer in the swarm, even when not warned, than if it is alone. Whereas a single fish can be easily fixated by a predator and then caught, it is much more difficult for the predator to focus and pursue one target out of many in a swarm. The aim of the predatory fish is confused and, of course, precise aiming is a prerequisite for successful hunting (I. Eibl-Eibesfeldt, 1962b; G. v. Wahlert, 1963). Predatory fish try to separate single fish from a swarm or lie in wait quietly until a fish comes within reach (Fig. 15-23). A falcon in pursuit of a flock of birds also tries to isolate an individual from the group by means of sham attacks (N. Tinbergen, 1951). When a hawk chases a flock of pigeons it usually aims for the bird that is farthest from the group. In a flock of white pigeons it chases the single black one, and in a black flock it chases the white one.

A large number of brood-care behavior patterns for care of the young can be understood as adaptations to predators, such as the **distraction display**, or the **mouth-breeding** of **cichlids** (Fig. 15-24). Black-headed gulls (*Larus ridibundus*) remove the egg shells from the vicinity of the

Figure 15-23 *Top:* Predatory *Mycteroperca olfax* about 1 meter in length chasing swarm fish *Xenocys jessiae* (Galápagos). The fish form a vacuole around the predators. *Bottom: Xenocys jessiae* in flight just above the sea bottom. (Photographs: I. Eibl-Eibesfeldt.)

Figure 15-24 *Tilapia nilotica* female taking young into her mouth during the brood-care stage. (Photograph: R. Apfelbach.)

nest after the young have hatched. If they fail to do this a predator can more easily detect the nest (N. Tinbergen et al., 1962). Adaptations to predators are also seen in the selection of the nesting place and the habit of breeding in colonies. Large gulls and crows can be driven off by many black-headed gulls. H. Kruuk (cited by N. Tinbergen, 1965) demonstrated that these attacks make it more difficult for predators to steal eggs from the center of the colony than from the periphery. The synchronization of egg laying in black-headed gulls is a protection against predators. By this simultaneous egg laying the "market is flooded," so to speak, and this overabundance of eggs then becomes a protection (I. J. Patterson, cited by N. Tinbergen, 1965).

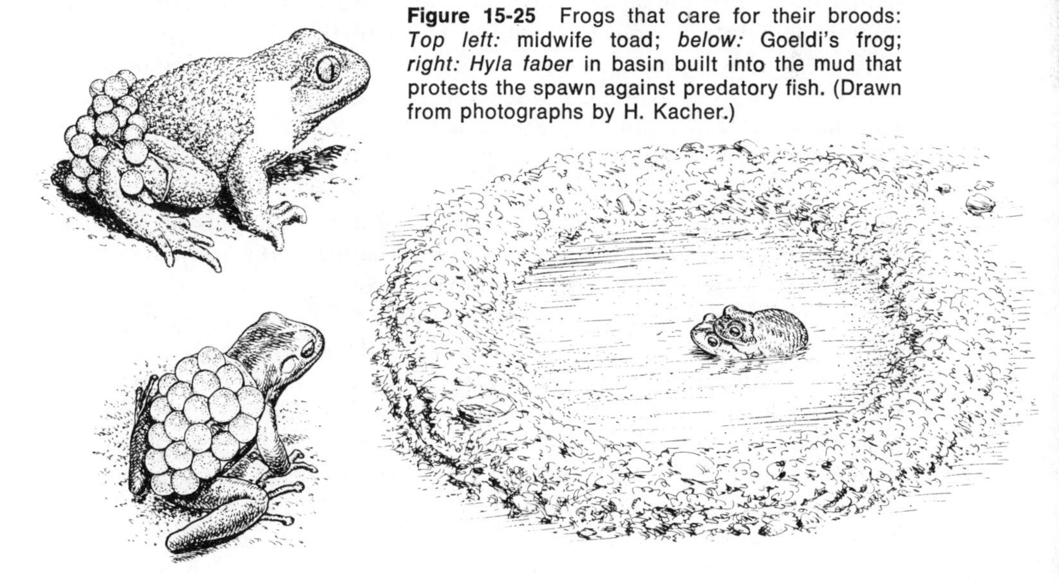

Figure 15-25 Frogs that care for their broods: *Top left:* midwife toad; *below:* Goeldi's frog; *right: Hyla faber* in basin built into the mud that protects the spawn against predatory fish. (Drawn from photographs by H. Kacher.)

To demonstrate the variety of brood-care patterns evolved even in the quite homogeneous group of anurans, some examples of frogs that care for their brood will be mentioned (Fig. 15-25). Additional literature is in W. Klingelhöffer (1956) and R. Mertens (1959). In the central European midwife toad (*Alytes obstetricans*), the male takes up the strands of spawn around its legs until the larvae wiggle within the gelatinous capsules. Then the toad seeks out a pond in which the larvae hatch. In this way the spawn is protected from the numerous predators in the water. The Central American poison arrow frogs (*Dendrobates auratus*) deposit their eggs on a leaf outside the water. The male guards them and sits next to the newly hatched tadpoles, which then climb the back of their father. There they adhere by suction and are thus transported to the nearest puddle. The frog (*Hyla faber*) of tropical South America builds small breeding basins at the edge of puddles by raising mud into a ringed wall. Spawning takes place in these small basins. Later the larvae are freed by the rising water. In the Chilean (Darwin's) frog (*Rhinoderma darwini*) the males guard the eggs, which are deposited on land. As soon as the larvae wiggle in the eggs they are snapped up by the males and are carried about in the vascularized throat pouch. In the South American toad (*Pipa americana*) the eggs develop in honeycomb-like pockets on the mother's back.

Marsupial frogs (*Nototrema*) protect their spawn in incubating pouches on their backs, and Goeldi's frog (*Hyla goeldii*) carries the spawn in a bowl-like depression on its back. Many frog species deposit their eggs in a frothy mass of air bubbles which they fasten to plants above the water, so that the hatching tadpoles fall into the water; for example, the female grey treefrog of Africa (*Chiromantis xerampelina*) clasps the foamy nest until the larvae have hatched. In this way the nest is protected against drying out too rapidly. The whistling frog (*Leptodactylus labialis*) builds its foam nests in caves it has dug into creek banks, and the larvae are freed when the water rises during the rainy season. A comparable multiplicity of brood-care behavior patterns can be observed in many other animal groups. We might think of the many fascinating adaptations of insects. Suffice it here to demonstrate this variability with the examples from the frogs. If so uniform a group of animals as frogs can show such a variety of brood-care behavior patterns, it is easy to imagine what diversity we can expect in other groups.

The nonspecies member may also have an impact as a competitor, forcing the most varied adaptations in animals which are so threatened. On the Galápagos Islands the frigate birds (especially *Fregata minor*) have specialized in taking food away from other sea birds. They circle above a bay until they see a fishing booby or other sea bird. Then they rush at it and peck it with their bills until it regurgitates its prey, which they then skillfully catch. I once saw a tropic bird killed by a frigate bird. This formidable predatory competition may have caused the swallow-tailed gull (*Creagrus furcatus*) to fish only at night. The dusky gull (*Larus fuliginosus*), which fishes during the day, has gray feathers and is well camouflaged, which I interpret as an adaptation to evade the competing frigate birds. In this connection it is remarkable that many adult red-footed boobies *(Sula piscator websteri)* on the Galápagos possess a brown-feathered plumage which resembles that of juveniles.

Finally, competition with other species demands behavior adaptations in the most varied areas. We already mentioned the food-getting behavior of the Darwin finches. As an example from another functional cycle: masked boobies *(Sula dactylatra granti)* and red-footed boobies *(Sula piscator websteri),* which occur sympatrically on the Galápagos Islands, have different breeding habits. The masked booby breeds on the ground, the red-footed booby in trees (Fig. 15-26).

Figure 15-26 Masked boobies (ground nesters) and red-footed boobies (tree nesters) (Galápagos). (Photograph: I. Eibl-Eibesfeldt.)

Special protective measures are required against parasites. Grass frogs, toads, and alpine salamanders free themselves of adhering leeches by sitting in the sun, which the leeches cannot tolerate. Fish that are plagued by parasites allow themselves to be cleaned by **cleaners.** In the leaf-cutting ants (*Atta cephalotes*) the small minima workers protect the larger workers against the attacks of parasitic flies of the Phoridae group. While the larger workers cut leaves and are defenseless, the minima workers position themselves with open mandibles around them and snap at the approaching flies. They also ride along on the cut leaves as guards on the way back to the nest (I. and E. Eibl-Eibesfeldt, 1967; see also Fig. 15-65).

Symbiotic Relationships

I want to conclude the discussion of interspecific relationships with some illustrations of symbioses, which are of especial interest to the student of behavior. As I said in Chapter 5, the problem of interspecific communication is raised in such partner relationships.

German biologists speak of symbiosis when two different species collaborate in some way to their mutual advantage. If only one species profits, without some disadvantage for the other, this is commensalism. In the English and American literature of marine biology, what we call symbiosis is often called mutualism; the concept of symbiosis is a larger category that includes parasitism, commensalism, and mutualism.

The point of departure for parasitism as well as symbiotic partnership is probably in most instances some kind of commensalism. Many fish of the high seas seek protection in the vicinity of larger fish. I have observed jackfish (*Caranx ruber*) accompanying barracudas (*Sphyraena barracuda*). They swam closely above their backs and performed each turn of the barracudas. Near the Cocos Islands I saw the jackfish (*Caranx chrysos*) accompanying sharks and rays, and in the Indian Ocean a closely related species of jackfish accompanied the large wrasse (*Cheilinus undulatus*). The blue runner (*Elagatis bipinnulatus*) sometimes swims with sharks and other large fish. They are safer there than

Figure 15-27 Mackerels (*Elagatis bipinnulatus*) rubbing themselves on the back of a shark (Maldive Islands). (Photograph: I. Eibl-Eibesfeldt.)

alone in the open water. Sometimes they rub themselves on the skin of the sharks (Fig. 15-27), which, in turn, seem to derive no advantage from their companions. From these kinds of loose associations obligatory relationships evolved. Pilot fish (*Naucrates ductor*, L.) are almost never seen alone. According to our observations they behave differently when

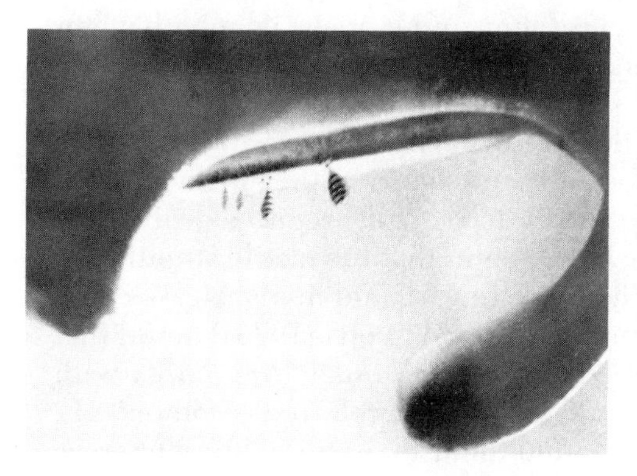

Figure 15-28 *Left:* Manta with pilot fish (Red Sea); *right:* shark with pilot fish (Azores). (Photographs: H. Hass.)

Figure 15-29 Cleaner fish (*Labroides dimidiatus*) cleaning *Odonus niger.* (Photograph: I. Eibl-Eibesfeldt.)

swimming with predatory sharks than with the giant mantas (*Manta*) and harmless whale sharks (*Rhincodon*). When accompanying predatory species, the pilot fish usually swam near the ventral and dorsal fins and farther behind, and only for a short period of time in front of the mouth, such as when they passed the shark and swam toward the divers. Whale sharks and mantas are surrounded by pilot fish which flee into their mouths when danger approaches (Fig. 15-28). H. Hass (1954) observed that the pilot fish cleaned the mouths of mantas. Here a partnership with mutual advantage for both has developed. *Echeneis* also clean sharks.

In true cases of mutualism simple signals develop between the partners in the service of interspecific interaction. These have been studied in some detail in the cleaning symbioses (I. Eibl-Eibesfeldt 1955a). I have referred to the "cleaner dance" as well as the "inviting" and "rejection" postures of the customers, which even change color when they want to be cleaned. *Naso tapeinosoma*, for instance, turns light blue when it is being cleaned. This makes the parasites visible against the contrasting background. The **cleaner** rapidly taps its host with its fibrillating ventral fins, keeping it informed as to where it is cleaning. The host acts accordingly: It stops moving the fins that are being cleaned (Fig. 15-29) and erects them, and it opens its mouth when the cleaner butts against its corners and allows the cleaner inside. If the hosts wants to breathe heavily, it signals its intention by partially closing its mouth and the cleaners leave the mouth cavity. Other warning signals are the movement of the operculae and the shaking of the entire body by the host fish. These behavior patterns have convergently evolved in various species of fish. G. Losey (1971) found that the tactile stimulation is most important for the host. It acts as a reward and in many cases conditions the host to the cleaner's presence. G. W. Potts (1973a) found that different species of hosts were cleaned by *Labroides dimidiatus* with different techniques and that certain hosts are preferred to others. He also found that young cleaners occupied themselves with different areas from those of the adults. Potts provides a detailed analysis of the stimulus–response relationship between the cleaner and its host.

Many different species of fish allow themselves to be cleaned by cleaners: predatory fish and peaceful ones, reef dwellers as well as fish of the open seas. Sometimes we saw swarms of fish appear out of the deep blue water and stand head down above a cleaner station as if in response to a single command. They waited for the cleaners, which soon were busy with their task. After several minutes the fish school again disappeared in the depths of the sea.

Even manta rays visit the cleaner stations in the coral reef to be cleaned (Fig. 15-30). In a reef canal of the Addu Atoll (Maldive Islands) we observed how four mantas slowly circled a coral clump 15 meters below the surface. During this time they were being cleaned by numerous cleaner wrasses (*Labroides dimidiatus* and *Thalassoma* species). They opened their large gill slits and permitted the cleaners to enter.

The cleaners play an important role in the life of the reef fishes, as has been demonstrated by C. Limbaugh (1961). He captured all cleaners from two reefs in the Bahamas. A large number of reef fish left the area, and those that remained showed many skin and fin ailments two weeks later. Not until new cleaners migrated to this area did customers return again. Limbaugh also observed that one cleaner was visited by more than 300 customers of various species during a period of 6 hours.

At times we have observed the fish actually crowding round cleaner stations. Various species waited their turn, and as agonistic as these fish are at other places in the reef, they were peaceful here. The cleaner station was, so to speak, a barber's shop in the reef, owned by all and therefore neutral ground.

Cleaning symbioses have been observed in various regions of the world, and it has been found that not all cleaners are specialized in the same way to carry out this "trade." In the Indo-Pacific area it is mostly the cleaner fish of the genus *Labroides* that are active. In the tropical Atlantic there are many species of fish that regularly or occasionally clean others. In 1955 we observed near Bonaire (Caribbean) the following cleaner fish: *Elacatinus oceanops* (Gobiidae), *Gramma hemichrysos* (Hemichromidae), *Thalassoma bifasciatum* and *Bodianus rufus* (Labridae), and *Anisotremus virginicus* (Haemulidae); in the Bermudas we found *Chaetodon striatus* (Chaetodontidae) and *Abudefduf saxatilis* (Pomacentridae). It appears here as if the ecological niche of the cleaner

Figure 15-30 Manta being cleaned by cleaner fish (*Thalassoma*). The manta opens the gill slits and permits the cleaners to enter. (Photograph: H. Hass.)

has been occupied by more than one specialized species. Various species are still competing for this position. Most specialized is the cleaner goby (*Elacatinus oceanops*), which has very similar colors to the cleaner wrasse (*Labroides dimidiatus*). A comprehensive summary of these cleaning symbioses can be found in H. M. Feder (1966) and G. W. Potts (1973a, 1973b). Near a Cuban island D. H. H. Kühlmann (1966) observed a toothed carp (*Gambusia*) cleaning the mouth of a crocodile (*Crocodylus acutus*).

Many shrimp of the genera *Stenopus* and *Periclimenes* are known to be cleaners of fish. By waving their long antennae they attract the attention of their customers. *Periclimenes pedersoni* climbs on the fish and crawls under its gill covers or into its mouth, with the fish exhibiting the appropriate "inviting" signals. When frightened the host fish spits out the cleaner prawns or warns them before swimming off. The cleaner prawns also remove parasites that are lodged below the skin (C. Limbaugh, H. Pederson, and F. A. Chace, 1961).

On land comparable cleaning symbioses exist. The crocodile bird (*Pluvianus aegyptius*) had been cited by Herodotus as entering the mouths of crocodiles and eating leeches. This symbiosis has not been studied in detail. Very little is also known about the oxpeckers (*Buphagus africanus*), the relatives of starlings, which climb about on larger animals, catching insects, larvae, ticks, and other parasites. They specialize in this and are sometimes parasites themselves. I have observed how they opened wounds on rhinoceroses and drank the blood. However, the benefits undoubtedly outweigh the disadvantages. Whether or not these species communicate with one another in the manner of the cleaner fish is not known. One of my own chance observations points in this direction. An oxpecker that was working on a part of the skin of a rhinoceros, which was infested by the larvae of the botfly, was repeatedly interrupted in its task when the animal rolled on the ground. When the rhino again stood on its feet, the bird uttered rapid calls before again alighting on it. The bird continued to call when it again perched on the rhinoceros before again approaching the wound. The rhinoceros did not roll on the ground after that. Zebras await oxpeckers by standing with spread legs, their tail raised and ears drooping, so that the birds can get at the ticks (H. Klingel, 1967).

Oxpeckers are completely dependent on their hosts. However, there are also a number of more casual relationships that may serve as models for the development of such cleaning symbioses. Starlings (*Sturnus vulgaris*) can frequently be seen near grazing cattle, where they catch insects. White wagtails are frequently seen chasing insects on pigs. Cattle egrets (*Ardeola ibis*), which often ride on elephants, catch the insects that have been scared up by the grazing animals. The small ground finch (*Geospiza fuliginosa*) of the Galápagos Islands searches for ticks on marine iguanas. A meeting of two species in this manner provides the prerequisite for the development of an actual symbiotic relationship.

In 1974 I filmed land iguanas *(Conolophus subcristatus)* from Galápagos being cleaned by small and medium ground finch *(Geospiza fuliginosa* and *G. fortis).* In this case a clear symbiotic relationship had developed. Whereas marine iguanas merely tolerate the activity of the finches, without showing any clear response, the land iguanas, at sight of a finch hopping in the vicinity on the ground, rise high on their legs and lift their tail so as to make the underside of the body available to the finch. C. MacFarlane and W. G. Reeder (1974) report similar posturing in the Galápagos tortoises, by which the folds of the soft skin become exposed to the cleaning finches.

On Wenman and Culpepper Island (Galápagos) the sharp-beaked ground finch *(Geospiza difficilis)* picks bird flies from boobies and has in doing so apparently learned in addition to bite open the skin surrounding the base of the feathers and drink the blood of its host. Thus from a symbiotic relationship a parasitic one has developed (R. I. Bowman and S. C. Billeb, 1965).

The partners may offer each other quite different advantages, and accordingly the symbioses vary. In the Red Sea and the tropical Indo-Pacific region live giant anemones (genera *Stoichactis, Radianthus, Discosoma*), between whose poisonous tentacles one often finds the anemone fish of the genera *Amphiprion* and *Premnas* (Fig. 15-31). The fish are hardly ever encountered without an anemone, and the advantage of living together can be readily seen: the fish are well protected between the tentacles of the anemones. No predators can catch them there without being caught themselves.

Figure 15-31 Anemone fish (*Amphiprion akallopisus*) between the tentacles of *Radianthus* anemone. (Photograph: I. Eibl-Eibesfeldt.)

In some instances it has been observed that the anemone fish clean their host: They carry off its waste products and swish the sand from its top side (I. Eibl-Eibesfeldt, 1960a). In the aquarium some anemone fish feed their anemones, but it is not known if they also do this in the wild.

In any case the anemone fish seem to have the decided advantage. How is it, then, that they are not stung by the poisonous tentacles of the anemone? Here some opposing points of view exist, because various authors have studied species that differ in their behavior.

We have experimented with *Amphiprion akallopisus, A. xanthurus,* and *A. percula* and found that the fish are covered by a protective substance on their skin (I. Eibl-Eibesfeldt, 1960a, 1964c). If one removes the mucus from the skin, the fish are stung and caught by the tentacles. An intact fish can be flung against the tentacles without harm resulting from this rough treatment. This holds true even when the fish is moved passively across the tentacles of the anemone in an atypical manner. This disproves the hypothesis of some who suggested that the anemone recognizes its fish by the type of movements they make. Investigations by D. Davenport and K. Norris (1958) and M. Blösch (1965) have also shown that anemone fish possess a protective substance. There exist anemones that tolerate all species of anemone fish, whereas others only tolerate a specific species of anemone fish and catch all other species. And finally Blösch found anemones that at first caught all kinds of anemone fish but gradually accepted them. But here, too, only anemone fish are able to accustom the anemone to their presence. Originally it was assumed that the protective mucus is secreted by the fish. D. Schlichter (1968), however, demonstrated that the anemone fish *Amphiprion bicinctus* impregnates itself actively with anemone mucus which makes the fish a part of the surface of the anemone itself. Whether this is true for all anemone fish needs investigation.

In the Mediterranean Sea E. Abel (1960a) observed the goby (*Gobius bucchichii*) as an "anemone fish" of *Anemonia sulcata*. Besides the typical anemone fish there exist a number of other fish that seek protection near the anemones but that avoid contact with the poisonous tentacles. Of special interest is the behavior of the damselfish (*Dascyllus trimaculatus*). Near the Maldive and Nicobar Islands we saw these fish, especially young ones, frequently near the giant anemones, but not touching their tentacles. In the Red Sea, on the other hand, we observed small schools of 1- to 2-cm-long damselfish between the tentacles of anemones. Here one can observe the development to an anemone fish within one particular species.

Other animals are also associated with anemones. It is well known that the hermit crabs of the genus *Eupagurus* have anemones on their snail shells in which they live and are protected (Fig. 15-32). It has been observed that an octopus which attempted to catch such a hermit crab was stung by the anemones and retreated. However, the anemones probably also derive some benefit from this union, for instance by partaking of the meals of the crab. In any case they seem to be well adapted to this life with the hermit crabs. It is even possible that the initiative to live on snail shells came from the anemones. Near England the anemone (*Calliactis*-[*Sagartia*] *parasitica*) climbs without the aid of the hermit crab (*Eupagurus bernhardus*) onto the snail shell (D. M. Ross, 1960). A detailed investigation showed that the classical conception of the stinging cells as independent effectors is not quite correct. If the anemone sits on the sea bottom and touches the horny outer surface of a snail shell, then the tentacles actively explore the shell, and many adhere to it by a discharge of the nettling cells. The anemone then releases its footing from the substratum and mounts the snail shell. Once in place its behavior changes. The tentacles no longer stick to the shell when they are

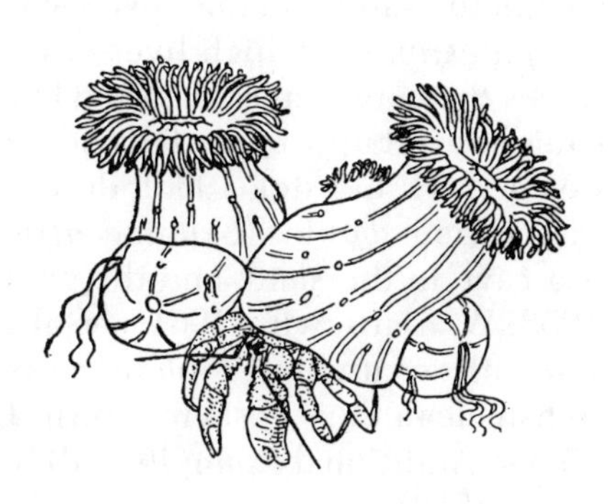

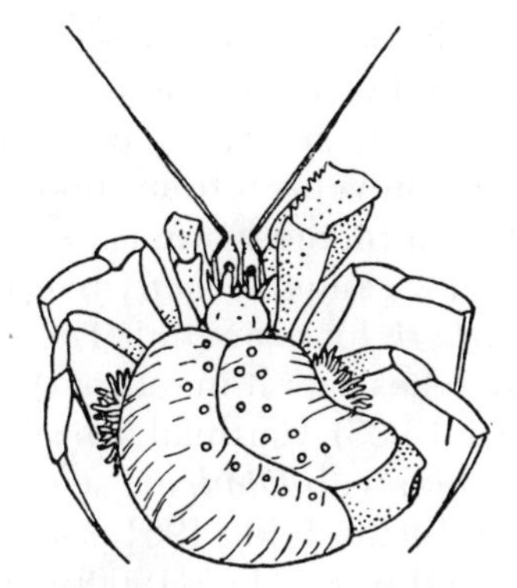

Figure 15-32 *Left:* Hermit crab (*Pagurus arrosor*) in symbiosis with *Calliactis parasitica; right:* hermit crab (*Eupagurus prideaux*) in a snail shell that is grown over with *Adamsia palliata.* (After L. Faurot from H. Fuller, 1958.)

touched with a piece of another shell. The threshold for the release of the nettling cells changes depending on the substratum on which the anemone sits (D. Davenport, D. M. Ross, and L. Sutton, 1961).

In the Mediterranean the hermit crab (*Pagurus arrosor*) helps its anemone (*Calliactis parasitica*) to mount its shell. When the crab encounters an anemone fastened to the bottom, it at first attempts to remove it from the substratum by tapping it with its claws and the first pair of walking legs. The contracted anemone opens up as a result of this and releases its hold on the bottom. Without this "cooperation" of the anemone such a removal from the bottom is not possible, because the foot of the anemone adheres tightly. The released anemone then sticks to the snail shell with its tentacles and bends its body in a U shape, thus bringing its foot onto the shell. When the hermit crab chooses a larger shell, it transfers its anemones to the new one (F. Brock, 1927).

These two examples illustrate two developmental stages of this interspecific relationship, because in the first example the crab is relatively uninvolved, whereas in the second one a clear adaptation is present in the behavior of the two symbiotic partners. It is possible, however, for each to get along without the other. In the crab *Eupagurus prideauxi* and the anemone *Adamsia palliata,* however, the relationship is so close that the anemone cannot live without the crab. The adult partners are never found alone. The anemone fastens itself below the mouth of the crab on the snail shell, which it gradually surrounds with its foot. Then it secretes a horny substance above the opening of the snail shell, and in this way achieves an enlargement of the snail shell. In this way it prevents frequent changes on the part of the crab, on which it depends as a supplier of food. Additional examples of symbioses with stinging animals can be found in H. Füller (1958).

In this way many protective alliances of various kinds are formed. A small coral, *Heteropsammia,* which lives in the sand, can only exist with the aid of a small sipunculoid worm (*Aspidosiphon*), which lives at the base of its lime skeleton. The worm moves the coral across the sand bottom, prevents it from sinking in, and straightens it up when it falls down. In return, the worm is protected by the coral (H. Feustel, 1966). Prawns of the genus *Alpheus* live together with various gobies (*Cryptocentrus lutheri,* for example). The prawn digs a cave in the sand, and the goby watches over it in return (W. Luther, 1958; W. Klausewitz, 1961; see also Fig 15-33). Cardinal fish seek protection with sea urchins, *Siphamia versicolor* with *Diadema* sea urchins. The fish clean their hosts in return (I. Eibl-Eibesfeldt, 1961c; see also Fig. 15-34). Additional examples will be found in E. Abel (1960b) and D. Magnus (1954).

The various symbioses known in insects are most fascinating. As an example I mention only those between ants and plant lice (*Aphids*). The plant lice excrete large amounts of a sugary substance and are visited by the ants for this reason. The ants tap the aphids with their feelers and thus stimulate them to excrete. The behavior of the ants looks very similar to that which is shown when one ant begs for food from another, and it is thought that the plant lice imitate the head of an ant with their abdomen, especially because they raise their hind legs into the air, resembling feelers (Fig. 15-35). The relationship between the plant lice and the ants can be very close. The plant lice *Lachnus taeniatoides, Anuraphis farfarae, Pemphigus caerulescens,* and the *Stomachis* species can no longer remove their excrements from their bodies because hairs surround the anal region, which retains the excrements for the ants. These plant lice are not only protected against enemies by the ants, but they are cared for like the way man keeps valuable cattle. The ants make roofs for them out of earth and they bring the winter eggs into the lower part of their hills for safekeeping over the winter. In the spring the hatching larvae are taken to the plants for feeding, and on cold nights they are re-

Figure 15-33 While the goby (*Cryptocentrus lutheri*) watches at the entrance, the prawn *Alpheus djiboutensis* busily digs out sand from the cave. (Photographs: W. Luther.)

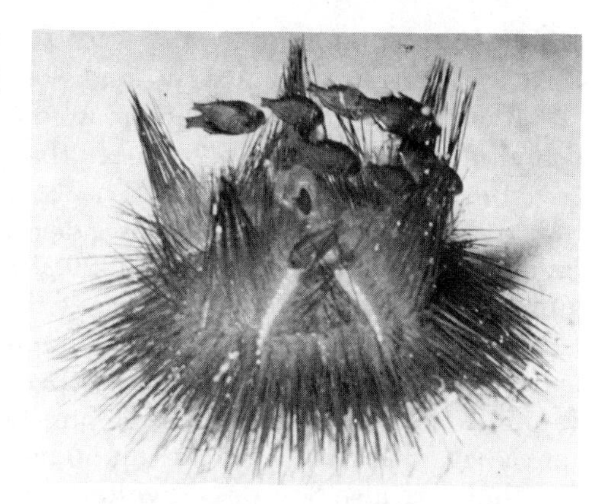

Figure 15-34 Cardinal fish (*Siphamia versicolor*) seek protection between the spines of a *Diadema* sea urchin (Nicobar Islands). (Photograph: I. Eibl-Eibesfeldt.)

turned to the ant hill. Sometimes the ants depend completely on plant lice for food. *Lasius brunneus* lives exclusively from the excrements of the *Stomachis* species. The symbioses between insects and flowers are mentioned at this time but will not be discussed further.

All symbiotic relationships raise interesting questions for the ethologist in respect to their ecological significance, the way they originated, the methods of interspecific communication, and the development of signals.

Parasitism

The various relationships between parasites and their hosts raise a number of fascinating problems for the student of behavior, such as selection of the host, finding the host, and the defensive reaction of the host (G. Osche, 1962, 1966). I have already described how the parasitic

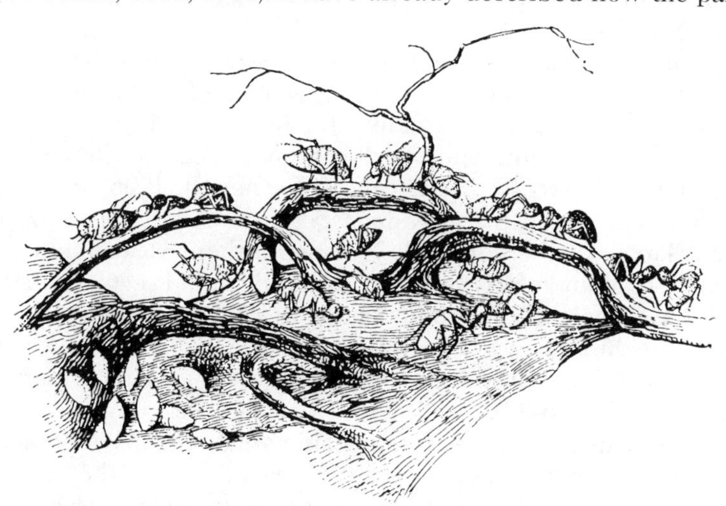

Figure 15-35 Husbandry of the plant louse *Tramaradicis* on the roots of *Artemisia* by the ant *Lasius umbrabus*. (After A. Forel.)

widow birds mimic the host species and how the larvae of certain **liver flukes** reach their final host by means of deceptive signals. D. Davenport (1966) investigated the specific reaction of polychaetes that live on starfish and hermit crabs to substances that are secreted by their hosts.

Between symbionts, commensals, predators, and parasites there are transitional forms. This can be traced in ants and their guests (K. Escherich, 1906; E. Wasmann, 1920, 1925; H. Bischoff, 1927; R. Hesse and F. Doflein, 1943). A number of these "guests" are true predators (synechtrans) that damage the ant population by eating the brood and avoid pursuit by the ants by retreating into small crevices. Their hosts cannot follow them there. Others are heavily armored and can roll up into a ball so that the ants can get no hold on them. Besides these predators there are also harmless dwellers (synoecetes) in an ant colony. They only eat waste products and are therefore tolerated. However, some of these manage to obtain food during the social feeding among ants. In this way the silverfish *Atelura* (Fig. 15-36) obtains nourishment. The ant cricket (*Myrmecophila acervorum*) robs food-carrying worker ants as well as larvae that have already been fed. The guest ant *Formicoxenus* begs from workers and is fed by them. True guests (symphiles) provide special aromatic secretions for the host ants in return for food and protection. As long as no disadvantage derives from this alliance for the hosts one could speak of a symbiosis; frequently, however, such symphiles cause extensive damage to their hosts. The rove beetles *Lomechusa strumosa* live with the red wood ants *Formica rufa,* which care for them, feed them, and in turn receive the aromatic gland secretion of the rove beetle (Fig. 15-36). While caring for their guests, however, the red wood ants neglect their own brood, which is lost if there are too many symphiles. Furthermore, the guests and their larvae eat a large number of their hosts' larvae.

The mite *Antennophorus* becomes a parasite in a different way. The mites attach themselves to the undersides of the ants' heads and stroke the throat region and the side of the head with the first pair of legs so that the host ant regurgitates a drop of food, which is taken up by this ectoparasite (C. Janet, 1897, cited by K. Escherich, 1906; see also Fig. 15-37). The ants often try, in vain, to remove this living muzzle. On many ant larvae of the genus *Pachycondyla* there is wrapped around their necks, as a kind of living collar, a phorid larva. It eats when the ant larvae is fed. When all the food is eaten, the fly larvae manipulates the behavior of its host larva by pinching its skin so that it becomes restless, which in turn arouses the attention of feeding worker ants, which return to feed them (Fig. 15-37).

Very unusual kinds of interspecific relationships are found in the slave-making ants, where one ant species becomes the parasite on another ant. The females of the slaveowners no longer establish their own colonies. If the female of the red ant *Formica rufa* does not find a

Figure 15-36 *Left:* Silverfish *Atelura* takes part in the feeding of an ant; *right:* the parasitic ant guest *Atemeles* is fed by an ant. *Left:* after C. Janet from K. Escherich, 1906; *right:* after A. Forel from K. Escherich, 1906.)

nest of its own species it will enter one of the related *Formica fusca.* There it will be adopted if no other queen is present. In the course of time this nest will become a *Formica rufa* state. The fertilized queen of *Formica sanguinea* always enters the nest of *Formica fusca* and robs some pupae, which it raises and defends against the host ants. The hatching workers care for the nonspecific queen and her brood, so that within the host nest a *Formica sanguinea* state develops. The host ants eventually kill their own queen, so the host population eventually dies

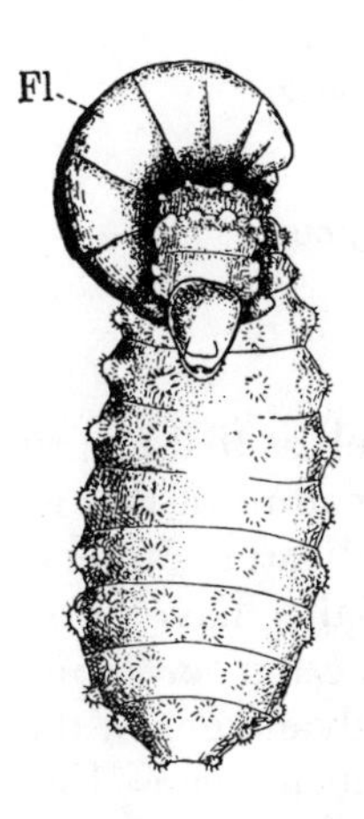

Figure 15-37 *Left:* A phorid larva (*Fl*) is wrapped around the neck of the ant larva (*Pachycondyla*) like a living collar. The phorid larva is able to control to some extent the behavior of the larva (see text). *Right:* The parasitic mite *Antennophorus,* which is attached to the underside of the ant's head (*Lasius*), stimulates the ant with its legs and thus releases feeding. (After C. Janet from K. Escherich, 1906.)

out. When there is a lack of workers, the *Formica sanguinea* workers rob workers from nearby *Formica fusca* nests. The American *Formica rubicunda* ant, which lives in *Formicà subserica* nests, collects new slaves when they are needed. The Amazon ants (*Polyergus*) are completely dependent on their slaves, because they are unable to break down food with their saberlike mandibles. They must be fed by their slaves and their main occupation is to go on slave-making raids. The small ant *Solenopsis fugax* builds its tunnels in the nests of *Formica rufa* so that it can rob their food. Similar prey robbers are also found among higher animals. I refer to the frigate birds mentioned earlier (for example, *Fregata minor* of the Galápagos Islands, which chases other birds and takes their food [I. Eibl-Eibesfeldt, 1964b]). The European cuckoo possesses a supranormal gape releaser and throws its nest mates, the young of the host species, out of the nest. This behavior wanes after several days.

Many species live on others either for the duration of their life or for a shorter period only. Some species regularly use others as a means of transportation, for which the term *phoresis* has been coined. The marine snail *(Janthina)* is often used by the crab *(Planes minutus)* as a means of transportation. The book scorpion sometimes clasps the legs of flies and is transported to other localities in this way. Such phoresic relationships may lead to a symbiotic as well as a parasitic relationship. The larvae of *Meloë* climb the flowers of anemones and dandelions and wait there for bees, which then carry them to their nests. There they eat the larvae and food stores.

BEHAVIOR TOWARD SPECIES MEMBERS

> In some aspects there exists a surprising agreement in the social behavior of animal and human groups, so that one may be encouraged to hope that animal psychology could be useful in discovering laws that also govern the social life of human groups (D. Katz, 1926:448).

The conspecific often plays the role of a partner as well as that of a rival in the ecology of an animal. As a result of the dual nature of this relationship, the species member frequently becomes the bearer of rejecting as well as attracting signals; some aspects of this duality have already been discussed under the heading of courtship and greeting ceremonies. In this section I want to demonstrate the selective advantages of these opposing contact-seeking and distance-maintaining mechanisms. I want to discuss the advantages of forming groups with few or many individuals and the advantages of social intolerance. Only a few species are socially indifferent in the sense that they seek no contact. Such animals do not even meet for reproduction. Many marine animals discharge their sexual products into the surrounding water. Some of them do this in response to a chemical stimulus that results in synchronization, and when they signal to each other in this way, we are already dealing with contact behavior of the most simple kind. Many ground-dwelling male arthropods deposit spermatophores which are accidentally discovered by the females. The males of *Polyxenus* build a signal structure in the

PLATE V

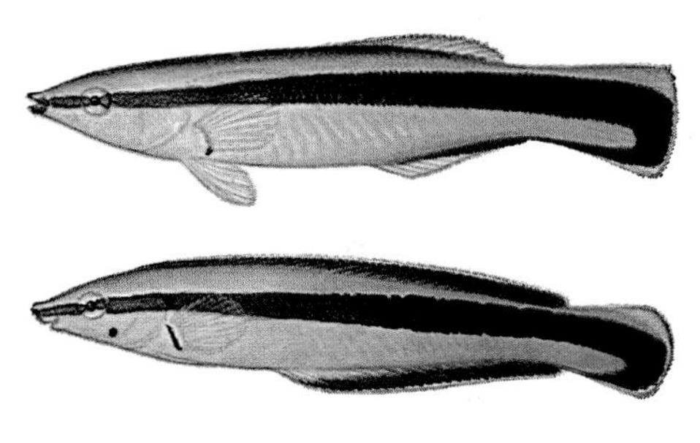

Top right: Plectorhynchus diagrammus being cleaned by cleaners. A cleaner mimic attacks the tail of the fish in the foreground. Two places where chunks were bitten from the edge of the caudal fin are clearly visible. (H. Kacher, artist.) *Bottom right:* the cleaner *Labroides dimidiatus* and below its mimic *Aspidontus taeniatus*. (H. Kacher, artist.) *Left:* egg-spot cichlids *(Haplochromis burtoni)*. Top: The male presents the egg spots on his ventral fin to the female during courtship. Below: The milting male exposes his egg models, which the female tries to pick up. (From W. Wickler [1964a]; H. Kacher, artist.)

PLATE VI

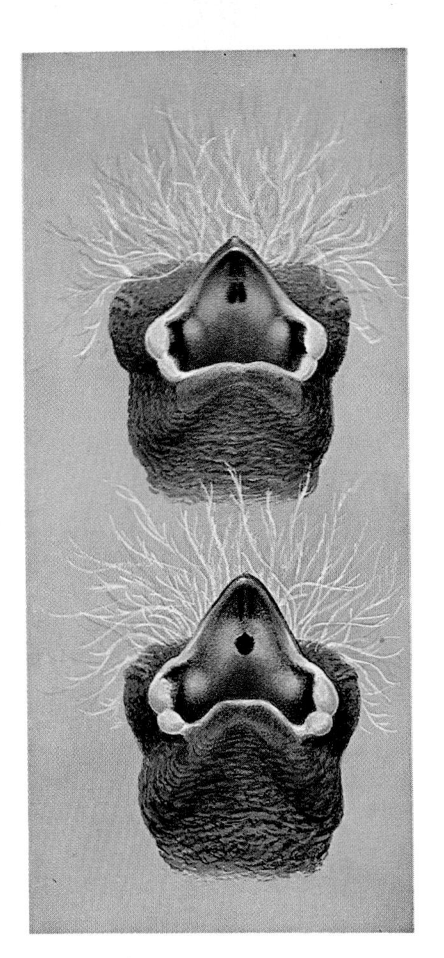

Top right: the Estrildid finch *(Pytilia melba),* to its left the paradise widow *(Steganura paradisea)*. Below them: 30-day-old young (right: Estrildid finch; left: paradise widow); on the left the gape markings of 1-day-old young; top, paradise widow; below, Estrildid finch. Whereas the adults look very much different, the gape markings of the young are almost identical, and the young fledglings are quite similar. (After J. Nicolai [1964]; H. Kacher, artist.)

PLATE VII

Right: chimpanzee paintings (Tierpark Hellabrunn, Munich). Top: painting of a high-ranking female. The sequence in which the colors were offered was red, blue, and green. The animal filled the available space. Bottom: painting of a low-ranking female. Color sequence: red, black, blue, and yellow. The animal painted one color over the next. (Photograph: I. Eibl-Eibesfeldt.) *Bottom:* protective coloration and behavior. The somatolytically colored perch *(Cephalopholis miniatus)* remains still when the diver approaches. The colorful *Paracanthurus theutis* tries to hide and offers the observer only a view of the narrow back. (Photograph: I. Eibl-Eibesfeldt.)

Vervet monkey *(Cercopithecus aethiops)* displaying the colorful genitalia when sitting (Tsavo National Park, Kenya). (Photograph: I. Eibl-Eibesfeldt.)

form of a double track made of filaments that leads the females to the spermatophores (F. Schaller, 1962).

Special internal motivating mechanisms ensure that an animal will seek social contact with members of its species. Distance-maintaining behavior is often based on special drive mechanisms (**aggressive drives**) that lead some animals to agonistic encounters with conspecifics. In the same way there exist drive mechanisms which are the basis for grouping behavior. The drives to sexual and parental-care behaviors are examples of this. For the graylag goose H. Fischer (1965) demonstrated an additional drive that is independent of sexual behavior and caring for the young. Such drives to maintain bonds probably exist in other animals as well. A young fish that is separated from the swarm exhibits appetitive behavior in search of a swarm. It has not been possible to demonstrate an underlying sexual motivation. What the animal seeks is just to be in the proximity of a conspecific, which constitutes the consummatory situation. The mechanisms that motivate an animal to bond with a conspecific are numerous, and they often develop convergently in different groups of animals.

Intraspecific Aggression

We use the terms *aggression* and *aggressive behavior* as synonyms. Aggression has been defined in a variety of ways. Psychologists usually define the term by the intent to hurt as contrasted to accidental hurting or hurting with intent to heal. In animals, however, such a definition would not apply. We may, of course, assume that higher animals do at times act "by intent," but for reasons explained in the introductory chapters we cannot operate with subjective terms. Nor is "hurting" a helpful characteristic. Often fights are ritualized, so as to avoid damage and destruction. Should not such a ritualized fight, a tournament of two antelopes for example, be labeled as an aggressive encounter? We would indeed draw an artificial distinction if we were to label as aggression only those encounters that lead to damage, in particular since we know that the physiological system underlying both types of intraspecific fighting is often the same, and since it can furthermore be proved that patterns of ritualized fighting often evolved from patterns of damaging fighting.

What we can clearly observe is that a range of behavior patterns convey hostility toward a conspecific. Two or more members of a species often compete, using either force or signals of hostility, with the effect that one finally gives in, either leaving the place or merely avoiding further approach, or, in the case of gregarious animals, subordinating himself, leaving the dominant winner with prior access to resources.

Behavior that leads to those observable results (spacing, subordination) we call aggressive. The term is thus functionally defined and does not imply that the mechanisms underlying aggression in different animal groups developed as homologies. There are, indeed, strong indications that the mechanisms underlying aggression evolved repeatedly in analogy.

One might propose labeling as aggressive only those encounters in which physical force is employed. But this does not seem justified. Threat displays regularly evolved by the ritualizations of acts of fighting and as ritualized intention movements of attack. There are, of course, cases where this is not so evident. The territorial song of some male birds serves as a spacing device against other males. The intensity of singing is increased if the rival approaches, and if he does not withdraw fighting starts. Females, however, are attracted. We know too little about the source of the territorial song. But territorial singing is certainly positively correlated with acts of aggression. Birds displaying territorial songs are also ready to attack a rival. Some insects "aggress" by means of pheromones. In the polistine wasp the highest-ranking female physically subjugates auxiliaries; in the honeybee the same is achieved by a pheromone. It seems that the latter behavior phylogenetically replaced actual fighting and may therefore be considered as ritualized aggression. Rats *(Rattus novegicus)* and many other rodents engage in threat duels by directing at each other pulses of ultrasonic sound. This strong sensory impact acts as a hostile gesture. Evidently it is a form of fighting, positively correlated with other acts of threat and physical aggression.

In fact of all these facts it seems advisable to stick to the definition given above and label all acts by means of which an animal drives off or dominates another conspecific as aggressive, whether physical force is used or not.

Territoriality

According to Rousseau the builder of the first fence was the founder of civilization. Since the work of H. E. Howard (1920) we know, however, that many animals defend a certain area of their habitat as a territory against members of their species and often mark it in a specific manner. The territory may be the possession of one individual that repels all conspecifics or merely of all those of the same sex, but it can also be the possession of a group that repels only conspecifics that are not members of the group.

Ethologically a territory is defined as a space in which one animal or a group generally dominates others, which in turn may become dominant elsewhere (E. O. Willis, 1967). Domination can be achieved by diverse means, for example by fighting threat, territorial songs, and olfactory marking. By these means the territory owners usually banish those that do not belong to the group or any conspecific if it is solitary.

In hamsters males and females live solitarily, and they share a den only temporarily during the reproductive season. The females live with their young for only a relatively short time. In many birds and some mammals (gibbon, *Hylobates lar;* see J. O. Ellefson, 1968) the pair defends a territory, but many animals live in larger units (packs, herds, or clans) that occupy an area they defend against conspecifics from different groups. This is the case with wolves, hamadryas baboons, and rats, to give only a few examples. House mice and house and Norway rats (*Mus musculus, Rattus rattus,* and *R. norvegicus*) live in groups that develop out of the family unit as succeeding generations remain together. These animals defend their territories against strangers from other groups of their species. This intolerance, which is tied to a specific area, has received much attention recently, because certain parallels exist to the human attitudes and behavior with respect to **property.**

Territorial behavior ensures a certain amount of living space or hiding places for an individual or a group of animals. Thus it is important if a songbird is to find sufficient food for its brood that no other species member breeds in the immediate vicinity of its nest. There may be competition for a suitable hiding place or nesting sites. Anemone fish do not defend their anemones as a feeding ground but as a hiding place, and the same is true for many other reef-dwelling fish. Animals distribute themselves equally through territorial behavior. Pressure is exerted on the neighbors and in the final analysis this results in an increase in the range of the animals. Finally, one of the results of territorial behavior is to prevent an overexploitation of the living space, for example through overgrazing (M. M. Nice, 1941; N. Tinbergen, 1957; F. S. Tompa, 1962; V. C. Wynne-Edwards, 1962). This principle holds whether individual animals, pairs, or larger groups oppose each other as intolerant units. When groups exert pressure on each other, then this also results in their dispersal. In free-ranging monkeys group territoriality is a widespread characteristic, and the analogies to human behavior are obvious according to C. R. Carpenter (1942). Members of different groups threaten and fight one another and actual fights between groups can develop. A. P. Wilson (1968) studied such fights in rhesus monkeys on the island of Cayo Santiago. In these conflicts the groups are lined up opposite each other. Several females of one group may rush forward in an attack; they fight briefly and retreat to their own line, while females that have waited until now rush forward and continue the fight. Fights between groups of Norway rats were described by F. Steiniger (1951).

Males and females may share equally in the defense of territory. Often, it is primarily if not exclusively the male that occupies the territory. This usually occurs during the reproductive period; at other times the animals may be quite peaceful. Here possession of the female is the goal. The selective advantage of such rival fights is that the stronger and healthier animal will breed, and in some animals the stronger also assumes the role of protector of the brood. In the Uganda kob (*Adenota kob thomasi*) there exist selected mating places or arenas that consist of a number of adjoining territories. Each territory is occupied by a male. In the center of these arenas, which have a diameter of 300 to 400 meters, the territories are most closely packed, 10 to 20 in number. The females seek out for mating the males that hold these center territories. There are also males in single, more widely dispersed, territories. The actual advantage of such arenas is not known. They have only been studied in this antelope (H. K. Buechner, 1961; W. Leuthold, 1966). F. R. Walther (1966) is of the opinion that this could be an adaptation against predators. When many neighbors are close together they can detect danger more easily. In addition, these otherwise very sociable animals can maintain social, if only visual, contact with their companions in spite of territorial separation.

To avoid misunderstanding it should be pointed out that territorial species do not defend all areas they may visit against members of their own kind. In areas that are frequented regularly by an animal neutral areas exist. An area that is not defended by an animal may be called its home range. In the Galápagos sea lion (*Zalophus wollebaeki*) males defend a specific coastal strip on land as well as the water close to it. The fishing grounds in the sea, however, are not defended. The hamster

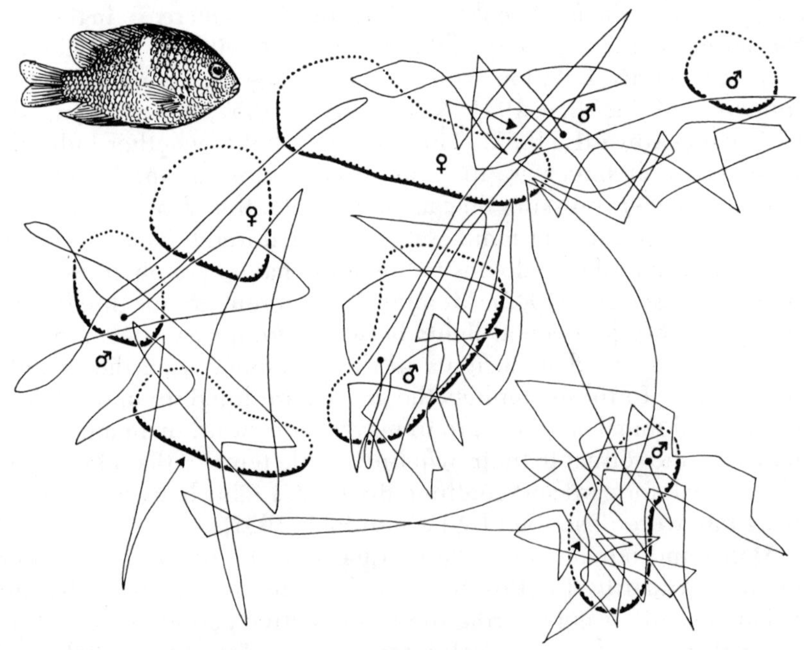

Figure 15-38 Territories of four male damselfishes (*Abudefduf leucozona*) can be recognized by the marked paths over which the animal swims. It can be seen that the fish remain in a small area. Longer excursions came about during pursuit of entering neighbors. Each fish was observed for 5 minutes. The area shown in the drawing covers approximately 5 × 6 meters. The fish measures about 12 cm. (From I. Eibl-Eibesfeldt, 1964c.)

(*Cricetus cricetus*) defends and marks its den and the immediate environment but retreats from other hamsters it may meet in the fields. There are then neutral areas that are not defended. In baboons (*Papio ursinus*) the troops each evening seek out specific sleeping places in trees, which are defended against other troops. Each troop also has its own feeding ground, where contact with other groups is avoided. At waterholes, however, they come into contact with other groups without conflicts resulting (I. DeVore, 1965). Sometimes an animal defends its entire living area, which then is usually small (Fig. 15-38). The territory is not necessarily a contained area with rigid boundaries. It can also be a system of paths with fixed points. Norway rats pursue strange rats from other territories only on paths they have marked; house rats (*Rattus rattus*), on the other hand, defend the entire area that is crossed by their paths (H. J. Telle, 1966).

J. H. Mackintosh (1970, 1973) has described in some detail territory formation in domestic mice (*Mus musculus*). Territories were initially formed by placing a barrier between two groups and then removing it. The territorial boundary, initially formed along the site of the screen or other natural barrier, could be shifted by moving visual clues, such as position of water bottles or small plastic rectangles placed on the side of the enclosure. A hierarchy of cues to the boundary was shown to exist, the most important being visual (large nearby objects being more important than distant small ones) with olfactory cues secondary. Mackintosh also showed that the movement of subordinate males is highly restricted, whereas females and juveniles can roam throughout both territories.

Both P. E. Maxim and J. Buettner-Janusch (1963) and J. D. Patterson (1973) have reported cases of baboon territoriality, involving intertroop fighting, during times of scarce resources (drought and an unusually long dry season respectively). The patterns observed were of "chase/counterchase between the adult and large juvenile males of both troops, culminating in out-and-out fights between opposing adult males" (Patterson, 1973:645).

Larger birds often have separate breeding and feeding territories. While the tree-nesting herons, such as the silk egret (*Egretta garzetta*), the rail heron (*Ardeola ralloides*), the night heron (*Nycticorax nycticorax*), and the gray heron (*Ardea cinerea*) breed in dense heronries in small forests on the bank of the Danube and Theiss, with individual distances reduced to pecking distances, the birds distribute themselves for feeding on stagnant sidestreams and rice fields over wider areas, where they often hold larger, individual territories. The daily flights (feeding) between the two separate areas are still performed in dense masses; only on the feeding ground itself are distances kept, owing to increased aggression (A. Festetics, 1959).

It is an error to believe that territory-owning animals are in a state of continuous fighting with their neighbors. Such a misconception seems to have led R. Schenkel (1966) to conclude that the black rhinoceros is not territorial because he never observed fights between neighboring animals. He failed to consider that animals usually fight when the territories are first established and occasionally with a trespasser, but not with their neighbors. These know one another and respect each other's territories. One rarely observes fights between two neighboring Galápagos sea lions, but intense fights with intruders take place.

Schenkel seems to restrict the term territory to an area that the animal can monitor continuously and where there is a possibility of an intruder being confronted and driven off within a short time. In line with this reasoning a wolverine would be highly intolerant toward conspecifics of the same sex but would be unable to protect its large home range from intruders. Therefore, it would be wrong to speak of its territoriality. I propose that any space-associated intolerance be called *territoriality*, where a "territory owner" is that animal before which another conspecific must retreat at a given time. In this connection P. Leyhausen's (1965b) observation that cats possess temporally defined territories deserves notice. Many male cats can use the same area, but at different, well-established times, and each is only a temporary owner of the territory—and is retreated from—during this time. This does not mean an equalization of territoriality with relative intolerance such as is observed in rank disputes. Intolerance connected with rank may lead to spatial avoidance on the part of the inferior but rarely leads to a spacing out. In territorial defense the competitors for rank often unite against foreign intruders. A low-ranking member of a group thus demonstrates that he is a member of the group, occupying and defending the group territory. Rank disputes are not linked to territorial claims; they might be observed even in a migrating flock, and the rank relationship does not change regularly with the time of the day, whereas in temporally defined territories the temporary owner gains superiority.

With this proviso I basically agree with Schenkel that a territory is

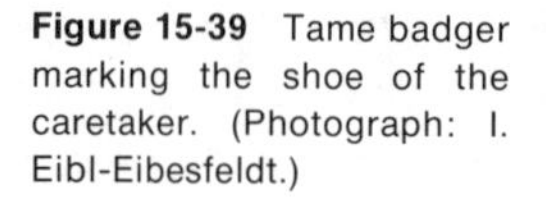

Figure 15-39 Tame badger marking the shoe of the caretaker. (Photograph: I. Eibl-Eibesfeldt.)

Figure 15-40 Olfactory marking of the tenrec (*Echinops telfairi*). The animal is marking the head of the caretaker. (Photographs: I. Eibl-Eibesfeldt.)

an area in which an individual or group does not tolerate particular members of the same species, either repulsing all strangers or only those of the same sex. Ownership of the territory may be restricted to established periods of time only. There are of course cases in which it is not easy to establish whether we are confronted with territoriality or not. Animals living in closed groups with nomadic habits (for example, on migration) defend their group at any place, but show no particular attachment to site. Such intolerance, however, would in effect automatically lead to territoriality, if the animals became resident in a particular area.

Natural landmarks are often used as territory boundaries. In the three-spined stickleback one can experimentally move the territorial boundary. A newly planted row of *Elodea* will be accepted by a cichlid fish as its territorial boundary, even if it reduces the size of the original territory. Likewise, a row of bicycle spokes, 3 to 4 cm apart, will be accepted (J. v. Iersel, 1958). However, the new border is only accepted if it also borders the territory of a neighbor and when it is not closer than 30 cm from the nest.

An area that is occupied by an animal or a group is often marked. Many mammals place scent marks by depositing gland excretions, urine, or feces at certain places around the territory. The method of marking differs from species to species. The hamster smears the excretion of its flank glands onto the walls of its den and on clumps of grass and stones in the vicinity. Badgers and martens mark objects with an excretion from a gland pocket under the base of their tails (Fig. 15-39). Antelopes place excretions from their preorbital glands on bushes and tips of grasses (H. Hediger, 1949; F. R. Walther, 1965). The tenrec (*Echinops telfairi*) puts saliva on the object to be marked and transfers its body odor by alternately scratching itself with a foot and rubbing it in the saliva (Fig. 15-40). The giant galago and the Senegal galago (*Galago crassicaudatus* and *G.*

senegalensis) urinate on the palms of their hands and rub it into the soles of their feet. When climbing about they leave behind obvious scent marks that are visible as dark spots. House mice and rats mark their paths with urine and follow these trails like trains on a track (I. Eibl-Eibesfeldt, 1950c, 1953c, 1965c). These odor trails can also be used by strange mice and rats. When one mouse population, which lived in a wooden barracks, was exchanged with another, the new mice quickly found their way about by utilizing the scent-marked paths of the previous owners (I. Eibl-Eibesfeldt, 1950c). H. J. Telle (1966) poisoned one of two adjoining rat populations, each of which had its own marked trails that were separated only by some narrow objects. When he introduced new rats in the freed area the newcomers used all the available trails, including those of the adjoining population. They soon learned that they would be attacked there, however, and thereafter they restricted their activities to the area that became available when the previous population had been poisoned. Male rabbits mark their territory with chin and anal glands. The chin glands in males are larger than in females, and this difference in size increases with the approach of sexual maturity. They are more strongly developed in higher-ranking males but have no direct relationship to body size; sometimes a smaller but more dominant male can have a larger chin gland than a heavier, sexually inactive animal. The secretion of these chin glands is rubbed on the ground, branches, and stones, and on females. High-ranking animals mark more frequently than low-ranking ones (R. Mykytowycz, 1955). The secretion of the anal glands adds a particular smell to the dung pellets. Those dung pellets that are deposited for the purpose of territory markings on especially excavated earth mounds have a stronger odor than those scattered all over the ground during feeding (R. Mykytowycz, 1966).

The scent marks are chemical property signs (F. Goethe, 1938). They aid the territory owner, first of all, as signs of recognition. They help in orientation and make the area familiar. A badger that becomes agitated or frightened in a strange environment can be calmed by letting it sniff an object that it had previously marked (I. Eibl-Eibesfeldt, 1950a). A male hamster that enters the territory of a female during the mating season will mark this strange territory before it begins to court. It is probable that this has also a repelling function for others. Strange scent marks have an aggression-releasing effect in hamsters, which show threat behavior when sniffing them (I. Eibl-Eibesfeldt, 1953a). Rabbit dung that is introduced into the territory of an established animal is highly arousing. A male rabbit marks with its chin glands and defecates significantly more than when some fresh earth is placed into its territory. Aroused in such a way male rabbits will also begin to attack members of their group, but will stop when they have come close enough to recognize them. It appears as if a rabbit in this mood considers everyone a potential intruder (R. Mykytowycz, 1966). The scent marks of the flying marsupials (*Petaurista*) have no repelling effect on strangers but they increase the aggressiveness of the territory owners while they lower that of the strangers (T. Schultze-Westrum and B. Braun, 1967).

In addition to scent markings there are other ways of indicating territorial ownership. For example, it can be advertised by calls and conspicuous behavior. The male sea lion calls continuously when swimming back and forth before its part of the beach. Near the territorial boundary it occasionally climbs out of the water and calls toward its neighbor, who reacts in the same way, without fighting. Male fur seals (*Callorhinus ursinus*) who own territories move toward the neighbor, throw themselves on their bellies, and glide toward each other until they bump noses near the territorial boundary. In this manner they indicate their boundaries without engaging in fights (G. A. Bartholomew, 1953). Groups of howling monkeys (*Alouatta palliata*) mark their territories by howling displays, especially during the morning hours (C. R. Carpenter, 1965). The territorial song of many songbirds is well known.

Many animals display themselves conspicuously within their territory and are frequently conspicuously colored. **Genital displays of some primates** can be interpreted as a visual marking of territory.

Ownership of territory is frequently a prerequisite for the occurrence of aggressive behavior. Sticklebacks swim peacefully together in a swarm without reproductive coloration until they have found a suitable location for the establishment of a territory. As soon as one fish has found a place, its belly turns red; other males are attacked when they approach too closely. Its readiness to attack is reduced the farther it gets from its own territory. This can be demonstrated easily if two neighbors are each placed in small glass containers. They can then be moved about at various distances from each other. If male *a* remains in its territory and *b* is placed close to it, it will attempt to attack *b* through the glass wall, while *b* attempts to flee. If both are placed into the territory of *b*, then the change in behavior can be observed: *b* attacks and *a* tries to escape (Fig. 15-41).

More aggressive individuals generally conquer more favorable and larger territories. If a territory-owning red grouse (*Lagopus*) is implanted with a pellet of male sex hormone under its skin, then the aggressivity of the male is increased and it expands its territory substantially at the expense of its neighbors (A. Watson, 1966).

Sometimes the defended areas are quite small. Many birds that breed in colonies nest just out of reach (pecking distance) of their neighbors (Fig. 15-42).

Outside their territory many animals are willing to attack a conspecific if it comes too close. They are surrounded with a small inviolable space (Fig. 15-43). This "individual distance," which if crossed by a member of the same species will release fighting, has been measured precisely by P. R. Marler (1956a) for the chaffinch. Males permit females

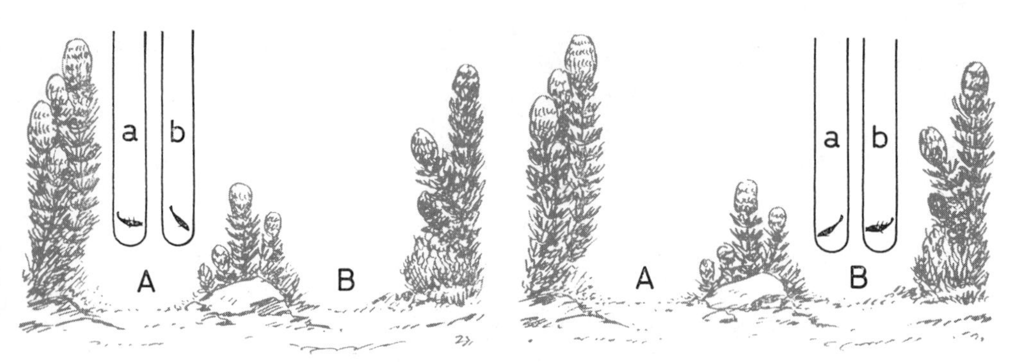

Figure 15-41 Two stickleback males (*a*) and (*b*), which live in two territories *A* and *B* respectively, are moved from one territory to another in glass tubes. When *a* is in its territory it attacks *b*, which tries to escape. If both are moved into *b*'s territory, he attacks and *a* tries to escape. (From N. Tinbergen, 1951.)

to approach closer than they do other males, where the males' plumage is the distinguishing mark. Females whose undersides have been artificially dyed red are not permitted as close as normal-colored ones and are attacked at a greater distance, as if they were males.

Figure 15-42 Flightless cormorants nesting side by side just out of reach (pecking distance) of the next nest. (Photograph: I. Eibl-Eibesfeldt.)

Figure 15-43 Individual distance between gulls on Lake Zurich. (Photograph: H. Hediger.)

Intraspecific fighting behavior

In rival fights and in fighting for territories we observe attack behavior, which is exclusively released by the appearance of conspecifics even before any physical contact has been made. This aggressive behavior has been the subject of many discussions and there are many contradictory views. These contradictions are especially prone to arise in discussion of the degree to which phylogenetic adaptations determine the behavior, and especially whether internal drive mechanisms are the cause for a spontaneous aggressive drive.

Intraspecific fighting behavior has several remarkable characteristics and in most species easily can be distinguished from interspecific fighting behavior. An oryx antelope will never use its horns to gore another oryx but fights according to strictly observed rules (Fig. 15-44).

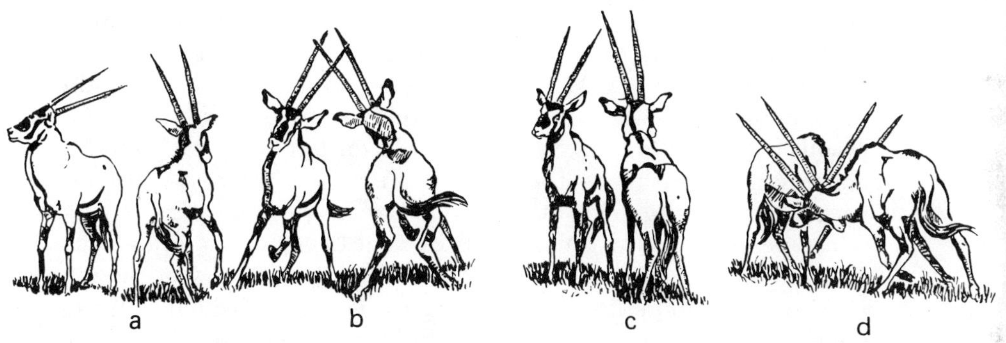

Figure 15-44 Fight between two bulls of *Oryx gazella beisa*. (*a*): Initial posture (head-up threats); (*b*): the first clash, where the horns touch in the upper third; (*c*): pause; (*d*): second thrust, which leads to the head-to-head pushing contest. (After F. Walther, 1958.)

It does, however, stab lions in this manner (F. R. Walther, 1958). A giraffe uses its short horns to fight rivals, but uses its hoofs in defense against predators (D. Backhaus, 1961). A predator fights differently with a species member than it does with a prey, and by electrical brain stimulation it can be shown in cats that these two types of behavior have different neural substrates in the central nervous system. Stimulation of areas in the lateral hypothalamus evokes eating and prey-killing responses. Stimulation of areas in the ventral and medial hypothalamus elicits intraspecific ("emotional type") aggression (B. Kaada, 1967). These differences between inter- and intraspecific aggression must be emphasized, because they are not always clearly recognized. Thus, R. Ardrey (1962) traces the aggressive behavior of man back to the predatory ways of his australopithecine ancestors. He overlooks the fact that there is no necessary connection between aggression and predation. After all, plant eaters are no more peaceful against their own kind. Bulls fight no less intensely than do rabbits, sparrows, hamsters, or cats. Z. Y. Kuo (1960–1961) also treats intra- and interspecific aggression as if it were the same. We do not want to assert that there is absolutely no connection between aggression and food getting, but only that a predatory way of life does not necessarily lead to an increase of intraspecific aggression. W. Wickler (1961a) and H. Albrecht (1966a) have shown that fighting behavior can frequently be derived from feeding behavior. Fish that feed on algae which they pull from the bottom fight and threaten with the same behavior patterns. Others, which capture larger prey, threaten with movements and postures that they assume before striking at prey. Furthermore, the readiness to feed and to fight are often positively correlated: stimuli that release fighting facilitate eating movements; eating increases the readiness to fight. The phenomenon of competition for food, which we can also observe in mammals, is facilitated in this way.

In this connection I should point out that the manner of fighting of a species is also naturally determined by a number of other factors which in themselves have little to do with aggressive behavior.

R. Apfelbach's (1967a) investigations on mouth- and bottom-breeding cichlids of the genus *Tilapia* revealed obvious differences in the mouth fights. Bottom-breeding fish hold on tightly to each other's lips and engage in pushing-pulling contests until a winner emerges. Mouthbreeders never bite each other in this way; instead they butt each other with opened mouths. This mouth butting becomes more and more ritualized as the mouthbreeding itself becomes more specialized. Apparently mouthbreeding requires a sensitive mouth, which is incompatible with the rough ways of push-pull fighting (Fig. 15-45).

According to a widely held opinion aggressive behavior ultimately aims at the destruction of the opponent. This can easily be disproved. Where one species possesses very dangerous weapons such as teeth or claws, which could easily kill an opponent if they were used, special inhibiting mechanisms have usually evolved that prevent killing of the species member; often the entire fight has become transformed into a

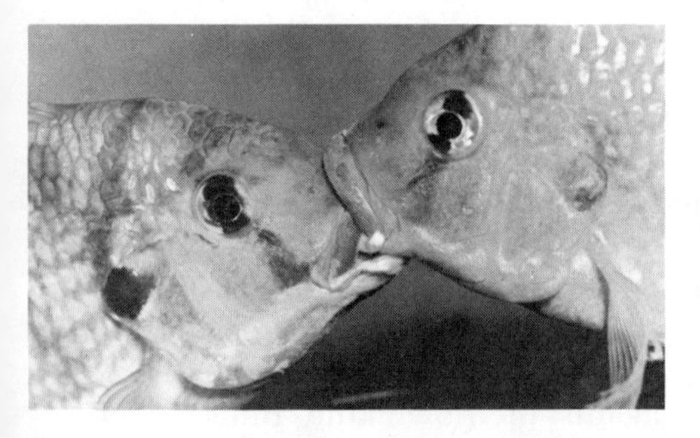

Figure 15-45 *Left:* The cichlid *Tilapia zilli* is a bottom breeder and grasps the opponent firmly with the jaws during fighting; *right:* the mouthbreeding *Tilapia nilotica,* on the other hand, fights by pushing with the mouth. (Photographs: R. Apfelbach.)

tournament (K. Lorenz, 1943, 1963a).[2] Only rarely do well-armed animals use their weapons against a conspecific without inhibition. This is true for some rodents, hamsters (*Cricetus*), for example, which can get away from one another quickly after a short exchange of bites. This ability to escape protects the pursued, and under natural conditions one hamster rarely kills another. Why lions in some areas of East Africa kill animals belonging to other prides without inhibition (R. Schenkel, 1966) remains to be studied in more detail.

There are certainly a number of animal species, particularly among invertebrates, in which aggression is destructive. But this fact does not invalidate the finding that aggression is very often ritualized. The rule remains unchallenged: where physical aggression can easily cause damage to a conspecific aggression is ritualized instead of dropping out altogether. In the crickets *Acheta domesticus* and *Gryllus pennsilvanicus*, for example, singing is positively correlated with aggression, and it can be considered as its ritualized form, since it suppresses aggression in subordinate animals. If one destroys the tympanal organs of a subordinate cricket, it can no longer hear the singing, is freed from the inhibition, and in consequence attacks (L. H. Phillips and M. Konishi, 1972).

A large number of invertebrates fight without doing damage to each other. The best-studied forms are the tournaments of some fiddler crabs (R. Altevogt, 1957; J. Crane, 1966). During low-intensity fighting the animals merely butt each other with their large claws, which are only slightly opened. Crane reports that small protuberances on the front side of the claws prevent them from slipping off one another. Only with an

[2] In German this is called *kommentkampf* (*comment* = French, how) and refers to the rules of duelling by students. In English: tournament or ritualized fight.

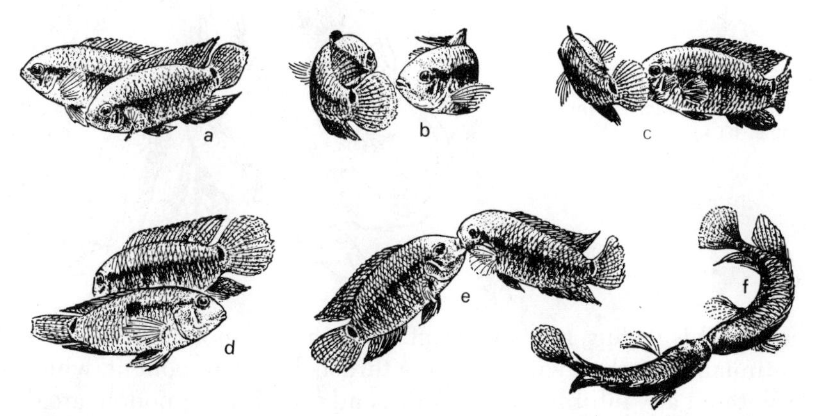

Figure 15-46 Fighting scenes from *Aequidens pulcher* (= *A. latifrons*). (*a*): Lateral display threats and light tailbeats, uneven fin spread; (*b*): circling, threatening swimming about, lower part of the mouth lowered; (*c*): tailbeat against the head of the opponent; (*d*): the front animal gives up (coloration, fin position); (*e*): rearing up in front of the other and grasping the mouth; (*f*): mouth pulling. (From W. Wickler, 1962b.)

increase in the intensity of fighting do the crabs hold on with their claws, but here again other morphological adaptations ensure that the animals grasp each other in a particular way.

Ritualized fights can be observed in many fish. In cichlids (Cichlidae) rivals threaten each other either frontally or by lateral displays, in which they spread the fins and especially the gill covers and gill membranes (Fig. 15-46*a–f*). At the same time the combatants display a colorful appearance. Before the fight begins they may circle each other.

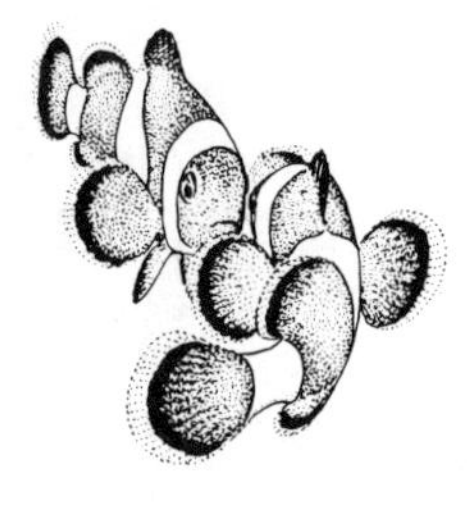

Figure 15-47 Sparring technique of fighting anemone fish (*Amphiprion percula*). (From I. Eibl-Eibesfeldt, 1960a.)

Then one animal beats its tail against the other, and the strength of the created pressure wave gives an indication to the other of the strength of its opponent. The cichlid *Apistogramma wickleri* specializes in tail beating. The tail beat aims below the opponent and creates a drag so that the other fish is pulled downward (W. Wickler, 1962b). After an exchange of tail beating most species begin mouth pulling or mouth pushing by grasping each other at the upper or lower mandible. Finally one of them

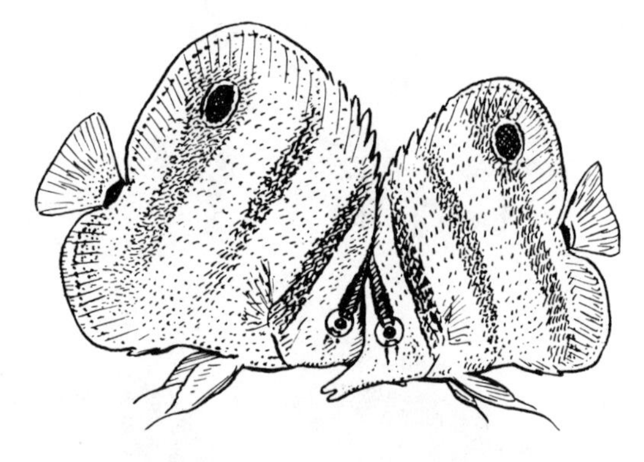

Figure 15-48 Fighting pipefishes (*Chelmon rostratus*). (After D. Zumpe, 1964.)

gives up, folds its fins, and swims off. If the fish cannot withdraw, then it is continually attacked with ramming thrusts by the opponent, which attacks without any inhibition. The sides and fins of the opponent are damaged and it is quickly killed. This occurs, however, only when the fish are kept in an aquarium.

Grunts (*Haemulon*) fight by attempting to push each other mouth-to-mouth from their places. Anemone fish (*Amphiprion percula*) also have tournament fights in which they parry thrusts of the opponent with their pectoral fins (I. Eibl-Eibesfeldt, 1960a, 1965a; see also Fig. 15-47). Butterfly fishes (*Chaetodon, Chelmon, Heniochus*) fight by head butting (D. Zumpe, 1965; see also Fig. 15-48). The clinid fish *Emblemaria pandionis* grasps its opponent by the head after an initial threat display and attempts to retreat into its own cave while holding on to the other. If this succeeds, the other will give up soon, because it is exposed helplessly before the entrance (W. Wickler, 1964e). Piranhas (*Serrasalmus*) fight by tailbeating. The loser flees to the surface and is left there unmolested. Only rarely do these well-armed fish bite each other (H. Markl, 1972).

In poisonous snakes rival males fight according to strict rules that

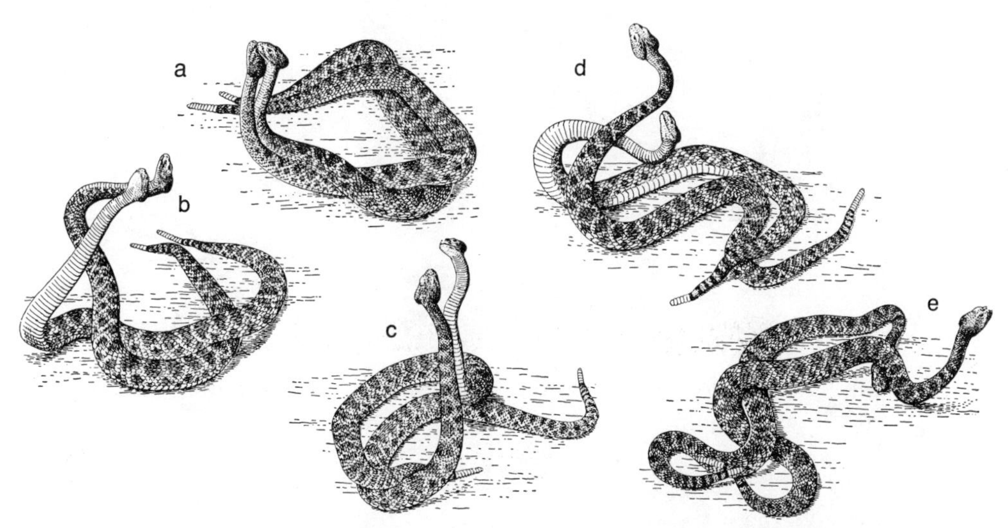

Figure 15-49 Fighting male rattlesnakes (*Crotalus ruber*): The rivals beat one another with their heads. The loser is pushed against the ground with the body. The rivals never bite one another. (After C. E. Shaw, 1948.)

differ slightly from species to species (C. E. Shaw, 1948; E. Thomas, 1961). Rattlesnakes (*Crotalus ruber*) wind themselves around each other's tails and raise the anterior third of their bodies. In this position each tries to hit the head of the other with its own (Fig. 15-49). This they do alternately until one of them is so fatigued that it gives up. In the marine iguanas (*Amblyrhynchus cristatus*) the males fight by rushing at each other after an initial threat display, and butting the tops of their heads together. Each attempts to dislodge the other. If one of them realizes that it is losing, it will give up and lie down flat on the ground before the victor, who in turn respects this submissive gesture of the loser, and waits in a threat posture until the other retreats (Fig. 15-50). In this way the stronger males do not kill their weaker, often younger rivals with

Figure 15-50 Ritualized fight of the marine iguana. *Top:* Territory of a male (center) with several females; *center:* two fighting males (head pushing); *bottom:* the weaker yields and lies flat before the victor, who ceases to fight as a result. (Photographs: I. Eibl-Eibesfeldt.)

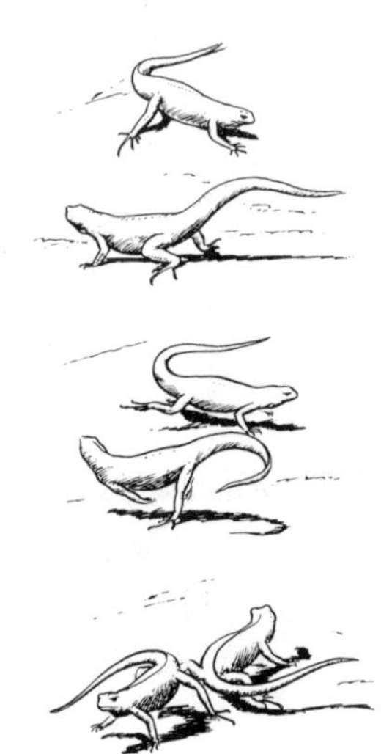

Figure 15-51 Lava lizards fighting by beating with their tails (*Tropidurus delanonis*). (From I. Eibl-Eibesfeldt, 1964a, 1966c.)

their powerful teeth and jaws, and the species does not lose its reserve of growing males (I. Eibl-Eibesfeldt, 1955d).

Fighting males of the lava lizards (*Tropidurus*) whip each other with their tails (Fig. 15-51). In special circumstances damaging fights will occur in these lizards. One *Tropidurus* male which had lost part of its tail in some way tried at first in vain to defend itself with tail whipping against its opponent. It seemed to be unaware that it did not possess a complete tail. Finally it successfully defended itself with biting (I. Eibl-Eibesfeldt 1966c; see also Fig. 15-52). Male marine iguanas bite and shake a rival if another animal is abruptly placed into their territories. The unwilling intruder does not have an opportunity in this case to show the threat display which initiates the usual encounters. This seems to be the reason an immediate attack occurs.

In sand lizards (*Lacerta agilis*) one of the combatants will permit the other to grasp its neck. This they do alternately until one seems to realize that the other is stronger. This it can recognize by the strength of the grip of its opponent. Sometimes a smaller animal will give up when it realizes, while biting the other, that it has encountered an especially large opponent. The loser prostrates itself on the ground, treads, and runs off (G. Kitzler, 1942).

Turkeys try to drive off rivals with threatening calls or by jumping at them or beating them with their legs, which have large spurs. If one of the opponents stands up to fight, a tournamentlike struggle ensues wherein each bird attempts to grasp the other by the conspicuously colored red neck and head skin, to push or pull the other and to press it to the ground. The strong skin resists this rough treatment much better than feathers could, and according to W. M. Schleidt (1966) the significance of the red signal color is that it draws the pecks of the opponent so that the feathers remain undamaged. Hens and vanquished males that show no red skin are not engaged in fights. Domestic roosters peck and kick each other after an initial display. The loser finally submits.

Many mammals change to a damaging fight after an initial threat display. Attacking Norway rats erect their hair, hump their backs, gnash their teeth, and approach their opponent broadside (Fig. 15-53). They avoid all sudden movements that could release defensive biting and begin to push their opponents from their places. So the fight has not resulted in damage, and it can end if the threatened animal gives up and leaves by jumping up and running away. After this beginning rats often box each other with their front paws while standing on their hind legs opposite each other, and they may kick with a hind leg. If the opponent falls down and the attacker lies on top of it, both become rigid and they make threats by gnashing their teeth and by squeaking. Finally, the

Figure 15-52 Damaging fight of lava lizards (*Tropidurus albemarlensis*) after one lost its tail. The animal at first tried unsuccessfully to fight the attacker with tail beats. This did not work, so at first it fled from its territory when the neighbor appeared. After several days the animal defended itself by grasping the opponent at its tail and winning the fight. (Photographs: I. Eibl-Eibesfeldt.)

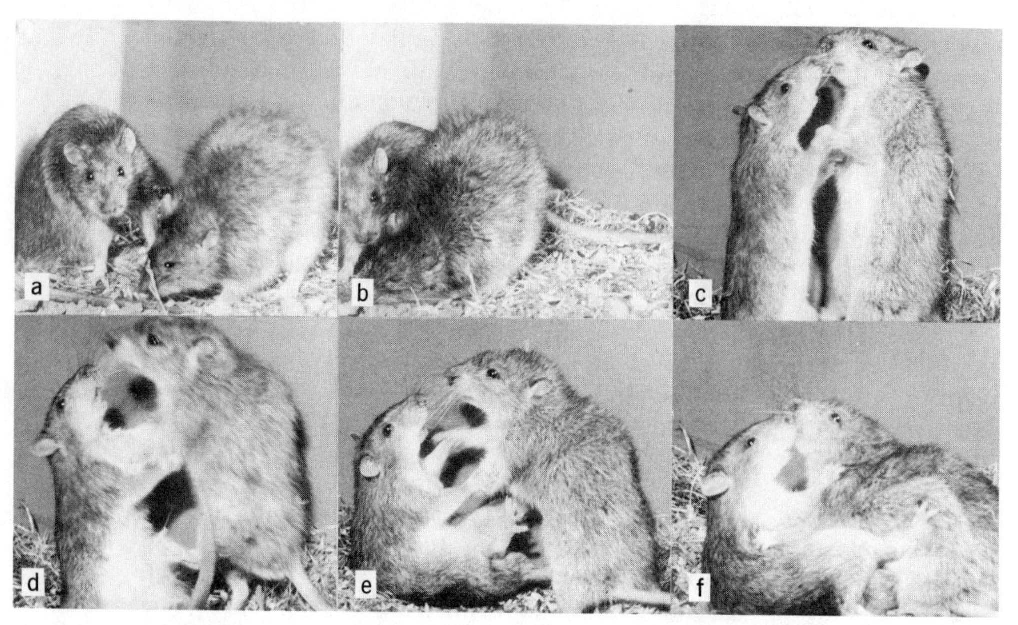

Figure 15-53 Fight of two male Norway rats (wild form). (*a*) and (*b*): The attacker approaches by exposing its lateral side to the opponent; (*c*): the opponents fight standing up in front of each other and the stronger attacker throws its opponent on its back and kicks with its hind leg; (*d*), (*e*), and (*f*): the opponents rigidly maintain their "wrestling" position. (From Scientific Film E131, I. Eibl-Eibesfeldt, 1957b.)

Figure 15-54 Damaging fight of Norway rats. The animals are biting each other and holding on.

Figure 15-55 Submissive behavior in wolves. *Left:* Appeasing by food begging; *right:* passive submission by rolling on the back. (After R. Schenkel, 1967.)

damaging part of the fight begins, when both animals bite each other and hold on (Fig 15-54). After a few seconds one of them will give up and flee (I. Eibl-Eibesfeldt, 1958a, 1963).

Wolves also bite each other in a fight until one becomes submissive either by behavior identical to the food begging of a puppy or by rolling on its back and remaining still (Fig. 15-55). The latter posture seems to be derived from the behavior repertoire of the puppy and may be interpreted as a ritualized presentation for cleaning of the anal region. Quite often the submissive wolf urinates, thus releasing actual cleaning by its opponent. Dogs behave similarly (K. Lorenz, 1963a; R. Schenkel, 1967). Submissive postures by presentation of the neck region has also been described by D. Backhaus (1960) in the zebra and by H. Kummer (1968) in the baboon.

The fights of horn- and antler-bearing hoofed animals are extensively ritualized. Bighorn sheep walk toward each other raised up on their hind legs and beat their heads together from above (H. Bruhin, 1953). F. R. Walther (1958) described a ritualized neck fight in nilgai bulls (*Boselaphus tragocamelus*) that occurs in addition to the usual head-against-head butting (Fig. 15-56). Hornlike organs have developed repeatedly and convergently in the service of intraspecific aggression (V.

Figure 15-56 Fighting rituals of various antelopes. *Top:* Two nilgai bulls (*Boselaphus tragocamelus*) during the neck fight. The animals try to press one another to the ground. This species also fights by thrusts with the horns. *Bottom:* In *Taurotragus oryx* the thrusts are always directed against the head of the opponent. Other antelopes with well-developed horns do the same.

Geist, 1966a). The multiplicity of forms in horn development within horn-bearing hoofed animals shows clearly that these organs primarily function in the service of intraspecific conflicts and that they are adapted to the specific style of fighting of a particular species. If they were, in fact, weapons against predators, then most of them would probably be dagger- or saberlike and certainly not coiled backward as in bighorn sheep. In those animals that ram their opponents the forehead is provided with a massive base for the horns, reminiscent of armor plate, as, for example, in the buffalo. In "wrestlers" the horns are shaped in such a way that they can hook or fork into each other (kudu and impala). "Fencers" beat each others' horns with the long side, where the middle part of the horns is especially subject to wear. In this case they are curved, and cross-ridges of horn prevent the horns from sliding off each other (*Capra ibex* and *Hippotragus niger*) (F. R. Walther, 1966).

In species with sideways-protruding horns the flanks are covered, preventing side attacks. The opponent is only able to attack from the front. In such cases the rules of tournament are determined by the very structure of the weapons; however this is rarely the case. The Thomson's gazelle has no protruding horns; a side attack would be easy and, indeed, the rivals sometimes stand so to each other that the observer would expect it. In spite of this, in more than a thousand fights F. R. Walther nerver saw a single flank attack. The rivals used exclusively frontal attacks, directed against the head, especially the horns. The rule and regulations of the fight must be based upon specific central nervous system structures in such cases.

Among the antelopes we find, besides these ritualized fighting methods, submissive, attack-inhibiting postures whereby the head is turned, the horns held backward, or the animal turns aside completely. In all cases the horns are turned from the partner and the appeasing animal makes itself appear smaller (F. R. Walther, 1966).

In a comparable manner we find that antlers are adapted to the style of fighting of their bearers. Here we also find a change in the style of fighting because each year the antlers are dropped. After the antlers have been dropped and the new ones are still covered with velvet, elk fight with their hoofs, without attempting to use their new horns.

Antlers and horns constitute visual signs of **rank** in some species.

The forehead-butting technique of the horn- and antler-bearing hoofed animals probably evolved out of biting (G. Tembrock, 1960). One could suppose that the bite attack became inhibited in social encounters, where the animals then lowered their head and butted together with their heads as they were carried forward by the initial momentum of the thrust. This hypothesis is supported by the observation that hornless females still perform snapping movements when they butt their heads together, as well as by the **biting-threat posture of elk.** The hornless females butt their opponents in the sides in 50 percent of the cases, something that horn-bearing males and females never do (F. R. Walther, 1961; see also Fig. 15-57). The tournaments of marine iguanas could also have evolved in this way.

Howling monkeys defend the territory of their troop by loud calls. C. R. Carpenter (1942) and C. H. Southwick (1963) talk of actual vocal battles that prevent more bloody encounters. In man we find song duels. For additional examples of tournament fights the reader is referred to B. Oehlert (1958), I. Eibl-Eibesfeldt (1961b), W. M. Schleidt and M. Schleidt (1962), K. Fiedler (1964), D. Ohm (1964), and J. Crane (1966).

In submissive postures, which terminate a fight, fight-releasing signals are turned away, as was discussed in respect to appeasing greeting gestures. The exact opposite of threat behavior is shown by making oneself smaller, the principle of antithesis, which was discussed in detail by C. Darwin (1872; Fig. 15-58).

Figure 15-57 Hornless females—here nilgai antelopes—butt in 50 percent of the cases with the forehead against the flank of the opponent, an act that the males and females of horn-bearing species rarely do. (After F. Walther, 1961.)

Many fish collapse their fins and change their coloration (Fig. 15-59). The cichlid *Tilapia mariae,* for example, changes from conspicuous coloration into the juvenile colors. The cross-bars provide excellent camouflage between plants (H. Albrecht, 1966a). Often infantile behavior will appease another animal. A dog that lies on its back and urinates frequently may be licked by the attacker as a mother would her young. We have already pointed out that many animals (for example, **hamsters**) appease during courtship by showing **infantile behavior.** Sometimes the **female presenting movement (baboons)** is used as a submissive posture. The aggression-inhibiting behavior patterns of man will be discussed in Chapter 18.

The existence of tournament fights indicates how strong a selection pressure in favor of aggressive behavior exists. Otherwise counterselection would have bred out aggressive behavior in species that can do damage to a conspecific. Instead, the most complex fighting techniques have evolved in order to allow fighting to occur as a spacing-out mechanism.

Fighting behavior evolved in different animal groups as a means of maintaining spacing. Although common selective pressures shaped the

Figure 15-58 Principle of antithesis discovered by Darwin. *Left:* Threat posture of the dog; *right:* submissive posture. (From C. Darwin, 1872.)

Figure 15-59 Threat posture (*above*) and submissive posture (*below*) of *Blennius fluviatilis*. (Photographs: W. Wickler.)

patterns along similar lines in different species, differences in the underlying mechanisms are to be expected.

The Dynamic-Instinct Concept of Aggression

The last few years have witnessed a lively discussion on the determinants of aggression. Several explanatory models have been advanced

1. The proponents of *learning theory* assume that aggressive behavior is learned. And indeed a decisive influence of learning has been demonstrated—for example in studies with mice, which can be trained to be aggressive by allowing them to win fights repeatedly. Losses dampen their aggressiveness and so does daily stroking (J. Uhrich, 1938; M. W. Kahn, 1951; J. P. Scott and E. Fredericson, 1951; J. P. Scott, 1960). Aggressive puppies that were lifted off the ground, so that they lost their footing, also became docile.

   J. P. Scott therefore came to the general conclusion that aggressive behavior is learned. The habit of attack is said to emerge gradually during the course of development, when the animals experienced pain in the competition for food and during play fighting.

> From a more general viewpoint the experiments with mice show us that aggression has to be learned. Defensive fighting can be stimulated by the pain of an attack, but aggression in the strict sense of an unprovoked attack can only be produced by training (J. P. Scott 1960:20).

Z. Y. Kuo (1960–1961) also is of the opinion that aggressive behavior is an acquired habit, and a bad one.

The learning theory of aggression plays a particular role in human psychology. The root of aggression is said to lie in early childhood, when the individual's attempts to satisfy his demands by aggressive means lead to success and are thus reinforced. In addition, social learning from a model plays an important part in the acquisition of aggressive behavior patterns (A. Bandura and H. Walters, 1963).

2. According to the *frustration-aggression hypothesis* aggressive behavior is the result of deprivation experiences in early childhood, frustration being defined as any impediment or external influence that interrupts striving toward a goal. Since in practice such experiences never can be totally avoided, the development of aggressive behavior is to all intents inevitable (J. Dollard et al., 1939; L. Berkowitz, 1962). Aggression in this view is primarily of a reactive nature.
3. The *dynamic-instinct concept* of aggression by S. Freud and K. Lorenz (see also A. Mitscherlich, 1957, 1959) assumes an innate drive toward aggression. This theory is also referred to as the instinct theory of aggression (K. Lorenz, 1963a). The dynamics of aggression were first explored by Freud, who, however, did not appreciate the importance of biological adaptation and postulated a mystical death wish, a concept from which many modern psychoanalysts have departed (H. Hartmann et al., 1949).
4. The biological concept of *phylogenetic adaptation* in a more specific way insists that phylogenetic adaptations preprogram aggressive behavior in specifiable ways, inborn motivating mechanisms being only one way in which the behavior might be preprogrammed (I. Eibl-Eibesfeldt, 1973a).

All these models are built on observations and experiments and we have therefore to assume that in most cases all these factors interact and that the different models are not incompatible. Since the concept of phylogenetic adaptations as determinants of aggressive behavior has been refuted so often, and since in particular the drive concept of aggression has been received with great skepticism, I would like to bring to the fore the facts in favor of this interpretation.

That motor patterns of fighting are inborn in animals can be demonstrated by deprivation experiments. Siamese fighting fish (*Betta splendens*) raised in isolation attack conspecifics or their own mirror image with species-specific motor patterns (H. Laudien, 1966), and so do cichlid fishes. Marine iguanas deprived of social experience fight by head-butting, whereas lava lizards that are raised in isolation to adulthood slash tails when they meet a conspecific. Jungle fowl cocks raised in isolation prove to be more aggressive than those raised in a group and they exhibited the typical motor patterns of threat and fighting (J. P. Kruijt, 1964, 1971).

Rats and mice that were raised in isolation attacked conspecifics that were introduced into their cages, and exhibited all the species-typical behavior patterns of threat and fighting (J. A. King and N. L. Gurney, 1954; E. M. Banks, 1962; I. Eibl-Eibesfeldt, 1963). Mice that were raised in isolation by Banks were even more aggressive than those raised with companions. Mice that were raised in isolation by King and Gurney attacked other mice less readily, perhaps because their innate aggressive behavior had been suppressed by their stimulus-poor environment compared to animals raised with others. In contrast, G. A. Hudgens, V. H. Denenberg, and M. X. Zarrow (1968) report that mice who were deprived of the opportunity to play with their litter mates from weaning on were more ready to fight than mice that grew up together. The authors presume that the mice learn during play to live peacefully together. Their fighting plays do not lead to injuries because their jaws and teeth are not fully developed. According to K. Lagerspetz and S. Talo (1967) aggressive behavior in the albino mice matures spontaneously around day 28. This maturation can be postponed by punishing them with painful stimuli and can be facilitated by withholding food. Experience with other mice in a group is not necessary. Genetic control of aggressive behavior was demonstrated by K. Lagerspetz and K. Worinen (1965) when they exchanged litters of mothers from aggressive and nonaggressive mouse populations. The young from the aggressive line that were raised by the docile mothers were clearly more aggressive than the animals that came from the docile strain and were raised by aggressive mothers (K. Lagerspetz, 1964).

Of course, various experiences with conspecifics have an effect on aggressiveness (E. McNeil, 1959; F. Merz, 1965). Rhesus monkeys that grow up in the exclusive company of their mothers are later more withdrawn as well as more aggressive when placed with playmates of the same age. In general, they adjust readily, however. The monkeys that were raised in this way are quite normal with respect to all other behaviors (B. K. Alexander, 1966). J. P. Scott (1963, 1964) repeatedly pointed out the importance of early social experiences for the development of friendly social relations in mammals. Nevertheless, the fact that aggressive behavior can be so greatly influenced does not justify the conclusion that it is completely learned.

For aggressive behavior to be shown an animal must generally be in its own territory, or it will flee rather than attack (N. Tinbergen, 1951; see also Fig. 15-41). An animal trainer who enters the cage first and only then allows the lions to enter utilizes this knowledge. In this way he is the territory owner and the lions are inhibited in their aggression from the start.

We have already discussed those **key stimuli** that release fighting behavior. These are often of a simple nature. Fence lizard males (*Sceloporus undulatus*) attack females whose abdominal sides have been painted blue, and they ignore males whose abdomens are painted grey (G. K. Noble and H. T. Bradley, 1933). Similarly, inexperienced male sticklebacks attack simple decoys with red undersides.

It is evident that fixed action patterns, releasing mechanisms, and releasing signals have evolved in the "service" of intraspecific aggression. The neural substrate for aggressive behavior is fairly well known in a number of instances (W. R. Hess, 1954; B. Kaada, 1967; J. M. K. Delgado, 1967). In their readiness to react aggressively animals show obvious fluctuations that are not necessarily related to corresponding fluctuations of the environment. In vertebrates the male sexual hormone plays a decisive role in inducing the specific readiness for aggression in the adult animal (T. G. Vandenberg, 1971), as well as in the organization of the neural structures during early ontogeny (A. B. Rothballer, 1967). Appetitive behavior for fighting develops in isolates, as was demonstrated with cocks in the experiments I have mentioned (J. P. Kruijt, 1964, 1971). There are strong indications, therefore, that inborn drive mechanisms underlie aggression.

For invertebrates this was proved to be the case in the hermit crab (*Pagurus samuelis*). They fight more after social isolation. Not all the patterns of aggressive behavior are equally influenced by isolation. Those labeled "low-ranking" remain unaffected, while the patterns of actual combat ("high-ranking" aggression) increase with isolation. Since the locomotor activity remains unaffected the increase in fighting cannot be attributed to an increased chance of meeting a conspecific. It must be considered as a genuine expression of the damming-up phenomenon of aggression (E. Courchesne and G. W. Barlow, 1971).

By means of electrical brain stimulation it is possible to release a true appetitive behavior for fighting in chickens (E. v. Holst and U. v. Saint-Paul, 1960). The previously discussed experiments of O. A. E. Rasa (Fig. 4-11) prove finally that aggression can be both dammed up and discharged.

In subsequent studies this author also demonstrated an propensity for aggression and the phenomenon of aggression storing in the damselfish (*Microspathodon chrysurus*). The fish learned a simple L maze when offered a conspecific to threaten at and fight through the glass pane of a goal box. Given freedom to leave the goal box, the animal's length of stay depended on the drive for aggression, which increased with isolation. O. A. E. Rasa (1971) obtained these results by testing the same five fish for fighting appetency after varying periods of isolation from conspecifics (Fig. 15-60). In criticizing this experiment, W. Wickler (1969, 1971) argued that the fish might have associated the entrance of the maze with the rival on the other side. Thus he would be aroused by the conditioned signal and not be acting in spontaneous appetence. However, according to A. Rasa the fish did not display any aggressive behavior at perceiving the maze entrance, which makes it very improbable that it considered it as a conditioned signal for the rival. Sexual motivation can be ruled out, because the fish were tested in the juvenile stage. The fact that the time spent in the goal box increased with the time of isolation is good evidence for a damming-up phenomenon. An alternative assumption could be that increased length of stay in the goal box after increased isolation was effected by a storing of curiosity. This possibility, however, was eliminated by a further experiment. The conditions were reproduced, with the exception that a silver star shape was presented instead of a fight-releasing rival. Here, the time spent in the goal box remained at the same low level.

But the physiology of aggression is by no means the same with all territorial fish. The cichlid *Haplachromis burtoni* needs certain releasing stimuli to maintain his readiness for aggression. In the absence of these stimuli, readiness to attack drops to a very low level within a few weeks (W. Heiligenberg and U. Kramer, 1972).

In the swordtail (*Xiphophorus helleri*) males that are kept for 14–56 days in social isolation show an increasing duration and intensity of fighting. "These results support the idea of a spontaneous accumulation of attack-readiness and are in contrast to the results with cichlid fishes. [Here the authors refer to the studies of Heiligenberg I have just cited.] It is supposed that the differential influence of social isolation on aggressive behavior corresponds to different types of social organization and ecological adaptations" (D. Franck and U. Wilhelmi, 1973:897).

Increased length of isolation leads to increased aggressiveness in the house mouse (L. Valzelli, 1969). T. I. Thompson (1963, 1964, 1969)

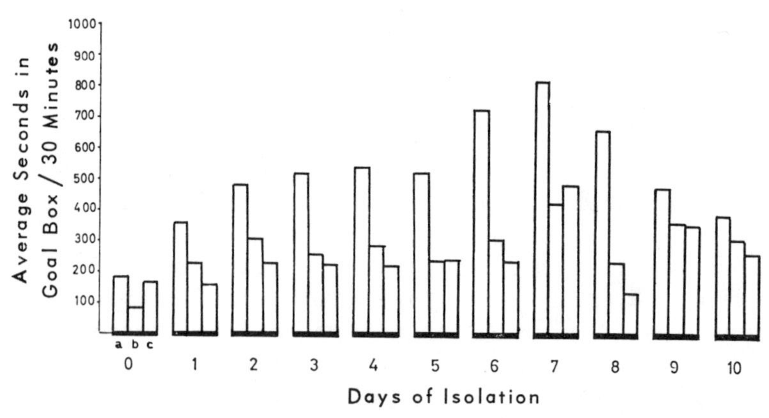

Figure 15-60 Increase in time spent in goal box after varying periods of isolation. The same five fish are kept in isolation 0, 1, 2, 3, etc., days and then tested with a rival seen through a glass pane from inside a goal box. The increase in length of stay with increasing isolation can be seen clearly from the diagram. On days of experimentation Rasa tested three times at different hours. Length of stay always decreased markedly at the second and third sessions. (From O. A. E. Rasa, 1971.)

found that fighting cocks and fighting fish (*Betta splendens*) learn a task when as a reward they are confronted with a stimulus situation releasing fighting and threat behavior. According to K. M. Lagerspetz (1964, 1969) and A. Tellegen, J. M. Horn, and R. G. Legrand (1969) the opportunity to fight acts as a learning incentive for mice. They learn a maze to get at a conspecific to fight with. Aggressively aroused mice even cross an electrified metal grill to reach an opponent.

During so-called paradoxical sleep the electrical activity of certain areas of the cat's brain is significantly increased. At the same time the animal performs movements of the eyes, ears, whiskers, and paws, as if it were dreaming. If one destroys a small area in the myelencephalon, spontaneous outbreaks of motor patterns of aggression can be observed during paradoxical sleep. This finding indicates that these motor patterns are served by spontaneously active groups of neurons, which are normally held in check by inhibitory centers in the myelencephalon. Upon their destruction the uninhibited spontaneity manifests itself in motor action (M. Jouvet, 1972). N. H. Azrin, R. R. Hutchinson, and R. McLaughlin (1965) released aggression by electric punishing stimuli in squirrel monkeys. The animals learned a task following a shock, when they were rewarded by being allowed to attack a ball for a short time. In this case we are not dealing with spontaneous aggression. The experiments demonstrate, however, that by provocation a physiological state is achieved that results in an appetite for attacking that is probably similar to the physiological state responsible for the spontaneous urge to fight, as observed in the isolated jungle fowl cocks and cichlids mentioned above. That this internal aggressive urge built up by provocation can be discharged, leading to a reduction of tension, has been demonstrated in experiments with humans. J. E. Hokanson and S. Shetler (1961) had an experimenter induce anger in student subjects and as a result their blood pressure rose. One group of angry subjects was then given a chance to administer electric shock to this experimenter, whenever he made an error in his task. Another group could inform him of his errors by flashing a light. In those who believed they were shocking the experimenter the blood pressure dropped rapidly, while it remained much higher in the other group. The possibility of administering verbal insults also resulted in a lowering of tension (J. E. Hokanson and M. Burgess, 1962). The experiments of J. W. Thibaut and J. T. Cowles (1952) and of S. Feshbach (1961) show that aggressive impulses can be discharged. However, the release of tension is only of short duration, as is the case with other instinctive behavior patterns. In the long run, the possibility of discharging aggressive impulses constitutes a kind of training for aggression. The individual becomes more aggressive. In the same way, an aggressive drive can atrophy when an animal does not have an opportunity to discharge it for some time (W. Heiligenberg, 1964). We want to emphasize this point because sometimes the view is expressed that a child should have the opportunity to discharge its aggressive impulses so that it will be all the more peaceful as an adult. This possibility needs to be investigated in humans directly. There is no evidence available, to my knowledge, of such a long-lasting cathartic effect.

There are many facts that argue for the dynamic-instinct concept of aggression. D. E. Davis (1962), who investigated the behavior of gangs, pointed out that rank and territory are the objects of aggression and he concludes from his observations on humans:

> A wide variety of observations suggest that fighting for rank and territory has innate features. . . . Thus contrary to the conclusion of some authors, it seems that aggression is heavily dependent on genetics. Probably only the means of fighting and the objects of attack are learned.

Quite certainly even man has innate aggressive dispositions that can be encouraged or suppressed by individual experience (D. Freeman, 1971). The readiness for aggression in man seems to be increased by the male sex hormone, as it is with many other vertebrates. Furthermore, neurogenic fits of temper were observed with human beings, originating in the spontaneous firing of cells in the temporal lobe and brain stem (F. A. Gibbs, 1951; K. E. Moyer, 1969, 1972; W. H. Sweet et al., 1969; V. H. Mark and F. R. Ervin, 1970). In such pathological cases the occurrence of spontaneous aggression has been demonstrated. Admittedly, this phenomenon does not constitute proof of the existence of a primary drive for aggression in healthy individuals. Yet it may allow for the assumption that a healthy individual is to a certain degree activated by these neuronal structures, which, after all, he shares with all humans. Considering these findings K. E. Moyer (1972:50) concludes:

> The Lorenzian hydraulic model for aggressive behavior is based on a number of physiological facts. When the neural systems underlying aggressive behavior get sensitized by changes in the chemical composition of the blood, the "pressure" toward aggressive behavior increases. The individual in consequence is more and more inclined to show hostile behavior. The idea however, that this "pressure" for aggression can only be diminished by acting out aggression is a little bit too simple.

I may add that such a simple solution was certainly not proposed by ethologists. Aggressive acts, like many other activities, do not as a rule come to an end by action-specific exhaustion of dammed-up energy. The consummatory situation "enemy not there" switches off the aggression as well, and the removal of aggression-eliciting stimuli therefore is a means of aggression control, although probably not the only one. The activation of our inborn dispositions to bond also seems to offer itself in this context (I. Eibl-Eibesfeldt, 1972a, 1972b, 1973b). Energy models and control-theory models, both cover an aspect of nature and they supplement each other.

Resistance to the instinct concept of aggression is primarily based on philosophical convictions. Thus L. Berkowitz (1962:4) writes:

> But aside from its theoretical significance Freud's hypothesis has some important implications for human conduct. An innate aggressive drive cannot be abolished by social reforms or the alleviation of frustration. Neither complete parental permissiveness nor the fulfillment of every desire will eliminate interpersonal conflict entirely, according to this view. Its lessons for social policy are obvious: Civilization and moral order ultimately must be based upon force, not love and charity.

Is this conclusion actually compelling? We are in a position to deny this. The observation of aggressive animal species shows that the gregarious forms are definitely able to **neutralize** their **aggression,** which in fact is a prerequisite for the formation of groups. Individual acquaintance generally inhibits aggression. The lions, which according to R. Schenkel (1966) show no social inhibitions about killing other lions as such, have an absolute inhibition against biting their own pride members. This is true for many other animals and on the whole also for man, as shown by the necessity for forbidding fraternization with the enemy during war. When the Patasiwa tribe in western Seran still engaged in headhunting activities, it was their custom to attack their victims from behind in order to kill them. To attack a man from the front to take his head was considered murder. Only as long as the headhunter cannot look his victim in the eye can he be considered prey with whom there exists no personal bond. This bond is at once established, however, when one man looks the other in the eye. To kill in that circumstance was considered a crime (O. D. Tauern, 1918).

In other cases the bond is established by sharing a meal. In certain New Guinea headhunting tribes even strangers cannot then be killed. (See also discussion in Chapter 18.)

Inhibitions against killing are nevertheless graded. They are stronger toward members of the individualized group than toward strangers. Women and children, especially small children, are more protected than men. Although verified reports exist that children have been killed in wartime, it is described as an outrage, a deviation from the normal. The strong inhibition against killing children is exemplified by the custom in several cultures of using a child for the establishment of bonds with strangers. The Massai of East Africa often push a little child to the fore, hands held open, to beg for sweets. H. Basedow (1906) reports how aborigines in Australia approached Europeans in a formal manner, with one or two high-ranking men pushing a little child in front of them, their hands on his shoulders. They were sure one would not harm a child. The same author relates how in Central Australia a woman who was suddenly surprised gripped her breasts to spray the intruders with milk. Later asked why, she explained that she did it to show that she was a mother, hoping they would then leave her unharmed. Thus the existence of an innate inhibition against killing may be presumed, especially as we find, as a worldwide subjective correlate, the emotion of pity. In this sense, innate and thus binding norms of ethical behavior seem to be programmed into man.

As an additional safeguard against the possible release of aggression in a member of the group, animals and humans possess a repertoire of behavior patterns that buffer aggression (**greeting ceremonies and other appeasement gestures**). When animals form larger groups, whose members can no longer recognize each other individually, they develop signals that unite them—for example, group odors. The familiar odor of the group inhibits aggression against a member. Man is also equipped with this additional capacity to identify with someone with whom one is not personally acquainted, and as this often involves abstract ideas and uniting symbols, it would be in the realm of the possible to create **symbols that unite** all mankind.

To control aggression one must promote those behavior mechanisms in man that appease aggression and facilitate the formation of bonds between members of a group—an idea advanced by S. Freud. Freud also considered it hopeless to attempt to abolish aggression, but he thought that the disruptions of life in human societies might be overcome by the promotion of "libidinal" forces, by the activation of all forces that are capable of facilitating emotional ties among people. Freud wrote in 1932:

> If the readiness to make war is a channel for the discharge of a drive towards destruction, then it seems logical to call upon its great opponent eros to curb it. Everything that establishes emotional ties between people must work against war. These ties can be of two kinds: first, the relationship towards a love object, albeit without sexual goals. Psychoanalysis need not be ashamed to speak of love in this context, because religion says the same: Love thy neighbor as thyself. This is easy to demand but difficult to fulfill.
>
> The other kind of emotional tie is by identification. Everything that establishes meaningful similarities between people calls forth such feelings of communality identifications. A good part of the structure of human societies is based on them (S. Freud, 1950, vol. 16:20).

Sometimes the objection is raised that man needs an enemy to discharge his aggressive drive, but experiments which show that **aggression** can be **discharged** without the performance of aggressive acts indicate that this is not so. It is not even necessary that verbal insults and similar behavior be engaged in. Athletic competition is just as helpful as the passive viewing of a motion picture. S. Feshbach (1961) presented angry and nonangry college students with either a 10-minute boxing film or a 10-minute neutral film. The angry students were less angry after viewing the boxing film than after viewing the neutral film. There was no significant difference between the groups among the nonangry students. The film experiment indicates that the viewers can benefit in the sense of a release of tension by seeing films with aggressive content. This needs to be studied further.

The appeal of television aggression in part is based on this cathartic principle (S. Feshbach and R. Singer, 1971). In our search for harmless outlets we should be aware, however, of the fact that repeated activation of physiological mechanisms trains the system, even though the short-term effect is cathartic. Furthermore, the violence shown in television often provides a model for social learning and therefore must be considered unfavorably (A. Bandura and R. H. Walters, 1963).

In primitive people customs that serve as a kind of safety valve have been described. It has been reported that some Australian tribes come together at certain times to insult one another and to fight according to specific rules. Eskimos settle many of their disputes by song duels. Additional examples of these customs are given by P. Bohannan (1966). K. Lorenz (1943) proposes that certain combative sports may be possible safety valves for aggression.

Another possible way to control aggression would be by radical counterconditioning. E. McNeil (1959), however, raised the question as to the degree to which this kind of training would result in a loss of general initiative. It is most certainly dangerous to conduct educational experiments before the characteristics that are correlated with aggression are known. We speak of attacking a problem, and there are many other indications that a general tendency to explore is positively correlated with aggression. This needs to be studied before one thinks of curing man of his aggression. The same applies to eugenic attempts to eradicate aggression.

Aggression definitely has positive aspects, which K. Lorenz has emphatically pointed out. Bloodless competition is an important driving force toward cultural development. I. Kant (1784) also recognized this positive aspect of aggression:

> Without those not especially loveable characteristics of unsociability, from which resistance is derived, which each person must necessarily encounter in his own striving, all talent would forever remain hidden in its buds as in the arcadian life of the shepherd, in complete harmony, satisfaction, and mutual love. People, docile as the sheep they herd, would impart no greater value to their lives than their domesticated animals possess: they would not fill the empty spaces left for them by creation in line with their purpose as rational creatures. Therefore we owe our gratitude to nature for the quarrelsomeness, for the envious, competitive vanity, for the never-satisfied desire to possess or even to rule. Without these all the excellent native potential of humanity would slumber eternally. Man desires harmony, but nature knows better what is good for his kind: it wants discord. He wants to

> live at ease and happy; nature, however, desires to stir him from his laxity and unproductive contentedness into work and hardship, so that he may again discover means to extricate himself cleverly from them. The natural drives to accomplish this are unsociability and universal opposition, from which much unhappiness comes, but which also motivate to new efforts and hence to a further development of natural potential. Thus they betray, it seems, the plan of a wise creator and not the hand of an evil spirit who meddled in his majestic plan or spoiled it because of envy (I. Kant, 1960, vol. 6:30).

However, we must be aware that aggression can be trained to be excessive and destructive, so that the natural counterforces of love fail to curb it. We should furthermore not take the one-sided view of accepting our aggressive disposition as an excuse for reckless competition. To a great extent the struggle for life consists of cooperation. Our aggressive impulses are counteracted by our bonding impulses and they indeed are so strong that, for example, in trench warfare soldiers have to be shifted from time to time in order to prevent bonding over the lines by exchanging cigarettes. Indeed, any war propaganda has to build up artificial barriers against communication and bonding. And all efforts aim at making the members of their own group believe that the others are not real human beings. It is my personal feeling that this capacity of man made him more murderous than did the invention of weapons. Control of aggression requires the diligent pursuit of friendly, altruistic behavior. Only then is peaceful competition possible. If uncontrolled, aggression will lead to further murderous strife between peoples, which will endanger our very existence. The understanding of the causal relationships involved should help us in the control of our aggressive impulses.

A criticism of the Freudian-Lorenzian instinct theory of aggression which is sometimes heard—that it is designed to exonerate us, and should therefore be rejected—is nonsense. Whoever argues along these lines has not read carefully what these authors have said. Actually there are surprisingly numerous misinterpretations presented in the discussion of human aggression. A. Plack (1968) imputes to Lorenz the statement that aggression is the basic drive of all life, although he never said such a thing. Sociability, readiness to cooperate, and altruism are as much part of human nature as man's occasional incompatibility.

One remarkable form of aggression that has been little studied is the expulsion reaction, which is directed not against strangers but against a fellow group member. T. Schjelderup-Ebbe (1922a, 1922b) found that chickens attack one member of the group and even kill it, if it deviates from the norm, whether it be different because of weakness or a physical handicap. He could release this reaction when he marked the comb of a chicken with paint or tied it down in an unnatural direction. C. Kearton (1935) described how three slightly different colored penguins were continuously attacked by their own kind. Young herring gulls attacked a sibling that had a caked cloaca (F. Goethe, 1939). J. van Lawick-Goodall

(1971) observed that chimpanzees feared and abandoned fellow group members whose behavior had been changed by polio.

The changes in their behavior shown toward their deviant group members was most dramatic. When Pepe dragged himself along the ground to the feeding place, the other chimpanzees who were already there touched and embraced each other for reassurance, with the fear grin showing on their faces, staring at unhappy Pepe, who did not know that he was the cause of their fear but looked around alertly to discover the cause of their excitement. The paralyzed McGregor, who moved in a highly aberrant fashion, released aggressive display in his healthy male companions, and Goliath even attacked him, hammering on his back. A second male was about to join in, when Hugo van Lawick intervened for his protection.

Later his companions got used to him, but they excluded him from all social interactions. They avoided all his approaches and invitations to groom him and it was indeed dramatic to watch the vain efforts of the sick animal to establish contact with his former friends.

Humans also tend to expel group members who deviate from the norm (K. Schlosser, 1952a–c). In a milder form this behavior can be seen in school classes or in the military. Someone who is fat, cross-eyed, or has some deviant habit is teased, laughed at, or even mistreated. This aggression against deviating group members undoubtedly results in its preservation of homogeneity, which under the archaic conditions of life in very small groups may be of selective value. In man a mild form of this expulsion reaction, teasing, can be considered as a kind of educational devise to bring the outsider into line, in that deviating "asocial" habits are suppressed. Where this is not possible, a more radical expulsion reaction may result. In these instances aggression tends to be more cruel and stronger than when it is directed against enemies who are less well known individually. This is perhaps because in addition to everything else, those common characteristics that unite the group with the outsider must also be destroyed. This norm-preserving function of the expulsion reaction is today not advantageous in human society; "outsiders" are frequently highly talented and valuable persons. Here a phylogenetic behavioral adaptation proves to be a historic burden, a handicap comparable to the appendix. We need by the use of insight to curb our intolerant impulses.

Living in Groups (Contact Behavior)

The Selective Advantages of Living in Groups

Most animals – but by no means all of them – come together, at least temporarily, with another species member for the purpose of mating. This is apparently the best method to ensure fertilization and an exchange of genetic codes on which further evolution is based. Animals

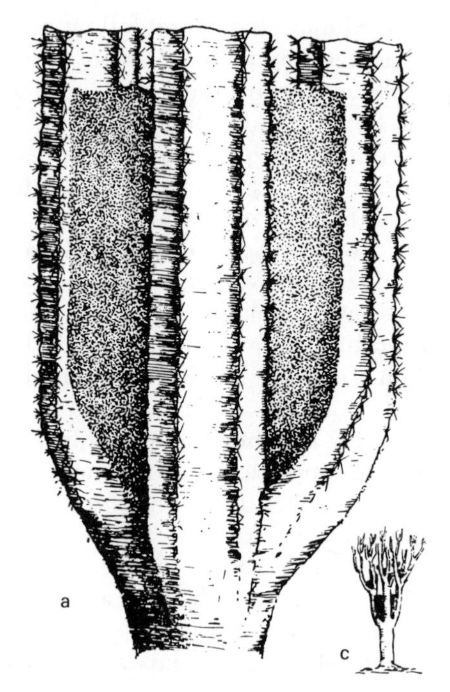

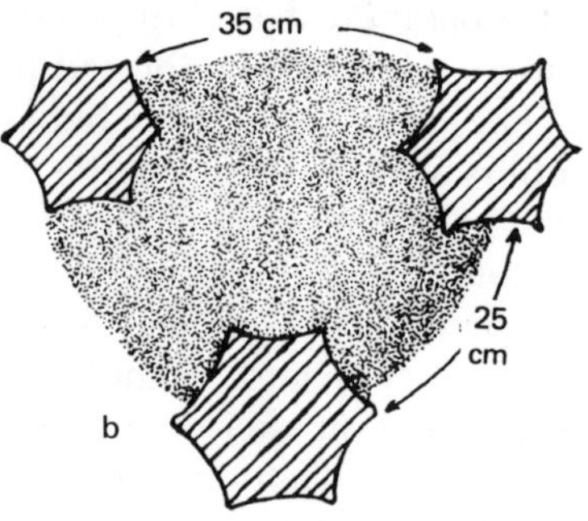

Figure 15-61 Collection of approximately 70,000 harvestmen (*Leiobunum cactorum*) in the branches of a large Mexican candelabra cactus. (*a*) Side view; (*b*) cross-section through the cactus and the collection; (*c*) view of the whole cactus. (From H. O. Wagner, 1954.)

also come together for other reasons to form permanent or temporary groups. Alpine salamanders (*Salamandra atra*) gather in the fall in cavities under rocks and sleep there during the winter. Land isopods, which are normally solitary, bunch up into balls during drought and thus protect themselves against excessive loss of moisture (W. C. Allee, 1926). Some species of harvestmen (Phalangidae) in Mexico gather in tight clumps in favorable locations during the dry period and so prevent desiccation. Such an aggregation of harvestmen (*Leiobunum cactorum*) was discovered by H. O. Wagner (1954) in the lowest fork of the branches of a candelabra cactus (Fig. 15-61). He estimated that approximately 70,000 animals had gathered there. The legs of these animals were folded over their backs and pointing outward, giving the mass the appearance of a piece of fur. Like a pelt it retained the moisture given off by the cactus. A pheromone that is secreted from a pair of glands at the edge of the head and thoracic segment attracts additional species members. Animals that were forcibly removed attempted to reach the aggregation from as far as 30 meters away. Here the aggregation is a protection against climatic conditions. The animals are not only attracted by a favorable location but also by each other and coordinate their behavior to a certain extent.

W. D. Hamilton (1964) has emphasized that the evolution of altruistic behavior can take place only in closed groups. If one member of a group sacrifices himself for the good of others, he is at the same time promoting the distribution of his genome. The probability of survival of his group members is increased by his altruistic behavior and these group members, as close relatives, harbor in all likelihood many of his genes. Though R. L. Trivers (1971) has pointed out that altruistic behavior can be selected even if kin selection can be ruled out, it is evident that the encapsulation of groups of fairly close relations strongly promotes the evolution of high-level altruism. In addition evolutionary processes in general will be speeded up by group inbreeding.

For protection against predators many fish and birds collect in swarms or flocks. Large mammals living in savannah, which affords little protection, gather in herds. These protective aggregations can be temporary, such as migratory groups and swarms of young fish, or they can be lifelong, as with herring swarms. In fish swarms the individual is protected by the confusion effect. Birds of a breeding colony actively assist each other in mutual protection when they are threatened by a predator in the same way as members of families or herds. Rhesus monkeys will even attack their keeper when he catches an animal from the group and it utters the alarm call. Jackdaws attack anyone who holds a conspecific in his hand, including a caretaker when he was holding a tame jackdaw (K. Lorenz, 1935). The reaction is released whenever something black is dangling, even a black pair of swimming trunks. Porpoises aid wounded species members and raise them to the surface so that they can breathe. They circle females giving birth and thus protect them against sharks (J.B. Siebenaler and D. K. Caldwell, 1956; see also Fig. 15-62).

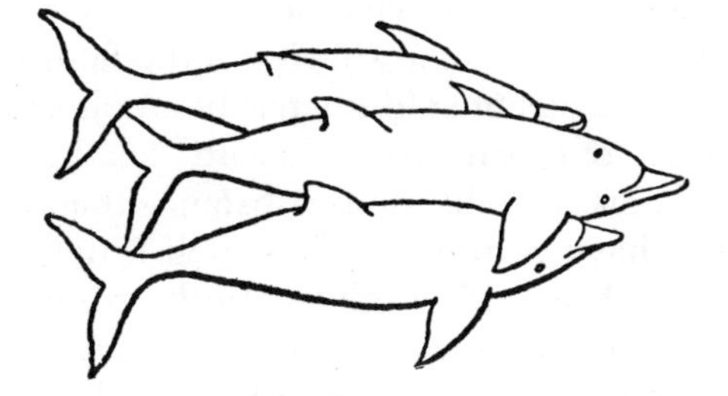

Figure 15-62 Porpoises carry an injured conspecific to the surface. (After J. B. Siebenaler and D. K. Caldwell, 1956.)

Various predators form hunting packs. Jackfishes (*Caranx*) circle swarms of fish, wolves run down game in packs; some pass an intended victim and try to cut it off, while the other pack members chase it (A. Murie, 1944). Cape hunting dogs (*Lycaon pictus*) hunt their prey in groups. At first each hunting dog chases the gazelle nearest it, but continually watches the other pack members. If it observes that another has a better chance of catching a gazelle than it has, it will come to the aid of its fellow (W. Kühme, 1965).

Family groups are often formed to care for young. If only the father rears the young, paternal families result, as in sticklebacks, pipefishes, sea horses, labyrinth fishes, and some birds, such as phalaropes (*Pha-*

laropus) and quail (*Turnices*) (N. Tinbergen, 1951; D. Morris, 1954; K. Fiedler, 1954; A. Remane, 1960). In maternal families the females alone take care of the young. This is true in many cichlids (*Tilapia macrochir*) and other mouthbreeders, wolfspiders, reeves, hummingbirds, and many mammals (polar bear, hamster, squirrels) to give only a few examples. Families in which males and females both care for the young are gibbons (*Hylobates*), crested bull-faced tamarins (*Oedipomidas*), songbirds, and many cichlids (*Hemichromis*). In such instances there is a certain division of labor in that the male usually undertakes the defense of territory, and sometimes even carries the young, as in *Oedipomidas* (H. Wendt, 1964). This is usually done by females, who in mammals also take care of the feeding and cleaning, but male wolves and foxes also bring food. In many birds both parents usually incubate and feed.

Frequently one male protects several females and young, for instance in sea lions. In the Congo cichlid (*Lamprologus congolensis*) each male has a large territory with several hiding places into which it entices females, one after another. Each of these females has her own subterritory and defends it against others. First, each newly acquired female is attacked by all the others, but because the male sides with her she eventually obtains a part of the territory. In this way an upper limit to the size of the male's harem is established because he can protect only a certain number of females (W. Wickler, 1965d).

The division of labor becomes possible only when animals come together. In insect states this division of labor has reached its ultimate form; we are reminded of the various castes in termites, which in addition to sexual animals include workers and soldiers with highly specialized tasks. In some species of the family Termitidae the anterior portion of the heads of soldiers has been elongated into a long proboscis. From the tip of this nose the "nasuti" (as these ants are called) secrete a sticky and perhaps poisonous substance that they use for defense. Their mandibles have retrogressed to such a degree that they can no longer feed without the aid of others. In the tropical ant *Colobopsis* one caste protects the entrances. Their heads are flattened and they possess a head protrusion that is colored like the bark of the plant into which they dig their tunnels. By placing their heads into the openings to this tunnel they can effectively close them. If a worker ant wants to leave or return to the nest, the guard must be notified by special signals (A. Forel, cited by K. Escherich, 1906; see also Figs. 15-63 and 15-64). The various castes of the leaf-cutting ants each have specific tasks to perform (Fig. 15-65).

In workers of the honeybee the activities differ at various ages. From days 1 to 10 after hatching bees clean the hive and keep the brood cells warm. After several days they develop nurse glands and the bees take care of the larvae. Toward the end of this first stage of its life the bee undertakes short exploratory flights outside the hive. In the second stage (10 to 20 days of life) the nurse glands regress and the wax

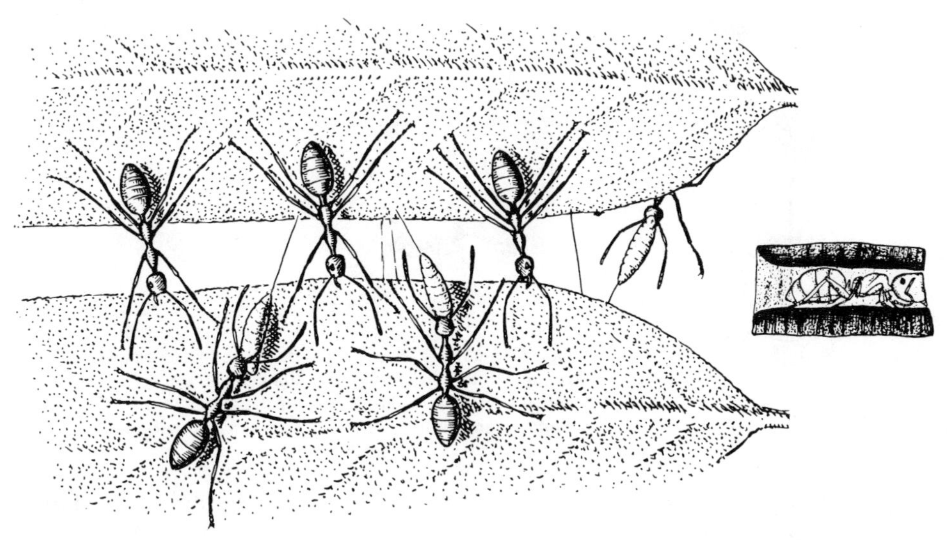

Figure 15-63 Examples of specializations of various caste-forming insects. *Left:* Division of labor and cooperation in the weaver ant (*Oecophylla longinoda*): while one group of workers pulls the edges of the leaves together, another group sews them together by pressing the spinning glands of their larvae against the edges and weaving back and forth with them. *Right:* Guard of *Colobopsis* closing the entrance with its head. (*Left* after F. Doflein, 1905; *right* after A. Forel from K. Escherich, 1906.)

glands develop strongly. The bee is now a building bee and receives nectar from other workers and stores it in the storage cells and also cleans the hive. Toward the end of the second stage some of these bees are guards near the entrance. From day 20 until death a bee is active as a pollen and nectar collector. These dates, which were determined by G. A. Rösch (1925), are accurate on the whole. The thorough investigations of M. Lindauer (1952) show, however, that most of the activities described above are not strictly separated in the sequence; there is considerable overlap between them, and nurse and wax glands are often simul-

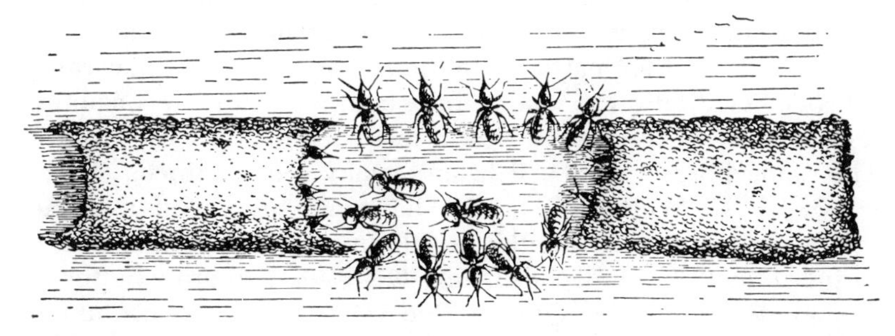

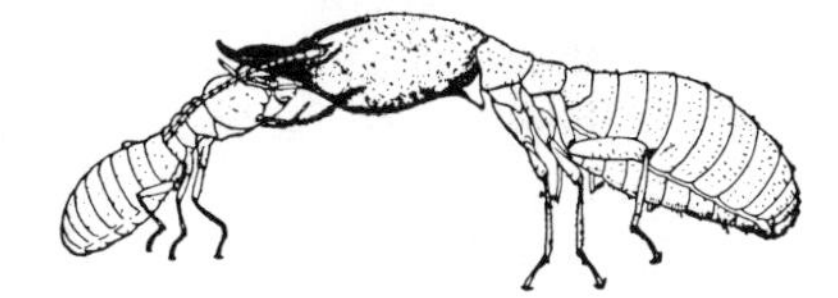

Figure 15-64 *Top:* Termite soldiers (*Eutermes*), the so-called "nasuti," guard the opening of a damaged passageway, while workers begin with repairs. *Bottom:* Workers of the termite *Bellicositermes natalensis* feed soldiers. (*Top* after P. Grasse, 1949; *bottom* after Beaumont from E. Hegh, 1922.)

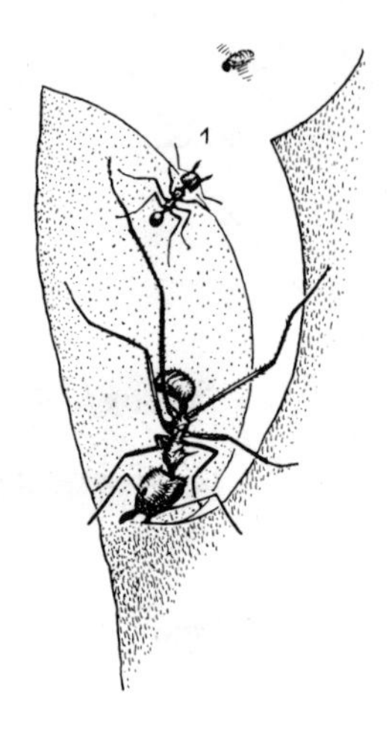

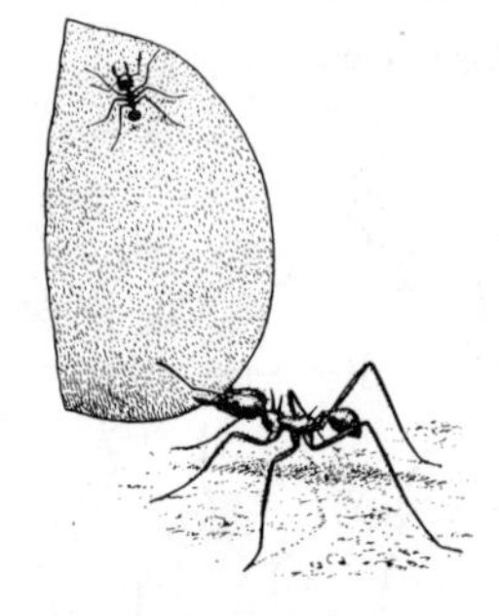

Figure 15-65 Defense against parasites by the caste of minima workers in the leafcutting ant (*Atta cephalotes*). *Left:* Leafcutting ant: the minima worker wards off the attack of a fly (Phoridae). *Right:* Transporting the leaf: the minima worker guards the carrier. (From I. Eibl-Eibesfeldt, 1967.)

taneously functional in one bee. Only the transition to foraging is fairly close to day 21.

The castes of many animals have —up to a point—their parallels in the physical constitutions of man. I. Schwidetzky (1950), among others, has pointed out that in humans there exist inherited body forms which must be considered to be adaptations to the environment. Herdsmen and warrior-herdsmen are long-legged and tall; planter types are more stocky and short-legged. It is noticeable that in sedentary, city-building peoples several constitutional types occur simultaneously and that preferences for various jobs seem to be correlated with certain constitutional types (examples can be found in Schwidetzky). Perhaps this rather conspicuous polymorphism is of selective advantage in providing a kind of predestined suitability for a diversity of jobs. An additional advantage of living in groups, especially in the higher vertebrates, is the possibility that experiences and inventions of individuals can be passed on to other members. They can then spread faster than if transmission from one generation to the next were the only available means (see **Japanese macaques**).

When living together partners influence one another in various ways. The social facilitation of moods, for example, results in consumption of more food in chicken that are kept in a flock than if they are kept singly. The same holds for rats, fish, and many other animals (H. F. Harlow, 1932; J. C. Welty, 1934). We have already noted that the presence of a male can stimulate the development of the gonads in females (**partner effect**). Cockroaches (*Blatella germanica*) grow better in groups than alone. The effect is transmitted via the sense organs on the antennae, because animals in which these have been amputated grow like those kept in isolation. In addition, there exists a nutritional effect, because the addition of pulverized feces to the food of isolated animals results initially in a higher growth rate and, with high concentrations, in an inhibition of growth (R. Chauvin, 1952). House mice females show a regular estrous cycle when excrements of familiar males are present. The odor of strange males, on the other hand, results in resorption of embryos or premature birth (H. M. Bruce, 1961). One result of population density is reduced fertility; for instance, flower beetles (*Tribolium confusum*) eat their eggs when the population density is too high.

In mammals overpopulation results in a kind of stress that eventually leads to a sudden population decline long before there is a food shortage. Fourteen miles from Cambridge, Maryland, is an island of 280 acres (James Island) where, in 1916, 4 or 5 sika deer (*Cervus hippon*) were released. In 1955 there were 300 healthy animals. In 1958 about half of them died, although the food supply was adequate, and the population continued to decrease to 80 animals during the following years. The animals that were studied during the years of the decline showed histological changes in the adrenal glands which indicated that the stress caused by the overpopulation had led to the decline (J. J. Christian, 1959, 1963).

In the tree shrew (*Tupaja belangeri*) density-dependent stress symptoms have been studied in detail (H. Autrum and D. v. Holst, 1968; D. v. Holst, 1969). Stress causes a delay in the development of the young and numerous changes in the behavior and physiology of the adult. Females under stress produce less milk or none at all. The sternal gland ceases secretion and the females therefore cannot mark their young olfactorily as usual. Without this protection the young get eaten by the cage mates or even by their mother. Under severe stress females cease reproducing and they show masculine behavior, by mounting cage mates. In young males the *descensus testiculorum* is delayed, and under extreme stress the testes, even of older males, recede into the body cavity. Stress is mainly but not solely caused by aggressive interactions. This leads to an activation of the sympathetic nervous system and the adrenal cortex. Tree shrews under stress fluff the hairs of their tails conspicuously. The time of tail-hair fluffing, expressed as a percentage of the total daily activity time, gives us a means to quantify the degree of stress. The interdependence of tail-hair fluffing and other changes from crowding can be seen in Fig. 15-66. Renal failure finally causes death when the social stress persists (D. v. Holst, 1972).

If a group of rats is kept in a limited space with adequate food supplies, their number will increase up to a certain point, and then they develop abnormal behavior. They no longer care properly for their young and do not build adequate nests; as a result the death rate among young increases to such a level that no additional increase occurs in the population, although theoretically there would have been more space available for them. The animals continuously disrupt each other's activities (J. B. Calhoun, 1962). F. Frank (1953) found similar conditions in field mice. The animals increase in number when the food supply is adequate until the optimal density has been passed. As a result of continuous conflicts with conspecifics various disturbances occur, including those of an endocrine nature, which eventually result in death of the animals. Under unfavorable social conditions the embryos of the wild rabbit are reabsorbed, and more so in low- than in high-ranking animals (R. Mykytowicz, 1960). Lemmings, on the other hand, reproduce without inhibition during summers in which the food supply is ample. They are finally forced into mass migrations, which for most of the animals ends in catastrophe. Here there is no mechanism of limiting births. In most instances, however, adaptations in the social behavior have been found which serve to prevent overpopulation of a given area (V. C. Wynne-Edwards, 1962). This is one of the functions of territorial behavior. In Scotland defeated willow grouse males (*Lagopus lagopus scoticus*) survive for a time hidden in the territories of the victors. However, they obtain only barely enough food and usually die in the later months of winter. If death of a territory owner opens up a territory one of these animals can take over and survive. On the average about 60 percent of the male offspring die as a result of this social mechanism. They are the price the species pays for the preservation of its kind. Catastrophic population declines are prevented (A. Watson, 1966; D. Jenkins et al., 1967). In some mammals, these mechanisms, which would lead to a density-dependent reduction in the rate of increase and so prevent a breakdown of the social system, do not function adequately. This most likely also applies to man, as H. Autrum (1966), E. T. Hall (1966), and T. Schultze-Westrum (1967) emphasize. "It is not the danger of hunger but the danger of a breakdown of the supporting and order-preserving social structures as a result of overpopulation which threatens our future" (H. Autrum, 1966).

I have discussed in this chapter a number of behavior patterns that are of advantage to the group but not to the individual. The jackdaw that

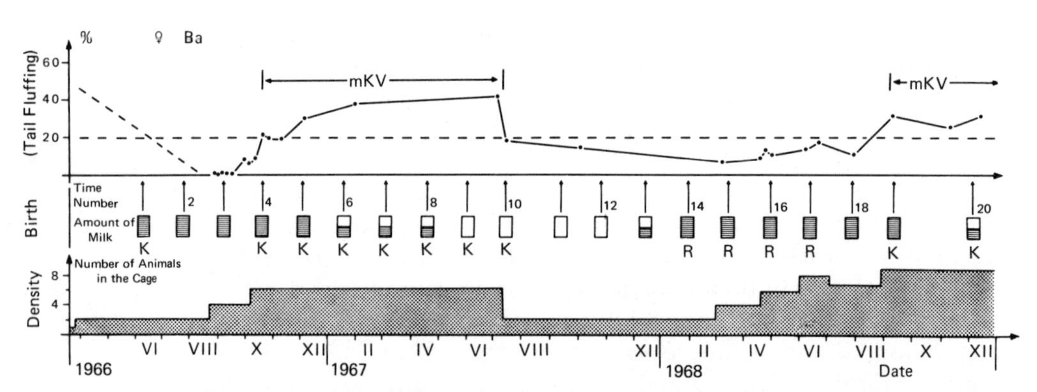

Figure 15-66 Relation between the number of animals present in one cage, tail fluffing value as expressed in the percentage of total activity time, propagation, and male copulatory behavior. Abbreviations: *mKv*, male copulatory behavior; *K*, cannabalism (eating of the young); *R*, rhythm of nursing disturbed, *hatched squares*, young fed normally; *half-hatched squares*, young inadequately fed; *empty squares*, young not fed at all. (After D. v. Holst, 1969.)

attacks an enemy is endangered, and superficially it might appear as if an animal that does not behave so altruistically would have a better chance to pass on its hereditary material. W. Wickler (1967b) has discussed the phylogenesis of **altruistic behavior** in great detail and has shown that it is possible to discuss it in terms of an evolutionary point of view. The fact that a genome prevails in a population means that its bearer distributes it better within a population than a competitor which has different characteristics. Even if the carrier of a certain genotype of parental care and defense of siblings should die, the population in which this altruistic behavior is contained survives better than one in which it is lacking. The individual that regularly kills group members of its species certainly prevails quickly; however, compared with a population with inhibitions against killing the former is clearly at a disadvantage. If one bird throws its siblings out of the nest it survives alone. However, if a mutation would occur so that a bird would not eject its nest mates, the female would raise correspondingly more young, and the population with an "altruistic" inhibition against throwing out siblings would prevail. From what has been said it is clear that mutations for altruistic behavior can best prevail in closed groups such as clans—a decided advantage for the formation of these groups (W. D. Hamilton, 1964; R. L. Trivers, 1971).

Mechanisms of group cohesion

Socially indifferent species sometimes form aggregations when they are brought together by a common goal such as a favorable roosting place for the night. However, not until they are socially attracted by other species members can they be said to constitute a true association. In general, it is enough that they possess one signal that attracts the partner, such as an odor, song, or a visual releaser. Such simple signals keep the **fish swarm** together, bring **harvestmen** to a gathering place, or **attract the sexual partner,** to recall only a few examples.

In some duet-singing birds the drive to sing can be satisfied only if a partner is present and sings its stanzas in turn. This sometimes serves, seemingly, as the only means of bonding (W. Wickler, 1972).

In some cases where the parents form a pair, cohesion may come about through the young. The partners of the cichlid *Tilapia mariae* at first attract each other. Then, after spawning, the mutual affect becomes one of repulsion. Yet there is no spatial separation, because both are attracted to the eggs and to the young. Thus the bond results from the brood. The partners no longer follow each other as frequently as would be expected in a bonded pair. In fact, the statistical incidence of following falls below that of chance, which would indicate avoidance. If we remove the brood, the parents' swimming activity increases markedly. The adults are obviously disturbed and make searching movements. However, if we remove its partner but leave the young untouched the fish will become even more peaceful than before. It does not seem to miss its partner; on the contrary, the other had been a source of disturbance (J. Lamprecht, 1972).

In most cases we can observe a clear disposition to bond. Phylogenetically it can be traced to various motivations. Flight, for example, leads an animal to seek the vicinity of a conspecific, as does sex or aggression. Sometimes, however autochthonous motivations can be found. Thus W. Wickler and U. Seibt (1972) observed that in the shrimp *Hymenocera picta* the male often sits in close proximity to a particular female, but without bodily contact. If one removes the female, the male searches for her, and no other female is accepted as a substitute. This search is not sexually motivated, since under sexual motivation the male mounts any female. Sitting close to the mate is not correlated with any other activity and it has to be assumed therefore that it has a motivation of its own. Wickler and Seibt speak of a bonding drive. An analogous **bonding drive** has also evolved in **geese.**

If the species is not especially aggressive by nature, no further obstacle exists to bonding. It is different with aggressive animals, but even they have evolved ways to form groups. This can come about, for example, by a periodic suppression of aggressive impulses. In the spring the three-spined stickleback moves peacefully in a swarm to the shallow breeding grounds. There it establishes a territory and only then assumes sexual colors and becomes pugnacious. In other species the aggressiveness may be restricted to a specific category of conspecifics. **Male fence lizards** (*Sceloporus undulatus*) attack only males of their own species, which they recognize by the blue stripes on their flanks. The females lack this inciting signal and they are tolerated by the males unless one paints blue stripes on their sides. Many young animals are especially protected by infantile characteristics, but this is not always so. If one hamster (*Cricetus cricetus*) meets another in its territory, they will attack each other regardless of sex. Only when a female is in estrus will it temporarily tolerate a male. What then inhibits her aggression in this case? The house mouse (*Mus musculus*), which lives in clans, attacks any mouse not belonging to its group that strays into its territory. Group members, however, get along very well with each other. They clean one another and do not even compete for the favor of an estrous female. Why then does not aggression break through within the group? Graylag geese and herring gulls allow contacts by their mates and young, but strange young or adults are attacked (N. Tinbergen, 1963). A sea lion female takes care of her own young; the same is true for the domesticated sheep or the mountain sheep (*Ovis aries musimon*) (B. Tschanz, 1962).

In the cases I have mentioned aggression by the female is always inhibited by individual acquaintanceship with the young. This close bond is established immediately after birth or hatching of the young, and during this time it is possible to have strange young adopted. Later this is only rarely possible (H. Blauvelt, 1964; P. H. Klopfer and J. Gamble,

1966; W. Leuthold, 1967). The bonds of individual acquaintanceship also unite siblings. This can also be established artificially between members of different species. I raised polecats with young rats. Both species got along well until the rats died a natural death when they were two years old. They cleaned one another and engaged in play fighting. If the polecats became too rough, the rats squeaked and this inhibited further attacks.

Strange rats, which were later introduced to the polecats, were sniffed thoroughly, and were not harmed. The deciding factor permitting this peaceful cohabitation of rats and polecats was probably the possession of similar bonding and aggression-inhibiting mechanisms so that they "understood" each other's expressive behavior. Polecats squeak when a conspecific nips them, and rats do likewise; and in both species further aggression is inhibited in this way. Both species understand grooming as a friendly gesture. When animals know one another individually, the aggression that is released by the partner frequently becomes redirected against other objects and this may even become transformed into a group-uniting ritual (K. Lorenz, 1963a).

Within such a group aggression is often neutralized by the establishment of a rank order. This prevents continuous conflict among members of the group. Not until there is a disruption of this order can one observe at times intense outbreaks of aggression within a previously harmonious group. This also holds, as revolutions teach us, for human societies. J. P. Scott has often stated that "social disorganization" is the cause of aggressive behavior. This point could be made more precise by saying that social disorder releases aggression against members of the group. On the other hand, we know that the group consciousness of a well-organized group increases aggressiveness against strangers (H. D. Schmidt, 1960).

If a group grows into a larger association, so that individual recognition of members is no longer possible, then the recognition of members (the familiarity effect) is brought about by other means. Rats and mice **mark** each other with odors, and they recognize group members by this odor. If a rat is removed for only a few days from the group, it loses the group odor and it will be attacked by all former group members (I. Eibl-Eibesfeldt, 1950c). **Symbol** identification in **man** is discussed later.

In addition to these cues for recognition that tie together aggressive animals living in groups, they possess a number of appeasement ceremonies to serve as buffers against aggression. I discussed these behavior patterns in Chapter 6 and refer here only to the **greeting ceremonies of the flightless cormorant,** whose appeasing function has been experimentally verified.

In the parrot *Agapornis personata* a brief interlocking of beaks is used as a gesture of appeasement, in greeting, to strengthen the bond, when danger threatens, and as an expression of tenderness. A similarity with preening and feeding behaviors suggests possible origins of this pattern, but the question remains open until we know more about related species (R. A. Stamm, 1960, 1962). The appeasement gestures do not in themselves keep the group together, but they enable animals to remain together in groups.

Display behavior patterns in which the **weapon** is turned away have an appeasing function and so do a number of behavior patterns of young animals, such as begging movements and other **infantilisms,** especially behavior patterns of **care for young, ritualized feeding, social grooming,** which appease as well as establish ties because they are rewarding. The significance of **sexual behavior patterns** as **appeasement gestures** have been discussed in more detail by W. Wickler (1965e; 1967a, 1967b). The equivalent **group-uniting mechanisms in man** will be discussed later.

In groups whose members know each other individually, the highest ranking animal often has an important group-uniting function. **Sea lion** bulls settle disputes between females with special **appeasement ceremonies.** In some animals the young inadvertently promote the cohesiveness of the group, as in lemurs (*Propithecus verreauxi* and *Lemur catta*), where the attraction of the young keeps the adults together. When an infant is born in a group of *Propithecus,* the adults groom each other four times as much as they normally would. They discharge the activated brood-care behavior of social grooming on adult group members (A. Jolly, 1966).

The social-bonding effect of the infant monkey rests on the friendly reactions released by its signals. Male Barbary macaques (*Macaca sylvana*) make use of this in their interactions with peers. When a subordinate male wishes to approach one of higher rank safely, he will "borrow" a young monkey and present it to the dominant animal (J. M. Deag and J. H. Crook, 1971). For a comparison with human behavior see Chapter 18.

All gregarious animals display an obvious desire for contact. Separated from their group and kept forcibly in isolation they do not do well. A fish that has been separated from its group swims back and forth rapidly and seeks contact with the group. Gregarious mammals become apathetic when they are kept alone; they suffer from "loneliness" (M. Meyer-Holzapfel, 1958). In higher animals group members attempt to bring back into the group those members that are in danger of becoming separated. This has been observed by K. Lorenz (1931) in jackdaws. In 1929 a large flock of migrating crows and jackdaws settled near his colony of tame jackdaws. The young birds of that and the previous year had mingled with the strangers and the possibility existed they would be carried away in the excitement when the flock took wing to migrate—which is a powerful flight-releasing stimulus. This undoubtedly would have taken place had not two old and experienced males of the tame colony brought the youngsters back individually. They flew from the house to the flock, searched out the young of their flock, and called them away from the strangers by flying closely above them from behind with their tails spread and by giving the flight-call note. In this way they returned all but two of the young during the course of two hours.

In cichlids the cohesion of the family is achieved by the following reaction of the young as well as by the brood-care behavior of the parents. The mother leads the young as soon as they are free swimming, takes into her mouth those who swim too far away from the swarm, and spits them back among the others (E. Kuenzer and P. Kuenzer, 1962).

In spite of this desire to be with others an individual animal may avoid bodily contact. This can be seen in many social birds. Starlings (*Sturnus vulgaris*) have great attraction for each other—they form large flocks outside the reproductive season—but they avoid body contact with flock members. When they perch on electric wires they maintain a certain distance between each other. Swallows (*Hirundo rustica*) behave similarly. The long-tailed tit (*Aegithalos caudatus*) and the gold crest (*Regulus regulus*), on the other hand, keep together in a family after the young have become fledged and sit huddled together at night in the closest possible contact. Alone they would probably freeze. Animals can then be divided into gregarious and solitary animals, and the gregarious ones can be grouped into *contact* and *distance* animals. It should be obvious that the nongregarious animals are also extreme distance animals.

A contact animal displays a definite appetitive behavior for bodily contact. This is true in chimpanzees, gorillas, and many other primates, which if kept alone often deteriorate, unless their keeper permits them to make contact, plays with them, and scratches or strokes them. This need for contact seems to have its roots in the drive for close contact on the part of young animals, and the behavior of contact-seeking adults shows clear similarities to parental care and infantile behavior (Fig. 15-67). Chimpanzees put their arms around each other, and even high-ranking animals, when frightened, will clasp a lower-ranking animal for reassurance. In general, however, other animals flee to the high-ranking one for protection. In *Papio hamadryas* females even flee to the high-ranking males who have just mistreated them. Bodily contact has a calming effect, and low-ranking chimpanzees beg for it (Fig. 15-68). Young gorillas and adult females seek body contact with old males when they rest (G. B. Schaller, 1963; see also Fig. 15-69) just as the young of most mammals seek contact with their mother.

In contact animals social grooming is a widespread behavior. We find it, for example, in Norway rats, agoutis, house mice, vervet monkeys, chimpanzees, and many other primates (Figs. 15-70 and 15-71). This behavior certainly seems to have its own strong motivation, because caged animals invite the keeper to groom them, for example,

Figure 15-67 The need for contact possibly has its roots in the drive to be with the mother. *Left:* Rhesus mother with an infant; *top right:* with an older infant (Cayo Santiago, Puerto Rico); *bottom:* two Sonjo children clasping each other in fright. (Photographs: I. Eibl-Eibesfeldt.)

agoutis by licking and combing the skin with their teeth. Tame vervet and other monkeys "delouse" the hair of their keeper and demand in return that the same be done to them, even when they are free of parasites. In this way one can make friends with shy **contact animals.** Such behavior patterns facilitate group ties via the reinforcement mechanism.

A bond can also be established and strengthened via aggression. A pair of **graylag geese** is bonded as a combat unit, and their greeting rituals derive from aggressive threat (K. Lorenz, 1963a). Fighting together establishes a bond in the rhesus monkey and other macaques, and this is also true for man. In everyday life one can observe how **laughing** at someone unites those that join in. Lorenz even expresses the opinion that love – defined as the personal bond – evolved in many instances from intraspecific aggression. This thesis I am reluctant to fully accept and will discuss it further on page 388.

Figure 15-68. For contact animals bodily contact has a calming effect. *Left:* An approximately 4-year-old female under the hand of an old chimpanzee male; *right:* a human couple. (Photographs: *Left:* Baron and Baroness H. van Lawick-Goodall, with permission of the National Geographic Society; *right:* I. Eibl-Eibesfeldt.)

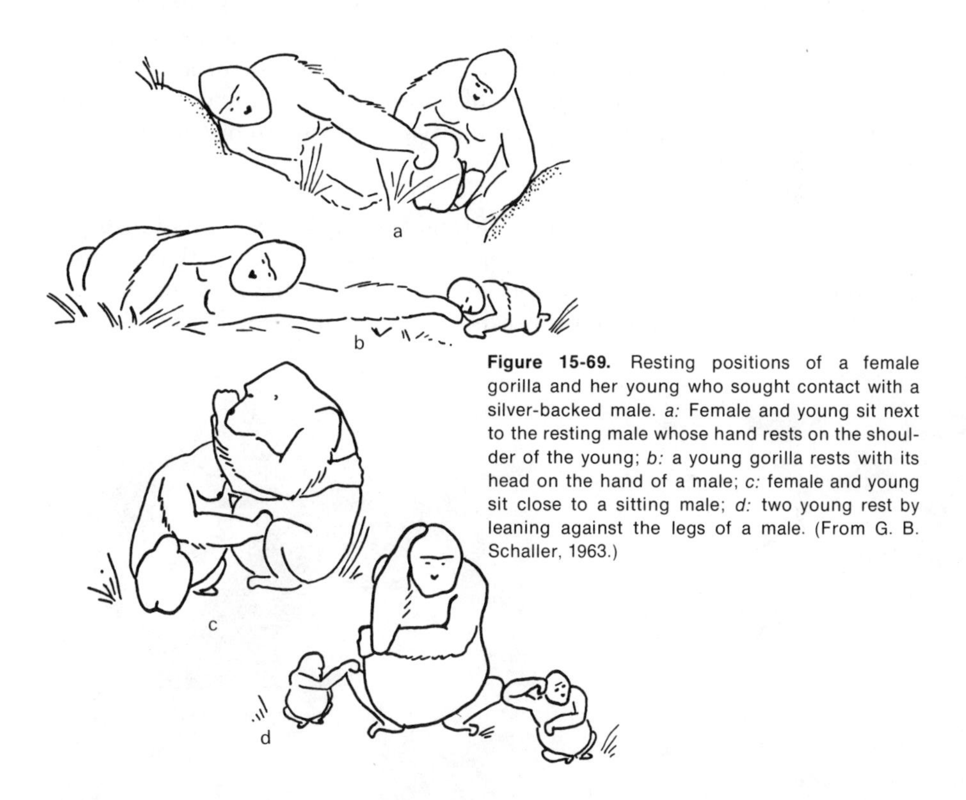

Figure 15-69. Resting positions of a female gorilla and her young who sought contact with a silver-backed male. *a:* Female and young sit next to the resting male whose hand rests on the shoulder of the young; *b:* a young gorilla rests with its head on the hand of a male; *c:* female and young sit close to a sitting male; *d:* two young rest by leaning against the legs of a male. (From G. B. Schaller, 1963.)

On the Phylogenetic Development of Group Bonding The many different bonding mechanisms discussed in this chapter point to several different origins of sociability. Comparative studies do, in fact, give evidence of this diversity.

Bonding through Fear Several species seek out conspecifics when in danger. I have already described this behavior in schooling fish.

Figure 15-70 Female lions licking each other. The animals clean those parts of the body the other cannot reach. (Photographs: W. Kühme, 1966.)

This tendency can be observed in man as well. When experiencing great distress we will join even with strangers. The child always flees to its mother. The flight goal is often provided by the **dominant animal in a group.** In such cases we may speak of bonding through fear, a very old phenomenon that plays a significant motivational role in organisms from fish to man (I. Eibl-Eibesfeldt, 1970).

Sexual Bonding As we know, Freud traced all social relationships in man to sexual origins. I have already expressed my view that he had the order of development reversed when he stated that the affectionate behavior patterns of a mother towards her child (hugging, kissing, strok-

Figure 15-71 Social grooming in the vervet monkey (*Cercopithecus aethiops*) and in humans (Bali). (Photographs: I. Eibl-Eibesfeldt.)

ing) are of a sexual nature. These behaviors belong primarily to the **mother-child interaction** and only secondarily to adult bonding. Although we might expect that the sexual drive – the most frequent basis of relationship with a partner – would also form a basis for bonding, it is surprisingly true that this drive is rarely employed in establishing a partner bond. In several species of social monkeys some forms of appeasement gestures (**presenting,** for example) are derived from the repertoire of sexual behaviors. Copulatory behavior is used as a greeting ritual, and the hamadryas baboon performs a bonding copulation without ejaculation. Finally, sexual bonding plays a prominent part in human social life.

Figure 15-72 *Left:* Aggregation of marine iguanas on Narborough (Galápagos); *right:* despite close body contact these animals show no altruistic behavior patterns. (Photographs: I. Eibl-Eibesfeldt.)

Bonding through Care of Young Various groups of animals differ markedly in their social potential. In reptiles one does not find groups whose members collaborate in any way. Marine iguanas seem to be gregarious; one sees them rest on rocks on the shore by the hundreds, crowded together and sometimes one on top of the other (Fig. 15-72). They tolerate each other but exhibit no altruistic "friendly" behavior patterns. They do not clean or feed one another. Their social behavior is limited to the repertoire of fighting and threat behavior, with which even their courtship behavior has much in common. This is conspicuously different from most birds, mammals, and bony fishes, which often form associations in which they cooperate in an altruistic manner; they may take different roles in hunting or defending a territory together. These animals also have a rich repertoire of friendly gestures such as grooming behavior, feeding ceremonies, and greeting ceremonies derived from them. These behavior patterns usually evolved from behavior associated with care of young, and some are derived from **infantile patterns.**

Apparently, many behavior patterns became available during the evolution of parental behavior that were also very suitable as indicators of readiness for social contact as well as for interaction between adult animals, all of which make it possible for altruistic cooperation to occur. W. Wickler (1967b) points out in this connection that only those insects that have highly developed brood-care behavior form states, and that their brood-care behavior also ties the adults together. Bees and ants feed not only their own young but also each other.

Feeding rituals play an important role in all higher vertebrates. Man and the chimpanzee use sharing of food as a means to create or reinforce bonds and to achieve or keep a high position in the **rank order.**

Individualized bonding and the associated inhibition of aggression also seem to have developed in connection with parental behavior. Mothers who look after their young and nurture them over a long period of time surely cannot accept indiscriminately any young of their species. Otherwise there would be the danger that females would take young away from one another and thus collect more than they could provide for. An individualized bond guarantees the opposite. Frequently we even observe a hostile attitude toward strange young. Female sea lions attack strange pups, seize them roughly, and toss them aside when the young attempt to suckle (I. Eibl-Eibesfeldt, 1955b). Herring gulls even kill strange young (N. Tinbergen, 1963). The pattern familiar = friend, strange = foe determines adult behavior in many other animals as well.

Bonding through Aggression K. Lorenz (1963a) suggests that the friendly union of two or more animals for the purpose of defense forms the basis for the development of individualized relationships. "The personal bond of love," he writes "has undoubtedly often originated in intraspecific aggression, frequently between acquaintances, by way of ritualization of response to a newly perceived attack or threat." I have already given examples of **bond-creating rituals** that have been derived from **threat behavior.** Aggression also has a much longer phylogenetic history than the individualized bond, and so Lorenz hypothesizes that the latter is a child of the former: "Through many long periods in the history of the earth there lived animals that were surely very ill-tempered and aggressive. All reptiles known today display these characteristics. And there is no reason to assume that those of prehistoric times displayed them less. The personal bond, however, occurs only in teleost fish, birds, and mammals, that is in groups which appeared without exception after the beginning of the late mesozoic. Thus we can observe intraspecific aggression without its opposite love, but not love without aggression."

Certainly this statement is true, yet there is much evidence for the assumption that the bonding effect of aggression developed only in association with defense of the young. All species that are bonded through aggression also display parental behavior. There is no friendship without care of young, with the exception of a few cases where we may assume that the species lost its brood-care behavior secondarily. There is no known case where animals are bonded solely on the basis of aggression, yet the opposite would be expected, should aggression be one of the main roots of sociability (I. Eibl-Eibesfeldt, 1970).

Types of groups

Aggregations At times aggregations of animals of one or several species occur where the sole reason for coming together lies in the attraction of some environmental factor. Butterflies may congregate at water places. If no social attraction exists, one speaks of *aggregations.*

Anonymous groups When animals are brought together by social attraction for one another, but subsequently do not develop a bond based on individual recognition, one speaks of *anonymous groups* (G. Kramer, 1950). They can be open or closed to others.

In an *open anonymous group* new species members may join at will. Individual animals are freely interchangeable as far as the group is concerned. One such example is a swarm of fish (I. Eibl-Eibesfeldt, 1962b). The swarm is kept together by means of simple species-specific signals. Swarms of minnows (*Phoxinus*) accept new conspecifics only if they do not deviate more than 1 cm from the average length of the swarm members (Berwein, cited by A. Remane, 1960). Individuals that have become separated from the swarm show a definite appetitive behavior to rejoin a swarm of their own species.

Anonymous groups can consist of subgroups in which members know one another individually. This is true in the breeding colonies of many bird species, where many breeding pairs form a larger association. They collectively attack predators and are clearly grouped together because of social attraction and not because of environmental factors. The mates know each other individually; each pair marks off its own nesting area and tolerates only its partner and later its own young in the immediate vicinity.

In *closed anonymous groups* individual members do not know each other individually, but they recognize from other cues whether or not an animal belongs. Only group members are tolerated; strangers are attacked vigorously. Rats and mice fall into this category, as I have already mentioned. The bond of individual familiarity is lacking, but a collective odor based on a mutual marking with urine identifies members of the group. They do not know one another, so there is no rank order. Males mate with females without any rivalry among the males. Conflicts about food are not bloody. The animals groom one another; sometimes an especially large individual may dominate others.

If one of two males of the house mouse, which have until then lived together peacefully, is marked with the urine of a strange mouse, aggressive behavior in the other will be released. On the other hand one can reduce the intensity of conflict between two strange males by rubbing one of the males with the urine of a mouse that is known to the other (J. H. Mackintosh and E. C. Grant, 1966). The members of a bee colony also know one another by a hive-specific odor, but they recognize each other as belonging to one colony only after they have exchanged food. If worker bees of one hive have been separated by a double screen, they fight each other even though they have been exposed to the same hive odor (J. Lecomte, 1961).

Individualized Groups: Rank Order If a group of animals is kept together by bonds of individual acquaintanceships, we speak of an *individualized group*. Its social organization can be quite complicated, with the establishment of a *social hierarchy* that develops as a result of occasional fights. Each group member learns from the repeated conflicts who is superior and who is inferior and behaves accordingly. Once the matter of rank has been settled, fights are rare, and usually a brief threat by a high-ranking animal is sufficient to keep a lower-ranking one in its place. The high-ranking animal not only has a number of special advantages, such as being the first at the feeding place or obtaining the best sleeping place, but it may also assume the responsibility for protecting the group against predators or one group member against another. The dominant animal may ensure the cohesion of the group by breaking up fights; it may assume the function of leadership in certain respects, such as determining the time for moving on and by giving the direction during migrations. The role of the protector makes the highest-ranking animal a focus around which the group gathers.

A social hierarchy or rank order presumes not only that some members of the group will seek authority either by fighting for it or by some other special achievement, but also that the lower-ranking animals will accept this order. Only the capacity and readiness to submit makes the formation of stable societies possible. This often does not become apparent until one attempts to raise a higher, solitary mammal. My own quite intelligent badger completely lacked the ability to submit. He remained self-willed and accepted no reprimands. If punished for some misdeed by as much as a light slap, he at once became seriously aggressive. A dog, however, will readily submit and adjust his behavior accordingly. It is a group animal by nature.

The presence of a high-ranking animal influences a low-ranking one in many ways. E. Diebschlag (1940) reported that low-ranking pigeons had more difficulty in learning a color and position discrimination in the presence of a higher-ranking animal. He also kept these low-ranking pigeons in individual cages, where they became used to models that retreated whenever the birds showed aggressive display. They were finally able to defeat the previously dominant bird when they were again placed into the old environment. The same bird also clearly improved its learning performance, when both were present in a training session.

The phenomenon of rank order was first studied by T. Schjelderup-Ebbe (1922a, 1922b, 1935) in chickens. At a particular feeding place some hens have certain privileges. They are first at the feeding place and peck at other lower-ranking hens that are already present or that come too close to them. Who may peck whom is well established. Chicken A may peck chickens B, C, D, and E; chicken B pecks all others except A; C all others except A and B; and so on. The lowest-ranking hen is pecked by all the others, but is generally left alone by the higher-ranking ones, because they are more attentive to the hens ranking immediately below them, as they are their most serious rivals. When strange chickens are put together, they will at first fight intensively. Each animal will fight every other animal and victory or loss will determine its future standing. A chicken that has lost a fight will remember the victor and avoid it in the future. The victor usually is the stronger animal, but agility, perseverance, and aggressiveness are also of importance. It is also possible that a high-ranking hen A, which had been victorious over B and C, will lose to D, which had lost to both B and C – perhaps because hen A had just been weakened in a fight or had been frightened by something. Then A is still dominant over B and C, but is below D in the rank order, although D is subordinate to B and C. Thus there are, in addition to simple linear rank orders, more complex *triangular* relationships as well.

The subordinate animal will be pursued by the winner for only a few days; after that it will usually be left alone. Once the peck order of a flock of chickens has been established, everything is peaceful. If necessary, the higher-ranking animal will assert itself with a short threat. Roosters are in general dominant over hens, but they must fight their way up through the ranks, including the hens. Several roosters within a flock also have their own peck order. In jackdaw colonies there is also a strict rank order. Very high ranking animals are quite peaceful toward very low ranking ones, but they are very aggressive toward those who are just below them. They may also become involved in fights between two lower-ranking birds, and then they always attack the lower ranking of the two. Male jackdaws will only mate with females that are of lower rank than they (K. Lorenz, 1931, 1935).

The colorful plumage of males plays an important role in the fights for rank. Female chaffinches whose undersides had been colored red to resemble those of males dominated the social hierarchy when they were placed together with normal-colored females. They also dominated over the others in fights and usually won, which further demonstrates the intimidating function on the other sex of the colorful male plumage. Very low ranking females from a group that had lived together for a long time could be raised in rank by artificially coloring them. Even hand-raised females that had never seen a male retreated before red-breasted females. This reaction, then, is not acquired as a result of social experiences (P. R. Marler, 1955a, 1955b). It was not possible to teach the females to avoid green-breasted birds. Even when they had been placed together for a long time with green-breasted males, they later did not avoid artificially green-colored females.

In many hoofed animals that bear antlers or horns, conspecifics judge the fighting qualities of other species members by the size of the antlers or horns. In the red deer only animals with approximately equally well developed antlers will fight together. After dropping their antlers high-ranking deer rapidly drop in rank, and the attacks of lower-ranking animals always take place immediately after the others lose their antlers. This occurs even though the antlers, just prior to being dropped, had already become useless for fighting because of the onset of osteolysis. This shows that their possessors were protected only by the symbolic significance of their antlers (H. Hediger, 1954; see also A. Bubenik, 1968). Wild sheep estimate the strength of conspecifics according to the size of their horns, and strange animals that join the group can readily fit into the existing rank order without fights (V. Geist, 1966b).

Rank order is by no means stable. Small changes occur continuously. A young hen with chicks advances in her standing; even higher ranking hens will tolerate her. Low-ranking baboon females rise in rank when they are in estrus or when they have small young (I. DeVore, 1965), and the same is true for many other animals. In animals that have permanent mates the rank position of the female may change dramatically as a pair bond is established. A low-ranking jackdaw female at once advances in the hierarchy when she mates with a high-ranking male and she changes her behavior accordingly (K. Lorenz, 1935). She is aware of her new position, which is based on protection by her higher-ranking mate.

One of the free-living chimpanzees that was observed by J. van Lawick-Goodall (1965) utilized an accidental discovery and substantially improved his standing within the group. Chimpanzees are afraid of loud noises. The low-ranking Mike had discovered that one can make loud noises with empty kerosene cans by dragging them over the ground or throwing them.

> Mike often walked to the tent while a group of chimps was resting peacefully nearby, selected a can from the veranda, and carried it outside. Suddenly he would begin to rock slightly from side to side, uttering low hoots. As soon as the hooting rose to a crescendo, he was off, hurling his can in front of him. He could keep as many as three cans in play, one after the other (p. 813).

J. van Lawick-Goodall later hid the cans, but by then Mike was no longer in need of them. Whenever he approached other chimps of his group, they bowed low to the ground, acknowledging his dominance.

It is therefore not physical strength alone, but intelligence as well, which contributes to the rank of an individual animal. In free-living rhesus monkeys the rank of a male does not depend only on his physical strength but also on his ability to form friendships (alliances) with others. In so doing males sometimes change groups. An animal that seeks contact with another male stays near him and tries to entice him to join him in a common aggressive action by sham attacks against other males or the observer, on the principle that he who fights alongside another is also his friend (see **graylag goose**). Females, on the other hand, remain in their groups (A. P. Wilson, personal communication). In the Japanese monkey low-ranking individuals seek the friendship of the high-ranking by assisting them in attacks. When the superior makes an attack the inferior individual joints it from the side, leading the attack and roaring. These "attacks for flattery" help to strengthen the bond and raise the confederate's status," for to be the friend of the most dominant animal is the surest and quickest way of promoting oneself (S. Kawamura, 1963:203).

In baboons the strongest male is usually the highest-ranking one. Sometimes, however, two or three older males join together, where each alone would be subdued by single, younger males in the group. By being allied they rule the troop. Within the central hierarchical group there is a certain ranking. A male outside this central group may be the highest-ranking individual animal. In one of the instances described by I. DeVore (1965) the central hierarchy consisted of the males Dano, Pua, and Kovu. The highest-ranking individual, however, was the fourth male, Kula, who was dominant over each individual of the group when they were encountered alone. Kula had to move aside only when he encountered all three together, who then usually acted in concert. With respect to the choice of sleeping place, the direction of migration, and in situations of danger the high-ranking animals determine the behavior of the group. The rank of an individual animal is also determined by his relationship to the highest-ranking animal. If a baboon secures the friendship of a high-ranking animal, he raises his status in relation to others. Lorenz found similar conditions in the graylag geese.

Most revealing observations about the function of high-ranking animals were made by L. Williams (1967). When he released his colony of woolly monkeys (*Lagothrix*) for the first time into an open area, only the alpha male climbed into the trees and carefully inspected all possible routes for climbing. He broke off dead branches. Group members that attempted to follow him were at first chased off. He permitted them access to the trees only after two days of careful exploration.

There is a study by D. W. Ploog et al. (1963; see also Fig. 15-73) of the development of social relationships within a group of squirrel monkeys during the course of one year. The fusion of two groups of squirrel monkeys began in a collective discharge of aggression, in the course of which the highest-ranking females engaged in screaming and biting fights. The males performed display duels. The vanquished group finally withdrew into a corner, and a new rank order then developed among the males, with one male of the losing group becoming the whipping boy. The aggression of all others was directed toward this one male, and this seemed to release sexual behavior that had previously been inhibited by the collective display of aggression. As social contacts increased, the groups fused more and more into one and the aggression against the whipping boy decreased gradually (R. Castell and D. W. Ploog, 1967). R. M. Yerkes (1948) described cyclical changes in the ranking of chimpanzee females that paralleled their estrous cycle. During the estrous period the females are superior to the males in rank order, but they are subordinate to them at other times.

The rank position of an animal is sometimes determined by the standing of its mother. Then it is transmitted by tradition. This has been reported in rhesus monkeys and Japanese macaques (M. Kawai, 1958; C. B. Koford, 1963a; D. S. Sade, 1967; S. A. Kaufmann, 1967).

Newborn young are introduced to other animals in an individualized group, a practice that undoubtedly protects them against aggres-

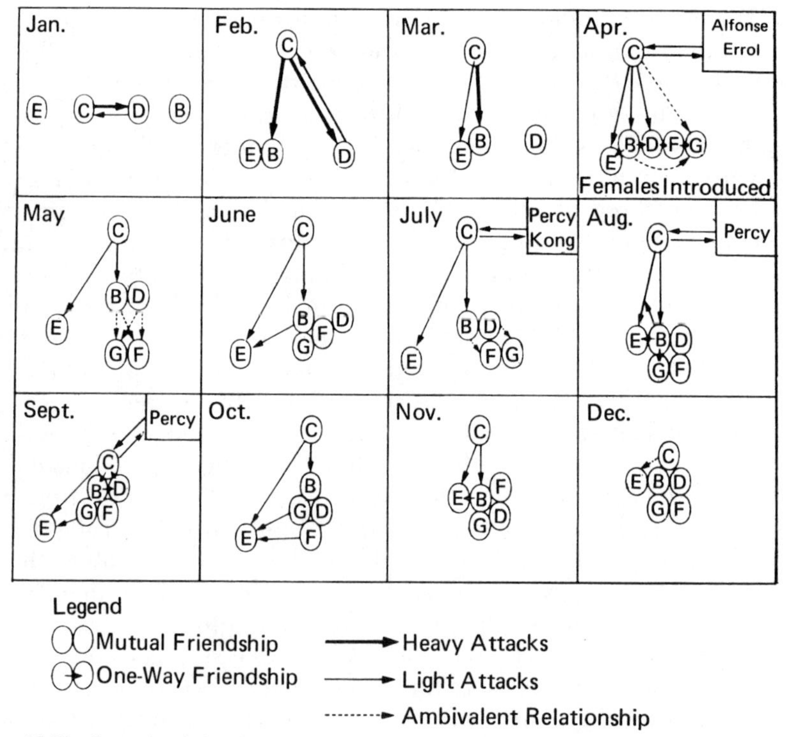

Figure 15-73 Example of the development of social relationships in a group of squirrel monkeys. (From D. W. Ploog, J. Blitz, and F. Ploog, 1963.)

sion. The chimpanzee mother is alone when she gives birth to her young. When she returns to the group she approaches other group members with an open hand, palm up, and shows her child at the same time. She behaves in an obviously apprehensive manner but is at once reassured when the other animal reaches out with its hand (J. v. Lawick-Goodall, 1965). The gorilla mother Achilla of the Basel Zoo presented her newborn young within the first few days to familiar persons when they came near her cage.

> She pressed "Jambo" against her body with her left arm, held his left arm with her right hand and stuck it through the cage bars while holding her own right hand next to his. She had learned to present her hand, and it seemed as if she also wanted to include her infant into this manner of indicating contact readiness. She seemed quite contented after one had given one's hand to her and her child (R. Schenkel, 1964:243).

It is remarkable that she reached out her hand in the manner of a chimpanzee mother, a gesture which Schenkel considered learned but which is possibly homologous to that of the chimpanzee. Achilla occasionally tried to push her child under the bars to the outside. She was taught to refrain from this simply by breaking off contact with her. Lion mothers also introduce their young to other members of the pride (R. Schenkel, 1966). In social carnivores (wolves, hunting dogs) and in the chimpanzees low-ranking animals that have established prior possesion of meat at a kill may retain exclusive possession even in the presence of high-ranking individuals. This rule of priority of possession prevents a monopolization of food by the most dominant males (J. van Lawick-Goodall, 1968, 1971; L. D. Mech, 1970; O. Aldis, on press). In the chimpanzee the owner of a kill shares its meat in an elaborate pattern with other members of the group, which in some ways reminds one of the sharing rituals of hunters and gatherers. High-ranking males in particular are skillful in sharing with nearly every group member and in not giving away everything at once, but rather doing so piecemeal over a period of time, thus staying for a while in the focus of attention and reinforcing their position. G. Teleki (1973) observed that the high-ranking wild chimpanzee Mike shared his prey with 13 of the 16 adult and subadult group members present.[3] He was busy from 8:00 A.M. to 5:30 P.M. and thus was the focus of attention (Fig. 15-74). The skill of the owner consists in not giving too much away to a few, but in sharing morsels while eating, and in keeping enough for further distribution. (See M. R. A. Chance, 1967, for a further description of the mechanisms of attention attracting.)

In some carnivores and primates a male with prior possession of a female in estrus may retain exclusive possession in the presence of

[3] However, it has been observed that a lower-ranking chimpanzee may refuse to share with a higher-ranking individual.

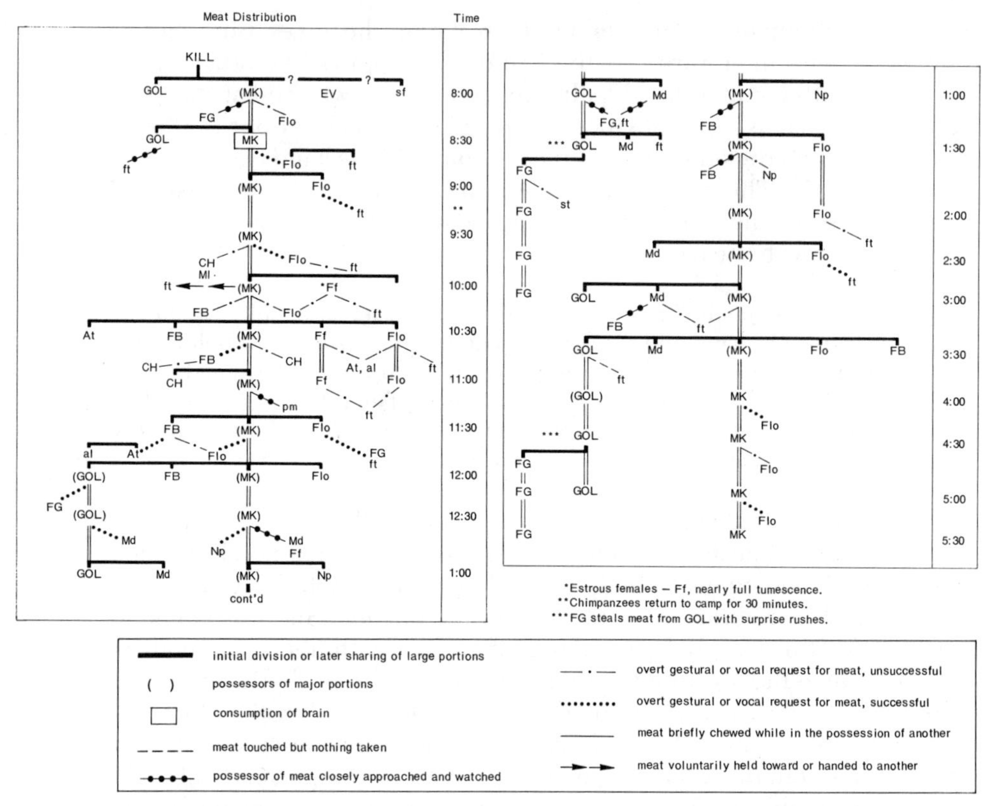

Figure 15-74 Diagram showing the pattern of meat distribution by the high-ranking male Mike. (From G. Teleki, 1973.)

higher-ranking males. Again, this is a measure to avoid conflict and it ensures subordinate males staying with the group (O. Aldis, in print).

Examples of Individualized Groups Individualized groups can be found expressing various degrees of mutual recognition. A pair of animals frequently know one another individually (K. Lorenz, 1963a). This is true for many fishes, especially in the cichlid group (Cichlidae) and damselfishes (Pomacentridae), where the parents know each other individually but the young are known only as a group (W. Wickler, 1967b). In many birds, graylag geese, for example, the parents know each other as well as their young. In mammals a monogamous, individualized family is rarer, but it does occur (for example, in the gibbon). Usually mammals form maternal families or larger packs or clans that are maintained on the basis of individual acquaintanceship.

The flying marsupial (*Petaurus breviceps*) lives in such clans and it is possible to demonstrate the existence of a clan odor as well as individual odors (T. Schultze-Westrum, 1965). These animals keep together in a group and have rank orders within the group. In this species group members mark one another and fight against outsiders. If one wants to get strange animals used to each other one can place them into a nest box together, separated by a screen.

The prairie dog (*Cynomys ludovicianus*) lives in exclusive family groups that consist of one male, several females, and the young of the present and previous year. The family members know one another individually (J. A. King, 1955). By constant control the dominant male makes certain that no stranger intrudes. During the breeding season the young males of these groups become quarrelsome and emigrate. They found new colonies and females at first move freely back and forth between them until they finally choose a specific male.

Cape hunting dogs (*Lycaon pictus*) live in packs with a highly developed social life. All pack members are of equal rank and feed each other; each may beg from any other and receive food. In this way all food is equally divided within the pack, so that the less skillful hunters also obtain food (W. Kühme, 1965). This lack of rank order in the Cape hunting dog is a case of specialization. During a hunt and when encountering a predator one animal may be seen in the forefront, but on the next occasion another animal may assume this position. If it were otherwise, according to Kühme, there would be a failure to adapt. If there were a division of labor according to rank, the pack would often be involved in internal rank-order fights, whenever an alpha animal became slightly injured; in Cape hunting dogs injuries are quite frequent.

In wolf packs that are otherwise quite similar there is a clear rank order (A. Murie, 1944; R. Schenkel, 1947; L. Crisler, 1962). In red deer and the wapiti we find groups of females that are led by an older female. The leading position is always held by a female with a calf; she is the most alert animal in the group. The other females with more mature young form subgroups that are driven off temporarily when a new calf is born. The adult males form loose groups outside the rutting season without any clear leadership in evidence, and they disperse with the onset of the breeding season. Then a male will join a female group, but will not assume leadership, which is still provided by the leading cow; she warns when danger approaches, and the male keeps the herd together by circling it (F. F. Darling, 1937; M. Altmann, 1952).

In horses one finds large herds that are led throughout the year by a stallion. He lives in the center of the herd and the younger stallions live on the periphery. When these are four to five years old they separate themselves from the group with a few mares each. Among the mares there is also a rank order. According to H. Ebhardt (1958) there are social organizations among horses in which a lead stallion lives with only a few mares and their foals in a family group. Zebras (*Equus quagga*) live in permanent families and in stallion groups. The families consist of an adult stallion and one to six mares and their foals. The old mares remain in the family for the remainder of their days, but sick and old stallions are replaced. During their first estrus young mares often leave the herd to accompany strange stallions. Between the first and fourth year of life stallions leave the herd on their own to associate with stallion groups. Within the zebra family an older mare leads the group. The highest-ranking stallion follows it (H. Klingel, 1967).

In eared seals the males rule over a stretch of coast as their territory where they gather their females. There they tolerate no other males. The Galápagos sea lion bulls participate in the care of young; they drive back young animals that swim too far out and protect them from sharks this way. The females in general are quite peaceable; any conflicts that arise are stopped by the bull.

Prosimians (*Lemur catta* and *Propithecus verreauxi*) live in exclusive groups. As with many primates, behavior patterns of social grooming and the attraction of the young are the most important binding forces within the group. There is a rank order among group members, and A. Jolly (1966) developed the interesting hypothesis that this complex social life facilitated the development of primate intelligence. During the mating season there are often great disruptions of the social structure, as intense fights break out among the males that often lead to injuries.

The group dynamics of rhesus monkeys are known from intensive observations made on the island of Cayo Santiago off Puerto Rico, where several groups live in freedom (A. P. Wilson, 1968). They, too, form exclusive groups with their own rank order. If one high-ranking group approaches a feeding place, the lower-ranking group will move away, even if the approaching animals, within their own group, have only low-ranking positions. The ranking of females within a group depends on ancestry. High-ranking females usually produce high-ranking daughters, and the animals that belong together as a result of common descent form permanent subgroups within the larger troops, consisting of several generations (D. S. Sade, 1967). Males, however, do on occasion move to other groups. They initially associate with the strangers by remaining near the periphery of the group they wish to join, and on occasion attempt to groom another male. They begin their life in the new troop with a low rank but they can rise higher. The rank of a male depends on his ability to establish an alliance (A. P. Wilson, personal communication). High-ranking animals settle disputes between group members. In group fights they leave the fighting to the next lower ranking animals, and if the strange group is victorious, they are the first to retreat.

The breeding season has no negative effect on the cohesion of the group. The animals are promiscuous, but usually only half the males mate. The lower-ranking males copulate rarely or not at all, apparently because of social inhibitions (C. B. Koford, 1963b). I. DeVore (1965) observed that low-ranking baboons do not mate. This is reminiscent of some instances of psychogenic impotence in humans. Otherwise, sexual activity seems to cement the rhesus group together, in contrast to the lemurs. According to C. R. Carpenter (1942) this is also the case in howling monkeys, baboons, chimpanzees, and gibbons. In rhesus monkeys the estrous period during which females permit copulation is relatively long. Estrus lasts 9.2 days per sexual cycle of the female, which is about one third of the total cycle. Carpenter interpreted this as an adaptation in the service of group cohesion.

In baboons (*Papio ursinus*) groups consist of numerous infants, juveniles, adult females, and males. One troop of 80 included 54 young, 18 adult females, and 8 adult males. Of these males the strongest is usually dominant, and between the others there is a graded hierarchy. Sometimes two or three older males unite to form a central group that rules the troop (I. DeVore, 1965).

The highest-ranking males advance toward predators and strange baboon males whenever there is danger. Otherwise they remain in the center of the troop and the females with very small young gather around them and are thus protected against the encroachment of other group members. These highest-ranking males copulate exclusively with females in full estrus, while the younger, lower-ranking males can mate only with females in partial estrus. They also have access to younger females. When two low-ranking animals fight, one of them may flee to the proximity of a high-ranking one, present his rear to him in an appeasing gesture, and threaten against the enemy. In these instances the high-ranking animal will take sides with the one that has fled to him and chase away the pursuer (H. Kummer, 1957). In migrating troops a certain marching order is evident. An advance group of strong adult males is followed by childless females and young males. Then follows a group of dominant males, including the highest-ranking ones and the females with small infants. Other young adult males bring up the rear (Fig. 15-75). Young baboons seek protection initially with their mother, later with the alpha male, even when he is the cause of the fear. Until they are two years old young baboons are not subject to the harsh rules of adult life. They become only gradually involved in the social tensions of the group (H. Kummer, 1957).

Chimpanzees in the wild live in large, loose groups of animals that know one another personally. Females with their young form subgroups, and young animals live in more or less close contact with their mother and siblings. Females with newborn infants introduce them to other group members, and the young are very interested in their newborn siblings and are permitted to hold them after several weeks have passed (J. Goodall, 1963, 1965; J. Van Lawick-Goodall, 1965). Among adult animals a rank order exists. The males frequently threaten other group members. During their occasional temper tantrums they will attack not only subordinate males but females as well. These behaviors seem to be assertions of dominance. But according to J. van Lawick-Goodall (1971), there is no fighting for possession of females, who when in estrus mate peacefully with all males present. J. van Lawick-Goodall (1971:116) refers to a chimpanzee community as "an extremely complex social organization."

> The members who comprise it move about in constantly changing associations and yet, though the society seems to be organized in such a casual manner, each individual knows his place in the social structure—knows his status in relation to any other chimpanzee he may chance upon during the day. Small wonder there is such a wide range of greeting gestures—and that most chimpanzees do greet each other when they meet after a separation.

Figure 15-75 Marching order of a troop of baboons. The dominant males accompany the females with infants in the center of the troop. A group of juveniles is seen at the lower part of the picture. Other males and females precede and follow the center group. Two females in estrus (swelling shown by dark markings) are each accompanied by a male. (From I. DeVore, 1965.)

The wealth of appeasing behavior patterns and their frequency of appearance indicate a large potential aggressivity that must be held in constant check. And in fact chimpanzees do show a great fear of strangers. Strange chimpanzees added to an enclosure by other observers fled into the water moat and would have drowned if no one had helped them. J. van Lawick-Goodall (1971:120) observed in the wild how two females attacked a strange female and chased her from the feeding station:

> Just as we were getting some bananas ready for her we noticed Flo and Olly staring fixedly at the stranger, every hair on their bodies bristling. It was Flo who took the first step forward, and Olly followed. They went

> quietly and slowly towards the tree, and their victim failed to notice them until they were quite close. Then, with pants and squeaks of fear, she climbed higher in the branches. Flo and Olly stood for a moment, looking up, and then Flo shot up the tree, seized the branch to which the now screaming female was clinging and, her lips bunched in fury, shook it violently with both hands. Soon the youngster, half shaken, half leaping, scrambled into a neighbouring tree with Flo hot on her heels and Olly uttering loud "waa" barks on the ground below. The chase went on until Flo forced the female to the ground, caught up with her, slammed down on her with both fists and then, stamping her feet and slapping the ground with her hands, she chased her victim from the vicinity. Olly, still barking, ran along behind.

The frequently disseminated opinion that chimpanzees are not at all aggressive is obviously false. Neither is it true that they live in open societies. I. Itani and A. Suzuki (1967) report groups of 30–50 animals living in West Tanzania which clearly distinguish themselves from other groups. The authors term such an organization a "preband." In the Mahali Mountains east of Lake Tanganyika T. Nishida and K. Kawanaka (1972) also identified distinct groups of chimpanzees with different home ranges.

G. B. Schaller has reported on clans of gorillas whose family life is more similar to that of humans. Here one group usually consists of several females and males. An especially large silver-backed male, as the dominant animal, determines the time and direction of moving and settles disputes between females. The males are tolerant toward others, but it is an exception to find two silver-backed males in one group. The bloodless rank-order fights are carried out by chest beating (G. B. Schaller, 1963). Members of different groups avoid contact. They threaten each other at a distance. Occasionally fights between males of different groups have been observed. M. Kawai and H. Mizuhara (1959) report a fight during which one male throttled another. In general, however, gorillas have peaceful habits in the wild. And this is true on the whole for primates, where violent aggression rarely occurs in nature except under situations of stress and overcrowding (C. Russell and W. W. S. Russell, 1968).

Incest Taboo, Dissolution of Family, Superfamilies Several vertebrates show a strong inhibition against mating with parents, offspring, or siblings. This applies, for example, to the graylag goose. Japanese and rhesus macaques have a mother–son incest taboo. J. van Lawick-Goodall (1968) reports that twice she observed an estrous female mate with all males of the group, except with her two sexually mature sons. A young female chimpanzee allowed her brother to mate with her when she showed her first incomplete swelling. But at her first complete swelling she refused him. Mating inhibitions seem to have appeared when members of a family group stay together for a long period of time, and probably work as precautions against detrimental interbreeding. This principle appears even among certain plants, which often develop rather complex mechanisms to prevent self-fertilization.

There is much dispute as to whether the incest taboo has a biological origin in man (F. David et al., 1963; K. Kortmulder, 1968; F. B. Livingstone, 1969). Studies by N. Bischof (1972a, 1972b) support the hypothesis of an innate base. The incest taboo is universal as a general rule. In rare cases special classes had exceptions. The inhibition against engaging in sexual activity with persons one has grown up with seems to mature without educational pressure. In the kibbutzim children are cared for from early childhood in small peer-groups of the same age. In early childhood sexual play is frequently observed, but by the age of ten sexual inhibitions appear and the relation between the sexes becomes tense. This tenseness vanishes when adolescence is reached and a strong affectional bond of a brother-sister type develops. Among 2769 marriages investigated by T. Shepher (1971) none occurred between members of the peer-group, although no social pressure was exerted in this direction. The avoidence was completely voluntary. In the group where Shepher worked no heterosexual activity was observed between peer-group members. In 13 cases where marriage between peers had occurred there had been an interruption in the peer-group membership before the age of 6. There seems to exist a critical period between birth and age 6 during which exposure to other children will define "with whom one will not fall in love" (T. Shepher, 1971).

Family groups frequently dissolve when the young grow up and intrafamilial hostility develops, as in polecats and hamsters, or because the mother drives off the young. Squirrel mothers become very unsociable shortly after their young are weaned and reject approaching young actively by threat calls and by biting or pushing with the feet. If such behavior mechanisms for breaking up the family group have not evolved, the groups grow into a larger clan association. If this group reaches a certain size, the members can no longer recognize each other individually and we have an **anonymous, closed group.**

The structure and the size of animal groups changes from species to species, or within a species according to the time of the year (compare with **stickleback**). Convergencies are numerous. There are monogamous birds and mammals as well as monogamous fish. Animals that live in open and exposed areas such as plains or the high seas tend to form large groups. Those who occupy biotopes with adequate cover live in smaller groups, but there are always exceptions – the house mouse, for example, which lives in superfamilies despite adequate cover. The multiplicity of social groupings has been discussed in detail by P. Deegener (1918). Excellent reviews can be found in E. Stresemann (1934), W. C. Allee (1938), W. Goetsch (1940), E. A. Armstrong (1947), G. P. Baerends (1950), F.

Bourliere (1950), G. le Masne (1950), A. Portmann (1953), N. Tinbergen (1953), H. M. Peters (1956), F. E. Lehmann (1958), A. Remane (1960), I. DeVore (1965), J. F. Eisenberg (1965), R. F. Ewer (1968), T. H. Crook (1970), E. O. Wilson (1971), and A. Jolly (1972).

SUMMARY

By means of morphological and behavioral adaptations animals cope on a variety of adaptational fronts with the demands of the environment. They protect themselves against climatic fluctuations – by constructing nests, for example. Numerous techniques of food gathering are found (tool use, luring devices, and so on), by which animals evolve to occupy sometimes extremely specialized ecological niches (for example, as a mimic of a cleaner fish). Predator defense provides examples of the numerous escape devices developed by animals, in their competition with predators. These adaptations include the sophisticated techniques of parental care behavior (such as the mouthbreeding of the cichlid fishes). Symbiotic relationships, too, have often evolved under the pressures exerted by predators, but also as a means of coping with other problems of survival. In the cleaning symbioses, for example, the host has the advantage of being cleaned and the cleaner has opened for himself a food source. All intermediate stages between symbiotic and parasitic relationships can be found.

Behavior toward a conspecific is often characterized by the counteracting tendencies of approach and contact-seeking on the one hand, and active repulsion, avoidance, and withdrawal on the other. Both the patterns of bonding and the patterns of spacing are adaptive, meaning that they contribute to the survival of the species.

Spacing prevents overcrowding and secures for the individual or group a territory. Territories are often marked (for example, by odors) and defended by ritualized displays or fighting. In addition to its territorial function aggression plays an important role in individual competition, such as for the mate. Intraspecific fighting is often ritualized to the extent that no physical harm results. The loser often ends a fight by adopting special submissive postures that inhibit further fighting by the victor. There is a definite selection pressure acting in favor of sparing the loser. At the same time a strong selection pressure favoring aggression must exist, otherwise tournament-like fighting would not have evolved. Damaging fights do occur, however, in a number of species, but these have developed other devices, such as an elaborate escape capacity, so that killing does not occur as a rule after each encounter.

Aggressive behavior is much influenced by individual experience but is not only learned. Phylogenetic adaptations in the form of motor patterns, releasers, releasing mechanisms, motivating mechanisms, and learning dispositions preprogram aggression in definable ways. Evidence in favor of an aggressive drive is available, but the motivational basis seems to differ even within the percid fish, according to the ecology of the species. Even though man's aggressive behavior is shaped to a great extent by cultural norms, phylogenetic adaptations play an important role as determinants, and, contrary to many claims, neurophysiological evidence demonstrates spontaneous creation of aggressive impulses within the human brain. We have therefore at least to consider the possibility of an innate aggressive drive in man.

Group formation provides a number of selective advantages. By schooling fish avoid predation. By forming aggregations some harvestmen resist desiccation during the dry season. In the higher organized groups division of labor becomes possible. There are many mechanisms of group cohesion. In open groups, as represented by a school of fish, the group member is recognized by simple sign stimuli but not as an individual. Furthermore, the individual shows clear appetitive behavior to seek the company of others characterized by these stimuli. With the development of parental behavior the capacity to develop individualized closed groups evolved. Patterns derived from parental care behavior as well as patterns derived from infantile signals are used by numerous gregarious insects and vertebrates as means to bond adult group members. The evolution of altruistic behavior was possible in closed groups, which can develop complicated social structures with rank order and role differentiation.

16

Orientation in Space

Every organism attempts to achieve continuous control of its environment through its sense organs, and each is programmed in such a way as to avoid unfavorable conditions and to stay in a favorable environment. A daphnia swims close to the water surface when the water contains much carbon dioxide, which is appropriate because the water contains more oxygen there. Two stimuli play a role in this reaction: carbon dioxide as a releasing stimulus and light as a directing stimulus. If one illluminates the water from below the daphnia will swim downward as soon as carbon dioxide is added (A. Kühn, 1919).

Yet this ability to stay in a favorable surrounding can be achieved by simpler means, such as by moving faster in an unfavorable environment and by slowing down in congenial surroundings, without necessarily altering the direction of movement. This principle, which is called kinesis, suffices to keep an animal for a longer period in an environment favorable to its survival.

In most animals of higher organization environmental stimuli are processed in such a way that an angle-controlled change of the direction of movement is achieved. Such topical reactions represent, of course, definite progress compared to the kinesic reactions. The orientation movement (taxes) and their mode of operation will now be considered. I want to restrict myself to a short summary of the attempts to classify orientation processes, concluding with a catalogue. This is followed by a more detailed discussion of selected examples, to demonstrate the main problems and the experimental approach involved.

As I have already pointed out, **orienting movements** are dependent on directing stimuli. According to the resulting position, or the involved mechanisms, taxes have been variously named. A. Kühn (1919) distinguished phototaxis or alarm movements and four kinds of topotaxes:

1. Tropotaxis: with the help of paired receptors the animal assumes a position in which each receptor is equally stimulated. If one receptor is destroyed, the animal turns in circles.
2. Menotaxis (compass orientations): nonsymmetric orientation toward an orienting stimulus, for example, by keeping a constant angle toward light rays.
3. Telotaxis: goal-directed orientation. During this process the goal is fixated.
4. Mnemotaxis: orientation based on memory.

In accordance with the kind of stimulus, one refers to phototaxis, rheotaxis, geotaxis, chemotaxis, and galvanotaxis. Helpful summaries are presented in O. Koehler (1950), G. S. Fraenkel and D. S. Gunn (1961), and M. Lindauer (1963).

One species may possess several orienting mechanisms. The grayling butterfly (*Eumenis semele*) will escape from an enemy by flying toward the sun. It flies in circles if blinded in one eye; this is a tropotaxis. The males fly toward passing females in response to optical stimuli; this they can do even when blinded in one eye, hence their orientation in this functional cycle is a telotaxis (N. Tinbergen et al., 1943).

Orienting processes are not governed by strict stimulus-response relationships. E. v. Holst (1950a) demonstrated that the specific physiological condition of an organism—its **drive state**—exerts a decisive influence. Many fishes orient simultaneously to gravity and to the light. In horizontal light from the side, the fish is turned 90° in response by the dorsal-light reaction, while the static receptors attempt to hold the fish in equilibrium in his normal position. The fish then comes to rest in an in-between position that can be measured precisely in high-backed forms (for example, *Pterophyllum*, angelfish). The stronger the intensity of the light, the more the fish will orient toward the light. If the weight is increased on the static organs by placing the entire experimental apparatus into a centrifuge, then the influence of the light is reduced. Up to this point it appears as if the fish is suspended between two arms of a scale—one representing the light components, the other the static components, but there is another factor at work: Should the hungry fish perceive a prey, the orientation to the light exerts more influence—that is, optical stimuli are now stronger. This example supports the contention that orienting reactions can be dependent upon an inner readiness to act, such as hunger (E. v. Holst, 1950a).

According to their function, orienting processes can be classified into three main groups:

1. Space orientation, which encompasses the orientation in spatially oriented reference systems, where the stimulus sources provide the coordinates.
2. Stabilization of posture and movement.
3. Object orientation, which encompasses all orientations toward directly perceived objects, which are at the same time goals.

Into these three main categories it is possible to fit all known orienting capacities, as can be seen from H. Schöne's (1965–1966) compilation:

- A. Orientation in fixed-spaced reference systems: Stimulus sources provide the coordinates and/or serve to maintain the normal position, and they are *not* the goal of the orientation.
  - 1. Orientation with gravity receptors.
    - a. Maintenance of physical balance against the pull of gravity and compensatory eye movements (postural reflexes).
    - b. Control of position and direction of action in space.
      - b_1. Of the body (directional control of free-swimming fishes, octopuses, crustaceans, and so on).
      - b_2. Of body appendages via proprioceptive control and of other objects by means of optical, tactile perception (recognition and distinction of horizontal and vertical structures).
  - 2. Orientation with the eyes.
    - a. Orientation related to light: orientation of many water animals; not quite correctly called "dorsal-light reaction."
    - b. Horizontal orientation: compass and sun-compass orientation, orientation to landmarks.
  - 3. Tactile, kinesthetic (registration and repetition of its own movements) orientation.
  - 4. Electrical and magnetic orientation (electric fishes).
  - 5. Orientation in currents (air, water).
- B. Stabilization of posture and movement (locomotion) independent of locality.
  - 1. Via mechanical rotational sense organs (semicircular canal structures).
  - 2. With the eyes (optomotoric [kinetic] control).
- C. Object orientation: Stimulus sources are the objects of orientation; frequently locating processes (goal *and* direction are considered).
  - 1. Optical target orientation (mantis, chameleon).
  - 2. Acoustical location (echolocation of bats, porpoises).
  - 3. Tactile location (dragonfly larvae, clawed frogs).
  - 4. Chemical orientation.
  - 5. Thermal orientation (infrared organs of pit vipers, temperature preferences).

In the following sections I want to demonstrate the way in which orienting mechanisms work by discussing some selected examples. We shall pay special attention also to the active role of the organism in the orienting process. Comprehensive reviews on this topic have been published by M. Lindauer (1963), S. Gerlach (1965), and B. Hassenstein (1966). The phylogeny of orienting capacities, especially the light orientations of arthropods, has been discussed by R. Jander (1966a. 1966b).

CONTROL OF BODY POSITION AND MOVEMENT IN SPACE

Many fishes and water beetle larvae (*Dytiscidae*) orient themselves with their eyes toward light, which can be easily demonstrated with an experiment (Fig. 16-1). Water beetle larvae swim up to the surface to get air. If an aquarium is illuminated from below, the animals will swim to the bottom and turn their backs toward it as if it were the surface. Unless the illumination is reversed the animals will suffocate there.

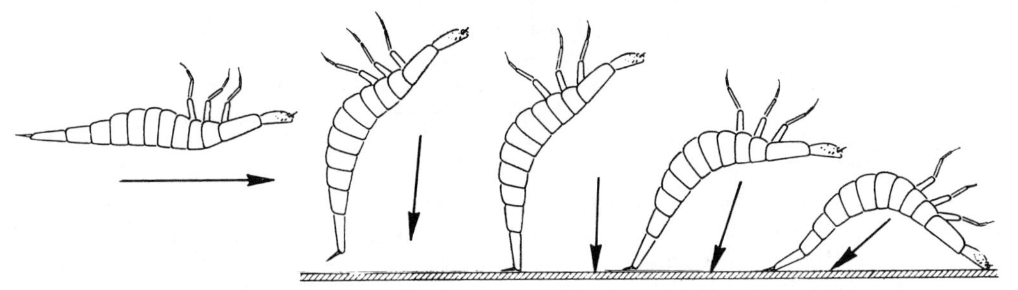

Figure 16-1 When the light comes from below, *Acilius* larvae swim with their backs down. Horizontal swimming and attempts to get air can be seen. (From H. Schöne, 1962.)

However, it is not as if the light stimulus is coupled, so to speak, with a rigid, obligatory reaction. The animal may orient its direction of movement toward the light, but which particular direction with respect to the light source is followed depends on its internal state. If the animal needs to get air it will swim upward, afterward it will swim away from it. Depending on the internal state different orders go to the orientating mechanism so that the animal can choose between various directions and orient itself accordingly. There is no fixed reflexlike stimulus–response relationship; the organism can actively change its direction with reference to the light. This was elegantly demonstrated by H. Schöne (1962).

Dytiscid larvae have six stemmata on each side from which information is integrated in such a way that the animal is able to move in a specific direction with respect to the angle of light incidence. The angles of light stimulation can be controlled by covering individual eyes and by diffusely illuminating the aquarium from all sides. The animal will then swim with its dorsal or ventral side up—in circles in the case of symmetric blinding—in an attempt to maintain a reference position with respect to the direction of illumination. If the angle of light incidence coincides with the angle for the reference position (*Soll-Lage*) the larva stops circling. The turning tendencies, which can be computed from the circles that the animal performs, can be represented on a curve where the orientation of the body depends on the angle of light incidence (produced by selective blinding of stemmata), so that one obtains a sinusoidal curve, from which the strength of the turning tendency for a given expected position can be read.

A normal larva shows a number of different behavior patterns in which it can assume various reference positions in space (Fig. 16-2, *1–8*). After breathing at the surface (*1*) it swims downward at an angle (*2, 3*) and continues horizontally when hunting (*4*), or it waits motionless for prey. Then it swims upward again, at an angle, toward the surface (*5, 6*).

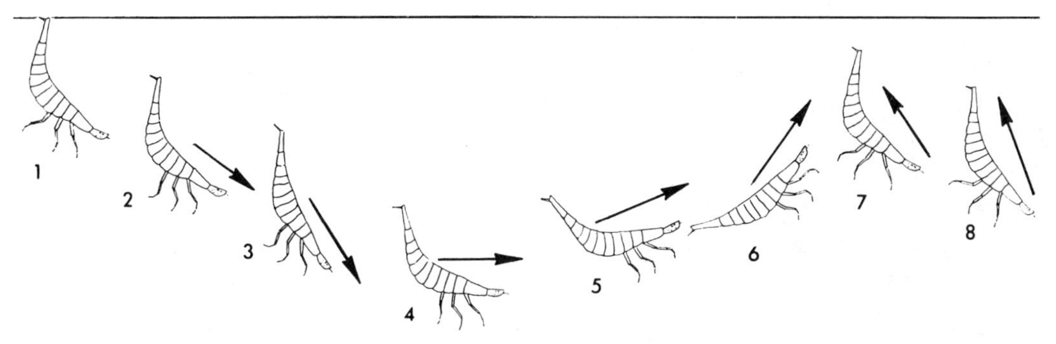

Figure 16-2 Characteristic positions of eight behavior patterns in *Acilius* larvae when the light comes from above (From H. Schöne, 1962.)

Shortly before reaching the surface the larva changes the direction of movement and swims backward until the tip of its abdomen breaks the surface (*7, 8*). For each of these behavior patterns there is a different reference value. H. Schöne (1962) determined the curves that indicate the turning movements in degrees of magnitude for various reference positions and found that with each change of the reference position the curve that represents the strength of the turning tendency is shifted up or down on the ordinate. The position of the upper and lower values on the curve remain unchanged (Fig. 16-3).

Postural orientation based on statolith organs has been studied in detail by E. v. Holst (1950b) and H. Schöne (1959). The gravity detectors of crustaceans and vertebrates consist of statocysts with a sensory epithelium. The statoliths rest on the hairs of the sensory cells. Pressure downward on the statoliths and pulling in the opposite direction elicit no excitation, but a parallel shearing force does. If the hair receptors of the left statocyst in a crustacean are bent outward, a turning tendency to the right with respect to the longitudinal axis is released. If bent inward they cause turning in the opposite direction. With unilateral removal of the entire statocyst, fish and crustaceans turn about their longitudinal axis in the direction of the injured side. E. v. Holst thought this was due to the continuous activity of the sensory cells in the statocysts. Normally these continuous discharges of impulses in the left and right statocysts cancel each other; after elimination of the statocyst on one side the turning tendency in the opposite direction becomes manifest. This interpretation is supported by the results of experiments in which sharks, whose

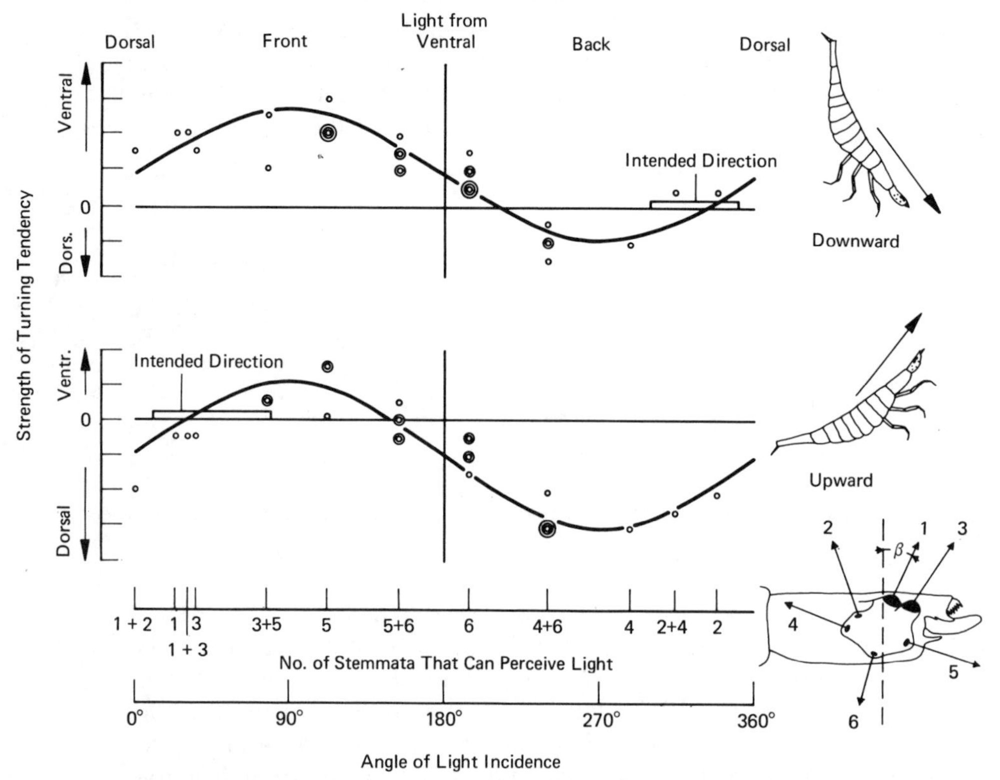

Figure 16-3 Example of an orientation with changeable reference (intended) positions: up-and-down swimming in larvae of the water beetle *Acilius sulcatus*. *Experimental technique:* The incidence of light in each case is coupled to the subject: all stemmata are covered, with the exception of those pointing in the desired direction toward the light. The larva is observed in a container that is illuminated equally from all sides. When the "set" direction of light incidence (dependent on which stemmata are left uncovered) coincides with the reference position, the animal swims straight; otherwise it turns toward its back or ventral side. The diameter of the performed circles serves as the measure for the turning tendency. *Result:* The values for the strength of the turning tendencies obtained in this way are plotted in relation to the direction of light incidence: in the upper curve for reference direction downward, in the lower curve for reference direction upward (swimming into the breathing position and leaving the breathing position). *Conclusion:* The curve for the downward direction is shifted upwards compared to the curve for the upward direction. The reference values for the positions are changed by central nervous system processes affecting the turning tendencies in correspondence to the direction of light incidence. All turning tendencies are increased by the same value in the same direction. (From H. Schöne, 1962.)

statoliths were removed without damaging the sensory epithelium, showed no tendency to turn (S. S. Maxwell, 1923). Crustaceans without statoliths, in which one of the empty statocysts had been removed, turned toward the injured side (H. Schöne, 1959).

The discharges of resting potentials of the sensory epithelium on both sides cause, according to these findings, opposing turning tendencies which cancel each other. The shearing forces in the statocysts, which emanate from the statoliths when the animal turns, increase the rate of discharge of impulses in one direction and decrease it in the other according to the principle represented in Fig. 16-4.

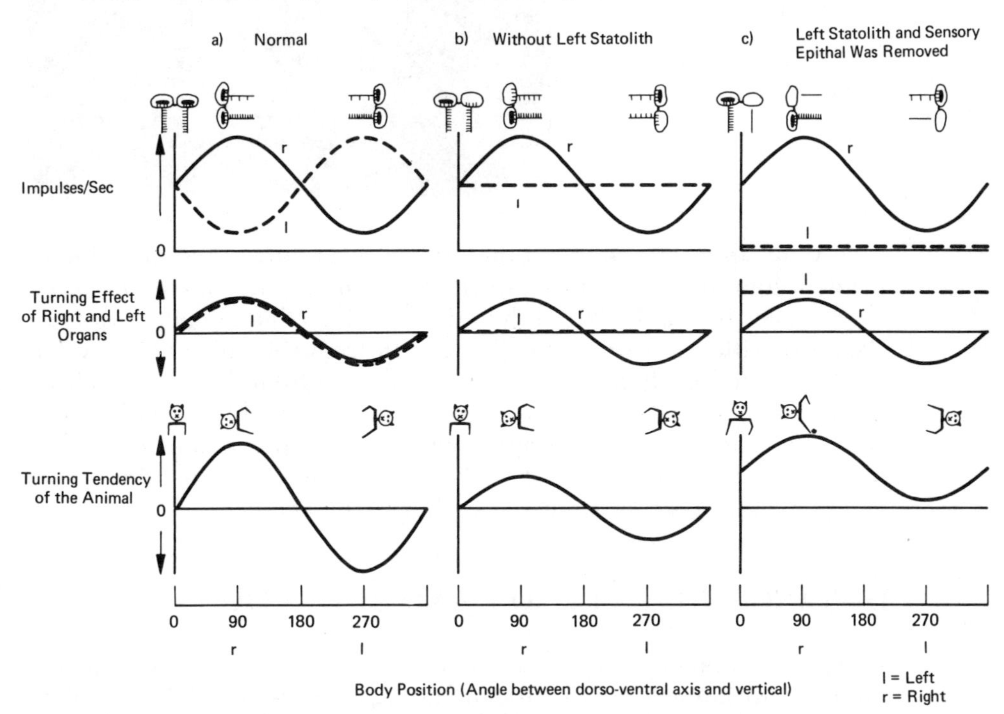

Figure 16-4 Example of a simple orientation process: the principle of postural orientation with the aid of the statolith apparatus in vertebrates. The normal position ($\alpha = 0°$) is the expected position; all turning tendencies seek to return to this zero position. The diagrams of the *top row* show the excitatory values in the statolith organs, as they can be determined in electrophysiological experiments from recordings of single fibers; the figures above the curves symbolize the two statolith organs and the sensory epithelia, statolith bodies, and the flow of excitatory currents. The shearing force of the statolith mass modifies the rate of discharge of the impulses that emanate from the sensory cells. In the *center row* the hypothetical possibility of the relationship between the excitation of both individual organs and the resulting turning tendency is portrayed. The turning tendencies of both sides add up to the turning tendency of the entire animal (*bottom row*). The figures above the curves illustrate the appropriate position reactions, which correspond to the turning tendencies in the example of a standing mammal. (*a*): A normal animal; (*b*) an animal that lacks its left statolith mass, but whose sensory epithelium is still intact; (*c*) an animal whose left sensory epithelium has also been removed. The curves show the relationships immediately after the operation. Later-appearing compensating processes result in a shifting of the turning tendency curve downward; it then cuts across the zero line. The initial continuously occurring rotations eventually come to rest in this position. (From H. Schöne, 1965-1966.)

Figure 16-5 *Myripristis murdjan* in a cave of the Miladummadulu atoll (depth approximately 30 meters). Some of the fish swim upside down. They orient themselves with the aid of the dorsal light reflex toward the light, which is reflected from the sandy bottom of the cave. (Photograph: I. Eibl-Eibesfeldt.)

Many animals also orient themselves according to the direction of incident light. E. v. Holst investigated this integration of two orienting mechanisms in fish. Fish turn their dorsal side to the light and assume a slanted position toward it when it comes from the side. This slanted position is the resultant of stimulation received from the eyes and from the statolith organs. Changes of the endogenous condition can lead to a changed evaluation of the optical components, which is then expressed in a different postural angle (Fig. 16-5). That the individual **drive state** enters the process was discussed earlier in the chapter.

DISTANCE ORIENTATION AND MIGRATION

Distance orientation has always been of special interest (G. Kramer, 1961). Many animals are capable of finding a goal which they cannot directly perceive, and their achievements during such migrations are at times astounding. The golden plover (*Pluvialis dominicus*), which breeds along the northern shore of Alaska, migrates during the fall via Labrador to Argentina with a flight over the open ocean from Nova Scotia to Guiana. The return flight is across land, over Central America, up along the Mississippi River toward the north. The long axis of this migratory ellipse measures 11,000 km. The Mongolian plover migrates from Siberia to Australia and South Africa. Figure 16-6 presents an overview of the extraordinary migratory capacities of some American migratory birds. E. Stresemann (1934) computed the work performed by such a migratory bird, on the assumption that a medium-sized wading bird such as the golden plover flies 26 meters per second and makes two wing beats per second. The distance of 3300 km, which is the shortest distance from

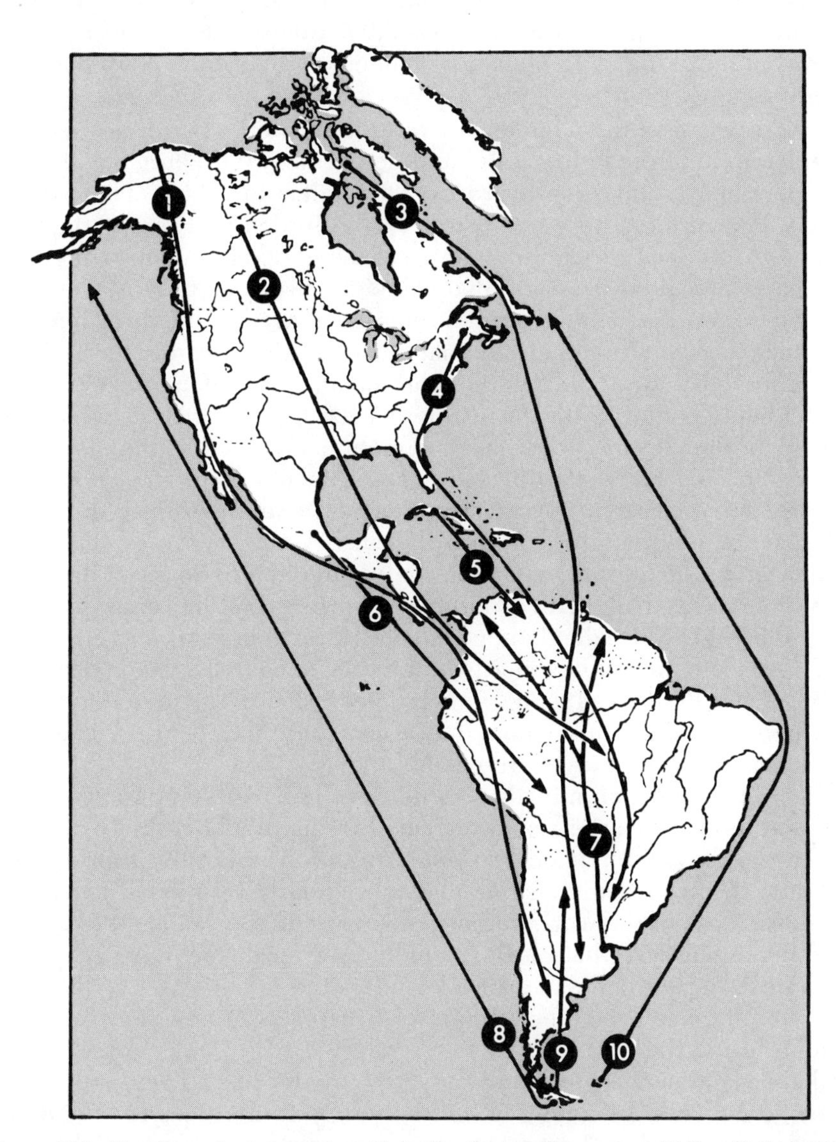

Figure 16-6 Migrations of some American birds after the breeding season. (*1*) Knot (*Calidris alba*), from northern Alaska to southern Argentina; (2) red-eyed vireo (*Vireo olivaceus*), from Mackenzie to Mato Grosso; (*3*) American golden plover (*Pluvialis dominicus*), from Melville Peninsula to Argentina; (*4*) bobolink (*Dolichnyx oryzivorus*), from Maine to Brazil; (*5*) gray kingbird (*Tyrannus dominicensis sequax*), from Cuba to Venezuela; (*6*) sulphur-bellied flycatcher (*Myiodynastis luteiventris luteiventris*), from southern Mexico to Bolivia; (*7*) swallow (*Phaeoprogne tapera fusca*), from Argentina to British Guiana; (*8*) sooty shearwater (*Puffinus griseus*), from the Magellan Islands to the coast of Alaska; (*9*) *Lessonia rufa,* from Tierra del Fuego Islands to New Argentina; (*10*) Wilson's petrel (*Oceanites oceanicus*), from the Falkland Islands to Newfoundland. (After van Tyne and Berger, from G. Diesselhorst, 1965.)

the Aleutians to Hawaii, would require 35 hours, during which the bird would move its wings up and down 252,000 times. The American golden plover requires about 48 hours for its nonstop flight from Nova Scotia to South America.

European storks migrate in two groups. Storks that live west of a line extending from Leiden (Holland) to Giessen, Würzburg, and Kempten (Germany) move westward over Gibraltar to Africa. The eastern storks fly across the Bosporus (Turkey), the Jordan Valley, and the Gulf of Suez to tropical Africa. Knowledge of the general migratory direction is innate; East Prussian storks that had been displaced to West Germany moved toward the southeast when released, that is, in the direction they would have taken from their original home location, in order to reach the Bosporus and East Africa. On the other hand, when young East Prussian storks made contact with the resident West German populations, they flew with them in a southwesterly direction. In this case they followed the group. The Baltic starlings winter in England and northern France. To reach these areas they must migrate in a southwesterly direction. This general direction is inborn, because young starlings that were displaced to the latitude of Genoa (Italy) migrated to Spain. If the same experiment was made with experienced starlings that had once migrated from the Baltic to England, they corrected their course to compensate for the displacement and flew northward from Genoa to England (E. Schüz, 1952; A. C. Perdeck, 1958b). Graylag geese learn the migratory route south from their parents, whom they accompany. Without their guidance they remain where they were raised.

Other groups of animals also perform remarkable migratory feats. Salmon return from the sea to the small rivers in which they spawned. To return they swim against strong current for hundreds of miles. The Atlantic green turtles (*Chelone mydas*) regularly visit small islands in various places of the world to bury their eggs in the sands. Marking experiments showed that turtles which feed near the Brazilian coast migrate to Ascension Island, 2000 km out in the Atlantic (A. Carr, 1965).

In all these examples the question arises as to how these animals orient during their travels. Some migratory birds orient themselves by utilizing landmarks, but many can do without these. They maintain a specific direction by orienting themselves with reference to stars or, as more recent work indicates, by using the earth's magnetic field. The various kinds of compass orientations will be discussed.

Many migrating animals use the sun as an aid in navigation. G. Kramer (1952, 1957, 1959) was the first to demonstrate the use of the sun as a compass by migratory birds. His caged birds, which showed migratory restlessness, fluttered in the same direction independently of optical landmarks or terrestial magnetic influences: in the fall toward the south and in the spring toward the north. They oriented themselves according to the position of the sun. When Kramer deflected the sun rays with mir-

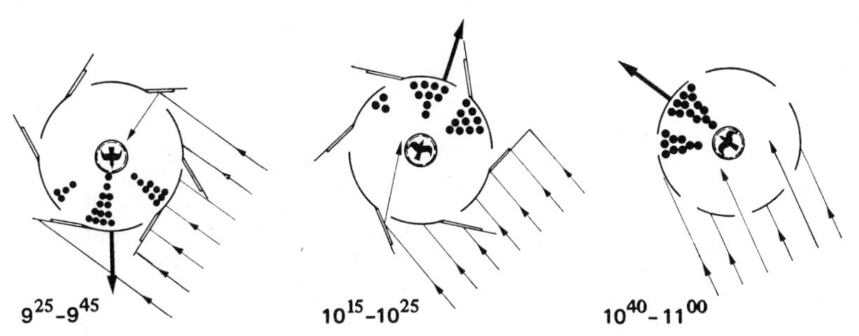

Figure 16-7 Mirror experiment of G. Kramer (1952). The dots represent individual observations that were made at equal time intervals.

rors by a specific angular value in the horizontal plane the birds compensated their directional tendency by the same value (Fig. 16-7). Starlings are also capable of compensating for the slow movement of the sun during the course of the day. Starlings that had learned to obtain food at the same time of day in a specific direction in a cage chose the same direction at other times of the day when the sun was in a different position. Three of these starlings that were trained to respond to a specific direction were tested by Kramer in a cellar under a fixed artificial sun. In these conditions the direction of choice changed in a lawful manner over time. The birds behaved in accordance with their central compensating mechanism as if the sun had moved by a specific amount (Fig. 16-8).

This ability presupposes a precise internal clock, and the experiments of K. Hoffmann (1954, 1960) show that the mechanism which underlies circadian periodicity is involved. This internal clock, which is so important for orientation, can be reset by an **artificial light–dark alternation schedule.** In a starling that had been trained to seek food in a specific direction, the choice direction changed accordingly. If a bird had been trained to a southerly direction before its internal clock was reset, then it chose an angle of about 45° to the right of the sun at 9:00 in the morning, and at 3:00 in the afternoon it selected the same angle to the left of the sun. When the day had been moved in such a way that it began 6 hours later than usual for the bird, then the early afternoon became the subjective morning (Fig. 16-9). The results of Hoffmann's experiments corresponded exactly to this prediction in experiments in which the 6-hour-deviation from the normal day was used.

When the artificial day is again brought in line with the actual time the starlings also made their choice according to the existing situation (Fig. 16-10). This demonstrates that the compensation for the movement of the sun depends upon an endogenous physiological clock which is synchronized with local time through the day–night cycle. Under constant conditions of continuous illumination this orienting mechanism continues to function. It is therefore not set in motion by an external "time giver" (***zeitgeber***) working on the hourglass principle. This physi-

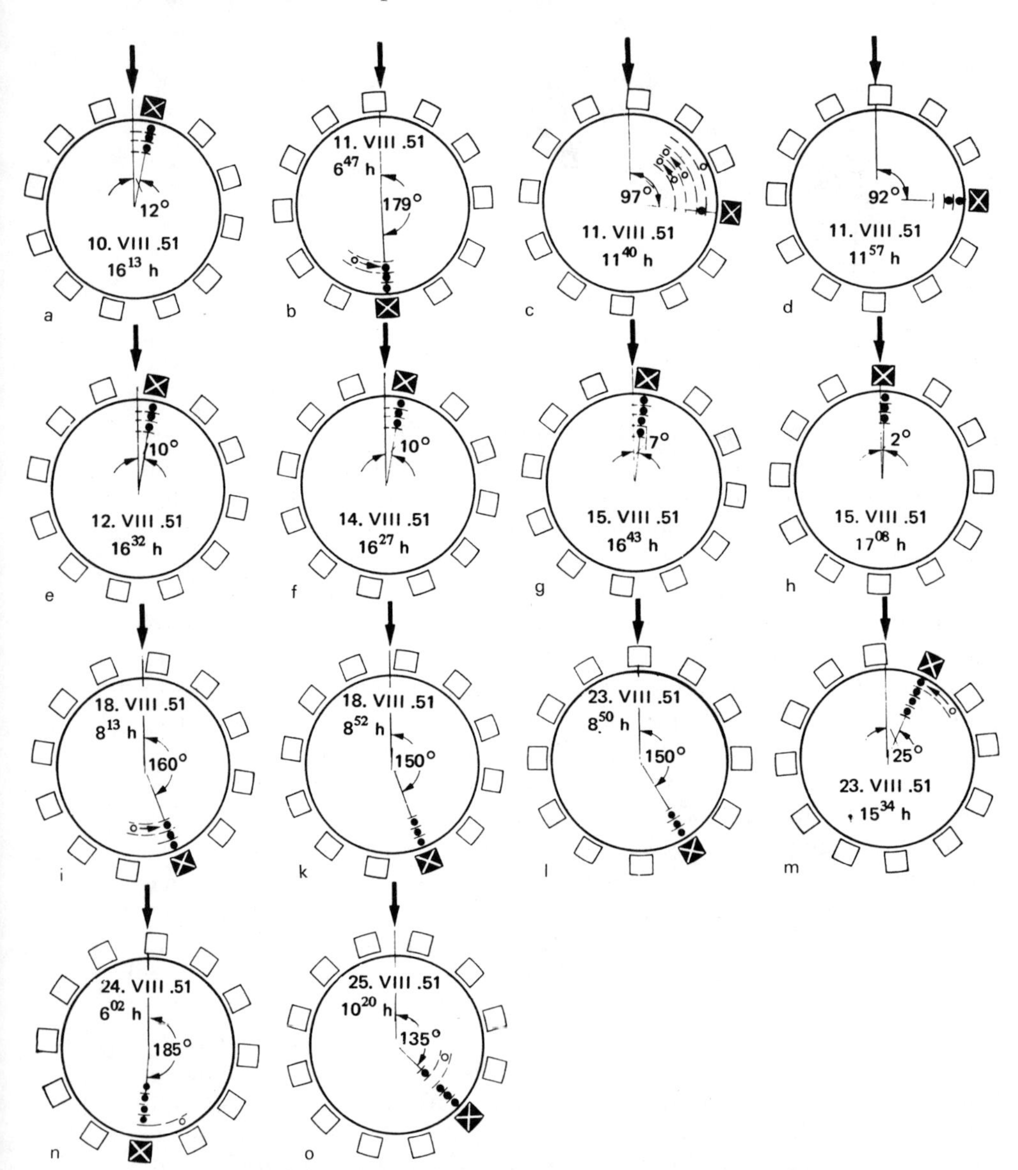

Figure 16-8 Series of choices made by a starling that was trained to the west with an artificial sun. *Arrows;* incidence of the light from the artificial sun; *filled dots:* choices that resulted in eating food; *empty circles:* choices without food intake; *blocks with cross near the periphery:* baited cork in the expected direction; *two points connected by an arrow:* choice corrected. (From G. Kramer, 1952.)

ological clock shows its own **circadian frequency** under constant conditions, which deviates somewhat from the time based on the rotation of the earth. By waiting long enough one obtains deviations that are similar to those achieved by artificial resetting of the day–night cycle (Fig. 16-11).

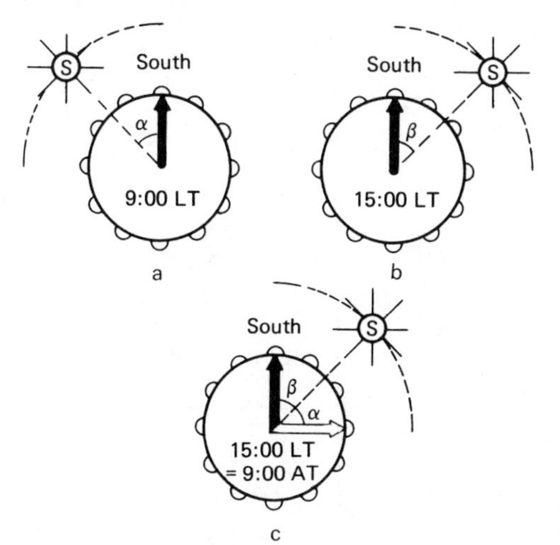

Figure 16-9 Expected results after changes in the internal clock. *S*, sun; *filled arrow*, direction of training; *empty arrow*, expected choice direction after the change. The *large circles* symbolize the choice presentation, the *half-circles around their diameters*, the feeding cups. *LT*, true local time; *AT*, artificial local time.

That birds utilize a sun compass during their migrations can be indirectly deduced from translocation experiments, which were successful only during clear weather. Under covered skies the birds were unable to orient themselves (G. V. T. Matthews 1955; J. D. Carthy 1956; G. Kramer 1961). Some observations have been made, however, of successful migrations under a cloudy sky (P. Steidinger, 1968). Perhaps these birds orient themselves by the earth's magnetic field (see below). According to the findings of A. D. Hasler and H. O. Schwassmann (1960), H. F. Winn et al. (1964), and W. Braemer (1960), fish are also capable of using the sun as a compass. Lizards (*Lacerta viridis*) (K. Fischer, 1961) and land and water turtles (E. Gould, 1957; K. Fischer, 1963) are also capable of this orientation. An interesting, innately programmed control of orienting behavior was described by C. Groot (1965), who investigated the migration of salmon (*Oncorhynchus nerka*) from their spawning grounds in Babine Lake (British Columbia) to the sea. The lake system has many branches so that the young which were born in various parts of the lake initially migrate in different directions until they reach the common river which leads to the sea (Fig. 16-12). They orient not by the currents but by the sun and other still-unknown cues when the sky is overcast. One group from one of the arms of the lake (Morrison Lake) has to migrate south-southeast to reach the exit from this lake, then turn 180° in a north-northwest direction, while other groups from the lake can simply maintain a general northwest migratory direction. Young fish that were caught in various parts of the lake at the beginning of migration and were kept in round or octagonal aquaria without a view of the horizon under the open sky oriented themselves according to the direction of migration their fellows showed in the various parts of the lake: for example, those from Morrison Lake in a southeasterly direction, those from the main lake in a northwesterly direction. As time passes the orienting direction of the

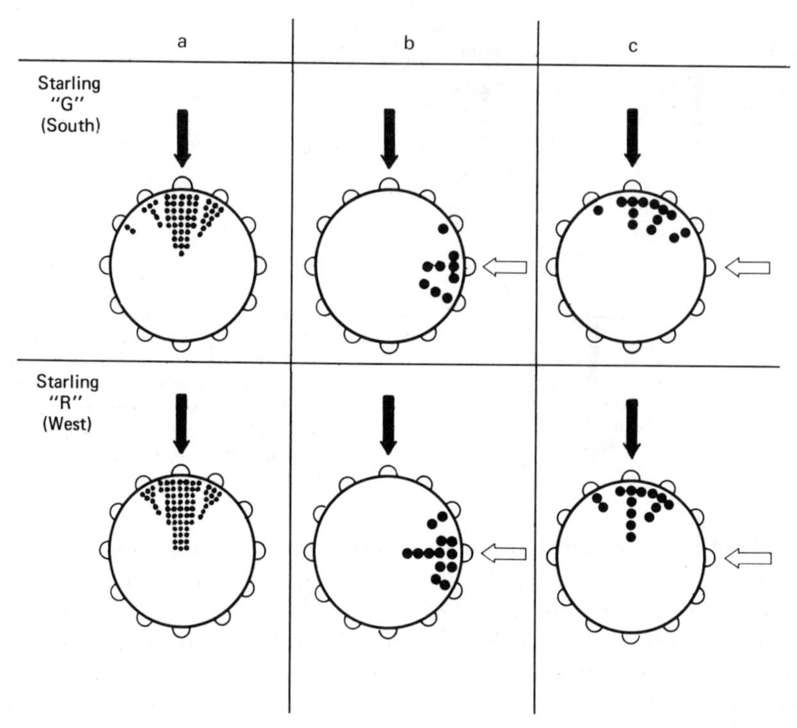

Figure 16-10 Result of experiments in which the internal clock was changed in starlings trained to respond in a certain direction. The starlings were trained to search for food in the south or west. (*a*): Choices during training under natural daylight conditions; (*b*): choices after 12 to 18 artificial days, which were shifted 6 hours from the actual time of day. The direction of choice shifted in (*b*) in the expected direction and in (*c*) returned again to the original direction. The *large circles* symbolize the choice situation, the *half-circles around their diameters* the food cups (which were empty during the tests). Each *dot* represents one choice. The *filled arrows* show the original training direction, the *empty arrows* show the expected training direction after the artificial shifting of the day. (From K. Hoffmann, 1965.)

fish from Morrison Lake changes. A similar change from the preferred direction was also observed by C. Groot in salmon from another lake as the migrating season progressed. Here there was also a correspondence of the change in direction to the natural migratory route, which none of the fish had previously traversed. The salmon find their way home by recognizing the odor of the waters in which they were born (A. D. Hasler, 1960).

Bees orient with the aid of the sun compass. They learn the direction of flight or obtain the information from conspecifics. In various species of wolf spiders, ants, and water boatmen a sun-compass reaction has also been demonstrated (G. Birukow and E. Busch, 1957b; F. Papi, 1959; R. Jander, 1966b).

When the sun is covered bees are able to orient themselves according to the pattern of polarization of a part of the exposed blue sky (K. v. Frisch, 1950), a capacity that has since been demonstrated in other arthropods as well (K. v. Frisch, M. Lindauer, and K. Daumer, 1960).

Quite remarkable is the orientation of the beach hopper (*Talitrus saltator*), which lives in and around the waterline of beaches. When displaced inland the animal escapes toward the sea using compass orientation with respect to the sun and compensating for the movement of the sun. The appropriate compass direction in which to escape is inborn in the respective populations that live in different locations (L. Pardi, 1960). Beach hoppers who were raised in the laboratory under artificial light and whose parents came originally from various different locations and therefore took different escape directions in their specific localities, spontaneously chose the flight direction of their parent populations when placed under the open sky. At night they oriented themselves by using the moon (F. Papi and L. Pardi, 1959; J. T. Enright, 1961). Some migratory birds that migrate at night, such as several species of warblers and lesser grey shrikes, orient themselves by the stars. They are disoriented when the sky is covered with clouds, but they are able to maintain their migratory direction in a planetarium under an artificial starry sky. They apparently orient themselves to the patterns of the fixed stars. Moonlight disrupts their orientation (F. Sauer and E. Sauer, 1955, 1960).

More recently, it has been discovered by planetarium experiments that the indigo bunting (*Passerina cyanea*) and the mallard duck are able to orient themselves by an artificial starry sky. The pattern of the ar-

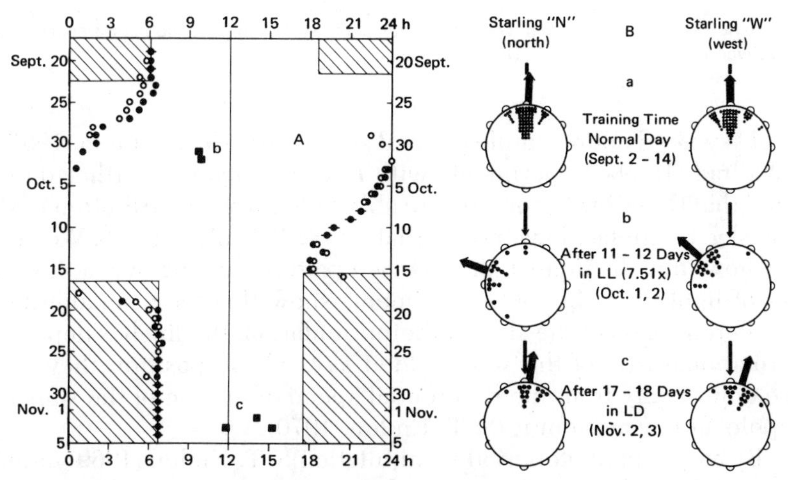

Figure 16-11 Comparison of the rhythms of locomotor activity (*left*) and finding of direction with the aid of the sun after maintenance under constant light and temperature conditions. (*a*): The starling *N* is represented by an *open circle*, starling *W* by a *dot* with respect to the onset of activity. When the onset of activity coincides for both, a line is drawn through the dot. *Crosshatched:* Times of darkness. Black squares indicate the time of the choice tests, which are shown at (*b*) and (*c*). (*b*): Choice test during the time of training in a normal day (*a*), after 10 to 11 days under constant conditions (*b*) and during an artificial day, which was synchronized with the natural day (*c*). *LL,* constant light; *LD,* artificial light–dark alternation, which corresponds to the normal day–night cycle. *Centripetal arrows* show the direction of training, *centrifugal arrows* the average direction of choices made. Additional explanations are in Fig. 16-10. It can be seen from this figure that the activity rhythm occurs one half hour sooner each day and that the direction of choice shifts accordingly. (From K. Hoffmann, 1965.)

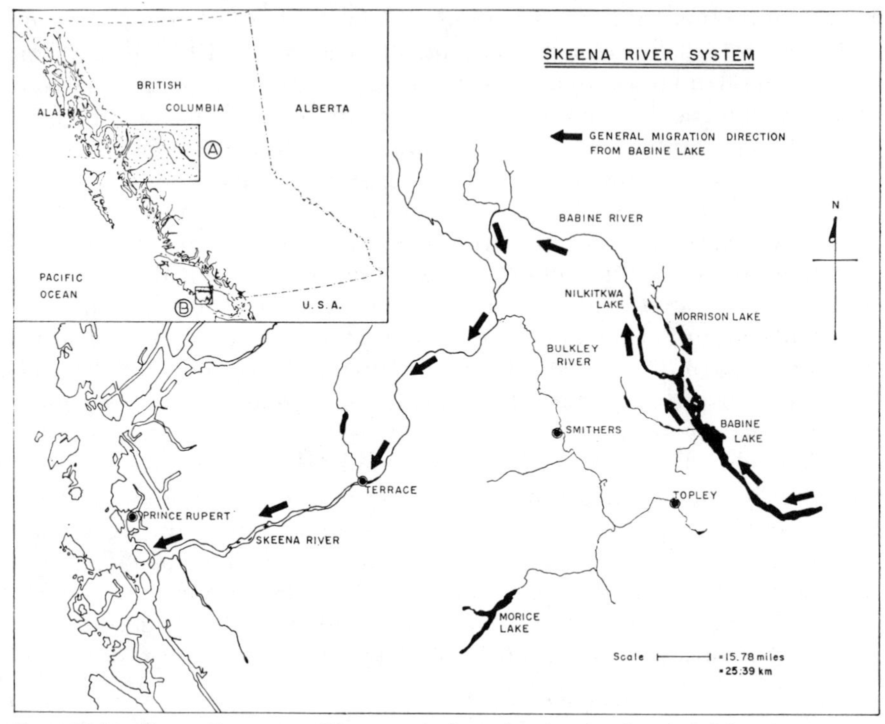

Figure 16-12 Skeena River system. The arrows indicate the migration direction of the salmon toward Babine River and the sea. (From C. Groot, 1965.)

tificial sky need not resemble the real star pattern (S. T. Emlen, 1967; H. G. Wallraff, 1969). Experiments with *Passerina cyanea* further demonstrated that these birds not only are able to learn the distribution of stars in the sky, but in fact have no alternative (S. T. Emlen, 1970). We do not know how they distinguish the relevant patterns among over a thousand spots of light. Blacking-out experiments show that they can alternately use different parts of the sky. In their orientation, the finches depend on the rotational axis of the star-studded sky, whose position they determine empirically from the differential speed of motion of the stars near the pole and away from it (S. T. Emlen, 1970).

By means of photoperiod manipulation S. T. Emlen (1969) induced the physiological states of spring and autumn migratory readiness in *Passerina cyanea*. He tested the birds simultaneously under an artificial spring planetarium sky and found that birds in spring condition oriented northward; those in autumnal condition, southward. The results suggest that the seasonal reversal of the orientative tendencies is brough about by changes of the physiological state of the birds rather than by differences in the external stimulus situation.

Some animals can perceive and utilize the earth's magnetic field for orientation. The direction of crawling of the snail (*Nassalia obsoleta*) can be influenced by the intensity of a magnetic field (A. W. A. Brown and H. M. Webb, 1960). By changing the alignment between a magnetic and an electrical field F. Schneider (1961) was able to influence the activity of the cockchafer (*Melolontha vulgaris*). According to G. Becker (1965) diptera reacted to changes in a magnetic field. Bees of one colony used to orient their combs in a specific direction. If the declination of the magnetic field is changed artificially, the direction of the combs was changed in the same way (M. Lindauer and H. Lindauer, 1972).

Some birds orient themselves by the earth's magnetic field when they migrate. F. W. Merkel and W. Wiltschko (1965) and W. Wiltschko (1968) were able to influence the direction of migration in the European robin in an artificial magnetic field.

Experienced pigeons find their way home even flying beneath a cloudy sky. If small magnets are atttached to their body, they become disoriented, whereas controls that carry an equally heavy brass weight show no such disorientation (W. T. Keeton, 1971).

Special problems are posed by true *navigation*, that is, the ability to find home from an unknown place. Homing pigeons that have been displaced for hundreds of miles and have no sensory contact with their home region find their way home to their loft, but only when the sun can be seen (G. Kramer, 1952, 1957; H. G. Wallraff, 1960b). Pigeons that have been raised in cages until they were tested can do this provided the cage is in an open area and is constructed of wire mesh (H. G. Wallraff, 1967). Just how the animals determine the geographic position of their home location and then compare it with the location at which they are released is unexplained to date. There are several hypotheses, but according to H. G. Wallraff (1959, 1960a, 1960b) they are not convincing. G. V. T. Matthews (1955) developed the sun-navigation hypothesis, according to which the bird knows the path of the sun's movement at its home location. When circling over the releasing point the bird is said to take cognizance of a part of the sun's path from which it can determine the apex of the sun in that locality. By comparing the highest position of the sun at both places the bird would then be able to compute the geographical latitude of its position. The bird is also said to remember the time at its home locality, so it could also compute the distance of the displacement. With some modifications C. J. Pennycuick (1960) supported this hypothesis, which presupposes the existence of an extremely precise internal clock; it would seem to be improbable for this reason.

According to F. Sauer (1956, 1957) lesser whitethroats, blackcaps, and garden warblers can navigate by the stars, a conclusion that Wallraff is unwilling to accept on the basis of a statistical analysis of Sauer's data. The later investigations of F. Sauer (1961), however, further support astronavigation. According to these findings star orientation involves more than a simple compass orientation. Although we are relatively well informed about the compass orientation of animals, we know very little about the mechanisms underlying true navigation.

Frequently more than one mechanism is involved in the orientation of an animal. Salmon can use a sun compass, but also stimuli from currents and finally the odor of the home river (A. D. Hasler, 1954, 1956, 1960). Eel larvae flow into river estuaries with the rising tides. With falling tides they sink to the bottom and are thus protected against being washed out to sea again. They react innately to the odor of specific substances in the inland water, and it has been demonstrated that the saline concentration is not a factor. This odor keeps them at the bottom until seawater returns, when they continue their movements (F. Creutzberg, 1961). A thorough discussion of animal orientation is presented by K. Hoffmann (1971).

PROXIMITY ORIENTATION DURING LOCOMOTION

During their daily excursions animals orient themselves by various means. Many insects and mammals leave **odor trails** when they move about, which serve as their means of orientation.

Acoustical orientation exists in the oilbirds (*Steatornis*), many bats, porpoises, and probably Weddell seals (*Leptonychotes*) (F. P. Möhres, 1953; W. E. Schevill, 1955; D. R. Griffin, 1958, 1962; J. Schwartzkopff, 1960; W. N. Kellogg, 1961; C. Ray, 1966). Electrical orientation with the aid of self-produced electrical fields is known to exist in some fish. The Nile pike (*Gymnarchus niloticus*), which lives in cloudy waters, sends out an almost continuous series of electrical impulses, approximately 300 per second with a voltage of 3 to 7 volts. The tail becomes negatively charged with respect to the head during a discharge. The fish are very sensitive to the changes in potential in the surrounding water. A drop in potential of as little as 0.04 millivolt per centimeter is responded to. The fish is able to detect and localize disruptions in the electrical field with the aid of its electrical sense (H. W. Lissmann and K. G. Machin, 1958; F. P. Möhres, 1961).

CHANGE IN THE REFERENCE VALUE DURING ACTIVE MOVEMENT ("REAFFERENCE PRINCIPLE")

Oriented behavior in space is possible only when an animal can actively take up a variety of postures. This requires special mechanisms. If a passive fish resting in its normal position is tilted, say, to the right, the increasing shearing force of the statolith on the sensory surface of the labyrinth on the same side results in increased activity in the postural center of the central nervous system, and this sets in motion the motor movements that bring the fish back to its normal position. Under such experimental conditions this postural reflex works untiringly. However, during the normal activity of the fish, as in the search for food, spontaneous tilting movements frequently take place. Why, in this case, does the postural reflex fail to pull the fish back automatically to its normal position? According to the reflex theory it has been assumed that the postural reflex is blocked during spontaneous movement. It can be shown by experiment that this is not the case. By placing the fish in a constant centrifugal field, it is possible to double the shearing force that the statolith exerts on the sensory surface of the labyrinth. Now, if one measures the frequent spontaneous tilting movements the free-swimming fish makes, one finds that they have decreased in magnitude and that the tilting movements decrease in proportion to the increase of weight of the statoliths. Thus the righting reflex is not blocked during spontaneous movements but is dependent upon or regulated by the afferent feedback that they cause (E. v. Holst et al., 1950).

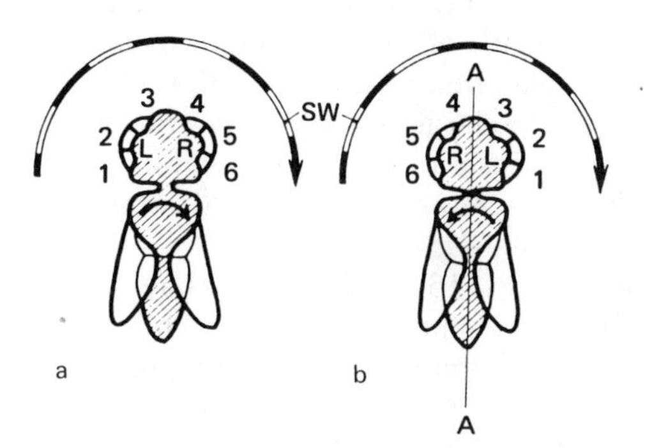

Figure 16-13 Schema of a fly with normal head position (*a*) and the head turned 180° about the longitudinal axis (*A–A*). *b:* Six sectors for the eyes are shown: (*1–3*) for the left eye (*L*) and (*4–6*) for the right eye (*R*). When a striped wall is turned to the right about the animal, it will turn to the right when its head is in the normal position (*a*). When the head is inverted it will turn to the left (*b*) with respect to the striped wall. (Additional explanation in the text.) (From E. v. Holst and H. Mittelstaedt, 1950.)

Likewise it was assumed that optomotor reflexes are blocked during active locomotion. Again it could be shown that this is not the case. The fly *Eristalis* orients optically by fixating its visual field when in a resting position. If a cylinder painted with vertical stripes is rotated around the fly, it will turn as a result of the optomotor reflex in the direction of the moving stripes. If the fly moves spontaneously in the resting cylinder it can make many turns without being forced to return to its original position, although there is a displacement of retinal images from the environment. The optomotor reflex in this case does not interfere with the fly's movements. H. Mittelstaedt (1954) turned the head of an *Eristalis* 180 degrees about the longitudinal axis and fixed it to the thorax in this position (Fig. 16-13). The sequence of visual elements was now reversed, so the animal turned to the left whenever the striped cylinder turned to the right. If active movement would result in inhibition of the optomotor reflexes the fly with the reversed head should be able to move in the resting stripe cylinder. Actually, however, when the fly begins to move spontaneously, it now moves alternately to the right and left in small circles and finally stops in a bent posture.

During spontaneous movements, therefore, the stimuli, which otherwise release postural reflexes, are not inactivated but must become neutralized in another way. Additional experiments and considerations lead to an hypothesis called the *principle of reafference* (E. v. Holst and H. Mittelstaedt, 1950).

This functional organization can be represented as a feedback loop. We differentiate in this schema *afferences*, which flow toward the central nervous system, and *efferences*, which lead from the central nervous system to the motor areas. Afferences in turn can be subdivided into receptor excitation caused by internal changes in the muscular system (*reafference*) and those produced passively by external energy changes from the external environment (*exafference*). Reafference and exafference are integrated in some higher center. E. v. Holst and H. Mittelstaedt postulate that with each voluntary movement of the organism a copy of the motoric efferent impulse branches off as an efferent copy which is stored in a subordinate lower center Z_1. The efferent impulse goes to the effector, and the sense organs report the result of the movement as a reafference. Centrally this reafference is matched against the efferent copy and is canceled. If the total afference is too large or too small as the result of external stimulation, then there remains a plus or minus value in the center Z_1. This is reported to the higher center and the initial command is correspondingly strengthened or weakened.

E. v. Holst and H. Mittelstaedt (Figs. 16-14 and 16-15) have illustrated this principle with the example of "space constancy." This is the observation that we recognize nonmoving objects as stationary and moving objects as being in motion regardless of whether or not we ourselves or parts of our bodies are moving. For example, if we look at a train that is beginning to move, its image passes across our retinas in the same manner as if we actively pass our glance along a standing train. In either case, however, we know when the train stands still and when it is moving. Three simple experiments help us to understand the mechanism that underlies this capacity.

If we fixate an object with one eye and then move the eye passively to the left by pressing lightly against the eyeball with a finger or by moving a ring placed on the eyeball (see Fig. 16-14*a*), we have the impression that the object moves to the right. In this case an intentional command is missing and as a result the efferent copy present during an active movement is lacking. The retinal displacement of the image is reported on to a higher center, and we draw the erroneous conclusion that the object is moving.

Next we may temporarily paralyze the eye muscles by a drug and ask the subject to look to the left. This movement cannot be carried out,

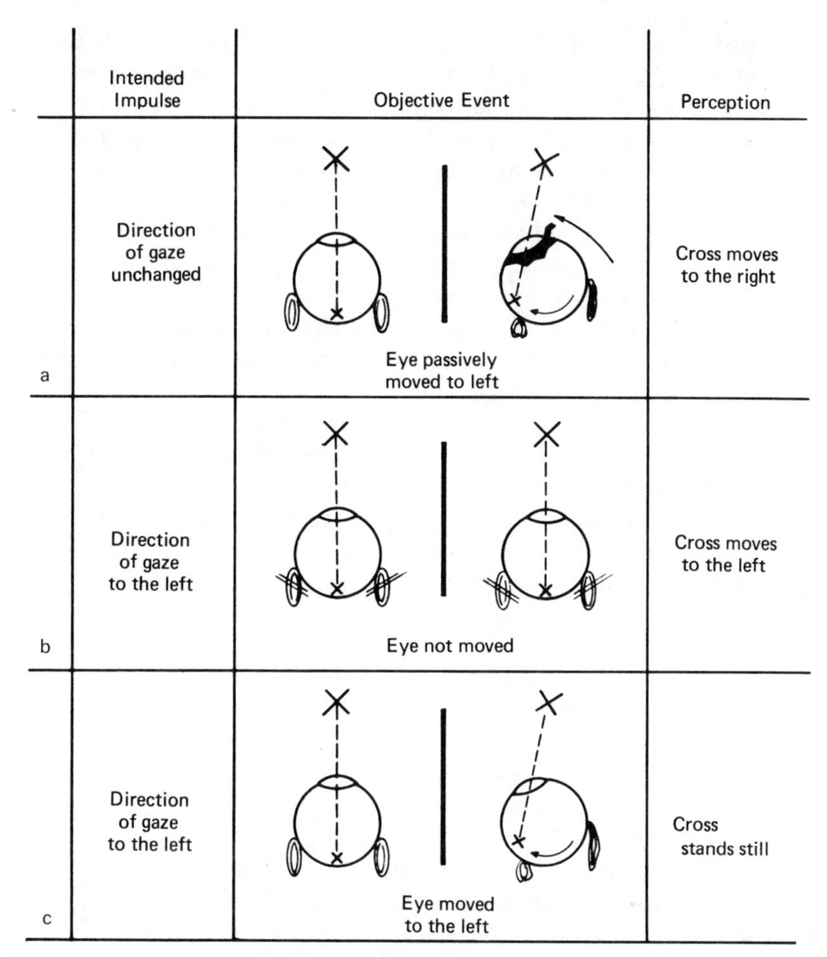

Figure 16-14 Eye movements and perception. Experiments for the determination of the functional schema (functional organization) of spatial constancy. (Explanation in the text.) (From E. v. Holst, 1956.)

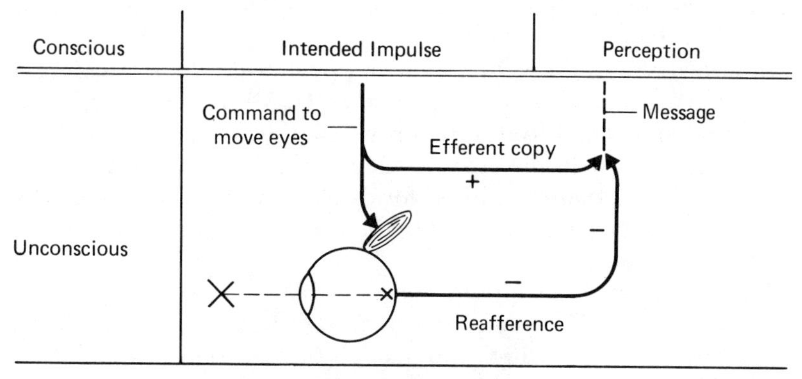

Figure 16-15 Functional schema of spatial constancy. (From E. v. Holst, 1956.)

but interestingly enough the subject experiences a movement. He sees the object moving to the left. In this case, then, a movement perception takes place although there is no retinal displacement of the image. This is again in agreement with the reafference principle, because the intentional command produces an efferent copy that passes to the higher centers without being canceled by afferent impulses.

Finally, we can combine both experiments by asking a subject to move the previously paralyzed eye that is focused on an object to the left. At the same time we move the eye passively in the same direction by means of a clamp ring. If this experiment is carried out properly, the subject will perceive no movement. Both commands, reafference and the efferent copy, which in the previous experiments led to erroneous perceptions, now cancel each other. This is the reason we perceive our environment as stationary although we actively look about. The functioning of the orienting mechanisms which we have discussed so far can be understood best in line with these kinds of theoretical considerations (see also the excellent discussions of this subject by B. Hassenstein, 1966, and N. Bischof, 1966a, 1966b).

ORIENTATION TOWARD OBJECTS

When we reach for an object, the grasping movement is controlled automatically by the eyes, which can detect each deviation to the left or right and which initiate the appropriate corrections of the hand movements by means of complex processes in the brain. Whereas here the grasping movement is under continuous corrective control of the eyes, in the very quick catching movements of the praying mantis the movement is not under the control of the eyes, as any corrective order given after the initial release of the action would be too late to have any effect, because of the speed with which the movement is carried out. The mechanism that regulates the orientation of the striking legs must be informed about the position of the fly in relation to the head as well as about the position of the head in relation to the body.

Before striking, the visual focusing on the target takes place in a very specific manner, as was established by H. Mittelstaedt (1953, 1954). The praying mantis first fixates the prey with the head and then strikes it by bringing forward its first pair of legs, which are folded under the prothorax and which are modified for catching prey. If necessary these forelegs can strike toward the side when the prey is not in the central plane of the prothorax.

The mantis has a cushion of sensory bristles ("neck organs") at the head joint (see Fig. 16-16) and the degree of bending of these bristles registers the degree by which the head is turned. If one cuts the nerve that comes from the left bristle pad, the animal will strike past the prey on the right for some time. Bilateral deafferentation does not produce a tendency to strike to the sides, but only prey is caught that is in a straight line with the prothorax. If the flies are situated to the right of this line the strike misses on the left and vice versa. Therefore the neck organs are involved in the aiming process.

It is possible that the information which determines the direction of the strike consists of the additive components of optical and proprioceptive information in approximately this manner: if the mantis fixates upon a fly positioned directly in front of it, the image of the fly is equally represented in both eyes, and the report from the neck organs is symmetrical; the catching legs strike out straight forward. If the fly is to the right of the mantis, the eyes again report symmetry, because the head is turned toward it, but in the neck organ the excitation coming from the right bristle pad is stronger. As a result the strike is directed to the right by a value corresponding to this difference in excitation (Fig. 16-16). If this hypothesis is correct, the orienting mechanism should function if

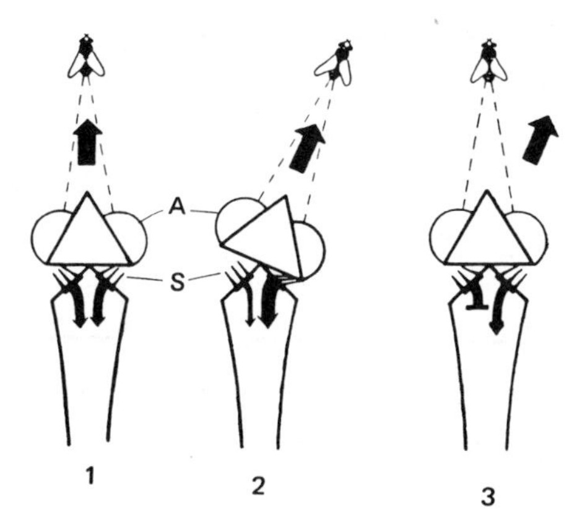

Figure 16-16 Schema to explain the functioning of the neck organs of a praying mantis. (*A*): eye; (*S*): sensory bristles; (*1*) and (*2*): intact animals; (*3*): left nerve cut; *arrow:* direction of strike. (From H. Mittelstaedt, 1953.)

the head of the mantis is fixed asymmetrically to the prothorax by a drop of glue. This forced position would be reported by the neck organs and the animal that was so treated could fixate the prey by compensating for the missing head movements by moving the body or the legs. From 60 to 80 percent of all strikes, however, miss in the opposite direction of the forced head position; it is as if the animal did not know that its head is at an angle to the body axis. Therefore this proposed hypothesis cannot be correct. It appears as if information coming from the receptors in the orienting mechanism is processed only when the head is freely moveable. Mantids whose entire proprioceptive afference of the neck region was removed fixated and struck at their prey in a well-coordinated fashion, so one can assume that the excitatory pattern of the optical center, which directs the neck muscles and thus determines the position of the head, also determines the direction of the strike. The bristle pads of the neck organs signal the actual head position, but this information does not seem to be used when aiming the striking legs but only in the control of the head-positioning musculature. The task of this mechanism is to make the head position, which has been effected by the eyes, independent of further external disturbances. The neck organs then seem to control the neck muscles (Fig. 16-17).

To illustrate this with one example: A mantis that fixates a fly positioned to its right orients its striking legs to the right according to the amount of effort that was required to obtain this head position, which is measured by the innervation on the right. To express it anthropomorphically: the praying mantis strikes in the direction in which it thinks it has turned its head. The knowledge of the actual head position, which is available through the neck organs, is not available to the localization apparatus but only to a lower motor center, which has the task of making the normal (zero) position of the head and the degree of its deviation independent of the mechanical strain

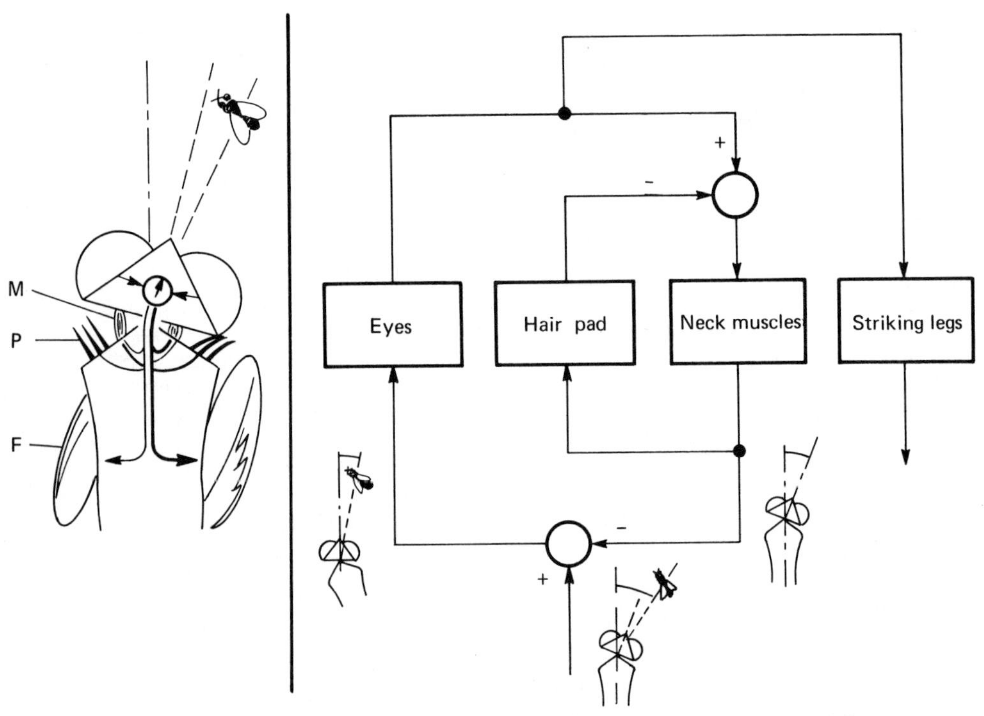

Figure 16-17 Functional organization of the aiming process in the praying mantis. *Left:* Overview of the organs involved: *M,* neck muscles; *P,* bristle pads; *F,* striking legs. *Right:* Abstracted schema. When aiming, the eyes give a turning command to the neck muscles, which turn the head and the fixed eyes in the direction of the target. The bristle pads, which are stimulated by this action, seem to counteract this fixating reaction, because they work toward a return of the head to its original position, which results in the head aiming a short distance behind the fly. The influence of these bristle pads is considered within the system, however. The task of the bristle-pad feedback system seems to be the nullification of external influences that might change the position of the head via an order to the neck muscles. (After H. Mittelstaedt from G. Wendler, 1966.)

on the neck muscles. This has been experimentally supported. Thus the head can be loaded with considerable rotational force (by attaching small weights) before the accuracy of the strikes is affected (H. Mittelstaedt, 1953:106).

SUMMARY

By means of their sensory organs animals maintain contact with their environment and orient themselves. They adopt oriented positions in space, orient toward directly perceived objects (such as their prey), and move in space using landmarks as reference points, or the sun (and for some even the earth's magnetic fields) as a compass.

The control of the body position and movement in space is by no means a passive affair—a rigid, solely stimulus-dependent, reaction—but the orienting mechanisms include active components, The reference value upon which body positions depend can be changed by the organism according to the motivational state. Movement in space could not take place at all if an animal were to be forced back into the normal position by its postural reflexes. A fish must also be able to tilt. It has been found that the postural reflexes are not simply blocked during active movement but regulated by a feedback control system which works on the principle of matching a copy of the efferent impulses, which is stored centrally, with the afferent feedback (reafference principle). By such feedback systems an animal can distinguish between sensory stimulation caused by its movement in the environment and the actual movements of its environment.

The mechanisms of distance orientation have been analyzed in a number of migratory animals. Salmon, for example, remember the odor of the river water in which they spent their larval stages and return to these places for spawning. Migratory birds orient by landmarks, the sun, the starry sky, and some even by help of the earth's magnetic field. The use of the sun as a compass in orientation demands as a prerequisite a central calculatory mechanism involving a internal clock that compensates for the movement of the sun. The general direction of the seasonal migration route is in part inborn in some species of birds. In other species this knowledge is passed on from the parent to the offspring by guidance during the first migration.

Some birds—for example, homing pigeons—demonstrate a capacity to return home from a place definitely unknown to them. This phenomenon of navigation is still an unresolved puzzle. It has not yet been discovered how the animals determine the geographical position of their home location and how they compare it with the location of release.

17

Temporal Factors in Behavior

Regularly recurring events, such as changes from day to night, high and low tides, changes of the moon and the seasons, and so on, are of the greatest importance to all organisms. Those who know animals are aware that different species are active at different times of the day or night. Some animals move about especially during the morning and evening hours, others are active during the day and sleep through the hours of darkness, and animals active at night rest during the day. This is as true for aquatic as it is for land animals (Fig. 17-1). Sometimes an animal changes from daytime to nighttime activity during the course of its development. The tortoises of the Galápagos Islands feed during the cooler evening and night hours when they are young. At that time they can eat even the dry grass because it is covered by dew. In contrast, young badgers play during the day in front of their den in the sun. Only gradually do they shift their activity toward dusk and night. This is accompanied by changes in their behavior. A young badger that until then has been trusting becomes shy during the day, but at night it is much less nervous (I. Eibl-Eibesfeldt, 1950a). W. Kühme (1966) observed a similar confidence in lions during the night in the wild; during the day they exhibited a much greater flight distance.

During the time of rest and sleep the locomotive activity generally comes to an end, the electric activity pattern of the brain changes, and the thresholds of the sensory organs are raised.

In man several levels of physiological activity, correlated with the depth of sleep, can be distinguished. The phases of shallow and half-deep sleep are normally ended within 30–60 minutes of the onset of sleep, to be followed by a phase of deep sleep that may last for another hour. During this phase the electroencephalographic pattern changes from the normal alpha-wave pattern to larger, slower wave patterns, interspersed with rapid pointed spikes. The period of deep sleep is followed by a dreaming phase, during which the brain-wave pattern returns to the alpha form and body movements and rapid eye movements

Figure 17-1 Fish fauna at a rocky shoreline in the southern Gulf of California (Baja California). Mexico. *Top:* at noon; *bottom:* around midnight on a moonless night. The species shown: (*1*), *Eupomacentrus rectifraenum:* (*2*), *Epinephelus labriformis;* (*3*), *Holocentrus suborbitalis;* (*4*), *Thalassoma lucasanum*, (5), *Abudefduf troscheli;* (6), *Runula azalea;* (7), *Myripristris occidentalis;* (8), *Microlepidotus inornatus;* (9), *Bodianus diplotaenia;* (*10*), *Scarus californiensis;* (*11*), *Balistes verres;* (*12*), *Rypticus bicolor;* (*13*), *Chromis atrilobata;* (*14*), *Prionurus punctatus;* (*15*), *Heniochus nigrirostris;* (*16*), *Pareques viola;* (*17*), *Apogon retrosella;* (*18*), *Lutianus argentiventris;* (*19*), *Anisotremus interruptus;* (*20*), *Haemulon sexfasciatum;* (*21*), *Mycteroperca rosacea.* (From E. S. Hobson, 1965.)

are observed. If we wake the sleeper during this period of REM (rapid eye movement) sleep, he may report dreaming. The dreaming phase may start 90 minutes after the onset of sleep and last for 30–60 minutes. The depth of sleep fluctuates following this pattern three to four times a night, and toward the morning hours the sleep becomes shallower. There are, however, individual differences. While most people reach the stage of deep sleep fairly soon after onset, some reach this phase much later toward morning.

Animals often adopt specific sleeping postures, and a number of activities continue during sleep—for example, ruminating. M. Holzapfel (1940) pointed to the fact that specific motivating mechanisms underlie sleep behavior, causing an appetitive behavior for sleeping. The sleepy animal searches for a sleeping place and performs a number of preparatory acts, specific for the species (E. S. Hobson, 1972). E. v. Holst and U. v. Saint Paul (1960) induced sleeping in the chicken through electrical stimulation of certain points in the hypothalamus. W. R. Hess (1954) in the same manner released sleeping in the cat. For further references on the physiology and ethology of sleep see G. Tembrock (1964), J. Altmann (1966), and W. Baust (1970).

Why animals sleep is a matter of dispute. It has been suggested that sleep is needed for regenerative processes to occur, especially within the central nervous system. It might well be, however, that other selective pressures were the prime cause for the development of sleeping. One could consider it as advantageous for a species to specialize with its sense organs and its behavioral repertoire in either day or night activity. Such a one-sided specialist is perhaps better kept passive during the time of the day for which it is not adapted. This would prevent unnecessary exposure to danger and spare energy.

By specializing in activity at different times, day-active and night-active animals fill various ecological niches (for example, birds of prey). Many animals that live in tidal zones must seek shelter for several hours before and during the time when their home range is without water to protect themselves against desiccation. The grunion of the California coast (*Leuresthes tenius*) must be ready to spawn at high tide, because they bury their eggs in the sand near the highest line reached by the

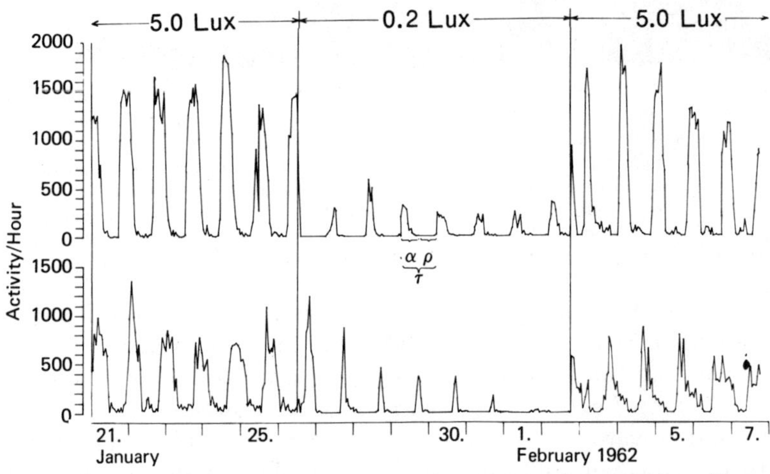

Figure 17-2 Activity periods of two chaffinches (*Fringilla coelebs*) under constant conditions during constant illuminations of 5, 0.2, and 5 lux. Period (activity time, α, plus resting time, ρ) is indicated by τ. (From J. Aschoff and R. Wever, 1962a.)

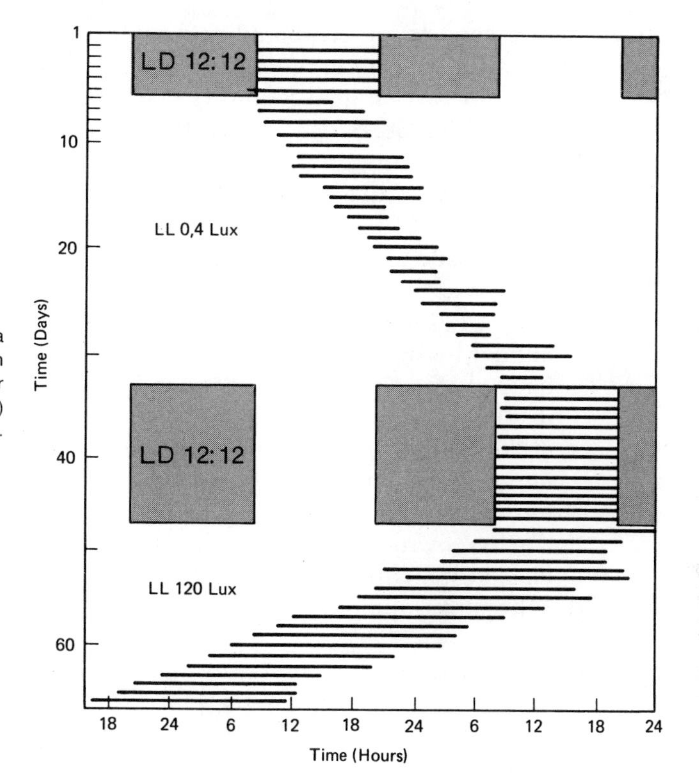

Figure 17-3 Activity periods of a chaffinch in constant illumination *LL* of different intensities and under periodic light–dark alternation (*LD*) of 12 hours light and 12 hours dark. (From J. Aschoff, 1965.)

water. Animals that are active during the day must go to their resting place for the night before it is too late; frogs must be ready to mate when the snows melt; in short, these changes must not find the animals unprepared. As we now know, many animal species are adapted in their endogenous activity cycles to the periodic changes in their environment. The 24-hour rhythms have been especially well studied (J. Aschoff, 1962, 1964, 1965; J. Aschoff and R. Wever, 1962a). If one registers the activity of animals in experimental cages one generally observes a distinct 24-hour periodicity. Usually it is accurately synchronized with the normal day–night cycle. If the animals are kept under constant conditions of continuous light or darkness at the same temperatures, they continue to show periodic activity, but the length of the periods deviate somewhat from the normal 24-hour periodicity. This proves that the periodicity is endogenous and is not induced by environmental factors.

An increase in the illumination determines not only the frequency but also the total amount of activity and the relationship of the activity time to the period of rest. With chaffinches, for example, the length of the periods decreases with increase in light intensity, but at the same time there is an increase in periods of activity and in the amount of activity (Figs. 17-2 and 17-3).

The circadian rhythm is apparently inborn in many species. Chicks that have been incubated and maintained under constant conditions show the rhythm in the same way as do lizards and mice that have been bred for several generations (J. Aschoff and J. Meyer-Lohmann, 1954; J. Aschoff, 1955b; K. Hoffmann, 1959; see also Fig. 17-4). Lizards that had been hatched in an incubator under temperature and light periods that corresponded to a day length of 16 and 36 hours, respectively, exhibited the normal 24-hour rhythm when they were tested under constant conditions, just as did control animals that had been kept under a 24-hour rhythm after they had hatched (K. Hoffmann, 1959).

An especially elegant piece of evidence for innate periodicity was offered by R. J. Konopka and S. Benzer (1971), who isolated three mutants of *Drosophila melanogaster.* One of these was arrhythmic, another displayed a short periodicity of 19 hours, and the third a long periodicity of 28 hours (Fig. 17-5). With special techniques it is possible to create mosaic individuals in *Drosophila,* which are composed of parts with different genetic material. A fly with a mutant head runs on a mutant rhythm, even if the rest of the body is normal. This indicates that the clock responsible for the circadian rhythm is associated with the head (R. J. Konopka, quoted by S. Benzer, 1971).

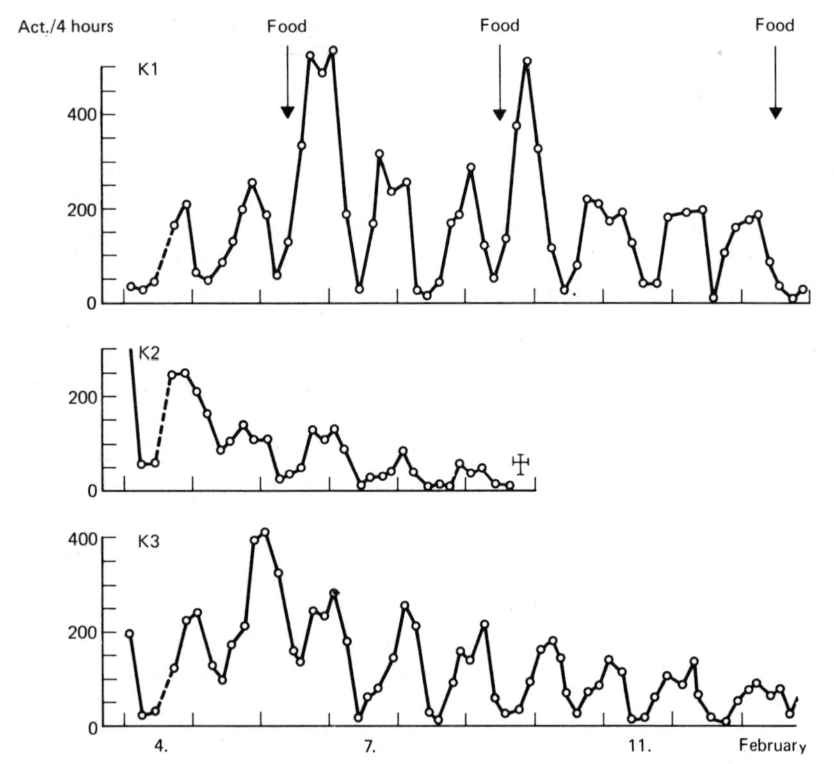

Figure 17-4 Activity periods of chicks under constant illumination. (From J. Aschoff and J. Meyer-Lohmann, 1954.)

Larvae of the marine fly *Clunio marinus* dwell in the lower tidal zones that lie free only during low spring tides, for about two hours each time. At this time the wingless female hatches from her pupa and is immediately fertilized by the male. These especially low tides occur shortly after full moon and new moon, and the pupae hatch only during springtide. Although during this period the lower tidal zone lies uncovered in the morning as well as evening, the pupae hatch only in the evening. This shows that the tides do not determine the time of hatching (Fig. 17-6). Larvae kept in a laboratory with an artificial day of light and dark periods always hatch towards the end of a light period. The rhythm is circadian and follows a shift in the light–dark cycle. The spring tide occurs

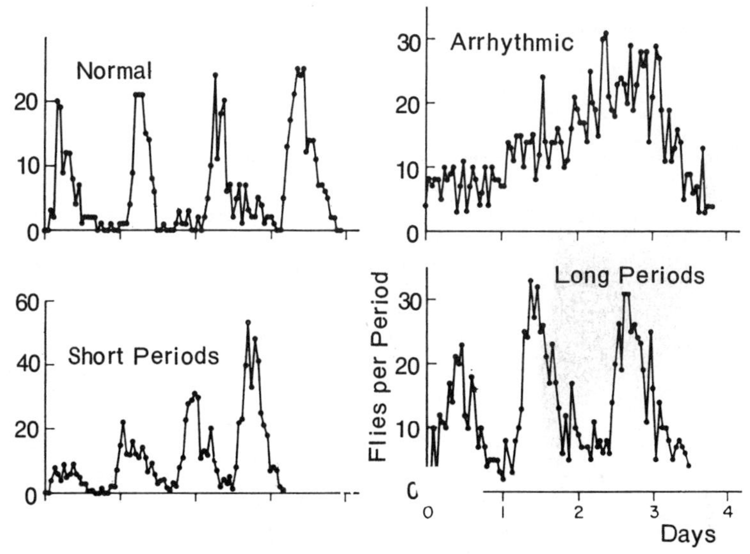

Figure 17-5 Hatching rhythm under conditions of constant darkness for normal *Drosophila* and their mutants. Before this observation, the populations were exposed to a periodic light–dark alternation of 12 hours light and 12 hours darkness. (From J. Konopka and S. Benzer, 1971.)

on the same date on both the North Sea and the North Atlantic coasts, yet the hour of the day at which ebb tide occurs differs from place to place and with it the hatching rhythm of the local populations. D. Neumann (1966) observed flies of different populations while holding conditions constant, and discovered that the variations were based on local genetic adaptations. When he crossbred populations whose time of hatching did not normally overlap, the resulting F_1 generation had an intermediate hatching time. With daily light–dark alternation, no half-monthly rhythm was evidenced. If, on the other hand, Neumann allowed a weak light of 0.4 lux on 4 to 6 nights in 30-day intervals, a semilunar rhythm was induced. A single period of artificial moonlight sufficed to keep such a

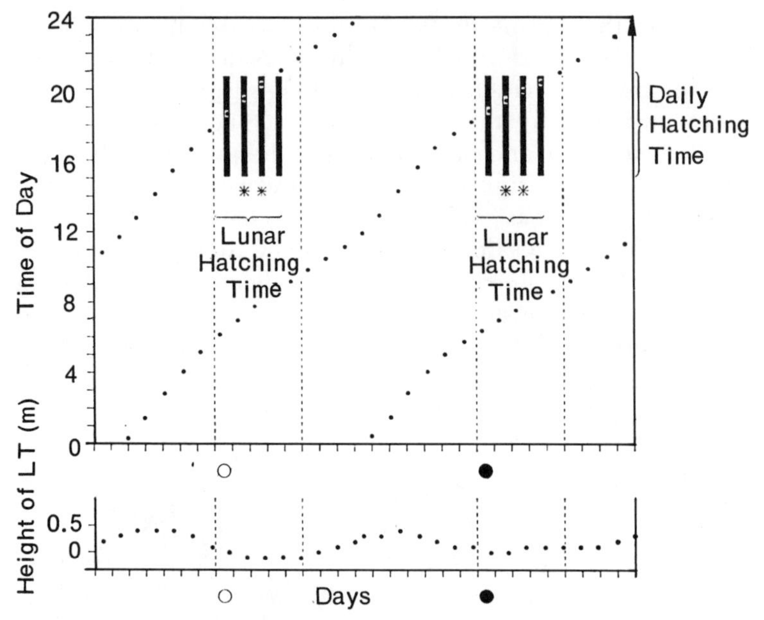

Figure 17-6 Hatching times of the fly *Clunio marinus* and conditions of low tide on the island of Helgoland in August 1960. *Above:* Times of low tide (*dots*) and hatching (*bars*). Days of increased hatching are indicated with asterisks. *Below:* Height of water above zero of the second of the daily low tides (*LT*) during the same month. (From D. Neumann, 1966.)

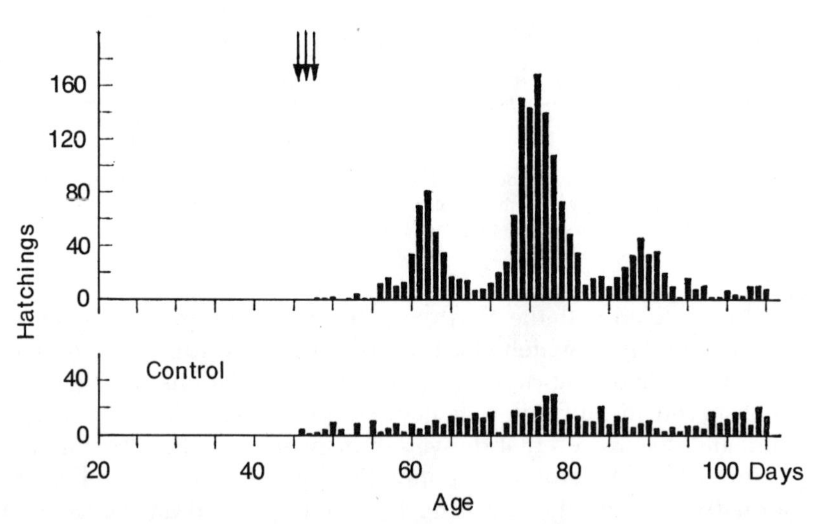

Figure 17-7 Hatching of the fly *Clun'ᵒ mannus* of the Por stock (Normandy) after a single treatment with artifical moonlight (3 successive nights, see arrows): summation diagram of six Petri dishes with cultures from three experimental series (light–dark 16:8, 20°C). *Below:* Control without artificial moonlight. (From D. Neumann, 1966.)

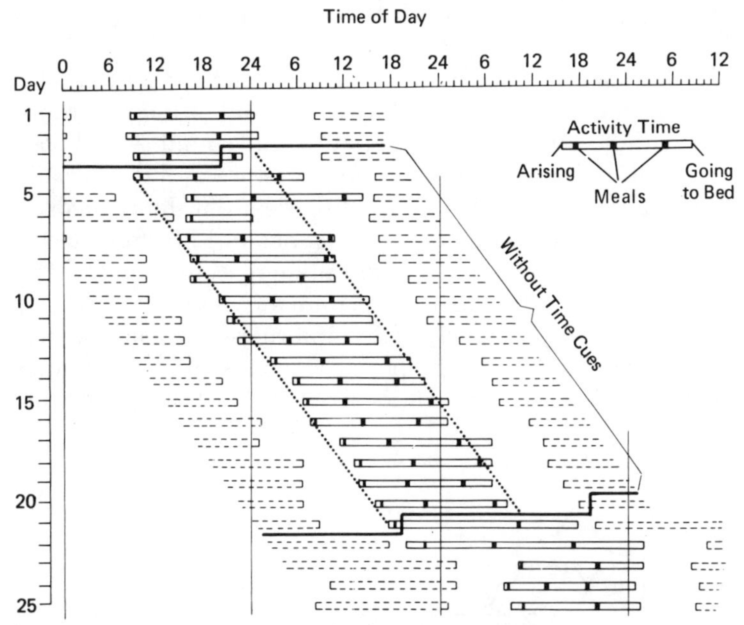

Figure 17-8 Periodic behavior of an experimental subject (human) in a bunker that was completely isolated from the outside world. The dates on the left indicate the beginning of a waking period (heavy horizontal bars). (From J. Aschoff and R. Wever, 1962b.)

cycle going for three periods, indicating that the induced rhythm was endogenous (Fig. 17-7).

If a human being is kept isolated from all environmental influences in a subterranean bunker, one can also observe a spontaneous rhythm (J. Aschoff and R. Wever, 1962b; J. Aschoff, 1966). Man's periodicity is also circadian—that is, it deviates only slightly from the normal 24-hour periodicity—which proves its endogenous origin (Fig. 17-8). It was also found that various physiological processes have their own distinct circadian rhythmicity, each with a slightly different frequency, which diverge from each other as time passes (Fig. 17-9). Beyond this 24-hour rhythm a 7-day rhythm (circasepton) was demonstrated (F. Halberg et al., 1965).

Babies exhibit a 4-hour rhythm, awaking approximately every fourth hour for feeding (Hellbrügge, 1967). The periods of sleep during the night lengthen as a part of maturation, by leaving out first one, and later two meals (M. Morath, according to B. Hassenstein, 1973).

At birth infants do not yet display a 24-hour periodicity; it develops during the first few weeks of life. Initially a circadian periodicity of 25 hours develops, which only gradually becomes synchronized to a 24-hour rhythm with maturation of the sense organs. This observation supports the assumption that here we are dealing with maturational processes, and that the periodicity must therefore be innate (T. Hellbrügge, 1967).

The internal circadian rhythm is synchronized by external stimuli termed *Zeitgeber,* time setters, with the rhythm of the environment at large such as light, humidity, temperature, and sound (Fig. 17-9). Some lizards can still be synchronized by temperature cycles of small deviations. If the deviation is 1.6° C, then 75 percent of the animals are still fully synchronized; at 0.9°C about 25 percent are (K. Hoffmann, 1968). In man even a weak electromagnetic field influences the circadian periodicity. When a field of 10 cycles per second was switched on, the period became shorter than previously. The "internal desynchronization phenomenon," in which the activity period becomes abnormally increased to 30 to 40 hours while the continuously monitored vegetative functions

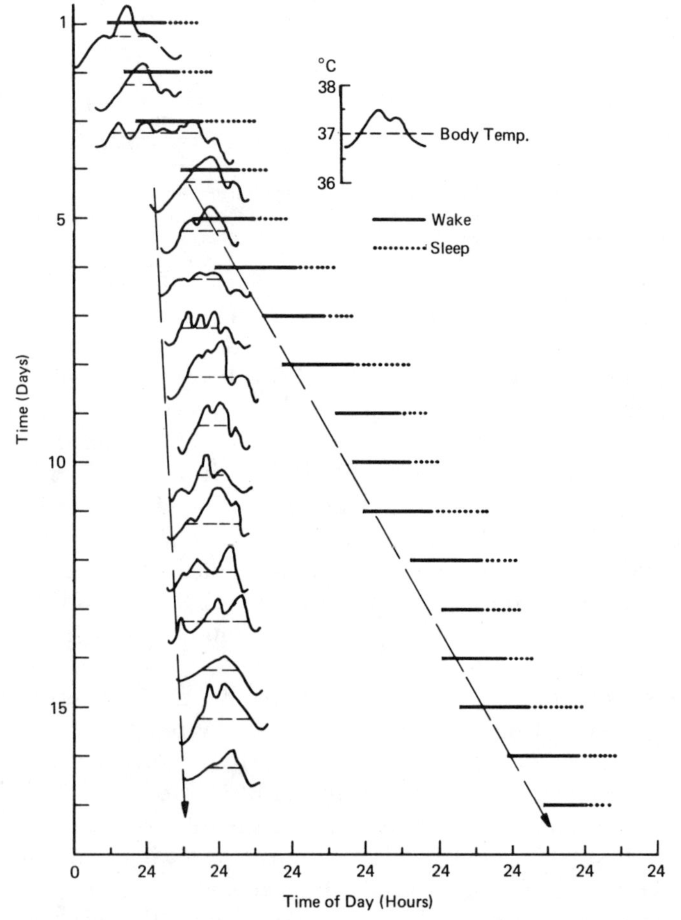

Figure 17-9 Desynchronization of the circadian rhythms of a subject maintained isolated and without a clock. Horizontal bars and dotted lines, waking and sleeping periods; dashed line, 37°C of body temperature. (From J. Aschoff, 1966.)

continue in a 25 to 26-hour periodicity, can be observed only when the magnetic field is switched off. Finally, it can be demonstrated that this field *Zeitgeber* is capable of maintaining the internal rhythm in nearly perfect synchrony as long as the timing of the *Zeitgeber* falls close to the time of spontaneous onset of the activity phase (R. Wever, 1968). In man reciprocal social effects lead to synchronization of circadian rhythms (E. Pöppel, 1968). Further references on *Zeitgeber* can be found in K. Hoffmann (1969, 1970).

Circadian periodicity has not only been described for intact organisms – from unicellular animals to man – but has also been demonstrated in isolated organs and tissues (E. Bünning, 1963).

Upon removal of the pineal organ the activity of sparrows that are kept under constant conditions becomes arrhythmic, but not if there is a change of day and night in illumination (S. Gaston and M. Menaker, 1968).

In marine animals a distinct lunar periodicity has been demonstrated in several instances. Monthly and 14-day reproductive cycles are known (P. Korringa, 1957). We already mentioned the grunion (*Leuresthes*), which spawns precisely at high tide along the California coast (B. W. Walker, 1952). Another well-known example is given by the palolo worm (*Eunice viridis*) (H. Caspers, 1961). In the closely related polychaete *Platynereis dumerilii* this monthly reproductive cycle is maintained under constant conditions for at least two cycles (C. Hauenschild, 1960). More recently the creek planarian, a freshwater animal, has been found to have lunar-periodic fluctuations in its light preference (E. May and G. Birukow, 1966).

Animals that live in the tidal zone show an activity rhythm that corresponds to the daily rhythms of the tides. The isopod *Excirolana chiltoni,* when exposed to constant conditions without a tidal rhythm, displays a periodicity in daily swinging activity which corresponds to the tidal rhythm. The free-running period lasted 24 hours and 55 min. and was roughly five minutes longer than the normal tidal rhythm. A monthly rhythm was superimposed on this rhythm, influencing the total activity for each activity phase (J. T. Enright, 1972). K. S. Rao (1954) observed in mussels (*Mytilus edulis* and *M. californicus*) rhythmic fluctuations in the rate of water propulsion, corresponding to the tides. The rhythms were maintained for weeks in the laboratory without deviating noticeably from the tidal rhythms. According to F. A. Brown, Jr. (1965) this points to the involvement of a still unknown *Zeitgeber,* because an endogenous rhythm of such precision is hard to imagine. J. T. Enright (1963) has expressed doubts about this interpretation. E. Naylor (1958) discovered a very precise tidal rhythm in the running activity of the crab *Carcinus maenas.* Under constant conditions this rhythm is lost by the sixth day. The sand beach hopper *Synchelidium* exhibits an activity rhythm of swimming and digging in at the sandy coasts of California that corre-

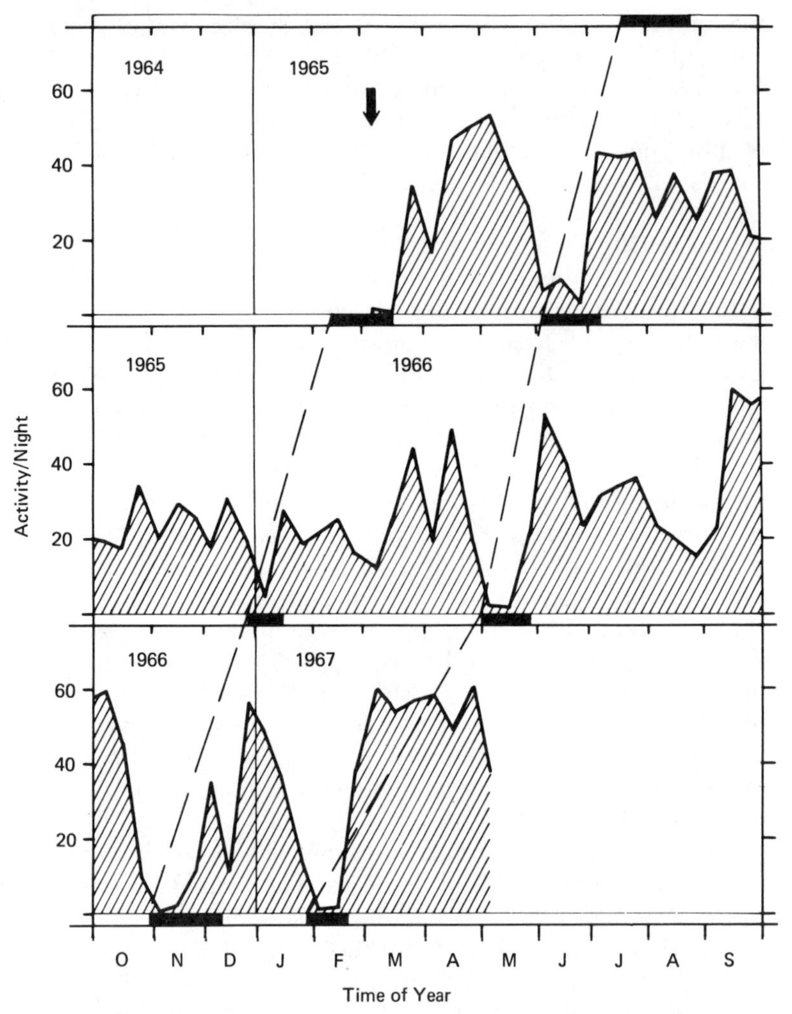

Figure 17-10 Molting time and nightly migratory restlessness of a willow warbler that lived for 27 months at 21°C (± 2°C) in an artificial 12:12 hour day (200:0.2 lux). Ordinate: number of 10-minute intervals in which the bird was active per night, averaged over one third of a month. *Black lines,* molt. After leaving the nest in June 1964 the bird lived first in a heated room with a natural light–dark cycle and then early in March was placed in the experimental conditions (*arrow*). (E. Gwinner, 1967.)

sponds exactly to the local tidal rhythm. In captivity the rhythm is maintained for a few days with decreasing precision. The turbulence of the water is thought to be the *Zeitgeber* (J. T. Enright, 1963).

The yearly rhythms in behavior are not exclusively exogenous. E. T. Pengelley and K. Fisher (1963) kept golden mantled ground squirrels (*Citellus lateralis tescorum*) under constant conditions. The animals still exhibited a physiological cycle of approximately one year, which manifested itself in the uptake of food and periodic onsets of hibernation.

Willow warblers (*Phylloscopus trochilus*) molt twice a year and exhibit migratory restlessness during the spring and fall. Willow warblers that were kept for 27 months on a 24-hour day (12 hours of light, 12 hours of darkness) retained this cycle; it was no longer synchronized exactly with the yearly seasons, but shifted from year to year, as can be expected from a free-running, endogenous, circannual rhythm (E. Gwinner, 1967; see also Fig. 17-10). We are indebted to J. Aschoff (1962) for a comprehensive review of the annual periodicity of activity. For a more general review of biological periodicity, see H. Remmert (1965). The comparison between animal and plant rhythms has been discussed by E. Bünning (1963).

SUMMARY

The activity of animals, organs, and even isolated cells follows a pattern of rhythmicity. The best-studied are the 24-hour rhythms. It has been found that many animals are driven by an inborn circadian (approximately 24-hour) rhythm which, by an external *Zeitgeber* (synchronizer), is kept in phase with the normal 24-hour day-and-night cycle. In man different physiological systems have been found to have their own circadian rhythms. Since their periods are not exactly of the same length they asynchronize if the subject is kept without external *Zeitgeber* under constant conditions.

Rhythms other than circadian are also found. Variations in tidal rhythms that accord with the location were found to be genetically determined in the marine fly *Clunio*. In some animals a yearly rhythmicity was exhibited by animals kept under constant conditions. The physiological nature of the internal clocks is not known. Experiments with *Drosophila* have shown that the clock is located in the head.

18

The Ethology of Man

Since the time of Darwin we have known that one key to the understanding of man's behavior lies in appreciation of the fact of his phylogenetic evolution. On various occasions I have mentioned that behavior mechanisms, owing their adaptiveness to evolutionary processes, play a part in human behavior as well as infrahuman behavior. This is acknowledged by many students of human behavior (R. Bilz, 1940, 1944; E. W. Count, 1959; B. Berelson and G. A. Steiner, 1964; L. Tiger, 1969; A. H. Esser, 1971; and P. Ekman, W. V. Friesen and P. Ellswerth, 1972, among others). In particular, ethological ideas have influenced the study of child behavior (J. Bowlby, 1969; S. J. Hutt and C. Hutt, 1970; N. Blurton Jones, 1972; and W. C. McGrew, 1972). E. A. Tinbergen and N. Tinbergen (1972) interpreted on ethological grounds the behavior of autistic children as a perpetuated fear response. These children are so sensitive that being looked at—which, as we know, contains a flight-provoking element—releases extreme fear, which cannot be overcome by the friendly smile that accompanies the looking. The authors propose ways to make contact with the children without visual encounter to overcome their fear. A very detailed ethology of the child has been presented by B. Hassenstein (1973). He discusses normal development and points out that emotional disturbances may result from neglect of certain aspects of human nature—for example, by lack of response to the interactors' aggression of the social-exploration type.

But at least an equal number of workers refuse—more or less emphatically—to accept ethology. E. H. Lenneberg's description (1967:393) of the situation is appropriate:

> There was a time when "innateness" was on the index of forbidden concepts. Much has changed in the official censorship of technical terms, but there are still many scientists who regard the postulation of anything innate as a clever parlor trick that alleviates the proponent from performing "truly scientific" investigations. This position is odd to say the least. Organisms are links in a chain of reaction called *life*. All living forms derive from this event and carry within them its principle; life itself is an innate principle of organisms. At present, biology does no more than discover how various forms are innately constituted, and this includes descriptions of a creature's reactions to environmental forces. Research into these reactions does not eventually free us from the postulation of innate features but merely elucidates the exact nature of innate constitutions. The discovery and description of innate mechanisms is a thoroughly empirical procedure and is an integral part of modern scientific inquiry.

And yet the rejection of ethological concepts is widespread even today, and occasionally manifests itself in rather extreme statements. An example is the opinion of M. F. A. Montagu (1968:11):

> There is, in fact, not the slightest evidence or ground for assuming that the alleged "phylogenetically adapted instinctive" behavior of other animals is in any way relevant to the discussion of the motive forces of human behavior. The fact is, that with the exception of the instinctoid reactions in infants to sudden withdrawals of support and to sudden loud noises, the human being is entirely instinctless.

Assertions of this kind are supported by general statements only (see, for example, the statement of W. LaBarre, p. 455). Lack of facts is compensated for by the pungency of the antibiological expression of opinion. Rather than being challenged by objective arguments we encounter emotional involvement of a kind hardly beneficial to the study of man. Frequently, the focus of attention is the man–beast comparison, which is attacked with an almost machinelike monotony. Biologists are accused of tabooed or premature equalization of animals and human beings. Especial doubts are expressed with regard to the basic comparability of human cultural achievements—such as custom—with animal behavior. Repeatedly we are reproached for a frivolous analogizing (for example, W. Schmidbauer, 1971a, b). On closer scrutiny, however, we often discover that our critics display an unfortunate lack of clarity in their notions of comparative ethological methods.

Although I have already discussed our methodology, it seems appropriate at this time to summarize briefly what is to be learned from the comparative study of behavior. First, since organisms are related at different levels, we have an opportunity for the meticulous use of homology in uncovering real traits of affinity. One example would be the clarification of the relationship of laughter and smiling with the "play face" or "**silent bared-teeth display**" observed in higher monkeys. It is an important task of comparative research to establish such phylogenetic connections; but it is far from the most important one. It is little known or understood that the study of analogies also provides us with a basic insight, and that this contributes to an understanding of human behavior. There is something to be learned, indeed, from fish and graylag geese—on some issues more than from our nearest relatives. Should we, for example, pose questions concerning conjugal behavior or the phenomenon of group aggression, we would be ill-advised to study our nearest relatives. They have become adapted to other environmental conditions and have failed to develop either conjugality or patterns of group aggression. The biological principles underlying such behavior could be discovered, however, through research with other species. In studying many different species that display conjugality or group aggression, we may learn the conditions under which these behavior patterns develop and with what other characteristics they are correlated. For this purpose we can compare inborn movement patterns or dispositions of animals with cultural ones in man. The study of **analogies** provides insight into the laws determined by function, and thus makes it possible to formulate principles or laws independent of phylogenetic relationships. The broader the basis for induction, the greater is the general validity of these laws. In this way, we derive from the study of animal behavior working hypotheses whose scientific value for the understanding of human behavior can of course be tested only in research on man himself.

In this book I have presented the facts that compel us to accept the existence of phylogenetic adaptations in the behavior of animals, and we have demonstrated that these adaptations are present in the form of fixed action patterns, internal motivating mechanisms, innate releasing mechanisms, releasers, and innate learning dispositions. We are now ready to examine to what extent similar adaptations also preprogram human behavior and to what degree they are still adaptive today, in the sense that they function in the service of the preservation of the species.

The quest for the phylogenetic determinants of human behavior is certainly not the sole task of human ethology. The problem of why we behave the way we do concerns cultural patterns too, and by asking this question as biologists we investigate the functional aspects of behavior. We want to learn in what ways a particular pattern contributes to the survival of the species and we want to learn the laws of function by which cultural patterns are shaped. Cultural evolution seems to follow the same basic rules that can be observed in biological evolution.

The ethological approach helps us to understand cultural rituals in a new way For example, my studies on greeting, festive behavior, courting, and flirtation reveal the existence of universal rules after which these rituals are shaped (I. Eibl-Eibesfeldt, 1973). One could speak of a grammar of ritualistic behavior which applies to all men. To study the rules by which certain behavioral elements are integrated in larger units is yet another main concern of human ethology. I want to emphasize this, since some authors seem to see human ethology as a discipline restricted to the study of the inborn in man. When we first used the term *Humanethologie* (I. Eibl-Eibesfeldt and H. Hass, 1966), we might have created that impression, since we started our program by searching for phylogenetic adaptations. But we have emphasized repeatedly that cultural behavior, including verbal behavior, is also the subject of our studies. I pointed, for example, to the fact that people certainly speak different languages, but that what they say in particular situations is in principle always the same, and I spoke of verbal clichés by which people appeal to their partner (I. Eibl-Eibesfeldt, 1973b). The rules by which they are applied and the rules by which a festival or other ritual is structured are the subjects of our studies as well. And human ethology certainly will not stop here. The recent monograph of K. Lorenz (1973) on the natural history of cognitive processes demonstrates the profound impact of ethology on epistemology. The achievements as well as the pitfalls of our ways of thinking appear indeed in a new perspective when we apply the phylogenetically oriented quest for the selective forces that have shaped the instrument of cognition.

FIXED ACTION PATTERNS AND THEIR RELEASE IN INFANTS

The newborn human being is equipped with a number of functional behavior patterns (A. Peiper, 1951, 1953, 1961; P. Mussen, 1970). In the main these are capacities located in the brain stem and the spinal cord; the cerebral cortex is at that time not functional in any real sense. Anencephalic children differ little in their behavior from that of healthy children during the first two months of their lives, although they lack a cerebral cortex (E. Gamper, 1926; M. Monnier and H. Willi, 1953).

Some behavior patterns, which serve the function of food intake, are phylogenetically quite old, and the human infant shares them with many other mammals. First I should mention the rhythmic searching movements for the nipple, a turning of the head left and right, which may occur spontaneously or following a touch on the mouth region (H. F. R. Prechtl and W. M. Schleidt, 1950). The seeking behavior ends when the infant gets the nipple into the mouth and when the lips close firmly around it. This rhythmic seeking for the breast is observed only during the first days after birth (Fig. 18-1). It is soon replaced by an oriented search for the breast: when the mouth region is touched the infant turns

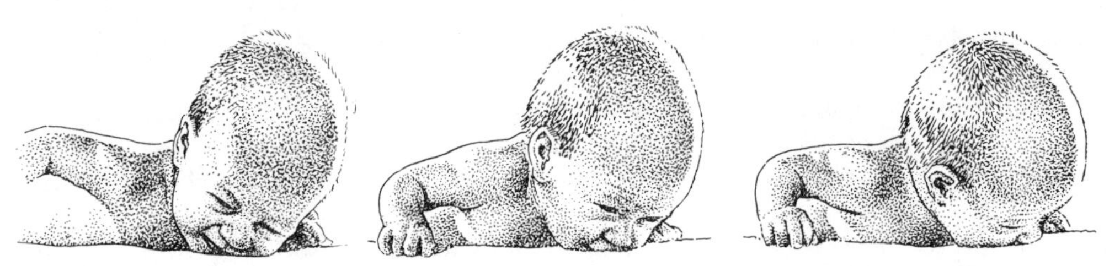

Figure 18-1 Rhythmic search for the breast (searching automatism). (After H. F. R. Prechtl, 1953c.)

toward the stimulus object, orienting in space so that he or she can get hold of it. This spatially oriented movement at first still has a rhythmic component that is soon lost (H. F. R. Prechtl, 1958). The motor patterns of sucking also change within the first few weeks of life. Initially the lips close firmly around the nipple area and suction is produced by a partial vacuum in the mouth cavity (pump sucking); later the tongue alone does the work involved in sucking by pressing the nipple against the roof of the mouth. During this lick sucking the corners of the mouth remain open.

A characteristic reaction of the newborn infant is the grasping reflex of the hand. If one touches the palm of the infant's hand the fingers close firmly around the object and, as H. F. R. Prechtl (1955) has shown by motion picture analysis, in an ordered sequence of finger movements (Fig. 18-2). This reflex grasping is especially strong during sucking. Quantitative investigations show that children react especially to hair. Undoubtedly the grasping reflex originally served the purpose of holding on to the mother's fur. This reflex is often considered a rudiment because man no longer possesses fur and therefore the reflex is thought to be no longer functional. The behavior does not seem to have completely lost its function, however; one can observe how small infants sleep close to their mother's body and how they hold on to her clothing. The hand-grasp reflex is so strongly developed in premature babies that they are able to hang on to a stretched-out clothesline. This capacity is lost later, which is an indication of the beginning of rudimentation (Fig. 18-3).

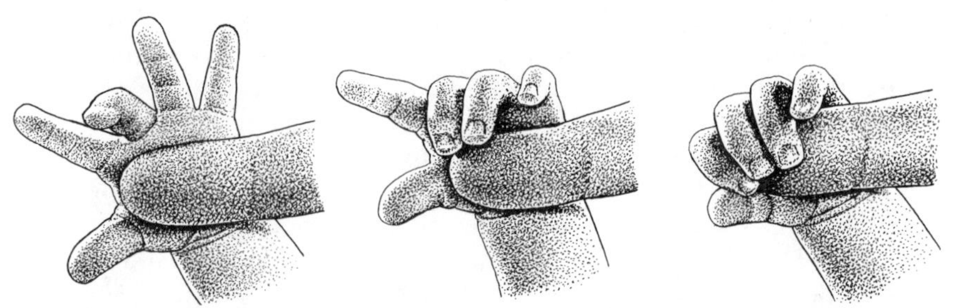

Figure 18-2 Grasping reflex of the human infant. At first the middle finger closes; the others follow, and the thumb is last. (After H. F. R. Prechtl, 1953c.)

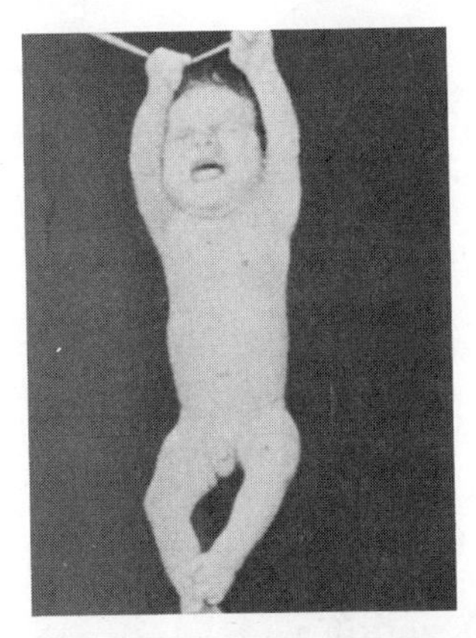

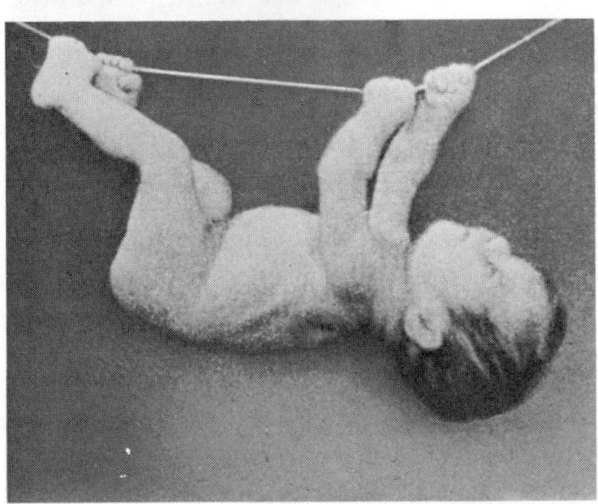

Figure 18-3 Hanging by the hands and with the toes in a premature human child (7 months). (Photographs: A. Peiper, 1961.)

Figure 18-4 Walking newborn baby. (Photograph: A. Peiper.)

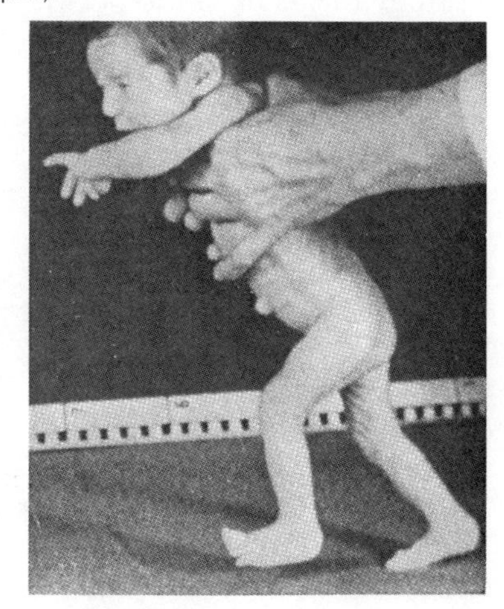

The climbing movements that can be seen in premature infants seem to be definite rudiments: placed on their backs they perform alternating, well-coordinated arm and hand movements. One arm is moved downward with a closed hand, the other moves upward while the hand opens slowly.

Swimming movements can be released in infants that are a few weeks old by placing them into the water in a prone position and merely holding them up at their chin. They paddle in a coordinated fashion with their hands and legs. The behavior disappears at three to four months.

One can also release walking and crawling behavior in the newborn child. A newborn infant on its stomach will commence to perform crawling movements by moving the diagonally opposed limbs (*Kreuzgang*). If one supports the infant and places the feet onto a firm surface it will begin to walk and place one leg before the other (A. Peiper, 1953; H. F. R. Prechtl, 1955; see also Fig. 18-4). In addition, one can observe a number of movements in infants that serve to protect the body. Evidence for innate releasing mechanisms in very small babies is presented on page 488.

As examples of expressive behavior in newborn infants I can offer crying and smiling. The first is a kind of "**lost call**"; a child can be easily quieted by picking it up or by imitating the presence of the mother with appropriate models. The primary function of the **smile** seems to be to appease. According to the legend, Cypselus, who later became the ruler of Corinth, was spared by those ordered to kill him while still a baby when he smiled at them. It is a fact that the smile releases delight in the mother, even those who initially were indifferent, and aids in the establishment of a strong emotional tie. The time of its first occurrence varies. Sometimes it can be observed in newborn or even prematurely born infants. The smile occurs spontaneously during sleep and also after drinking, diapering, and passing of wind (O. Koehler, 1954a; J. A. Ambrose, 1961). Laughing and cries of delight with a widely opened mouth mature around the fourth month of life.

> Laughing as well as joyful shouts appear at a time when the laughing of adults does not facilitate the same behavior in the baby but startles it more than anything else, or can even cause the baby to cry when it has been laughing. The old imitation hypothesis does not hold up very well here (D. W. Ploog, 1964a:321).

The baby, by smiling and fixating (unconsciously at first) strengthens the bond to the mother. K. S. Robson (1967) emphasized that visual contact is at the base of human sociability. Mothers react very strongly when the child starts to look at them at approximately four to five weeks of age. It seems rewarding to them when their babies smile.[1] It is quite remarkable in this context that the visual fixating process occurs even in **blind-born infants.**

During the first three months of life the children of deaf-mute parents are indistinguishable from those whose parents do not have this impediment (E. H. Lenneberg et al., 1965). We have already mentioned that **deaf-born children** begin to babble.

The initially spontaneous smile is later superseded by an answering smile. The former is frequently contrasted as a "grimace" to the later "genuine" smile. The latter is said to exist only when there exists a mutual relationship, that is, when the smile is a response to the smile of another person (A. Nitschke, 1953). This type of distinction is a rather ar-

[1] "The human mother is subject to an extended, exceedingly trying and often unrewarding period of caring for her infant. Her neonate has a remarkably limited repertoire with which to sustain her. Indeed, his total helplessness, crying, elimination behavior and physical appearance, frequently elicit aversive reactions. Thus, in dealing with the human species, nature has been wise in making both eye-to-eye contact and the social smile, that often releases in these early months, behaviors that at this stage of development generally foster positive maternal feelings and a sense of payment for services rendered. . . . Hence, though a mother's response to these achievements may be an illusion, from an evolutionary point of view it is an illusion with survival value" (K. S. Robson 1967:15).

tificial break in a continuous maturational process. This can be clearly demonstrated in the answering smile, which matures quite independently of the mimic expressions of the partner and becomes a personal greeting only very much later. R. A. Spitz and K. M. Wolf (1946) were able to release a smile in three- to six-month-old children by presenting them scarecrow faces and distorted grimaces as well as a normal human face. Within this wide spectrum everything was smiled at that was placed over the bed. R. Ahrens (1953) followed up the development of the recognition of mimic expressions. Until the onset of the second month eye-sized, well-defined, contrasting spots on a square or round two-dimensional plane, representing a cardboard model of a head, release smiling better than a painted face or a rectangular bar on the same background. It makes no difference whether the pair of dots is presented in a parallel or vertical position or whether three pairs of dots are shown. One dot alone, on the other hand, is ineffective.

Around the second month of life dots presented in a horizontal plane in front of the infant's face are more effective than when presented vertically, and soon the child pays attention to the entire area around the eyes but not to the lower part of the face. This is included gradually toward the third month. At four months of age the child reacts to the movements of the mouth, without differentiating all details; it is not until the fifth month that the broadening of the mouth specifically releases smiling, and this is especially true for the six-month-old child. The effectiveness of models then decreases. The child clearly distinguishes between models and faces of adults, but it does not understand the mimic expressions of smiling until it is seven to eight months old, when it reacts appropriately to a laughing person.

According to Ahrens the mimic expressions of the forehead were not adequately responded to until the children in the study were 14 months old. Vertical threat wrinkles frightened the institutionalized children; they turned away, moved away, and cried or screamed. Ahrens emphasized that the children had hardly ever seen threatening mimic expressions prior to these tests. The facial expressions following the perception of sweet, sour, and bitter tastes are present in the newborn. Similar gustofacial responses were observed in anencephalic neonates and the movement pattern therefore can be considered as being inborn (J. E. Steiner and R. Horner, 1972).

While the releasing mechanisms, expressive, and locomotor movements continue to mature and are increasingly integrated with individual experience, other behavior patterns become superimposed during the course of development and are exhibited only under special circumstances, for example, in **degenerative processes of the central nervous system.** Some early-childhood behavior patterns are also taken over into the repertoire of expressive movements of adults.

BEHAVIOR OF CHILDREN BORN BLIND OR DEAF-BLIND

With respect to the question of innate components in human behavior the behavior of children born blind or deaf-blind is most informative. We have here the accidental experiments of nature that can be considered as **deprivation** (Kaspar-Hauser) **experiments.** From this point of view J. Thompson (1941) studied the expressions of blind and blind-born children and compared them with those of seeing children. The results support and supplement the observations cited in the preceding section. Smiling, laughing, and crying, also the expressions of anger, pouting, fear, and sadness, looked the same in blind-born children, although they could not have imitated anyone. Blind-born children did, however, smile less as time went on in comparison with seeing children or those who had become blind after birth; no comparable decrease in crying was noted. In smiling a certain social feedback must play a role, which has yet to be investigated. When this feedback is missing the behavior atrophies somewhat. My own observations on a deaf-blind 7-year-old girl lead me to the conclusion that this feedback is of a general nature. After the girl's teacher and nurse intensively interacted with her and played with her, she laughed more frequently than before.

D. G. Freedman (1964) published a picture of a blind-born infant. She laughed when her mother spoke to her. The continuous nystagmus seen in the blind then ceased and the eyes fixated upon the source of the sound, although she could not see it (Fig. 18-5). This seems to be achieved by a central fixating process.

Differences in the expressions of blind and seeing children are to be found not in the basic pattern, but in the additional occurrences of uncontrolled, superfluous "grimacing" movements in the blind, which was pointed out by G. Mackensen (1965) among others.

Of special interest is the behavior of children born deaf and blind,

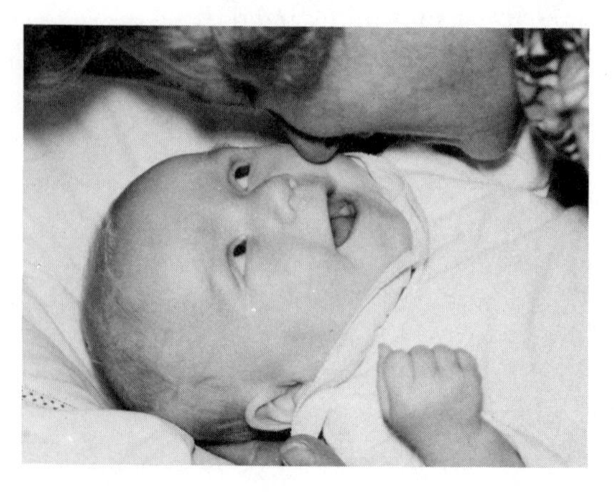

Figure 18-5 Two-month 20-day-old girl, blind from birth, smiling. Although the child cannot see, the eyes fixate, probably as the result of a central process. Normally the eyes move about restlessly. (Photograph: D. G. Freedman, 1964.)

about which very little is known. These unfortunate children grow up in eternal night and silence. They have no means for imitation and education is very difficult. In spite of the lack of formal training these children nevertheless show a number of well-coordinated motor patterns. F. L. Goodenough (1932) reported that a 10-year-old congenitally deaf and blind girl, who grew up without instruction under conditions of poverty, was able to laugh heartily when she found her lost doll. She laughed also when she danced while standing on her toes, which she had learned by herself. When angry, she turned her head away, furrowed her brow, and pouted her lips. When she was very angry she threw back her head, shook it violently, and showed her clenched teeth.

Figure 18-6 Deaf-blind 7-year-old girl, laughing. When fully laughing she throws her head back, opens her mouth, and laughs audibly, although in a restrained manner. (Photograph: I. Eibl-Eibesfeldt.)

As part of a continuing investigation[2] I filmed the laughing and smiling of a 7-year-old congenitally deaf and blind girl and a 5-year-old boy who otherwise had no mental impairment. The motor patterns of laughing corresponded in all details to those of normal children. These two deaf and blind children threw back their heads during high-intensity laughing in a fashion typical for normal children and they also opened their mouths (Figs. 18-6, 18-7, and 18-8). The rhythmic sounds

[2] I wish to thank Karl Heinz Baaske, teacher for the deaf-blind at the State Institution for the Blind in Hannover, for his cooperation in making these observations possible.

are very clear, but their laughing is somewhat restrained, more like a giggle. The girl also showed a number of typical expressive movements, for example, crying. When angry she stomped with her feet. She rejected by shaking her head or by pushing away with her hand, while she also shook her hand. If she stumbled she extended both hands forward. When taken on her caretaker's lap or shoulders, she liked to cuddle against him. This girl, who made a very alert impression, especially when actively exploring her environment with her hands, was able to distinguish strange persons from familiar ones by sniffing briefly at the presented hand. She showed obvious fear of strangers, although she had no reason to fear them, since she had never experienced any harm from strangers. On the contrary, every stranger behaved very tenderly, to give the poor girl a feeling of general security. Only the mother had sometimes inflicted mild punishment. Nevertheless the mother was the beloved reference person, and strangers were feared. Later this fear changed to a more aggressive aversion of strangers: she would push the stranger away and even slap him, before turning away.[3] A basic reaction of man—his distrust of strangers—seems to be rooted in this inborn disposition. Among the consequences of this reaction pattern is our living in closed groups and our hostility toward strangers.

In short, a whole array of even quite complex behavior patterns, which are typical for human beings, have also developed in the deaf and blind and are therefore present as phylogenetic adaptations. Some characteristics of social behavior developed even contrary to educational efforts, such as, for example, the fear of strangers. Similarly, in a boy of the same institution approaching puberty, certain aggressive inclinations developed and needed to be curbed by education (I. Eibl-Eibesfeldt, 1973).

The possible objection[4] that the deaf-blind child could have learned the complex movement coordinations of crying and laughing by a step-by-step reinforcement of these components can be answered as follows: if no phylogenetic adaptations existed, a large number of individual steps would be necessary to establish such complex movement coordinations. The mother would have to initially reward the child when she raised the corners of her mouth but not reward any other lip movement. She would have to shape the rhythmic vocalizations of laughing by step-by-step reinforcement of the appropriate inhaling and exhaling movements until finally the typical rhythm of typical laughing behavior was achieved. Again, different steps would have to be followed in the training of crying. The best Skinnerians would be confronted with great difficulties if they were asked to teach a higher mammal such movement

[3] Of 8 children born deaf and blind and studied so far, 6 showed clear evidence of stranger rejection.

[4] This objection was raised by R. L. Birdwhistell in a discussion in which I took part in Minneapolis.

coordination. The deaf and blind children, on the other hand, are expected to have learned all this by the accidentally adequate behavior of the mother, although these children as a rule have great difficulty in learning such relatively simple skills as holding a spoon and bringing it to the mouth. A heavily brain damaged 12-year-old girl in the same institution who was born deaf and blind was able to laugh and cry. She clasped her caretaker and tried to climb up on him, whimpering, in the manner of a small child. But she had been unable to learn simple acts such as bringing a spoon to her mouth, despite the greatest efforts to teach them. The hypothesis that the complex expressive movements of these children are learned thus lacks all foundation. If one still were to maintain this view, then one would have to assume special innate learning dispositions and would have to rely even more on the ethological

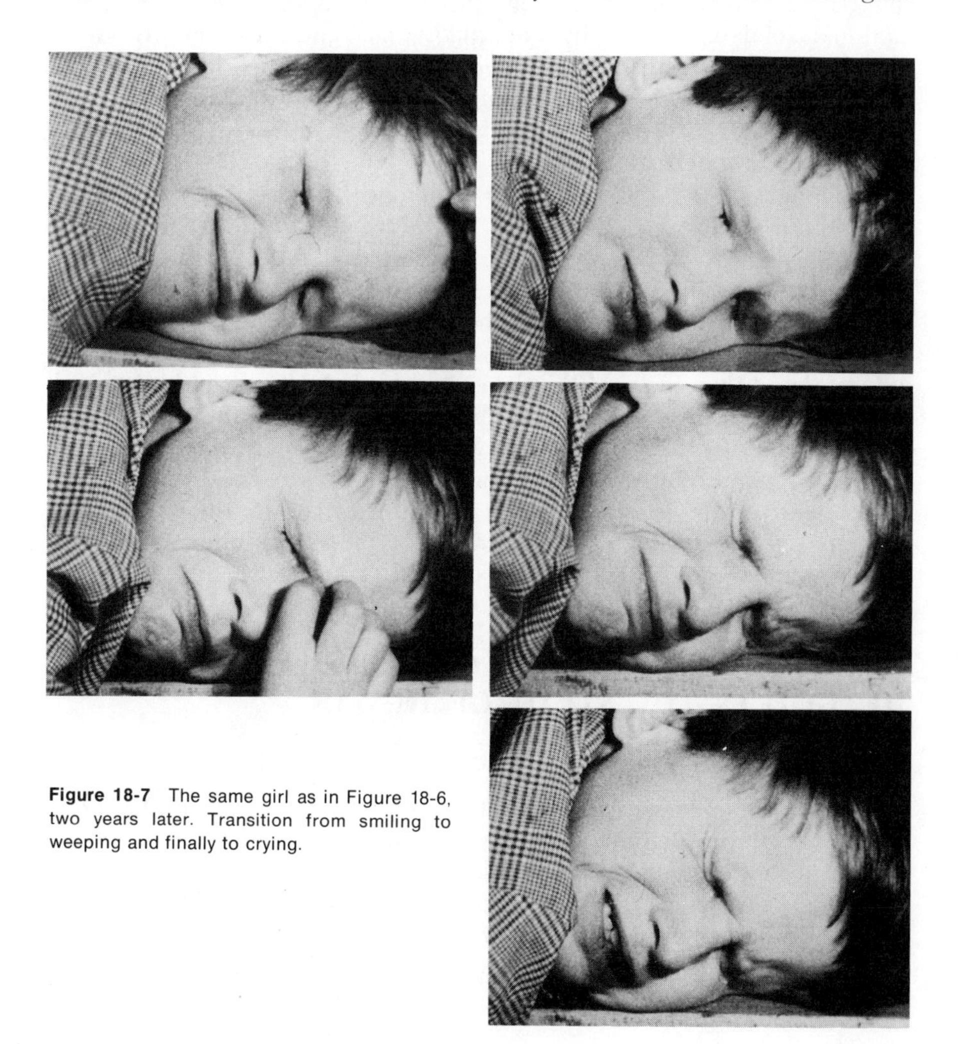

Figure 18-7 The same girl as in Figure 18-6, two years later. Transition from smiling to weeping and finally to crying.

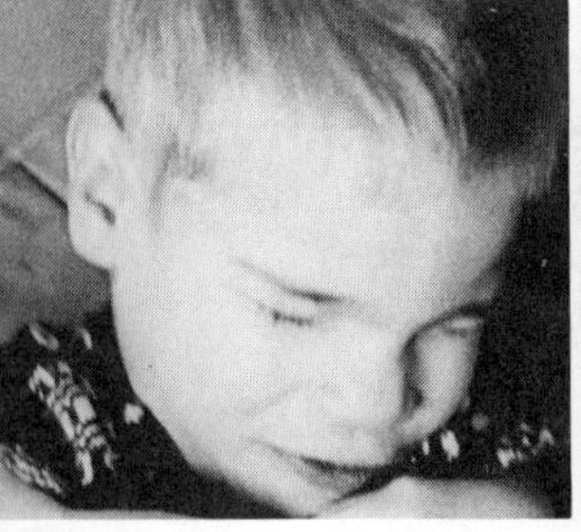
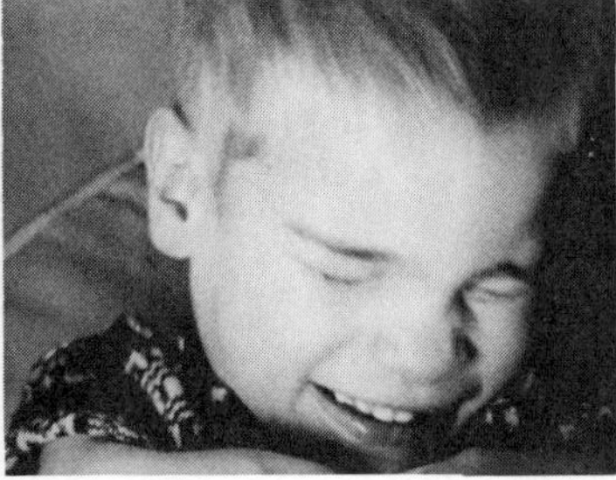
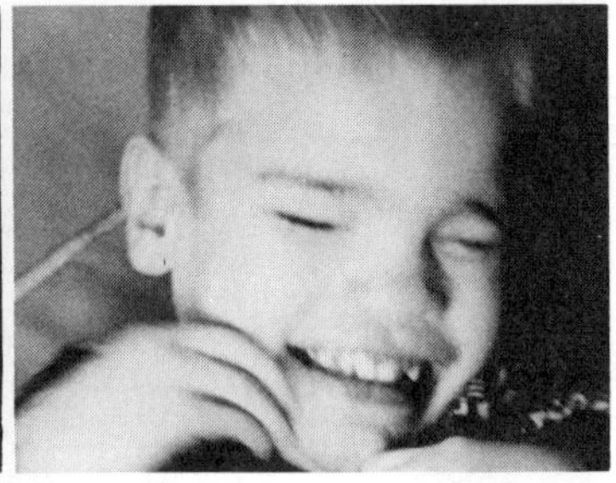

Figure 18-8 Deaf and blind 5-year-3-month-old boy (Thalidomide child), laughing. (Photographs: I. Eibl-Eibesfeldt.)

concept of phylogenetic adaptation in behavior. The possibility that a congenitally deaf and blind child acquires information about his mother's facial expressions by touching her face and consequently learning by imitation can also be excluded; I know of a congenitally deaf and blind boy who, in addition to this handicap, does not have arms to reach out but only very short stumps. Nevertheless he shows the basic facial expressions in a normal way.

A number of complex expressive behavior patterns, such as coquettish embarrassment, cannot be seen in the deaf and blind. This may be due to lack of relevant experience or to the fact that the channels which usually receive such perceptions are closed in these children. That the latter is at least partially true can be deduced from the observation of congenitally blind children, whose mimic expressions are more highly differentiated than those of the deaf and blind. A 10-year-old blind girl showed embarrassed smiling with flushing, a lowering of the head, and incipient head rocking movements when I praised her for her performance at the piano. Those that are born blind have fewer expressive gestures but are otherwise most similar to seeing persons in their spontaneous expressions. They are, however, able to act out mimic expressions only in an incomplete form (F. Dumas 1932; M. N. Mistschenko 1933).

SOME RESULTS OF THE COMPARATIVE METHOD IN THE STUDY OF HUMAN BEHAVIOR

The observations on congenitally blind and both deaf and blind people allow only limited statements to be made about human behavior. Such people lack the more complex behavior sequences that are normally released by visual and auditory signals. The question of if and how much of complex human behavior is inborn may be answered by the comparison of behavior in individuals of different cultures. If one can demonstrate communalities in expressions and gestures, then we may conclude that they derived from a common inherited root, especially when the behavior patterns concerned are specific and widespread in occurrence among people of different ecological, cultural, and racial backgrounds. This commonality of behavior patterns is particularly significant because the behavior of man is extremely susceptible to comparatively rapid molding by cultural forces, as the evolution of language clearly demonstrates. This line of thought was expressed by C. Darwin (1872), and many present-day psychologists are aware of the basic agreement in mimic expressions among different peoples:

> As far as the accuracy of the data allows, the mimic expressions and pantomime which correspond to the described states, seem to appear in each people and race in a similar context or with the same meaning (perhaps one laughs without joy, but also because of it). It is true that important cultural differences exist in expressive behavior, but this in no way detracts from the constancy of these primary expressions (N. H. Frijda, 1965:376).

S. Asch (1952) writes in a similar vein:

> The findings of the ethnologists agree that there are basic expressions which occur without exception in all human societies. Cries of pain and of grief are universally distributed: When frightened one becomes pale and trembles; laughing and smiling are quite generally an expression of joy and happiness. It is probable that the agreements are even more encompassing and that reactions such as surprise, boredom, and astonishment are included. We may therefore speak of certain invariables in our emotional expressions, even though they have not been described in sufficient detail (p. 195, retranslated from the author's translation).

One is then quite surprised when one reads the statements of others, such as A. Gehlen (1956), who writes:

> Inborn, instinctive behavior patterns are actually only demonstrable in very small children where they can hardly be distinguished from reflexes, such as sucking, grasping and holding-on movements. Otherwise and quite generally human motor patterns are bare of all instincts . . . they are learned in their totality and concreteness in the way they are performed, they are built up individually through the integration of external stimuli and experiences.

In a similar way W. LaBarre (1947:49) denies the universality of human expressive movements. He writes: "The anthropologist is wary of those who speak of an 'instinctive' gesture on the part of a human being," and he tries to back this statement with a few examples (p. 52):

> Smiling, indeed, I have found may almost be mapped after the fashion of any other culture trait; and laughter is in some senses a geographic vari-

able. On a map of the Southwest Pacific one could perhaps even draw lines between areas of "Papuan hilarity" and others where a Cobuan, Melanesian dourness reigned. In Africa Gorer noted that laughter is used by the Negro to express surprise, wonder, embarrassment, and even discomfiture; it is not necessarily, or even often a sign of amusement; the significance given to "black laughter" is due to a mistake of supposing that similar symbols have identical meanings. Thus it is that even if the physiological behaviour be present, its cultural and emotional functions may differ. Indeed, even within the same culture, the laughter of adolescent girls and the laughter of corporation presidents can be functionally different things. . . .

R. L. Birdwhistell (1963, 1968, 1970) states that no expressive movement has any universal meaning, that they are all the product of culture and are not inborn.

Beyond these anecdotal remarks, I have failed to find any references. LaBarre does not offer any correlational analyses of Papuan or Melanesian laughter. He did not collect numerous incidents of laughter in this or that culture, so that he could have correlated them to context, movement sequence, and elicited responses and make a cross-cultural comparison. Instead, LaBarre associates the different meanings of an expression with different cultures, without posing the question of whether the same or similar spectrum is to be found in every culture; nor does he look for a common denominator but presents in a fairly naive way his personal impressions. The complexity of associations will be demonstrated in a later section with the example of the **eyebrow flash,** on which I have done correlational analyses.

Recent investigations of P. Ekman and W. Friesen (1971) clearly demonstrate a cross-cultural consistency not only in the movement patterns of some facial expressions, but also in the interpretation of these expressions. Illiterate Papuans were, for example, able to select from a choice of photographs those showing the expression that was appropriate to the test story told to them before, even if the depicted persons belonged to another culture.

A sample of still photographs may serve for the moment to illustrate the similarity of facial expressions exhibited by people of different cultures in the same situation. Indeed, people smile everywhere in a friendly encounter, they show their anger in a similar fashion, and they cry in distress (Figs. 18-9–18-12). Of course, still photographs are inadequate to document movement patterns. What we need are motion pictures, but documents of this type are lacking. It is in part insufficient documentation and hence a lack of knowledge that has allowed statements as those quoted above to remain unchallenged. N. H. Fridja (1965:399) states this very plainly:

Let us repeat, in a particular instance the explanation (especially of facial

> expressions) is lacking, not only because of a failure to test hypotheses, but also because of insufficient or wrong hypotheses. Many of Darwin's explanations have been superseded or have been considered somewhat far-fetched. But no one has since given his attention so conscientiously and in such detail to the facial expressions of man. Most people seem to be satisfied with global half-objective, half-interpretive descriptions, as Kirchhoff (1960) critically remarked. Especially: What work other than that of Darwin contains such detailed, theoretically unbiased descriptions of facial expressions as were actually observed in particular situations? We certainly take the easy way out.

There are only a few isolated descriptive-analytic investigations of human facial expression. One example is a study of P. Lersch (1951), but there is hardly any comparative work. No one seems to have thought about documenting human behavior objectively, that is, through films taken without the subject's awareness. As unbelievable as it may seem, the ethogram of man has not yet been documented and recorded in a way that would permit one scientist to examine the data of another which are not colored by the interpretations of the observer.

However, E. C. Grant (1965, 1968, 1969, 1972) has compiled a list of human facial expressions and movements, which includes over a hundred elements. These are described as movement patterns of mouth, eyes, eyebrows, head, and so on. That these expressions carry affective meanings has been shown by sequence analysis; five main clusters of behavior have been isolated: Flight, Assertion, Relaxation, Contact, and Ambivalence. These results have been confirmed by T. Pitcairn and E. C. Grant (1972), who investigated the behavioral relations between patient and doctor in a psychiatric interview situation. The main clusters of behavior were again demonstrated, and also the fact that the behavior of the two interactors was always synchronized, in other words, they were both operating within the same category of behavior. The interpretation

Figure 18-9 Smiling people: *Left:* Schom-Pen man (Great Nicobar Island); *center:* Negro boy (near Bihamarulu, Tanzania); *right:* Balinese (Sanur). (Photographs: I. Eibl-Eibesfeldt.)

of this phenomenon is that each actor is using synchrony as a means of adjusting his behavior to the other, simultaneously with the postural synchrony demonstrated by W. S. Condon and W. D. Ogston (1967).

A search in the large film library of the Institute for Scientific Film in Göttingen, Germany, revealed numerous films of certain cultural activities, such as the weaving of mats, making of pots, tilling of fields, building of boats and houses, dances, and so on. Almost always it is a staged activity (though this is frequently not even mentioned), which reduces the value of these films considerably. On the other hand, there are many film documents made in the field in which one can find material that is revelant to the kinds of questions we are asking, but they are always incidental to some other topic (S. R. Sorenson and D. C. Gajdusek, 1966).

There do exist, in fact, some excellent ethnological monographs. I am referring, for example, to the film *Dead Birds* by R. G. Gardener (see also R. G. Gardener and K. G. Heider, 1968). It contains many documents of social interactions among the Dani (Papua), including battle episodes.

Figure 18-10 *Left:* Kabuki actor showing rage (Tokyo); *right:* expression of rage in a 4-year-old girl, whose motivation is probably jealousy. The father had photographed her sister repeatedly in her confirmation dress. Finally this little girl jumped forward and shouted "I want to have my picture taken too!" The child cried and was also very aggressive. Note the forward stance, which is an intention to attack, the clenched fists, and the expression of rage, especially around the corners of the mouth. (Photographs: *left:* I. Eibl-Eibesfeldt; *right:* from E. F. v. Eickstedt, 1963.)

Figure 18-11 *Left:* Crying during great depression and bodily pain. A girl in Vietnam (near Don Xai), who had lost her parents and been injured during the fighting in the city. *Right:* Indian baby crying (Pisac, Peru). (Photographs: *left:* Wide World Photos; *right:* I. Eibl-Eibesfeldt.)

But even in this film we encounter difficulties in localizing shots for any particular issue.

If one wants to know whether a Papuan, Bantu, Japanese, or Italian stomps his foot when he is angry, one will search in vain in the film archives for unstaged films of people in rage. The same is true if one searches for comparative pictures of flirting, laughing, crying, and gestures of disdain and other expressive movements. This surprising discrepancy in the documentation of cultural activities, on the one hand, and of expressive behavior, on the other, is partly rooted in the historical development of psychology and ethnography. There are also certain methodological difficulties. Making of pots, weaving of mats, and cultural activities are readily performed for observers. The documents, then, do not actually reflect reality, but in general it is possible to capture the process. Facial expressions and gestures, on the other hand, must be recorded without the subject's awareness. Even a learned activity changes markedly when it has an audience; this is even more true for emotional behavior. People are shy by nature and they do not like to be photographed. It is amazing over what great distances people perceive a camera that is pointing at them. Their behavior changes instantly. The facial expression becomes rigid, most people look restlessly toward the camera, smile in embarrassment, or exaggerate or overdo the behavior if they decide to continue it. Even in learned skills the smoothness of the performance is often lost.

A method that was developed by H. Hass and that we tested in various parts of the world in photographing people without their awareness overcomes all these difficulties (I. Eibl-Eibesfeldt and H. Hass 1966, 1967; H. Hass, 1968). An attachment that is mounted in front of the normal lens of the camera and contains a mirror prism makes it possible to film to the side (Fig. 18-13). With this technique it was possible to photograph people even from nearby without their awareness. They see, of course, that filming is going on, but the camera and the attention of the cameraman point in another direction, so they soon ignore it.

In these films we used the technique of time transformations (fast and slow motion) to make visible the lawfulness of the behavior sequences, which normally escapes an observer. The method of time acceleration (one to seven frames per second) has so far not been used in the study of the behavior of higher vertebrates, and I would like to discuss its special advantages.

First, by using the fast-motion technique one can obtain film protocols of long behavior sequences (for a small amount of film per unit of

Figure 18-12 International language of facial expressions (Highland Indian woman from Pisac, Peru): (*a*), feeding her child; (*b*), smiling at the child, who replies in kind; (*c*), obviously in thought; (*d*), a little later, smiling at her husband. (Photographs: I. Eibl-Eibesfeldt.)

Figure 18-13 *Left:* Bolex camera with a mirror lens. A prism built into the attachment permits filming to the side. In the front the aperture is covered by a mock lens. *Right:* H. Hass filming with the mirror lens in the marketplace at Arusha, Tanzania (Photographs: I. Eibl-Eibesfeldt.)

time), which document an entire sequence. For example, if we are interested in the technique of making a pot we find the films taken by ethnologists unsatisfactory, in that they never show the whole process. Single episodes have always been selected: how the clay was brought, how it was kneaded, and how the bottom was shaped. Then we may see several stages of shaping, smoothing of the walls, and so on. The entire sequence is chopped up in this way and so includes the photographer's interpretation of what is important. What occurs between the cuts one learns, if at all, from the accompanying publication. This has been accepted as a necessary part of the methodology, as we can see from the writing of G. Spannaus (1961), who published guidelines for the preparation of ethnological films. He emphasized that an ethnographer has to depend upon a "representative" collection of complete movement sequences taken from the overall behavior because it is as a rule not possible to record the entire event, such as making a pot or a religious festival. "It is enough when all parts that occur once are recorded once, and those that are repeated are recorded once or twice. . . ." (G. Spannaus, 1961:77). However, the example of making pots is one where we are dealing with a behavior sequence of a high level of integration, which should be recorded in its entirety, and can be by means of the fast-motion technique.

If one selects the right number of frames per second, the movements run off quite rapidly, but each individual action remains clearly visible. One can see how the product grows under the shaping hands and one can count later how many individual movements were necessary to produce the particular pot or vase or whatever—that is, what amount of effort went into its making. This in turn makes it possible to compute the relative efficiency of different techniques. With comparative fast-motion film records of, for example, the tilling of a field, one can determine the different amounts of work involved. The comparison of fast-motion films of trained and untrained workers can also be very useful.

Of course, the behavior patterns that belong to a lower level of integration, from which the more complex behavior sequence is made up, should also be recorded. These individual movements we film in slow motion and without the subjects' awareness.

The fast-motion technique opens up new paths of documentation and analysis to the ethnographer. One can think of the investigation of religious ceremonies and other rites, which can now be recorded in their entirety. An event that takes half an hour can be recorded with five minutes of film and four frames per second. If, later, someone should become interested in the study of cultural ritualization, then fast-motion films—for example, of a Catholic mass—would be most useful. If this event were filmed at regular time intervals, one could see the changes directly from the films.

In addition to these uses, the fast-motion film technique can make visible certain regularities in behavior that normally escape direct observation. A newspaper seller whom H. Hass filmed in Vienna proved to be a most rewarding subject for demonstrating the value of this technique. When he was observed normally nothing unusual was detected about his behavior. The fast-motion technique revealed, however, that the man ran back and forth in front of a 1.5-meter-wide part of the wall, which was bordered on both sides by large display windows. His behavior was so stereotyped that it appeared as if he were tied to this small spot before the walled portion of the building. It is possible that this was an indication of an inborn tendency to keep oneself covered from behind.

Pictures taken from a height show that people approach conspicuous landmarks—for example, flagpoles—without any special reason, perhaps because of an innate orienting mechanism. This happens even if they must deviate from a more direct path to their goal to accomplish it.

During the analysis of fast-motion film of persons eating we noticed that individuals who ate alone looked up and around into the distance after each bite or two; their gaze often swept automatically to the sides as if scanning the horizon (H. Hass, 1968). Baboons and chimpanzees show the same behavior. This seems to be an alert behavior of guarding against enemies, a behavior that is also a phylogenetically inherited part of man, although today there is very little danger to man when he is eating. The same phenomenon has attracted the attention of D. P. Barash (1972), who found that single persons look up significantly more often than those sitting in groups.

The fast-motion technique is of special value in the investigation of people in different sized groups. Normally even the observation of two persons is quite difficult, because the behavior of both cannot be recorded simultaneously. It is even more difficult when still more people are involved, as in a large family. On the other hand, if we have a fast-

a

b

Figure 18-14 Flirting Turkana woman (Lorukumu, Kenya) as an example of successive ambivalence. She makes contact with the eyes, laughs, lowers the head and eyelids (in embarrassment?), and repeats full contact with the eyes. From a slow-motion sequence (48 frames/second). The entire sequence *a–d* takes 6.04 seconds (290 frames). (*b*): Frame 40 [0.83 second after (*a*)]; (*c*): frame 177 [3.68 seconds after (*a*)]. (Photographs: H. Hass.)

c

d

motion protocol we can view the film as often as is necessary and we can recognize how the behavior patterns of individuals are attuned to one another and how they are grouped. With this technique we have filmed playing children, mothers with children, couples and larger congregations of people, and the behavior of people in cultural ceremonies. We have also found that the fast-motion technique can be used successfully for the investigation of human and animal behavior in other situations. I should add that for fast-motion filming one can mount a camera in an elevated place, such as the roof of a car, and let it run without being attended. To allay all suspicions we worked with the lens to the side even under these circumstances, although no one suspected that the camera standing by itself was running.

When we want to analyze facial expressions and gestures we film the subject in slow motion (50 frames per second) without his awareness. For later analysis it is especially important to have a record of what the subject did just prior to and after the filming. We strive, in other words, to understand the behavior within the context of the situation and in the sequence in which it takes place, in the same way as it is necessary in motivational analysis of animal behavior to avoid later subjective interpretations. In some cases it is possible to create the releasing stimulus situation. When we were filming, curious onlookers would often gather around us, and we experimented with them without their being aware of it. By handing such a person a small box out of which popped a cloth snake when it was opened, we obtained the unrehearsed expression of fright. By casually looking at a person one can release greeting and sometimes even flirting behavior.

In countries with a highly developed theater culture (Japan, Thailand, Europe) we had the actors act out certain expressions according to a list prepared by us, which we filmed with 50 frames per second. These scenes allow us to compare the natural expressions with the actor's portrayal of them.

Employing this method, we have already gathered film records of unstaged social interactions in many countries. They will be made available to the public through the Film Archive for Human Ethology of the Max Planck Society. For the technical details of data collecting see I. Eibl-Eibesfeldt (1972a).

Figure 18-15 Flir''ng Samburu girl. Eye contact, closing of the eyelids, looking away. The ritualized "flight" is limited to eye movements, and only very slightly indicated in movements of the head. Taken from a film by the author. (I. Eibl-Eibesfeldt, 1970.)

Concerning the selectivity of the sampling method employed when filming unstaged social interactions, I want to emphasize that we film whenever an interaction can be expected to occur. We deduce this from the orientation of the persons observed. We press the camera release, for example, when two persons approach each other, or turn toward each other, without knowing for sure whether an interaction will take place and of what type it will be.

Although the work is still in progress, we have filmed enough to say that some of the more complex human expressions can be traced back to the superposition of a few fixed action patterns which do not seem to be culturally determined. To give just one example, we found agreement in the smallest detail in the flirting behavior of girls from Samoa, Papua, France, Japan, Africa (Turcana and other Nilotohamite tribes, Himba, Bushmen) and South American Indians (Waika, Orinoko).

Figure 18-16 Simultaneous ambivalence in flirting behavior. A 10-year-old Samburu girl turning her eyes toward the observer while showing an intention to turn away. (Photograph: I. Eibl-Eibesfeldt, from 16-mm movie film, near Maralal, Kenya.)

The flirting girl at first smiles at her partner and lifts her eyebrows with a quick, jerky movement upward so that the eye slit is briefly enlarged. The most probably inborn greeting with the eyes is quite typical (Figs. 18-14–18-21). Flirting men show the same movement of the eyebrow, which also can be observed during a friendly greeting between members of the same sex. After this initial, obvious turning toward the person, in the flirt there follows a turning away. The head is turned to the side, sometimes bent toward the ground, the gaze is lowered, and the eyelids are dropped (Fig. 18-14). Frequently, but not always, the girl may cover her face with a hand and she may laugh or smile in embarrassment. She continues to look at the partner out of the corners of her eyes and sometimes vacillates between that and an embarrassed looking away.

There is, then, a fluctuation between turning towards and turning away, the latter sentiment probably originating in fear, and the whole

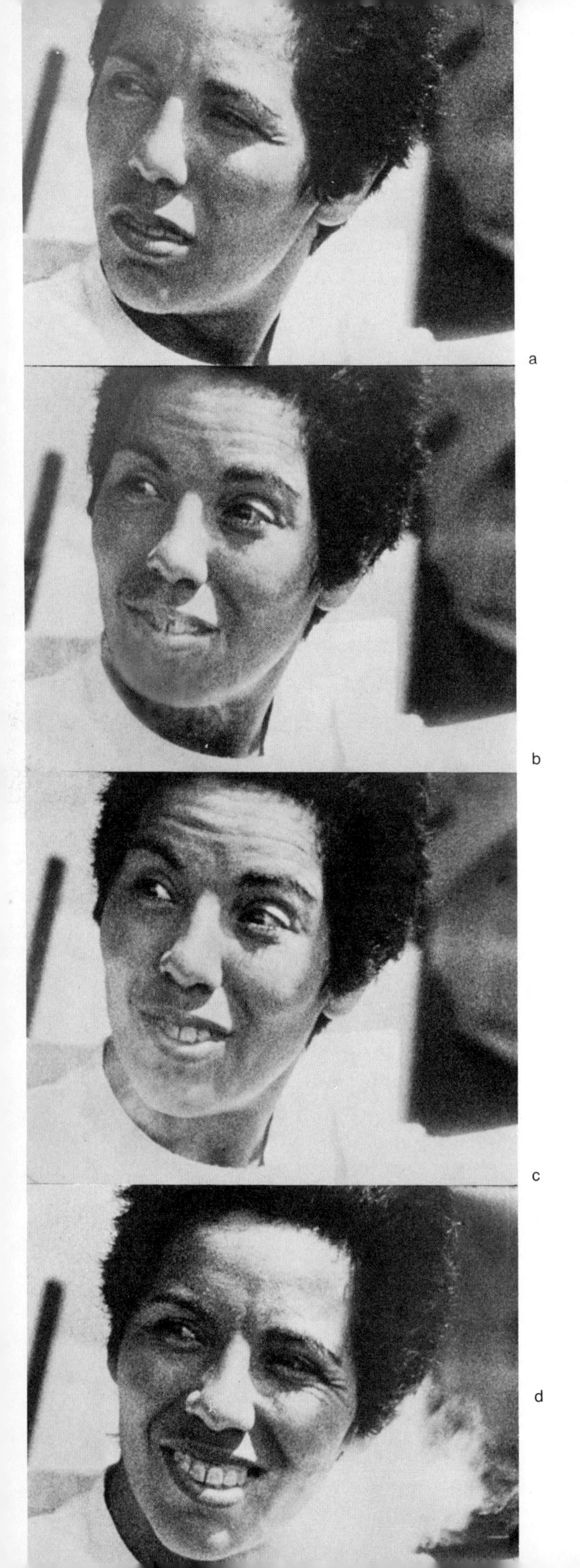

Figure 18-17 Greeting with the eyes by a French woman: *a*, neutral face, *b* and *c*, raising the eyebrows; *d*, a smile. Six frames after the first picture one notices a slight raising of the eyebrows. Slow-motion movie film (48 frames/second). The sequence *a–c* includes 41 frames (0.87 second). Between frames 19 and 26 the eyes are maximally raised. *b* shows frame 23 [0.047 second after *a*]. *c* is frame 41. The entire sequence of raising the brows and lowering them again includes 18 frames (0.37 second), and they are raised maximally only for 7 frames. (Photographs: H. Hass, from 16-mm movie film.)

procedure becoming ritualized into flirting behavior. Often this fluctuation is limited to eye movements (Fig. 18-15). In the case of a simultaneous ambivalence, turning the eyes towards the partner may be combined with a turning away of the body (Fig. 18-16; I. Eibl-Eibesfeldt, 1967, 1972b).

Here we find that the **superposition** of a few invariable components (intention movements of turning toward someone, responsiveness, and turning away) yields a relatively complex and variable expression. The assertion of R. L. Birdwhistell (1963, 1968) that there is no expressive behavior independent of culture and that everything is learned is disproved by these results. In particular, the principle of using antithetical movements in combination is a universal one.

The visual cut-off by looking away or hiding the face on the part of persons flirting or exhibiting coyness may, besides its expressive function, also serve to control the state of arousal. At least, it does so in infants, as was demonstrated by L. Matas, E. Waters, and L. Srouf (on press). While recording infants from 6 to 10 months of age on videotape, with a continuous heart rate record, the subjects were approached by an unfamiliar female adult. The infants were seated in high chairs, with their mother sitting beside them at a distance of 4 feet. By looking down or at their mothers the children exercised some control over their arousal. Heartbeat accelerations as the stranger approached were often followed by the infant looking away or at his mother – usually for less than 3 seconds. The heart rate dropped by 10–15 beats per minute during this period. Furthermore, the results of this study, indicate that infants who perform this cut-off maneuver do not become aroused enough to cry and are more likely to have pleasant reactions to subsequent approaches.

In friendly greeting over a distance, the eyebrow flash is transmitted in combination with smiling, headtossing, and nodding (I. Eibl-Eibesfeldt, 1968, 1971a, 1972b). This expressive pattern is universal and stereotyped: I have observed it in all cultures so far visited. However, differences did exist in the readiness with which this greeting was used. The Japanese show a considerable reserve and answered upon being questioned that the eyebrow flash greeting is not considered proper. Yet they were unaware of the generosity with which they used this signal when joking around with children. At the other extreme, Samoans regularly greeted everyone with the eyebrow flash, while in Central Europe only very close friends were received in this way. In Samoa, the eyebrow flash is also used to express a sober affirmation and accompanies the verbal "Yes," occasionally even replacing it. Europeans act this way only in enthusiastic affirmation. Here I observed a quick raising of the eyebrows also in thanking, flirting, joking with children, emphasizing a fact, and, more infrequently, in asking for information.

In 155 cases of quick eyebrow raising the following categories could be distinguished: 45 were for greeting and parting, 9 for flirting, 12 for women joking with babies, 4 for thanks, 31 for affirmation, 14 for agreeing with a statement ("Yes, that is so!") and 40 for emphasizing an assertion. Almost one half of all the recorded cases (70) involved a willingness for social contact (flirting, greeting, joking, thanking). Affirmation and agreement were involved in 45 cases. Raising the brows also plays an important part in emphasizing an assertion. Here, too, we are dealing with an attempt to establish social contact (seeking affirmation). There is a desire to convince the partner, expressed by the assurance that the statement is true. In essence, it is always just such an agreement to social contact (greeting, flirting) that is expressed. By way of this more general form of agreement, the Samoans made the raising of the eyebrows into a sign for the factual "Yes." This movement is used by Europeans for emphasizing a statement, and more rarely even for a negation. In the latter case, the negation is confirmed: "Yes, that is so!" In posing a question, Europeans frequently raise the eyebrows just as they do when demanding agreement to the verbal question "Yes?", often waiting for the reply with raised eyebrows as well. Yet I do not believe that the inquisitive raising of the brows derived from those forms meaning "Yes." If an individual has posed a question, he will listen attentively for the answer, opening the doors of perception as in curiosity.

If we look further at the various conditions under which the eyebrows are raised, we can find hints as to the phylogenetic development of this behavior. Europeans raise the eyebrows in curiosity—to this category we may add the just-mentioned inquisitive eyebrow raising—and often as a sign of surprise. In both cases, the eyebrows are kept raised a little longer than usual. This may constitute an opening of the perceptual doors, in which case the raising of the eyebrows accompanies the widening of the eyes. This process would then form the basis for a ritualization of the behavior to the eyebrow flash greeting. Even today, upon meeting a friend, Central Europeans will exclaim in surprise: "Aha, it's you!" The eyebrows are raised briefly in a flash greeting and at the same time other mimic signs (smiling) express joy at this surprise.

There is also another direction which a ritualization could take in this case: in Europe the eyebrows are also raised in displeasure: for example, when wishing to express disapproval at the bad manners of another person. This would constitute an unpleasant surprise, so to speak, very much of personal concern. The eyebrows remain raised and thus the gaze becomes threatening. Displeasure is further expressed through accompanying facial movements (vertical creases on the forehead). Finally, mention should be made of eyebrow raising in an expression of arrogance, representing a further ritualization of this behavior to a gesture of social rejection. Among some cultural groups, for example the Greeks, this gesture has developed into an expression of the factual "No."

Figure 18-18 Eyebrow flash of a !Kung Bushman woman (Southwest Africa) on answering a verbal greeting. Frames 1, 3, 7, and 9. (Taken from a 16-mm movie film, 50 frames/second. Recording: I. Eibl-Eibesfeldt.)

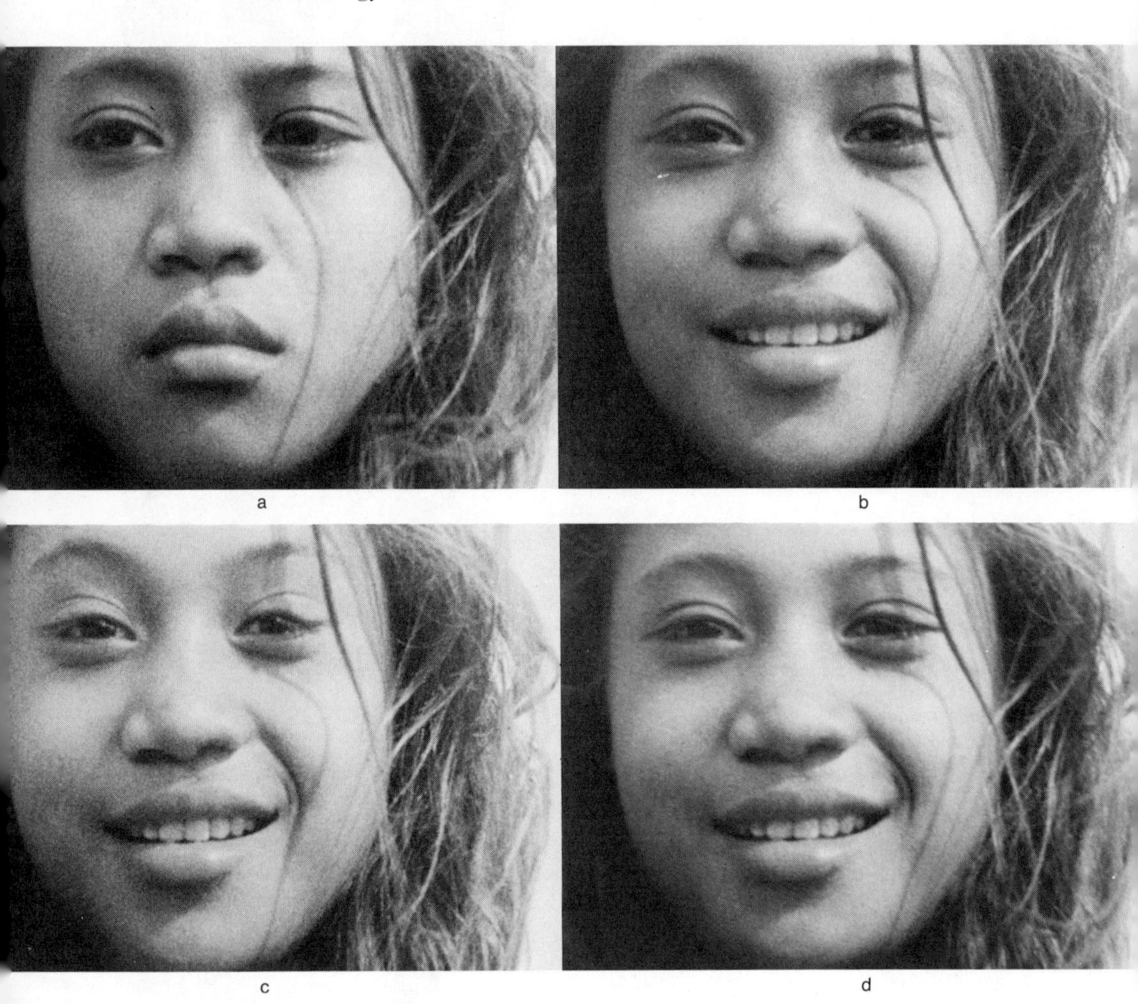

Figure 18-19 Greeting with the eyes by a flirting girl from Samoa, from a slow-motion film (48 frames/second). *a:* Neutral face; *b:* smiling at the partner (frame 41); *c:* sudden raising of the eyebrows (frame 107). The entire sequence runs very fast. The eyebrows are clearly raised for only 1/6 second. (Photographs: H. Hass, 16-mm movie film.)

The example of the eyebrow flash serves well to illustrate the care that must be taken in generalizing about the meaning of expressive movements. Table 18-1 shows a schema of the various forms of eyebrow flash.

Wide agreement is also found in many other expressive behaviors. Thus, arrogance and disdain are expressed by an upright posture, raising of the head, moving back, looking down, closed lips, exhaling through the nose – in other words through ritualized movements of turning away

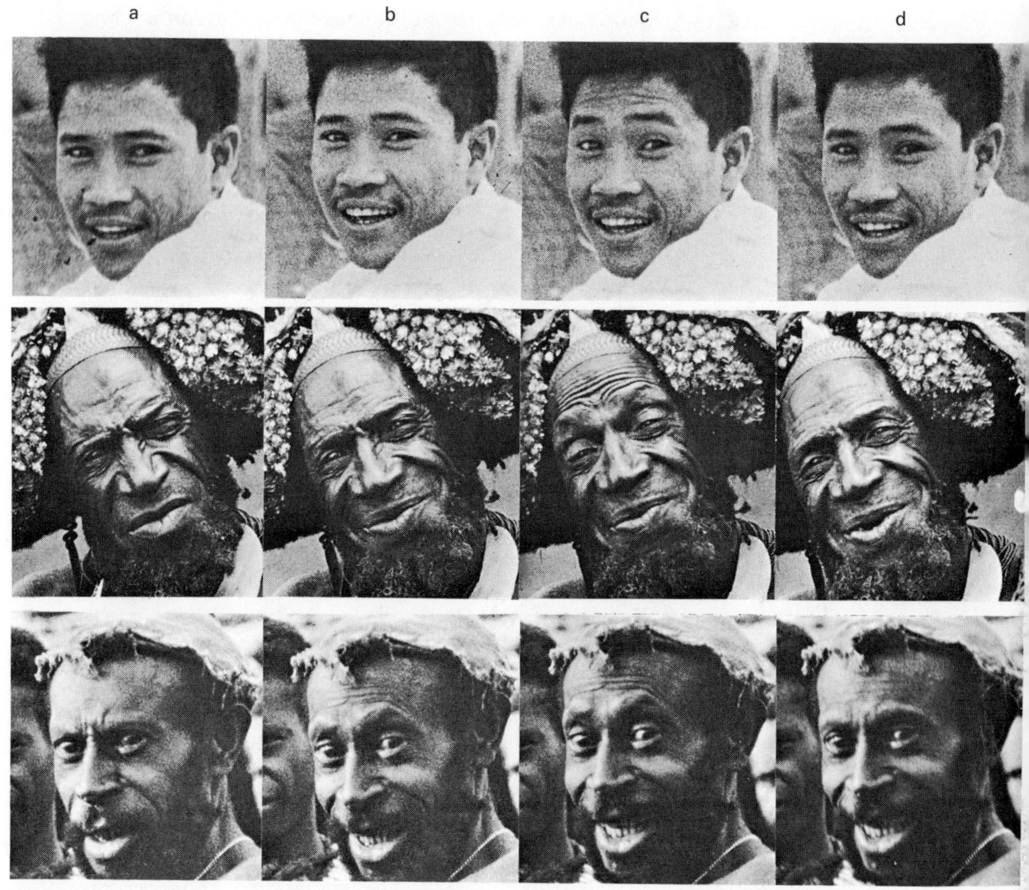

Figure 18-20 Eyebrow flash during greeting. *Upper row:* Balinese of the island Nusa Penida. Sequence (*a*)–(*d*) lasts 19 frames; (*b*) shows frame 6; (*c*) frame 11. With frame 6 raising of the eyebrows set in and they were maximally raised in frame 11. The lowering movement began with frame 17. *Middle row:* Papuan, Huri tribe near Tari (New Guinea). Sequence (*a*)–(*d*) lasts 45 frames; (*b*) shows frame 30; (*c*) frame 36. Twenty-six frames after he had started to smile he began to lift the eyebrows, and 4 frames later they were maximally raised and held so over 7 frames. *Lower row:* Papuan, Woitapmin tribe near Bimin (New Guinea). Sequence (*a*)–(*d*) lasts 85 frames; (*b*) shows frame 75; (*c*), frame 79. At frame 16 he starts to smile and at frame 75 raises the eyebrows. They are maximally raised during frames 78–84. All photographs are reproduced from 16-mm movie film taken at 48 frames/second. (Photographs: I. Eibl-Eibesfeldt.)

and rejection. When enraged, people bare their teeth at the corners of the mouth (see Fig 18-10).

With respect to gestures one also finds many agreements among peoples of different cultures. Bowing everywhere seems to be a gesture of submission—for example, during greeting or if one approaches a high-ranking person, or in praying (Fig. 18-22; T. Ohm, 1948). Differences are only quantitative: we may nod, while a Japanese bows very low. In triumph and when we are enthusiastic we throw up the arms (Fig. 18-

TABLE 18-1 *Schematic relationship of various forms of eyebrow lifting*

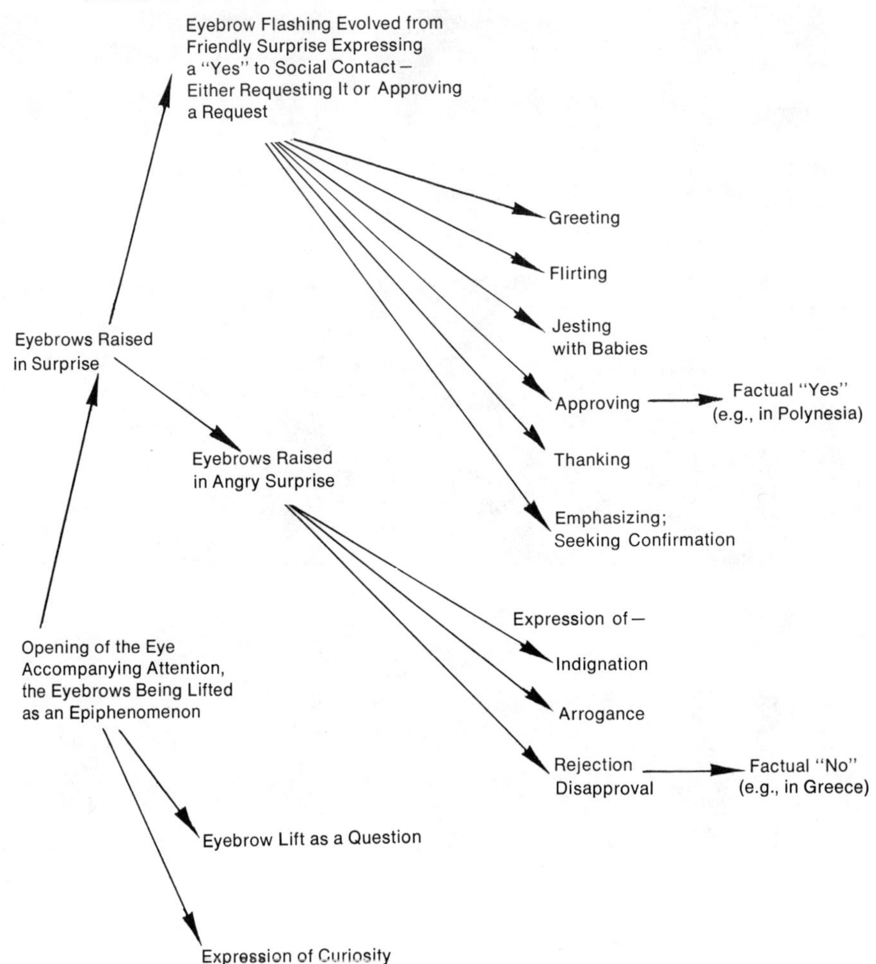

23). Members of the most varied cultures greet by raising the open hand (Fig. 18-24). If one man wants to impress another – to display – it is again done quite similarly in different peoples by an erect posture, stern facial expression, and frequently with an artificial enhancement of the body size and **width of the shoulders.** The only variable is in the means of achieving this expression in different cultures. Some men place feathered crowns on their heads, others fur caps made of bear hide, another displays with weapons and colorful dress – the principle remains the same. When we are angry, we jump up into an intention movement for attack, make fists, and may even bang the table, which is a **redirected attack behavior.** When angry we may stamp with a foot, an intention motion of attack which among Europeans is especially found in small, un-

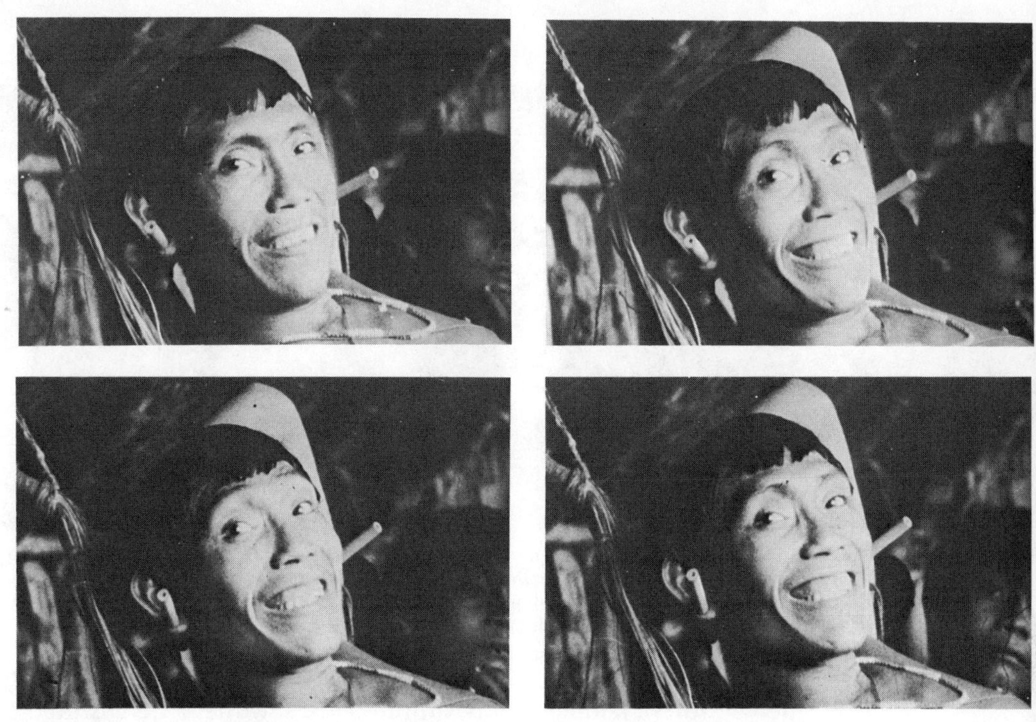

Figure 18-21 Greeting with the eyes by a Waika (upper Orinoco) man. The pictures were taken from a slow-motion (16-mm, 48 frames/second) film and show frames 1, 15, 33, and 76 of a sequence. (Photographs: I. Eibl-Eibesfeldt.)

Figure 18-22 *Left:* A praying Japanese woman in the temple; *right:* Portuguese before the King of the Congo. (*Right:* From Ordoado Lopez, 1957.)

Figure 18-23 Throwing up the hands in triumph: *Left:* An enthusiastic German sports fan during the 1966 World Championship soccer games in England, when Germany was playing Spain, after the first German goal; *right:* exuberant joy expressed at the carnival in Rio de Janeiro. (Photographs: *left:* Associated Press; *right:* I. Eibl-Eibesfeldt.)

controlled children; adults usually suppress it. I saw the same gesture in an angry Bantu boy.

It remains to be investigated to what degree gestures of approval or disapproval have an innate basis. Many races indicate a general "No" by shaking the head, closing the mouth, some by showing the tongue (ritualized spitting out; see Fig. 18-25), and they say "Yes" by nodding their heads. Darwin points out that the first act of saying "No" (disapproving) in children is the rejection of food, by turning the head to the side from the breast or a spoon. One might think of a shaking-off movement.

The **blind and deaf girl** who was described earlier shook her head when she did not want to eat, and also when she refused something—for example, an invitation to play. That people also say "No" with different gestures (for example, a Sicilian by bending his head back) does not argue against Darwin's interpretation. We know that innate behavior patterns can be suppressed by training. For example, one would like to know whether a rejecting shaking of the head is also used by Sicilian children.

It is possible that several primary forms of saying "No" exist and that people in different cultures accept one or the other by convention. One movement of rejection can be traced from the intention of turning away. In saying "No" a Greek, for example lifts his head with a jerk backward, at the same time lowering his eyelids and often raising one or both hands with the open palms showing to the opponent (Fig. 18-26). This behavior can be observed in northern Europe as a gesture of emphatic refusal ("For heaven's sake!"). It is also very similar to the posture of arrogance. Sometimes instead of lifting the head backward we can observe a turning-to-the-side movement. Another widespread gesture of refusal or "No" is head shaking, and sometimes one can observe a rejecting form of shaking the hand, which may be a ritualized shaking off.

In 1971 I worked among the unacculturated Ayoréo Indians (Paraguay), who combined two gestures of refusal. The lips were pointed and pushed outward, the eyes closed tight with a wrinkling of the nose, and occasionally the head was moved slightly towards the back. In cases of lesser intensity, the gesture consisted only of a closing of the eyes by wrinkling the nose (Fig. 18-27).

According to Darwin nodding was derived from an intention movement to eat. Another possible interpretation is given by H. Hass (1968), who says that nodding can be regarded as an intention movement to bow, as a ritualized gesture of submission, so to speak. When expressing agreement one does submit to the will of another. Much is to be said for this interpretation. Nodding is a widespread gesture of approval. Papuans nod and so do Waika Indians and Bantu. Like others, many Indians and Ceylonese also nod when saying, in effect, "Yes, this is so." However, when expressing their agreement to do something they have been asked to do, they sway their head in a peculiar sideways movement. If one asks a Ceylonese, "Do you drink coffee?" he will nod upon confirmation. If, however, we say to him, "Let us drink coffee," he sways his head in agreement. I have seen no nodding in the deaf and blind-born so far.

If accounts are correct, the kiss is not found everywhere. In spite of

Figure 18-24 *Left:* A Schom Pen (Great Nicobar Island) greets by raising the open hand; *right:* a woman from Karamojo (Africa) greeting in the same manner. (Photographs: I. Eibl-Eibesfeldt.)

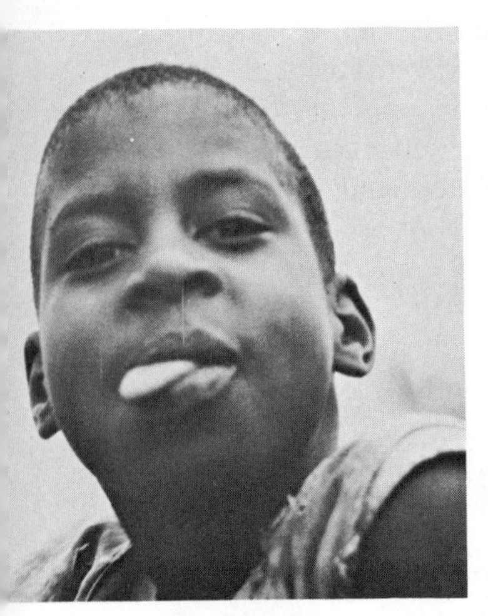

Figure 18-25 Showing the tongue, a widespread gesture of disdainful rejection. *Left:* A Negro boy from the Bihamarulu region (Tanzania); *right:* warriors from New Guinea ridicule one another. (Photographs: *left:* I. Eibl-Eibesfeldt; *right:* from J. Cook, 1784.)

this, however, one might think of it as a kind of ritualized feeding derived from the **care-of-young behavior system,** which has been taken over as one of the expressions of tenderness (Figs. 18-28 and 18-29). In this connection the accounts of L. v. Hörmann (1912) of the behavior of the inhabitants of the Hinterzillertal (a mountain valley in Austria) are of especial interest. It is the custom there to chew pine resin, which gradually changes into a viscous mass that is no longer sticky and is shifted from one cheek to the other, and sometimes is visible from the corners of the mouth. "When chewing pitch the same custom of exchanging wads prevails, as is also done with chewing tobacco. Among lovers this exchange plays an important role" (L. v. Hörmann, 1912:99).[5] The boy exposes a piece of pitch from between his teeth and invites the girl to pull it out with her teeth, an attempt that the boy tries to prolong as a kind of love play. When the dancing partner responds to this invitation of the boy, it is a sign of her interest and affection and even more.

Passing of food as a gesture of contact readiness can also be observed in small children. I recently observed this in a 3-year-old girl who was a guest in our house for the first time. The child watched her parents, who were engaged in friendly conversation with us, but she was at first shy. After lunch, while we were drinking coffee, the girl suddenly came up, took a cookie from a plate, and gave it to me, smiling somewhat embarrassedly. She repeated this with an obviously flirtatious behavior

[5] Hörmann previously reported that in the Zillertal, Pustertal, and Pinzgau regions of Austria the exchange of chewing tobacco is an expression of friendship between men. Acceptance of a chewed piece of tobacco by a girl is confirmation of love returned.

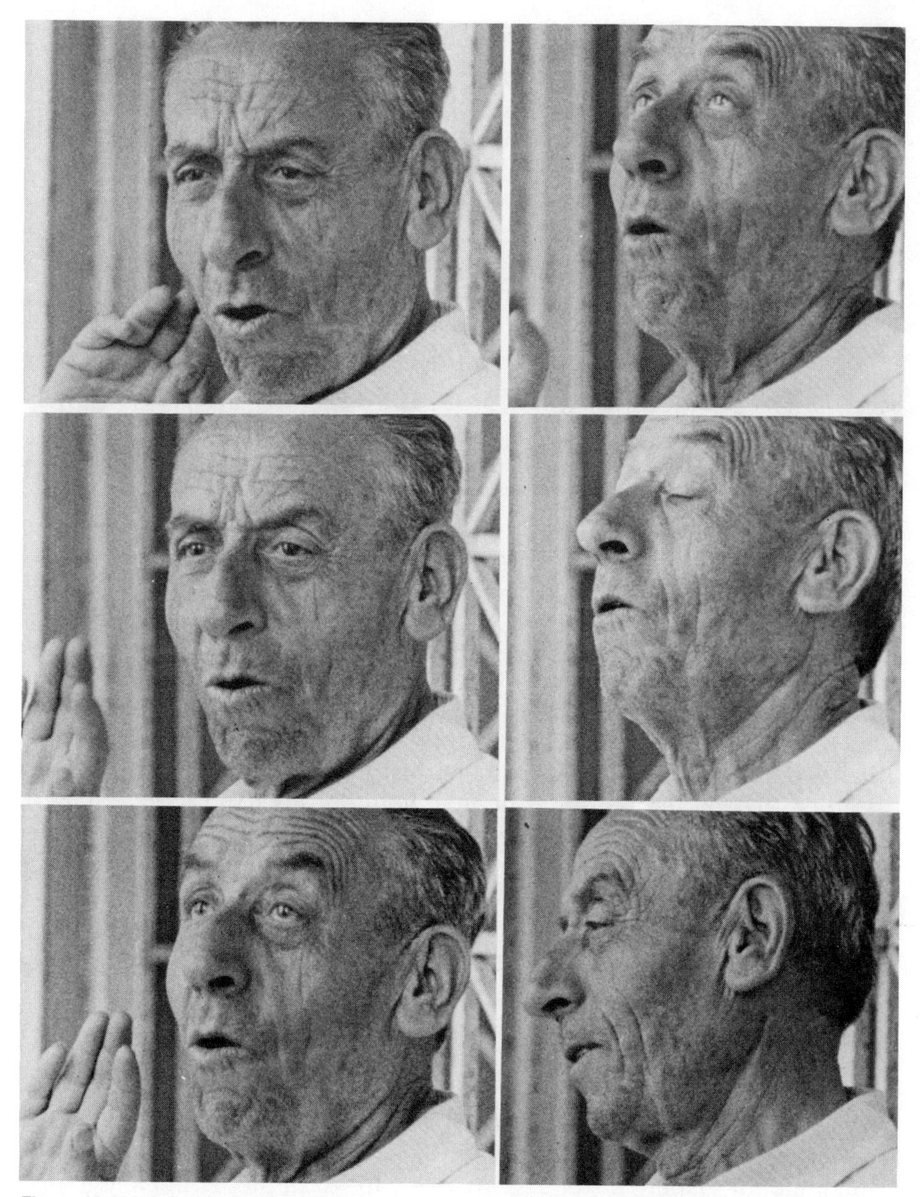

Figure 18-26 A Greek expressing negation. He raises the hand in refusal, raises the head in a backward movement, rolls the eyes upwards, closes the lids, and turns the head away. The **eyebrows** are **raised** as with indignant surprise. This behavioral sequence can be regarded as an expression of strong negation. Normally, the head is thrown back lightly and the eyes are briefly closed. Island of Aegina. Frames 1, 14, 20, 25, 34, 59 from a 16-mm movie film, 50 frames/second. (Recording: I. Eibl-Eibesfeldt.)

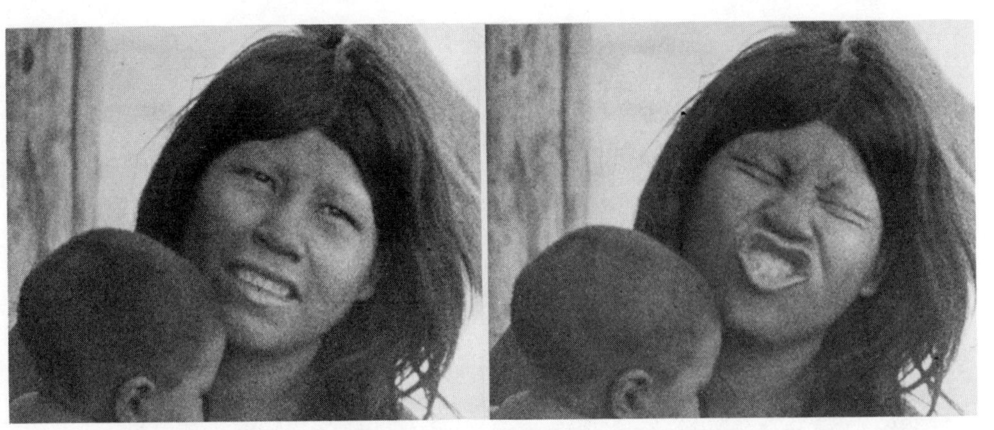

Figure 18-27 An Ayoréo Indian woman expressing negation. She points the lips as if pushing something away and closes the eyes with a wrinkling of the nose. Taken from a 16-mm movie film, 50 frames/second. (Recording: I. Eibl-Eibesfeldt.)

and was happy when I accepted and ate the cookie. From then on she felt completely at ease.

This gesture appeases even those who are enemies, as an acquaintance of mine found during World War II. He had been ordered to capture a prisoner from an enemy trench to obtain information from him, an act that he had carried out successfully on previous occasions. When he jumped into the trench with a drawn pistol and pointed it at the enemy soldier, the soldier, scared as he was, held out his hand with a piece of bread in it. This gesture so changed the mood of my acquaintance that he was unable to carry out his task and withdrew. After that he was unable to carry out similar missions. The food industry uses the function of forming bonds by means of food and drink in its advertising (Fig. 18-30).

In the cases just described the agreement lies in the principle, not in the formal pattern of movements. The motor patterns are not innate but certain inclinations are. It remains to be ascertained whether these are caused by innate releasing mechanisms or by specific drives.

The comparative study of animal behavior can be as revealing as that of people of different cultures. In addition to true homologies, there are many **analogies,** which I have discussed in earlier chapters. The open-mouth display of the chimpanzee, for example, is homologous to the human smile and the relaxed open-mouth display is homologous to laughing (Fig. 18-31; N. Kohts, 1935, and J. A. van Hooff, 1971).

A. Jolly (1972) traced the homologies of various facial expressions in primates. Her findings are summarized in Table 18-2. A. Kortlandt (1972) listed numerous homologous patterns in the agonistic signaling of apes and man. The aggressive displays of chimpanzees are quite human, as can be seen from the illustrations from J. van Lawick-Goodall (1974).

Figure 18-28 !Ko Bushwoman kiss-feeding a baby. Upon contact of the lips the baby opens its mouth and the woman pushes a small piece of melon toward the baby with her tongue. From a 16-mm movie film. (Photograph: I. Eibl-Eibesfeldt.)

Figure 18-29 !Ko Bushman (Kalahari) kissing an infant. (Photograph: I. Eibl-Eibesfeldt.)

TABLE 18-2 Primate Homologies in Facial Expressions

Name	Face	Situation	Bush-baby	Lemur	Cebus	Spider	Baboon	Guenon	Chim-panzee	Man	Human name
Relaxed face			Yes	Yes	Yes	Yes	Yes	Yes	Yes	Yes	
Alert face	Eyes wide, lips may be parted	Novelty, etc.	Yes	Yes	Yes	Yes	Yes	Yes	Yes	Yes	
Tense-mouth face	Eyes wide, mouth narrow slit	Confident threat or attack	No	No	Yes, brows down	?	Yes, brows down	Yes, brows normal	Yes, brows frown	Yes, brows frown	Silent glare
Staring open-mouth face	Eyes wide, mouth open, lips cover teeth	Inhibited threat or mobbing predator	Yes, intention bite	Yes	Yes, brows down	Yes	Brows up	Brows normal	Brows frown	Brows frown	Angry shout or scold
Staring bared-teeth scream face	Eyes wide, mouth corners drawn back, teeth and gums show	Terror, flight, temper tantrums	Yes	Yes	Yes	Yes	Brows up	Brows normal	Brows up	Brows up	Scream
Frowning bared-teeth scream face	Eyes narrow, brows down, mouth corners back, teeth show	Total submission, infant distress	No	No	?	?	Yes	Rare	Yes	Yes, eyes narrow	Intense crying

Silent bared-teeth face	Eyes stare or evade, brows relaxed or up, mouth corners back, teeth show	Social fear or submission, or friendly approach, or bad smell	Yes, with protective responses, not social	Yes	Yes	Yes, but also with attack	Yes	Yes	Yes, often greeting	Yes, eyes narrow	Polite smile
Bared-teeth gecker[a] face	Same with rapid noise	Subordinate flee–approach conflict, infant discomfort	Yes, with defensive threat calls and infant clicks	Yes	Yes, with defensive threat calls and infant squeaks	?	Yes	Yes	Yes	Yes	Begging whimper or laugh
Lip-smacking face	Sucking jaw movements and tongue protrusion, eyes wide	Greeting, sex grooming	No	Rare	Yes	No	Yes, brows up	Rare	Rare	No	
Pout face	Eyes wide, mouth corners forward, "o-mouth"	With contact calls, and especially infant begging	No	Yes	?	Yes, brows up	Yes	Infant only?	Yes	Infant mainly	Pout, begging
Hoot face	Mouth corners far forward, "trumpet-mouth"	With long-distance calls	No	Yes	? (Marked in howling howler)		No	?	Yes	Rare	Howling
Relaxed open-mouth face	Eyes normal or narrow, mouth wide with corners up, brows normal	Play, especially rough-and-tumble	No	No	No	Yes	Yes	Yes	Yes, and with panting grunts	Yes, and with laughing, eyes narrow	Laughter, play

[a] The word *gecker* to describe this expression was first used by Hinde and Rowell (1962) and adopted by Van Hooff (1971). It is Yorkshire (England) dialect, describing the indeterminate whimper made by a baby upon losing the nipple or being surprised, and is an excellent example of the way in which the observer's background shapes the language of science—Dr. Rowell comes from Yorkshire (A. Jolly, personal communication).

From A. Jolly, 1972. Reproduced by permission.
Classification from Van Hooff, with material from Andrew, Moynihan, Struhsaker, Van Lawick-Goodall, and Eisenberg.

Figure 18-30 Appeal to the bonding function of feeding in advertisements for food products.

Most of the depicted patterns can be readily observed during a child's temper tantrum (Fig. 18-32). The similarities found by cross-culture comparison are numerous indeed. Many are to found in **greeting behavior.** I have mentioned the eyebrow flash, nodding, kissing, clasping, and giving hands. The form of greeting that is found among many peoples, the rubbing of noses, is probably not derived from the kiss but has another origin. When Bali lovers greet each other in this way and they breathe in deeply, it is a kind of friendly sniffing. The sense of smell does play a larger role in the social relations between people than is generally realized. In the German language one speaks of not being able to stand another's odor (*man kann jemanden nicht riechen*) when one cannot stand another person. T. Schultze-Westrum (1968) cites a statement by K. Nevermann that among the Kanum-irebe tribe of southern New Guinea it is an expression of close friendship when one takes something of the odor of the person who leaves. The person who remains reaches under the arm pits of the one who is leaving, smells the hand, and rubs the odor over himself (Fig. 18-33).

It has frequently been stated that in cultures in which tenderness is expressed by rubbing noses—for example, among the Papuans, Polynesians, Indonesians, and Eskimos—no kissing exists. This statement is based, however, on imcomplete observation. In the first three cultural regions I observed that mothers hugged and kissed their children, even among stone age Papuans of a remote Kukukuku village who only seven months prior to my visit had had their first brief contact with a government patrol (I. Eibl-Eibesfeldt, 1968). It is very unlikely that these mothers learned this behavior from the patrol members. In the same Papuans I also saw a father kiss his son on the cheek when he greeted him.

Our threat posture can undoubtedly be traced back very far. It is expressed by rolling our arms inward in the shoulders, during which the hair erectors on the shoulders and back contract, although we no longer have any fur. We experience this contraction only as a shudder. In chimpanzees, which assume the same posture, the hair becomes erect and their outline is enlarged (K. Lorenz, 1943).

In response to strong acoustic stimuli people raise their shoulders, bend their head slightly forward, and close their eyes. Reactions homologous to this "neck-shoulder reaction" are known in other mammals (P. Spindler, 1958).

There are certainly numerous motor patterns inborn to man. The analysis of these motor patterns characterized by **form constancy** is made difficult by the fact that motor patterns can superimpose upon each other in a way K. Lorenz demonstrated, when he discussed the variety of **facial expressions in the dog** which derive from the superposition of the intention movements of attack and flight at different intensities.

Figure 18-31 Homologies: (*left*) to smiling (open-mouth display) and (*right*) to laughing (relaxed open-mouth display) in chimpanzees. (From J. A. van Hooff, 1971.)

Further complications are due to the fact that behavior patterns may substitute for each other. Coyness, which plays a significant role in flirting, can be expressed in various ways. A girl may look at you and at the same time show you her shoulder, she may look at you and alternatively look away. She may show tongue protrusion, a defensive reaction which has its origin in a movement of food rejection, and add a smile. Or the smile may be merely repressed by the activation of antagonistic facial muscles, which gives raise to a "coy smile." What remains constant is the combination of antithetical movements of friendly approach and contact readiness with patterns of rejection and withdrawal. In Table 18-3 I enumerate some of the most common patterns of approach and withdrawal which universally occur as elements of coyness behavior.

Figure 18-32 Aggressive displays by chimpanzees. (From J. van Lawick-Goodall, 1974; drawings from originals by David Bygott.)

TABLE 18-3 Elements of coyness behavior

Patterns expressing contact avoidance (mainly derived from intention movements of withdrawal, hiding and cut-off behavior)	*Patterns expressing contact readiness* (mainly derived of intention movements of approach, readiness to accept, and patterns of appeasement)
gaze avoidance (looking away)	turning body or head toward the partner
head lowering	looking at
turning away (head or whole body)	smiling
dropping the eyelid	eyebrow flash
raising the shoulders as to hide the head (a pattern also often seen in defense reactions)	
hiding the face including eyes behind the hands (also observed in blindborn)	
hiding lower face behind the hands—(-hiding a smile)	
clasping hands in front of pubic region or crossing them there in a protective way	
tongue protrusion	
inhibiting expressions of contact readiness, as by activating muscles antagonistic to the smile	
biting lips or fingers (patterns of aggression directed against oneself)	

This may provide an idea of the complexity which derives from the fact that patterns can substitute for one another within a given frame.

Each of the patterns listed under *contact avoidance* may be combined with each of the patterns of *contact readiness*, either simultaneously (for example the eyebrow flash and lowering of eyelid, looking at and turning away) or in succession (looking at and looking away) when coyness is expressed (Fig. 18-16; I. Eibl-Eibesfeldt, 1972b).

Further principles by which the antithetical movements are combined need still to be elucidated by careful correlational analyses. The basic principle of signaling contact readiness in combination with signals of flight and refusal can be discovered in the more elaborate cultural rituals of courtship dances, where approach and withdrawal are acted out, and the principle is probably an old one. In a number of mammals the females use ritualized flight behavior in combination with patterns of contact-seeking to entice males (*Sprödigkeitsverhalten*, I. Eibl-Eibesfeldt, 1957a).

Expressive patterns can be used in different contexts within the framework of a universal grammar analogous to the use of words. I have provided as an example the rapid eyebrow raising. The more generalized meaning of the signal, a "yes" to social contact, acquires a specific meaning according to the particular context in which it occurs.

A very curious display behavior of many primates, including man, was pointed out by D. W. Ploog et al. (1963) and W. Wickler (1966c). Squirrel monkeys display against conspecifics by presenting the erect

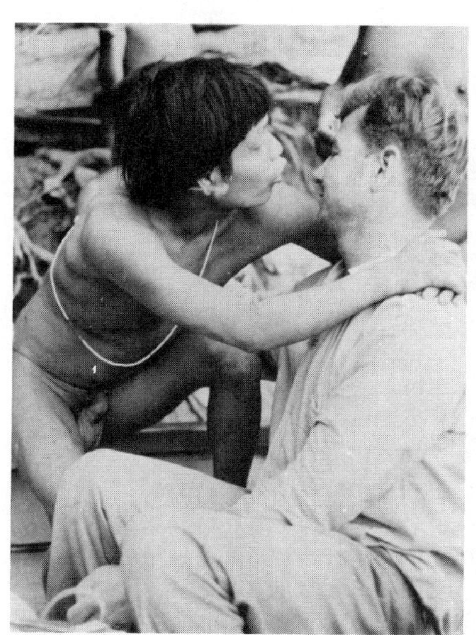

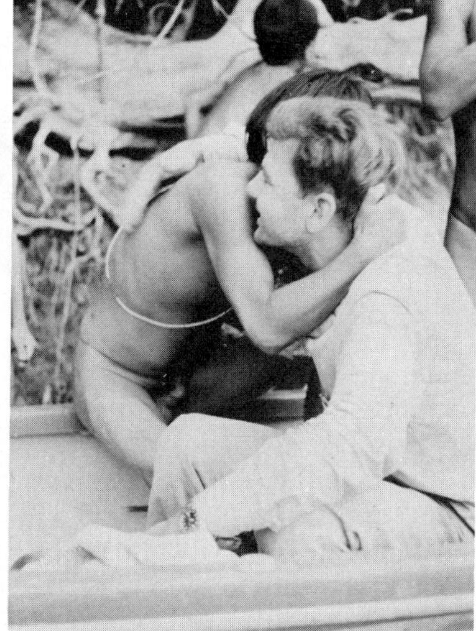

Figure 18-33 The author is greeted by a Waika (upper Orinoco) man with embracing and kissing. The Waika man took the initiative.

Figure 18-34 Genital displays of male primates. Left to right: squirrel monkey (*Saimiri*), vervet monkey (*Cercopithecus*), proboscis monkey (*Nasalis*), and baboon (*Papio*). (From W. Wickler, 1966c.)

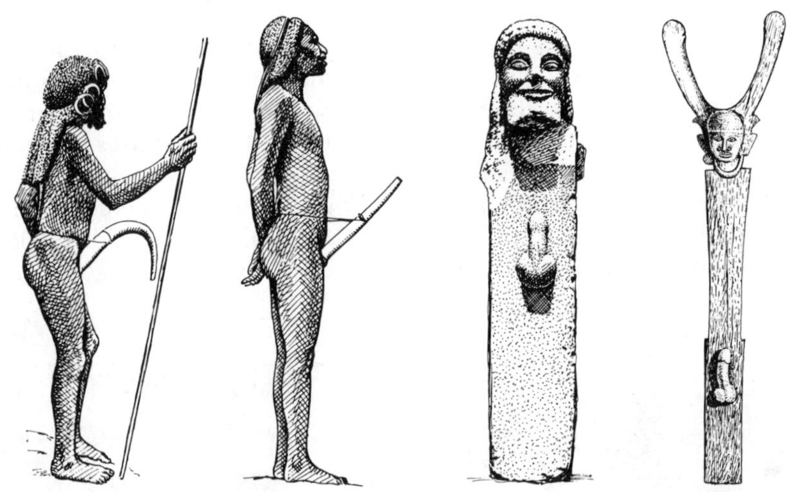

Figure 18-35 Genital displays in man. *Left:* two Papuans from Kogume on the Konca River; *to the right:* herma of Siphnos (490 B.C.), 66 cm high, Athens, National Museum; *right:* house guardian (Siraha) of the natives on the island of Nias. The man-high figures are still in use. In the Greek statue the beard is emphasized as a male symbol; in beardless peoples the male head ornaments are emphasized. (From W. Wickler, 1966c.)

penis when they meet. Even young animals show this. In the common marmoset (*Callithrix jacchus*) the male displays in defense of its family by raising its tail and exposing the rear to the opponent. The testes are pressed into the scrotum, an erection takes place, and the male urinates. After this display, it retreats to a marking place and marks it with urine. During the threat display it looks back at the opponent. The posture of showing the rear to the viewer is probably explained by the flight motivation of the animal. Females display in a similar posture, and not knowing the behavior of the males one could be misled to assume that the posture derived from a female sexual presentation. This is not the case, however. The females mimic the male posturing.

Vervet monkeys, baboons, and many other monkeys have been observed where several males sit at the periphery of their group "on guard." It was believed that they were watching out for predators. But this is precisely what they do not do. Instead they slink away as inconspicuously as possible in such cases. W. Wickler has discovered that this behavior is directed against neighboring troops. The "guards" always sit with their backs to their own group and display their male genitals prominently, which in these animals are very conspicuously colored (Plate VIII and Fig. 18-34). When a strange conspecific approaches the penis becomes erect and in some species it is moved rhythmically. This behavior is a display that serves to mark the territory. Interestingly enough, the same behavior could be demonstrated in man. Some Papuan tribes emphasize their masculinity by artificial means (Fig. 18-35). In some male dresses of Europe this region is still emphasized today by decorative embroideries.

On the Nicobar Islands and on Bali I saw fetishes with an erect penis which are used to ward off ghosts (I. Eibl-Eibesfeldt and W. Wickler, 1968). W. Wickler called attention to stone columns ("hermae") in ancient Greece with a man's head and a penis that were used as property markers. Phallic "guardians" carved in wood or stone can be discovered in Romanesque churches (for example, in Lorch, West Germany, and St. Remy, France). In modern Japan phallic amulets are still used, for example, to protect against car accidents (Fig. 18-36; I. Eibl-Eibesfeldt, 1970).In the Museum of Linz (Austria) one finds amulets that depict male sexual organs. It is possible that pathological exhibitionism can be traced back to a drive to display (Fig. 18-36). This hypothesis is supported by the observations of J. H. Schultz (1966). The sitting posi-

Figure 18-36 Two amulets the author acquired in Japan (Tagata Temple), which are intended to protect the wearer. One amulet shows a threatening face on its front. By removing a cover on the back a golden penis becomes visible. The cover is inscribed with the words: "To protect against traffic accidents." The threatening face and phallus are frequent elements in figures that are meant to offer protection against demons. (See also I. Eibl-Eibesfeldt and W. Wickler, 1968.)

tion of men differs clearly from that of women and is reminiscent of that of the monkeys we discussed (G. H. Hewes, 1957).

W. Wickler derives the genital displays of primates from **urine rituals** that contain elements of copulation behavior. In many mammals males mount conspecifics of the same sex during an aggressive assertion of rank. R. Schenkel (1947) described it in wolves, I. Eibl-Eibesfeldt (1950c) in house mice, S. Zuckermann (1932) in baboons, and C. B. Koford (1963b) in rhesus monkeys. In rhesus monkeys "rage copulations" were also observed during the course of aggressive conflicts, where the aggressively aroused individuals often mount a third one who is not involved (C. B Koford, 1963b). To what degree this occurs in humans is an interesting question. In a recent Polish novel, *The Painted Bird*,[6] I found the statement that young herdsmen rape strangers who enter their territory. A. P. Wilson (personal communication) told me that in some prisons in the United States new prisoners are occasionally beaten up by the other inmates. If the person does not fight back like a man, he is treated like a girl and raped. In short, mounting, in many primates, possibly including man, is a demonstration of rank of an aggressive nature. It appears warranted to me to interpret the "sitting on guard" of many primates as a further ritualization of this behavior (the threat to mount).

RELEASING MECHANISMS, KEY STIMULI, AND RELEASERS IN MAN

The experiments carried out on a large scale by industry and the arts using various models show, just as do certain miscarriages of our esthetic and ethical value judgments, that we react almost automatically and in a predictable manner to certain releasing stimulus situations. This is likely to be on an innate basis, although a definite proof cannot be obtained because no one has ever grown up without experience. We know, however, that **infants** react innately to certain stimulus configurations by **smiling.** Furthermore, R. L. Frantz (1967) has shown that newborns prefer a schematic representation of a human face over an array of other stimuli during their first week of life.

If infants 2–11 weeks old are fastened to a chair and are shown a projection of a symmetrically expanding silhouette, they react as if an object were heading for them on a collision course. They turn their head away, lift their hands protectively, and show a marked increase in pulse rate. They react in the same way to large objects that are moved towards them. If, on the other hand, the projected silhouettes are expanded asymmetrically, making it appear as if an object were moving past the in-

[6] J. Kosinski, 1966, New York (Pocket Books).

fant, no reaction is observed (W. Ball and E. Tronick, 1971). T. G. Bower (1971:32) makes the following comments on these experiments:

> The precocity of this expectation is quite surprising from the traditional point of view. Indeed, it seems to me that these findings are fatal to traditional theories of human development. In our culture it is unlikely that an infant less than two weeks old has been hit in the face by an approaching object, so that none of the infants in the study could have learned to fear an approaching object and expect it to have tactile qualities. We can only conclude that in man there is a primitive unity of senses, with visual variables specifying tactile consequences, and that this primitive unity is built into the structure of the human nervous system.

Other researchers have observed that infants "freeze" at the simulated edge of a visual cliff, indicating an innate fear of falling. Furthermore, infants as young as two months are able to recognize form constancies under varying transformations. Thus, for example, it was possible to condition infants to work an electric switch in their cushions with movements of the head. Reward was the appearance of a person smiling, the training signal a cube measuring 30 cm at the sides presented at a distance of 1 meter. No reaction was observed to a 90-cm cube at 3 meters' distance, although its image on the retina is the same size as that of a 30-cm cube at 1 meter (T. G. Bower, 1966).

The ability of children to combine visual and tactile impressions is also innate. When an object disappears behind a screen, we know it is still there. According to classical theory, a child comes to understand this by reaching behind the screen. On this point, T. G. Bower (1971) performed further experiments: he measured surprise reactions of infants (increase in pulse frequency) while testing them with various optical illusions. For example, he exposed his subjects to objects projected on a screen. The infant reached out to grasp the object, whereupon it failed and showed surprise by an increased pulse rate. If, on the other hand, it succeeded in grasping a sighted object, no change in pulse rate was recorded. Thus, the infant expects to be able to touch an object it sees. The fact that children as young as two weeks of age react this way to the experimental situation permits us to assume that expectation of tactile consequences from visual impressions is inborn. "These results were surprising and interesting. They show that at least one aspect of the eye and hand interaction is built into the nervous system." (T. G. Bower, 1971:33).

Next, Bower posed the question of whether more complex processes could also be preprogrammed. He allowed infants to witness the concealment of an object behind a screen, which was pulled away at various intervals. The children displayed no uneasiness as long as the object was there. But if the object had disappeared when the screen was removed, even subjects 20 days old reacted with an increase in pulse rate—provided that the interval was not too long. "It seems that even very young infants know that an object is still there after it has been hidden, but if the time of occlusion is prolonged they forget about the object altogether. The early age of the infants and the novelty of the testing situation make it unlikely that such a response has been learned" (p. 35). Bower discovered by means of further experiments that children of 8 weeks anticipate the reappearance of an object which had disappeared behind a screen. If this object does not reappear or does so too quickly, the subjects become uneasy. Yet they do not seem to care if instead of a disappearing ball a cube appears on the other side. Only the movement pattern has to fit, if a child is to follow visually. Thus, object identity has to be learned.

These experiments support the hypothesis of innate data-processing mechanisms in human beings and are thus of great theoretical significance. They confirm the view of K. Lorenz that innate releasing mechanisms probably underlie not only our perceptions but also our forms of thought. We do not know for certain how much of our social life is determined by innate releasing mechanisms. There is, however, some solid evidence for their influence.

K. Lorenz (1943) stated that the behavior patterns of caring for young and the affective responses which a person experiences when confronted with a human child are probably released on an innate basis by a number of cues that characterize infants. Specifically, the following characteristics are involved:

1. Head large in proportion to the body;
2. Protruding forehead large in proportion to the size of the rest of the face;
3. Large eyes below the midline of the total head;
4. Short, thick extremities;
5. Rounded body shape;
6. Soft, elastic body surfaces;
7. Round, protruding cheeks, which are probably genuine differentiations with a signal function. Sometimes it is said that in the corpus adiposum buccae we have a mechanical reinforcement of the sides of the mouth to aid in sucking, but this has not been proved. Such an additional function is feasible, of course, but we notice that monkeys and other mammals can get along without this formation. This fact argues for a specifically human organ that evolved in the service of signaling.

These physical attributes are further enhanced by behavioral ones, such as clumsiness. When an object possesses some of these characteristics it releases affects and behavior patterns typical of those toward young children. We find these objects "cute" and may want to pick them up—to cuddle them. B. Hückstedt (1965) and B. T. Gardener and L. Wallach (1965) demonstrated experimentally that the rounded forehead and the relatively large brain case are important characteristics of "cute-

Figure 18-37 "Baby schema" of man. *Left:* Head proportions that are generally considered to be "cute"; *right:* adult forms, which do not activate the drive to care for the young (brood care). (From K. Lorenz, 1943.)

ness" which can be exaggerated in an experiment. The doll and film industry utilizes this posssibility and constructs "**supernormal**" models to elicit behavior of caring for young. Animals are also considered cute if they have some of the childlike characteristics (Figs. 18-37 and 18-38). To be considered cute it is enough that the parakeet has a round head and that a young dog is clumsy and has feet much too big for his body. In Pekinese dogs breeders seem to have produced a perfect substitute object for the unfulfilled mothering reaction of older ladies. By offering cats to experimental subjects of different ages P. Spindler (1961) released the typical patterns of caring for. The reactions (affection, euphoria, patting, bending down the head, using pet names in a high-pitched voice) mature at an age of three.

It is also possible that the understanding of expressions is given a priori by innate releasing mechanisms, because we are easily deceived by simple models. A crying or laughing face can be depicted with a few strokes.[7] When we see such expressions in animals we consider them

[7] When we recognize the simplified expression it could be the result of a secondary process of abstraction, just as we recognize the caricature of a person. This needs to be tested in very small children. If it could be shown that at a certain age the sketchy representation is more effective than the natural object, one could suspect an innate releasing mechanism, because then the abstraction is available before any learning. Experiments are also needed on the congenitally blind who have later had their vision restored.

Figure 18-38 *Left:* Overemphasis of baby characteristics, from the December 1966 issue of *Ladies Home Journal*; *above:* Disney dog as an example of the exaggeration of baby characteristics. Note the rounded shape and the head-body relation.

friendly (Mandarin ducks), arrogant (camel), or daring (eagle), although this has nothing to do with the actual mood of the particular animal (Fig. 18-39). Finally, the automatic reactions to the expressions of another person argue for innate releasing mechanisms that determine a response to an expression. We already mentioned the disarming smile.

That we can respond innately to human expressions has been doubted, primarily on the basis of experiment. Subjects were presented with enlargements made from a film and other photographs of facial expressions for their evaluation (for example, B. M. Turhan, 1960). The subjects judged the pictures quite variably, something that should not be surprising. Expressions are sequential structures. If one wants to examine their releasing effects, one must present the film sequences to the subjects and not present only single frames. When a biologist wants to test the releasing function of a certain bird song he does not present only a single tone. Undoubtedly there are static expressions that can be recognized even on a still photograph, but usually the releasing effect comes about as a result of the entire sequence.

P. Ekman et al. (1969) found that observers in both literate and preliterate cultures (New Guinea, Borneo, United States, Brazil, and Japan) chose the predicted emotion from photographs of the face. The association between facial muscular movements and discrete primary emotions is evidently the same cross-culturally.

Eye spots catch the immediate attention of an observer. R. G. Coss (1965, 1968) measured the pupil reactions of persons who viewed eye spots that were presented singly, in pairs, or in a group of three, each consisting of concentric circles imitating pupil and iris. The strongest responses were obtained to the paired spots, and the responses were stronger when the inside of the eye spot was dark. The response to double pairs of eye spots varied, depending upon their spacing. Horizontal spacing caused a stronger dilation of the pupils than a vertical one.

In the most divergent cultures we find ornamental eye spots in connection with protective magic—for example, on uniforms, folk costumes, and amulets. Eyes adorn the bows of many Greek ships (O. Koenig, 1970). In human communication, staring at someone for any length of time is perceived as an unfriendly gesture, often done on purpose as a **threat stare** or to intimidate and **challenge.**

We judge our fellowmen on the basis of information that is most likely also inborn. The wide agreement of certain male and female ideals of beauty among people of different cultures points in this direction, as well as the fact that exaggeration of individual characteristics on models is so effective. In men broad shoulders are desirable, and rarely will we find a hero in art or literature who has narrow shoulders. The width of the shoulders in relation to the narrow hips is very effective, although it may be tremendously exaggerated,, as is the case with Greek vases and statues (Fig. 18-40). The shoulders are also frequently emphasized by men through clothing.

In human beings the growth line of the reduced hair covering is such that tufts of hair would grow on the shoulders, if more hair were present. P. Leyhausen (personal communication) interprets this trait by the assumption that our apelike ancestors developed a fur covering with tufts at the shoulders to serve as a display characteristic. This interpreta-

Figure 18-39 Many people "misunderstand" a camel's expression. Man has an innate releasing mechanism that responds to the relative position of the camel's eyes to his nose; only in man does this mean an "arrogant turning away." We therefore consider the camel to be an aloof animal. In the eagle the bony ridge above the eyes is seen as a wrinkling of the forehead. Together with the pulled-back corners of the mouth the expression is one of "proud decisiveness." (From K. Lorenz, 1965a.)

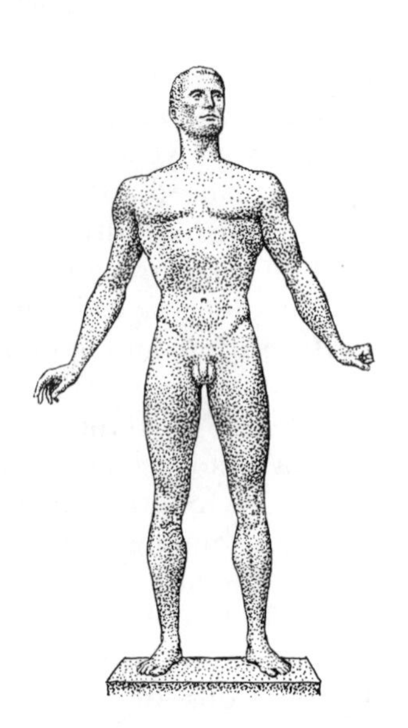

Figure 18-40 Exaggerations of the width of the shoulders in artistic male representations. *Left:* Early Greek bronze figure: Apollo, 20 cm high, formerly in the collection of Tyskiewicz, Boston Museum of Fine Arts; *right:* Dionysos of A. Breker. (Drawings: H. Kacher.)

tion throws new light on the fact that, in greatly divergent cultures, men emphasize their shoulders even today as embellishment and in particular fashions. It appears as if the receiver's selectivity, determined by an IRM, had outlived the reduction of the hair covering (Figs. 18-41 and 18-42). Furthermore, we consider long limbs and slenderness to be aristocratic and against all reason consider gazelles and other animals with such characteristics as possessing nobility, whereas the obese hippopotamus is considered the opposite, although the gazelle and the hippopotamus are each perfect adaptations to particular ecological niches. The ideal of feminine beauty seems to consist of characteristics that may be illustrated, according to K. Lorenz (1943), by the shape of the classical Venus and the prehistoric Venus of Willendorf (Fig. 18-43). To this day people exist whose ideal of beauty corresponds to that of the prehistoric Venus in the main proportions, which are characterized, among other things, by a conspicuous steatopygia (fat deposit on the buttocks), displayed to this day by Hottentot women.

Some of the important secondary sex characteristics that are indicators of normal sexual function in a woman are a slender waistline (lacking heavy adipose deposits), red cheeks and lips, as well as, perhaps, the distribution of the pubic hair. This is not a complete list nor does it necessarily apply to members of all races. We know that some of these characteristics are exaggerated in art and are emphasized by fashion. Fashion designers improve the releasing effect of the female breast by the use of padding, and in the nineteenth century the buttock region was especially enlarged (*cul de Paris*).

D. Morris (1968) interprets the breasts and lips of women as sexual signals that are projected to the front. Our apelike ancestors, argues Morris, mated by mounting from behind and reacted to releasers visible from behind (fleshy buttocks, red labia). Walking erect led to a change in the copulatory position and it became necessary to develop sexual releasers on the front of the body. According to Morris this came about by the evolution of the lips and breasts as the frontal equivalents of the labia and buttocks. The artificially up-lifted breast of a movie star may evoke such an association, but a normal breast is as dissimilar from a buttock as lips are from the labia. Morris also overlooks the fact that men also have red lips. His thesis is hard to support, especially because more likely in-

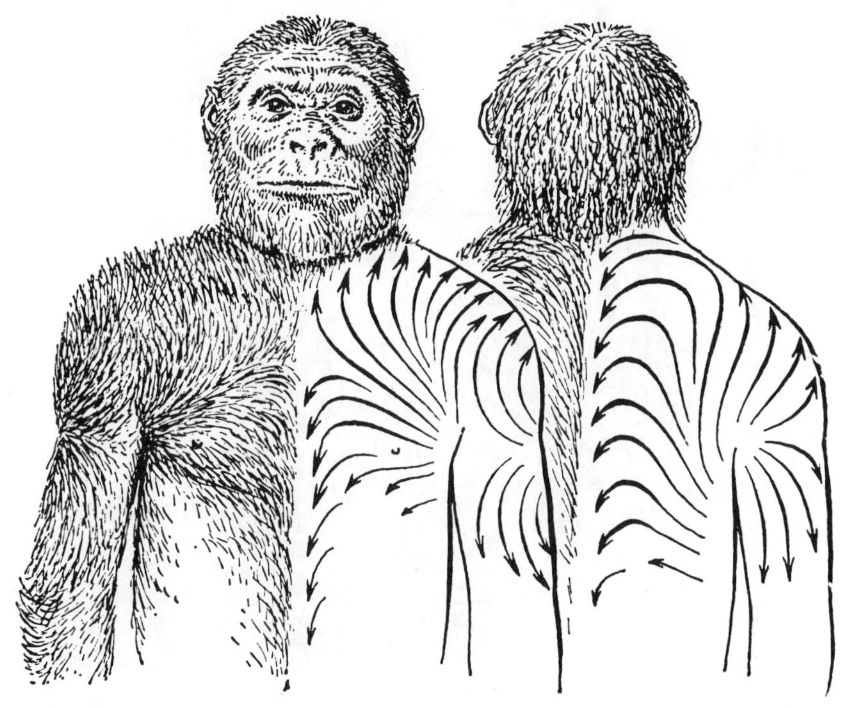

Figure 18-41 The hairline of modern man is such that by raising our hair we would enlarge the outline of our shoulders, if we still had a fur covering. According to the hypothesis of P. Leyhausen, this was probably the case with our apelike ancestors. In this reconstruction, direction of the hairline of modern man, front and back, is projected onto a hypothetical ancestor, showing the latter's probable appearance when the hair is raised. Humans with a strong growth of body hair even today sport tufts of hair on their shoulders. After reduction of the hair covering, the human male's preference for emphasizing his shoulders remained.

terpretations exist for the development of these releasers. In higher vertebrates behavior patterns of caring for the young have a calming effect, for example, cuddling, feeding, clasping, and social grooming. Frightened young mammals run toward their mother to nurse, and children can be calmed with a pacifier. In many higher mammals behavior patterns of caring for the young have been taken over as precopulatory behavior. In man, among other behaviors, this includes sucking (not to be confused with kissing, which is a form of ritualized feeding). The offering of the

Figure 18-42 In the most divergent cultures, the human male tends to emphasize his shoulders in fashion. *Above:* Waika Indian; *middle:* Kabuki actor (Japan) (both taken from photographs by the author); *below:* Alexander II of Russia (drawn by H. Kacher from a contemporary portrait).

breast as a female contact gesture could have been taken over into the sexual domain and in this way the breasts may have acquired their specific releasing function (see also W. Wickler, 1968b). The lips in turn acquired their signal function as a result of their role in kissing. Kissing is a mutual activity, so both sexes evolved lips with signal functions.

The advertising industry uses our readiness to respond to sexual releasers to attract attention and to direct it to the actual message (Fig. 18-44).

There is much evidence that other species besides ours are influenced in their esthetic judgments by innate releasing mechanisms. The **paintings of chimpanzees** indicate that certain basic esthetic perceptions are present in animals. This has been demonstrated by B. Rensch (1957, 1958, 1961) for other apes or monkeys and for some birds (carrion crows and jackdaws), who preferred regular geometric patterns over irregular ones.

We know only very little about releasing stimuli of other senses. With respect to odors I have cited the investigations of **J. LeMagnen,** who found that sexually mature women can smell musk substances which men cannot perceive, unless they have received estrogen injections. A. Comfort (1971) has drawn attention to the presence of pheromones in humans. He describes how women living in close proximity to each other (such as nurses or students living in the same dormitory) show a tendency toward simultaneous menstrual cycles. R. v. Krafft-Ebing (1924) reported the case of a young man who sexually aroused peasant girls by wiping their perspiring brows after a dance with a handkerchief that he had carried in his armpit. In Mediterranean countries forms of dancing exist in which men dance around their female partners while waving a kerchief. It is said that in some areas these have been carried in their armpits. It seems likely that certain pleasant as well as disgusting odors are reacted to innately; the same seems true for certain taste perceptions, although occasionally the key stimuli can be falsified, as shown by our reaction to saccharin. We seem to prefer sweet-tasting food. Normally such substances are rich in carbohydrates and hence in calories.

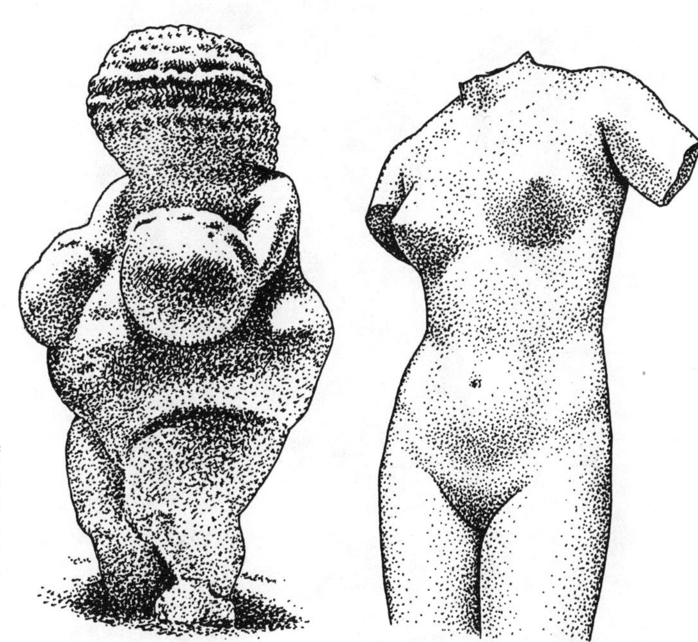

Figure 18-43 *Left:* Venus of Willendorf, limestone carving from Aurignac, France (Museum of Natural History of Vienna, Prehistoric Section); *right:* Aphrodite of Cyrene (Therme Museum, Rome).

Certain perceptions of tactile stimuli release specific defensive reactions. "Creeping things" on the back of the hand release a shaking movement of the hand, which K. Lorenz interprets as a defense against insects. Damage to the teeth is prevented by a reaction to acoustical stimuli. Sharp, screeching noises release this reaction whether we bite on a hard object or scrape a plate with a knife, which in some persons is felt as pain and projected into the teeth. The reaction consists of pulling the cheeks between the teeth and performing cleaning movements with the tongue.

Acoustical releasing stimuli have been little investigated. The crying of babies, the sobbing of another person, the desperate cry for help of a child or a woman move and alarm us so strongly that one may suspect an innate basis. A pilot study by E. H. Hess and B. Beck (1967, personal communication) supports this hypothesis. Using the pupillometric technique developed by Hess, they presented male and female adult subjects with a tape-recorded sounds of a baby (1) crying in pain, (2) crying in hunger, and (3) babbling. The greatest positive pupil reaction was recorded to the babbling. Negative pupil responses were recorded for the two types of crying. The reactions were not uniform in all subjects; that is, some showed clearer distinctions than others. Sex differences were not clear cut in the small sample of the study, and further work is being done. Undoubtedly the internal states of the subjects at the time of the study and their marital and parental status would have had an effect

Figure 18-44 "The legs of your car." Advertisement for Pirelli tires, which utilizes the attention-getting quality of sexual releasers.

on the responses. Soft or harsh conversation can also be recognized without specific knowledge of the particular language.

A basis of inborn releasing mechanisms may be responsible for the effectiveness and appeal of the highest artistic expressions. In music rhythm undoubtedly plays a large role and it has been shown that various physiological rhythms can be brought into phase with a metronome even in animals. I have pointed out that all our close primate relatives display by hitting resounding objects. Drums are among the oldest musical instruments of man, and as war drums they still serve the functions of threat display. This is true also of some other typical noise-producing instruments (horns), which are used to intimidate in place of shouting. There are strong indications favoring the hypothesis that we innately recognize particular melodies as touching, charming, soothing, and so on. We describe melodies by likening them to typically inborn vocalizations such as sobbing or joyous shouting. Tender or rumbling sounds stand for the linguistic expression of tender or angry words. Tender words resemble higher notes and we know that even little girls raise the pitch of their voices when they talk to a little baby. The shrill vocalizations of an angry person are universally understood and we find them uncomfortable.

J. Kneutgen (1970; G. Last and J. Kneutgen, 1970) studied the lullabies of various peoples and came to the conclusion that this is the most uniform musical expression throughout the world. A Chinese lullaby is just as soothing to a particular child as a German song or any other. When listening to lullabies, breathing becomes shallow and regular, like that of a sleeping person. The characteristics of this form of breathing are also in the structure of the lullaby. The regularity is reflected in the regular components of the song. When a lullaby is played on tape, the breathing rhythm of the listener adjusts to the melody; that is, the breaths become as long as the phrases of the song. The inhaling phases of subjects generally coincide with the gradual rises of the melody; the exhaling accompanies the gradually falling melody at the end of the phrase. The breathing seems to accompany the music. The shallowness of breathing is matched by the simplicity of the melody. There are no large intervals, which gives the impression of a gentle gliding. The listeners feel relaxed, the heart beat slows and the psychogalvanic skin response shows little change. In another experiment, when jazz was presented under identical conditions, the subjects became excited. Breathing became irregular and the psychogalvanic skin response showed irregular changes. After the subject did knee bends, heart and breathing became normal within 3 minutes when they listened to a lullaby, and within 6 minutes without it. When jazz was played it took 8–9 minutes until breathing and pulse returned to the base line.

Composers use these key stimuli intuitively to evoke various emotions in the listener—think for a moment of the rumbling drums of

Beethoven's Fifth Symphony. The releasing stimuli are artfully encoded and lose much of their flashy obtrusiveness, which is a characteristic of popular music produced largely for commercial purposes. Because of this coding of key stimuli it also takes some time before one is able "to listen one's self into the music," so to speak. By the artistic manipulation of the releasing stimuli the composer can create and dissolve tensions in the listener. The highs and lows of emotional experiences are touched in an ever-changing pattern that cannot be experienced in everyday life. The heightening of experiences is perhaps one of the most important effects of music. It is most certainly not the only component of musical creation, but it seems to be a substantial one. Pleasure in playful experimentation and in the construction of new and different structures and sequences is another factor.

Inborn releasing mechanisms also seem to determine our need for cover and unobstructed lines of sight. Persons who have had no aversive experiences with others or with predatory animals occupy corner and wall tables first in a restaurant, the tables in the center last. Children feel comfortable in niches and like to build such cover when they play. This has been investigated by D. P. Barash (1972), who found that people eating alone show a stronger tendency to seek cover than do people dining in groups.

Experiments with **infants,** mentioned earlier, have given support to our view that the processing of data in **spatial perception** is, to a significant degree, inborn.

Finally, I should note K. Lorenz's striking suggestion (1943) that a number of releasing stimulus situations which affect our ethical value judgments are governed by innate releasing mechanisms. In the art and literature of all peoples there are recurring themes, situational clichés: loyalty of friends, manly courage, love of homeland, love of wife or husband, love of children and parents—these are the noble basic motives of human actions that we follow because of an inner disposition. They are the basic themes of literature and the theater from the ancient world to this day. We are gripped by the account of the friend who sacrifices himself for his fellow men and we identify with the hero of the legend or the Western movie who liberates and protects the innocent girl or helpless child. "Ethical" norms function as safety devices ensuring species survival. Critical situations in social life where an animal could interact with a conspecific in a way that would very probably be detrimental to the species are controlled by inbuilt inhibitions. I have mentioned that aggressive encounters are often controlled so as to prevent damage to the conspecific. In some social animals the disruption of an existing pair bond by a third (rival) is prevented by inhibitions that prevent courting of a mated conspecific. In a similar way, rivalry over prey is prevented in some carnivores and in chimpanzees. The ownership of prey is, so to speak, respected.

W. Wickler (1969, 1971) has made some fascinating studies on the biological foundations of ethical norms. If there were no innate predispositions for ethical behavior in man, if there were no binding norms for what is essentially good or bad based on phylogenetically acquired adaptations, we would be in a dangerous position indeed. A cultural relativism is the logical consequence, and any cultural norm agreed upon by a society could exist rightfully. Environmentalism would provide the excuse, and I sometimes wonder why this danger is not seen.

I do not imply, however, that we have to follow every "inner" value judgment. I have pointed out that animals carry historic burdens—structures and behaviors that have evolved during phylogeny—which because of changing environmental situations have become maladaptive. The appendix in man, which caused many deaths in former centuries, is one example. In the same way we can consider some of our preprogrammed ethical "values" as being outdated. I mentioned, for example, that there is a strong conformity pressure in groups of men, and outsiders are reacted to with violence—a pattern found all over the world, even in recent times. Throughout history people have reacted emotionally against deviants and minorities, and demagogues have justified this as a "sound popular instinct." Needless to say, this is maladaptive. We have furthermore reached a level of consciousness that makes us realize that those different from us are nonetheless basically the same and that diversity constitutes the particular beauty of mankind. We have therefore to curb our archaic intolerance by encouraging this level of consciousness and taking advantage of our deeply rooted drive to bond (I. Eibl-Eibesfeldt, 1972b).

Our reactions to these situational clichés, on the one hand, ensure our correct behavior in crises—behavior in line with requirements for the preservation of the species. At the same time, however, these reactions also harbor great danger in the present age, which is why insight into the basic underlying mechanisms is needed. This applies especially to the social fighting reaction, whose function is to defend the group. The affective correlate of fighting, which is experienced as enthusiasm, is not released only by an actual threat. With our great readiness for aggression, even a reference to a possible threat is sufficient to release it—an observation that demagogues of all times have expertly used.

PRIMARY AND SECONDARY MEANS OF STABILIZING HUMAN SOCIAL BEHAVIOR

> One instance of the innate and ineradicable inequality of men is their tendency to fall into the two classes of leaders and followers. The latter constitute the vast majority; they stand in need of an authority which will make decisions for them and to which they for the most part offer an unqualified submission. (S. Freud, 1932, in a letter to A. Einstein [1950, Vol. 16:20].)

Are striving for rank and readiness to submit actually characteristic of our species, as S. Freud, among others, assumes, or are they the products of education? To what degree are we by nature social and in the final analysis political beings? Or are we, as Hobbes asserted, forced together against our will by authority? In this section I want to examine and trace the biological bases of our social behavior and pose the question of the extent to which this behavior is determined by phylogenetic adaptations and by what has been acquired.

It is easiest to disprove Hobbes' statement when we take the family as our point of departure. Mother and child are in a naturally well-adjusted relationship to one another on a mutual basis. J. Bowlby (1958) has enumerated in detail the ways in which the bond between mother and child is at first brought about by a number of inborn reactions (such as sucking, clasping, crying, smiling, following) and by the appropriate responses of the mother who loves her baby, in which she responds to certain releasing stimuli. Theories based on the proposition that the child is only secondarily tied to the mother because she fills its need for food and warmth have to date been no more supported by fact than the assertion that the child resents being born and attempts to return to his mother's womb (M. Klein, cited by J. Bowlby, 1958). The bond of the child to its mother is a primary one and does not develop because of the "self-love" of the child at being fed by the mother (K. M. Banham, 1950). Even the individualized relationships between mother and child, which develop gradually through learning processes, are, as I have pointed out, programmed by innate learning dispositions.

However, not only the mother–child bond but also the permanent bond of the parents to each other seem to depend in man on something more than mere tradition. Tradition does determine whether a man may have one or many wives. A permanent, long-lasting association of the partners, however, is generally the rule, and is necessitated by the slow development of the human child. Among the many mechanisms that tie people together, which I discussed earlier, we also find a form of sexual bonding that we do not know of in other mammals; the latter usually mate only during the short estrous periods of the females. Only in the chimpanzee have occasional copulations outside of these periods been observed. R. M. Yerkes (1948) writes that chimpanzee females sometimes presented themselves successfully outside the estrous period and gained certain advantages from a particular male, such as being first at the feeding place. In humans the limitation of the sexual drive and desire to specific cycles or seasons has been largely eliminated. A woman is physiologically ready to respond most of the time to the sexual desires of the man, although she is able to conceive only during a fraction of that time. This enables her also to maintain a tie with the man on the basis of a sexual reward, and this is probably the function of this unique physiological adaptation. Also in the service of maintaining a bond between partners is the ability of the woman to experience an orgasm comparable to that of the man. This increases her readiness to submit and, in addition, strengthens her emotional bond to the partner.

Because of this, the sexual act of humans has acquired a significance in the social life of man that goes beyond the need for reproduction. One argument of the Catholic Church against birth control by artificial means is the supposed unnaturalness of such a measure, which is based on the widespread assumption that the sexual act is in the service of reproduction only. This is so in animals. In man, in addition to this function, there is also the important one of maintaining the bond between partners. The sexual act enhances the relationship between people in a way not present in animals. The erroneous interpretation of this process has often resulted in calling immoral the specifically human aspect of this behavior by admitting only the animal aspects of reproduction, which in the long run results in a disruptive superficiality of the relationships between partners (W. Wickler, 1968c; I. Eibl-Eibesfeldt, 1966a, 1970). We have already seen other comparable extensions of functions and I want to refer here only to the ritualized feeding in the service of maintaining a **pair bond.** W. Wickler (1965b) showed that the presenting movement of female primates has often become ritualized into a greeting and has been used as such by males. In a similar manner courtship movements in fish have been ritualized into appeasement gestures. In all these instances sexual behavior became invested with a new meaning (W. Wickler, 1966e).

We humans do not live only in family units but also find ourselves in village communities, circles of friends, and so on, and in addition to all those whom we know individually, we associate with many people in an **anonymous group.** We are also predestined to this type of group formation. This was emphasized by C. Darwin: "Since man is a social animal it is quite certain that he has inherited a tendency to be loyal towards his fellows and obedient to the leader of his tribe; since these characteristics are common to almost all social animals" (*The Descent of Man*).

As a gregarious being man is characterized by a number of behavior patterns that serve in forming groups, and he is bonded by the same mechanisms. Fear strengthens the group bond—a fact exploited by demagogues throughout the ages. They encourage fear of enemies when desiring to consolidate group cohesion and loyalty to a protective ruler. Similarly, they exploit fear of chaos—for only order guarantees security. This order represents orientation in space and time, including conspecific phenomena. We wish to know how others of our kind will behave.

Furthermore, we are well aware of the intensity with which nurturing activities can motivate a bond. We can easily demonstrate the existence of a strong drive to establish bonds with fellow human beings through gifts and friendly acts (I. Eibl-Eibesfeldt, 1970, 1973b). This tendency goes so far as to motivate soldiers who have faced each other in warfare for several months to exchange cigarettes and to hold their fire. Such behavior is then termed "demoralization of the troops." And since man has so strong a tendency to fraternize, it becomes essential to suppress it in order to maintain hostilities. Even today we can observe the worldwide efforts of various power groups to erect barriers to communication—at an expense of millions of dollars.

It is common knowledge, too, that human beings can be bonded through aggression. K. Lorenz (1963a) has pointed out that individuals will manifest intense emotion (enthusiasm) when partaking in collective aggression against a common enemy. These emotions are accompanied by truly primitive expressive movements.

The sexual bond, as I have pointed out, serves only in the formation of heterosexual partnerships. On the other hand, the three motivational factors mentioned above provide the cement for larger associations. Male groupings are exclusive to varying extents. But exclusiveness is a universal feature and it is based on an innate disposition, which expresses itself at a very early age as "**fear of strangers.**"

We observe in man, as in most vertebrates, distinct territorial behavior. Individuals maintain specific distances between themselves and others. How close we are permitted to approach another person is determined by cultural patterns, but some generally valid basic outlines can readily be discerned (E. T. Hall, 1966). One can experimentally overstep the individual distance by casually sitting close to a person in a library. The behavior of persons subjected to this experiment has been described by N. J. Felipe and R. Sommer (1966). The "victims" at first try to move away from the intruder, and, failing that, they erect artificial barriers against him with books, rulers, and so on. If all these efforts to withdraw fail, they leave the table. The various forms of bodily contact, such as shaking hands, putting arms around another during greeting, or sitting in close contact, are restricted to certain situations and social circles. Children develop individual distances at the time they develop a feeling for property (D. W. Ploog, 1964a). The expression of both tendencies seems based on a common mechanism.

We must expect also that human beings have certain needs for space which are based on an innate disposition and whose fulfillment is necessary for our well-being. It is true that man largely creates his own environment, but its structure is surely in line with his biological constitution (R. Sommer, 1966). Even within a family each person has his own individual domain. The areas owned by each family collectively are more clearly marked. Apartments and gardens are areas to which we assert territorial rights, and this "natural right" is almost everywhere recognized by law makers. No one may enter another's dwelling without special permission; this is considered illegal entrance, "breaking and entering." Fences and signs designate our rightful ownership. Wickler's surprising interpretation of the territory-marking function of the priapic busts and other artifacts has been discussed. Each trespass across territorial boundaries must be accompanied by special ceremonies if it is to remain unpunished. Even when we visit friends we obey certain rituals that appease aggression, for example, giving presents, which have their parallel in the appeasing **greeting ceremonies of animals.**

In everyday life we can observe examples of territorial behavior on many occasions. If one wants to sit down at an already occupied table in a restaurant, it is proper to ask politely if one may do so. If one fails to observe this rule one releases anger in the other person. The same is true when one enters a partially occupied compartment in a train. If one does not greet the occupants in a friendly way one may experience an air of rejection. Mentally ill persons are especially territorial and aware of rank order (B. Staehelin, 1953, 1954; A. H. Esser, 1968, 1971). Only patients on a very regressed stage show no interest in establishing and defending a territory. A degree of aggressive behavior toward intruders is, therefore, considered a healthy sign (A. H. Esser et al., 1965; E. Hackett et al., 1966). These patients who could not acquire and maintain even a chair or a corner for themselves wandered around aimlessly, occasionally attempting—if only for a few moments—to occupy someone else's chair. Mentally retarded boys who stayed together in a fairly large room initially fought for certain territories. When at last everyone had claimed a spot, quarreling became rare. In this way—by means of their territorial behavior—the boys had created some order that was of benefit to all. Each had his place and knew that there he would be left in peace (R. J. Palluck and A. H. Esser, 1971a, 1971b).

In addition to family possessions humans also defend group territories. Basically, human territorial behavior has the same function as in other territorial animals. Anthropological investigations support this view. In the New Guinea edition of the *American Anthropologist* (1965) it was reported that before the intervention of the Australian government some highland tribes lived in areas in which survival was just barely possible. After the tribal wars had been outlawed the people moved into already settled and more suitable regions. Overpopulation and starvation were the result. Among these people, who cultivate tuberiferous plants, food production soon reached an upper limit and tribal wars served here as the spacing-out mechanism for the various groups.

A. Vayda (1970) emphasized the functional aspect of warfare for the Maoris, by which the population was distributed. This mechanism worked as long as the traditional armorment and strategies of warfare were used, but was seriously disturbed with the introduction of muskets by the Europeans.

Occasionally it is asserted that the statements of biologists are defeatist, as if the assertion that a behavioral characteristic is inborn would imply that nothing, therefore, can be done about it. This is a false assumption. Certainly many of our behavior patterns and motivations evolved as adaptations in the service of specific functions. It is as true, however, that changes in the environment can convert the adaptive value of a behavior into its opposite. This is certainly the case with territorial aggression in our overpopulated world bristling with arms. As I have emphasized, effective control of our aggressive impulses is needed for our survival and is certainly possible by environmental manipulation or other measures, provided we explore the causes of this phenomenon without any bias. Those that simply deny man's inclination to act aggressively take the easy road. H. Helmuth (1967), M. F. A. Montagu (1968), and W. Schmidbauer (1971b) have argued along that line, pointing to the existence of nonaggressive people—for example, the Eskimos, the Zunis, the Arapesh, the Hadzas, and the Bushmen (R. Benedict, 1934; M. Mead, 1935; K. Birket-Smith, 1948; R. Lee and I. DeVore, 1968). According to most reports these people lack territorial aggression, except some of the Eskimo tribes. But what has escaped their attention is the fact that they nonetheless show quite a number of aggressive acts within the group. The Eskimos engage in singing duels; they beat each other within the family (K. Rasmussen, 1908).

The myth of the nonaggressive hunter and gatherer is old indeed. Among other sources we must include Nansen, who wanted to represent the Eskimos as a friendly people. H. König (1925:294) has commented on this:

> Let us start with the latter assumption [of Eskimo peacefulness]. Here, his only source of information upon which he based his opinion of these people is none other than Nansen. This man, however, had only very superficial experience of the Eskimos in their natural state. His moral judgment, which he proclaims mainly in his *Eskimo Life,* is certainly biased, for

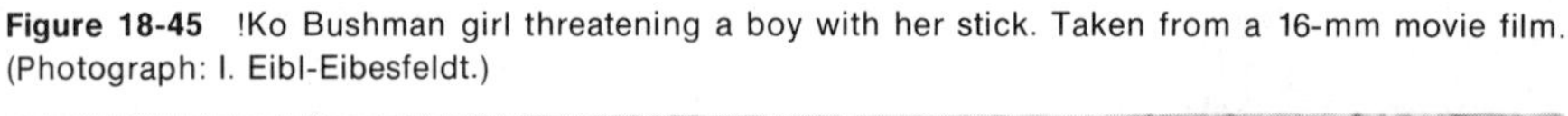

Figure 18-45 !Ko Bushman girl threatening a boy with her stick. Taken from a 16-mm movie film. (Photograph: I. Eibl-Eibesfeldt.)

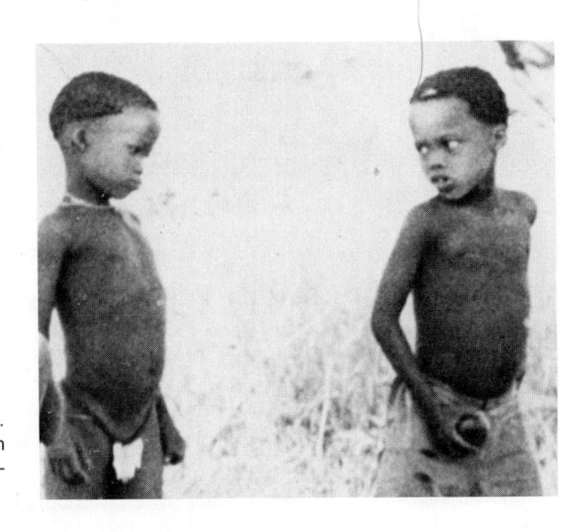

Figure 18-46 !Ko Bushman children (Kalahari). Boy performing threat stare, girl pouting. Taken from a 16-mm movie film. (Photograph: I. Eibl-Eibesfeldt.)

his purpose was to arouse pity. In particular, he draws conclusions for earlier times on the basis of his visit—160 years after the introduction of Christianity. Reports of the first visitors to Greenland do not support the claim that the inhabitants were especially peaceful. Knud Rasmussen reports incidents, especially on the southern east coast, that would have convinced even Steinmetz, had he received the information, of the fallacy of his views.

Yet scholars continue to disseminate this myth, spread by theorists who take their information from second-hand literature. Thus, W. Schmidbauer (1971b), who has never seen a hunter and gatherer even from a distance, claims that "most" hunters and gatherers are remarkably nonaggressive and, above all, do not defend territories. He draws this conclusion mainly from some recent publications on the Bushmen and Hadzas. It is all too obvious that Schmidbauer does not really know this literature, for otherwise he could not have missed the many reports of territoriality in these peoples (S. Passarge, 1907; B. v. Zastrow and H. Vedder, 1930; V. Lebzelter, 1934; I. Brownlee, 1943; H. Vedder, 1952; L. Kohl-Larsen, 1958; L. Marshall, 1961, 1965; P. V. Tobias, 1964; H. J. Heinz, 1966, 1967, 1972). In my monograph on the !Ko-Bushmen, I have cited these studies and have analyzed the issue (I. Eibl-Eibesfeldt, 1972a). My documentation on the !Ko Bushmen contains numerous examples of aggressive confrontations (Fig. 18-45). For example, I counted the altercations of children playing in groups; in one such play group of 9 children, I observed in 191 minutes a total of 116 aggressive acts (hitting with hand or fist, spitting, kicking with the foot, biting, and so on). Of these conflicts 10 ended with 1 partner crying. Only about one third of the scufflings could be classified, according to such criteria as laughing, as friendly and playful. Many of the patterns of threat and fight behavior seemed to be innate—for example, the threat stare and submissive pouting (Figs. 18-46 and 18-47); both of these behaviors are also found in

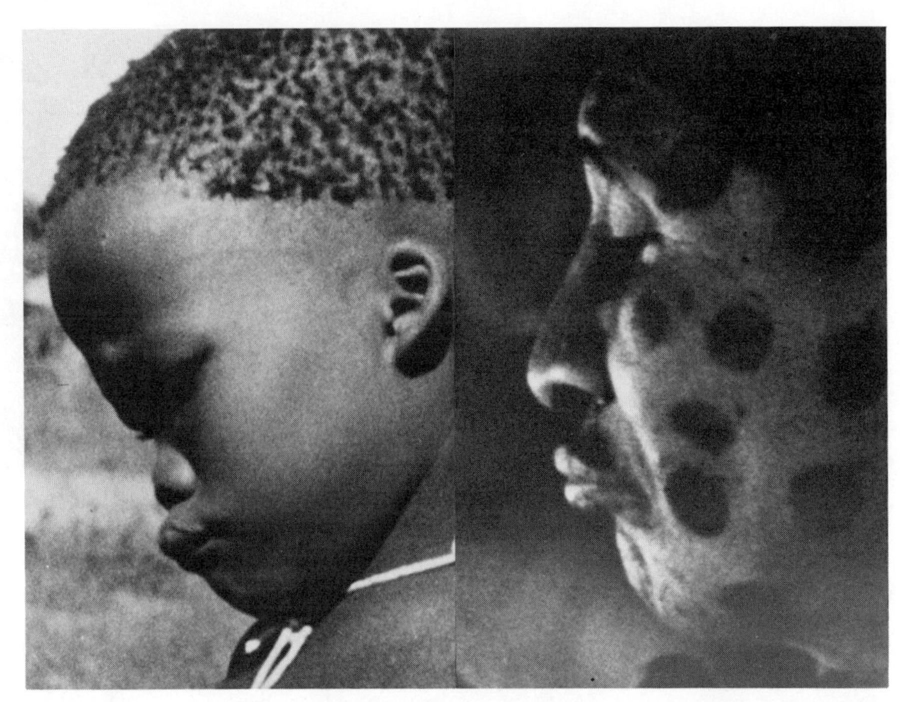

Figure 18-47 *Left:* pouting !Ko Bushman girl (Kalahari); *right:* pouting Waika Indian. The girl had been insulted by a playmate. The Waika man was insulted because, despite his pleas, we could not take him with us because of the illness of one of our boat passengers. Note the similarity of expression. Taken from 16 mm movie films. (Photographs: I. Eibl-Eibesfeldt.)

other cultures. Pouting stops the partner's aggression and often releases comforting behavior.

It is remarkable to observe how early Bushman infants show aggressiveness. They push down their peers, scratch each other, and hit each other from above with the hand (Fig. 18-48). It is therefore quite untrue that Bushmen live in a society without aggression. What is true is that outwardly they are rather peaceful, and manage to contain their aggression within the group as well, through emphasis on bonding behavior (gift-giving rituals, sharing, and so on). Aggression is especially socialized in the children's play groups, whose members learn to get along with each other from experience and through the appeasing influence of older children (I. Eibl-Eibesfeldt, 1972a). Aggression in early childhood is not only a result of frustrations, but is often exploratory. The child by its aggressive actions tests out the range of what is permitted. They therefore expect to get a feedback, and if the partner fails to give the response, they tend to escalate their behavior (B. Hassenstein, 1973).

R. Benedict reports quite aggressive initiation rites in the Zunis (1934; see also P. Weidkuhn, 1968–1969). Recent investigations on the Arapesh reveal evidence for warfare between groups of these allegedly peaceful people. Hunting territories and their defense have been described for the Pygmies, and war between bands for the Hadza. In fact our Stone Age ancestors were already engaging in warfare, as can be seen from cave paintings of war scenes, which date back to the upper Paleolithic. A book presenting the evidence on warfare and aggression in primitive societies is in preparation (I. Eibl-Eibesfeldt 1974, on press; Fig. 18-49).

Certainly we cannot conclude, on the basis of behavior observed with present-day hunters and gatherers, that primeval societies were peaceful. On the contrary, many of these modern hunters and gatherers are singularly aggressive – we need only look at the Andamanese. Reports on warfare among the Waika Indians (who are hunters, gatherers, and horticulturists) are offered by N. A. Chagnon (1968). According to his records, roughly 25 percent of the men die in battle. The victors show little mercy for the vanquished men, often killing even the latter's children and taking their women as spoil (E. Biocca, 1970). We can easily see how this procedure would encourage the distribution of the victor's genes, resulting in a constant, very strict selection process. R. S. Bigelow (1970, 1971) hypothesizes that the origin of man's rapid evolution lies in such warfare. The weight of man's brain tripled during the Pleistocene alone, about four million years ago! Man's tendency to cluster into small groups (**pseudospeciation**), and to compete aggressively with others certainly provided a motive force for this evolutionary development. In a tragic way we are indebted to aggression for the rapid development of our intellect. Yet this same aggression has encouraged our astonishing capacity for cooperation. In the course of this development, man has reached a level of consciousness that should enable him to replace this sanguinary blind alley with a more rational kind of evolution, for we are not "creatures of Cain" animated with nothing but murderous drives, as L. Szondi (1969) would have us believe.

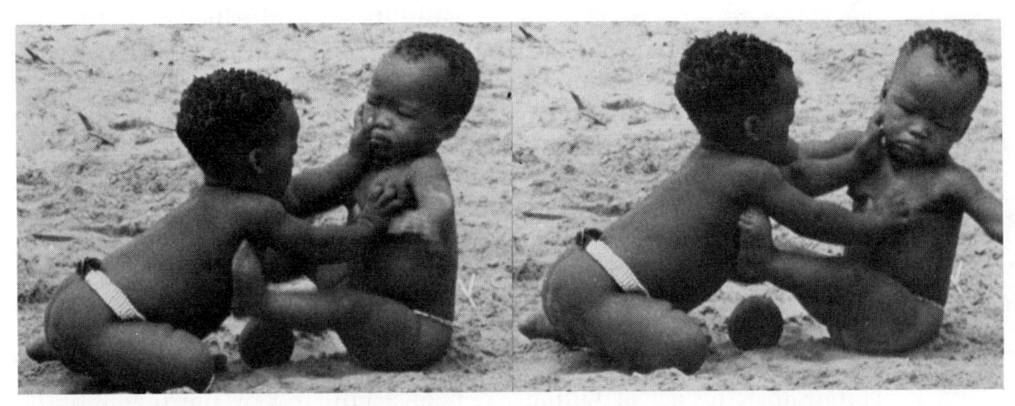

Figure 18-48 Male infant, approximately 10 months old, pushing another over. His fingers dig into the victim's skin and scratch him. Taken from a 16-mm movie film. (Photographs: I. Eibl-Eibesfeldt.)

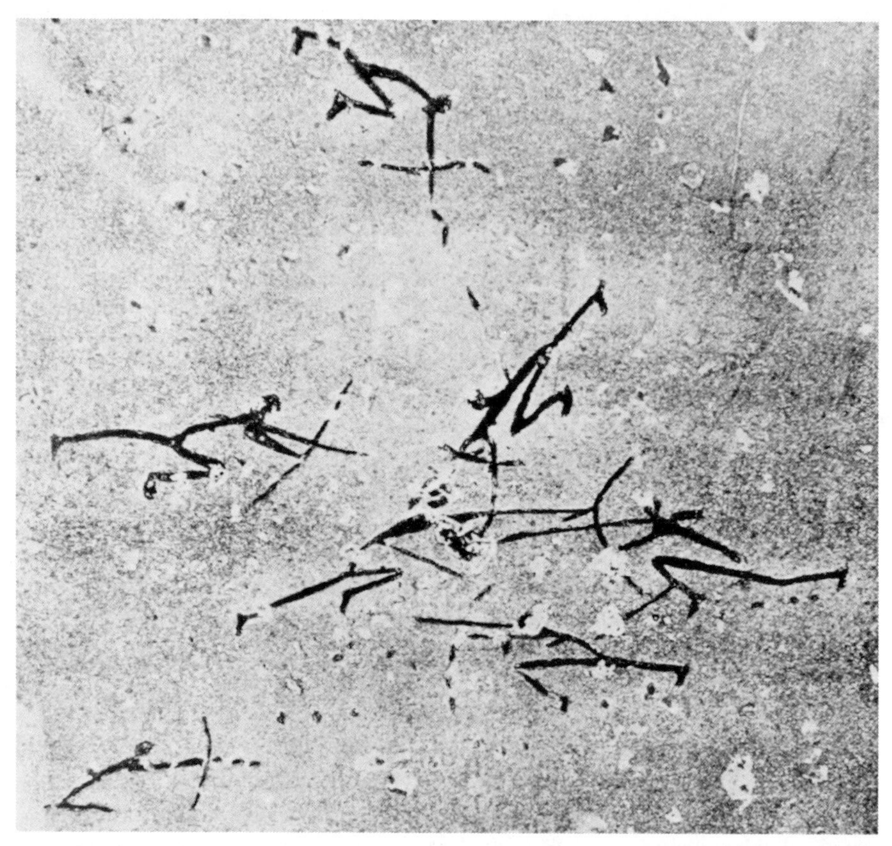

Figure 18-49 Fighting scene in Upper Paleolithic cave painting at Morella la Vella, Castile, Spain. (From H. Kühn, 1929.)

We are aggressive, but at the same time gregarious and endowed with altruistic tendencies—which are just as phylogenetically rooted in our nature as our potential incompatibility (I. Eibl-Eibesfeldt, 1972b). We can even demonstrate through experiment that an individual will learn a task, if as a reward another is freed from suffering (R. Weiss, 1971). If this is true, well may we ask how such a bloodstained history could have come to be? How can a species so endowed with **inhibitions against killing,** whose members show such a capacity for pity, be so murderous? Could it be that the appeasing influence of certain behaviors (smiling, crying, and so on) is effective only in a social group bonded by personal acquaintance? Has the stranger no way of releasing pity? Or has the individual here acquired special means of overcoming inhibitions, that is, of secondarily weakening his empathy?

Observation favors the latter alternative. To begin with, I am quite certain that the act of killing has been facilitated by the invention of weapons. An attacker is able to kill his victim with a weapon—be it only a club—so quickly that the other person has no opportunity to perform a ritual of submission. But this fact alone does not suffice to explain the merciless massacres, during which no victim is spared. Rather, we have here the working of another mechanism. A human being has the ability to convince himself that others are not fellow humans. Even jungle-dwelling Indians speak of their neighbors as though they were but prey for hunting, and members of civilized nations behave in much the same way. They dehumanize their enemy, brand him as subhuman, or make him an "animal." Communication barriers are raised to prevent individuals from becoming acquainted and discovering the truth (R. F. Murphy, 1957; L. Tiger, 1969). In my opinion one of man's most dangerous abilities is this of dehumanizing his fellows, for only through this mechanism can he become a merciless killer. He need only convince himself often enough that certain others are not human, and this indoctrination becomes rooted in structures of the brain. These new associations then grow into subjective reality. Yet it is precisely the fact that man needs to indoctrinate himself, and to create communication barriers so that he may fight, which demonstrates the real strength of his peaceful tendencies (I. Eibl-Eibesfeldt, 1970).[8] I would think that, with this marked drive to fraternize, man would be able to use his wealth of technological means of communication for the purpose of breaking through the mechanism of mutual isolation and persecution.

Studies of the fear-of-stranger response—which develops into stranger avoidance and stranger repulsion even in congenitally deaf and blind children who have never experienced any harm from strangers, and which can be found universally—demonstrate that the concept of "enemy" develops according to the pattern "familiar = friend; unfamiliar (stranger) = enemy." Man's urge to become acquainted and create bonds offers itself as the key for future conflict resolution.

Within the group the aggressive behavior of individuals often leads to the establishment of rank orders, which provide society with a certain amount of stability. The high-ranking persons usually assume some kind of leadership function. The development of such a ranking presupposes not only that some members of the group succeed in establishing their authority, either by fighting or special achievements, but also that the subordinates accept this **rank order.**

This readiness to accept subordination, the corollary of the quest for rank, is very striking and poses particular problems for us. Obedience to the father or some other "recognized personage of public life" is generally considered of some ethical value. In all forms of government there is a tendency to the cult of personality. If necessary people will create models to be honored, and they seem to have a need to follow them. Human beings fight against the rule of brute force, to be sure, but they also seem to have a distinct disposition to follow those whose authority

[8] In contrast to aggression between group members, war aims at destruction. But this destructiveness is a product of cultural evolution. I discuss this matter in more detail in my new book (on press).

they voluntarily accept. When one has voluntarily submitted oneself to authority, one is also at its mercy up to a certain point, as experiments by S. Milgram (1963, 1965a, 1966) have shown in a surprising manner. Milgram invited his subjects (men between 20 and 50 years of age) from various backgrounds (40 percent laborers, 40 percent white-collar workers, 20 percent professional men) to participate in a supposed learning experiment for a small honorarium. They were given the task of administering progressively stronger electrical shocks for each error another person made. The latter was supposed to be learning something, but in actuality was an accomplice of the experimenter. In one sequence of trials the "learner" was strapped to a chair in a separate room from the "teacher" (subject) and electrodes were fastened to his body. The teacher was helped in this by the experimenter. Then the experimenter explained to the teacher that he was to administer a shock for each erroneous answer, and that he should begin with a low voltage, which should be increased as more errors were made. In this way one would be able to study the effect of punishing stimuli on the learning process. The punishing stimuli were administered by the teacher by means of an apparatus that contained switches for 30 steps ranging from 30 to 450 volts. In addition to the voltage designation there were labels ranging from "low shock" to "danger: heavy shock." To test the effect of immediate or delayed feedback from the victim various conditions of feedback were investigated.[9] In the first group of trials the "learner" protested against the treatment by pounding against the wall when the 300-volt lever was pressed, and he ceased to respond at 315 volts and above. In the second series a tape recorder played back the protesting voice. From 75 volts on, each step had a specific response linked to the lever: at first mere mumbling, at 120 volts the report that the shocks were painful, finally protests with the demand to discontinue the test and to release the "learner." At 180 volts the victim cried that he could no longer stand it; from 315 volts on, he refused to answer but groaned in pain when he supposedly received a shock.

The third series was like the second, but here the teacher was in the room with the learner, about 2 feet from him.

The fourth series of trials was like the third, with the difference that the victim received a shock only when his hand rested on an electrified piece of metal. From 150 volts on, he refused to put his hand back on the shock device, and the experimenter then ordered the teacher (who was, in fact, unknowingly a subject) to force the hand of the victim onto the grid.

Forty persons were tested in each group. Under the condition of

[9] Pilot studies had shown that almost all experimental subjects went through the entire scale of punishing stimuli when they received no feedback from the subject.

weak feedback 34 percent of the subjects ("teachers") opposed the experimenter's instructions; when voice feedback was heard 37.5 percent did. When they were next to the subject 60 percent refused and when they had to touch him 70 percent refused. The less abstract and distant the suffering of the victims was, the more the pleading reactions were perceived by the subjects and the greater was the inhibition in the subject against administering shocks even under the authoritarian pressure of another person (the experimenter). However, 30 percent of the subjects still obeyed the instructions of the experimenter when they had to press down the victim's hands.

Frequently the subject would be unsure and ask the experimenter if they were to continue in view of the expressions of pain of the learner. He received the stereotyped answer: "You have no choice; you must continue." In these instances a divergence between what the subjects said and what they did appeared. Although they protested that they could not do such things to the poor fellow, they nevertheless continued to administer the shocks in obedience to the authority of the experimenter.

In another set of experiments in which the degree of supervision by the experimenter was varied it was found that the subjects more readily disobeyed when the experimenter was absent. The number of obedient subjects was three times as large when the experimenter was present as when he gave his instructions by telephone. In addition, many subjects did not increase the shock intensity when the experimenter was absent, although they claimed to have done so. When some of the subjects were allowed to observe a staged experiment in progress before they took part in it as "teachers," and where they could see the other subject refuse to obey the experimenter's order to apply further shocks, in 90 percent of the cases they also refused to obey his instructions (S. Milgram, 1965b).

The results of these experiments prove that a large number of persons had difficulty in opposing the authority of the person in charge of the experiment. Even under conditions of vocal feedback subjects administered shocks to the victims (which in reality would have killed or severely injured these people) in 62.5 percent of the cases. And this took place in the United States, a culture that educates its children against blind obedience. This result, in effect, contradicts all expectations one would have based on the cultural ideal. Of 40 leading psychiatrists who were asked to make predictions about the outcome of the experiments, most expected that the subjects would not go beyond 150 volts and that only 0.1 percent would follow the instructions through all the steps. Between the expectations and the reality there is a remarkable discrepancy. This points to innate tendencies that assert themselves against the cultural ideal.

The postscript that S. Milgram (1966:460f.) attached to his paper leaves one pondering (retranslated from the German):

> With a numbing regularity we saw good people submit to the demands of authority and commit actions that were without feeling and cruel. Persons who in their daily lives were responsible and decent were led to commit cruel acts by the pretension of authority and the uncritical acceptance of the experimenter's definition of the situation.
>
> Where is the boundary for such obedience? In many instances we tried to provide such limits. Screams of the victims were used; they were not enough. The victim complained about heart trouble; still some subjects shocked them when they were ordered to do so. The victim pleaded to be released and his answers were not registered by the signal apparatus; the subjects continued to shock them. Initially we had not expected that such drastic measures would be needed to obtain a refusal to cooperate, and each step in the experimental procedure was only added to the degree to which the ineffectiveness of the previous condition became revealed. The last attempt to erect a barrier was the condition in which the subject had to touch the victim. However, the very first subject already used force against the victim and proceeded to the highest shock intensity. One quarter of the subjects behaved similarly under these conditions.
>
> The results—as they were observed and felt in the laboratory—cause great concern to this author. They allow for the possibility that human nature or—more specifically—the kind of person produced by American society would not afford much protection to its citizens against brutal and inhuman treatment at the behest of an evil authority. A large percentage of these people did what they were told to do, irrespective of the nature of their activity and without conscientious objections, as long as they saw that the order came from a legitimate authority. When, as in this study, an anonymous experimenter could successfully order adults to force a 50-year-old man into submission and administer painful electric shocks to him despite his protests, then one can only be apprehensive about what a government—with much more authority—could order its subjects to do.

All of us know the example of the God-fearing Abraham, frequently glorified by artists, who was willing upon the command of God to sacrifice his own son (Fig. 18-50).

Abraham's sacrifice undoubtedly symbolizes one of the greatest human dilemmas. Obedience is an ethical value, just as is "Love thy neighbor," but when does it cease to be? When these values are in conflict, then obedience is often the stronger of the two, apparently because it is based on an inborn disposition whose roots probably reach back into the rank-order structure of our primate ancestors. Generally it is of advantage for a primate group when they follow the stronger and usually more intelligent alpha males.

From this insight it follows, however, that love of neighbor and the morality of the individual are often not enough to resist the contradictory

Figure 18-50 Abraham's sacrifice (etching by Rembrandt).

orders of strong authorities.[10] Mankind during peace accepts certain humanitarian norms. If it were possible to establish them by law on an international scale and spell them out in detail, this would indeed be a decided advance in humanitarian development. The individual could then call upon the protection of the abstract authority of law against the orders of an evil authority. He would no longer stand alone with his moral decision against one authority but would have another as his ally.

It is also important to educate people to be critical in their attitude toward authority, to avoid blind obedience. Any subordination should be based on reason. In this context the slogan "antiauthoritarian education" has recently been heard. But this seems to be but a phrase, because antiauthoritarian educators still take advantage of authorities. On the walls of antiauthoritatian kindergartens one may see pictures of politicians.

[10] But of the cities of these people, which the Lord thy God doth give thee for an inheritance, thou shalt save alive nothing that breatheth: But thou shalt utterly destroy them; namely, the Hittites, and the Amorites, the Canaanites, and the Perizzites, the Hivites, and the Jebusites; as the Lord thy God has commanded thee. Deut. 20:16–17.

And they utterly destroyed all that was in the city, both man and woman, young and old, and ox, and sheep, and ass, with the edge of the sword. Josh. 6:21.

Probably there is no other way. After all, society can only exist if individual egoism is, to a certain extent, curbed by society.

High rank in man is dependent to a certain point on age, and respect of the old seems to have some biological roots. In baboons we find the relatively high rank of older males, who then also aid the group by their experience even after their physical strength has waned. In man, old people play an important social role (council of elders in a senate) and in many peoples the old impress the group members by special attributes (white hair, bushy eyebrows, white beard), which could be considered as special display attributes that help to compensate for the waning bodily strength.

Membership in a particular group is learned. Some arguments can be made for suspecting imprinting-like fixations as factors in this process. The religious and political ideals of their youth are usually tenaciously adhered to by human beings. The same holds for the identification with an ethnic group. Once a young man has committed himself to a particular group, his attitude is probably determined for the rest of his life. Despite a similar genetic basis people can be very different because of this imprinting-like process: Germans or Russians, Frenchmen or Americans. We owe to this tenacious clinging to a once-acquired attitude the colorful multiplicity of human cultures. However, because groups always close themselves off *against* others, we find here also the root of all ethnocentrism, whose consequences are often a destructive intolerance that still remains to be overcome.

Darwin himself stated (1875:158) that it was man's task to spread his feelings of sympathy from the members of his own small group to all mankind.

> If man progresses in culture, and smaller tribes are unified into larger societies, then the simplest consideration on the part of every individual will tell him that he must extend his social instincts and sympathies to all members of the same nation, even if they are unknown to him personally. Once this point is reached there is only an artificial barrier preventing him from extending his sympathies to all people of all nations and all races. (Retranslated from the German.)

A. Gehlen (1969:121) also pursues this line of thought, speaking of an expanded clan ethos:

> The ethos of brotherly love is familial. It first comes to life within the extended family, yet is capable of being expanded until in principle it embraces all of humanity.

In these religious, political, and ethnic groupings we have already reached the level of anonymous groups. The members of one religious body or of one nation do not know each other individually. However, they are united by common ideas, common representatives (head of nation, head of church), and frequently by very simple common symbols. These symbols may be badges or flags or forms of dress. How important such symbol identification is for the cohesion of a group can be seen from the fact that new nations and political groups see it as one of their first tasks to build tremendous and expensive memorials and to exhibit their insignia of state and the pictures of their presidents everywhere. An important group-binding function is also found in festivals and national holidays.

Within the anonymous group, aggression is largely well buffered, although not as well as in the individualized group. We are on the whole less altruistic toward an unknown member of our anonymous group than toward our personal friend. The necessity of a morality based on reason (Kant) seems to be an inevitable consequence of this fact. We serve a friend without question because we are fond of him, but there are also situations in which people serve anonymous members of their group with a high emotional investment, especially when they direct their attention against members of another anonymous group who are adherents of other ideals. This aggressive characteristic of groups could possibly be overcome by symbolic identification if it were possible to create ideals and symbols that unite the whole of mankind. In this connection an idea of F. Fremont-Smith (1962:104) deserves special attention. He asked himself on the occasion of attending a conference in Russia what common interest could perhaps unite people of different political orientations, and he found an answer: the protection of the child. In an address before his Russian hosts, he said, among other things:

> We have now reached that point in our history when no nation can any longer protect its own children. No government, may it be ever so powerful, can any longer guarantee the safety of its most precious possession, the safety of its children. If a nuclear holocaust should come, New York, London and Moscow will perish and with it all its children. If the nations could, however, agree to protect the children of the others, then the children could be saved. If the USSR would guarantee the safety of all American children, then all children could be saved.

Out of this common desire a common symbol could be formed: the helpless child that needs protection—a symbol that, among others, has been used by the Christian religions as a uniting force. It is interesting that it is again the **care-of-young behavior** which even on this high plane is probably the most effective "cement" to keep a group together.

In all anonymous groups there exists a strong tendency toward conformism, an assimilation to the other members in appearance and behavior. Outsiders who resist this assimilation release aggression. This conformity in behavior becomes especially obvious in large-scale activities such as sports festivals and parades. The phenomenon is based on certain mood-facilitating mechanisms that are similar to those we find in animals. A significant means of inducing a similar mood among people to bring them into harmony is music. Marching music virtually forces everyone into step. In this connection it is interesting to note that, in animals, simple behavior patterns such as the breathing of fishes, songs of birds, and simple movement stereotypes (squirrels) one can be forced into the rhythm of the monotonous sound of a metronome (J. Kneutgen, 1964).

It has sometimes been stated that biologically man is adapted only for life in an individualized group, such as the family or a circle of friends. What I have said so far should have shown that man also possesses all the prerequisites for the formation of anonymous groupings that contain millions of members. He is adapted to both, and the step toward the anonymous group seems to be more recent, with the processes of adaptation still in progress.

We see, for example, that men, in an adaptation to the large society of millions, have shed their individualistic displays with respect to clothing and behavior. A Papuan may still develop a full masculine display within his small community. The members of such a group are so closely tied together by personal familiarity that this basically aggressive display is not disruptive. However, where this bond of personal friendship is lacking, such display behavior would be a cause for conflict. By giving up individualistic and challenging displays in dress and behavior, the members of large societies are adapted to the new conditions. Shaved and in the gray flannel suit we live with less friction in the crowded conditions of modern times. Women, on the other hand, are allowed to enhance their attractiveness and to appear colorful, because their "display" activates binding mechanisms.

In our feelings toward our fellow men we are in the process of developing new attitudes of social responsibility. Emotionally we are still better adapted to a life in individualized communities, the emotional ties with members of the anonymous group being less strong than those with our family and friends. The need is urgent, however, to develop a new social responsibility also toward members of the anonymous group, and ways to achieve this goal are offered by different ideologies. Both Christianity and Marxism propagate the identification of the individual with the anonymous group while at the same time opposing egoism and individualism. Because the source of individualism is the family, attempts have been made repeatedly in theory and in practice to dissolve this core group and also to attack the establishment of individual bonds. Thus, it was hoped, man would become one with the collective. These attempts have failed so far, owing to man's inborn inclination to form families and seek personal relationships. One will thus be compelled to look for methods that enhance the sense of social responsibility of individuals toward the anonymous community but also allow the formation of individualistic bonds. The one does not necessarily exclude the other.

In examining the possible innate dispositions of man, as far as his social behavior is concerned, special attention needs to be given to the question of whether there are any significant differences in the social dispositions of men and women. K. Lorenz (1956) and more recently L. Tiger and R. Fox (1966, 1969) point to the formation of male groups that exclude women, which are found in all cultures and which may be a direct adaptation to hunting and fighting behavior. Men also seem more ready to form anonymous groups than women, who are more oriented toward the family. This leads us to ask if there is a predisposition for the various roles in a society that are played by men and women.

This, however, was refuted by A. Booth (1972*a*, *b*), who showed that woman bond as well and certainly just as often with members of their own sex as men. Nor can it be said that the bond is less lasting. Tiger (1973) answered that his statements on male bonding were more concerned with the difference in functions. In this context a recent study of H. Sbrzesny (1974) is worthwhile mentioning. She reports that !Ko-Bushmen children show a clear preference for playing with members of their own sex. Of the 126 playing groups counted 47.77 percent consisted of boys only, 38.09 percent of girls only, and 13.49 percent were groups in which both boys and girls were members. There seems to be a clear tendency in the children to bond with their own sex. This preference may have stemmed from the obvious sex-specific play interests of the children. Playful exploration of technical equipment consumed 45 percent of the play activity for boys against 4.16 percent in girls. Rough and tumble play made up 16.66 percent in boys against 6.25 percent in girls. These results are remarkable, since Bushmen are in principle very liberal in their educational practices, and the children have free choice to play as they want.

Assertions that differences between boys and girls are the direct result of planned education by the parents are not lacking (M. Mead, 1965). The proof for these statements remains to be provided, and the fact that in almost all cultures, as far as is known, men are more aggressive and less passive than women should give us something to think about. Furthermore, male characteristics, such as a desire for higher rank and increased aggressiveness, are characteristics we share with other primates. For these reasons one should consider again the possibility that constitutional differences may exist (J. Kagan and H. A. Moss, 1962; D. G. Freedmann, 1967).[11] Physiologists indeed emphasize, in considering clinical and experimental data, that human psychosexuality and the derived sexual behavior are dependent upon genetic and endocrine fac-

[11] That women are inherently less curious than men must be doubted. It is more likely they are interested in different things. The possible sex-specific shift of interests should be studied in various cultures in a comparative way.

tors present prior to birth. Genetic factors structure the endocrine environment, and this in turn affects the psychosexual bias of the nervous system (M. Diamond, 1965, 1967).

A very thorough treatment of human sexual differentiation has been presented by J. Money and A. A. Ehrhardt (1972). The authors emphasize the importance of the interaction of genetics and environment in psychosexual differentiation.

> Interactionism as applied to the differentiation of gender identity can be best expressed by using the concept of a program. There are phylogenetically written parts of the program. They exert their determining influence particularly before birth, and leave a permanent imprimatur.[12] Even at that early time, however, the phyletic program may be altered by idiosyncrasies of personal history, such as the loss or gain of a chromosome during cell division, a deficiency or excess of maternal hormone, viral invasion, intrauterine trauma, nutritional deficiency or toxicity, and so forth. . . . Postnatally, the programing of psychosexual differentiation is, by phyletic decree, a function of biographical history, especially social biography. There is a close parallel here with the programing of language development. The social-biography program is not written independently of the phyletic program, but in conjunction with it, though on occasion there may be dysjunction between the two. Once written, the social-biography program leaves its imprimatur as surely as does the phyletic. The long-term effects of the two are equally fixed and enduring, and their different origins not easily recognizable. Aspects of human psychosexual differentiation attributable to the social-biography program are often mistakenly attributed to the phyletic program (1972:2).

Let us present and examine some of the evidence upon which these conclusions are based. The XX and XY chromosomal combinations determine the destiny of the undifferentiated gonad as testis (XY) or ovary (XX). The differentiated gonad in turn determines the further events. Secretions of the testes are a prerequisite for the development of male reproductive structures. In the absence of male hormones the fetus will always develop with a female anatomy. Ovarian hormones seem to play no decisive role at these early stages of development. It is important to remember that **male hormones** not only determine the shape of the external genitalia but also the organization of the brain, in particular the hypothalamic pathways. Without androgen influence the hypothalamic nuclei responsible for later pituitary gonadal cycling differentiate a permanently cyclic, feminine function. If during a critical period in fetal life an androgen influence is exerted, feminine cycling will as a result be abolished. The foundations for boyish and girlish behavior are also determined by the prenatal hormonal influence, but the social environ-

[12] The term is used in analogy to the permission to print something, to mean that a permitted or sanctioned part of the program of development expresses itself and leaves permanent sequelae.

ment from birth on will play a decisive role in shaping the gender role, within the framework provided by phylogenetic adaptations.

Normally the sexual dimorphism of the gonads goes parallel with a dimorphism of the rest of the anatomy. However, since these differences are caused by hormonal influences, incongruencies that constitute hermaphroditism can be brought about. If, for example, testosterone is added to the bloodstream of a genetic female fetus, the girl will be born with a enlarged clitoris or even with a normal-looking penis and an empty scrotum. This can be caused, for example, by abnormally functioning adrenocortical glands of the fetus or by the pregnant mother suffering from a tumor that produces male hormones. If the fetal testes fail to produce sufficient hormone this may result in incompletely masculinized external genitalia of genetic males.

In lower mammals the prenatal determination of behavior by androgenic influences is not influenced by later history. However, if human hermaphrodites of the same type are reared in opposite ways they differentiate in gender identity in agreement with their biography, irrespective of chromosomal, gonadal, or hormonal sex. Nevertheless they will share certain traits of temperament or personality, presumably as a result of the similar prenatal hormonal environment (W. C. Young, 1961). Thus, genetic females masculinized in utero but raised as girls will behave as "tomboys." According to J. Money and A. A. Ehrhardt (1972) the elements of tomboyishness are:

1. The ratio of athletic to sedentary energy expenditure is weighted in favor of vigorous activity. Tomboyish girls will be athletic and will prefer to join boys in outdoor sports, specifically ball games. "Groups of girls do not offer equivalent alternatives, nor do their toys. Tomboyish girls prefer the toys that boys usually play with" (J. Money and A. A. Ehrhardt, 1972:10).
2. Self-assertiveness in competition for position in the dominance hierarchy is strong enough to permit successful rivalry with boys.
3. Self-adornment is spurned in favor of functionalism and utility.
4. "Rehearsal of maternal behavior in doll play is negligable. Dolls get relegated to the permanent storage" (J. Money and A. A. Ehrhardt, 1972:10). In later life they do not show great enthusiasm for baby sitting and similar activities and consider motherhood as something to be postponed rather than hastened, giving preference in anticipation to one or two children, instead of a large family.
5. Romance and marriage are given second place to achievement and career. Priority of career is already evident in the childhood fantasies. The boyfriend stage in adolescence is reached later than in most of their compeers. "Priority assigned to career is typically based on high achievement in school and on high IQ. There is some preliminary evidence to suggest that an abnormally elevated prenatal androgen level, whether in genetic males or females, enhances IQ" (J. Money and A. A. Ehrhardt, 1972:10).
6. In adulthood, responsiveness to the visual (or narrative) erotic image resem-

bles that of men rather than women. The viewer objectifies the opposite-sexed figure in the picture as a sexual partner, as men typically do.

Should the prenatally androgenized girl be raised as a boy she would be assimilated into a purely boyish gender identity. The same individual if assigned as a girl but raised ambivalently might easily differentiate a gender identity as lesbian.

J. Money and A. A. Ehrhardt (1972) emphasize that we may expect prenatal hormonal influences which affect the central nervous system, but not the differentiation of the external sexual organs. The result in such a case would be either an androgenized genetic female with normal anatomy or conversely a genetic male with normal genitalia but a prenatal history of androgen deficit.[13]

As far as we know, pubertal hormones do not determine the relationship between erotic image and erotic arousal but only the degree of arousal to an image already predetermined to have some form of arousal power. The study of matched pairs of hermaphrodites concordant for diagnosis and hormonal output but disconcordant for sex of rearing, typically are also disconcordant for sex of erotic stimulus arousal. This is true even if the hormonal and bodily development of puberty has not been therapeutically corrected to agree with the sex of rearing. It seems thus that the social environment, probably in an imprinting-like fashion, imprints to the erotic image congruent to the sex of rearing.

It should be clear from these findings that the social environment plays an important role in the final gender identification. On the other hand the personality characteristics of maleness seem much less dependant on the social environment, and more determined by prenatal hormonal (androgenizing) influences. And these behavior characteristics that prove comparatively resistent to change are certainly not minor.

I have sketched a tentative framework into which human social behavior has been placed by phylogenetic adaptation. These adaptations consist less of rigid behavior patterns and more of innate motivations and **learning dispositions.** I mentioned drumming and the phallic display as examples. Both dispositions are probably basic drives and innate releasing mechanisms—adaptations on the receptor side that allow recognition of the biologically adequate signal and thus in principle shape the activity of people. The detachment of these territorial displays from rigid motor patterns allows, however, a greater range of expression. Man does not need to sit guard the way other primates do, but may carve statues instead and thus create symbols (see **hermae**). Learning dispositions allow a wider range of freedom than do innate behavior patterns. Despite a basic similarity, this leads to a multiplicity of cultural modifications of human social behavior, where each culture and subculture develop their

[13] In animal experiments it has been found that simultaneous administration of androgens and sleeping pill medication cancelled the androgen effect.

rites in diverging ways. Once formed, they are as rigid as phylogenetically developed rites. Just as the phylogenetically evolved rites of animals control the inborn motivations, so cultural rites do this in man, and for this reason they are just as important for an orderly life together in groups (K. Lorenz, 1966). To gloss over them as just so much "cultural whitewash"—as a sort of superficial varnish—is basically wrong. Our inborn mechanisms are insufficient to control our drives. They became secondarily reduced during the course of phylogenesis and were replaced by cultural control patterns. This is a gain in adaptive modifiability, because various patterns of culture could be developed that made possible the exploitation of various habitats. An Eskimo requires different patterns for the control of his sexual or aggressive impulses than does a modern city dweller of central Europe or the United States. Cultural control patterns can also be changed more quickly following a change in living conditions, but in all cases they are indispensable for social communal life, and man is, as A. Gehlen aptly remarked in this case, a cultural creature by nature.

Everywhere, the disciplining of drives, the ability to maintain control—though perhaps manifesting itself in varied ways—seems to represent an ethical value. Everywhere men strive to refine their customs, and we could almost speak of an inborn appetency for culture. Beyond this general appetency, the various cultural rituals seem to develop on the basis of diverse and rather specific innate learning dispositions (I. Eibl-Eibesfeldt, 1970, 1971b).

To examine these questions along ethological lines is a most attractive task for us in the future. Dance forms, for example, show culturally determined differences, but also some universal traits. In display dances men show off their physical prowess by high jumps, clapping their hands, stomping their feet, showing their weapons, and so on. This is true in the Cossack dances, the Scottish Highlander's dances, or those of the Nilotohamites (Fig. 18-51).

The human desire to impress leads to parallel symptoms in different cultures. Ethnologists have coined the term "prestige economy"

Figure 18-51 Dancing Samburu warriors as an example of human display activity (near Maralal, Kenya). *Left:* The young warriors show off their strength by jumping high up into the air, one after another, to the rhythm of songs and the clapping of hands. We find parallels in many exhibition dances of men (Cossack dances, Scottish Highlander's dances, and so on). Girls sitting nearby seemingly ignore the proceedings. *Right:* Somewhat later the girls join the warriors in the dance. In some phases of the dance the warriors also display their weapons by throwing their spears high and spinning them around the long axis so that the blade glitters in the sun. The dance is not arranged and has been filmed without the awareness of the dancers. (Photographs: I. Eibl-Eibesfeldt.)

(J. Faublée, 1968). This includes expenses for official robes as well as the annual trading in of automobiles, unnecessary from a technical point of view; the herds of the Masai, which are often impractically large; and the huge and expensive stone covers on the graves of some shepherd tribes in Southern Madagascar, erected to demonstrate distinction and power. The costs of these power demonstrations are always considered necessary expenditures.

This prestige display takes the most fantastic form in the Kwakiutl of the Vancouver Islands (R. Boas, 1895; R. Benedict, 1934, 1955). There rival chiefs compete in the destruction of property. The rivals are invited and in front of them valuables are broken in pieces and large amounts of highly prized oil poured into the fire. The guests have to answer in turn by inviting their hosts and joining in this destructive competition by destroying even more. The higher the value of the destroyed property, the higher the prestige achieved. In some Papuan tribes pig feasts tend to escalate in a similar way, since every host in turn does his best to outdo his predecessor (A. Strathern, 1971). And certainly our modern societies follow the same pattern, the Olympic games being only one example. So too certain aspects of the race to the moon, both in the East and West, seem motivated by prestige thinking.

Below their diversity, cultural rituals often share the same basic structure and the same structural elements. We may recall **greeting ceremonies,** which certainly show great cultural variation but in which basic elements (movement patterns) occur as universals. And on a closer look the whole pattern of these rituals of encounter is structured along similar lines, determined by the shared universal functions of appeasement and bonding displays. Power and strength may be expressed by a firm hand shake or a punctilious salute.

So too, during greeting individuals may express their peaceful intentions in various—but certainly not unlimited—ways. A Masai will thrust his spear into the ground in front of him; a Kalahari Bushman may lay his bow and arrow on the ground; our soldiers may present arms when greeting a visitor of state; in principle they are thus all doing the same. Similarly, a friendly bond is usually established by the exchange of food and gifts, and the rituals connected with this are numerous indeed. The basic pattern, however, remains the same. Indeed, so restricted are our means that we even fall back upon these patterns when we deal with revered or feared supernatural powers. Man seeks to appease them by giving presents of various kinds, including ritualized feeding. On Bali there are daily offerings of rice and flowers at designated places, and during large festivals abundant food sacrifices are offered. Such remnants of old customs are also found in the Enns Valley in Austria. A special food made of milk and bread (*Peschtmilch*) is eaten on Epiphany, and the remainder of the food is left on the table with a spoon in it. The *Pescht* (a fairytale character) will come and eat it (V. R. v.

Geramb, 1918). In the United States children leave food for the Easter Bunny and Santa Claus.

In the same way, festivals—according to my observations—follow a similar pattern cross-culturally, whether it be a palmfruit festival of the Waika Indians, a Bavarian rifle club meeting, or a mourning festival in the highlands of New Guinea (I. Eibl-Eibesfeldt, 1973b). Everywhere the occasion is used as a stage on which to show off, in order to gain prestige. This does not lead to hostility, however, but only to mutual respect. The basic object of a feast is, after all, to establish and strengthen a bond. Therefore hand-in-hand with the displays of wealth, power, or merely personal skill goes the demonstration of peaceful intent. The exchange of presents by actual giving and sharing, by mutual dining or by verbal promises strengthens the bond. Sympathy and concern are also demonstrated, for example by expressly remembering the dead and by public grieving. Finally, a number of rituals (special dances and so on) bond by joint action, often with two groups of actors (singers and dancers) interacting in a way resembling the duets of songbirds. However, these rituals are also very culture-specific and thus serve to separate one group from another, by allowing only those to whom the rules are familiar to join in. The group-separating function of rites can be so absolute that such groups separate as if they were different species. Cultures at this level, indeed, behave like ethospecies (see also **pseudospeciation**).

The palmfruit festival of the Waika Indians, which I have studied in some detail (I. Eibl-Eibesfeldt, 1972a, 1973b), may serve as an example to illustrate what I have just presented in more general terms. Among these very bellicose people, village communities form alliances that are initiated and cemented through feasts. Men of another village are invited; they arrive with wife and children. The men are splendidly adorned and dance into the village, at first alone, then in groups. They display themselves imposingly, swinging their weapons. Yet at the same time they appease their hosts by being accompanied by children who dance along and wave green palm fronds (Fig. 18-52). Afterwards the hosts offer their guests plantain soup, and finally the men sniff a drug together. The dead, too, are mourned together and the men will jointly drink the ground-up ashes of the deceased. Gifts are then exchanged. Of special interest are the alternating duet songs of the men, where demands and promises are exchanged. While singing, the partners squat facing each other. At first their presentation is quite intelligible, but the duet alternations become progressively faster and the partners begin to call mere cue words to each other. The roots of this ritual of synchronization appear to lie in the bonding dialogue (I. Eibl-Eibesfeldt, 1972a).

Such dialogues also play an important part in our own daily life. Informational content is often quite banal ("Nice weather today"; "Yes, very nice weather"; and so on) but the social communication is very important; we are expressing our willingness to converse with each other. We communicate that the channels are open and we express agreement. If discord enters the dialogue the conversation is supplemented with specific verbal appeals—for example, infantilisms, if a partner feels neglected and wishes to elicit nurturance. Probably the original form of the bonding conversation is the lulling dialogue between mother and child.[14] From our comparative analyses of festivals and greeting encounters it has become evident that both share the same *basic structures* or *grammar*. One could call a feast a greeting encounter of a group and unite both under the heading "rituals of meeting." Display, appeasing patterns, and signals of bonding are exchanged at encounter in both occasions. And the patterns fulfill the same function of opening the channels for communication without submitting and thus giving up identity. A handshake signals firmness as do the various methods of greeting with the arms—for example by saluting. Friendly signals at the same time bridge the barriers between the individuals or groups.

The opening phase is followed by a phase of communication. The two persons or groups that have met or come together exchange information—one important functional aspect of any encounter ritual—and reassure each other that the channels are open. If no factual information is there to be exchanged, phrases at least are used to demonstrate the readiness to communicate. Thus information of social importance—"I am open to you"—is exchanged. And on both occasions mutual concern is also expressed ("How are you?" or shared mourning for the deceased). And the meeting ends with formal rituals of termination (*Abschied*) including the exchange of presents, which in greeting encounters is verbalized by exchanging good wishes. This grammar of meeting encounters is indeed universal. The complete ritual consists everywhere of an *opening phrase*, a *phase of communication* and a *phase of termination*.

The basic features observed in these rituals of meeting can be observed in a condensed form even in children between one and two years of age. When they establish contact with a visitor they first put themselves in the focus of attention by display. Soon they offer presents (toy or morsels of food) and thus open up for communication. A phase of intensive interaction (for example, by playing together) follows and at the end the child expects to exchange farewells and not just to end abruptly. Indeed, rituals of leavetaking help to relieve the tension otherwise brought about by parting. A child who has learned to wave goodbye to the mother or another reference person is less apt to cry. It overcomes by this ritual the fear of separation. The parting rituals at the occasion of a burial may in a similar way help the distressed.

[14] The social function of these dialogues is well expressed by the name given to them by some psychologists: "stroking."

Figure 18-52 Display and appeasement, two basic components of human expressive behavior: During the palm-fruit festivities different Waika villages (upper Orinoco) extend invitations to one another in order to reaffirm alliances. They exchange, among other things, gifts and promises. The visitors greet their hosts by a dance through the village, with the men displaying themselves in full regalia and war-like postures, waving their weapons and stamping on the ground with fierce expressions on their faces. As with all threat-greeting (**salute, handshake**) the individual thus demonstrates his worth as a warrior and at the same time his value as an ally. The aggressive (and as a result possibly fight-releasing) demeanor is counteracted by the presence of the dancing child (see also p. 367) who follows behind with two ragged palm fronds in his hands. The rituals of man are not arbitrarily invented; they are also shaped by phylogenetically acquired dispositions. (Photograph: I. Eibl-Eibesfeldt.)

There is certainly a great cultural diversity to be seen in the rituals of man. But on a closer look striking similarities in the basic structures as well as in the elements are to be found. Some similarities concern the formal patterns and may even concern the minute details, while other relationships concern functional principles. Some can be explained on the basis of phylogenetic adaptations (for example, by a bias in the perceptual field, which allows the ritual to be formed only along specific lines). And all cultural inventions are subject to selection and are thus shaped according to laws determined by function. In order to understand human behavior, we have to disentangle the determining factors.

The comparative study of customs along ethological lines has hardly begun. Here human ethology as a "biology of cultural achievements" opens up a wide and interesting area of research.

INTELLIGENCE TO USE TOOLS AND LANGUAGE

In the same spirit as J. G. Herder, A. Gehlen (1940) called man a "deficient creature" (*Mängelwesen*), referring to the nakedness of man and his lack of natural defenses. He is more helpless than any animal in this world and would hardly have a chance of surviving if he did not have the aid of technology. This view of man as a deficient being continues to be held in the anthropological literature and has more recently found its way into the popular literature. It is, however, very one-sided. It overlooks the fact that there is no such thing as a perfect organism. Each specialization in one area means a loss in another. A mammal such as a seal, which has been adapted to life in the sea, can move only with difficulty on land. In addition, each presently living organism is the result of countless restructuring processes. The fact that all land vertebrates have descended from fishes is evident as an historical burden in the circulatory system of these vertebrates. The still incompletely separated circulatory system of amphibians and many reptiles can be considered a deficiency from the point of view of an engineer. The step-by-step transformation of the fish circulatory system to include circulation through the lungs as an adaptation to life on land at first leads to an incomplete separation of venous and arterial blood. Frogs, salamanders, and even the lively lizards, as "vertebrates with mixed blood," do not have the endurance to reach peak performance during, say, an escape, as fish, birds, or mammals easily do (G. Kramer, 1949). That embryonic baleen whales form tooth rudiments, only to resorb them again, and that we in an early developmental stage develop branchial arches, can only be interpreted as historical burdens. Whenever a way of life changes, the morphological and physiological adaptations limp after them. The tendency of man to develop fallen arches and varicose veins in the legs shows that these systems still lack some of the necessary adaptations to upright walking (K. Saller, 1963). These things are not, however, specific deficiencies of man, rather they are the expression of evolution in progress.

Furthermore, many characteristics of man considered "faults" prove upon close observation to be true adaptive characteristics. This is true for hairlessness, which, along with the numerous sweat glands, makes it possible to chase prey animals with perseverance in warm climates. Bushmen pursue antelopes until their prey collapses from fatigue. Mammals that have fur often suffer from overheating.

Finally, it can hardly be considered a deficiency that man is not specialized in a particular way. He is, in K. Lorenz's (1959) words "a specialist in not being specialized." To this fact man owes his worldwide distribution. His sense organs are excellent and his physical capacities were illustrated by K. Lorenz (1959:154) by comparing them with those of other mammals.

> If one were to set the three tasks: to march 35 km in one day, to climb up a 5-meter-long rope, and to swim a distance of 15 meters in 4 meters depth and pick up a number of objects in a certain order from the bottom, all activities which a highly nonathletic person who sits much of his time behind a desk—like myself—can do without difficulty, then there is no single other mammal which can duplicate this feat.

This universality in the physical realm is matched by a surprising individual adaptability. Man is, as K. Lorenz said, a "curious being open toward the world." Whereas most mammals are curious only in their youth, this characteristic of youth is retained by man for the rest of his life. He is forever ready to actively explore new things and to experiment with things in his environment (see **play**). From his tree-dwelling apelike ancestors he inherited a few adaptations that were originally developed for climbing: the hand for grasping, binocular vision, and spatial intelligence—the ability to comprehend spatial relationships "centrally." He who climbs about in trees by the use of hands must be able to judge distances well and to integrate the observed relationships centrally. Even to this day our thinking is organized in terms of space; we translate all obscure relationships into concepts that we can centrally "grasp."

> We gain insight into an intertwined relationship—like an ape in a confusion of branches—but we have really comprehended the object only when we have completely "grasped" it. In this last expression the age-old haptic primate reveals himself as a predecessor to the optical one (K. Lorenz, 1959:153).

All these adaptations to life in the trees—the central representation, binocular vision, the grasping hand—became useful in a new way when our ancestors were forced, with an increasing change from jungle to savannah, to move to the ground. At first they probably moved from one group of trees to another, but even this requires erection of the body to look out for enemies above the grass. The hands could become free for carrying objects. Chimpanzees carry fruit in their hands while walking erect.

With the continued spread of the grasslands, the need to hunt prey probably arose and this produced a strong selection pressure toward the use of tools, which eventually led to the making of them. The australopithecines of South Africa and the Olduvai Gorge made tools (R. A. Dart, 1957; L. S. B. Leakey, 1963; G. Heberer, 1965). With these instruments they killed their prey.

Simple chipped-stone tools for a long time determined the state of human culture, and the explosive development of tool-producing cultures came about only in relatively recent times. Why the development stagnated for hundreds of thousands of years after the invention of the first stone implements, and then suddenly evolved into a tool-producing culture, we do not know. Perhaps this was directly tied up with the development of language. Language facilitates the communication of individual experience. In a brilliant report H. Hass (1968) has discussed the various selective advantages of tool using. Tools serve the function of organs; they are "artificial organs," so to speak. They do not need to be nourished, however, and one does not need to carry them around all day. It is also possible to exchange them for others and thus change one's specialization. Different people may use the same artifacts and cooperate in their production. Not only do such instruments as a fork or knife have to be considered artificial, but also trucks, planes, or bridges.

The human hand shows a number of adaptations that enable it to be used in making and using tools, adaptations which are already present in rudimentary form in other primates. Thus, all primates can grasp an object with a hand (adaptation for climbing). The thumb becomes more and more specialized in the primates and finally can be independently moved and opposed to the other fingers (Fig. 18-53). This development is furthest advanced in man; we can hold an object with one or several fingers and the opposing thumb (precision grip). This firm thumb–finger grip is further improved by the broadening of the terminal phalanges. The thumb is long in relation to the index finger and it is moved by strong muscles. The joints connecting the thumb to the metacarpal and the trapezoid bones enable man to rotate it 45° about the long axis so that it can be opposed to all other fingers (J. Napier, 1962).

Higher mammals have the capacity to solve problems without actually trying out all possibilities. A chimpanzee, confronted with the task of getting a banana attached out of reach to the roof of the cage, may sit down quietly and, by looking around, spot a box and finally, without moving, discover the solution of placing the box beneath the banana. In man the capacity to experiment in imagination has advanced so far that we may rightly call him the "phantasy being" (A. Gehlen, 1940). We combine our engrams and thoughts in ever-new ways, and not only when a concrete task demands a solution. We play with these contents of our mind, build castles in the air, devise plans for our actions, and thus dissolve old habits. This capacity enables us to stay open to adaptive modifications. Sometimes, however, we create images in our phantasy that determine as imaginary forces our future behavior (H. Hass, 1967). This is likely to happen when our phantasy constructs originate under the influence of strong motivations (aggression, sex). To a degree man lives a second life in phantasy, and this may have cathartic effects when real life denies the role a man would like to play.

Man's ability to learn new motor patterns purely centrally without

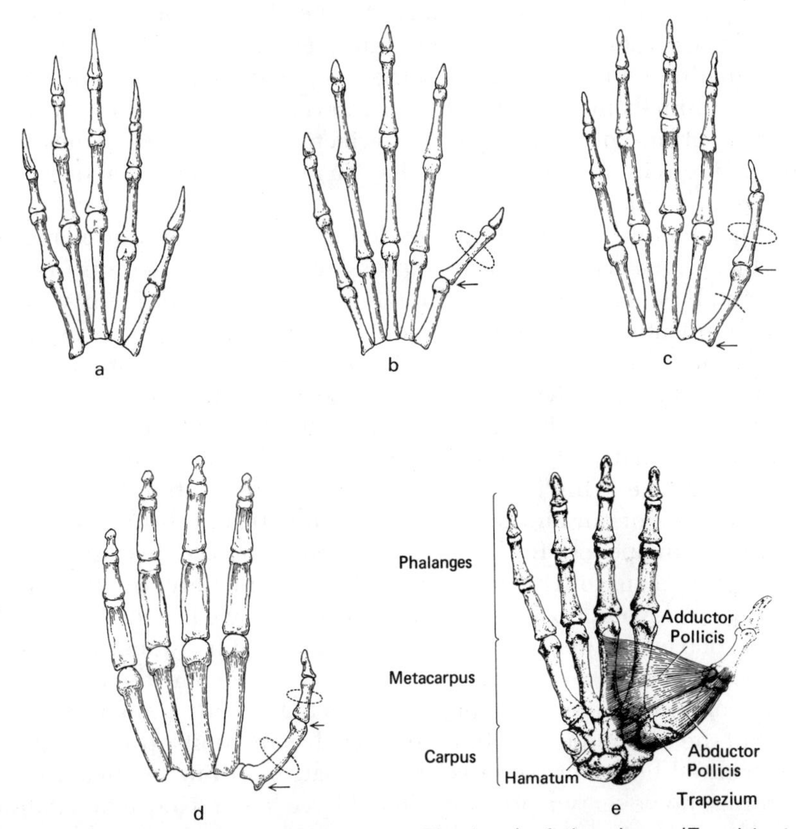

Figure 18-53 Hands of living primates. *a:* The hand of the shrew (*Tupaia*) shows a beginning of the specialization of the thumb that is typical for primates. *b:* In the tarsiers (*Tarsius*) the thumb has already become separated from the other fingers and is able to rotate about the joint it forms with the carpal bone. *c:* In the capuchin monkey (*Cebus*) the angle between the thumb and other fingers is still wider and movement can occur independently at the base of the first carpal joint. *d:* In the Old World gorilla the first carpal is connected by a saddle joint with the first trapezium. This makes possible a rotating motion of the first carpal bone. *e:* The hand of modern man; the development shown in *a–d* has continued. The thumb is quite long in relation to the index finger and curves inward considerably. Strong muscles move the thumb to and away from the palm of the hand. Saddle joints between thumb and carpal bones and carpal bone and trapezium make it possible to rotate the thumb 45° about its long axis so that it can be opposed to all other fingers. The flattening of the distal phalanges makes possible the firm grip between the thumb and fingers. (Napier, J. "The Evolution of the Hand." Copyright © 1962 by Scientific American, Inc. All rights reserved.)

exercise is remarkable. We are not only capable of reproducing an oral or written word in an instant, but we can also invent new movement patterns in our phantasy and reproduce them in behavior according to this phantasy.

The necessity of communal hunting enforced the evolution of a highly differentiated communications system. Our repertoire of expressive movements is very rich: we can signal even with our eyes (I. Eibl-Eibesfeldt, 1968; M. v. Cranach, 1971). It is a reasonable guess that the white eyeballs in man developed to allow the perception of minute eye movements.

Typical for humans is language. I have already pointed out that almost all vocalizations of animals are interjections; only rarely can we observe anything like naming or other language-like communications. Whereas animals merely reflect their internal state with their primarily innate repertoire of **noninsightful interjections,** man gives names to objects in his environment and hence he can make statements about it (J. B. Lancaster, 1966).

To communicate emotions we do not necessarily require language even today, because our innate expressive behavior repertoire is quite sufficient. It is possible that the function of language originally was only to communicate certain environmental contingencies, such as during collaboration for the hunt. Children also use their first sentences for communicating about the environment and only much later express their emotions by the use of language. R. J. Andrew (1963a) calls the emancipation of vocalizations from emotions an important prerequisite for human language. This is in accord with the observation that the most sensitive parts of our auditory range, around 3000 cycles per second, is not utilized in speech. In this frequency range lie the distress calls of a child or a woman, to which we probably react innately; in other words, this range is devoted to the emotions. For speech we use the frequencies around 1000 cycles per second. It is in speech that freedom from instinct develops, and only because of this can language become the basis for objective communication. This objectivity is a distinctly human characteristic, but the independence from instincts has its limits. It is true that, in man, learned behavior patterns predominate over innate ones. In absolute terms, however, we probably do not have fewer fixed action patterns than other primates; it is more likely that we have more. In addition, we have inborn motivations such as play, hunting, and gathering drives, striving for rank and status, which have no definite correlated motor patterns and which are in part recent acquisitions in our phylogeny; we may think here of our **drive to speak.** Such primary motivations are probably the cause of the principal analogies in various cultures, because they have a determining influence as learning dispositions. In the course of this learning, man learns by innate mechanisms when his behavior is drive-reducing or when a terminal, drive-rewarding situation has been met. This liberation of drives from the bonds of strictly programmed courses of motor behavior allows for the wide range of adaptive modifiability of our behavior and is one more of man's distinct behavioral characteristics.

SUMMARY

The sciences of man have for a long time been dominated by the environmentalist assumption, according to which all human behavior, with the exception of some basic reflexes, is learned. As the study of animal behavior has revealed that it is preprogrammed in well-defined ways by phylogenetic adaptations the question has been asked whether human behavior might not be structured in similar ways.

Recent investigations have shown this to be the case. Human beings have motor patterns that developed their adaptations during phylogeny. Some are already functional at birth, others mature during ontogeny, as can be demonstrated by the study of those born deaf and blind, as well as by cross-cultural investigations. Human beings, furthermore, possess inborn detector devices that are tuned to certain environmental stimuli, to which they react in specific ways. Some are functional shortly after birth, others seem to appear during maturation. Social releasers (such as facial expressions) evolved as devices for communication. We are also motivated by inbuilt physiological processes (drives) some of which seem very specific to man, such as the drive we call curiosity, which is shared by only a few of the higher vertebrates. Hunger, sex, and aggression follow patterns we can observe in other mammals. The question whether aggression is based upon inborn motivating mechanisms is much disputed. Neurophysiological evidence is in favor of the inborn-aggression-drive hypothesis. Learning also is preprogrammed by inborn learning dispositions, the most striking one being our capacity and urge to learn language. Many observations allow us to assume the existence of sensitive periods during which certain social attitudes and responses are acquired.

The value of the comparisons between animals and man consists partly in tracing homologies. Analogies, however, are not of lesser importance, as their study informs us about the laws dependent on function.

Cultural rituals can be studied in the same way as phylogenetically evolved ones. They are shaped by selection along basically similar lines, and we should, as the starting point of our investigation, search for the functions of rituals in order to understand the selection pressures that have caused them to evolve. The comparative approach also reveals the existence of universal rules according to which cultural rituals are structured.

In human rituals phylogenetic and cultural adaptations interact in an intricate way, which can be elucidated by cross-cultural comparisons.

Phylogenetic adaptations influence our everyday life at different levels, and this needs to be understood for very practical reasons. We experience serious disturbances in our social behavior and in order to cope with them we have to learn about the determinants of such behavior. It is occasionally argued that by dealing with the innate ethologists surrender to a conservative doctrine, implying that the finding of genetically determined patterns saddles us with accepting them as inevitable. Ethologists therefore emphasize that man is by nature a cultural being, capable of controlling all his inborn reactions. For this reason, if some of them prove to be malfunctional in modern times—a burden of history, so to speak—we can, of course, control them; indeed, it is our intent to help man this way. On this point we would agree with Skinner, who proposes to shape man for his survival. We only take into consideration the nature of man, being aware that man is not equally easily moulded in every direction by the environment. He resists modification in certain areas more than in others and to remember this may prove more humane than negligence on the basis of an extreme environmentalism. Cultural relativism, which derives from environmentalism, could even be held responsible for authoritarian behavior control. If no binding ethical norms are assumed for man, radically different norms could be derived functionally, and nothing could be said against any of them. Fortunately man has been given some universal guide lines of how to behave. His inborn nature is the bench mark by which he can, after all, communicate as a human being across the barriers set up by culture.

References

ABEL, E. (1960a) Liaison facultative d'un poisson (*Gobius bucchichii* Steindachner) et d'une Anémone (*Anemonia sulcata*) en Mediterranée. *Vie et Milieu, 11:*518–531.

——— (1960b) Fische zwischen Seeigel-Stacheln. *Natur u. Volk, 90:*33–37.

ADAMSON, J. (1960) *Born Free.* New York (Pantheon) and London (Collins).

ADLER, J., LINN, G., and MOORE, A. V. (1958): Pushing in Cattle: Its Relations to Instinctive Grasping in Humans. *Animal Behaviour, 6:*85–86.

ADRIAN, E. D., and BUYTENDIJK, F. J. J. (1931) Potential Changes in the Isolated Brainstem of the Goldfish. *J. Physiol., 71:*121–135.

AGRANOFF, B. W. (1967) Memory and Protein Synthesis. *Sci. Am., 216*(6):115–123.

AHRENS, R. (1953) Beitrag zur Entwicklung des Physiognomie- und Mimikerkennens. *Z. exp. angew. Psychol., 2:*412–454; 599–633.

ALBRECHT, H. (1966a) *Tilapia mariae (Cichlidae): Kampf zweier Männchen.* Encycl. cinem. E 603. Göttingen (Inst. wiss. Film).

——— (1966b) Zur Stammesgeschichte einiger Bewegungsweisen bei Fischen: Untersuch. am Verhalten von *Haplochromis (Pisces, Cichlidae). Z. Tierpsychol., 23:*270–302.

ALBRECHT, H., and DUNNETT, S. C. (1971) *Chimpanzees in Western Africa.* München (Piper).

ALBRECHT, H., and WICKLER, W. (1968) Freilandbeobachtungen zur "Begrüßungszeremonie" des Schmuckbartvogels *Trachyphonus d'arnaudii* (Prevost u. Des Murs). *J. Ornith., 109:*255–263.

ALDIS, O. (on press) Priority of Possession as a Determinant of Dominance in Competition for Food and Mates: A Broadening of the Concept of Territoriality.

ALEXANDER, B. K. (1966) The Effects of Early Peer Deprivation on Juvenile Behavior of Rhesus Monkeys. Doctoral Dissertation, Univ. Wisconsin. Summary in *Am. Zool., 6:*560.

ALLEE, W. C. (1926) Studies in Animal Aggregations: Causes and Effects of Bunching in Land Isopods. *J. Exp. Zool., 45:*255–277.

——— (1938) *The Social Life of Animals.* London

ALLEMANN, C. (1951) *Die Spieltheorien: Menschenspiel und Tierspiel.* Zürich.

ALTEVOGT, R. (1955) Beobachtungen und Untersuchungen an indischen Winkerkrabben. *Z. Morph. Ökol. Tiere, 43:*501–522.

——— (1957) Untersuchungen zur Biologie, Ökologie und Physiologie indischer Winkerkrabben. Z. *Morph. Ökol. Tiere, 46:*1–110.

ALTMAN, J. (1966) *Organic Foundations of Animal Behavior.* New York (Holt, Rinehart and Winston).

ALTMANN, M. (1952) Social Behavior of Elk, *Cervus canadensis nelsoni* in the Jackson-Hole Area of Wyoming. *Behaviour, 4:*116–143.

ALTMANN, S. A. (1962) A Field Study of the Sociobiology of Rhesus Monkeys, *Macaca mulatta. Ann. N.Y. Acad. Sci., 102:*338–435.

ALTUM, B. (1868) *Der Vogel und sein Leben.* Münster.

AMBROSE, J. A. (1960) The Smiling and Related Responses in Early Human Infancy: An Experimental and Theoretical Study of Their Course and Significance. Univ. London, Ph.D. Dissertation.

——— (1961) The Development of the Smiling Response in Early Infancy. In FOSS, B. M. (ed.), *Determinants of Infant Behavior.* London (Methuen).

——— (1963) The Age of Onset of Ambivalence in Early Infancy: Indications from the Study of Laughing. *J. Child Psychol. Psychiat., 4:*167–181.

ANDREW, R. J. (1963a) The Origin and Evolution of the Calls and Facial Expressions of the Primates. *Behaviour, 20:*1–109.

——— (1963b) Evolution of Facial Expression. *Science, 142:*1034–1041.

ANTHONEY, T. R. (1968) The Ontogeny of Greeting, Grooming and Sexual Motor Patterns in Captive Baboons (Superspecies *Papio cynocephalus). Behaviour, 31:*358–372.

ANTONIUS, O. (1939) Über Symbolhandlungen und Verwandtes bei Säugetieren. Z. *Tierpsychol., 3:*263–278.

——— (1947) Beobachtungen an einem Onagerhengst. *Umwelt, 1:*299–300.

APFELBACH, R. (1967a) Kampfverhalten und Brutpflegeform bei *Tilapia. Naturwiss., 54:*72.

——— (1967b) *Tilapia macrochir (Cichlidae), Laichablage.* Wiss. Film E1019, Göttingen, Publikationen zu Wiss. Filmen 1A, pp. 63–67.

ARDREY, R. (1962) *African Genesis.* New York (Atheneum) and London (Collins).

——— (1966) *The Territorial Imperative.* New York (Atheneum).

ARMSTRONG, E. A. (1947) *Bird Display and Behavior.* 2nd ed., London (Lindsay and Drummond).

ARONSON, L. R. (1949) An Analysis of the Reproductive Behavior of the Mouth-Breeding Cichlid Fish *Tilapia macrocephala* Bleeker. *Zoologica, 34:*133–157.

——— (1951) Orientation and Jumping Behavior in the Gobiid Fish *Bathygobius soporator. Am. Museum Novitates,* 1486.

——— (1956) Further Studies on Orientation and Jumping Behavior in the Gobyfish *Bathygobius soporator. Anat. Rec., 125:*606.

ASCH, S. (1952) *Social Psychology.* Englewood Cliffs, N.J. (Prentice-Hall).

ASCHOFF, J. (1955a) Jahresperiodik der Fortpflanzung bei Warmblütlern. *Studium Gen., 8:*742–776.

——— (1955b) Tagesperiodik bei Mäusestämmen unter konstanten Umgebungsbedingungen. *Arch. Ges. Physiol., 262:*51–59.

——— (1960) Exogenous and Endogenous Components in Circadian Rhythms. *Cold Spring Harbor Symp. Quant. Biol., 25:*11–26.

——— (1962) Spontane lokomotorische Aktivität. In KÜKENTHAL, *Handb. d. Zool., 8,* 11(4), 1–74.

——— (1964) Die Tagesperiodik licht- und dunkelaktiver Tiere. *Rev. Suisse Zool., 71:* 528–558.
——— (1965) Circadian Clocks. *Proc. Feldafing Summer School, Sept. 1964.* Amsterdam (North-Holland).
——— (1966) Tagesrhythmus des Menschen bei völliger Isolation. *Umschau, 12:*378–383.
——— (1967) Human Circadian Rhythms in Activity, Body Temperature and Other Functions. *Life Sciences and Space Research.* Amsterdam (North Holland), 159–173.
ASCHOFF, J., GERECKE, U., and WEVER, R. (1967) Phasenbeziehungen zwischen den circadianen Perioden der Aktivität und der Kerntemperatur beim Menschen. *Arch. Ges. Physiol., 295:*173–183.
ASCHOFF, J., and MEYER-LOHMANN, J. (1954) Angeborene 24-Stunden-Periodik bei Küken. *Arch. Ges. Physiol., 260:*170–176.
ASCHOFF, J., and WEVER, R. (1962a) Aktivitätsmenge und $a{:}\alpha$-Verhältnis als Meßgröße der Tagesperiodik. *Z. vergl. Physiol., 46:*88–101.
——— (1962b) Spontanperiodik des Menschen bei Ausschluß aller Zeitgeber. *Naturwiss., 49:*337–342.
AUTRUM, H. (1943) Über kleinste Reize bei Sinnesorganen. *Biol. Zbl. 63:*209–236.
——— (1948) Über Energie- und Zeitgrenzen der Sinnesempfindungen. *Naturwiss. 35:*361–369.
——— (1952) Nerven- und Sinnesphysiologie. *Fortschr. Zool., 9:*537–604.
——— (1958) Electrophysiological Analysis of the Visual Systems in Insects. *Exptl. Cell. Res. Suppl., 5:*436–439.
——— (1962) Die Sinnesorgane und ihre Arbeitsweise. *Der große Herder, Ergeb., 1:*519–534. Freiburg (Herder).
——— (1966) Tier und Mensch in der Masse. *Festrede. Munich (Verl. Bayr. Akad. Wiss.).*
AUTRUM, H., and HOLST, D. v. (1968) Sozialer "Stress" bei Tupajas (*Tupaia glis*) und seine Wirkung auf Wachstum, Körpergewicht und Fortpflanzung. *Z. vergl. Physiol., 58:*347–355.
AZRIN, N. H., HUTCHINSON, R. R., and MCLAUGHLIN, R. (1965) The Opportunity for Aggression as an Operant Reinforcer during Aversive Stimulation. *J. Exptl. Anal. Behavior, 8:*171–180.
BABICH, F. R., JACOBSON, A. L., BUBASH, S., and JACOBSON, A. (1965) Transfer of a Response to Naive Rats by Injection of Ribonucleid Acid Extracted from Trained Rats. *Science 149:*656–657.
BACKHAUS, D. (1960) Über das Kampfverhalten beim Steppenzebra. *Z. Tierpsychol., 17:*345–350.
——— (1961) Giraffen in zoologischen Gärten und freier Wildbahn. *Inst. des Parcs Nat. du Congo et du Ruanda-Urundi.* Brussels.
BAERENDS, G. P. (1941) Fortpflanzungsverhalten und Orientierung der Grabwespe, *Ammophila campestris. Tijdschr. Entomol., 84:*68–275.
——— (1950) Les Sociétés et les Familles de Poissons. *Colloq. Intern. Centre Natl. Rech. Sci., 34:*207–219 (Paris).
——— (1956) Aufbau tierischen Verhaltens. In KÜKENTHAL, *Handb. d. Zool., 8,* 10 (3), 1–32.
——— (1957) Behavior: The Ethological Analysis of Fish Behavior. In BROWN, M. (ed.), *Physiology of Fishes II.* New York (Academic Press), 229–269.
——— (1958) Comparative Methods and the Concept of Homology in the Study of Behaviour. *Arch. Neerl. Zool., 13:*401–417.
BAERENDS, G. P., and BAERENDS-VAN ROON, J. M. (1950) An Introduction to the Study of the Ethology of Cichlid Fishes. *Behaviour, Suppl.; 1:*1–243.
BAERENDS, G. P., BRIL, K. A., and BULT, P. (1965) Versuche zur Analyse einer erlernten Reizsituation bei einem Schweinsaffen. *Z. Tierpsychol., 22:*394–411.
BAERENDS, G. P., BROWER, R., and WATERBOLK, H. T. (1955) Ethological Studies on

Lebistes reticulatus Peter: I. Analysis of the Male Courtship Pattern. *Behaviour, 8:*249–334.

BAERENDS, G. P., and DRENT, R. H. (1970) The Herring Gull's Egg. *Behaviour, 17:* (Suppl).

BAEUMER, E. (1955) Lebensart des Haushuhns. Z. *Tierpsychol., 12:*387–401.

BAK, I. J. (1965) Electron Microscopic Observations in the Substantia nigra of Mouse during Reserpine Administration. *Experientia, 21:*568.

BALL, W., and TRONICK, E. (1971) Infant Responses to Impending Collision: Optical and Real. *Science, 171:*818–821.

BALLY, G. (1945) *Vom Ursprung und von den Grenzen der Freiheit, eine Deutung des Spieles bei Tier und Mensch.* Basel (Birkhäuser).

BANDURA, A., and WALTERS, R. H. (1963) *Social Learning and Personality Development.* New York (Holt, Rinehart and Winston).

BANHAM, K. M. (1950) Development of Affectionate Behavior in Infancy. *J. Gen. Psychol., 76:*283–289.

BANKS, E. M. (1962) A Time and Motion Study of Prefighting Behavior in Mice. *J. Gen. Psychol., 101:*165–183.

BARASH, D. P. (1972) Human Ethology: The Snack-Bar Security Syndrome. *Psychol. Reports, 31:*577–578.

BARDACH, J. E., WINN, H. E., and MENZEL, D. W. (1959) The Role of the Senses in the Feeding of Nocturnal Reef Predators *Gymnothorax moringa* and *G. vicinus. Copeia, 2:*133–139.

BARLOW, G. W. (1962) Ethology of the Asian Teleost *Badis badis:* IV. Sexual Behavior. *Copeia,* 346–360.

BARLOW, H. B., HILL, R. M., and LEVICK, W. R. (1964) Retinal Ganglion Cells Responding Selectively to Direction and Speed of Image Motion in the Rabbit. *J. Physiol. (Cambridge), 173:*377–407.

BARTHOLOMEW, G. A. (1953) Behavioral Factors Affecting Social Structures in the Alaska Fur Seal. *Trans. 18th North Am. Wildlife Conf.,* Wildlife Management Inst., Washington.

BASEDOW, H. (1906) Anthropological Notes on the Western Coastal Tribes of the Northern Territory of South Australia. *Trans. Roy. Soc. South Australia, 31:*1–62.

BASTOCK, M. (1956) A Gene Mutation Which Changes a Behavior Pattern. *Evolution, 10:*421–439.

BASTOCK, M., MORRIS, D., and MOYNIHAN, M. (1953) Some Comments on Conflict and Thwarting in Animals. *Behaviour, 6:*66–84.

BATESON, G., and MEAD, M. (1942) Balinese Character: A Photographic Analysis. *Special Publ. N. Y. Acad. Sci.,* 2.

BATESON, P. P. B. (1966) The Characteristics and Context of Imprinting. *Biol. Rev., 41:*177–220.

BATHAM, E. J., and PANTIN, C. F. A. (1950) Inherent Activity in the Sea-Anemone *Metridium senile. J. Exptl. Biol., 27:*290–301.

BAUMANN, H., and HALX, G. (1972) Ophrys: die Pflanze mit Sex. *Kosmos, 68:*78–80.

BAUST, W. (1970) *Ermüdung, Schlaf und Traum.* Stuttgart (Wiss. Verl. Ges.).

BEACH, F. A. (1937) The Neural Basis of Innate Behavior: I. Effects of Cortical Lesions upon the Maternal Behavior Pattern in the Rat. *J. Comp. Physiol. Psychol., 24:*393–439.

——— (1938) The Neural Basis of Innate Behavior: II. Relative Effects of Partial Decortication in Adulthood and Infancy upon the Maternal Behavior of the Primiparous Rat. *J. Gen. Psychol., 53:*108–148.

——— (1940) Effects of Cortical Lesions upon the Copulatory Behavior of Male Rats. *J. Comp. Physiol. Psychol., 29:*193–245.

——— (1942) Comparison of Copulatory Behavior of Male Rats Raised in Isolation, Cohabitation and Segregation. *J. Gen. Psychol., 60:*121–136.

——— (1947) Evolutionary Changes in the Physiological Control of Mating Behavior in Mammals. *Psychol. Rev., 54:*297–315.
——— (1948) *Hormones and Behavior.* New York (Harper & Row).
——— (1958) Normal Sexual Behavior in Male Rats Isolated at Fourteen Days of Age. *J. Comp. Physiol. Psychol., 51:*37–42.
BECHTEREW, W. (1913) *Objektive Psychologie oder Psychoreflexologie.* Leipzig.
——— (1926) *Reflexologie des Menschen.* Leipzig.
BECKER, G. (1965) Zur Magnetfeldorientierung von Dipteren. *Z. vergl. Physiol., 51:*135–150.
BEEBE, W. (1953) A Contribution to the Life History of the Euchromid Moth *Aethria carnicauda* Butler. *Zoologica, 38:*155–160.
BELLOWS, R. T. (1939) Time Factors in Water Drinking in Dogs. *Am. J. Physiol., 125:*87–97.
BENEDICT, R. (1934) *Patterns of Culture.* Boston (Houghton Mifflin).
——— (1955) Urformen der Kultur. *Rowohlts Deut. Enzykl., 7.*
BENSON, C. W. (1948) Geographical Voice Variation in African Birds. *Ibis, 90:*48–71.
BENTLEY, D. R. (1971) Genetic Control of an Insect Neuronal Network. *Science, 174:* 1139–1141.
BENTLEY, D. R., and HOY, R. R. (1972) Genetic Control of the Neuronal Network Generating Cricket (*Telegryllus gryllus*) Song Patterns. *Animal Behaviour, 20:*478–492.
BENZER, S. (1971) From the Gene to Behavior. *J. Am. Med. Ass., 218:*1015–1022.
BERELSON, B., and STEINER, G. A. (1964) *Human Behavior.* New York (Harcourt).
BERKOWITZ, L. (1962) *Aggression: A Social Psychological Analysis.* New York (McGraw-Hill).
BERNICK, N., KLING, A., and BOROWITZ, G. (1971) Physiologic Differentiation of Sexual Arousal and Anxiety. *Psychosom. Med., 33:*341–352.
BEST, J. B. (1963) Protopsychology. *Sci. Am., 208(2):*55–62.
BETHE, A. (1898) Dürfen wir den Ameisen und Bienen psychische Qualitäten zuschreiben? *Arch. Ges. Physiol. 70:*15–100.
BIERENS DE HAAN, J. A. (1940) *Die tierischen Instinkte und ihr Umbau durch Erfahrung.* Leiden.
BIGELOW, R. S. (1970) *The Dawn Warriors: Man's Evolution toward Peace.* Boston (Little, Brown) and London (Hutchinson).
——— (1971) Relevance of Ethology to Human Aggressiveness. *Int. Soc. Sci. J., 23:*19–26.
BILZ, R. (1940) *Pars pro toto: Ein Beitrag zur Pathologie menschlicher Affekte.* Leipzig (Thieme).
——— (1944) Zur Grundlegung einer Paläopsychologie. I. Paläophysiologie. II. Paläopsychologie. *Schweiz. Z. Psychol., 3:*202–212, 272–280.
BIOCCA, E. (1970) *Yanoama: The Narrative of a White Girl Kidnapped by Amazonian Indians.* New York (Dutton).
BIRCH, H. G. (1945) The Relation of Previous Experience to Insightful Problem-Solving. *J. Comp. Physiol. Psychol., 38:*367–383.
BIRDWHISTELL, R. L. (1963) The Kinesis Level in the Investigation of the Emotions. In KNAPP, P. H. (ed.), *Expressions of the Emotions in Man.* New York (International Univ. Press).
——— (1968) Communication without Words. In ALEXANDRE, P. (ed.) *L'Aventure Humaine.* Encycl. Sci. de l'Homme (Paris), *5:*157–166.
——— (1970) *Kinesics and Context.* Philadelphia (Univ. Pennsylv. Press).
BIRKET-SMITH, K. (1948) *Die Eskimos.* Zürich (Orell Füssli).
BIRUKOW, G. (1953) Photogeomenotaxische Transpositionen bei *Geotrupes sylvaticus. Rev. Suisse Zool., 60:*534–540.
——— (1956) Angeborene und erworbene Anteile relativ einfacher Verhaltenseinheiten. *Zool. Anz., Suppl., 19:*32–48.

BIRUKOW, G., and BUSCH, E. (1957) Lichtkompaßorientierung beim Wasserläufer *Velia currens* am Tage und zur Nachtzeit. Z. *Tierpsychol., 14:*184–203.

BISCHOF, N. (1966a) Psychophysik der Raumwahrnehmung. *Handb. d. Psychol.,* 1(1), Göttingen, 307–408.

——— (1966b) Stellungs-, Spannungs- und Lagewahrnehmung. *Handb. d. Psychol.,* 1(1), Göttingen, 409–497.

——— (1972a) The Biological Foundations of the Incest Taboo. *Soc. Sci. Inform., 11* (6):7–36.

——— (1972b) Inzuchtbarrieren in Säugetiersozietäten. *Homo, 23:* 330–351.

BISCHOFF, H. (1927) Biologie der Hymenopteren. *Biol. Stud. bücher.* Berlin (Springer).

BITTERMANN, M. E. (1965) The Evolution of Intelligence. *Sci. Am., 212:*92–100.

BLAIR, W. F. (1957a) Mating Call and Relationship of *Bufo hemiophrys. Texas J. Sci.,* 9:99–108.

——— (1957b) Structure of the Call and Relationship of *Bufo microscaphus. Copeia.,* 208–212.

——— (1958) Structure and Species Groups in U.S. Tree Frogs (*Hyla*). *Southwestern Naturalist, 3:*77–89.

BLAKEMORE, C., and COOPER, G. F. (1970) Development of the Brain Depends on the Visual Environment. *Nature, 228:*477.

BLAKEMORE, C., and MITCHELL, D. E. (1973) Environmental Modification of the Visual Cortex and the Neural Basis of Learning and Memory. *Nature, 241:*467–468.

BLAUVELT, H. (1964) Dynamics of the Mother-Newborn Relationship in Goats. *Group Proc. Josiah Macy Jr. Found.* New York, 221–258.

BLEST, A. D. (1957) The Function of Eye-Spot Patterns in the Lepidoptera. *Behaviour, 11:*209–255.

——— (1960) The Evolution, Ontogeny and Quantitative Control of Settling Movements of Some New World Saturnid Moths, with Some Comments on Distance Communication by Honey-Bees. *Behaviour, 16:*188–253.

——— (1966) Learning, Instinct and Evolution. *Nature, 212:*564.

BLODGETT, H. C. (1929) The Effect of the Introduction of Reward upon Maze Performance of Rats. *Univ. Calif. Publ. Psychol., 4:*113–134.

BLÖSCH, M. (1965) Untersuchungen über das Zusammenleben von Korallenfischen (*Amphiprion*) mit Seeanemonen. Dissertation, Univ. Erlangen.

BLURTON-JONES, N. (1972) *Ethological Studies of Child Behaviour.* Cambridge (Univ. Press).

BOAS, F. (1895): The Social Organization and the Secret Societies of the Kwakiutl Indians. *Report of the US National Museum. Reprint:* New York 1970 (Johnson Reprint Corp.).

BODMER, W. F., and CAVALLI-SFORZA, L. L. (1970) Intelligence and Race. *Sci. Am., 223* (4):19–29.

BOGERT, C. M. (1961) The Influence of Sound on the Behavior of Amphibians and Reptiles. In LANYON, W. E., and TAVOLGA, W. N. (eds.), *Animal Sounds and Communication.* Washington.

BOHANNAN, P. (1966) *Law and Warfare.* New York (Natural History Press).

BOLWIG, N. (1964) Facial Expression in Primates with Remarks on Parallel Development in Certain Carnivores. *Behaviour, 22:*167–192.

BONNER, J. (1964) The Molecular Biology of Memory. *Summary of the Symposium on the Role of Macromolecules in Complex Behavior,* Kansas State University, 89–95.

BOOTH, A. (1972a) Sex and Social Participation. *Am. Soc. Rev., 37:*183–192.

——— (1972b) Reply to Tiger. *Am. Soc. Rev., 37*(5):637.

BOPP, P. (1954) Schwanzfunktionen bei Wirbeltieren. *Rev. Suisse Zool., 61:*83–151.

BORELL DU VERNAY, W. v. (1942) Assoziationsbildung und Sensibilisierung bei *Tenebrio.* Z. *vergl. Physiol., 30:*84–116.

BOURLIERE, F. (1950) Classification et Caractéristiques des Principaux types de groupements sociaux chez les Vertébres sauvages. *Colloq. Intern. Centre Natl. Rech. Sci. (Paris), 34:*71–79.

——— (1955) *The Natural History of Mammals.* London (Harrap).

BOWER, T. G. (1966) Slant Perception and Shape Constancy in Infants. *Science, 151:*832–834.

——— (1971) The Object in the World of the Infant. *Sci. Am., 225:*30–38.

BOWLBY, J. (1952) Maternal Care and Mental Health. *World Health Organization Monogr. Ser., 2.*

——— (1958) The Nature of the Child's Tie to His Mother. *Intern. J. Psychoanal., 39:*350–373.

——— (1969) *Attachment and Loss. Vol. 1. Attachment.* The Int. Psycho-Analytical Library No. 79. New York (Basic) and London (Hogarth Press).

BOWMAN, R. I., and S. C. BILLEB (1965) Blood Eating in a Galápagos Finch. *The Living Bird, 10:*243–270.

BOYCOTT, B. (1965) Learning in the Octopus. *Sci. Am., 212* (3):42–50.

BRAEMER, W. (1960) A Critical Review of the Sun-Azimuth Hypothesis. *Cold Spring Harbor Symp. Quant. Biol., 25:*413–427.

BRAUM, E. (1963) Die ersten Beutefanghandlungen junger Blaufelchen (*Coregonus wartmanni* Bloch) und Hechte (*Esox lucius* L.). *Z. Tierpsychol., 20:*257–266.

BRELAND, K., and BRELAND, M. (1966) *Animal Behavior.* New York (Macmillan).

BROCK, F. (1927) Das Verhalten des Einsiedlerkrebses *Pagurus arrosor* während des Aufsuchens, Ablösens und Aufpflanzens seiner Seerose *Sagartia parasitica. Roux Arch. Entwicklungsmech., 112:*204–238.

BROWER, L. P., and ZANDT-BROWER, J. v. (1962) Investigations into Mimicry. *Natural Hist., 71:*8–19.

BROWN, A. W. A., and WEBB, H. M. (1960) A "Compass-Direction Effect" for Snails in Constant Conditions and Its Lunar Modulation. *Biol. Bull., 119:*307.

BROWN, F. A., Jr. (1965) A Unified Theory for Biological Rhythms: Rhythmic Duplicity and Genesis of "Circa" Periodisms. In ASCHOFF, J. (ed.), *Circadian Clocks.* Amsterdam (North-Holland), 231–261.

BROWN, R. (1970) The First Sentences of Child and Chimpanzee. In BROWN, R. (ed.) *Selected Psycholinguistic Papers.* New York (Macmillan).

BROWNLEE, F. (1943) The Social Organization of the Kung (!Un) Bushmen of the North-Western Kalahari. *Africa, 14:*124–129.

BRUCE, H. M. (1961) Time Relations in the Pregnancy-Block Induced in Mice by Strange Males. *J. Reprod. Fertility, 2:*138.

BRUHIN, H. (1953) Zur Biologie der Stirnaufsätze bei Huftieren. *Physiol. Comp. Oecol., 3:*63–92, 93–127.

BUBENIK, A. B. (1968) The Significance of the Antlers in the Social Life of the Cervidae. *Deer, 1:*208–214.

BUCHHOLZ, K. F. (1957) Das Sitzverhalten einiger *Orthretum*-Arten (Odonata) *Bonner Zool. Beitr., 9:*296–301.

BUDDENBROCK, W. v. (1952) *Vergleichende Physiologie, 1.* Basel (Birkhäuser).

BUECHNER, H. K. (1961) Territorial Behavior in the Uganda Kob. *Science, 133:*698–699.

BULLOCK, T. H. (1953) Predator Recognition and Escape Responses of Some Intertidal Gastropods in the Presence of Starfish. *Behaviour, 5:*130–140.

——— (1961) The Origins of Patterned Nervous Discharge. *Behaviour, 17:* 48–59.

——— (1962) Integration and Rhythmicity in Neural Systems. *Am. Zoologist, 2:*97–104.

BULLOCK, T. H., and HORRIDGE, G. A. (1965) *Structure and Function in the Nervous System of Invertebrates. I. and II.* San Francisco (Freeman).

BÜNNING, E. (1963) *Die physiologische Uhr.* 2nd ed. Heidelberg (Springer).

BÜRGER, M. (1959) Eine vergleichende Untersuchung über Putzbewegungen bei *Lagomorpha* und *Rodentia. Zool. Garten, 24:*434–506.

BURGHARDT, G. M. (1966) Stimulus Control of the Prey Attack Response in Naive Garter Snakes. *Psychol. Sci., 4:*37–38.

BURGHARDT, G. M., and HESS, E. H. (1966) Food Imprinting in the Snapping Turtle *Chelydra serpentina. Science, 151:*108–109.

BURKHARDT, D. (1958) Kindliches Verhalten als Ausdrucksbewegungen im Fortpflanzungszeremoniell einiger Wiederkäuer. *Rev. Suisse Zool., 65:*311–316.

——— (1960) Die Eigenschaften und Funktionstypen der Sinnesorgane. *Ergeb. Biol., 22:*226–267.

——— (1961) Allgemeine Sinnesphysiologie und Elektrophysiologie der Rezeptoren. *Fortschr. Zool., 13:*146–189.

BURKHARDT, D., SCHLEIDT, W. M., and ALTNER, H. (1966) *Signale in der Tierwelt: Vom Vorsprung der Natur.* Munich (Moos).

BUSNEL, R. G. (1964) *Acoustic Behaviour of Animals.* Amsterdam (Elsevier).

BUTENANDT, A. (1955) Über Wirkstoffe des Insektenreiches. II. Zur Kenntnis der Sexual-Lockstoffe. *Naturw. Rundschau, 12:*457–464.

BUTENANDT, A., BECKMANN, R., STAMM, D., and HECKER, E. (1959) Über den Sexuallockstoff des Seidenspinners *Bombyx mori.* Reindarstellung und Konstitution. Z. *Naturforsch., 143:*283–284.

BUTLER, R. A. (1953) Discrimination Learning by Rhesus Monkeys to Visual Exploration Motivation. *J. Comp. Physiol. Psychol., 46:*95–98.

BUYTENDIJK, F. J. J. (1933) *Wesen und Sinn des Spieles.* Berlin.

——— (1940) *Wege zum Verständnis der Tiere.* Zürich.

BYRNE, W. L., et al. (1966) Memory Transfer. *Science, 153:*658–659.

CALHOUN, J. B. (1962) Population Density and Social Pathology. *Sci. Am., 206(2):*139–148.

CARMICHAEL, L. (1926) The Development of Behavior in Vertebrates Experimentally Removed from the Influence of External Stimulation. *Psychol. Rev., 33:*51–58.

——— (1927) A Further Study of the Development of the Behavior of Vertebrates Experimentally Removed from the Influence of External Stimulation. *Psychol., Rev., 34:*34–47.

——— (1928) A Further Experimental Study of the Development of Behavior. *Psychol. Rev., 35:*253–260.

CARPENTER, C. R. (1942) Societies of Monkeys and Apes. *Biol. Symp., 8:*177–204.

——— (1965) The Howlers of Barro Colorado Island. In DEVORE, I. (ed.), *Primate Behavior.* New York (Holt, Rinehart and Winston), 250–291.

CARR, A. (1965) The Navigation of the Green Turtle. *Sci. Am., 212(5):*78–86.

CARTHY, J. D. (1956) *Animal Navigation: How Animals Find Their Way about.* London (G. Allen).

CASPARI, E. (1964) Behavior Genetics. *Am. Zoologist, 4:*95–99.

CASPERS, H. (1961) Beobachtungen über Lebensraum und Schwärmperiodizität des Palolowurmes *Eunice viridis. Intern. Rev. Ges. Hydrobiol., 46:*175–183.

CASTELL, R., and PLOOG, D. W. (1967) Zum Sozialverhalten der Totenkopf-Affen *(Saimiri sciureus):* Auseinandersetzung zwischen zwei Kolonien. Z. *Tierpsychol., 24:*625–641.

CAVILL, G. W. K., and ROBERTSON, P. L. (1965) Ant Venoms: Attractants and Repellents. *Science, 149:*1337.

CHAGNON, N. A. (1968) *Yanomamö: The Fierce People.* New York (Holt, Rinehart and Winston).

CHANCE, M. R. A. (1956) Social Structure of a Colony of *Macaca mulatta. Brit. J. Anim. Behav., 4:*1–13.

——— (1963) The Social Bond in Primates. *Primates, 4:*1–22.

——— (1967) Attention Structure as the Basis of Primate Rank Orders. *Man,* N. S., *2:*503–518.

CHANCE, M. R. A., and JOLLY, C. J. (1970) *Social Groups of Monkeys, Apes and Men.* New York (Dutton) and London (Cape).

CHANCE, M. R. A., and RUSSELL, W. M. S. (1959) Protean Displays: A Form of Allaesthetic Behaviour. *Proc. Zool. Soc. London, 132:*65–70.

CHARLESWORTH, W. R. (on press) Limits to the Application of Ethology to Problems in Human Development: A Real Problem?

CHAUVIN, R. (1952) L'effet de groupe. Structure et physiologie des sociétés animales. *Colloq. Intern. Centre Natl. Rech. Sci. (Paris), 34:*81–97.

CHMURZYNSKI, I. A. (1967) On the Role of Relations between Landmarks and the Nest-Hole in the Proximate Orientation of Female *Bembex rostrata* (L.). *Acta Biol. Exptl. (Warsaw), 27:*221–254.

CHOMSKY, N. (1970) *Sprache und Geist.* Frankfurt (Suhrkamp) (1972 *Language and Mind.* New York [Harcourt]).

CHRISTIAN, J. J. (1959) The Roles of Endocrine and Behavioral Factors in the Growth of Mammalian Populations. In GORBMAN, A. (ed), *Comparative Endocrinology.* New York (Wiley), 71–97.

——— (1960) Factors in Mass Mortality of a Herd of Sika Deer (*Cervus hippon*). *Chesapeake Sci. 1:*79–95.

——— (1963) Endocrine Adaptive Mechanisms and the Physiologic Regulation of Population Growth. In MEYER, M. V., and GELDER, R. van (eds.), *Physiological Mammalogy. 1.* New York (Academic Press), 189–353.

CLARK, E., ARONSON, L. R., and GORDON, M. (1954) Mating Behavior Patterns in Two Sympatric Species of Xiphophorin Fishes: Their Inheritance and Significance in Sexual Isolation. *Bull. Am. Mus. Nat. Hist., 103:*135–226.

COGHILL, G. E. (1929) *Anatomy and the Problem of Behaviour.* Cambridge (Univ. Press).

COMFORT, A. (1971) Likelihood of Human Pheromones. *Nature, 230:*432.

CONDON, W. S., and OGSTON, W. D. (1967) A Segmentation of Behaviour. *J. Psychiat. Res., 5:*221–235.

COOK, J. (1784) *A Voyage to the Pacific Ocean (1776–1780):* London (C. J. KING, ed.).

CORNING, W. C., and JOHN, E. R. (1961) Effect of Ribonuclease on Retention of Conditioned Response in Regenerating Planaria. *Science, 134:*1363–1365.

COSS, R. G. (1965) Mood Provoking Visual Stimuli, their Origins and Applications. *Industrial Design Graduate Program, Los Angeles,* The Regents of the University of California.

——— (1968) The Ethological Command in Art. *Leonardo I.* London and New York (Pergamon Press), 273–287.

——— The Perceptional Aspects of Eye-Spot. In: HUTT, S. J., and HUTT, C. (eds.): *Behavior Studies in Psychiatry.* New York and Oxford (Pergamon Press), 121–147.

——— (1972) Eye-Like Schemata: Their Effect on Behavior. Thesis Dept. Psychology Univ. of Reading.

COTT, H. B. (1957) *Adaptive Coloration in Animals.* London (Methuen).

COUNT, E. W. (1959) Eine biologische Entwicklungsgeschichte der menschlichen Sozialität. *Homo, 10:*1–35.

COURCHESNE, E., and BARLOW, G. W. (1971) Effect of Isolation on Components of Aggressive and Other Behavior in the Hermit Crab, *Pagurus samuelis. Z. vergl. Physiol., 75:*32–48.

COWLES, J. T. (1937) Food-Tokens as Incentives for Learning by Chimpanzees. *Comp. Psychol. Monogr., 14:*1–96.

CRAIG, W. (1918) Appetites and Aversions as Constituents of Instincts. *Biol. Bull Woods Hole, 34:*91–107.

——— (1928) Why do Animals Fight? *Intern. J. Ethics, 31:*264–278.

CRANACH, M. v. (1971) Über die Signalfunktion des Blickes. *Mannheimer Sozialwiss. Stud. 3:*201–224.

CRANE, J. (1943) Display, Breeding and Relationship of the Fiddler Crabs (*Brachyura Genus Uca*). *Zoologica, 28:*217–223.

——— (1949) Comparative Biology of Salticid Spiders at Rancho Grande, Venezuela: IV. An Analysis of Display. *Zoologica, 34:*159–214.
——— (1952) A Comparative Study of Innate Defensive Behavior of Trinidad Mantids. *Zoologica, 37:*259–293.
——— (1957) Basic Patterns of Display in Fiddler Crabs. *Zoologica, 42:*69–82.
——— (1966) Combat, Display and Ritualisation in Fiddler Crabs. *Phil. Trans. Roy. Soc. London B 251:*459–472.
CRAWFORD, M. P. (1937) The Cooperative Solving of Problems by Young Chimpanzees. *Comp. Psychol. Monogr., 14*(2).
CREUTZBERG, F. (1961) On the Orientation of Migrating Elvers (*Anguilla vulgaris*) in a Tidal Area. *Neth. J. Sea Res., 1:*257–338.
CRISLER, L. (1962) *Wir heulten mit den Wölfen.* Munich (dtv), 75.
CROOK, J. H. (1970) *Social Behaviour in Birds and Mammals.* London and New York (Academic Press).
CULLEN, E. (1957) Adaptations in the Kittiwake to Cliffnesting. *Ibis. 99:*275–302.
——— (1960) Experiments on the Effects of Social Isolation on Reproductive Behaviour in the Three-Spined Stickleback. *Animal Behaviour, 8:*235.
CUNNING, D. C. (1968) Warning Sounds of Moths. *Z. Tierpsychol., 25:*129–138.
CURIO, E. (1960) Ontogenese und Phylogenese einiger Tribäußerungen von Fliegenschnäppern. *J. Ornithol., 101:*291–309.
——— (1961) Versuche zur Spezifität des Feinderkennens durch Trauerschnäpper. *Experientia, 17:*516.
——— (1963) Probleme des Feinderkennens bei Vögeln. *Proc. 13th Intern. Ornithol. Congr.*, 206–239.
——— (1965a) Die Schlangenmimikry einer südamerikanischen Schwärmerraupe. *Natur u. Mus., 95:*207–211.
——— (1965b) Zur geographischen Variation des Feinderkennens einiger Darwinfinken (*Geospizidae*). *Zool. Anz., Suppl., 28:*466–492.
——— (1966a) Die Schutzanpassungen dreier Raupen eines Schwärmers (*Lepidopt., Sphingidae*) auf Galapagos. *Zool. Jb. System., 91:*1–29.
——— (1966b) Färbung und Ruheverhalten dreier Raupenformen eines Schwärmers. *Umschau, 14:*475.
——— (1966c) Wie Insekten ihre Feinde abwehren. *Naturwiss. Med., 11:*3–21.
CUSHING, J. E. (1941) An Experiment on Olfactory Conditioning in *Drosophila guttifera. Proc. Natl. Acad. Sci. U.S., 27:*296–299.
DAANJE, A. (1950) On the Locomotory Movements in Birds and the Intention Movements Derived from it. *Behaviour, 3:*48–98.
DANE, B., and van der KLOOT, W. G. (1964) An Analysis of the Display of the Goldeneye Duck (*Bucephala clangula* L.) *Behaviour, 22:*282–328.
DARLING, F. FRASER (1937) *A Herd of Red Deer.* New York and Oxford (Univ. Press).
DART, R. A. (1957) The Osteodontokeric Culture of *Australopithecus prometheus. Transvaal. Mus. Mem., 10.*
DARWIN, C. (1872) *The Expression of Emotions in Man and Animals.* London.
——— (1875) *Die Abstammung des Menschen (The Descent of Man), Bd. 1* (Ausgabe V. v. Carus, *Gesammelte Werke,* B. 5) Stuttgart (Schweizerbart).
——— (1881) *Das Bewegungsvermögen der Pflanzen.* Translated by J. V. CARUS, Stuttgart.
DATHE, H. (1964) Zur Körperpflege der Tiere in freier Wildbahn und Gefangenschaft. *Wiss. kult. Mitt. Tierpark Berlin. 1:*349–383.
DAVENPORT, D. (1955) Specificity and Behavior in Symbioses. *Quart. Rev. Biol., 30:*29–46.
——— (1966) The Experimental Analysis of Behavior in Symbiosis. In HENRY, S. M. (ed.), *Symbiosis I.*, New York (Academic Press), 381–429.
DAVENPORT, D., and NORRIS, K. (1958) Observations on the Symbiosis of the Sea Anemone *Stoichactis* and the Pomacentrid Fish *Amphiprion percula. Biol. Bull., 115:*397–410.

DAVENPORT, D., ROSS, D. M., and SUTTON, L. (1961) The Remote Control of Nematocyst Discharge in the Attachment of *Calliactis parasitica* to Shells of Hermit Crabs. *Vie et Milieu, 12:*197:209.

DAVID, F. A., BRONFENBRENNER, U., HESS, E. H., MILLER, D. R. SCHNEIDER, D. M., and SPUHLER, J. N. (1963) The Incest Taboo and the Mating Patterns of Animals. *American Anthropologist 65:*253–265.

DAVIS D. E. (1962) The Phylogeny of Gangs. In BLISS, E. (ed.), *Roots of Behavior.* New York (Harper & Row).

DAWKINS, R. (1968) The Ontogeny of a Pecking Preference in Domestic Chicks. Z. *Tierpsychol., 25:*170–186.

DEAG, J. M., and CROOK, J. H. (1971) Social Behaviour and "Agonistic Buffering" in the Wild Barbary Macaque *Macaca sylvana L. Folia primat., 15:*183–200.

DEEGENER, P. (1918) *Die Formen der Vergesellschaftung im Tierreiche.* Leipzig.

DELGADO, J. M. R. (1967) Aggression and Defense under Cerebral Radio Control. *UCLA Forum in Medical Science, 7:*171–193. Reprinted in CLEMENTE, C. D., and LINDSLEY, D. B. (eds.), *Aggression and Defense.* Berkeley (Univ. California Press).

DEMBOWSKI, J. (1955) *Tierpsychologie.* Berlin (Akademie-Verlag).

DENNIS, W. (1960) Causes of Retardation among Institutional Children. *Iran. J. Genet. Psychol., 96:*47–59.

DETHIER, V. G. (1957) Communication by Insects: Physiology of Dancing. *Science, 125:*331–336.

DETHIER, V. G., and BODENSTEIN, D. (1958) Hunger in the Blowfly. Z. *Tierpsychol., 15:*129–140.

DEVORE, I. (1965) *Primate Behavior: Field Studies of Monkeys and Apes.* New York (Holt, Rinehart and Winston).

DIAMOND, M. (1965) A Critical Evaluation of the Ontogeny of Human Sexual Behavior. *Quart. Rev. Biol., 40:*147–175.

——— (1967) Genetic-Endocrine Interactions and Human Psychosexuality. In: DIAMOND, M. (ed.) *Reproduction and Sexual Behavior.* Bloomington (Indiana Univ. Press, 417–443.

DIEBSCHLAG, E. (1940) Psychologische Beobachtungen über die Rangordnung bei der Haustaube. Z. *Tierpsychol., 4:*173–188.

DIESSELHORST, G. (1965) Klasse *Aves,* Vögel. *Handb. d. Biol., 6,* 2. Frankfurt (Athenaion), 745–866.

DIETERLEN, F. (1959) Das Verhalten des syrischen Goldhamsters *(Mesocricetus auratus).* Z. *Tierpsychol., 16:*47–103.

DILGER, W. C. (1960) The Comparative Ethology of the African Parrot Genus *Agapornis.* Z. *Tierpsychol., 17:*649–685.

——— (1962) The Behavior of Lovebirds. *Sci. Am., 206(1):*88–98.

DOFLEIN, F. (1905) Beobachtungen an Weberameisen. *Biol. Zbl., 25:*497–507.

DOLLARD, J., DOOB, L., MILLER, N., MOWRER, O., and SEARS, R. (1939) *Frustration and Aggression.* New Haven (Yale Univ. Press).

DOLLO, L. (1895) Sur la Phylogénie des Dipneustes. *Bull. Soc. Belge de Géol., 9:*79–128.

——— (1909) Poissons Voiliers. *Zool. Jahrb. System, 27:*419.

DREES, O. (1952) Untersuchungen über die angeborenen Verhaltensweisen bei Springspinnen *(Salticida).* Z. *Tierpsychol., 9:*169–207.

DÜHRSSEN, A. (1960) *Psychogene Erkrankungen bei Kindern und Jugendlichen.* Göttingen (Verlag f. medizin. Psychologie).

DUMAS, F. (1932) La mimique des aveugles. *Bull. Acad. Med., 107:*607–610.

DUPEYRAT, A. (1963) *Papua—Beast and Men.* London (MacGibbon and Kee).

EBHARDT, H. (1958) Verhaltensweisen verschiedener Pferdeformen. *Säugetierkdl. Mitt., 6:*1–9.

ECCLES, J. C. (1953) *The Neurophysiological Basis of Mind: The Principles of Neurophysiology.* New York and London (Oxford Univ. Press).

EDWARDS, D. (1968) Mice: Fighting by Neonatally Androgenized Females. *Science, 161:*127–128.

EIBL-EIBESFELDT, I. (1949) Über das Vorkommen von Schreckstoffen bei Erdkrötenquappen. *Experientia, 5:*236.

—— (1950a) Über die Jugendentwicklung des Verhaltens eines männlichen Dachses (*Meles meles* L.) unter besonderer Berücksichtigung des Spieles. *Z. Tierpsychol., 7:*327–355.

—— (1950b) Ein Beitrag zur Paarungsbiologie der Erdröte (*Bufo bufo* L.). *Behaviour, 2:*217–236.

—— (1950c) Beiträge zur Biologie der Haus- und der Ährenmaus nebst einigen Beobachtungen an anderen Nagern. *Z. Tierpsychol., 7:*558–587.

—— (1951a) Nahrungserwerb und Beuteschema der Erdkröte (*Bufo bufo* L.). *Behaviour, 4:*1–35.

—— (1951b) Gefangenschaftsbeobachtungen an der persischen Wüstenmaus (*Meriones persicus* Blanford): Ein Beitrag zur vergleichenden Ethologie der Nager. *Z. Tierpsychol., 8:*400–423.

—— (1951c) Zur Fortpflanzüngsbiologie und Jugendentwicklung des Eichhörnchens. *Z. Tierpsychol., 8:*370–400.

—— (1953a) Zur Ethologie des Hamsters (*Cricetus cricetus* L.). *Z. Tierpsychol., 10:*204–254.

—— (1953b) Vergleichende Studien an Ratten und Mäusen. *Prakt. Desinfektor, 45:*166–168.

—— (1953c) Eine besondere Form des Duftmarkierens beim Riesengalago, *Galago crassicaudatus. Säugetierkdl. Mitt., 1:*171–173.

—— (1953d) Ethologische Unterschiede zwischen Hausratte und Wanderratte. *Zool. Anz., Suppl., 16:*169–180.

—— (1954) *Paarungsbiologie der Anuren.* Wiss. Film, C628. Göttingen (Inst. wiss. Film).

—— (1955a) Über Symbiosen, Parasitismus und andere zwischenarliche Beziehungen bei tropischen Meeresfischen. *Z. Tierpsychol., 12:*203–219.

—— (1955b) Ethologische Studien am Galapagos-Seelöwen *Zalophus wollebaeki* Sivertsen. *Z. Tierpsychol., 12:*286–303.

—— (1955c) *Sexualverhalten und Eiablage beim Alpenmolch.* Wiss. Film, C698. Göttingen (Inst. wiss. Film).

—— (1955d) Der Kommentkampf der Meerechse (*Amblyrhynchus cristatus* Bell) nebst einigen Notizen zur Biologie dieser Art. *Z. Tierpsychol., 12:*49–62 (siehe auch wiss. Film der Encycl. cinem., E591. Göttingen [Inst. wiss. Film] 1964).

—— (1955e) *Biologie des Iltisses (Putorius putorius).* Wiss. Film C 697, Göttingen (Inst. wiss. Film).

—— (1956a) Vergleichende Verhaltensstudien an Anuren. 2. Zur Paarungsbiologie der Gattungen *Bufo, Hyla, Rana* und *Pelobates. Zool. Anz., Suppl., 19:*315–323.

—— (1956b) Einige Bemerkungen über den Ursprung von Ausdrucksbewegungen bei Säugetieren. *Z. Säugetierkde., 21:*29–43.

—— (1956c) Zur Biologie des Iltis (*Putorius putorius* L.). *Zool. Anz., Suppl., 19:*304–314.

—— (1957a) Ausdrucksformen der Säugetiere. In KÜKENTHAL, *Handb. d. Zool., 8*(6), 1–26.

—— (1957b) *Rattus norvegicus. Kampf I (erfahrener Männchen).* Encycl. cinem., E131. *Kampf II (unerfahrener Männchen).* Encycl. cinem., E132. Göttingen (Inst. wiss. Film).

—— (1958a) Das Verhalten der Nagetiere. In KÜKENTHAL., *Handb. d. Zool., 8*(10), 13, 1–88.

—— (1958b) *Versuche über den Nestbau erfahrungsloser Ratten.* Wiss. Film B 757, Göttingen (Inst. wiss. Film).

——— (1959) Der Fisch *Aspidontus taeniatus* als Nachahmer des Putzers *Labroides dimidiatus. Z. Tierpsychol., 16:*19–25.

——— (1960a) Beobachtungen und Versuche an Anemonenfischen der Malediven und der Nicobaren. Z. *Tierpsychol., 17:*1–10.

——— (1960b) Naturschutzprobleme auf den Galapagos-Inseln. *Acta Trop., 17:*97–137.

——— (1961a) The Interactions of Unlearned Behavior Patterns and Learning in Mammals. In DELAFRESNAYE, J. F. (ed.). *Brain Mechanisms and Learning.* Oxford (Blackwell), 53–73.

——— (1961b) The Fighting Behavior of Animals. *Sci. Am., 205*(6):112–121.

——— (1961c) Eine Symbiose von Fischen *(Siphamia versicolor)* und Seeigeln. Z. *Tierpsychol., 18:*56–59.

——— (1962a) Die Verhaltensentwicklung des Krallenfrosches *(Xenopus laevis)* und des Scheibenzünglers *(Discoglossus pictus)* unter besonderer Berücksichtigung der Beutefanghandlungen. Z. *Tierpsychol., 19:*385–393.

——— (1962b) Freiwasserbeobachtungen zur Beutung des Schwarmverhaltens verschiedener Fische. Z. *Tierpsychol., 19:*165–182.

——— (1963) Angeborenes und Erworbenes im Verhalten einiger Säuger. Z. *Tierpsychol., 20:*705–754.

——— (1964a) *Tropidurus delanonis (Iguanidae). Kommentkampf der Männchen.* Encycl. cinem., E609. Göttingen (Inst. wiss. Film).

——— (1964b) *Galapagos, die Arche Noah im Pazifik.* 3rd ed. Munich (Piper).

——— (1964c) *Im Reich der tausend Atolle.* Munich (Piper).

——— (1965a) *Amphiprion percula: Kampfverhalten.* Encycl. cinem., E752. Publ. zu wiss. Filmen, 2 A. Göttingen (Inst. wiss. Film).

——— (1965b) *Nannopterum harrisi (Phalacrocoracidae): Brutablösung.* Encycl. cinem., E596. Publ. zu wiss. Filmen, 1 A, 303–306 Göttingen (Inst. wiss. Film).

——— (1965c) Das Duftmarkieren des Igeltanreks (*Echinops telfairi* Martin). Z. *Tierpsychol., 22:*810–812.

——— (1966a) Ethologie, die Biologie des Verhaltens. In GESSNER, F., and BERTALANFFY, L. v. (eds.), *Handbuch der Biologie 2.* Frankfurt (Athenaion), 341–559.

——— (1966b) *Dicranura vinula–Feindabwehr.* Encycl. cinem., E973. Göttingen (Inst. wiss. Film).

——— (1966c) Beobachtungen über das innerartliche Kampfverhalten der Kielschwanzleguane *(Tropidorus)* der Galapagos-Inseln. Z. *Tierpsychol., 23:*672–676.

——— (1967) Concepts of Ethology and Their Significance for the Study of Human Behavior. In STEVENSON, H. W. (ed.), *Early Behavior: Comparative and Developmental Approaches.* New York (Wiley), 127–146.

——— (1968). Zur Ethologie des menschlichen Grußverhaltens. I. Beobachtungen an Balinesen, Papuas und Samoanern nebst vergleichenden Bemerkungen. Z. *Tierpsychol., 25:*727–744.

——— (1970) Männliche und weibliche Schutzamulette im modernen Japan. *Homo, 21:* 175–188.

——— (1971a) Zur Ethologie menschlichen Grußverhaltens. II. Das Grußverhalten und einige andere Muster freundlicher Kontaktaufnahme der Waika-Indianer (Yanoama). Z. *Tierpsychol., 29:*196–213.

——— (1971b) Eine ethologische Interpretation des Palmfruchtfestes der Waika-Indianer (Yanoama) nebst Bemerkungen über die bindende Funktion des Zwiegespräches. *Anthropos, 66:*767–778.

——— (1971c) Das Humanethologische Filmarchiv der Max-Planck-Gesellschaft. *Homo, 22:*252–256.

——— (1972a) *Die !Ko-Buschmanngesellschaft: Aggressionskontrolle und Gruppenbindung.* Monographien zur Humanethologie 1. München (Piper).

——— (1972b) *Love and Hate: The Natural History of Behavior Patterns.* New York (Holt, Rinehart and Winston).
——— (1972c) Similarities and Differences between Cultures in Expressive Movements. In ARGYLE, J. M., and HINDE, R. (eds.) *Nonverbal Communication.* London (Cambridge Univ. Press), 297–314.
——— (1973a) The Expressive Behavior of the Deaf-and-Blind Born. In CRANACH, M. VON, and VINE, I. (eds.), *Social Communication and Movement.* London (Academic Press), 1963–193.
——— (1973b) *Der Vorprogrammierte Mensch: Das Ererbte als bestimmender Faktor im menschlichen Verhalten.* Wien (Molden).
——— (1974) The Myth of the Aggression-free Hunter and Gatherer Society. In: R. HOLLOWAY (ed); *Primate Aggression, Territoriality and Xenophobia:* A *Comparative Perspective.* London (Academic Press).
——— (on press) *Krieg und Frieden aus der Sicht der Verhaltensforschung.* München (Piper) (to be published 1975).
EIBL-EIBESFELDT, I., and EIBL-EIBESFELDT, E. (1967) Die Parasitenabwehr der Minima-Arbeiterinnen der Blattschneiderameise *Atta cephalotes. Z. Tierpsychol., 24:*279–281.
EIBL-EIBESFELDT, I., and HASS, H. (1959) Erfahrungen mit Haien. Z. *Tierpsychol., 16:* 733–746.
——— (1966) Zum Projekt einer ethologisch orientierten Untersuchung menschlichen Verhaltens. *Mitt. Max-Planck-Ges., 6:*383–396.
——— (1967) Neue Wege der Humanethologie. *Homo, 18:*13–23.
EIBL-EIBESFELDT, I., and KLAUSEWITZ, W. (1961) *Gnathypops rosenbergi annulata* n. ssp. von den Nikobaren. *Senckenbergiana Biol., 42:*421–426.
EIBL-EIBESFELDT, I., and SIELMANN, H. (1962) Beobachtungen am Spechtfinken *Cactospiza pallida. J. Ornithol., 103:*92–101.
——— (1965) *Cactospiza pallida (Fringillidae): Werkzeuggebrauch beim Nahrungserwerb.* Encycl. cinem., E597. Publ. zu wiss. Filmen, 1 A, 385–390. Göttingen (Inst. wiss. Film).
EIBL-EIBESFELDT, I., and WICKLER, W. (1962) Ontogenese und Organisation von Verhaltensweisen. *Fortschr. Zool., 15:*354–377.
——— (1968) Die ethologische Deutung einiger Wächterfiguren auf Bali. Z. *Tierpsychol., 25:*719–726.
EICKSTEDT, E. Fr. v. (1963) *Die Forschung am Menschen. Teil 3: Psychologische und philosophische Anthropologie.* Stuttgart (Enke), 1513–2645.
EISENBERG, J. F. (1965) The Social Organisations of Mammals. In KÜKENTHAL, *Handb. d. Zool., 8*(10), 7, 1–92.
EISENBERG, J. F., and KLEIMAN, D. G. (1972) Olfactory Communication in Mammals. *Ann. Rev. Ecol. System., 3:*1–32.
EISNER, T., and MEINWALD, J. (1966) Defensive Secretion of Arthropods. *Science, 153:* 1341–1350.
EKMAN, P. (1972) *Darwin and the Human Face.* New York (Academic Press).
EKMAN, P., and FRIESEN, W. (1971) Constants across Cultures in the Face and Emotions. *J. Person. Soc. Structure, 17:*124–129.
EKMAN, P., FRIESEN, W., and ELLSWERTH, P. (1972) *Emotions in the Human Face.* New York (Pergamon).
EKMAN, P., SORENSON, E. R., and FRIESEN, W. V. (1969) Pan-Cultural Elements in Facial Displays of Emotion. *Science 164:*86–88.
ELLEFSON, J. O. (1968) Territorial Behavior in the Common White-Handed Gibbon *Hylobates lar.* In JAY, P. C. (ed.): *Primates.* New York (Holt, Rinehart and Winston).
ELLIOTT, M. H. (1930) Some Determining Factors in Maze Performance. *J. Psychol., 42:*315–317.
EMERSON, A. E. (1956) Ethospecies, Ethotypes, and the Evolution of *Apicotermes* and *Allognathotermes* (Isoptera, Termitidae). *Am. Museum Novitates, 1771:*1–31.

EMLEN, S. T. (1967): Migratory Orientation in the Indigo Bunting. *Auk., 84:*309–342, 463–489.

——— (1969) Bird Migration: Influence of Physiological State upon Celestial Orientation. *Science, 165:*664–672.

——— (1970) Celestial Rotation: Its Importance in the Development of Migratory Orientation. *Science, 170:*1198–1201.

ENRIGHT, J. T. (1961) Lunar Orientation of *Orchestia corniculata. Biol. Bull., 120:*148–156.

——— (1963) The Tidal Rhythm of Activity of a Sand-Beach Amphipod. *Z. vergl. Physiol., 46:*276–313.

——— (1972) A Virtuoso Isopod: Circa-Lunar Rhythms and Their Tidal Fine Structure. *J. Comp. Physiol., 77:*141–162.

ERICKSON, C. J., and LEHRMAN, D. S. (1964) Effect of Castration of Male Ring Doves upon Ovarian Activity of Females. *J. Comp. Physiol., 58:*164–166.

ERIKSON, E. H. (1953) *Wachstum und Krisen der gesunden Persönlichkeit.* Stuttgart (Klett).

——— (1966) Ontogeny of Ritualisation in Man. *Phil. Trans. Roy. Soc. London, B 251:* 337–349.

ESCH, H. (1967) The Evolution of Bee Language. *Sci. Am., 216*(4):97-104.

ESCHERICH, K. (1906) *Die Ameise.* Braunschweig (Vieweg).

ESSER, A. H. (1968) Dominance Hierarchy and Clinical Course of Psychiatrically Hospitalized Boys. *Child Develop., 39:*147–157.

——— (1970) Interactional Hierarchy and Power Structure on a Psychiatric Ward. In HUTT, S. J., and HUTT, C. (eds.). *Behaviour Studies in Psychiatry.* New York and Oxford (Pergamon Press), 25–59.

——— (1971) *Behavior and Environment: The Use of Space by Animals and Men.* New York and London (Plenum Press).

ESSER, A. H., CHAMBERLAIN, A. S., CHAPPLE, E. D., and KLINE, N. S. (1965) Territoriality of Patients on a Research Ward. In WORTIS, J. (ed.), *Recent Advances in Biological Psychiatry, 7.* New York (Plenum Press).

EULER, H. A. (on press) Der Effekt von aggressionsabhängiger Strafreizung (Elektroschock) auf das Kampfverhalten von Leghorn Hähnen. *28. Kongress Deutschen Ges. Psychol. Okt.,* 1072.

——— (in preparation) Effect of Contingent Electric Shock on Submissive Responses in White Leghorn Cockerels.

EVERETT, G. M. (1961) Some Electrophysiological and Biochemical Correlates of Motor Activity and Aggressive Behavior. *Neuropharmacology, 2:*479–484.

EVERETT, G. M., and WIEGAND, R. G. (1962) Central Amines and Behavioral States: A Critique and New Data. Proc. 1st Intern. Pharmacol. Meeting, 8: *Pharmacol. Analysis of Central Nervous Action.* New York (Macmillan), 85-92.

EWER, R. F. (1963) The Behaviour of the Meerkat *Suricata suricatta. Z. Tierpsychol., 20:*570–607.

——— (1968) *Ethology of Mammals.* New York *(Plenum Press)* and London (Logos Press).

EWERT, J. P. (1972) Zentralnervöse Analyse und Verarbeitung visueller Sinnesreize. *Naturwiss. Rundschau, 25:*1–11.

——— (1973) Lokalisation und Identifikation im visuellen System der Wirbeltiere. *Fortschr. Zool., 21:*307–333.

——— (1974a) Neurobiologie und System-Theorie eines visuellen Muster-Erkennungsmechanismus bei Kröten. *Kybernetik, 14:*167–183.

——— (1974b) The Neural Basis of Visually Guided Behavior. *Sci. Am., 230(3):*34–42.

FABER, A. (1953a) Ausdrucksbewegungen und besondere Lautäußerungen bei Insekten als Beispiel für eine vergleichend-morphologische Betrachtung der Zeitgestalten. *Zool. Anz., Suppl., 16:*106–115.

——— (1953b) Laut- und Gebärdensprache bei Insekten. *Orthoptera I.* 278. Mitt. Mus. Naturkde, Stuttgart.

FABRÉ, J. H. (1879–1910) *Souvenier entomologique.* Paris (Delagrave), 1–10.

FABRICIUS, E. (1951) Zur Ethologie junger Anatiden. *Acta Zool. Fenn., 68:*1–178.

FANTZ, R. L. (1967) Visual Perception and Experience in Infancy. In STEVENSON, H. W. (ed.), *Early Behavior.* New York (Wiley), 181–224.

FARRIS, H. E. (1967) Classical Cinditioning of Courting Behavior in the Japanese Quail *Coturnix coturnix japonica. J. Exptl. Anal. Behavior, 10:*213–217.

FAUBLÉE, J. (1968) Note sur l'economie ostentatoire. *Rev. Tiers-Monde, 9:*17–23.

FEDER, H. M. (1966) Cleaning Symbiosis in the Marine Environment. In HENRY, S. M. (ed.), *Symbiosis I.* New York (Academic Press), 327–380.

FELIPE, N. J., and SOMMER, R. (1966) Invasions of Personal Space. *Social Prob., 14:*206–214.

FESHBACH, S. (1961) The Stimulating Versus Cathartic Effects of a Vicarious Aggressive Activity. *J. Abnorm. Soc. Psychol., 63:*381–385.

FESHBACH, S., and SINGER, R. D. (1971) *Television and Aggression.* San Francisco (Jossey Bass).

FESTETICS, A. (1959) Ökologische Untersuchungen an Brutvögeln des Saser. *Vogelwelt, 80:*1–21.

——— (1961) Ährenmaushügel in Österreich. *Z. Säugetierkde., 26:*1–14.

FEUSTEL, H. (1966) Anatomische Untersuchungen zum Problem der *Aspidosiphon-Heterocyathus*-Symbiose. *Zool. Anz., Suppl., 29:*131–143.

FIEDLER, K. (1954) Vergleichende Verhaltensstudien an Seenadeln, Schlangennadeln und Seepferdchen. *Z. Tierpsychol., 11:*358–416.

——— (1964) Verhaltensstudien an Lippfischen der Gattung *Crenilabrus. Z. Tierpsychol., 21:*521–591.

FISCHER, H. (1965) Das Triumphgeschrei der Graugans *(Anser anser). Z. Tierpsychol., 22:*247–304.

FISCHER, K. (1961) Untersuchungen über die Sonnenkompaßorientierung und Laufaktivität von Smaragdeidechsen (*Lacerta viridis* Laur). *Z. Tierpsychol., 18:*450–470.

——— (1963) Spontanes Richtungsfinden nach dem Sonnenstand bei *Chelonia mydas* (Suppenschildkröte). *Naturwiss., 51:*203.

FISHER, J., and HINDE, R. A. (1949) The Opening of Milk Bottles by Birds. *Brit. Birds, 42:*347–358.

FJERDINGSTAD, E. J. (ed.) (1971) *Chemical Transfer and Learned Information.* New York (Elsevier) and Amsterdam (North Holland).

FLETCHER, R. (1948) *Instinct in Man.* Aberdeen (Univ. Press).

FOLLEY, S. J., and KNAGGS, G. S. (1965) Levels of Oxytocin in the Jugular Vein Blood of Goats During Parturition. *J. Endocrinol., 33:*301–315.

FOREL, A. (1892) *Die Nester der Ameisen.* Zürich.

FOSSEY, D. (1970) Making Friends with Mountain Gorillas. *Natl. Geogr. Mag., 137:*48–67.

FOUTS, R. S. (1974) Communication with Chimpanzees. In EIBL-EIBESFELDT, I., and KURTH, G. (eds.) *Hominisation und Verhalten.* Heidelberg (Fischer).

FOX, M. W. (1969) The Anatomy of Aggression and its Ritualization in *Canidae:* A Developmental and Comparative Study. *Behaviour, 35:*242–258.

FRAENKEL, G. S., and GUNN, D. S. (1961) *The Orientation of Animals.* Oxford (Clarendon Press).

FRANCK, D. (1966) Möglichkeiten zur vergleichenden Analyse auslösender und richtender Reize mit Hilfe des Attrappenversuches, ein Vergleich der Successiv- und Simultanmethode. *Behaviour, 27:*150–159.

FRANCK, D., and WILHELMI, U. (1973) Veränderungen der aggressiven Handlungsbereitschaft männlicher Schwertträger, *Xiphophorus helleri,* nach sozialer Isolierung. *Experimentia, 29:*896–897.

FRANK, F. (1953) Über den Zusammenbruch von Feldmausplagen. *Zool. Jahrb. Abt. System, 82:*1–156.

FRANZISKET, L. (1955) Die Bildung einer bedingten Hemmung bei Rückenmarksfröschen. *Z. vergl. Physiol., 37:*161–168.

FREEDMAN, D. G. (1964) Smiling in Blind Infants and the Issue of Innate vs. Acquired. *J. Child Psychol. Psychiat., 5:*171–184.

——— (1965a) Hereditary Control of Early Social Behavior. In FOSS, B. M. (ed.), *Determinants of Infant Behavior. 3.* London (Methuen).

——— (1965b) An Ethological Approach to the Genetic Study of Human Behavior. In VANDENBERG, S. G., *Methods and Goals in Human Behavior Genetics.* New York (Academic Press), 141–161.

——— (1967) A Biological View of Man's Social Behavior. In ETKIN, W., and FREEDMAN, D. G. (eds.), *Social Behavior from Fish to Man.* Chicago (Phoenix Books).

FREEDMAN, D. G., and FREEDMAN, N. C. (1969) Behavioural Differences between Chinese-American and European-American Newborn. *Nature, 224:*1227–1235.

FREEDMAN, D. X., and GIARMAN, N. J. (1963) Brain Amines, Electrical Activity and Behavior. In GLASER, G. H. (ed.), *EEG and Behavior.* New York (Basic Books), 198–243.

FREEMAN, D. (1971) Aggression: Instinct or Symptom? *Australian N. Zealand J. Psychiatry, 5:*66–73.

FREMONT-SMITH, F. (1962) Saving the Children Can Save Us. *Saturday Rev.,* Aug. 11.

FREUD, S. (1950) *Gesammelte Werke.* 18 vols., London (Imago Publ.).

FREUD, J., and UYLERT, J. E. (1948) Micturation and Copulation Behavior in Dogs. *Acta Brevia, 16:*49–53.

FREYE, H. A., and GEISSLER, H. (1966) Das Ohrenspiel der Ungulaten als Ausdrucksform. *Wiss. Z. Univ. Halle, 15:*893–915.

FRIJDA, N. H. (1965) Mimik und Pantomimik. In KIRCHHOFF, R. (ed.), *Handb. d. Psychol., 5, Ausdruckspsychol.,* 351–421.

FRINGS, H., and FRINGS, M. (1959) Reactions of American and French Species of *Corvus* and *Larus* to Recorded Communication Signals Tested Reciprocally. *Ecology, 39:* 126–131.

FRISCH, K. v. (1914) Der Farben- und Formensinn der Biene. *Zool. Jahrb. Allgem. Zool. Physiol., 40:*1–186.

——— (1923) Ein Zwergwels, der kommt, wenn man pfeift. *Biol. Zbl., 43:*439–446.

——— (1941) Über einen Schreckstoff der Fischhaut und seine biologische Bedeutung. *Z. vergl. Physiol., 29:*46–145.

——— (1950) Die Sonne als Kompaß im Leben der Bienen. *Experientia, 6:*210–221.

——— (1959) Aus dem Leben der Bienen. *Verständl. Wissenschaft, I,* 6th ed.

——— (1965) *Die Tanzsprache und Orientierung der Bienen.* Berlin (Springer).

——— (1968) Honeybees: Do They Use Direction and Distance Information Provided by Their Dancers? *Science, 158:*1072–1076.

FRISCH, K. v., LINDAUER, M., and DAUMER, K. (1960) Über die Wahrnehmeng polarisierten Lichtes durch das Bienenauge. *Experientia, 16:*289–302.

FRISCH, O. v. (1958) Die Bedeutung der elterlichen Warnrufe für Brachvogel und Limicolenkücken. Z. *Tierpsychol., 15:*381–382.

——— (1962) Zur Biologie des Zwergchamäleons. *Z. Tierpsychol., 19:*276–289.

FROMME, A. (1941) An Experimental Study of the Factors of Maturation and Practise in Behavioral Development of the Embryo of the Frog *Rana pipiens. Genet. Psychol. Monogr., 24:*219–261.

FUCHS, P. (1967) Tatauierung in Afrika. *Bild d. Wiss., 2:*109–117.

FÜLLER, H. (1958) *Symbiose im Tierreich.* Wittenberg (Ziemsen).

FULLER, J. L., and THOMPSON, W. R. (1960) *Behavior Genetics.* New York (Wiley).

FUNKE, W. (1965) Untersuchungen zum Heimfindeverhalten und zur Ortstreue von *Patella* L. *Zool. Anz., Suppl., 28:*411–418.

GAITO, J. (1964) Nucleic Acids and Brain Function. *Symp. on the Role of Macromolecules in Complex Behavior.* Kansas State University, Manhattan, 68–75.

—— (1966) *Macromolecules and Behavior.* Amsterdam (North-Holland).

GAMPER, E. (1926) Bau- und Leistungen eines menschlichen Mittelhirnwesens (Arhinencephalie mit Encephalocelie). *Z. ges. Neurol. Psychiat., 102:*154–235, *104:*49–120.

GARCIA, J., and ERVIN, F. R. (1968) Gustatory-visceral and Telereceptorcutaneous Conditioning-Adaptation in Internal and External Milieus. *Communications in Behavioral Biol., Part A, 1:*389–415.

GARCIA, J., MCGOWAN, B. K., ERVIN, F. R., and KOELLING, R. A. (1968) Cues: Their Relative Effectiveness as a Function of the Reinforcer. *Science, 160,* 794–795.

GARCIA, J., MCGOWAN, B. K., and GREEN, K. F. (1969) Sensory Quality and Integration: Constraints on Conditioning. *Conference on Conditioning at Dept. Psychol. Ontario* (McMaster University).

GARDENER, R. G., and HEIDER, K. G. (1968) *Gardens of War: Life and Death in the New Guinea Stone Age.* New York (Random House).

GARDNER, B. T., and GARDNER, R. A. (on press) Comparing the Early Utterances of Child and Chimpanzee. In PICK, A. (ed.) *Minnesota Symposium on Child Psychology.*

GARDNER, B. T., and WALLACH, L. (1965) Shapes of Figures Identified as a Baby's Head. *Perceptual and Motor Skills, 20:*135–142.

GARDNER, R. A., and GARDNER, B. T. (1967) Acquisition of Sign Language in the Chimpanzee. *Univ. Nevada Progr. Report* (Ms.)

—— (1969) Teaching Sign Language to a Chimpanzee. *Science, 165:*664–672.

—— (1971) Two-Way Communication with an Infant Chimpanzee. In SCHRIER, A., and STOLLNITZ, F. (eds.) *Behavior of Nonhuman Primates, 4:* 117–184. New York and London (Academic Press).

GASTON, S., and MENAKER, M. (1968) Pineal Function: The Biological Clock in the Sparrow. *Science, 160:*1125–1127.

GATENBY, J. B. (1960) The New Zealand Glow-Worm. *Tuatera, 8:*86–92.

GATENBY, J. B., and COTTON, S. (1960) Snare Building and Pupation in *Bolitophila luminosa. Trans. Roy. Soc. New Zealand, 88:*149–156.

GAUL, A. T. (1952) Audiomimicry: An Adjunct to Color Mimicry. *Psyche, 59:*82–83.

GAZZANIGA, M. S. (1967) The Split Brain in Man. *Sci. Am., 217*(2):24–29.

GEBER, M. (1958) The Psycho-Motor Development of African Children in the First Year and the Influence of Maternal Behavior. *J. Soc. Psychol., 47:*185–195.

GEHLEN, A. (1940) *Der Mensch, seine Natur und seine Stellung in der Welt.* Berlin.

—— (1956) *Urmensch und Spätkultur.* Bonn.

—— (1961) Anthropologische Forschung. *Rowohlts Deut. Enzycl.,* 138.

—— (1969) *Moral und Hypermoral: Eine pluralistische Ethik.* Frankfurt (Athenaion).

GEIST, V. (1966a) The Evolution of Horn-Like Organs. *Behaviour, 27:*175–214.

—— (1966b) The Evolutionary Significance of Mountain-Sheep Horns. *Evolution, 20:*558–566.

—— (1966c) Ethological Observations on Some West-African Cervids. *Zool. Beitr., 12:*219–250.

—— (1971) *Mountain Sheep: A Study in Behavior and Evolution.* Chicago (Univ. Press).

GELBER, B. (1965) Studies on the Behavior of *Paramaecium aurelia. Animal Behaviour, Suppl., 1:*21–29.

GENTZ, K. (1935) Zur Brutpflege des Wespenbussards. *J. Ornithol., 83:*105–115.

GERAMB, V. R. v. (1918) Zur Volkskunde des Gesäusegebietes. *Z. Deut. Oesterr. Alpenver., 49:*33–66.

GERARD, R. W. (1961) The Fixation of Experience. In DELAFRESNAYE, J. F. (ed.), *Brain Mechanisms and Learning.* Oxford (Blackwell), 21–35.

GERLACH, S. (1965) Tierwanderungen. *Handb. d. Biol., 5.* Frankfurt (Athenaion), 413–472.

GESSNER, F. (1942) Die Leistungen des pflanzlichen Organismus. *Handb. d. Biol., 4:*34–187. Frankfurt (Athenaion).

GIARMAN, N. J., and FREEDMAN, D. X. (1965) Biochemical Aspects of the Actions of Psychomimetic Drugs. *Pharmacol. Rev., 17:*1–25.

GIBBS, F. A. (1951) Ictal and Non-ictal Psychiatric Disorders in Temporal Lobe Epilepsy. *J. Nerv. Ment. Dis., 113:*522–528.

GIBSON, E. J., and WALK, R. D. (1960) The Visual Cliff. *Sci. Am., 202*(4):64–71.

GILLIARD, E. T. (1963) The Evolution of Bowerbirds. *Sci. Am., 209*(2):38–46.

GINSBURGH, B., and ALLEE, W. (1942) Some Effects of Conditioning on Social Dominance and Subordination in Inbred Strains of Mice. *Physiol. Zool., 15:*485–506.

GLICKMANN, S. E., and SROGES, R. W. (1966) Curiosity in Zoo Animals. *Behaviour, 24:*151–188.

GNADENBERG, W. (1962) Erlebnisse mit Hunden. *Z. Tierpsychol., 19:*586–596.

GOETHE, F. (1938) Beobachtungen über das Absetzen von Witterungsmarken beim Baummarder. *Deut. Jäger, 13.*

——— (1939) Über das "Anstoß-Nehmen" bei Vögeln. *Z. Tierpsychol., 3:*371–374.

——— (1955) Beobachtungen bei der Aufzucht junger Silbermöwen. *Z. Tierpsychol., 12:*402–433.

GOETSCH, W. (1940) *Vergleichende Biologie der Insektenstaaten.* Leipzig.

GOLDFARB, W. (1943) The Effects of Early Institutional Care on Adolescent Personality. *J. Exptl. Educ., 12:*106–129.

GOLDSTEIN, K. (1939) *The Organism.* New York (American Book).

GOODALL, J. (1963) My Life among Wild Chimpanzees. *Natl. Geogr. Mag., 125*(8):272–308.

——— (1965) Chimpanzees of the Gombe Stream Reserve. In DEVORE, I. (ed.), *Primate Behavior.* New York (Holt, Rinehart and Winston), 425–473.

GOODENOUGH, F. L. (1932) Expressions of the Emotions in a Blind-Deaf Child. *J. Abnorm. Soc. Psychol., 27:*328–333.

GOTTLIEB, G. (1965a) Imprinting in Relation to Parental and Species Identification by Avian Neonates. *J. Comp. Physiol. Psychol., 59:*345–356.

——— (1965b) Prenatal Auditory Sensitivity in Chickens and Ducks. *Science, 147:*1596–1598.

——— (1966) Species Identification by Avian Neonates: Contributory Effects of Perinatal Auditory Stimulation. *Animal Behaviour, 14:*282–290.

GOTTLIEB, G., and KUO, Z. Y. (1965) Development of Behavior in the Duck Embryo. *J. Comp. Physiol. Psychol., 59:*183–188.

GOULD, E. (1957) Orientation in Box Turtles *Terrapene c. carolina. Biol. Bull., 112:*336–348.

GOULD, J. L., HENEREY, M., and MACLEOD, M. S. (1970) Communication of Direction by the Honey Bee. *Science, 169:*554.

GRABOWSKI, U. (1941) Prägung eines Jungschafes auf den Menschen. *Z. Tierpsychol., 4:*326–329.

GRAHAM BROWN, T. (1911) The Intrinsic Factors in the Act of Progression in the Mammal. *Proc. Roy. Soc. London B84:*308–320.

——— (1912) The Factors in Rhythmic Activity of the Nervous System. *Proc. Roy. Soc. London B85:*278–289.

GRANIT, R. (1955) *Receptors and Sensory Perception.* New Haven (Yale Univ. Press).

GRANT, E. C. (1965) An Ethological Description of some Schizophrenic Patterns of Behaviour. *Proc. Leeds Symp. Behav. Disorders,* Chapter 12.

——— (1968) An Ethological Description of Non-verbal Behaviour During Interviews. *British J. Med. Psychol, 41:*177–184.

——— (1969) Human Facial Expression. *Man, N.S., 4:*525–536.

——— (1972) Communication in the Mentally Ill. In HINDE, R. A. (ed.) *Non-Verbal Communication.* Cambridge (Univ. Press), 349–358.

GRASSÉ, P. P. (1949) Ordre des Isopteres ou Termites. *Traité de Zoologie 9:*408–544.

GRAY, J. (1950) The Role of Peripheral Sense Organs during the Locomotion of Vertebrates. Physiological Mechanisms in Animal Behavior. *Symp. Soc. Exptl. Biol.* 112–126 (Cambridge Univ. Press).

GRAY, J., and LISSMANN, H. W. (1946a) Further Observations on the Effect of Deafferentiation on the Locomotory Activity of Amphibian Limbs. *J. Exptl. Biol., 23:121-132.*

——— (1946b) The Coordination of Limb Movements in the Amphibia. *J. Exptl. Biol., 23:*133–142.

GRAY, P. H. (1958) Theory and Evidence of Imprinting in Human Infants. *J. Psychol., 46:*155–160.

GRETHER, W. F. (1939) Color Vision and Color Blindness in Monkeys. *Comp. Psychol. Monogr., 15:*1–38.

GRIFFIN, D. R. (1958) *Listening in the Dark.* New Haven (Yale Univ. Press).

——— (1962) Echo-Ortung der Fledermäuse, insbesondere beim Fangen fliegender Insekten. *Naturwiss. Rundschau, 15:*169–173.

GRIFFITH-SMITH, N. (1966) Evolution of Some Arctic Gulls *(Larus):* An Experimental Study of Isolating Mechanisms. *Ornithol. Monogr., 4.*

GROHMANN, J. (1939) Modifikation oder Funktionsreifung? Z. *Tierpsychol., 2:*132–144.

GROOS, K. (1933) *Die Spiele der Tiere.* 3rd ed. Jena.

GROOT, C. (1965) On the Orientation of Young Sockeye Salmon *(Oncorhynchus nerka)* During Their Seaward Migration out of Lakes. *Behaviour, Suppl., 14.*

GROSSMANN, K. E. (1967) Behavioral Difference between Rabbits and Cats. *J. Gen. Psychol., 3:*171–182.

GRZIMEK, B. (1949a) Die "Radfahrer-Reaktion." Z. *Tierpsychol., 6:*41–44.

——— (1949b) Ein Fohlen, das kein Pferd kannte. Z. *Tierpsychol., 6:*391–405.

——— (1951) *Affen im Haus.* Stuttgart.

——— (1954) Beobachtungen an Schimpansen, *Pan tr. troglodytes* (Blumenbach, 1775) in den Nimbabergen. *Säugetierkdl. Mitt., 1:*1–5.

GUITON, P. (1960) On the Control of Behavior during the Reproductive Cycle of *Gasterosteus aculeatus. Behaviour, 15:*163–184.

GÜNTHER, K. (1956) Systematik und Stammesgeschichte der Tiere 1939–1953. *Fortschr. d. Zool., 10:*33–278.

GUTHRIE, E. R. (1952) *The Psychology of Learning.* New York (Harper).

GÜTTINGER, H. R. (1970) Zur Evolution von Verhaltensweisen und Lautäußerungen bei Prachtfinken (Estrilidae). Z. *Tierpsychol., 27:*1011–1075.

GWINNER, E. (1961) Über die Entstachelungshandlung des Neuntöters (*Lanius collurio*). *Vogelwarte, 21:*36–47.

——— (1964) Untersuchungen über das Ausdrucks- und Sozialverhalten des Kolkraben (*Corvus corax*). Z. *Tierpsychol., 21:*657–748.

——— (1966) Über einige Bewegungsspiele des Kolkraben. Z. *Tierpsychol., 23:*28–36.

——— (1967) Circannuale Periodik der Mauser und Zugunruhe bei einigen Vögeln. *Naturwiss., 54:*447.

GWINNER, E., and KNEUTGEN, J. (1962) Über die biologische Bedeutung der "zweckdienlichen" Anwendung erlernter Laute bei Vögeln. Z. *Tierpsychol., 19:*692–696.

HAAS, A. (1962) Phylogenetisch bedeutungsvolle Verhaltensänderungen bei Hummeln. Z. *Tierpsychol., 19:*356–370.

——— (1965) Weitere Beobachtungen zum "generischen Verhalten" bei Hummeln. Z. *Tierpsychol., 22:*305–320.

HACKETT, E., ESSER, A. H., and KLINE, N. S. (1966) *Heterogeneous Dimensions of Chronic Schizophrenic Behavior.* Manuscript, Research Center, Rockland Hospital, Orangeburg, N.Y.

HÄDECKE, W. (1969) Verhaltensforschung – Resultate und Perspektiven. *Neue Rundschau, 80:*447–462.

HAILMAN, J. P. (1967) The Ontogeny of an Instinct. *Behaviour, Suppl. 15.*

HALBERG, F., ENGELI, M., HAMBURGER, C., and HILLMAN, D. (1965) Spectral Resolution of Low-Frequency Small-Amplitude Rhythms in Excreted 17-Ketosteroids: Probable Androgen-Induced Circasepton Desynchronisation. *Acta Endocrinol., Suppl., 103:*1–54.

HALL, E. T. (1966) *The Hidden Dimension.* New York (Doubleday).
HALL, K. R. L. (1965) Social Learning in Monkeys. *J. Zool. 148:*15–87.
HALL, K. R. L., and DEVORE, I. (1965) Baboon Social Behavior. In DEVORE, I. (ed.) *Primate Behavior.* New York (Holt, Rinehart and Winston).
HALL, K. R. L., and SCHALLER, G. B. (1964) Tool Using Behavior of the California Sea Otter. *J. Mammal., 45:*287–298.
HALL, M. F. (1962) Evolutionary Aspects of Estrildid Song. *Symp. Zool. Soc. London, 8:*37–55.
HAMBURGER, V. (1963) Some Aspects of the Embryology of Behavior. *Quart. Rev. Biol., 38:*342–365.
HAMBURGER, V., and OPPENHEIM, R. (1967) Prehatching Motility and Hatching Behavior in the Chick. *J. Exptl. Zool., 166:*171–203.
HAMBURGER, V., WENGER, R. E., and OPPENHEIM, R. (1966) Motility in the Chick-Embryo in Absence of Sensory Input. *J. Exptl. Zool., 162:*133–160.
HAMILTON, W. D. (1964) The Genetical Evolution of Social Behavior. *J. Theoret. Biol. 7:*1–52.
HARLOW, H. F. (1932) Social Facilitation of Feeding in the Albino Rat. *J. Gen. Psychol., 41:*211–221.
——— (1953) Higher Functions of the Nervous System. *Ann. Rev. Physiol., 15:*493–514.
HARLOW, H. F., and HARLOW, M. K. (1962a) The Effect of Rearing Conditions on Behavior. *Bull. Menninger Clin., 26:*213–224.
——— (1962b) Social Deprivation in Monkeys. *Sci. Am., 207:*137–146.
HARLOW, H. F., HARLOW, M. K., and MEYER, D. R. (1950) Learning Motivated by a Manipulation Drive. *J. Exptl. Psychol., 40:*228–234.
HARRIS, G. W. (1964) Female Cycles of Gonadotrophic Secretion and Female Sexual Behavior in Adult Male Rats Castrated at Birth. *J. Physiol. (London), 175:*75–76.
HARRIS, T. v. (1950) Habitat Selection of *Peromyscus.* Ph.D. Thesis Univ. Michigan, Ann Arbor.
HARRISON, C. J. O. (1965) Allopreening as Agonistic Behavior. *Behaviour, 24:*161–209.
HARTMANN, H., KRIS, E., and LOEWENSTEIN, R. M. (1949) Notes on the Theory of Aggression. *Psychoanalytic Study of the Child 3–4.* New York (International Univ. Press).
HARTMANN, M. (1956) *Die Sexualität.* Stuttgart (G. Fischer).
HARTRY, A. L., KEITH-LEE, P., and MORTON, W. D. (1964) Planaria: Memory Transfer through Cannibalism Reexamined. *Science, 146:*274–275.
HASKINS, C. P., and HASKINS, E. F. (1958) Note on the Inheritance of Behavior Patterns for Food Selection and Cocoon Spinning in F_1-hybrids of *Callosamia promethea* and *C. angulifera. Behaviour, 13:*89–95.
HASLER, A. D. (1954) Odour Perception and Orientation in Fishes. *J. Fisheries Res. Board Can., 11:*107–129.
——— (1956) Perception of Pathways by Fishes in Migration. *Quart. Rev. Biol., 31:*200–209.
——— (1960) Guideposts of Migrating Fishes. *Science, 131:*785–792.
HASLER, A. D., and SCHWASSMANN, H. O. (1960) Sun Orientation of Fish at Different Latitudes. *Cold Spring Harbor Symp. Quant. Biol., 25:*429–441.
HASS, H. (1951) *Drei Jäger auf dem Meeresgrund.* Zürich (Füssli).
——— (1957) *Wir kommen aus dem Meer.* Berlin (Ullstein).
——— (1968) *Wir Menschen.* Wien (Molden).
HASS, H., and EIBL-EIBESFELDT, I. (1964) *Dotilla sulcata (Brachyura): Fressen und Graben.* Encycl. cinem., E538. Publ. zu wiss. Filmen, 1 A, 165–168. Göttingen (Inst. wiss. Film).
HASSENSTEIN, B. (1955) Abbildende Begriffe. *Zool. Anz., Suppl., 18:*197–202.
——— (1965) *Biologische Kybernetik.* Heidelberg (Quelle u. Meyer).
——— (1966) Kybernetik und biologische Forschung. *Handb. d. Biol., 1,* 631–719. Frankfurt (Athenaion).

—— (1973) *Verhaltensbiologie des Kindes.* München (Piper).

HASSLER, R., and BAK, I. J. (1966) Submikroskopische Catecholaminspeicher als Angriffspunkte der Psychopharmaka Reserpin und Mono-Amino-Oxydase-Hemmer. *Nervenarzt, 37:*493–498.

HAUENSCHILD, C. (1960) Lunar Periodicity. *Cold Spring Harbor Symp. Quant. Biol., 25:*73–86.

HAYES, C. (1951) *The Ape in Our House.* New York (Harper & Row).

HEBB, D. O. (1949) *The Organization of Behaviour.* New York (Wiley).

—— (1953) Heredity and Environment in Mammalian Behaviour. *Brit. J. Animal Behaviour, 1:*43–47.

HEBERER, G. (1965) Über den systematischen Ort und den psychisch-physischen Status der Australopithecinen. In HEBERER, G. (ed.), *Menschliche Abstammungslehre.* Stuttgart (G. Fischer), 310–356.

HEDIGER, H. (1933) Beobachtungen an der marokkanischen Winkerkrabbe (*Uca tangeri*). *Verhandl. Schweiz. Naturforsch. Ges., 114:*388–389.

—— (1934) Zur Biologie und Psychologie der Flucht bei Tieren. *Biol. Zbl., 54:*21–40.

—— (1942) *Wildtiere in Gefangenschaft.* Basel.

—— (1949) Säugetierterritorien und ihre Markierung. *Bijdr. tot de Dierkde, 28:*172–184.

—— (1954) *Skizzen zu einer Tierpsychologie im Zoo und im Zirkus.* Zürich (Gutenberg).

—— (1961) The Evolution of Territorial Behavior. In WASHBURN, S. L. (ed.), *Social Life of Early Man.* Chicago (Aldine), 34–57.

—— (1963) Weitere Dressurversuche mit Delphinen und anderen Walen. *Z. Tierpsychol., 20:*487–497.

HEGH, E. (1922) *Les Termites.* Brussels (L. Desmet).

HEILIGENBERG, W. (1963) Ursachen für das Auftreten von Instinktsbewegungen bei einem Fische (*Pelmatochromis subocellatus kribensis*). *Z. vergl. Physiol., 47:*339–380.

—— (1964) Ein Versuch zur ganzheitsbezogenen Analyse des Instinktverhaltens eines Fisches (*Pelmatochromis subocellatus kribensis*). *Z. Tierpsychol., 21:*1–52.

—— (1965) A Quantitative Analysis of Digging Movements and Their Relationship to Aggressive Behaviour in Cichlids. *Animal Behaviour, 13:*163–170.

HEILIGENBERG, W., and KRAMER, U. (1972) Aggressiveness as a Function of External Stimulation. *J. Comp. Physiol., 77:*332–340.

HEILIGENBERG, W., KRAMER, U., and SCHULZ, V. (1972) The Angular Orientation of the Black Eye-Bar in *Haplochromis Burtoni* (Cichlidae, Pisces) and its Relevance to Aggressivity. *Z. vergl. Physiol., 76:*168–176.

HEINROTH, O. (1910) Beiträge zur Biologie, insbesondere Psychologie und Ethologie der Anatiden. *Verh. 5. Int. Ornith. Kongr.,* Berlin, 589–702.

HEINROTH, O., and HEINROTH, M. (1928) *Die Vögel Mitteleuropas.* Berlin-Lichterfelde (Bermühler).

HEINROTH-BERGER, K. (1965) Über Geburt und Aufzucht eines männlichen Schimpansen im Zool. Garten Berlin. *Z. Tierpsychol., 22:*15–35.

HEINZ, H. J. (1949) Vergleichende Beobachtungen über die Putzhandlungen bei Dipteren und bei *Sarcophaga carnaria* L. im besonderen. *Z. Tierpsychol., 6:*330–371.

—— (1966) The Social Organization of the !Ko-Bushmen. Master's Thesis, Dept. Anthropology, Univ. of South Africa, Johannesburg.

—— (1967) Conflicts, Tensions and Release of Tensions in a Bushmen Society. *The Institute for the Study of Man in Africa, Isma Papers,* 23.

—— (1972) Territoriality among the Bushmen in General and the !Ko in Particular. *Anthropos, 67:*405–416.

HELD, R., and HEIN, A. (1963) Movement-Produced Stimulation in the Development of Visually Guided Behavior. *J. Comp. Physiol. Psychol., 56:*872–876.

HELLBRÜGGE, T. (1967) Chronophysiologie des Kindes. *Verh. dtsch. Ges. innere Med., 73:* 895.

HELLMUTH, H. (1967) Zum Verhalten des Menschen: die Aggression. *Z. Ethnol.*, *92*(2): 265–273.

HERAN, H. (1966) Sinnesphysiologie. *Handb. d. Biol.*, *5*, 473. Frankfurt (Athenaion).

HERING, H. E. (1896) Über Bewegungsstereotypien nach centripetaler Lähmung. *Arch. Exptl. Pathol. Pharmakol.*, *38:*266–283.

HERING, E. (1921) *Über das Gedächtnis als eine allgemeine Funktion der organisierten Materie.* 3rd ed. Leipzig (Akadem. Verl. Ges.). (First edition, 1870.)

HERNANDEZ-PEON, R., and BRUST-CARMONA, H. (1961) Functional Role of Subcortical Structure in Habituation and Conditioning. In DELAFRESNAYE, J. F. (ed.), *Brain Mechanisms and Learning.* Oxford (Blackwell), 393–412.

HERREBOUT, W. M., KUYTEN, P. J., and RUITER, L. DE (1963) Observations on Color Patterns and Behavior of Caterpillars Feeding on Scots Pine. *Arch. Neerl. Zool.*, *15*(3):315–357.

HERTER, K. (1943) Beziehungen zwischen der Ökologie und der Thermotaxis der Tiere. *Biol. Gen.*, *17:*243–309.

——— (1952) Der Temperatursinn der Säugetiere. *Beitr. Tierkde–Tierzucht, 3:*1–171.

——— (1953) *Der Temperatursinn der Insekten.* Berlin (Dunker u. Humbolt).

HERTER, K., and SGONINA, K. (1938) Vorzugstemperatur und Hautbeschaffenheit bei Mäusen. *Z. vergl. Physiol.*, *26:*366–415.

HESS, C. v. (1913) Experimentelle Untersuchungen über den angeblichen Farbensinn der Bienen. *Zool. Jahrb. Allgem. Zool. Physiol.*, *34:*81–106.

HESS, E. H. (1956) Space Perception in the Chick. *Sci. Am.*, *195:*71–80.

——— (1959) Imprinting: An Effect of Early Experience. *Science*, *130:*133–141.

——— (1965) Attitude and Pupil Size. *Sci. Am.*, *212*(4):46–54.

——— (1973) *Imprinting: Early Experience and the Developmental Psychobiology of Attachment.* New York (Van Nostrand).

HESS, E. H., SELTZER, A. L., and SHLIEN, J. M. (1965) Pupil Response of Hetero- and Homosexual Males to Pictures of Men and Women: A Pilot Study. *J. Abnorm. Psychol.*, *70:*165–168.

HESS, W. R. (1954) *Das Zwischenhirn.* 2nd ed. Basel (Schwabe).

——— (1957) Die Formatio reticularis des Hirnstammes im verhaltensphysiologischen Aspekt. *Arch. Psychiat. Nervenkr.*, *196:*329–336.

HESSE, R., and DOFLEIN, F. (1943) *Tierbau und Tierleben. 2. Das Tier als Glied des Naturganzen.* Jena (G. Fischer).

HEUSSER, H. (1960) Über die Beziehungen der Erdkröte (*Bufo bufo* L.) zu ihrem Laichplatz II. *Behaviour*, *16:*93–109.

HEWES, G. H. (1957) The Anthropology of Posture. *Sci. Am.* *196*(2):123–132.

HILGARD, E. R. (1956) *Theories of Learning.* New York (Appleton).

HINDE, R. A. (1953) Appetitive Behaviour, Consummatory Act and the Hierarchical Organization of Behaviour: With Special Reference to the Great Tit (*Parus major*). *Behaviour*, *5:*189–224.

——— (1956) The Behaviour of Certain Cardueline F_1 Interspecies Hybrids. *Behaviour*, *9:*202–213.

——— (1958) The Nestbuilding Behaviour of Domesticated Canaries. *Proc. Zool. Soc. London*, *131:*1–48.

——— (1959) Some Recent Trends in Ethology. In KOCH, S. (ed.), *Psychology: A Study of a Science.* New York (McGraw-Hill) Vol. 2:561–610.

——— (1965) Interaction of Internal and External Factors in Integrations of Canary Reproduction. In BEACH, F. A. (ed.), *Sex and Behavior.* New York (Wiley), 381–415.

——— (1966) *Animal Behavior: A Synthesis of Ethology and Comparative Psychology.* New York (McGraw-Hill).

HINZE, G. (1950) *Der Biber.* Berlin (Akademie-Verlag).

HIRSCH, J. (1967) *Behavior: Genetic Analysis.* New York (McGraw-Hill).

HIRSCH, J., and BOUDREAU, J. C. (1958) Studies in Experimental Behavior Genetics: I. The Heritability of Phototaxis in a Population of *Drosophila melanogaster. J. Comp. Psychol., 51:*647–651.
HOBSON, E. S. (1965) Diurnal-Nocturnal Activity of Some Inshore Fishes in the Gulf of California. *Copeia, 3:*291–302.
——— (1972) Activity of Hawaiian Reef Fishes during the Evening and the Morning Transitions between Daylight and Darkness. *Fishery Bull. 70:*715–740.
HOCKETT, C. F. (1960) Logical Considerations in the Study of Animal Communication. *Am. Inst. Biol. Sci. Publ., 7:*392–430. Also *Z. Tierpsychol., 23:*250–254.
HOEBEL, B. G., and TEITELBAUM, P. (1962) Hypothalamic Control of Feedings and Self-Stimulation. *Science, 135:*375–377.
HOFFMANN, K. (1954) Versuche zu der im Richtungsfinden der Vögel enthaltenen Zeitschätzung. *Z. Tierpsychol., 11:*453–475.
——— (1955) Aktivitätsregistrierungen bei frisch geschlüpften Eidechsen. *Z. vergl. Physiol., 37:*253–262.
——— (1959) Die Aktivitätsperiodik von im 18- und 36-Stundentag erbrüteten Eidechsen. *Z. vergl. Physiol., 42:*422–432.
——— (1960) Experimental Manipulation of the Orientational Clock in Birds. *Cold Spring Harbor Symp. Quant. Biol., 25:*379–387.
——— (1965) Overt Circadian Frequencies and Circadian Rule. In ASCHOFF, J. (ed.), *Circadian Clocks.* Amsterdam (North-Holland), 87–94.
——— (1968) Synchronisation der circadianen Aktivitätsperiodik von Eidechsen durch Temperaturcyclen verschiedener Amplitude. *Z. vergl. Physiol., 58:*225–228.
——— (1969) Die relative Wirksamkeit von Zeit. *Oecologia (Berl.), 3:*184–206.
——— (1970) Zur Sychronisation biologischer Rhythmen. *Verh. Dtsch. Zool. Ges., 64. Tagung.* Stuttgart (G. Fischer), 266–273.
——— (1971) Biological Clocks in Animal Orientation and in other Functions. *Proc. International Symposium Circadian Rhythmicity* (Wageningen), 170–200.
HOFSTÄTTER, P. R. (1959) *Psychologie.* 3rd ed. Frankfurt (Fischer-Bücherei).
HOHORST, W., and GRAEFE, G. (1961) Ameisen – obligatorische Zwischenwirte des Lanzettegels *(Dicrocoelium dendriticum). Naturwiss., 48:*229–230.
HOKANSON, J. E., and BURGESS, M. (1962) The Effects of Three Types of Aggression on Vascular Processes. *J. Abnorm. Soc. Psychol., 64:*446–449.
HOKANSON, J. E., and SHETLER, S. (1961) The Effect of Overt Aggression on Physiological Tension Level. *J. Abnorm. Soc. Psychol., 63:*446–448.
HÖLLDOBLER, B. (1967) Zur Physiologie der Gast-Wirt-Beziehungen (Myrmecophilie) bei Ameisen. *Z. vergl. Physiol., 56:*1–21.
HOLST, D. v. (1969) Sozialer Stress bei Tupajas *(Tupaia belangeri) Z. vergl. Physiol., 63:*1–58.
——— (1972) Renal Failure as the Cause of Death in *Tapaia belangeri* Exposed to Persistent Social Stress. *J. Comp. Physiol., 78:*236–273.
HOLST, E. v. (1932) Untersuchungen über die Funktionen des Zentralnervensystems beim Regenwurm *(Lumbricus terrestris* L.). *Zool. Jb. (Physiol.), 51,* 4:547–588.
——— (1933) Weitere Versuche zum nervösen Mechanismus der Bewegungen beim Regenwurm *(Lumbricus terrestris* L.). *Zool. Jb. (Physiol.), 53,* 1:67–100.
——— (1935) Über den Prozeß der zentralen Koordination. *Arch. Ges. Physiol., 236:* 149–158.
——— (1936) Versuche zur Theorie der relativen Koordination. *Arch. Ges. Physiol., 237:* 93–121.
——— (1937) Baustein zu einer vergleichenden Physiologie der lokomotorischen Reflexe bei Fischen II. *Z. vergl. Physiol., 24:*532–562.
——— (1938) Neuere Versuche zur Deutung der relativen Koordination bei Fischen. *Arch. Ges. Physiol., 240:*1–43.

——— (1939) Die relative Koordination als Phänomen und als Methode zentralnervöser Funktionsanalyse. *Ergeb. Physiol., 42:*228–306.

——— (1943) Über die relative Koordination bei Arthropoden. *Pflüg. Arch., 246:*847–865.

——— (1950a) Quantitative Messung von Stimmungen im Verhalten der Fische. In *Physiological Mechanisms in Animal Behavior.* Symp. Soc. Exp. Biol., 4. Cambridge (Univ. Press), 143–172.

——— (1950b) Die Tätigkeit des Statolithenapparates der Wirbeltiere. *Naturwiss., 12:* 265–272.

——— (1955) Regelvorgänge in der optischen Wahrnehmung. *5th Conf. Soc. Biol. Rhythm,* Stockholm, 26–34.

——— (1956) Optische Wahrnehmungen, die wir selbst erzeugen, und ihre Bedeutung für unser Dasein. *Jahrb. Max-Planck-Ges.,* 121–149.

——— (1957) Die Auslösung von Stimmungen bei Wirbeltieren durch "punktförmige" elektrische Erregung des Stammhirns. *Naturwiss., 44:*549–551.

——— (1969) *Zur Verhaltensphysiologie bei Tieren und Menschen. I. und II.* Munich (Piper).

HOLST, E. v., KAISER, M., RÖBIG, G., and GÖLDNER, G. (1950) Die Arbeitsweise des Statolithen-Apparates bei Tieren. *Z. verg. Physiol., 32:*60–120.

HOLST, E. v., and MITTELSTAEDT, H. (1950) Das Reafferenz-Prinzip. *Naturwiss., 37:*464–476.

HOLST, E. v., and SAINT-PAUL, U. v. (1960) Vom Wirkungsgefüge der Triebe. *Naturwiss., 18:*409–422.

HOLZAPFEL, M. (1938) Über Bewegungsstereotypien bei gehaltenen Säugern. I. und II. *Z. Tierpsychol., 2:*46–72.

——— (1939) Über Bewegungsstereotypien bei gehaltenen Säugern. III. Analyse der Bewegungsstereotypie eines Gürteltieres. *Zool. Garten, 10:*184–193.

——— (1940) Triebbedingte Ruhezustände als Ziel von Appetenzhandlungen. *Naturwiss., 28:*273–280.

——— (1949) Die Beziehungen zwischen den Trieben junger und erwachsener Tiere. *Schweiz. Z. Psychol., 8:*32–60.

HOOFF, J. A. R. A. M. van (1971) *Aspecten van Het Sociale Gedrag En De Communicatie Bij Humane En Hogere Niet-Humane Primaten. (Aspects of the Social Behaviour and Communication in Human and Higher Nonhuman Primates).* Rotterdam (Bronder Offset).

HOOGLAND, R., MORRIS, D., and TINBERGEN, N. (1957) The Spines of the Sticklebacks *(Gasterosteus* and *Pygosteus)* as Means of Defence against Predators *(Perca* and *Esox). Behaviour, 10:*205–236.

HOPPENHEIT, M. (1964) Beobachtungen zum Beutefangverhalten der Larve von *Aeschna cyanea* Müll. *(Odonata). Zool. Anz., 172:*216–232.

HÖRMANN, L. v. (1912) Genuß- und Reizmittel in den Ostalpen, eine volkskundliche Skizze. *Z. Deut. Oesterr. Alpenver., 43:*78–100.

HÖRMANN-HECK, S. v. (1957) Untersuchungen über den Erbgang einiger Verhaltensweisen bei Grillenbastarden *(Gryllus campestris* x *Gryllus bimaculatus). Z. Tierpsychol., 14:*137–183.

HOWARD, H. E. (1920) *Territory in Bird Life.* New York (Dutton).

HSIAO, H. H. (1929) An Experimental Study of the Rat's "Insight" within a Spatial Complex. *Univ. Calif. Publ. Psychol., 4:*57–70.

HUBEL, D. H., and WIESEL, T. N. (1959) Receptive Fields of Single Neurons in the Cat's Striate Cortex. *J. Physiol. (London), 148:*574–591.

——— (1962) Receptive Fields, Binocular Interactions and Functional Architecture in the Cat's Visual Cortex. *J. Physiol. (London), 160:*106–154.

——— (1963) Receptive Fields of Cells in Striate Cortex of Very Young, Visually Inexperienced Kittens. *J. Neurophysiol., 24:*994–1002.

HUBER, F. (1955) Sitz und Bedeutung nervöser Zentren für Instinkthandlungen beim Männchen von *Gryllus campestris*. *Z. Tierpsychol., 12:*12–48.

HÜCKSTEDT, B. (1965) Experimentelle Untersuchungen zum "Kindchenschema." *Z. Exptl. Angew. Psychol., 12:*421–450.

HUDGENS, G. A., DENENBERG, V. H., and ZARROW, M. X. (1968) Mice Reared with Rats: Effects of Preweaning and Postweaning Social Interaction upon Adult Behaviour. *Behaviour, 30:*259–274.

HUET, M. (1952) Dix années de pisciculture aux Congo Belge et aux Ruanda. *Traité de pisciculture*. Brussels.

HUIZINGA, J. (1956) *Homo ludens*. Rowohlts Deut. Encycl., 21.

HULL, C. L. (1943) *Principles of Behaviour*. New York.

HUNSAKER, D. (1962) Ethological Isolating Mechanisms in the *Sceloporus torquatus* Group of Lizards. *Evolution, 16:*62–74.

HUNTER, W. S. (1913) The Delayed Reaction in Animals and Children. *Behavior Monogr., 2:*21–30.

HUTT, S. J., and HUTT, C. (1970) *Behaviour Studies in Psychiatry*. Oxford (Pergamon Press).

HUXLEY, J. S. (1923) Courtship Activities in the Red-Throated Diver (*Colymbus stellatus* Pontopp): Together with a Discussion of the Evolution of Courtship in Birds. *J. Linnean Soc. London Zool., 53:*253–292.

HYDÉN, H. (1961) Satellite Cells in the Nervous System. *Sci. Am., 205*(6):62–70.

HYDÉN, H., and EGYHAZI, E. (1962) Nuclear RNA Changes of Nerve Cells during a Learning Experiment in Rats. *Proc. Natl. Acad. Sci. U.S., 48:*1366–1373.

IERSEL, J. v. (1953) An Analysis of the Parental Behavior of the Three-Spined Stickleback *(Gasterosteus aculeatus). Behaviour, Suppl., 3.*

——— (1958) Some Aspects of Territorial Behavior of the Male Three-Spined Stickleback. *Arch. Neerl. Zool., 13:*383–400.

IMANISHI, K. (1957) Social Behaviour in Japanese Monkeys, *Macaca fuscata. Psychologia, 1:*47–54.

IMMELMANN, K. (1959) Experimentelle Untersuchungen über die biologische Bedeutung artspezifischer Merkmale beim Zebrafinken (*Taeniopygia castanotis* Gould). *Zool. Jahrb. Abt. System, 86:*438–592.

——— (1961) Beitrag zur Biologie und Ethologie australischer Honigfresser *(Meliphagidae). J. Ornithol., 102:*164–207.

——— (1962a) Vergleichende Beobachtungen über das Verhalten domestizierter Zebrafinken in Europa und ihrer wilden Stammform in Australien. *Z. Tierzücht., 77:*198.

——— (1962b) Beiträge zu einer vergleichenden Biologie australischer Prachtfinken *(Spermestidae). Zool. Jahrb. System, 90:*1–196.

——— (1965) Prägungserscheinungen in der Gesangsentwicklung junger Zebrafinken. *Naturwiss., 52:*169–170.

——— (1966) Zur Irreversibilität der Prägung. *Naturwiss., 53:*209.

——— (1967) Zur ontogenetischen Gesangsentwicklung bei Prachtfinken. *Zool. Anz., Suppl., 30:*320–332.

——— (1970) Zur ökologischen Bedeutung prägungsbedingter Isolationsmechanismen. *Verh. Dtsch. Zool. Ges., 64. Tagung*. Stuttgart (G. Fischer), 304–314.

INHELDER, E. (1955) Zur Psychologie einiger Verhaltensweisen, besonders des Spiels von Zootieren. *Z. Tierpsychol., 12:*88–144.

ITANI, J. (1958) On the Acquisition and Propagation of a New Food Habit in the Troop of Japanese Monkeys at Takasakiyama. *Primates, 1:*84–98 (in Japanese with English summary).

ITANI, J., and SUZUKI, A. (1967) The Social Unit of Chimpanzees. *Primates, 8:*355–381.

JACOBS, W. (1953a) Vergleichende Verhaltensstudien an Feldheuschrecken *(Orthoptera, Acrididae)* und einiger anderer Insekten. *Zool. Anz., Suppl., 19:*115–138.

——— (1953b) Verhaltensbiologische Studien an Feldheuschrecken. Z. *Tierpsychol., Suppl., 1.*

——— (1966) Die Gesänge der Heuschrecken. In BURCKHARDT, D., SCHLEIDT, W., and ALTNER, H. (eds.): *Düfte, Farben und Signale.* Munich (Moos).

JACOBS-JESSEN, U. F. (1959) Zur Orientierung der Hummeln und einiger anderer Hymenopteren. *Z vergl. Physiol., 41:*597–641.

JACOBSON, A. L., BABICH, F. R., BUBASH, S., and JACOBSON, A. (1965) Differential-Approach Tendencies Produced by Injection of RNA from Trained Rats. *Science, 150:* 636–637.

JAMES, W. (1890) *Principles of Psychology.* New York (Holt, Rinehart and Winston).

JANDER, R. (1966a) Die Phylogenie von Orientierungsmechanismen der Arthropoden. *Zool. Anz., Suppl., 29:*266–306.

——— (1966b) Die Hauptentwicklungen der Lichtorientierung bei den tierischen Organismen. *Verh. Verb. Deut. Biol., 3:*28–34.

JANISSE, M. P. (1973) Pupil Size and Affect: A Critical View of the Literature Since 1960. *Canad. Psychologist, 14:*311–329.

JANISSE, M. P., and PFAVLER, W. S. (1974) Pupillary Research Today: Emotion in the Eye. *Psychology Today* (February).

JANTSCHKE, F. (1972) *Orang-Utans in Zoologischen Gärten.* München (Piper).

JARVIK, L., KLODIN, V., and MATSUYAMA, S. S. (1973) Human Aggression and the Extra Y Chromosome: Fact or Fantasy? *Am. Psychologist, 28:*674–682.

JAY, P. (ed.) (1968) *Primates: Studies in Adaptation and Variability.* New York (Holt, Rinehart and Winston).

JENKINS, D., WATSON, A., and MILLER, G. R. (1967) Population Fluctuations in the Red Grouse *Lagopus lagopus scoticus. J. Animal Ecol., 36:*97–122.

JENNINGS, H. S. (1906) *The Behavior of the Lower Organisms.* New York.

JENSEN, A. R. (1969) How Much Can We Boost IQ and Scholastic Achievement? *Harvard Educational Review, 39* (1):1–123.

JESPERSEN, O. (1925) *Die Sprache.* Heidelberg.

JOHNSON, D. L. (1967) Honeybees: Do They Use the Direction Information Contained in Their Dance Maneuver? *Science, 155:*844–847.

JOLLY, A. (1966) Lemur Social Behavior and Primate Intelligence. *Science, 153:*501–506.

——— (1972) *The Evolution of Primate Behavior.* New York (Macmillan).

JOUVET, M. (1972) Le Discours Biologique. *Rev. Medécine, 16:*1003–1063.

KAADA, B. (1967) Brain Mechanisms Related to Aggressive Behaviour. *UCLA Forum in Medical Science, 7:*95–133. Reprinted in CLEMENTE, C. D., and LINDSLEY, D. B. (eds.), *Aggression and Defense: Neural Mechanisms and Social Pattern.* Berkeley (Univ. California Press).

KAESTNER, A. (1965) *Lehrbuch der speziellen Zoologie. I. Wirbellose.* 2nd ed. Jena (G. Fischer).

KAGAN, J., and MOSS, H. A. (1962) *Birth to Maturity: A Study in Psychological Development.* New York (Wiley).

KAHN, M. W. (1951) The Effect of Severe Defeat at Various Levels on the Aggressive Behavior of Mice. *J. Gen. Psychol., 79:*117–130.

——— (1954) Infantile Experience and Mature Aggressive Behavior of Mice: Some Maternal Influences. *J. Gen. Psychol., 84:*65–75.

KAISSLING, K. E., and PRIESNER, E. (1970) Die Riechschwelle des Seidenspinners. *Naturwiss., 57:*23–28.

KANT, I. (1960) *Werke.* Wiesbaden.

KAPUNE, T. (1966) Untersuchungen zur Bildung eines "Wertbegriffes" bei niederen Primaten. Z. *Tierpsychol., 23:*324–363.

KÄSTLE, W. (1963) Zur Ethologie des Grasanolis *(Norops auratus). Z. Tierpsychol., 20:* 16–33.

——— (1965) Zur Ethologie des Anden-Anolis *Phenacosaurus richteri. Z. Tierpsychol., 22:*751–769.

KATZ, D. (1926) Sozialpsychologie der Vögel. *Ergeb. Biol., 1:*447–478.

KAUFMANN, S. A. (1967) Social Relations of Adult Males in a Free-Ranging Band of Rhesus Monkeys. In ALTMAN, S. A. (ed.) *Social Communication among Primates.* Chicago (Univ. Press), 73–98.

KAWAI, M. (1958) On the Rank System in a Natural Group of Japanese Monkeys. *Primates, 1:*84–98 (in Japanese with English summary).

——— (1965) Newly Acquired Pre-Cultural Behavior of the Natural Troop of Japanese Monkeys on Koshima Island. *Primates, 6:*1–30.

KAWAI, M., and MIZUHARA, H. (1959) An Ecological Study on the Wild Mountain Gorilla *(Gorilla gorilla beringei). Primates, 2:*1–42.

KAWAMURA, S. (1963) The Process of Sub-Cultural Propagation among Japanese Macaques. In SOUTHWICK, C. H. (ed.), *Primate Social Behavior.* Princeton, N.J. (Van Nostrand), 82–90.

KEARTON, C. (1935) *Die Insel der fünf Millionen Pinguine.* Stuttgart.

KEENLEYSIDE, M. H. A. (1955) Some Aspects of Schooling of Fish. *Behaviour, 8:*183–248.

KEETON, W. T. (1971) Magnet Interference with Pigeons. *Proc. Nat. Acad. Sci., 68:*102–106.

KELLOGG, W. N. (1961) *Porpoises and Sonar.* Chicago (Univ. Press).

——— (1968) Communication and Language in the Home-Raised Chimpanzee. *Science, 165:*423–427.

KEMPENDORFF, W. (1942) Über das Fluchtphänomen und die Chemorezeption von *Heliosoma nigricans. Arch. Molluskenk., 74.*

KENNEDY, J. S. (1951) The Migration of the Desert Locust *(Schistocerca gregaria Forsk). Phil. Trans. Roy. Soc. London, B235:*163–290.

KING, J. A. (1955) Social Behavior, Social Organization and Population Dynamics in a Black-Tailed Prairiedog Town in the Black Hills of South Dakota. *Contrib. Lab. Vert. Biol. Univ. Mich., 67.*

——— (1957) Relationships between Early Social Experience and Adult Aggressive Behavior in Inbred Mice. *J. Gen. Psychol., 90:*151–166.

KING, J. A., and GURNEY, N. L. (1954) Effect of Early Social Experience on Adult Aggressive Behavior in C 57 BL/10 Mice *J. Comp. Physiol. Psychol., 47:*326–336.

KIRCHSHOFER, R. (1960) Über das "Harnspritzen" des Großen Mara. *Z. Säugetierkde., 25:*112–127.

KISLAK, J. W., and BEACH, F. A. (1955) Inhibition of Aggression by Ovarian Hormones. *Endocrinology, 56:*684–692.

KITZLER, G. (1942) Die Paarungsbiologie einiger Eidechsen. *Z. Tierpsychol., 4:*353–402.

KLAUSEWITZ, W. (1961) Einige systematisch und ökologisch bemerkenswerte Meergrundeln *(Pisces, Gobiidae). Senckenbergiana Biol., 41:*149–162.

——— (1965) Osteichthyes, Knochenfische. *Handb. d. Biol., Frankfurt, 6* (2):542–628.

KLAUSEWITZ, W., and EIBL-EIBESFELDT, I. (1959) Neue Röhrenaale von den Malediven und Nikobaren *(Pisces, Apodes, Heterocongridae). Senckenbergiana Biol., 40:*135–153.

KLINGEL, H. (1967) Soziale Organisation und Verhalten freilebender Steppenzebras. *Z. Tierpsychol., 24:*580–624.

KLINGELHÖFFER, W. (1956) *Terrarienkunde. 2.: Lurche.* SCHERPNER, C. (ed.), 2nd ed. Stuttgart (Kernen).

KLINGHAMMER, E. (1967) Factors Influencing Choice of Mate in Altricial Birds. In STEVENSON, H. W. (ed.), *Early Behavior.* New York (Wiley), 5–42.

KLINGHAMMER, E., and HESS, E. H. (1964) Parental Feeding in Ring Doves *(Streptopelia roseogrisea):* Innate or Learned? *Z. Tierpsychol., 21:*338–347.

KLOOT, W. G. van der, and WILLIAMS, C. M. (1953) Cocoon Construction by the Cecropia Silkworm: II. The Role of the Internal Environment. *Behaviour, 5:*157–174.

KLOPFER, P. H. (1957) An Experiment on Emphatic Learning in Ducks. *Am. Naturalist, 91*:61–63.

——— (1962) *Aspects of Ecology.* Englewood Cliffs, N.J. (Prentice-Hall).

——— (1963) Behavioral Aspects of Habitat Selection. *Wilson Bull., 75*:15–22.

——— (1971) Mother Love: What Turns It on? *Am. Sci., 59*:404–407.

KLOPFER, P. H., and GAMBLE, J. (1966) Maternal "Imprinting" in Goats: The Role of Chemical Senses. *Z. Tierpsychol., 23*:588–592.

KLOPFER, P. H., and HAILMAN, J. P. (1965) Habitat Selection in Birds. *Advan. Study Behavior, 1*:279–303.

KLUYVER, H. N. (1947) Over het gedrag van een jonge Grauwe Vliegenvanger en van een troep Pestvogels in de winter. *Ardea, 35*:131–135.

KNEUTGEN, J. (1964) Beobachtungen über die Anpassung von Verhaltensweisen an gleichförmige akustische Reize. *Z. Tierpsychol., 21*:763–779.

——— (1970) Eine Musikform und ihre biologische Funktion: Über die Wirkungsweise der Wiegenlieder. *Z. f. experiment. u. angew. Psychol., 17*(2):245–265.

KOEHLER, O. (1943) "Zähl"–Versuche an einem Kolkraben und Vergleichsversuche an Menschen. *Z. Tierpsychol., 5*:575–712.

——— (1949) "Zählende" Vögel und vorsprachliches Denken. *Zool. Anz., Suppl., 13*: 129–238.

——— (1950) Die Analyse der Taxisanteile instinktartigen Verhaltens. *Symp. Soc. Exptl. Biol. Cambridge, 4*:269–302.

——— (1952) Vom unbenannten Denken. *Zool. Anz. Suppl., 16*:202–211.

——— (1953) *Orientierungsvermögen von Mäusen: Versuche im Hochlabyrinth.* Wiss. Film B635. Göttingen (Inst. wiss. Film).

——— (1954a) Das Lächeln als angeborene Ausdrucksbewegung. *Z. menschl. Vererb. u. Konst. lehre, 32*:330–334.

——— (1954b) Vorbedingungen und Vorstufen unserer Sprache bei Tieren. *Zool. Anz., Suppl., 18*:327–341.

——— (1955) Zählende Vögel und vergleichende Verhaltensforschung. *Acta 11, Congr. Intern. Ornith.*, Basel, 588–598.

——— (1966) Vom Spiel bie Tieren. *Freiburger Dies Universitatis, 13*:1–32.

KOEHLER, O., and ZAGARUS, A. (1937) Beiträge zum Brutverhalten des Halsbandregenpfeifers (*Charadrius hiaticula* L.). *Beitr. Fortpfl. Biol. Vögel, 13*:1–9.

KOENIG, L. (1951) Beiträge zu einem Aktionssystem des Bienenfressers (*Merops apiaster* L.). *Z. Tierpsychol., 8*:169–210.

——— (1953) Beobachtungen am afrikanischen Blauwangenspint (*Merops superciliosus chrysocercus*) in freier Wildbahn und Gefangenschaft, mit Vergleichen zum Bienenfresser (*Merops apiaster* L.). *Z. Tierpsychol., 10*:180–204.

KOENIG, O. (1951a) Das Aktionssystem der Bartmeise (*Panurus biarmicus* L.). *Oesterr. Zool. Z., 1*:1–82.

——— (1951b) Das Aktionssystem der Bartmeise (*Panurus biarmicus* L.). *Oesterr. Zool. Z., 3*:247–325.

——— (1968) Biologie der Uniform. *Naturwiss. Medizin., 5*(22):3–19 and 5(23):40–50.

——— (1970) *Kultur und Verhaltensforschung.* München (dtv).

KOFORD, C. B. (1963a) Rank of Mothers and Sons in Bands of Rhesus Monkeys. *Science, 141*:356–357.

——— (1963b) Group Relations in an Island Colony of Rhesus Monkeys. In SOUTHWICK, C. (ed.), *Primate Social Behavior.* Princeton, N.J. (Van Nostrand), 136–152.

——— (1966) Population Changes in Rhesus Monkeys: Cayo Santiago 1960–1964. *Tulane Stud. Zool., 13*:1–7.

KÖHLER, W. (1921) *Intelligensprüfungen an Menschenaffen.* Berlin.

KOHL-LARSEN, L. (1958) *Wildbeuter in Ostafrika: Die Tindiga, ein Jäger- und Sammlervolk.* Berlin (Reimer).

KOHTS, N. (1935) Infant Ape and Human Child (Instincts, Emotions, Play, Habits). *Sci. Mem. Mus. Darwinianum, 3* (with Russian and English summaries).

KOLLER, G. (1955) Hormonale und psychische Steuerung beim Nestbau weisser Mäuse. *Zool. Anz., Suppl., 19:*123–132.

KOMISARUK, B. R., and OLDS, J. (1968) Neuronal Correlates of Behaviour in Freely Moving Rats. *Science, 161:*810–812.

KÖNIG, H. (1925) Der Rechtsbruch und sein Ausgleich bei den Eskimo. *Anthropos, 20:* 276–315.

KONISHI, M. (1963) The Role of Auditory Feedback in the Vocal Behavior in the Domestic Fowl. *Z. Tierpsychol., 20:*349–367.

——— (1964) Effects of Deafening on Song Development in Two Species of Juncos. *Condor, 66:*85–102.

——— (1965a) Effects of Deafening on Song Development of American Robins and Black-Headed Grosbeaks. Z. *Tierpsychol., 22:*584–599.

——— (1965b) The Role of Auditory Feedback in the Control of Vocalization in the White-Crowned Sparrow. Z. *Tierpsychol., 22:*770–783.

——— (1966) The Attributes of Instinct. *Behaviour, 27:*316–328.

KONOPKA, R. J., and BENZER, S. (1971) Clock Mutants of *Drosophila melanogaster. Proc. Nat. Acad. Sc. U.S., 68:*2112–2116.

KORRINGA, P. (1957) Lunar Periodicity. In HEDGEPETH, J. W. (ed.), *Treatise on Marine Ecology and Palaeoecology: 1 Ecology.* Memoire 67, Geol. Soc. Am. Baltimore (Waverly Press), 917–934.

KORTLANDT, A. (1940) Eine Übersicht über die angeborenen Verhaltensweisen des mit-eleuropäischen Kormorans. *Arch. Neerl. Zool., 4:*401–442.

——— (1955) Aspects and Prospects of the Concept of Instinct. *Arch. Neerl. Zool., 11:* 155–284.

——— (1962) Chimpanzees in the Wild. *Sci. Am., 206:*128–138.

——— (1965) How Do Chimpanzees Use Weapons When Fighting Leopards? *Yearbook Am. Phil. Soc.,* 327–332. Philadelphia.

——— (1967a) Experimentation with Chimpanzees in the Wild. In STARCK, D., SCHNEIDER, R., and KUHN, H. J. (eds.): *Neue Ergebnisse der Primatologie.* Stuttgart (Fischer) 208–224.

——— (1967b) Handgebrauch bei freilebenden Schimpansen. In RENSCH, B. (ed.) *Handgebrauch und Verständigung bei Affen und Frühmenschen.* Bern (Huber), 59–102.

——— (1972) New Perspectives on Ape and Human Evolution. *Stichting voor Psychobiologie, Zoolog. Lab.* Amsterdam.

KORTLANDT, A., and KOOIJ, M. (1963) Prohominid Behavior in Primates. *Symp. Zool. Soc. London, 10:*61–88.

KORTMULDER, K. (1968) An Ethological Theory of the Incest Taboo and Exogamy. *Current Anthropology 9:*437–449.

KOVACH, J. K. (1971) Ethology in the Soviet Union. *Behaviour, 38:*237–265.

KRAFFT-EBING, R. v. (1924) *Psychopathia sexualis.* 17th ed. Stuttgart.

KRAMER, G. (1933) Untersuchungen über die Sinnesleistungen und das Orientierungsverhalten von *Xenopus laevis. Zool. Jahrb. Abt. Phys., 52:*629–676.

——— (1949) Macht die Natur Konstruktionsfehler? *Wilhelmshavener Vorträge, Schriftenreihe d. Nordwestdtsch. Universitätsges, 1:*1–19.

——— (1950) Über individuell und anonym gebunden Gemeinschaften der Tiere und Menschen. *Studium Gen., 3:*564–572.

——— (1952) Die Sonnenorientierung der Vögel. *Zool. Anz., Suppl., 16:*72–84.

——— (1957) Experiments on Birds' Orientation and Their Interpretation. *Ibis., 96:*173–185.

——— (1959) Recent Experiments on Bird Orientation. *Ibis., 101:*399–416.

——— (1961) Long-Distance Orientation. In MARSHALL, A. J. (ed.), *Biology and Comparative Physiology of Birds, 2.* New York (Academic Press), 341–371.

KRECHEVSKI, I. (1932) "Hypotheses" in Rats. *Psychol. Rev., 39:*516–532.

KROTT, P., and KROTT, K. (1963) Zum Verhalten der Braunbären *(Ursus arctos)* in den Alpen. Z. *Tierpsychol., 20:*160–206.

KRUEGER, F. (1948) *Lehre vom Ganzen.* Bern (Huber).

KRUIJT, J. P. (1958) Speckling of the Herring Gull Egg in Relation to Brooding Behavior. *Compt. Rend. Soc. Neerl. Zool., 12:*565–567.

——— (1964) Ontogeny of Social Behavior in Burmese Red Jungle Fowl *(Gallus gallus spadiceus). Behaviour, Suppl., 12.*

——— (1971) Early Experience and the Development of Social Behaviour in Jungle Fowl. *Psychiatr. Neurol. Neurochir., 74:*7–20.

KRUMBIEGEL, I. (1940) Die Persistenz physiologischer Eigenschaften in der Stammesgeschichte. Z. *Tierpsychol., 4:*249–258.

KRUSHINSKII, L. V. (1962) Animal Behavior, Its Normal and Abnormal Development. In WORTIS, J. (ed.), *The International Behavioral Sciences Series.* New York (Consultants Bureau).

KRUUK, H. (1972) *The Spotted Hyena: A Study of Predation and Social Behavior.* Chicago (Univ. Press).

KUENZER, E., and KUENZER, P. (1962) Untersuchungen zur Brutpflege der Zwergcichliden. Z. *Tierpsychol., 19:*56–83.

KUENZER, P. (1968) Die Auslösung der Nachfolgereaktion bei erfahrungslosen Jungfischen von *Nannacara anomala* (Cichlidae). Z. *Tierpsychol., 25:*257–314.

KÜHLMANN, D. H. H. (1966) Putzerfische säubern Krokodile. Z. *Tierpsychol., 23:*853–854.

KÜHME, W. (1965) Freilandstudien zur Soziologie des Hyänenhundes. Z. *Tierpsychol., 22:*495–541.

——— (1966) Beobachtungen zur Soziologie des Löwen in der Serengeti-Steppe. Z. *Säugetierkde, 31:*205–213.

KÜHN, A. (1919) *Die Orientierung der Tiere im Raum.* Jena (G. Fischer).

——— (1955) *Vorlesungen über Entwicklungsphysiologie.* Heidelberg (Springer).

KÜHN, H. (1929) *Kunst und Kultur der Vorzeit: Das Paläolithikum.* Berlin (De Gruyter).

KÜHNELT, W. (1965) *Grundriß der Ökologie unter besonderer Berücksichtigung der Tierwelt.* Jena (G. Fischer).

KULLENBERG, B. (1956) Field Experiments with Chemical Sexual Attractants on Aculeate Hymenoptera Males. *Zool. Bidr. Uppsala, 31:*253–354.

KULZER, E. (1954) Untersuchungen über die Schreckreaktion bei Erdkrötenquappen. Z. *vergl. Physiol., 36:*443–463.

KUMMER, H. (1957) Soziales Verhalten einer Mantelpavian-Gruppe. *Schweiz.* Z. *Psychol., Suppl., 33.* Bern (Huber).

——— (1968) Social Organization in Hamadryas Baboons: A Field Study. *Bibliotheca Primat.,* Basel (Karger).

KUMMER, H., and KURT, F. (1965) A Comparison of Social Behavior in Captive and Wild Hamadryas Baboons. In VOGTBERG, H. (ed.), *The Baboon in Medical Research.* Austin (Univ. Texas Press), 1–16.

KUNKEL, P. (1959) Zum Verhalten einiger Prachtfinken. Z. *Tierpsychol. 16:*302–350.

KUO, Z. Y. (1930) The Genesis of the Cat's Responses to the Rat. *J. Comp. Psychol., 11:*1–35.

——— (1932) Ontogeny of Embryonic Behavior in Aves. *J. Exptl. Biol., 61:*395–430, 453–489.

——— (1960–1961) Studies on the Basic Factors in Animal Fighting *J. Genet. Psychol., 96:*201–239 and 97:181–209.

——— (1967) *The Dynamics of Behavior Development: An Epigenetic View.* New York (Random House).

——— (1970) The Need for Coordinated Efforts in Developmental Studies. In ARONSON, L. R., TOBACH, E., LEHRMAN, D. S., and ROSENBLATT, J. S. (eds.) *Development and Evolution of Behavior.* San Francisco (Freeman), 181–193.

LABARRE, W. (1947) The Cultural Basis of Emotions and Gestures. *J. Person. 16:*49–68.
LACK, D. (1943) *The Life of the Robin.* London (Cambridge Univ. Press).
——— (1947) *Darwin's Finches.* London (Cambridge Univ. Press).
LAGERSPETZ, K. (1964) Studies on the Aggressive Behaviour of Mice. Suomalaisen Tiedeakatemian Toimituksia. *Ann. Acad. Sci. Fennice, B131* Helsinki.
——— (1969) Aggression and Aggressiveness in Laboratory Mice. In GARATTINI, S., and SIGG, E. B. (eds.) *Aggressive Behaviour.* Amsterdam (Excerpta Medica Foundation): 77–85.
LAGERSPETZ, K., and TALO, S. (1967) Maturation of Aggressive Behaviour in Young Mice. *Rept. Inst. Psychol. Univ. of Turku, 28:*1–9.
LAGERSPETZ, K., and WORINEN, K. (1965) A Cross-fostering Experiment with Mice Selectively Bred for Aggressiveness and Nonaggressiveness. *Rept. Inst. Psychol. Univ. of Turku, 17,* 1–6.
LAMPRECHT, J. (1973) Mechanismen des Paarzusammenhaltes beim Cichliden *Tilapia mariae.* Boulenger 1899 (Cichlidae, Teleostei). *Z. Tierpsychol., 32:*10–61.
LANCASTER, J. B. (1968) Primate Communication Systems and the Emergence of Human Language. In JAY, P. (ed.), *Primates: Studies in Adaptation and Variability.* New York (Holt, Rinehart and Winston).
LANG, E. M. (1961) *Goma, das Gorillakind.* Zürich (A. Müller).
——— (1964) Jambo: First Gorilla Raised by Its Mother in Captivity. *Natl. Geogr. Mag., 125*(3):446–453.
LARSSON, K. (1959) Experience and Maturation in the Development of Sexual Behavior in Male Puberty Rats. *Behaviour, 14:*101–107.
LASHLEY, K. S. (1915) Notes on the Nesting Activity of the Noddy and Sooty Terns. *Pap. Dept. Marine Biol. Carnegie Inst. Wash., 7:*61–84.
——— (1929) *Brain Mechanisms and Intelligence: A Quantitative Study of Injuries to the Brain.* Chicago.
——— (1931) Mass Action in Cerebral Function. *Science, 73:*245–254.
——— (1935) The Behavior of Rats in Latch Box Situations. *Comp. Psychol. Monogr., 11*(2):1–42.
——— (1938) Experimental Analysis of Instinctive Behavior. *Psychol. Rev., 45:*445–471.
LAST, G., and KNEUTGEN, J. (1970) Schlafmusik. *Münchner Mediz. Wochschr., 44.*
LAUDIEN, H. (1966) Untersuchungen über das Kampfverhalten der Männchen von *Betta splendens. Z. wiss. Zool., 172:*134–178.
LAWICK-GOODALL, J. van (1965) New Discoveries among Africa's Chimpanzees. *Natl. Geogr. Mag., 128*(6):802–831.
——— (1968) The Behavior of Free-living Chimpanzees in the Gombe Stream Reserve. *Animal Behaviour Monographs 1*(3):161–311.
——— (1970) Tool-Using in Primates and Other Vertebrates. In LEHRMAN, D. S., HINDE, R. A., and SHAW, E. (eds.) *Advances in the Study of Behavior.* New York and London (Academic Press) 195–249.
——— (1971) *In the Shadow of Man.* Boston (Houghton Mifflin) and London (Collins).
——— (1974) The Behaviour of the Chimpanzee. In KURTH, G., and EIBL-EIBESFELDT, I (eds.), *Hominisation und Verhalten.* Stuttgart (Fischer), 56–104.
LAWICK, H. van, and LAWICK-GOODALL, J. van (1971) *Innocent Killers.* Boston (Houghton Mifflin).
LAWICK-GOODALL, J. van, and LAWICK, H. van (1966) Use of Tools by the Egyptian Vulture *Neophron percnopterus. Nature, 212:*1468–1469.
LEAKEY, L. S. B. (1963) Adventures in the Search of Man. *Natl. Geogr. Mag., 123:*132–152.
LEBZELTER, V. (1934) *Eingeborenenkulturen von Süd- und Südwestafrika.* Leipzig (Hiersemann).
LECOMTE, J. (1961) Le comportement agressif des ouvrières d'*Apis mellifica* L. *Ann. Abeille, 4:*165–275.

LEE, R. B., and DEVORE, I. (1968) *Man the Hunter.* Chicago (Aldine).
LEHMANN, F. E. (1958) *Gestaltung sozialen Lebens bei Tier und Mensch.* Bern (Francke).
LEHR, E. (1967) Experimentelle Untersuchungen an Affen und Halbaffen über Generalisation von Insekten- und Blütenabbildungen. *Z. Tierpsychol., 24:*208–244.
LEHRMAN, D. S. (1953) A Critique of Konrad Lorenz's Theory of Instinctive Behavior. *Quart. Rev. Biol., 28:*337–363.
——— (1955) The Physiological Basis of Parental Feeding Behavior in the Ring Dove *(Streptopelia risoria). Behaviour, 7:*241–286.
——— (1961) The Presence of the Mate and of Nesting Material as Stimuli for the Development of Incubation Behavior and for Gonadotropic Secretion in the Ring Dove. *Endocrinology, 68:*507–516.
——— (1970) Semantic and Conceptual Issues in the Nature–Nurture Problem. In ARONSON, L. R., TOBACH, E., LEHRMAN, D. S., and ROSENBLATT, J. S. (eds.) *Development and Evolution of Behavior.* San Francisco (Freeman), 17–52.
LEMAGNEN, J. (1952) Les phénomenes olfacto-sexuels chez l'homme. *Arch. Sci. Physiol., 6:* 125–160.
LENNEBERG, E. H. (1964) A Biological Perspective of Language. In LENNEBERG, E. H. (ed.), *New Directions in the Study of Language.* Cambridge, Mass. (M. I. T. Press), 65–88.
——— (1967) *Biological Foundations of Language.* New York and London (John Wiley).
LENNEBERG, E. H., REBELSKY, F. G., and NICHOLS, J. A. (1965) The Vocalization of Infants to Deaf and to Hearing Parents. *Human Develop., 8:*23–27.
LEON, M., and MOLTZ, H. (1971) Maternal Pheromone: Discrimination by Pre-Weanling Albino Rats. *Physiol. Behav., 7:*265–267.
——— (1972) The Development of the Pheromonal Bond in the Albino Rat. *Physiol. Behav., 8*(4):683–686.
——— (1973) Endocrine Control of the Maternal Pheromone in the Postpartum Female Rat. *Physiol. Behav., 10*(1):65–67.
LEONG, C. Y. (1969) The Quantitative Effect of Releasers on the Attack Readiness of the Fish *Haplochromis burtoni* (Cichlidae). *Z. vergl. Physiol., 65:*29–50.
LERSCH, P. (1951) *Gesicht und Seele.* 2nd ed. Munich.
LEUTHOLD, W. (1966) Variations in Territorial Behavior of Uganda Kob, *Adenota kob thomasi. Behaviour, 27:*215–258.
——— (1967) Beobachtungen zum Jugendverhalten von Kob-Antilopen. Z. *Säugetierkde, 32:*59–63.
LEVY, D. M. (1934) Experiments on the Suckling Reflex and Social Behavior of Dogs. *Am. J. Orthopsychol., 4:*203.
LEYHAUSEN, P. (1954a) Die Entdeckung der relativen Koordination. *Studium Gen., 7:*45–60.
——— (1954b) Vergleichendes über die Territorialität bei Tieren und den Raumanspruch des Menschen. *Homo, 5:*68–76.
——— (1956a) Das Verhalten der Katzen *(Felidae).* In KUKENTHAL, *Handb. d. Zool.,* 8(10), Berlin (de Gruyter).
——— (1956b) Verhaltensstudien an Katzen. Z. *Tierpsychol., Suppl., 2.*
——— (1965a) Über die Funktion der relativen Stimmungshierarchie (dargestellt am Beispiel der phylogenetischen und ontogenetischen Entwicklung des Beutefangs von Raubtieren). Z. *Tierpsychol., 22:*412–494.
——— (1965b) The Communal Organisation of Solitary Mammals. *Symp. Zool. Soc. London, 14:*249–263.
LIMBAUGH, C. (1961) Cleaning Symbiosis. *Sci. Am., 205*(8):42–49.
LIMBAUGH, C., PEDERSON, H., and CHACE, F. A. (1961) Shrimps that Clean Fishes. *Bull. Marine Sci. Gulf Caribbean, 11:*237–257.
LINDAUER, M. (1952) Ein Beitrag zur Frage der Arbeitsteilung im Bienenstaat. Z. *vergl. Physiol., 34:*299–345.

——— (1957) Sonnenorientierung der Bienen unter der Äquatorsonne und zur Nachtzeit. *Naturwiss., 44:*1–6.

——— (1961) *Communication among Social Bees.* Cambridge, Mass. (Harvard Univ. Press).

——— (1963) Allgemeine Sinnesphysiologie, Orientierung im Raum. *Fortschr. Zool., 16:*58–140.

LINDAUER, M., and LINDAUER, H. (1972) *Magnetic Effect on Dancing Bees.* Symposium NASA sp. 262 Animal Orientation and Navigation. Washington D.C. (Govt. Printing Office).

LISSMANN, H. W. (1946) The Neurological Basis of the Locomotory Rhythm in the Spinal Dogfish. *J. Exptl. Biol., 23:*143–176.

LISSMANN, H. W., and MACHIN, K. G. (1958) The Mechanism of Object Location in *Gymnarchus niloticus* and Similar Fish. *J. Exptl. Biol., 35:*451.

LIVINGSTONE, F. B. (1969) Genetics, Ecology and the Origins of Incest and Exogamy. *Current Anthropology 10:*45–61.

LLOYD, J. E. (1965) Aggressive Mimicry in Photuris: Fireflies Femmes Fatales. *Science, 149:*653–654.

LOEB, J. (1913) Die Tropismen. *Handb. vergl. Physiol., 4.*

LORENZ, K. (1931) Beiträge zur Ethologie sozialer Corviden. *J. Ornithol., 79:*67–127.

——— (1935) Der Kumpan in der Umwelt des Vogels. *J. Ornithol., 83:*137–413.

——— (1937) Über die Bildung des Instinktbegriffes. *Naturwiss., 25:*289–300, 307–318, 325–331.

——— (1939) Vergleichende Verhaltensforschung. *Zool. Anz., Suppl., 12:*69–102.

——— (1940) Durch Domestikation verursachte Störungen arteigenen Verhaltens. *Z. Angew. Psychol. Charakt.kde, 59:*2–81.

——— (1941) Vergleichende Bewegungsstudien bei Anatiden. *J. Ornithol., 89:*194–294.

——— (1943) Die angeborenen Formen möglicher Erfahrung. *Z. Tierpsychol., 5:*235–409.

——— (1949) *Er redete mit dem Vieh, den Vögeln und den Fischen.* Vienna (Borotha-Schoeler).

——— (1950a) Ganzheit und Teil in der tierischen und menschlichen Gemeinschaft. *Studium Gen., 3:*455–499.

——— (1950b) The Comparative Method in Studying Innate Behaviour Patterns. *Symp. Soc. Exptl. Biol., 4:*221–268.

——— (1950c) *So kam der Mensch auf den Hund.* Vienna (Borotha-Schoeler). (*Man Meets Dog.* New York and London [Penguin].)

——— (1951) Ausdrucksbewegungen höherer Tiere. *Naturwiss. 38:*113–116.

——— (1953) Die Entwicklung der vergleichenden Verhaltensforschung in den letzten 12 Jahren. *Zool. Anz., Suppl., 16:*36–58.

——— (1954a) Das angeborene Erkennen. *Natur u. Volk, 84:*285–295.

——— (1954b) Morphology and Behavior Patterns in Allied Species. *1st Conf. on Group Proc. Josiah Macy Jr. Found.,* New York, 168–220.

——— (1956) The Objectivistic Theory of Instinct. In GRASSE, P. P. (ed.), *L'instinct dans le comportement des animaux et de l'homme.* Paris (Fondation Singer-Polignac), 51–76.

——— (1957) Methoden der Verhaltensforschung. In KÜKENTHAL, *Handb. d. Zool, 8,* 10(1), 1–22.

——— (1958) The Evolution of Behaviour. *Sci. Am., 199*(6):67–78.

——— (1959) Psychologie und Stammesgeschichte. In HEBERER, G. (ed.), *Evolution der Organismen.* Stuttgart (G. Fischer).

——— (1961) Phylogenetische Anpassung und adaptive Modifikation des Verhaltens. *Z. Tierpsychol., 18:*139–187.

——— (1962) Naturschönheit und Daseinskampf. *Kosmos, 58:*340–348.

——— (1963a) *Das sogenannte Böse.* Vienna (Borotha-Schoeler).

——— (1963b) Die "Erfindung" von Flugmaschinen in der Evolution der Wirbeltiere. *Therap. Monats, 13:*138–148. Mannheim (Boehringer).

——— (1965a) *Über tierisches und menschliches Verhalten: Aus dem Werdegang der Verhaltenslehre (Ges. Abhandl.). I u. II.* Munich (Piper).

——— (1965b) *Evolution and Modification of Behavior.* Chicago (Univ. Press).

——— (1966) Stammes- und kulturgeschichtliche Ritenbildung. *Mitt. Max-Planck-Ges., 1:*3–30 and *Naturwiss. Rundschau, 19:*361–370.

——— (1969) Innate Basis of Learning. In: PRIBRAM, H. (ed.) *On the Biology of Learning.* New York (Harcourt).

——— (1971) Der Mensch, biologisch gesehen: Eine Antwort an Wolfang Schmidbauer. *Studium Gen., 24:*495–515.

——— (1973) *Die Rückseite des Spiegels.* München (Piper).

LORENZ, K., and SAINT PAUL, U. v. (1968) Die Entwicklung des Spießens und Würgens bei den drei Würgerarten *Lanius collurio, L. senator* u. *L. excubitor. J. Ornithol., 109:*137–156.

LORENZ, K., and TINBERGEN, N. (1938) Taxis und Instinkthandlung in der Eirollbewegung der Graugans. *Z. Tierpsychol., 2:*1–29.

LOSEY, G. (1971) Communication between Fishes in Cleaning Symbioses. In: CHENG, T. C. (ed.) *Aspects of the Biology of Symbioses.* Baltimore (Univ. Park Press).

LOEWENFELD, I. E. (1966) Comment on Hess' Findings. *Survey of Ophthalmology, 11:*293–294.

LUDWIG, J. (1965) Beobachtungen über das Spiel von Boxern. *Z. Tierpsychol., 22:*813–838.

LÜLING, V. H. (1958) Morphologisch-anatomische und histologische Untersuchungen am Auge des Schützenfisches *Toxotes jaculatrix,* nebst Bemerkungen zum Spuckgehaben. *Z. Morphol. Ökol. Tiere, 47:*529–610.

LUTHER, W. (1958) Symbiose von Fischen *(Gobiidae)* mit einem Krebs *(Alpheus)* im Roten Meer. *Z. Tierpsychol., 15:*175–177.

MCCONNELL, J. V. (1962) Memory Transfer through Cannibalism. *J. Neuropsychiat., 3:*542–548.

MCCONNELL, J. V., JACOBSON, A. L., and KIMBLE, D. P. (1959) The Effects of Regeneration upon Retention of a Conditioned Response in the Planarian. *J. Comp. Physiol. Psychol., 52:*1–5.

MCDERMOTT, F. A. (1917) Observations on the Light Emission of American Lampyridae. *Can. Entomol., 49:*53–61.

MCDOUGALL, W. (1936) *An Outline of Psychology.* 7th ed. London.

MACFARLAND, C., and REEDER, W. G. (1974) Cleaning Symbiosis Involving Galápagos Tortoises and Two Species of Darwin's Finches. *Z. Tierpsychol., 34:*464–483.

MACFARLANE, D. A. (1930) The Role of Kinaesthesis in Maze Learning. *Univ. Calif. Publ. Psychol., 4:*277–305.

MCGILL, T. (1965) *Readings in Animal Behavior.* New York (Holt, Rinehart and Winston).

MCGREW, W. C. (1972) *An Ethological Study of Children's Behavior.* New York and London (Academic Press).

MCILWAIN, J. T. (1972) Central Vision. *Ann. Rev. Physiol., 34:*291–314.

MCKINNEY, F. (1965) The Comfort Movements of *Anatidae. Behaviour, 25:*120–220.

MCNEIL, E. (1959) Psychology and Aggression. *J. Conflict Resolution, 3:*195–293.

MCPHALL, J. D. (1969) Predation and the Evolution of a Stickleback *(Gasterosteus). J. Fish. Res. Bd. Canada, 26:*3183–3208.

MACHEMER, H. (1966) Versuche zur Frage nach der Dressierbarkeit hypotricher Ciliaten unter Einsatz hoher Individuenzahlen. *Z. Tierpsychol., 23:*641–654.

MACINTOSH, J. H. (1970) Territory Formation by Laboratory Mice. *Animal Behaviour, 18:*177–183.

——— (1973) Factors Affecting the Recognition of Territory Boundaries by Mice *(Mus musculus). Animal Behaviour, 21:*464–470.

MACKENSEN, G. (1965) Zur Verhaltensweise blinder Kinder. *Studium Gen., 18:*9–14.
MACKINTOSH, J. H., and GRANT, E. C. (1966) The Effect of Olfactory Stimuli on the Agonistic Behaviour of Laboratory Mice. Z. *Tierpsychol., 23:*584–587.
MACLENNAN, R. R., and BAILEY, E. D. (1972) Role of Sexual Experience and Breeding Behavior of Male Ranch Mink. *J. Mammol., 53:*380–382.
MAGNUS, D. (1954) Zum Problem der "überoptimalen" Schlüsselreize. *Zool. Anz., Suppl., 18:*317–325.
——— (1958) Experimentelle Untersuchungen zur Bionomie und Ethologie des Kaisermantels *Argynnis paphia* L. I.: Über optische Auslöser von Auffliegereaktionen und ihre Bedeutung für das Sichfinden der Geschlechter. Z. *Tierpsychol., 15:*397–426.
——— (1964) Zum Problem der Partnerschaften mit Diademseeigeln. *Zool. Anz., Suppl., 27:*404–417.
MAIER, N. R. F., and SCHNEIRLA, T. C. (1935) *Principles of Animal Psychology.* New York.
MAKKINK, G. F. (1936) An Attempt at an Ethogram of the European Avocet (*Recurvirostra avosetta* L.) with Ethological and Psychological Remarks. *Ardea, 25:*1–60.
MANLEY, G. (1960) *The Agonistic Behavior of the Black-Headed Gull.* (Dissertation, Oxford, quoted by R. A. STAMM, 1964).
MANNING, A. (1956) Some Aspects of the Foraging Behaviour of Bumblebees. *Behaviour, 9:*164–201.
——— (1961) The Effects of Artificial Selection for Mating Speed in *Drosophila melanogaster. Animal Behaviour, 9:*82–92.
——— (1965) *Drosophila* and the Evolution of Behaviour. *Viewpoints Biol., 4:*125–169.
——— (1967) *An Introduction to Animal Behaviour.* London (E. Arnold).
MARK, V. H., and ERVIN, F. R. (1970) *Violence and the Brain.* New York (Harper & Row).
MARKL, H. (1972) Aggression und Beuteverhalten bei Piranhas *(Serrasalminae).* Z. *Tierpsychol., 30:*190–216.
MARLER, P. R. (1955a) Studies of Fighting in Chaffinches: (1) Behaviour in Relation to the Social Hierarchy. *Brit. J. Animal Behaviour, 3:*111–117.
——— (1955b) Studies on Fighting in Chaffinches: (2) The Effect on Dominance Relations of Disguising Females as Males. *Brit. J. Animal Behaviour 3:*137–146.
——— (1956a) Studies of Fighting in Chaffinches: (3) Proximity as a Cause of Aggression. *Brit. J. Animal Behaviour, 4:*23–30.
——— (1956b) Über einige Eigenschaften tierlicher Rufe. *J. Ornithol., 97:*220–227.
——— (1957a) Specific Distinctness in the Communication Signals of Birds. *Behaviour, 11:*13–39.
——— (1957b) Studies of Fighting in Chaffinches: (4). Appetitive and Consummatory Behaviour. *Brit. J. Animal Behaviour, 5:*29–37.
——— (1959) Developments in the Study of Animal Communication. In BELL, P. R. (ed.), *Darwin's Biological Work.* London (Cambridge Univ. Press), 150–206.
MARLER, P. R., and BOATSMAN, D. J. (1951) Observations on the Birds of Pico, Azores. *Ibis, 93:*90–99.
MARLER, P. R., and HAMILTON, W. J. (1966) *Mechanisms of Animal Behavior.* New York (Wiley).
MARLER, P. R., and TAMURA, M. (1964) Culturally Transitted Patterns of Vocal Behavior in Sparrows. *Science, 146:*1483–1486.
MARQUENIE, J. (1950) De balts van de kleine Watersalamander. *Levende Natuur, 53:*147–155.
MARSHALL, L. (1961) Sharing, Talking, and Giving: Relief of Social Tensions Among !Kung Bushmen. *Africa, 31:*231–249.
——— (1965) The !Kung Bushmen of the Kalahari Desert. In: GIBBS, J. L. (ed.) *Peoples of Africa.* New York (Holt, Rinehart and Winston).
MARTIN, R. D. (1966a) Tree Shrews: Unique Reproductive Mechanism of Systematic Importance. *Science, 152:*1402–1404.

——— (1966b) Sind Spitzhörnchen wirklich Vorfahren der Affen? *Umschau, 13:*437–438.

——— (1968) Reproduction and Ontogeny in Tree-shrews *(Tupaia belangeri),* with Reference to Their General Behaviour and Taxonomic Relationships. *Z. Tierpsychol., 25:*409–495, 505–532.

MARTINS, T., and VALLE, J. R. (1948) Hormonal Regulation of Micturition Behavior. *J. Comp. Physiol. Psychol., 41:*301–311.

MASNE, G. le (1950) Classification et caractéristiques des principaux types de groupements sociaux réalisés chez les invertébrés. *Colloq. Intern. Centre Natl. Rech. Sci. (Paris), 34:*19–70.

MASON, W. A. (1965) The Social Development of Monkeys and Apes. In DEVORE, I. (ed.), *Primate Behavior.* New York (Holt, Rinehart and Winston), 514–543.

MATAS, L., WATERS, E., and SROUFE, L. (on press) The Description and Function of Wariness in Infants. *Child Development.*

MATTHEWS, G. V. T. (1955) *Bird Navigation.* New York (Cambridge Univ. Press).

MATURANA, H. R., LETTVIN, J. Y., MCCULLOCH, W. S., and PITTS, W. H. (1960) Anatomy and Physiology of Vision in the Frog *(Rana pipiens). J. Gen. Physiol., Suppl., 6:*129–175.

MAXIM, P. E., and BUETTNER-JANUSCH, J. (1963) A Field Study of the Kenya Baboon. *Am. J. Phys. Anthrop., 21:*165–180.

MAXWELL, S. S. (1923) *Labyrinth and Equilibrium.* Philadelphia.

MAY, E., and BIRUKOW, G. (1966) Lunar-periodische Schwankungen des Lichtpreferendums bei der Bachplanarie. *Naturwiss., 53:*182.

MAYER, G. (1952) Untersuchungen über Herstellung und Struktur des Radnetzes von *Aranea diadema* und *Zilla* x- *notata* mit besonderer Berücksichtigung des Unterschiedes von Jugend- und Altersnetzen. *Z. Tierpsychol., 9:*337–362.

MAYER, J., and THOMAS, D. W. (1967) Regulation of Food Intake and Obesity. *Science, 156:*328–337.

MAYR, E. (1950) Ecological Factors in Speciation. *Evolution, 1:*263–288.

——— (1958) Behavior and Systematics. In ROE, A., and SIMPSON, G. (eds.), *Behavior and Evolution.* New Haven (Yale Univ. Press), 341–366.

——— (1970) Evolution und Verhalten. *Verh. d. Dtsch. Zool. Ges., 64:*322–336.

MEAD, M. (1935) *Sex and Temperament in Three Primitive Societies.* New York (Morrow).

——— (1956) Birth of a Baby in New Guinea. In SODDY, K. (ed.), *Mental Health and Infant Development. I.* New York (Basic Books).

——— (1965) *Leben in der Südsee.* Munich (Szczesny).

MECH, L. D. (1970) *The Wolf.* Garden City (Natural History Press).

MEISENHEIMER, J. (1921) *Geschlecht und Geschlechter im Tierreich.* Jena (G. Fischer).

MELCHERS, M. (1960) *Cupiennius salei (Ctenidae): Kokonbau und Eiablage.* Encycl. cinem., E363, Göttingen (Inst. wiss. Film).

——— (1963) Zur Biologie und zum Verhalten von *Cupiennius salei* (Keyserling), einer amerikanischen *Ctenidae. Zool. Jahrb. System., 91:*1–90.

——— (1964) *Cupiennius salei (Ctenidae): Spinnhemmung beim Kokonbau.* Encycl. cinem., E364. Publ. zu wiss. Filmen, 1A, 21–24. Göttingen (Inst. wiss. Film).

MERKEL, F. W., and WILTSCHKO, W. (1965) Magnetismus und Richtungsfinden zugunruhiger Rotkehlchen *(Erithacus rubecula). Vogelwarte, 23:*71–77.

MERTENS, R. (1959) *La Vie des Amphibiens et Reptiles.* Paris (Horizons de France).

MERZ, F. (1965) Aggression und Aggressionsantrieb. In THOMAE, H. (ed.), *Handb. d. Psychol., 2:*569–600.

MEYER-HOLZAPFEL, M. (1956) Das Spiel bei Säugetieren. In KUKENTHAL, *Handb. d. Zool., 8*(10):1–36.

——— (1958) Sozial Beziehungen bei Säugetieren. In LEHMANN, F. E. (ed.), *Gestaltungen sozialen Lebens bei Tier und Mensch.* Bern (Francke), 86–109.

MILGRAM, S. (1963) Behavioral Study of Obedience. *J. Abnorm. Soc. Psychol., 67:*372–378.

——— (1965a) Liberating Effects of Group Pressure. *J. Person. Soc. Psychol., 1:*127–134.
——— (1965b) Some Conditioning Obedience and Disobedience. *Human Relations, 18:*57–76.
——— (1966) Einige Bedingungen von Autoritätsgehorsam und seiner Verweigerung. *Z. Exptl. Angew. Psychol., 13:*433–463.
MILL, J. S. (1843) *A System of Logic, II.* London (Parker).
MILNE, L. J., and MILNE, M. (1963) *Die Sinneswelt der Tiere und Menschen.* Hamburg (Parey).
MISTSCHENKO, M. N. (1933) Über die mimische Gesichtsmotorik der Blinden. *Folia Neuropathol. Estonia, 13:*24–43.
MITSCHERLICH, A. (1957) Aggression und Anpassung. I. *Psyche, 10:*177–193.
——— (1959) Aggression und Anpassung. II. *Psyche, 12:*523–537.
MITTELSTAEDT, H. (1953) Über den Beutefangmechanismus der Mantiden. *Zool. Anz., Suppl., 16:*102–106.
——— (1954) Regelung und Steuerung bei der Orientierung der Lebewesen. *Regelungstechnik, 10:*226–232.
MIYADI, D. (1965) Social Life of Japanese Monkeys. *Science in Japan.* American Association for the Advancement of Science.
——— (1967) Differences in Social Behavior among Japanese Macaque Troops. *Neue Ergebnisse der Primatologie (First Congress Int. Primatol. Soc.)* 228–231. Stuttgart (G. Fischer).
MIZE, R. R., and MURPHY, E. H. (1973) Selective Visual Experience Fails to Modify Receptive Field Properties of Rabbit Striate Cortex Neurons. *Science, 180*(4083):320–322.
MÖHRES, F. P. (1953) Über die Ultraschallorientierung der Hufeisennasen. *Z. vergl. Physiol., 34:*547–588.
——— (1961) Die elektrischen Fische. *Natur u. Volk, 91:*1–12.
MOLTZ, H., and LEON, M. (1973) Stimulus Control of the Maternal Pheromone in the Lactating Rat. *Physiol. and Behav., 10*(1):69–70.
MONEY, J., and EBERHARDT, A. A. (1972) *Man and Woman, Boy and Girl: The Differentiation and Dimorphism of Gender Identity from Conception to Maturity.* Baltimore (Johns Hopkins Univ. Press).
MONNIER, M., and WILLI, H. (1953) Die integrative Tätigkeit des Nervensystems beim meso-rhombo-spinalen Anencephalus (Mittelhirnwesen). *Monatschr. Psychiat. Neurol., 126:*239–258.
MONTAGU, M. F. A. (1962) *Culture and the Evolution of Man.* New York (Oxford Univ. Press).
——— (1968) *Man and Aggression.* New York (Oxford Univ. Press).
MONTESSORI, M. (1952) *Kinder sind anders.* 5th ed. Stuttgart (Klett).
MORGAN, C. LLOYD (1894) *Introduction to Comparative Psychology.* London.
——— (1900) *Animal Behaviour.* London.
MORRIS, D. (1954) The Reproductive Behaviour of the River-Bullhead *(Cottus gobio* L.), with Special Reference to the Fanning Activity. *Behaviour, 7:*1–32.
——— (1956) The Feather Postures of Birds and the Problem of the Origin of Social Signals. *Behaviour, 9:*75–113.
——— (1957) "Typical Intensity" and Its Relation to the Problem of Ritualization. *Behaviour, 11:*1–12.
——— (1958) The Reproductive Behaviour of the Ten-Spined Stickleback (*Pygosteus pungitius* L.). *Behaviour, Suppl., 6.*
——— (1963) *Biologie der Kunst.* Düsseldorf (Rauch) *(The Biology of Art.* New York [Knopf].)
——— (1968) *Der nackte Affe.* Munich (Droemer). (*The Naked Ape.* New York [McGraw-Hill].)

MORSE, D. H. (1968) The Use of Tools by Brown-headed Nuthatches. *Wilson Bull., 80:* 220–224.

MOSS, A. M. (1920) Sphingidae of Para, Brazil. *Nov. Zool., 27:*333–424.

MOYER, K. E. (1969) Internal Impulses to Aggression. *Trans. New York Acad. Sci., Ser. II, 31:*104–114.

——— (1972) *A Physiological Model of Aggression: Does It Have Different Implications?* Symposium on Neural Basis of Violence and Aggression March 9–11, 1972, Houston.

MOYNIHAN, M. (1955) Some Aspects of Reproductive Behavior in the Blackheaded Gull (*Larus ridibundus ridibundus* L.) and Related Species. *Behaviour, Suppl. 4.*

——— (1964) Some Behavior Patterns of Platyrrhine Monkeys. I. The Night Monkey *(Aotes trivirgatus). Smithsonian Misc. Collections, 146*(5):1–84.

MUIR, D. W., and MITCHELL, D. E. (1973) Visual Resolution and Experience: Acuity Deficits in Cats Following Early Selective Visual Deprivation. *Science, 180*(4084): 420–421.

MÜLLER, D. (1961) Quantitative Luftfeind-Attrappenversuche bei Auer- und Birkhühnern *(Tetrao urogallus* L. und *Lyrurus tetrix* L.). *Naturforsch., 16b:*551–553.

MÜLLER-USING, D. (1952) Über einige bisher unbeachtete Übersprunghandlungen bei höheren Säugern. *Z. Tierpsychol., 9:*479–481.

MUNN, N. L. (1950) *Handbook of Psychological Research on the Rat.* New York (Houghton-Mifflin).

MURIE, A. (1944) The Wolves of Mount McKinley: Fauna of the National Parks of the U.S.A. *Fauna Ser., 5.*

MURPHY, R. F. (1957) Intergroup Hostility and Social Cohesion. *Am. Anthropol., 59:*1028.

MUSSEN, P. (1970) *Carmichael's Manual of Child Psychology. 1.* New York (John Wiley).

MYERS, R. E. (1956) Functions of Corpus Callosum in Interocular Transfer. *Brain, 79:* 358–363.

MYKYTOWYCZ, R. (1955) Further Observations on the Territorial Function and Histology of the Submandibular Cutaneous (Chin) Glands in the Rabbit *(Oryctolagus cuniculus). Animal Behaviour, 13:*400–412.

——— (1960) Social Behaviour of an Experimental Colony of Wild Rabbits. *CSIRO Wildlife Res. (Canberra), 5*(1):1–20.

——— (1966) Observations on Odoriferous and Other Glands in the Australian Wild Rabbit, *Oryctolagus cuniculus* L., and the Hare, *Lepus europaeus* P. I. The Anal Gland. II. The Inguinal Gland. III. Harders Lacrimal and Submandibular Glands. *CSIRO Wildlife Res. (Canberra), 11:*11–29, 49–90.

MYKYTOWYCZ, R., and DUDZINSKI, M. L. (1966) A Study of the Weight of Odoriferous and other Glands in Relation to Social Status and Degree of Sexual Activity in the Wild Rabbit, *Oryctolagus cuniculus. CSIRO Wildlife Res. (Canberra), 11:*31–47.

MYRBERG, A. A. (1964) An Analysis of Preferential Care of Eggs and Young by Adult Cichlid Fishes. *Z. Tierpsychol., 21:*53–98.

——— (1965) Sound Production by Cichlid Fishes. *Science, 149:*555–558.

NAPIER, J. (1962) The Evolution of the Hand. *Sci. Am., 207*(6):56–63.

NAYLOR, E. (1958) Tidal and Diurnal Rhythms of Locomotor Activity in *Carcinus maenas. J. Exptl. Biol., 35:*602–610.

NELSON, K. (1964) Behavior and Morphology in the Glandulocaudine Fishes (*Osteriophysi, Characidae*). *Univ. Calif. Publ. Zool., 75:*59–152.

NEUMANN, D. (1966) Die lunare und tägliche Schlüpfperiodik der Mücke *Clunio:* Steuerung und Abstimmung auf die Gezeitenperiodik. *Z. vergl. Physiol., 53:*1–61.

NEUMANN, F., and STEINBECK, H. (1972) Influence of Sexual Hormones on the Differentiation of Neural Centers. *Archives sex. Behav., 2*(2):147–162.

NEVERMANN, H. (1941) *Ein Besuch bei Steinzeitmenschen.* Stuttgart (Frankh).

NICE, M. M. (1941) The Role of Territory in Bird Life. *Am. Midland Naturalist, 26:*441–487.

——— (1962) Development of Behavior in Precocial Birds. *Trans. Linnean Soc. New York*, 8.

Nicolai, J. (1956) Zur Biologie und Ethologie des Gimpels. *Z. Tierpsychol., 13:*93–132.

——— (1959a) Familientradition in der Gesangstradition des Gimpels (*Pyrrhula pyrrhula* L.). *J. Ornithol., 100:*39–46.

——— (1959b) Verhaltensstudien an einigen afrikanischen und paläarktischen Girlitzen. *Zool. Jahrb. System., 87:*317–362.

——— (1964) Der Brutparasitismus der *Viduinae* als ethologisches Problem. Prägungsphänomene als Faktoren der Rassen- und Artbildung. *Z. Tierpsychol., 21:*129–204.

——— (1965a) *Vogelhaltung und Vogelpflege.* Das Vivarium. Stuttgart (Franckh).

——— (1965b) *Columba livia: Flug von Haustaubenrassen. I. Der Flug des Birmingham Rollers. II. Der Flug des Steller-Kröpfers.* Filme, Göttingen (Inst. wiss, Film).

——— (1965c) Der Brutparasitismus der Witwenvögel. *Naturwiss. Med.* 2(7):3–15.

Nieboer, H. J. (1960) Ethological Observations on the Ant Lion (*Euroleon nostras* Fourcroy). *Arch. Neerl. Zool., 13:*609–611.

Nitschke, A. (1953) Über Eigenart und Ausdrucksgehalt frühkindlicher Motorik. *Dent. Med. Wochschr., 78:*1787–1792.

Noble, G. K. (1927) The Value of Life History Data in the Study of the Evolution of Amphibia. *Ann. N.Y. Acad. Sci., 30.*

——— (1931) *Biology of Amphibia.* New York.

——— (1934) Experimenting with the Courtship of Lizards. *Nat. Hist., N.Y. 34:*3–15.

Noble, G. K., and Bradley, H. T. (1933) The Mating Behavior of Lizards. *Ann. N.Y. Acad. Sci., 35:*25–100.

Noble, G. K., and Curtis, B. (1939) The Social Behaviour of the Jewel Fish, *Hemichromis bimaculatus* Gill. *Bull. Am. Mus. Nat. Hist., 76:*1–46.

Oehlert, B. (1958) Kampf und Paarbildung einiger Cichliden. *Z. Tierpsychol., 15:*141–174.

Ohm, D. (1964) Die Entwicklung des Kommentkampfverhaltens bei Jungcichliden. *Z. Tierpsychol., 21:*308–325.

Ohm, T. (1948) *Die Gebetsgebärden der Völker und das Christentum.* Münster W.

Olds, J. (1958) Self-Stimulation of the Brain. *Science, 127:*315–324.

Oppenheim, R. (1966) Amniotic Contractions and Embryonic Motility in the Chick Embryo. *Science, 152:*528–529.

Osche, G. (1952) Die Bedeutung der Osmoregulation und des Winkverhaltens für freilebende Nematoden. *Z. Morphol. Ökol. Tiere, 41:*54–77.

——— (1962) Ökologie des Parasitismus und der Symbiose. *Fortschr. Zool., 15:*125–164.

——— (1966) Die Welt der Parasiten. *Verständl. Wiss., 89.* Heidelberg (Springer).

Palluck, B. J., and Esser, A. H. (1971a) Controlled Experimental Modification of Aggressive Behavior in Territories of Severely Retarded Boys. *Am. J. Mental Deficiency, 76:*23–29.

——— (1971b) Territorial Behavior as an Indicator of Changes in Clinical Behavioral Condition of Severely Retarded Boys. *Am. J. Mental Deficiency, 76:*284–290.

Papi, F. (1959) Sull'orientamento astronomico in specie del gen. *Arctosia. Z. vergl. Physiol., 41:*481–489.

Papi, F., and Pardi, L. (1959) Nuovi reperti sull'orientamento lunare die *Talitrus saltator. Z. vergl. Physiol., 41:*583–596.

Pardi, L. (1960) Innate Components in the Solar Orientation of Littoral Amphipods. *Cold Spring Harbor Symp. Quant. Biol., 25:*395–401.

Passarge, S. (1907) *Die Buschmänner der Kalahari.* Berlin (D. Reimer).

Patterson, J. D. (1973) Ecologically Differentiated Patterns of Aggressive and Sexual Behavior in Two Troops of Ugandan Baboons, *Papio anubis. Am. J. Phys. Anthrop., 38:*641–647.

Pavlov, I. P. (1927) *Conditioned Reflexes* (Oxford).

PECKHAM, G., and PECKHAM, E. (1904) *Instinkt und Gewohnheiten der solitären Wespen.* Berlin.

PEIPER, A. (1951) Instinkt und angeborenes Schema beim Säugling. *Z. Tierpsychol., 8:*449–456.

——— (1953) Schreit- und Steigbewegungen beim Neugeborenen. *Arch. Kinderheilkde., 147:*135.

——— (1961) *Die Eigenart der kindlichen Hirntätigkeit.* 3rd ed. Leipzig.

PEIPONEN, V. A. (1960) Verhaltensstudien am Blaukehlchen. *Ornis Fennica, 37:*69–83.

PENFIELD, W. (1952) Memory Mechanisms. *Arch. Neurol. Psychiat.* (Chicago), *67:*178–191.

PENGELLEY, E. T., and FISHER, K. C. (1963) The Effect of Temperature and Photoperiod on the Yearly Hibernating Behavior of Captive Golden-Mantled Ground Squirrels *(Citellus lateralis tescorum). Can. J. Zool., 41:*1103–1120.

PENNYCUICK, C. J. (1960) The Physical Basis of Astronavigation in Birds: Theoretical Considerations. *J. Exptl. Biol., 37:*573–593.

PERDECK, A. C. (1958a) The Isolating Patterns in Two Sibling Species of Grasshoppers *(Chorthippus brunneus* and *Chorthippus biguttulus). Behaviour, 12:*11–75.

——— (1958b) Two Types of Orientation in Migrating Starlings, *Sturnus vulgaris* and Chaffinches, *Fringilla coelebs,* as Revealed by Displacement Experiments. *Ardrea, 46:*1–37.

PERNAU, F. A. v. (1716) *Unterricht, was mit dem lieblichen Geschöpff denen Vögeln, auch außer dem Fang, nur durch die Ergründung deren Eigenschafften, und Zahmmachung, oder anderer Abrichtung, Man sich vor Lust und Zeit-Vertreib machen können.* Nürnberg.

PEETERS, G., DEBACKERE, M., LAURYSSENS, M., and KUHN, E. (1965) Studies on the Release of Oxytocin in Domestic Animals. In PINKERTON, J. H. M. (ed.), *Symposium on Advances in Oxytocin Research.* New York (Pergamon).

PETERS, H. M. (1937a) Experimentelle Untersuchungen über die Brutpflege von *Haplochromis multicolor,* einem maulbrütenden Knochenfisch. *Z. Tierpsychol., 1:*200–218.

——— (1937b) Studien am Netz der Kreuzspinne. *Z. Morphol. Ökol. Tiere, 32:*613–649 and *33:*128–150.

——— (1939) Die Probleme des Kreuzspinnennetzes. *Z. Morphol. Ökol. Tiere, 36:* 179–266.

——— (1953) Weitere Untersuchungen über den strukturellen Aufbau des Radnetzes der Spinnen. *Z. Naturforsch., 8b:*355–370.

——— (1956) Gesellungsformen der Tiere. In *Handb. d. Soziologie, II.* Stuttgart.

PETERSEN, B., LUNDGREN, L., and WILSON, L. (1957) The Development of Flight Capacity in a Butterfly. *Behaviour, 10:*324–339.

PETERSEN, E. (1965) Biologische Beobachtungen über Verhaltensweisen einiger einheimischer Nager beim Öffnen von Nüssen und Kernen. *Z. Säugetierkde, 30:*156–162.

PFEIFFER, W. (1960) Über die Schreckreaktion bei Fischen und die Herkunft des Schreckstoffes. *Z. vergl. Physiol., 43:*578–614.

——— (1963) Vergleichende Untersuchungen über die Schreckreaktion und den Schreckstoff der Ostariophysen. *Z. vergl. Physiol., 47:*111–147.

PHILLIPS, L. H., and KONISHI M. (1972) Control of Aggression by Singing in Crickets. *Nature 241:*64–65.

PILLERI, G. (1960) Über das Auftreten von "Kletterbewegungen" im Endstadium eines Falles von Morbus Alzheimer. *Arch. Psychiat. Nervenkde, 200:*455–461.

——— (1960b) Kopfpendeln ("Leerlaufendes Brustsuchen") bei einem Fall von Pickscher Krankheit. *Arch. Psychiat. Nervenkde, 200:*603–611.

——— (1961) Orale Einstellung nach Art des Klüver-Bucy-Syndroms bei hirnatrophischen Prozessen. *Schweiz. Arch. Neurol. Neurochirg. Psychiat, 87:*286–298.

PITCAIRN, T. K., and GRANT, E. C. (1972) A Dyadic Analysis of Human Behaviour. Paper to Assoc. Study Anim. Behav., London, Nov. 1972.

PLACK, A. (1968) *Die Gesellschaft und das Böse.* 2nd ed. Munich (List).

PLOOG, D. W. (1964a) Verhaltensforschung und Psychiatrie. In GRUHLE, H. W., JUNG, R., MAYER-GROSS, W., and MÜLLER, M. (eds.), *Psychiatrie der Gegenwart. 1.* Berlin (Springer), 291–443.

——— (1964b) Verhaltensforschung als Grundlagenwissenschaft für die Psychiatrie. 6. Psychiatertagg. d. Landschaftsverb. *Rheinland—123. wiss. Vers. d. Rhein. Ver. f. Psychiatrie,* 1–23.

——— (1964c) Über experimentelle Grundlagen der Gedächtnisforschung. *Nervenarzt, 35:*377–386.

——— (1972) Kommunikation in Affengesellschaften. In GADAMER, H. G., and VOGLER, P. (eds.) *Neue Anthropologie, 2:*98–178, Stuttgart (Thieme).

PLOOG, D. W., BLITZ, J., and PLOOG, F. (1963) Studies on Social and Sexual Behavior of the Squirrel Monkey *(Saimiri sciureus). Folia Primat.,* 29–66.

PLOOG, D., and MELNECHUK, T. (1971) Are Apes Capable of Language? *Neurosciences Res. Prog. Bull. 9:*600–700.

POTTS, G. W. (1973a) The Ethology of *Labroides dimidiatus* (Cuv. and Val.) on Aldabra. *Animal Behaviour, 21:*250–291.

——— (1973b) Cleaning Symbiosis among British Fish with Special Reference to *Crenilabrus melops* (Labridae). *J. Mar. Biol. Ass. U.K., 53:*1–10.

PÖPPEL, E. (1968) Desynchronisation circadianer Rhythmen innerhalb einer isolierten Gruppe. *Arch. Ges. Physiol., 299:*364–370.

PORTMANN, A. (1953) *Das Tier als soziales Wesen.* 2nd ed. Zürich (Rhein).

POULSEN, H. (1953) A Study of Incubation Responses and Some Behavior Patterns in Birds. *Vidensk. Medd. Danks. Naturh. Foren.*

PRECHTL, H. F. R. (1951) Zur Paarungsbiologie einiger Molcharten. *Z. Tierpsychol., 8:*337–348.

——— (1953a) Zur Physiologie des Angeborenen Auslösemechanismus. *Behaviour, 5:* 32–50.

——— (1953b) Die Kletterbewegungen beim Säugling. *Monatsschr. Kinderheilkde, 12:* 519–521.

——— (1953c) Stammesgeschichtliche Reste im Verhalten des Säuglings. *Umschau, 21:* 656–658.

——— (1955) *Die Entwicklung der frühkindlichen Motorik. I-III.* Wiss. Filme, C 651, C 652, C 653. Göttingen (Inst. Wiss. Film).

——— (1956) Neurophysiologische Mechanismen des formstarren Verhaltens. *Behaviour, 9:*243–319.

——— (1958) The Directed Head-Turning Response and Allied Movements of the Human Baby. *Behaviour, 13:*212–242.

PRECHTL, H. F. R., and KNOL, A. R. (1958) Fußsohlenreflexe beim neugeborenen Kind. *Arch. Psychiat. Z. Ges. Neurol., 196:*542–553.

PRECHTL, H. F. R., and SCHLEIDT, W. M. (1950) Auslösende und steuernde Mechanismen des Saugaktes. I. *Z. vergl. Physiol., 32:*252–262.

——— (1951) Auslösende und steuernde Mechanismen des Saugaktes. II. *Z. vergl. Physiol., 33:*53–62.

PREMACK, D. (1971) Language in the Chimpanzee? *Science, 172:*808–822.

PRÉVOST, J. (1961) Ecologie du Manchot empereur: *Aptenodytes forsteri* Gray. Expeditions polaires Françaises. *Miss. Paul-Emile Victor. Publ. 222. Actualités Sci. Ind. No. 1291.* Paris (Hermann).

PREYER, W. (1885) *Spezielle Physiologie des Embryo.* Leipzig (Grieben).

RABER, H. (1948) Analyse des Balzverhaltens eines männlichen Truthahns (*Meleagris*). *Behaviour, 1:*237–266.

RANDALL, J. E. (1958) A Review of the Labrid Fish Genus Labroides with Descriptions of Two Species and Notes on Ecology. *Pacific Sci., 12:*327–347.

RANDALL, J. E., and RANDALL, H. E. (1960) Examples of Mimicry and Protective Resemblance in Tropical Marine Fishes. *Bull. Marine Sci. Gulf Caribbean, 10:*444–480.

RAO, K. S. (1954) Tidal Rhythmicity of Rate of Water Propulsion in *Mytilus* and Its Modificability by Transplantation. *Biol. Bull., 106:*353–359.

RASA, O. A. E. (1969) The Effect of Pair Isolation on Reproductive Success in *Etroplus maculatus* (Cichlidae). Z. *Tierpsychol., 26:*846–852.

——— (1971) Appetence for Aggression in Juvenile Damsel Fish. *Beiheft 7 zur Z. Tierpsychol.,* Berlin (Parey).

RASMUSSEN, K. (1908) *People of the Polar North.* London.

RATTNER, J. (1970) *Aggression und menschliche Natur.* Olten/Switzerland (Walter).

RAY, C. (1966) Stalking Seals under Antarctic Ice. *Natl. Geogr., 129:*54–65.

RAZRAN, G. (1971) *Mind in Evolution: An East-West Synthesis of Learned Behavior and Cognition.* Boston (Houghton Mifflin).

REAUMUR, R. A. F. (1734–1742) *Mémoires pour servir à l'histoire des insectes.* Paris (Impr. Royale), 1–6.

REESE, E. S. (1962a) Submissive Posture as an Adaptation to Aggressive Behavior in Hermit Crabs. Z. *Tierpsychol., 19:*645–651.

——— (1962b) Shell Selection Behavior of Hermit Crabs. *Animal Behaviour, 10:*347–360.

——— (1963a) The Behavioral Mechanisms Underlying Shell Selection by Hermit Crabs. *Behaviour, 21,* 78–126.

——— (1963b) A Mechanism Underlying Selection or Choice Behavior Which Is Not Based on Previous Experience. *Am. Zool., 3:*508.

——— (1968) Shell Use: An Adaptation for Emigration from the Sea by the Coconut Crab. *Science, 161:*385–386.

REGEN, J. (1924) Über die Orientierung des Grillenweibchens nach dem Stridulationschall des Männchens. *Sitz. Ber. Akad. Wiss. Wien, Math. Nat. Kl., 132.*

REIMARUS, H. S. (1762, 1973) *Allgemeine Betrachtungen über Triebe der Tiere hauptsächlich über ihre Kunsttriebe.* Hamburg.

REMANE, A. (1952) *Die Grundlagen des natürlichen Systems der vergleichen den Anatomie und der Phylogenetik.* Leipzig.

——— (1960) Das soziale Leben der Tiere. *Rowohlts Deut. Enzykl.,* 97. Hamburg (Rowohlt).

——— (1971) *Sozialleben der Tiere.* Stuttgart (G. Fischer).

REMMERT, H. (1965) Biologische Periodik. *Handb. d. Biol.,* 5. Frankfurt (Athenaion), 335–411.

RENSCH, B. (1957) Ästhetische Faktoren bei Farb- und Formbevorzugungen von Affen. Z. *Tierpsychol., 14:*71–99.

——— (1958) Die Wirksamkeit ästhetischer Faktoren bei Wirbeltieren. Z. *Tierpsychol., 15:*447–461.

——— (1961) Malversuche mit Affen. Z. *Tierpsychol., 18:*347–364.

——— (1962) Gedächtnis, Abstraktion und Generalisation bei Tieren. *Arbeitgemeinschaft Forsch. Landes Nordrhein-Westfalen, Köln* (Westdeutscher Verlag).

——— (1963) Versuche über menschliche "Auslöser-Merkmale" beider Geschlechter. Z. *Morphol. Anthropol., 53:*139–164.

——— (1965) Die höchsten Lernleistungen der Tiere. *Naturwiss. Rundschau, 18:*91–101.

RENSCH, B., and DÖHL, J. (1968) Wahlen zwischen zwei überschaubaren Labyrinthwegen durch einen Schimpansen. Z. *Tierpsychol. 25:*216–231.

RENSCH, B., and DÜCKER, G. (1959) Versuche über visuelle Generalisation bei einer Schleichkatze. Z. *Tierpsychol., 16:*671–692.

REUTER, O. M. (1913) *Lebensgewohnheiten und Instinkte der Insekten.* Berlin.

RICHARD, P. B. (1955) Bièvres constructeurs de barrages. *Mammalia. 19:*293–301.

——— (1964) Les matériaux de construction du Castor (*Castor fiber*), leur signification pour ce rongeur. Z. *Tierpsychol., 21:*592–601.

RICHTER, R. (1927) Die fossilen Fährten und Bauten der Würmer. *Palaeont. Z., 9:*193–240.
RIESEN, A. H. (1960) Effects of Stimulus Deprivation on the Development and Atrophy of the Visual Sensory System. *J. Orthopsychiat., 30:*23–26.
RIESS, B. F. (1954) The Effect of Altered Environment and of Age on the Mother-Young Relationships among Animals. *Ann. N.Y. Acad. Sci., 57:*606–610.
RISLER, H. (1953) Das Gehörorgan des Männchens von *Anopheles stephensi* Liston (Culicidae). *Zool. Jahrb. (Anat.), 73:*165–186.
——— (1955) Das Gehörorgan der Stechmücken. *Mikrokosmos, 44:*217–220.
RITTINGHAUS, H. (1963) *Sterna hirundo (Laridae): Balz und Kopulation.* Encycl. cinem., E 659, Göttingen (Inst. wiss. Film).
ROBERTS, W. W., and BERGQUIST, E. H. (1968) Attack Elicited by Hypothalamic Stimulation in Cats Raised in Social Isolation. *J. Comp. Physiol. Psychol. 66:*590–595.
ROBERTS, W. W., and CAREY, R. J. (1965) Rewarding Effect of Performance of Gnawing Aroused by Hypothalamic Stimulation in the Rat. *J. Comp. Physiol. Psychol., 59:* 317–324.
ROBERTS, W. W., and KIESS, H. O. (1964) Motivational Properties of Hypothalamic Aggression in Cats. *J. Comp. Physiol. Psychol., 58:*187–193.
ROBERTS, W. W., STEINBERG, M. L., and MEANS, L. W. (1967) Hypothalamic Mechanisms for Sexual, Aggressive, and other Motivational Behaviors in the Opossum, *Didelphis virginiana. J. Comp. Physiol. Psychol. 64:*1–15.
ROBINSON, J. (1962) *Pilobolus* sp. and the Translocation of Infective Larvae of *Dictyocaulus viviparus* from Faeces to Pastures. *Nature, 193:*353–354.
ROBSON, K. S. (1967) The Role of Eye-to-Eye Contact in Maternal-Infant Attachment. *J. Child Psychol. 8:*13–25.
ROEDER, K. D. (1935) An Experimental Analysis of the Sexual Behavior of the Praying Mantis, *Biol. Bull. 69:*203–220.
——— (1937) The Control of Tonus and Locomotory Activity in the Praying Mantis (*Mantis religiosa* L). *J. Exptl. Zool., 76:*353–374.
——— (1955) Spontaneous Activity and Behavior. *Sci. Monthly. Wash. 80:*362–370.
——— (1963a) *Nerve Cells and Insect Behavior.* Cambridge, Mass. (Harvard Univ. Press).
——— (1963b) Ethology and Neurophysiology. *Z. Tierpsychol., 20:*434–440.
ROEDER, K. D., and TREAT, E. A. (1961) The Reception of Bat Cries by the Tympanic Organ of Noctuid Moths. In ROSENBLITH (ed.), *Sensory Communication.* Cambridge, Mass. (M. I. T. Press) and New York (Wiley).
ROPER, M. K. (1969) A Survey of Evidence for Intrahuman Killing in the Pleistocene. *Current Anthropology, 10:*427–459.
RÖSCH, G. A. (1925) Untersuchungen über die Arbeitsteilung im Bienenstaat. *Z. vergl. Physiol., 6:*571–631.
ROSE, S. P. R. (1970) Neurochemical Correlates of Learning and Environmental Change. In: HORN, G., and HINDE, R. A. (eds.) *Short-Term Changes in Neural Activity and Behavior.* Cambridge (Univ. Press).
RÖSEL v. ROSENHOF, A. J. (1746–1761) *Insekten-Belustigungen, I–IV.* Nuremberg (Fleischmann).
——— (1758) *Die natürliche Historie der Frösche hiesigen Landes, etc.* Nuremberg (Fleischmann).
ROSENKÖTTER, L. (1966) Auf Excursion in die Menschenkunde. *Frankfurter Hefte 21.*
ROSS, D. M. (1960) The Association between the Hermit Crab *Eupagurus bernhardus* L. and the Sea Anemone *Calliactis parasitica. Proc. Zool. Soc. London, 134:*43–47.
ROSS, S. (1951) Sucking Behavior in Neonate Dogs. *J. Abnorm. Soc. Psychol., 46:*142–149.
ROTH, L. M. (1948) An Experimental Laboratory Study of the Sexual Behaviour of *Aedes aegypti. Am. Midland Naturalist., 40:*265–352.
ROTHBALLER, A. B. (1967) Aggression, Defense and Neurohumors. *U.C.L.A. Forum*

*Medical Science, 7:*135–170. Reprinted in CLEMENTE, C. D., and LINDSLEY, D. B. (eds.), *Aggression and Defense.* Berkeley (Univ. Calif. Press).

ROTHENBUHLER, W. C. (1964) Behavior Genetics of Nest Cleaning in Honeybees: IV. Responses of F_1 and Backcross Generations to Disease Killed Brood. *Am. Zoologist, 4:*111–123.

ROTHMANN, M., and TEUBER, E. (1915) Einzelausgabe der Anthropoidenstation auf Teneriffa: I. Ziele und Aufgaben der Station sowie erste Beobachtungen an den auf ihr gehaltenen Schimpansen. *Abhandl. Preuss. Akad. Wiss. Berlin,* 1–20.

ROUBAUD, E. (1916) Recherches biologiques sur les guepes solitaires et sociales d'Afrique: La genese de la vie sociale et l'evolution de l'instinct maternal chez les vespids. *Ann. Sci. Nit., 1:*1–160.

ROWELL, T. E. (1967) Female Reproductive Cycles and the Behavior of Baboons and Rhesus Macaques. In ALTMAN, S. A. (ed.), *Social Communication among Primates.* Chicago (Univ. Chicago Press).

RUITER, L. DE (1952) Some Experiments on the Camouflage of Stick-Caterpillars. *Behaviour, 4:*222–232.

——— (1955) Countershading in Caterpillars. *Arch. Neerl. Zool., 11:*1–57.

——— (1963) The Physiology of Vertebrate Feeding Behaviour: Towards a Synthesis of the Ethological and Physiological Approaches to Problems of Behaviour. *Z. Tierpsychol., 20:*498–516.

RUSSELL, C., and RUSSELL, W. W. S. (1968) *Violence, Monkeys and Man.* New York and London (Macmillan).

RUSSELL, E. S. (1938) *The Behaviour of Animals: An Introduction to Its Study.* London.

SACKETT, G. P. (1966) Monkeys Reared in Isolation with Pictures as Visual Input: Evidence for an Innate Releasing Mechanism. *Science, 154:*1468–1473.

SADE, D. S. (1967) Determinants of Dominance in a Group of Freeranging Rhesus Monkeys. In ALTMAN, S. A. (ed.), *Social Communication among Primates.* Chicago (Univ. Chicago Press), 99–114.

SAINT PAUL, U. v. (1967) *Lanius collurio: Ontogense des Beutespießens.* Encycl. cinem., E 1241. Göttingen (Inst. wiss. Film).

SALLER, K. (1963) Die Aufrichtung des Menschen und ihre Folgen. Z. *Morphol. Anthropol., 54:*82–111.

SAUER, F. (1954) Die Entwicklung der Lautäußerungen vom Ei ab schalldicht gehaltener Dorngrasmücken (*Sylvia c. communis* Latham). *Z. Tierpsychol., 11:*1–93.

——— (1956) Zugorientierung einer Mönchsgrasmücke *(Sylvia atricapilla)* unter künstlichem Sternenhimmel. *Naturwiss., 43:*231–232.

——— (1957) Die Sternenorientierung nächtlich ziehender Grasmücken *(Sylvia atricapilla, borin* und *curruca). Z. Tierpsychol., 14:*29–70.

——— (1961) Further Studies on the Stellar Orientation of Nocturnally Migrating Birds. *Psychol. Forsch., 26:*224–244.

SAUER, F., and SAUER, E. (1955) Zur Frage der nächtlichen Zugorientierung von Grasmücken. *Rev. Suisse Zool., 62:*250–259.

——— (1960) Star Navigation of Nocturnal Migrating Birds. *Cold Spring Harbor Symp. Quant. Biol., 25:*463–473.

SBRZESNY, H. (1973) Die Spiele der !Ko-Buschleute unter besonderer Berücksichtigung ihrer sozialisierenden und gruppenbindenden Funktion. *Diplomarbeit Zoolog. Inst. München.*

——— (1974) !Ko-Buschleute (Kalahari): Das von Kindern gespielte Mädchen-Initiations-Ritual (Eland Tanz). *Homo.*

SCHAFER, W. (1965) Aktualpaläontologische Beobachtungen. *Natur Mus., 95:*83–90.

SCHALLER, F. (1962) Die Unterwelt des Tierreiches. *Verständl. Wiss., 78.* Berlin (Springer).

SCHALLER, F., and SCHWALB, H. (1961) Attrappenversuche mit Larven und Imagines einheimischer Leuchtkäfer. *Zool. Anz., Suppl., 24:*154–166.

SCHALLER, G. B. (1963) *The Mountain Gorilla.* Chicago (University Press).

——— (1972) *The Serengeti Lion: A Study of Predator-Prey Relations.* Chicago (University Press).

SCHEIN, W. M. (1963) On the Irreversibility of Imprinting. *Z. Tierpsychol., 20:*462–467.

SCHENKEL, R. (1947) Ausdrucksstudien an Wölfen. *Behaviour, 1:*81–129.

——— (1956) Zur Deutung der Phasianidenbalz. *Ornithol. Beobacht., 53:*182.

——— (1958) Zur Deutung der Balzleistungen einger Phasianiden und Tetraoniden. *Ornithol. Beobacht., 55:*65–95.

——— (1964) Zur Ontogenese des Verhaltens bei Gorilla und Mensch. *Z. Morphol. Anthropol., 54:*233–259.

——— (1966) Zum Problem der Territorialität und des Markierens bei Säugern—am Beispiel des Schwarzen Nashorns und des Löwen. *Z. Tierpsychol., 23:*593–626.

——— (1967) Submission: Its Features and Function in the Wolf and Dog. *Am. Zoologist, 7:*319–329.

SCHEVILL, W. E. (1955) Evidence for Echolocation by Cetaceans. *Deep Sea Res., 3:*153.

SCHIFTER, H. (1965) Beobachtungen am Großmaulwels, *Chaca chaca. Natur Mus., 95:* 465–469.

SCHILDKRAUT, J. J. (1965) The Catecholamine Hypothesis of Affective Disorders: A Review of Supporting Evidence. *Am. J. Psychiat., 122:*509–522.

SCHILDKRAUT, J. J., and KETY, S. S. (1967) Biogenic Amines and Emotion. *Science, 156:* 21–30.

SCHJELDERUP-EBBE, T. (1922a) Soziale Verhältnisse bei Vögeln. *Z. Psychol., 90:*106–107.

——— (1922b) Beiträge zur Sozialpsychologie des Haushuhns. *Z. Psychol., 88:*225–252.

——— (1935) Social Behavior of Birds. In MURCHISON, A. (ed.), *A Handbook of Social Psychology,* 947–972.

SCHLEIDT, M. (1954) Untersuchungen über die Auslösung des Kollerns beim Truthahn. *Z. Tierpsychol., 11:*417–435.

SCHLEIDT, W. M. (1961a) Über die Auslösung der Flucht vor Raubvögeln bei Truthühnern. *Naturwiss., 48:*141–142.

——— (1961b) Reaktionen von Truthühnern auf fliegende Raubvögel und Versuche zur Analyse ihrer AAMs. *Z. Tierpsychol., 18:*534–560.

——— (1962) Die historische Entwicklung der Begriffe "Angeborenes auslösendes Schema" und "Angeborener Auslösemechanismus." *Z. Tierpsychol., 19:*697–722.

——— (1964a) Über die Spontaneität von Erbkoordinationen. *Z. Tierpsychol., 21:*235–256.

——— (1964b) Wirkungen äußerer Faktoren auf das Verhalten. *Fortschr. Zool., 16:*469–499.

——— (1966) Aus dem Signal-Inventar der Truthühner. In BURKHARDT, D., SCHLEIDT, W. M., and ALTNER, H. (eds.), *Signale in der Tierwelt.* Munich (H. MOOS), 130–134.

SCHLEIDT, W. M., and SCHLEIDT, M. (1962) *Meleagris gallapavo silvestris (Meleagrididae): Kampfverhalten der Hähne.* Wiss. Film E 487, Göttingen (Inst. wiss. Film).

SCHLEIDT, W. M., SCHLEIDT, M., and MAGG, M. (1960) Störungen der Mutter-Kind-Beziehung bei Truthühnern durch Gehörverlust. *Behaviour, 16:*254–260.

SCHLICHTER, D. (1968) Das Zusammenleben von Riffanemonen und Anemonenfischen. *Z. Tierpsychol., 25:*933–954.

SCHLOSSER, K. (1952a) Körperliche Anomalien als Ursache sozialer Ausstoßung bei Naturvölkern. *Z. Morphol. Anthropol., 44.*

——— (1952b) Der Signalismus in der Kunst der Naturvölker. Biologischpsychologische Gesetzlichkeiten in den Abweichungen von der Norm des Vorbildes. *Arbeiten a. d. Mus. f. Völkerkde Univ. Kiel, I.* Kiel (Mühlau).

——— (1952c) Der Rangkampf biologisch und ethnologisch gesehen. *Act. IV Congr. Intern. Sci. Anthropol. Ethnol. Vienna, 2:*43–50.

SCHMIDBAUER, W. (1971a) Methodenprobleme der Human-Ethologie. *Studium Gen., 24:*462–522.

——— (1971b) Zur Anthropologie der Aggression. *Dynamische Psychiatrie, 4:*36–50.

SCHMIDT, H. D. (1960) Bigotry in School Children. *Commentary, 29:*253–257.

SCHMIDT, L. (1952) Gestaltheiligkeit im bäuerlichen Arbeitsmythos. *Veröffentl. Öst. Mus. Volkskde, 1:*1–240.

SCHMIDT, R. S. (1957) The Evolution of Nestbuilding Behavior in Apicotermes. *Evolution, 9:*157–181.

——— (1958) The Nests of *Apicotermes trägardhi:* New Evidence on the Evolution of Nest-Building. *Behaviour, 12:*76–94.

SCHNEIDER, D. (1962) Electrophysiological Investigation on the Olfactory Specificity of Sexual Attracting Substances in Different Species of Moths. *J. Insect. Physiol., 8:*15–30.

——— (1966) Vergleichende Neurophysiologie. *Nervenarzt, 37:*454–457.

——— (1967) Wie arbeitet der Geruchsinn bei Mensch und Tier? *Mitt Max-Planck-Ges.,* 294–314.

SCHNEIDER, F. (1961) Beeinflussung der Aktivität des Maikäfers durch Veränderung der gegenseitigen Lage magnetischer und elektrischer Reize. *Mitt. Schweiz. Entomol. Ges., 33:*223–237.

SCHNEIDER, H. (1963) Bioakustische Untersuchungen an Anemonenfischen der Gattung *Amphiprion. Z. Morphol. Ökol. Tierre, 53:*453–474.

SCHNEIRLA, T. C. (1946) Problems of Biopsychology and Social Organization. *J. Abnorm. Soc. Psychol., 41:*385–402.

——— (1956) Interrelationships of the "Innate" and the "Acquired" in Instinctive Behavior. In GRASSÉ, P. P. (ed.), *L'Instinct dans le Comportement des Animaux.* Paris. 387–452.

——— (1959) An Evolutionary and Developmental Theory of Biphasic Processes Underlying Approach and Withdrawal. *Nebraska Symp. Motivation.* Lincoln (Univ. Nebraska Press), 1–41.

——— (1965) Aspects of Stimulation and Organisation in Approach Withdrawal Processes Underlying Vertebrate Behavioral Development. *Advan. Animal Behavior, 1:*1–74.

——— (1966) Behavioral Development and Comparative Psychology. *Quart. Rev. Biol., 41:*283–302.

SCHÖNE, H. (1951) Die Lichtorientierung der Larven von *Acilius sulcatus* und *Dytiscus marginalis. Z. vergl. Physiol., 33:*63–98.

——— (1959) Die Lageorientierung mit Statolithenorganen und Augen. *Ergeb. Biol., 21:* 161–209.

——— (1962) Optisch gesteuerte Lageänderungen (Versuche an Dytiscidenlarven Vertikalorientierung). *Z. vergl. Physiol., 45:*590–604.

——— (1965–1966) *Vorlesung über Orientierungsvorgänge, gehalten im Wintersem, 1965–1966.* Univ. Munich.

——— (1973) Raumorientierung, Begriffe und Mechanismen. *Fortschr. Zool., 21*(2/3).

SCHÖNE, H., and EIBL-EIBESFELDT, I. (1965) *Grapsus grapsus (Brachyura): Drohen.* Encycl. cinem., E599. Publ. z. wiss. Filmen, 1 A, 391–396. Göttingen (Inst. wiss. Film).

SCHÖNE, H, and SCHÖNE, H. (1963) Balz und andere Verhaltensweisen der Mangrovekrabbe, *Goniopsis cruentata,* und das Winkverhalten der eulitoralen Brachyuren. *Z. Tierpsychol., 20:*641–656.

SCHREMMER, F. (1960) Beobachtungen über die Bestäubung der Blüten von *Ophrys fuciflora* durch Männchen der Bienenart *Eucera nigrilabris. Österr. Botan. Z., 107:* 6–17.

SCHULTZ, J. H. (1966) *Organstörungen und Perversionen im Liebesleben.* Munich (E. Reinhardt).

SCHULTZE-WESTRUM, T. (1965) Innerartliche Verständigung durch Düfte beim Gleitbeutler *Petaurus breviceps papuanus* Thomas (Marsupialia, Phalangeridae). *Z. vergl. Physiol., 50:*151–220.

——— (1967) Biologische Grundlagen zur Populationsphysiologie der Wirbeltiere. *Naturwiss.*, *54*:576–579.

——— (1968) Ergebnisse einer zoologisch-völkerkundlichen Expedition zu den Papuas. *Umschau*, *68*, 295–300.

——— (1974) *Biologie des Friedens.* München (Kindler).

SCHUTZ, F. (1956) Vergleichende Untersuchungen über die Schreckreaktion bei Fischen und deren Verbreitung. *Z. vergl. Physiol.*, *38*:84–135.

——— (1964) Über geschlechtlich unterschiedliche Objektfixierung sexueller Reaktionen bei Enten im Zusammenhang mit dem Prachtkleid des Männchens. *Zool. Anz.*, *Suppl.*, *27*:282–287.

——— (1965a) Sexuelle Prägung bei Anatiden. *Z. Tierpsychol.*, *22*:50–103.

——— (1965b) Homosexualität und Prägung bei Enten. *Psychol. Forsch.*, *28*:439–463.

——— (1968) Sexuelle Prägungserscheinungen bei Tieren. In GIESE, H. (ed.), *Die Sexualität des Menschen.* Handb. d. Med. Sexualforschung. Stuttgart (Enke), 284–317.

SCHÜZ, E. (1952) *Vom Vogelzug.* Frankfurt (Schöps-Verlag).

SCHWARTZKOPFF, J. (1960) Vergleichende Physiologie des Gehörs. *Fortschr. Zool.*, *12*: 206–264.

——— (1962) Vergleichende Physiologie des Gehörs und der Lautäußerungen. *Fortschr. Zool.*, *15*:214–336.

SCHWIDETZKY, I. (1950) *Grundzüge der Völkerbiologie.* Stuttgart (Enke).

SCHWINCK, I. (1955) Weitere Untersuchungen zur Frage der Geruchsorientierung der Nachtschmetterlinge: partielle Fühleramputation bei Spinnermännchen, insbesondere des Seidenspinners. *Z. vergl. Physiol.*, *37*:439–458.

SCOTT, J. P. (1960) *Aggression.* Chicago (Univ. Press).

——— (1962) Critical Periods in Behavioral Development. *Science*, *138*:949–95.

——— (1963) The Process of Primary Socialization in Canine and Human Infants. *Monogr. Soc. Res. Child Develop.*, *28*.

——— (1964) The Effects of Early Experience on Social Behavior and Organization. In ETKIN, W. (ed.), *Social Behavior and Organization among Vertebrates.* Chicago (Univ. Press).

SCOTT, J. P., and FREDERICSON, E. (1951) The Causes of Fighting in Mice. *Physiol. Zool.*, *24*:273–309.

SCOTT, J. P., and FULLER, J. L. (1965) *Genetics and Social Behavior of the Dog.* Chicago (Univ. Press).

SEILACHER, A. (1967) Fossil Behavior. *Sci. Am.*, *217*(2):72–80.

SEISS, R. (1965) Beobachtungen zur Frage der Übersprungbewegungen im menschlichen Verhalten. *Psychol. Beitr.*, *8*:1–97.

SEITZ, A. (1940) Die Paarbildung bei einigen Zichliden I. *Z. Tierpsychol.*, *4*:40–84.

——— (1941) Die Paarbildung bei einigen Zichliden. II. *Z. Tierpsychol.*, *5*:74–101.

SEVENSTER, P. (1961) A Causal Analysis of a Displacement Activity: Fanning in *Gasterosteus aculeatus. Behaviour*, *Suppl.*, *9*.

——— (1968) Motivation and Learning in Sticklebacks (*Gasterosteus aculeatus* L.). In INGLE, D. (ed.), *The Central Nervous System and Fish Behavior.* Chicago (Univ. Press), 233–245.

SEVENSTER-BOL, A. C. A. (1962) On the Causation of Drive Reduction after a Consummatory Act. *Arch. Neerl. Zool.*, *15*:175–236.

SEXTON, O. J. (1960) Experimental Studies on Artificial Batesian Mimics. *Behaviour*, *15*: 244–252.

SHARPE, R. S., and JOHNSGARD, P. A. (1966) Inheritance of Behavioral Characters in F_2 Mallard × Pintail (*Anas platyrhynchos* L. × *Anas acuta* L.) Hybrids. *Behaviour*, *37*: 259–272.

SHAW, C. E. (1948) The Male Combat "Dance" of Some Crotalid Snakes. *Herpetologica*, *4*:137–145.

SHEFFIELD, F. D., and ROBY, T. B. (1950) Reward Value of a Non-Nutritive Sweet Taste. *J. Comp. Physiol. Psychol., 43:*471–481.

SHEPHER, J. (1971) Mate Selection among Second Generation Kibbutz Adolescents and Adults: Incest Avoidence and Negative Imprinting. *Arch. Sec. Behavior, 1:*293–307.

SHERRINGTON, C. S. (1931) Quantitative Management of Contraction in Lowest Level Coordinations. *Brain, 54:*1–28.

SHIELDS, J. (1962) *Monozygotic Twins Brought Up Apart and Brought Up Together: An Investigation into the Genetic and Environmental Causes of Variation in Personality.* Oxford (Univ. Press).

SHINKMAN, P. G. (1963) Visual Depth Discrimination in Day Old Chick. *J. Comp. Physiol. Psychol., 56:*410–414.

SIEBENALER, J. B., and CALDWELL, D. K. (1956) Cooperation among Adult Dolphins. *J. Mammal., 37:*126–128.

SIELMANN, H. (1955) *Brutbiologie des Schwarzspechtes.* Wiss. Film, C695. Göttingen (Inst. wiss. Film).

——— (1958) *Das Jahr mit den Spechten.* Berlin (Ullstein).

——— (1967) *Ptilonorhynchus violaceus–Bauen an der Laube,* E1075. *Ptilonorhynchus violaceus–Balz und Kopulation,* E1076. *Chlamydera nuchalis–Verteilen von Sammlungsstücken an der Laube,* E1077. *Chlamydera nuchalis–Balz und Kopulation,* E1078. *Chlamydera lauterbachi–Bauen an der Laube,* E1080. *Chlamydera lauterbachi–Balz,* E1081. *Prionodura newtonia–Behängen der Reisigtürme,* E1082. *Amblyornis macgregoriae–Behängen des "Maibaumes" und Balz,* E1083. *Amblyornis subalaris–Schmücken der Laube,* E1084. Encycl. cinem. Göttingen (Inst. wiss. Film).

SKINNER, B. F. (1938) *The Behavior of Organisms.* New York.

——— (1953) *Science and Human Behavior.* New York (Macmillan).

——— (1966) Phylogeny and Ontogeny of Behavior Contiguencies of Reinforcement Throw Light on Contiguencies of Survival in the Evolution of Behavior. *Science, 153:*1203–1213.

——— (1971) *Beyond Freedom and Dignity.* New York (Knopf).

SLUCKIN, W. (1965) *Imprinting and Early Learning.* Chicago (Aldine).

SMALL, W. S. (1900) An Experimental Study of the Mental Processes of the Rat. *Am. J. Psychol., 11:*133–165.

SMITH, R. I. (1958) On Reproductive Patterns as a Specific Characteristic among Nereid Polychaetes. *Systematic Zool., 7:*60–73.

SOMMER, R. (1966) Man's Proximate Environment. *J. Social Issues, 22:*59–70.

SORENSON, E. R., and GAJDUSEK, D. C. (1966) The Study of Child Behavior and Development in Primitive Cultures. *Pediatrics, Suppl., 37:*149–243.

SOUTHWICK, C. H. (1963) Challenging Aspects of the Behavioral Ecology of Howling Monkeys. In SOUTHWICK, C. H. (ed.), *Primitive Social Behavior.* Princeton, N.J. (Van Nostrand), 185–191.

SPALDING, D. A. (1873) Instinct with Original Observation on Young Animals. *MacMillans Mag., 27:*282–283 (reprinted in *Brit. J. Animal Behaviour, 2:*1–11, 1954).

SPANNAUS, G. (1961) Der wissenschaftliche Film als Forschungsmittel der Völkerkunde. Der Film im Dienste der Wissenschaft. *Festschr. z. Einweihung d. Neubaus f. d. Inst. wiss. Film, Göttingen,* 67–83.

SPERRY, R. W. (1940) The Functional Results of Muscle Transposition in the Hind Limb of the Rat. *J. Comp. Neurol., 73:*379–404.

——— (1943a) Functional Results of Crossing Sensory Nerves in the Rat. *J. Comp. Neurol. 78:*59–90.

——— (1943b) Effect of 180 Degree Rotation of the Retinal Field on Visuomotor Coordination. *J. Exptl. Zool. 92:*263–279.

——— (1945a) The Problem of Central Nervous Reorganization after Nerve Regeneration and Muscle Transposition. *Quart. Rev. Biol. 20:*311–369.

——— (1945b) Restoration of Vision after Crossing of Optic Nerves and after Contralateral Transplantation of Eye. *J. Neurophysiol. 8:*15–28.

——— (1951a) Mechanisms of Neural Maturation. In S. S. STEVENS (ed.), *Handbook of Experimental Psychology.* New York (Wiley), 236–280.

——— (1951b) Regulative Factors in the Orderly Growth of Neural Circuits. *Symp. Soc. Study Develop. Growth, 10:*63–87.

——— (1958) Physiological Plasticity and Brain Circuit Theory. In HARLOW, H. F., and WOOLSEY, C. C. N. (eds.), *Biological and Biochemical Bases of Behavior.* Madison (Univ. Wisconsin Press), 401–424.

——— (1963) Chemoaffinity in the Orderly Growth of Nerve Fiber Patterns and Connections. *Proc. Nat. Acad. Sci. U.S. 50:*703–710.

——— (1964) The Great Cerebral Commissure. *Sci. Am., 210*(1):42–52.

——— (1965) Selective Communication in Nerve Nets: Impulse Specificity vs. Connection Specificity. *Neurosci. Res. Program, Bull. 3:*37–43.

——— (1971) How a Brain Gets Wired for Adaptive Function. In: TOBACH, E., ARONSON, L. R., and SHAW, E. (eds.). *The Biopsychology of Development.* New York (Academic Press), 27–44.

SPERRY, R. W., and PREILOWSKI, B. (1972) Die beiden Gehirne des Menschen. *Bild der Wissenschaft:* 920–928.

SPINDLER, M., and BLUHM, E. (1934) Kleine Beiträge zur Psychologie des Seelöwen *(Eumetopias calif.). Z. vergl. Physiol., 21:*616–631.

SPINDLER, P. (1958) Studien zur Vererbung von Verhaltensweisen. 1. Verhalten auf einen starken akustischen Reiz. *Anthropol. Anz., 22*(2):137–155.

——— (1961) Studien zur Vererbung von Verhaltensweisen. 3. Verhalten gegenüber jungen Katzen. *Anthropol. Anz., 25*(1):60–80.

SPITZ, R. A. (1945) Hospitalism. In *The Psychoanalytical Study of the Child, 1.* New York (International Univ. Press), 53–74.

——— (1946) Anaclitic Depression: An Inquiry into the Genesis of Psychiatric Conditions in Early Childhood. In *The Psychoanalytical Study of the Child. 2.* New York (International Univ. Press), 313–342.

——— (1951) Psychogenic Diseases in Infancy. In *The Psychoanalytical Study of the Child. 6.* New York (International Univ. Press), 255–275.

——— (1957) *Die Entstehung der ersten Objektbeziehungen.* Stuttgart (Klett).

——— (1965) *The First Year of Life.* New York (International Univ. Press).

SPITZ, R. A., and WOLF, K. M. (1946) The Smiling Response: A Contribution to the Ontogenesis of Social Relations. *Gen. Psychol. Monogr., 34:*57–125.

STAEHELIN, B. (1953) Gesetzmäßigkeiten im Gemeinschaftsleben schwer Geisteskranker. *Schweiz. Arch. Neurol. Psychiat., 72:*277–298.

——— (1954) Gesetzmäßigkeiten im Gemeinschaftsleben Geisteskranker, verglichen mit tierpsychologischen Ergebnissen. *Homo, 5:*113–116.

STAMM, J. S. (1954) Control of Hoarding Activity in Rats by the Median Cerebral Cortex. *J. Comp. Physiol. Psychol., 47:*21–27.

——— (1955) The Function of the Median Cerebral Cortex in Maternal Behaviour of Rats. *J. Comp. Physiol. Psychol., 48:*347–356.

STAMM, R. A. (1960) Paarintimität und Streitigkeiten bei *Agapornis personata. Verhandl. Naturforsch. Ges. Basel, 71:*1–14.

——— (1962) Aspekte des Paarverhaltens von *Agapornis personata* Reichenow. *Behaviour, 29:*1–56.

——— (1964) Perspektiven zu einer vergleichenden Ausdrucksforschung. In KIRSCHHOFF, R. (ed.), *Handb. d. Psychol., 5:*255–288.

STARCK, D. (1959) Neuere Ergebnisse der vergleichenden Anatomie und ihre Bedeutung für die Taxonomie. *J. Ornithol., 100:*47–59.

STEINEN, K. von den (1894) *Unter den Naturvölkern Zentralbrasiliens. Reiseschilderungen und Ergebnisse der zweiten Schingu-Expedition 1887–1888.* Berlin (Geogr. Verlagshandlg., D. Reimer). Reprinted in BÖLSCHE, W., *Neue Welten.* Berlin (Deutsche Bibliothek), 1917.

STEIDINGER, P. (1968) Radarbeobachtungen über die Richtung und deren Streuung beim nächtlichen Vogelzug im Schweizerischen Mittelland. *Ornithol. Beob., 65:*197–226.

STEINER, J. E., and HORNER, R. (1972) The Human Gustoficial Response. *Israel J. Med. Sci.,* 8(4).

STEINIGER, F. (1950) Zur Soziologie und sonstigen Biologie der Wanderratte. *Z. Tierpsychol., 7:*356–379.

——— (1951) Revier- und Aktionsraum bei der Wanderratte. *Z. Hyg. Zool., 39:*33–51.

STOKES, B. (1955) Behaviour as a Means of Identifying Two Closely-Allied Species of Gall Midges. *Brit. J. Animal Behaviour, 9:*154–157.

STRASSEN, O. ZUR (1952) Zweckdienliches Sprechen beim Graupapagei. *Verh. Dt. Zool. Ges., Freiburg.,* 84–89.

STRESEMANN, E. (1934) Aves. In KÜKENTHAL, *Handb. d. Zool.,* 7(2). Berlin (de Gruyter).

STRUHSAKER, T. T. (1967) Auditory Communication among Vervet Monkeys, *Cercopithecus aethiops.* In ALTMAN, S. A. (ed.), *Social Communication in Primates.* Chicago (Univ. Press), 281–334.

STRUHSAKER, T. T., and HUNKELER, P. (1971) Evidence of Tool-Using by Chimpanzees of the Ivory Coast. *Folia primat., 15:*212–219.

SWEET, W. H., ERVIN, F., and MARK, V. H. (1969) The Relationship of Violent Behaviour to Focal Cerebral Disease. In: GARATTINI, S., and SIGG, E. B. (eds.) *Aggressive Behaviour.* Amsterdam (Excerpta Medica Foundation): 77–85.

SZONDI, L. (1969) *Gestalten des Bösen.* Bern (Huber).

TAUB, E., and BERMAN, A. J. (1964) *The Effect of Massive Somatic Deafferentiation on Behavior and Wakefulness in Monkeys.* Papers presented at Psychonomic Science Meeting, Niagara, Ont., October.

TAUB, E., ELLMAN, S. J., and BERMAN, A. J. (1965) Deafferentiation in Monkeys. Effects on Conditioned Grasp Response. *Science, 151:*593–594.

TAUB, E., PERELLA, P., and BARRO, G. (1973) Behavioral Development after Forelimb Deafferentiation on Day of Birth in Monkeys with and without Blinding. *Science, 181:*959–960.

TAUERN, O. D. (1918) *Patasiwa und Patalima, vom Molukkeneiland Seran und seinen Bewohnern.* Leipzig (Voigtländer).

TIETELBAUM, P. (1961) Disturbances in Feeding and Drinking Behavior after Hypothalamic Lesions. *Nebraska Symp. Motivation, 9:*39–65.

TELEKI, G. (1973) *The Predatory Behavior of the Chimpanzees.* Lewisburg (Bucknell Univ. Press).

TELLE, H. J. (1966) Beitrag zur Kenntnis der Verhaltensweise bei Ratten, vergleichend dargestellt bei *Rattus norvegicus* und *Rattus rattus. Angew. Zool., 9:*129–196.

TELLEGEN, A., HORN, J. M., and LEGRAND, R. G. (1969) Opportunity for Aggression as a Reinforcer in Mice. *Psycho. Sci., 14:*104–105.

TEMBROCK, G. (1954) Rotfuchs und Wolf. *Z. Säugetierkde, 19:*152–159.

——— (1960) Spielverhalten und vergleichende Ethologie. *Z. Säugetierkde, 25:*1–14.

——— (1964) *Verhaltensforschung.* 2nd ed. Jena (G. Fischer).

TETS, G. P. v. (1965) A Comparative Study of Some Social Communication Patterns in the *Pelecaniformes. Am. Ornithol. Union, Ornithol. Monogr., 2.*

THIBAUT, J. W., and COWLES, J. T. (1952) The Role of Communication in the Reduction of Interpersonal Hostility. *J. Abnorm. Soc. Psychol., 47:*770–777.

THIELCKE, G. (1961) Stammesgeschichte und geographische Variation des Gesanges unserer Baumläufer (*Certhidea familiaris* L. u. *C. brachydactyla* Brehm). *Z. Tierpsychol., 18:*188–204.

——— (1965) Gesangsgeographische Variation des Gartenbaumläufers *(Certhidea brachydactyla)* in Hinblick auf das Artbildungsproblem. *Z. Tierpsychol., 22:*542–566.

THOMAS, E. (1961) Fortpflanzungskämpfe bei Sandottern *(Vipera ammodytes). Zool. Anz., Suppl., 24:*502–505.

THOMPSON, J. (1941) Development of Facial Expression of Emotion in Blind and Seeing Children. *Arch. Psychol. N.Y., 264:*1–47.

THOMPSON, T. I. (1963) Visual Reinforcement in Siamese Fighting Fish. *Science, 141:* 55–57.

——— (1964) Visual Reinforcement in Fighting Cocks. *J. Exptl. Anal. Behavior, 7:*45–49.

——— (1969) Aggressive Behaviour of Siamese Fighting Fish. In GARATTINI, S., and SIGG, E. B. (eds.), *Aggressive Behaviour.* Amsterdam (Excerpta Medica Foundation): 15–31.

THORNDIKE, E. L. (1911) *Animal Intelligence.* New York (Macmillan).

THORPE, W. H. (1938) Further Experiments on Pre-Imaginal Conditioning in Insects. *Proc. Roy. Soc. London, B126:*370–397.

——— (1939) Further Studies on Pre-Imaginal Olfactory Conditioning in Insects. *Proc. Roy. Soc. London, B127:*424–433.

——— (1951) The Definition of Some Terms Used in Animal Behaviour Studies. *Animal Behaviour, 9:*1–7.

——— (1954) The Process of Song Learning in the Chaffinch as Studied by Means of the Sound Spectograph. *Nature, 173:*465.

——— (1958a) The Learning of Song Patterns by Birds, with Special Reference to the Song of the Chaffinch. *Ibis, 100:*535–570.

——— (1958b) Further Studies on the Process of Song Learning in the Chaffinch *(Fringilla coelebs). Nature, 182:*554–557.

——— (1961a) Sensitive Periods in the Learning of Animals and Men: A Study of Imprinting with Special Reference to the Introduction of Cyclic Behavior. In THORPE, W. H., and ZANGWILL, O. L. (eds.), *Current Problems in Animal Behaviour.* Cambridge (Univ. Press), 194–224.

——— (1961b) Bird Song: The Biology of Vocal Communication and Expression in Birds. *Cambridge Monogr. Exptl. Biol., 12.*

——— (1963) *Learning and Instinct in Animals.* London (Methuen).

——— (1966) Ritualisation in the Individual Development of Bird Song. *Phil. Trans. Roy. Soc. London, 551*(772):351–358.

THORPE, W. H., and JONES, F. H. W. (1937) Olfactory Conditioning in a Parasitic Insect and Its Relation to the Problem of Host Selection. *Proc. Roy. Soc. London, B124:* 56–81.

THORPE, W. H., and NORTH, M. E. W. (1965) Origin and Significance of the Power of Vocal Imitation: With Special Reference to the Antiphonal Singing of Birds. *Nature, 208:*219–223.

TIGER, L. (1969) *Men in Groups.* New York (Random House).

——— (1972) Comment on "Sex and Social Participation." *Am. Sociolog. Rev., 37*(5):634–637.

TIGER, L., and FOX, R. (1966) The Zoological Perspective in Social Science. *Man, 1:*75–81.

TINBERGEN, N. (1935) Über die Orientierung des Bienenwolfes: II. Die Bienenjagd. *Z. vergl. Physiol., 21:*699–716.

——— (1940) Die Übersprungbewegung. *Z. Tierpsychol., 4:*1–40.

——— (1948) Social Releasers and the Experimental Method Required for Their Study. *Wilson Bull., 60:*6–52.

——— (1951) *The Study of Instinct.* New York and London (Oxford Univ. Press).

——— (1952) "Derived" Activities, Their Causation, Biological Significance and Emancipation during Evolution. *Quart. Rev. Biol., 27:*1–32.

——— (1953) *Social Behaviour in Animals, with Special Reference to Vertebrates.* New York (Barnes and Noble) and London (Methuen).

——— (1955) *Tiere untereinander.* Berlin (Parey).

——— (1957) The Functions of Territory. *Bird Study, 5:*14–27.
——— (1959) Einige Gedanken über "Beschwichtigungsgebärden." *Z. Tierpsychol., 16:*651–665.
——— (1963) *The Herring Gull's World.* New York (Basic) and London (Collins).
——— (1965) Behavior and Natural Selection. In MOORE, J. A. (ed.), *Ideas in Modern Biology. Proc. 16th Inst. Zool. Congr. Washington, 1963, 6:*521–542.
TINBERGEN, N., BROEKHUYSEN, G. J., FEEKES, F., HOUGHTON, J. C. W., KRUUK, H., and SZULC, E. (1962) Egg-Shell Removal by the Black-Headed Gull *Larus ridibundus* L.: A Behaviour Component of Camouflage. *Behaviour, 19:*74–118.
TINBERGEN, N., and KRUYT, W. (1938) Über die Orientierung des Bienenwolfes: III. Die Bevorzugung bestimmter Wegmarken. *Z. vergl. Physiol., 25:*292–334.
TINBERGEN, N., and KUENEN, D. J. (1939) Über die auslösenden und richtunggebenden Reizsituationen der Sperrbewegung von jungen Drosseln (*Turdus m. merula* L. und *T. e. ericetorum* Turton). *Z. Tierpsychol., 3:*37–60.
TINBERGEN, N., MEEUSE, B. J. D., BOEREMA, L. K., and VAROSSIEAU, W. W. (1943) Die Balz des Samtfalters (*Eumenis semele* L.). *Z. Tierpsychol., 5:*182–226.
TINBERGEN, N., and PERDECK, A. C. (1950) On the Stimulus Situation Releasing the Begging Response in the Newly-Hatched Herring Gull Chick (*Larus argentatus*). *Behaviour, 3:*1–38.
TINBERGEN, E. A., and TINBERGEN, N. (1972) Early Childhood Autism: An Ethological Approach. *Beihefte zur Z. Tierpsychol., 10.*
TOBIAS, P. v. (1964) Bushmen-Hunter-Gatherers: A Study in Human Ecology. In: DAVIS, D. H. S. (ed.) *Ecological Studies in Southern Africa.* The Hague (W. Junk). Reprinted in COHEN, Y. A. (ed.) (1969) *Man in Adaptation.* Chicago (Aldine), 196–208.
TOLMAN, E. C. (1932) *Purposive Behaviour in Animals and Men.* New York (Appleton).
TOLMAN, E. C., and HONZIK, C. H. (1930a) Insight in Rats. *Univ. Calif. Publ. Psychol., 4:* 215–232
——— (1930b) Introduction and Removal of Reward, and Maze Performance in Rats. *Univ. Calif. Publ. Psychol., 4:*257–275.
TOMPA, F. S. (1962) Territorial Behavior: The Main Factor Controlling a Local Song Sparrow Population. *Auk, 79:*687–697.
TOWBIN, E. J. (1949) Gastric Distension as a Factor in the Satiation of Thirst in Esophagostomised Dogs. *Am. J. Physiol., 159:*533–541.
TRACY, H. C. (1926) The Development of Motility and Behavior in the Toad-Fish (*Opsanus tau*). *J. Comp. Neurol., 40:*253–369.
TRIVERS, R. L. (1971) The Evolution of Reciprocal Altruism. *Quart. Rev. Biol., 46:*35–57.
TRUMAN, J. W., and SOKOLOVE, P. G. (1972) Silk Moth Eclosion: Hormonal Triggering of a Centrally Programmed Pattern of Behavior. *Science, 175:*1491–1493.
TRUMLER, E. (1959) Das "Rossigkeitsgesicht" und ähnliches Ausdrucksverhalten bei Einhufern. *Z. Tierpsychol., 16:*478–488.
TRYON, R. C. (1940) Genetic Differences in Maze Learning in Rats. In *39th Yearbook National Society for the Study of Education.* Bloomington, Ill. Publ. School, *1:*111–119.
TSCHANZ, B. (1962) Über die Beziehungen zwischen Muttertier und Jungen beim Mufflon. *Experientia, 18:*187.
——— (1965) Beobachtungen und Experimente zur Entstehung der "persönlichen" Beziehung zwischen Jungvogel und Elterntier bei Trottellummen. *Verhandl. Schweiz. Naturforsch. Ges. Zürich,* 211–216.
——— (1968) Trottellummen: Die Entstehung der persönlichen Beziehungen zwischen Jungvogel und Eltern. *Beih. Z. Tierpsychol.,* 103.
TURHAN, B. M. (1960) Über die Bedeutung des Gesichtsausdrucks. *Psychol. Beitr., 5:*440–454.
TWEEDIE, M. (1966) Butterfly Mimics. *Animals, 8:*318–321.
UEXKÜLL, J. v. (1921) *Umwelt und Innenwelt der Tiere.* 2nd ed. Berlin.

——— (1937) Umweltsforschung. *Z. Tierpsychol., 1:*33–34.

UEXKÜLL, J. v., and KRISZAT, G. (1934) Streifzüge durch die Umwelten von Tieren und Menschen: Ein Bilderbuch unsichtbarer Welten. *Verständl. Wiss.;* new ed., 1963.

UHRICH, J. (1938) The Social Hierarchy in Albino Mice. *J. Comp. Physiol. Psychol., 25:*373–413.

VALENSTEIN, E. S., RISS, W., and YOUNG, W. C. (1955) Experimental and Genetic Factors in the Organisation of Sexual Behavior in Male Guinea Pigs. *J. Comp. Physiol. Psychol., 48:*397–403.

VALZELLI, L. (1969) Aggressive Behaviour Induced by Isolation. In GARATTINI, S., and SIGG, E. B. (eds.) *Aggressive Behaviour.* Amsterdam (Excerpta Medica Foundation): 70–76.

VANDENBERG, J. G. (1971) The Effects of Gonadal Hormones on the Aggressive Behaviour of Adult Golden Hamsters *(Mesocricetus auratus). Animal Behaviour, 19:*589–594.

VAYDA, A. (1970) Maoris and Muskets in New Zealand: Disruption of a War System. *Political Sci. Quart., 85:*560–584.

VEDDER, H. (1952–53) Über die Vorgeschichte der Völkerschaften von Südwestafrika. *J. South West Africa Sci. Soc., 9:*45–56.

VERWEY, J. (1930) Die Paarungsbiologie des Fischreihers. *Zool. Jahrb. Allgem. Zool. Physiol., 48:*1–120.

WAGNER, H. O. (1938) Beobachtungen über die Balz des Paradiesvogels *Paradisea guilielmi. J. Ornithol., 86:*550–553.

——— (1954) Massenansammlungen von Weberknechten. *Z. Tierpsychol., 11:*348–352.

WAHLERT, G. v. (1957) Weitere Untersuchungen zur Phylogenie der Schwanzlurche. *Zool. Anz., Suppl., 20:*347–352.

——— (1962) Beobachtungen und Bemerkungen zum Putzverhalten von Mittelmeerfischen. *Veröffentl. Inst. Meeresforsch., 7:*71–78.

——— (1963) Die ökologische und evolutorische Bedeutung der Fischschwärme. *Veroeffentl. Inst. Meeresforsch. Meeresbiol. Symp., 3:*197–213.

WALK, R. D. (1966) The Development of Depth Perception in Animals and Human Infants. *Monographs of the Society for Research in Child Development, 31:*82–108.

WALKER, B. W. (1952) A Guide to the Grunion. *Calif. Fish and Game, 38:*409–420.

WALLRAFF, H. G. (1959) Örtlich und zeitlich bedingte Variabilität des Heimkehrverhaltens von Brieftauben *Z. Tierpsychol., 16:*513–544.

——— (1960a) Können Grasmücken mit Hilfe des Sternenhimmels navigieren? *Z. Tierpsychol., 17:*165–177.

——— (1960b) Über Zusammenhänge des Heimkehrverhaltens von Brieftauben mit meteorologischen und geophysikalischen Faktoren. *Z. Tierpsychol., 17:*82–114.

——— (1967) The Present Status of Our Knowledge about Pigeon Homing. *Proc. 14th Intern. Ornithol. Congr. Oxford.* (Blackwell), 331–358.

——— (1969) Über das Orientierungsvermögen von Vögeln unter natürlichen und künstlichen Sternenmustern. Dressurversuche mit Stockenten. *Verh. Dtsch. Zool. Ges., 1968 (Zool. Anz. Suppl. 32):*348–357.

WALTHER, F. R. (1958) Zum Kampf- und Paarungsverhalten einiger Antilopen. *Z. Tierpsychol., 15:*340–380.

——— (1961) Entwicklungszüge im Kampf- und Paarungsverhalten der Horntiere. *Jahrb. G. v. Opel—Freigehege Tierforsch., 3:*90–115.

——— (1964) Verhaltensbeobachtungen an Thomsongazellen. *Z. Tierpsychol., 21:*871–890.

——— (1965) Verhaltensstudien an der Grant Gazelle (*Gazella granti* Brooke) im Ngorongoro Krater. *Z. Tierpsychol., 22:*166–208.

——— (1966) *Mit Horn und Huf.* Berlin (Parey).

WASHBURN, S. L., and DEVORE, I. (1961) The Social Life of Baboons. *Sci. Am., 204:*62–71.

WASMANN, E. (1920) Die Gastpfledge der Ameisen. *Abhandl. Theoret. Biol., 2:*4. Berlin.
——— (1925) Ameisenmimikry. *Abhandl. Theoret. Biol., 19:*1–164.
WATSON, A. (1966) Social Status and Population Regulation in the Red Grouse (*Lagopus lagopus scoticus*). *The Royal Soc. Pop. Study Group, Proc. 2.* Royal Society, London, 22–30.
WATSON, J. B. (1919) *Psychology from the Standpoint of a Behaviorist.* Philadelphia (Lippincott).
——— (1930) *Der Behaviorismus.* Stuttgart.
WEHMER, F. (1965) Effects of Prior Experience with Objects on Maternal Behaviors in the Rat. *J. Comp. Physiol. Psychol., 60:*294–296.
WEIDKUHN, P. (1968/69) Aggressivität und Normativität. Über die Vermittlerrolle der Religion zwischen Herrschaft und Freiheit. Ansätze zu einer kulturanthropologischen Theorie der sozialen Norm. *Anthropos, 63/64.*
WEIDMANN, U. (1951) Über den systematischen Wert von Balzhandlungen bei Drosophila. *Rev. Suisse Zool., 54:*502–511.
——— (1955) Some Reproductive Activities of the Common Gull *Larus canus* L. *Ardea, 43:*85–132.
——— (1956) Verhaltensstudien an der Stockente. *Z. Tierpsychol., 13:*209–271.
——— (1959) The Begging Response of the Black-Headed Gull Chick. *Bericht vorgetr. 6. Intern. Ethologenkonf.,* Cambridge.
——— (1965) Colour and Behaviour. In *Colour and Life.* Inst. Biol. London, 79–100.
WEIH, A. S. (1951) Untersuchungen über das Wechselsingen (Anaphonie) und über das angeborene Lautschema einiger Feldheuschrecken. *Z. Tierpsychol., 8:*1–41.
WEISS, P. (1936) Selectivity Controlling the Central Peripheral Relations in the Nervous System. *Biol. Rev. 11:*494–531.
——— (1937) Further Experimental Investigation on the Phenomenon of Homologous Response in Transplanted Amphibian Limbs. *J. Comp. Neurol. 66:*481–535.
——— (1939) *Principles of Development.* New York (Holt, Rinehart and Winston).
——— (1941a) Self-Differentiation of the Basic Patterns of Coordination. *Comp. Psychol. Monogr., 17:*1–96.
——— (1941b) Autonomous versus Reflexogenous Activity of the Central Nervous System. *Proc. Am. Phil. Soc., 84:*53–64.
WEISS, R. F. (1971) Altruism is Rewarding. *Science, 171,* No. 3977.
WELTY, J. C. (1934) Experiments on Group Behaviour of Fishes. *Physiol. Zool., 7:*85–127.
WENDLER, G. (1964) Laufen und Stehen der Stabheuschrecke *Carausius morosus:* Sinusborstenfelder in den Beinen als Glieder von Regelkreisen. *Z. vergl. Physiol., 48:*198–250.
——— (1965) The Coordination of Walking Movements in Arthropods. *Nerv. and Horm. Mech. of Integration, 20th Symp. Soc. Exptl. Biol.,* 229–249.
——— (1966) Der Regelkreis gezielter Bewegungen. In BURKHARDT, D., SCHLEIDT, W. M., and ALTNER, H. (eds.), *Düfte, Farben und Signale.* Munich (Moos).
——— (1968) Ein Analogmodell der Beinbewegungen eines laufenden Insekts. *Kybernetik, Beih. zu "Elektron. Rechenanlagen." Munich (Oldenbourg), 18,* 67–74.
WENDT, H. (1964) Erfolgreiche Zucht des Baumwollköpfchens oder Pincheäffchens *Leontocebus (Oedipomidas) oedipus. Säugetierkdl. Mitt., 12:*49–52.
WENNER, A. M. (1967) Honey Bees: Do They Use the Distance Information Contained in Their Dance Maneuver? *Science, 155:*847–849.
WESENBERG-LUND, C. (1939) *Biologie der Süßwassertiere.* Wien (Springer).
——— (1943) *Biologie der Süßwasser-Insekten.* Berlin (Springer).
WEVER, R. (1968) Einfluß schwacher elektromagnetischer Felder auf die circadiane Periodik der Menschen. *Naturwiss; 55:*29–32.
——— (1973) Hat der Mensch nur eine "innere Uhr"? *Umschau, 18:*551–558.
WHEELER, W. M. (1928) *Social Life among the Insects.* New York.

WHITMAN, C. O. (1899) Animal Behavior. *Biol. Lect. Marine Biol. Lab., Woods Hole,* 285–338.

——— (1919) The Behavior of Pigeons. *Publ. Carnegie Inst., 257:*1–161.

WICKLER, W. (1957) Vergleichende Verhaltensstudien an Grundfischen. I. Beiträge zur Biologie, besonders zur Ethologie von *Blennius fluviatilis* Asso im Vergleich zu einigen anderen Bodenfischen. *Z. Tierpsychol., 14:*393–428.

——— (1958) Vergleichende Verhaltensstudien an Grundfischen. II. Die Spezialisierung des *Steatocranus. Z. Tierpsychol., 15:*427–446.

——— (1959) Vergleichende Verhaltensstudien an Grundfischen. III. Die Umspezialisierung von *Noemacheilus kuiperi. Z. Tierpsychol., 16:*410–423.

——— (1960a) Belegexemplare zu Ethogrammen. *Z. Tierpsychol., 17:*141–142.

——— (1960b) Die Stammesgeschichte typischer Bewegungsformen der Fisch-Brustflosse. *Z. Tierpsychol., 17:*31–66.

——— (1961a) Ökologie und Stammesgeschichte von Verhaltensweisen. *Fortschr. Zool., 13:*303–365.

——— (1961b) Über das Verhalten der Blenniiden *Runula* und *Aspidontus. Z. Tierpsychol., 18:*421–440.

——— (1961c) Über die Stammesgeschichte und den ökologischen Wert einiger Verhaltensweisen der Vögel. *Z. Tierpsychol., 18:*320–342.

——— (1962a) Ei-Attrappen und Maulbrüten bei afrikanischen Cichliden. *Z. Tierpsychol., 18:*129–164.

——— (1962b) *Das Züchten von Aquarienfischen. Das Vivarium.* Stuttgart (Franckh). (1966 *Breeding Aquarium Fish.* New York [Van Nostrand].)

——— (1963) Zum Problem der Signalbildung, am Beispiel der Verhaltens-Mimikry zwischen *Aspidontus* und *Labroides. Z. Tierpsychol., 20:*657–679.

——— (1964a) Das Problem der stammesgeschichtlichen Sackgassen. *Naturwiss. Med., 1*(2):6–29.

——— (1964b) Phylogenetisch-vergleichende Verhaltensforschung mit Hilfe von Enzyklopädie-Einheiten. *Res. Film, 5:*109–118.

——— (1964c) Signalfälschung, natürliche Attrappen und Mimikry. *Umschau, 64:*581–585.

——— (1964d) *Antennarius nummifer (Antennariidae): Beutefang.* Encycl. cinem., E141, Publ. zu wiss. Filmen, 1A, 41–47, Göttingen (Inst. wiss. Film).

——— (1964e) *Emblemaria pandionis: Kampfverhalten.* Encycl. cinem., E517, Publ. zu wiss. Filmen, 1A, 176–180. Göttingen (Inst. wiss. Film).

——— (1965a) Über den taxonomischen Wert homologer Verhaltensmerkmale. *Naturwiss., 52:*441–444.

——— (1965b) Die äußeren Genitalien als soziale Signale bei einigen Primaten. *Naturwiss., 52:*269–270.

——— (1965c) *Gastromyzon borneensis (Gastromyzonidae): Kriechen und Schwimmen.* Encycl. cinem., E611, Publ. zu wiss. Filmen, 1A, 421–426. Göttingen (Inst. wiss. Film).

——— (1965d) Neue Varianten des Fortpflanzungsverhaltens afrikanischer Cichliden *(Pisces, Perciformes). Naturwiss., 52:*219.

——— (1965e) Die Evolution von Mustern der Zeichnung und des Verhaltens. *Naturwiss., 52:*335–341.

——— (1965f) Signal Value of the Genital Tassel in the Male *Tilapia macrochir* Blgr. *(Pisces, Cichlidae). Nature, 208:*595–596.

——— (1965g) Mimicry and the Evolution of Animal Communication. *Nature, 208:*519–521.

——— (1966a) Specialization of Organs Having a Signal Function in Some Marine Fish. *Intern. Conf. Tropical Oceanogr.,* Miami Lab. 1965.

——— (1966b) Über die biologische Bedeutung des Genitalanhanges der männlichen *Tilapia macrochir. Senckenbergiana Biol., 47:*419–427.

——— (1966c) Ursprung und biologische Deutung des Genitalpräsentierens männlicher Primaten. *Z. Tierpsychol., 23:*422–437.

——— (1966d) Orchideen und Mimikry. In BURKHARDT, D., et al. (eds.), *Signale in der Tierwelt.* Munich (H. Moos).

——— (1966e) Natürliche "Übersexualisierung" des Soziallebens beim Brabantbuntbarsch. *Umschau, 17:*571–572.

——— (1967a) Socio-Sexual Signals and Their Intraspecific Imitation among Primates. In MORRIS, D. (ed.), *Primate Ethology.* London (Weidenfeld and Nicolson), 69–147 (1967, Chicago [Aldine]).

——— (1967b) Vergleichende Verhaltensforschung und Phylogenetik. In HEBERER, G. (ed.), *Die Evolution der Organismen. I.* 3rd ed. Jena (G. Fischer), 420–508.

——— (1968a) *Mimikry-Signalfälschung in der Natur.* Munich (Kindler).

——— (1968b) Mutter-Kind-Signale- Ursprung und Bedeutungswandel. *Umschau, 23:* 718–719.

——— (1968c) Das Mißverständnis der Natur des ehelichen Aktes in der Moraltheologie. *Stimmen d. Zeit, 182:*289–303.

——— (1969) *Sind wir Sünder? Naturgesetze der Ehe.* München (Droemer).

——— (1971) *Die Biologie der zehn Gebote.* München (Piper).

——— (1972) *Ökologie und Verhalten.* München (Piper).

WICKLER, W., and SEIBT, U (1972) Über den Zusammenhang des Paarsitzens mit anderen Verhaltensweisen bei *Hymenocera picta* Dana. *Z. Tierpsychol., 31:*163–170.

WICKLER, W., and UHRIG, D. (1969) Bettelrufe, Antwortszeit und Rassenunterschiede im Begrüßungsduett des Schmuckbart-Vogels *Trachyphonus d'arnaudii. Z. Tierpsychol., 26:*651–661.

WIEPKEMA, P. R. (1961) An Ethological Analysis of the Reproductive Behaviour of the Bitterling (*Rhodeus amarus* Bloch). *Arch. Neerl. Zool., 14:*103–199.

WIESEL, T. N., and HUBEL, D. H. (1963a) Effects of Visual Deprivation on Morphology and Physiology of Cells in the Cat's Lateral Geniculate Body. *J. Neurophysiol., 24:* 978–993.

——— (1963b) Single-Cell Responses in Striate Cortex of Kittens Deprived of Vision in One Eye. *J. Neurophysiol., 24:*1003–1017.

——— (1965) Comparison of the Effects of Unilateral and Bilateral Eye Closure on Cortical Unit Responses in Kittens. *J. Neurophysiol., 28:*1029–1040.

WIESER, S. (1955) Die motorischen Schablonen des Oralsinnes. *Fortschr. Neurol. Psychiat., 23:*94–184.

WIESER, S., and ITIL, T. (1954) Die Aufbaustufen der primitiven Motorik. *Arch. Psychiat. Nervenkr., 191:*450–462.

WIGGLESWORTH, V. B. (1964) *The Life of Insects.* London (Weidenfeld and Nicolson).

WILEY, R. H. (1973) Territoriality and Non-random Mating in Sage Grouse, *Centrocercus urophasianus. Animal Behaviour Monogr.* 6(2).

WILLIAMS, L. (1967) *Man and Monkey.* London (Deutsch).

WILLIES, E. O. (1967) The Behavior of Bicolored Antbirds. *Univ. Calif. Publ. Zool.* 79. Berkeley (Univ. Calif. Press).

WILLOWS, A. O. D. (1971) Giant Brain Cells in Mollusks. *Sci. Am., 224:*69–75.

WILSON, A. P. (1968) Social Behavior of Free-ranging Rhesus Monkeys with an Emphasis on Aggression. Dissertation. Berkeley (Univ. of California).

WILSON, D. M. (1961) The Central Nervous Control of Flight in Locust. *J. Exptl. Biol., 38:*471–490.

——— (1964) Relative Refractoriness and Patterned Discharge of Locust Flight Motor Neurones. *J. Exptl. Biol., 41:*191–205.

——— (1965) Motor Output Patterns during Random and Rhythmic Stimulation of Locust Thoracic Ganglia. *Biophys. J., 5:*121.

——— (1966) Insect Walking. *Ann. Rev. Entomol., 11:*103–123.

——— (1968) The Flight Control System of the Locust. *Sci. Am., 218*(5):83–90.

WILSON, E. O. (1963) Pheromones. *Sci. Am., 208*(5):100–114.

——— (1965) Chemical Communication in Social Insects. *Science, 149:*1064–1071.

——— (1971) *The Insect Societies.* Cambridge, Mass. (Belknap Press of Harv. Univ.).

WILSONCROFT, W. E., and SHUPE, D. U. (1965) Tail, Paw and Pup Retrieving in the Rat. *Psychon. Sci., 3:*494.

WILTSCHKO, W. (1968) Über den Einfluß statischer Magnetfelder auf die Zugorientierung der Rotkehlchen *(Erithacus rubecula). Z. Tierpsychol., 25:*537–558.

WINDLE, W. F. (1940) *Physiology of the Fetus.* Philadelphia (Saunders).

——— (1944) Genesis of Somatic Motor Function in Mammalian Embryo: A Synthesizing Article. *Physiol. Zool., 17:*247–260.

WINKELSTRÄTER, K. H. (1960) Das Betteln der Zoo-Tiere. *Beih. Schweiz. Z. Psychol. Suppl., 39.*

WINN, H. F., SALMON, M., and ROBERTS, N. (1964) Sun-Compass Orientation by Parrot Fishes. *Z. Tierpsychol., 21:*798–812.

WINTER, P., PLOOG, D. W., and LATTA, J. (1966) Vocal Repertoire of the Squirrel Monkey *(Saimiri sciureus):* Its Analysis and Significance. *Exptl. Brain Res., 1:*359–384.

WOLF, A. V. (1958) *Thirst: Physiology of the Urge to Drink.* Springfield, Ill. (Thomas).

WOLF, G. (1957a) Der wissenschaftliche Film. *Naturwiss., 44:*477–482.

——— (1957b) Encyclopaedia Cinematographica. *Forschungsfilm, 2:*304–310.

WOLFE, J. B. (1936) Effectiveness of Token-Rewards in Chimpanzees. *Comp. Psychol. Monogr., 12:*1–72.

WOODMANSEE, J. J. (1966) Methodological Problems in Pupillographic Experiments. *Proc. 74th Ann. Convention, Am. Psychol. Ass.,* 133–134.

WUNSCHMANN, A. (1963) Quantitative Untersuchungen zum Neugierverhalten von Wirbeltieren. *Z. Tierpsychol., 20:*80–109.

WYNNE-EDWARDS, V. C. (1962) *Animal Dispersion in Relation to Social Behaviour.* London (Oliver & Boyd).

YERKES, R. M. (1948) *Chimpanzees: A Laboratory Colony.* 4th ed. New Haven (Yale Univ. Press).

YERKES, R. M., and ELDER, J. H. (1936) Oestrus, Receptivity, and Mating in Chimpanzees. *Comp. Psychol. Monogr., 13*(5):1–39.

YOUNG, J. Z. (1961) Learning and Discrimination in the Octopus. *Biol. Rev., 36:*32–96.

——— (1965) The Organization of Memory System. *Proc. Roy. Soc., B159:*565–588.

YOUNG, W. C. (1961) The Hormones and Mating Behavior. In W. C. YOUNG (ed.), *Sex and Internal Secretions.* 3rd ed. Baltimore (Williams and Wilkins).

YOUNG, W. C. (1965) The Organisation of Sexual Behavior by Hormonal Action during Prenatal and Larval Periods in Vertebrates. In BEACH, F. A. (ed.), *Sex and Behavior.* New York (Wiley).

ZANGWILL, O. L. (1961) Lashley's Concept of Cerebral Mass Action. In THORPE, W. H., and ZANGWILL, O. L. (ed.), *Current Problems in Animal Behaviour.* London (Cambridge Univ. Press), 59–87.

ZASTROW, B. v., and VEDDER, H. (1930) Die Buschmänner. In SCHULTZ-EWERTH, E., and ADAM, L. (eds.) *Das Eigeborenenrecht: Togo, Kamerun, Südwestafrika, die Südseekolonien.* Stuttgart (Strecker und Schröder).

ZEEB, K. (1964) Zirkusdressur und Tierpsychologie. *Mitt. Naturforsch. Ges. Bern., 21.*

ZIEGLER, H. E. (1920) *Der Begriff des Instinkts einst und jetzt.* Jena (G. Fischer).

ZIPPELIUS, H. M. (1949) Untersuchungen über das Balzverhalten heimischer Molche. *Zool. Anz. Suppl., 12:*127–130.

ZUCKERMANN, S. (1932) *The Social Life of Monkeys and Apes.* London.

ZUMPE, D. (1964) *Chelmon rostratus: Kampfverhalten.* Encycl. cinem, E207, Publ. wiss. Film 1 A, 335–339, Göttingen (Inst. Wiss. Film).

——— (1965) Laboratory Observations on the Aggressive Behaviour of Some Butterfly Fishes. *Z. Tierpsychol., 22:*226–236.

A SELECT LIST OF ETHOLOGICAL FILMS FOR RENT

The following 16-mm silent films (and many more) may be rented from Audio-Visual Services of The Pennsylvania State University, 17 Willard Building, University Park, Pa., 16802 (814-865-6315). A complete catalog is available upon request.

The first and larger group is from the Institut für den Wissenschaftlichen Film, Göttingen, West Germany. A small second group, also available from Audio-Visual Services, is added because of its relevance to topics discussed in the text. Films are listed alphabetically by the common name of the animal. Figures in parentheses indicate the running time in minutes and the order number. Except where indicated the rental fee is $2.80 per day of actual use. Higher fees usually indicate color.

Adder, puff (*Bitis arietans*): Males, ritualized fight (10, E269)
Ant, leaf-cutting (*Atta cephalotes*): Cutting and carrying leaves; defense (7, E1407)
Beaver, Canadian (*Castor canadensis*): Filling dam gap (4, E1474, $4.30)
Blenny (*Aspidontus taeniatus*): Fighting behavior (7, E123)
Ingestion of food (4, E140)
Bower bird (*Ptilonorhynchidae*): Bower decoration and courtship (various times, E1077–E1084, $4.30 each)
Butterfly fish (*Chaetodon melanotus*): Displacement eating (3, E1581)
(*Chaetodon xanthocephalus*): Pair formation (3½, E1503)
(*Chaetodon melanotus*): Threatening and fighting (3½, E755)
Cardinal fish (*Siphamia versicolor*): Symbiosis with sea urchins (3½, E755)
Caterpillar, fork-tail (*Dicranura vinula*): Threatening behavior (7, E973)
Cichlid (*Haplochromis burtoni*): Courtship and spawning (11, E470)
Care of the fry (6, E1206)
Egyptian mouthbreeder (*Hemihaplochromis multicolor*): Care of the fry (4½, E1137)
(*Talapia grahami*): Territorial behavior (9, E1257)
(*Talapia nilotica*): Care of the fry (10½, E1158)
Clown fish (*Amphiprion melanopus*): Behavior toward giant sea anemone (3, E291) (2½, E357) (2, E358)
Comorant, flightless (*Nannopterum harrisi*): Return of mate to nest (4½, E596, $4.30)
Crab, hermit (*Coenobita scaevola*): Locomotion and change of habitat (8½, E489, $4.30)
Duck, mandarin (*Aix galericulata*): Mating behavior (4½, E340)
Goby (*Elacatinus oceanops*): Cleaning (3, E515)
Goose, graylag (*Anser answer*): Mating behavior (7, E1389)
Threatening and fighting (2½, E1388)

Gull, black-headed (*Larus ridibundus*): Agonistic displays (2, E336)
Pair formation (2, E335)
Hamster, common (*Cricetus cricetus*): Fight between rivals (2, E97)
Pair formation (1½, 4, 2, E98–E100)
Setting scent marks (2, E96)
Horse, domestic (*Equus caballus*): Social behavior during pair formation (10½, E508)
Iguana, sea (*Amblyrhynchus cristatus*): Ritualized fight (5, E591, $4.30)
Macaque, Japanese (*Macaca fiscata*): Food preparation (4, E1466, $4.30)
Oyster catcher (*Haematopus ostralegus*): Territorial defense (7½, E354)
Plover, Kentish (*Charadrius alexandrinus*): Distracting display (6½, E192)
Pole cat (*Putorius putorius*): Killing brown rats (11, E106)
Rat, brown (*Rattus norwegicus*): Fight, experienced males (8½, E131)
inexperienced males (8½, E132)
Sea lion, Steller's (*Eumetopias jubata*): Behavior patterns in colony (7, E1498, $4.30)
Sheep, bighorn (*Ovis canadensis*): Fighting males (10, E1334)
Social behavior of males (6½, E1333)
Shrew, common tree (*Tupaia glis*): Laying scent trail (3, E299)
Spider (*Cupiennius salei*): Failure of spinning mechanism (4, E364)
Squirrel (*Sciurus vulgaris*): Hiding food, inexperienced (7, E143)
Stickleback, three-spined (*Gasterosteus aculeatus*): Courtship and spawning (6½, E721, $4.30)
Tenrec (*Echinops telfairi*): Setting scent marks (5, E972)
Tree creeper (*Certhia familiaris*): "Showing" nest hole by male (3, E1051)
Viper, sand (*Vipera ammodytes montandoni*): Ritualized fight–males (5½, E329)
Wrasse (*Labroides dimidiatus*): Cleaning with different fish (9, E127) (8½, E754)

Baboon behavior (Washburn and DeTore) (31, PCR-2107K, $11.60)
Baboon troop, dynamics of male dominance in (Devore) (30, 31292, $10.60)
Bees, dance of the (v. Frisch) (33, PCR-103, $5.40)
Ducklings, imprinted, social reaction in (Hoffman) (21, PCR-2180K, $7.60)
Macaques of Japan, behavior of (Carpenter) (28, PCR-2184K, $10.60)
Rhesus monkeys, free-ranging, behavior of on Cayo Santiago: Mother-offspring interaction (Miller and Kling) (16, PCR-2243, $6.90)

Author Index

Subject Index